Theilheimer's Synthetic Methods of Organic Chemistry

Vol. 37

Vol. 37 Yearbook 1983

Theilheimer's Synthetic Methods of Organic Chemistry

Synthetische Methoden
der Organischen Chemie

Jahrbuch mit deutschem Registerschlüssel

Editor Alan F. Finch, London

Editorial Consultant William Theilheimer, Nutley, N.J.
Assistant Editor David G. Hawkins, London

KARGER

S. Karger · Basel · München · Paris · London · New York · Tokyo · Sydney

Deutsche Ausgaben

Vol. 1	1946	1. Auflage
	1948	2., unveränderte Auflage
	1950	3., unveränderte Auflage
Vol. 2	1948	
Vol. 3	1949	with English Index Key
	1953	2., unveränderte Auflage
	1966	3., unveränderte Auflage
	1975	4., unveränderte Auflage
Vol. 4	1950	with English Index Key
	1966	2., unveränderte Auflage

English Editions

(ab Vol. 5 deutscher Registerschlüssel)

Vol. 1	1948	Interscience Publishers
	1975	(Karger) Second Edition
Vol. 2	1949	Interscience Publishers
	1975	(Karger) Second Edition
Vol. 5	1951	with Reaction Titles Vol. 1–5 and Cumulative Index
	1966	Second Edition
Vol. 6	1952	
	1975	Second Edition
Vol. 7	1953	
	1975	Second Edition
Vol. 8	1954	
	1975	Second Edition
Vol. 9	1955	
Vol. 10	1956	with Reaction Titles Vol. 6–10 and Cumulative Index
	1975	Second Edition
Vol. 11	1957	
	1975	Second Edition
Vol. 12	1958	
	1975	Second Edition
Vol. 13	1959	
	1975	Second Edition
Vol. 14	1960	
	1975	Second Edition
Vol. 15	1961	with Reaction Titles Vol. 11–15 and Cumulative Index
Vol. 16	1962	
Vol. 17	1963	
Vol. 18	1964	
Vol. 19	1965	
Vol. 20	1966	with Reaction Titles Vol. 16–20 and Cumulative Index
Vol. 21	1967	
Vol. 22	1968	
Vol. 23	1969	
Vol. 24	1970	
Vol. 25	1971	with Reaction Titles Vol. 21–25 and Cumulative Index
Vol. 26	1972	
Vol. 27	1973	
Vol. 28	1974	
Vol. 29	1975	
Vol. 30	1976	with Reaction Titles Vol. 26–30 and Cumulative Index
Vol. 31	1977	
Vol. 32	1978	
Vol. 33	1979	
Vol. 34	1980	
Vol. 35	1981	with Reaction Titles Vol. 31–35 and Cumulative Index
Vol. 36	1982	

National Library of Medicine, Cataloging in Publication

Theilheimer's synthetic methods of organic chemistry = Synthetische Methoden der organischen Chemie. – Vol. 37 (1983) – Basel; New York: Karger, © 1982–
v.
Continues: Synthetic methods of organic chemistry.
Editor: Alan F. Finch.
Text in English; table of contents in English and German.
1. Chemistry, Organic – yearbooks I. Finch, Alan F. II. Theilheimer, William, 1914–
ISBN 3-8055-3600-3

Distributed by S. Karger AG, Allschwilerstrasse 10, P.O. Box, CH-4009 Basel (Switzerland)
Printed in Switzerland by Graphische Anstalt Schüler AG, Biel
ISBN 3-8055-3600-3

Contents
Inhalt

Preface

This is the second volume of the eighth series and contains abstracts and supplementary references from papers published in 1981 and the first half of 1982. As in the past few volumes, the information has been derived from Derwent's Chemical Reactions Documentation Service.

In these yearbooks, the policy has always been to provide the organic chemist with information which is difficult to access by other means. For this reason, continued emphasis is laid on functional group chemistry and ring reactions of a general type. For casual and frequent users alike, the Subject Index – in this case cumulative for Volumes 36 and 37 – is the primary aids to retrieving information, whereas the systematic (symbol) classification is largely for scanning purposes.

I much appreciate the valuable advice and cooperation of my colleagues in the Chemical Reactions Documentation Service and technical advisors in Basel. My thanks are also extended to Dr. Theilheimer in his capacity of editorial consultant.

Derwent Publications Ltd., London — A.F. Finch
May 1983

From the Prefaces to the Preceding Volumes

New methods for the synthesis of organic compounds and improvements of known methods are being recorded continuously in this series.

Reactions are classified on a simple though purely formal basis by symbols, which can be arranged systematically. Thus searches can be performed without knowledge of the current trivial or author names (e.g., "Oxidation" and "Friedel-Crafts reaction").

Users accustomed to the common notations will find these in the subject index. By consulting this index, use of the classification system may be avoided. It is thought that the volumes should be kept close at hand. The books should provide a quick survey, and obviate the immediate need for an elaborate library search. Syntheses are therefore recorded in the index by starting materials and end products, along with the systematic arrangement for the methods. This makes possible a sub-classification within the reaction symbols by reagents, a further methodical criterion. Complex compounds are indexed with cross reference under the related simpler compounds. General terms, such as synthesis, replacement, heterocyclics, may also be brought to the attention of the reader.

A brief review, *Trends in Synthetic Organic Chemistry,* stresses highlights of general interest and calls attention to developments too recent to be included in the body of the text.

The abstracts are limited to the information needed for an appraisal of the applicability of a desired synthesis. In order to carry out a particular synthesis it is therefore advisable to have recourse to the original papers or, at least, to an abstract journal. In order to avoid repetition, selections are made on the basis of most detailed description and best yields, whenever the same method is used in similar cases. Continuations of papers already included will not be abstracted, unless they contain essentially new information. They may, however, be quoted at the place corresponding to the abstracted papers. These supplementary references (see page 557) make it possible to keep abstracts of previous volumes up-to-date.

Syntheses that are divided into their various steps and recorded in different places can be followed with the help of the notations *startg. m. f.* (starting material for the preparation of...) and *prepn. s.* (preparation, see).

Method of Classification

1. Reaction Symbols

As summarized in the Systematic Survey (p. XX), reactions are classified firstly according to the bond formed in the synthesis, secondly according to the reaction type, and thirdly according to the bond broken or the element eliminated. This classification is summarized in the reaction symbol, e.g.

OC ⇑ N

↗	↑	↖
Bond formed	Reaction type	Bond broken or element eliminated

The first part of the symbol refers to the chemical bond formed during the reaction, expressed as a combination of the symbols for the two elements bonded together, e.g. HN, NC, CC. The order of the elements is as follows:

H, O, N, Hal (Halogen), S, Rem (Remaining elements), and C

Thus, for the formation of a hydrogen-nitrogen bond, the notation is HN, not NH.

If two or more bonds are formed in a reaction, the 'principle of the latest position' applies. Thus, for the reduction

$$RCH{=}O + H_2 \rightarrow RCH(H){-}OH$$

in which both hydrogen-oxygen and hydrogen-carbon bonds are formed, the symbol is HC ⇓ OC and not HO ⇓ OC.

The second part of the symbol refers to the reaction type. Four types are distinguished: addition (⇓), rearrangement (∩), exchange (↓↑), and elimination (⇑), e.g.

$$RCH{=}CH_2 + H_2O \rightarrow RCH(OH){-}CH_3$$ OC ⇓ CC

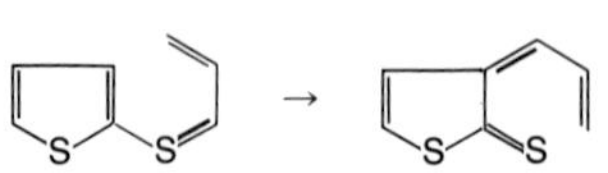

CC ∩ SC

$R{-}Cl + CN^- \rightarrow R{-}CN \quad [+ Cl^-]$ CC↓↑Hal

$RCH(Br){-}CH_3 \rightarrow RCH{=}CH_2 \quad [+ HBr]$ CC ⇑ Hal

Monomolecular reactions are either rearrangements (∩), where the molecular weight of the starting material and product are the same, or eliminations (⇑), where an organic or inorganic fragment is lost; bimolecular and multicomponent reactions are either additions (⇓), where the combined molecular weight of the starting materials is the same as that of the product,[1] or exchanges (↓↑), such as substitutions and condensations, where an organic or inorganic fragment is lost.

The last part of the symbol refers to the essential bond broken or, in the case of exchange reactions and eliminations, to a characteristic fragment which is lost. While the addition symbol is normally followed by the two elements denoting the bond broken, in the case of valency expansion, where no bonds are broken, the last part of the symbol indicates the atom at which the addition occurs, e. g.

$R_2S \rightarrow R_2SO$ OS ⇓ S

$RONO \rightarrow RONO_2$ ON ⇓ N

For additions, exchanges, and eliminations, the 'principle of the latest position' again applies if more than one bond is broken. However, for rearrangements, the most descriptive bond-breakage is used instead.[2] Thus, for the thio-Claisen rearrangement depicted above, the symbol is CC∩SC, and not CC∩CC.

Deoxygenations, quaternizations, stable radical formations, and certain rare reaction types are included as the last few methods in the yearbook. The reaction symbols for these incorporates the special symbols El (electron pair), Het (heteropolar bond), Rad (radical), Res (resolutions), and Oth (other reaction types), e.g.

$R_2S{=}O \rightarrow R_2S{:}$ ElS ⇑ O

$R_3N + R'Cl \rightarrow R_3\overset{+}{N}R'\ Cl^-$ Het ⇓ N

---

[1] Exceptions being additions of organometallics, e.g.

$RCH{=}O + R'MgBr \rightarrow RCH(R'){-}OH$ CC OC

$RCH{=}CHCOOR + LiCuR'_2 \rightarrow RCH(R'){-}CH_2COOR$ CC ⇓ CC

[2] Similarly, the formation of a peroxy function is classified under OO ⇓ CC, instead of OC ⇓ CC.

The following rules simplify the use of the reaction symbols:

1. The chemical bond is rigidly classified according to the structural formula without taking the reaction mechanism into consideration.
2. Double or triple bonds are treated as being equivalent to two or three single bonds, respectively.
3. Only stable organic compounds are usually considered: intermediates such as Grignard compounds and sodiomalonic esters, and inorganic reactants, such as nitric acid, are therefore not expressed in the reaction symbols.

2. *Reagents*

A further subdivision, not included in the reaction symbols, is based on the reagents used. The sequence of the reagents usually follows that of the periodic system. Reagents made up of several components are arranged according to the element significant for the reaction (e.g., $KMnO_4$ under Mn, NaClO under Cl). When a constituent of the reagent forms part of the product, the remainder of the reagent, which acts as a carrier of this constituent, is the criterion for the classification; for example, phophorus is the carrier in a chlorination with PCl_5 and sodium in a nitrosation with $NaNO_2$.

Systematik

1. Reaktionszeichen

Wie aus der systematischen Übersicht (S. XX) ersichtlich, erfolgt die Einordnung der Reaktionen primär nach den Bindungen, die bei der Synthese entstehen, zweitens nach dem Reaktionstyp, drittens nach den bei der Reaktion gelösten Bindungen oder nach einem charakteristischen Element, das eliminiert wird. Diese Einordnung ist kurz ausgedrückt im Reaktionszeichen, z. B.

OC ⇑ N

↗ ↑ ↖

Neue Bindung — Reaktionstyp — Gelöste Bindung oder eliminiertes Element

Der erste Teil des Reaktionszeichens bezeichnet die während der Reaktion entstehende Bindung, ausgedrückt durch die zwei neu gebundenen Elemente, z.B. HN, NC, CC. Die Reihenfolge der Elemente ist hierbei folgende:

H, O, N, Hal (Halogen), S, Rem (Übrige Elemente), C.

Die Bezeichnung für die Bildung einer Wasserstoff-Stickstoff-Bindung ist also HN, nicht NH.

Werden bei einer Reaktion zwei oder mehr Bindungen gebildet, so wird das «Prinzip der letzten Stelle» angewendet. So ist das Reaktionszeichen für die Reduktion

$$RCH{=}O \quad + \quad H_2 \quad \rightarrow \quad RCH(H){-}OH$$

bei welcher eine Wasserstoff-Sauerstoff- und eine Wasserstoff-Stickstoff-Bindung gebildet wird, HC ⇓ OC, und nicht HO ⇓ OC.

Der zweite Teil des Reaktionszeichens bezeichnet den Reaktionstyp. Es werden vier Typen unterschieden: Aufnahme (⇓), Umlagerung (∩), Austausch (↓↑) und Abgabe (Abspaltung) (⇑), z.B.

$$RCH{=}CH_2 \quad + \quad H_2O \quad \rightarrow \quad RCH(OH){-}CH_3 \qquad OC \Downarrow CC$$

CC∩SC

$R{-}Cl + CN^- \rightarrow R{-}CN \quad [+ Cl^-]$ CC↓↑Hal

$RCH(Br){-}CH_3 \rightarrow RCH{=}CH_2 \quad [+ HBr]$ CC⇑Hal

Monomolekulare Reaktionen sind entweder Umlagerungen (∩), bei welchen Ausgangs- und Endprodukt das gleiche Molekulargewicht besitzen, oder Abspaltungen (⇑) eines organischen oder anorganischen Fragments. Bimolekulare und höhere Reaktionen sind entweder Aufnahmen (Additionsreaktionen) (⇓), wo das Molekulargewicht des Produkts die Summe der Molekulargewichte der Ausgangsmaterialien ist[1], oder Austauschreaktionen wie Substitutionen und Kondensationen, bei welchen ein organisches oder anorganisches Fragment abgespalten wird.

Der letzte Teil des Reaktionszeichens bezeichnet die wichtigste gelöste Bindung oder, im Fall von Austausch- oder Abspaltungsreaktionen, ein charakteristisches Element, das abgespalten wird. Auf das Zeichen für Aufnahme (⇓) folgen normalerweise die zwei Elemente der gelösten Bindung; im Falle einer Vergrösserung der Wertigkeit, wenn keine Bindung gelöst wird, jedoch das Symbol des Atoms, an welchem die Aufnahme stattfindet, z.B.

$R_2S \rightarrow R_2SO$ OS⇓S

$RONO \rightarrow RONO_2$ ON⇓N

Bei Aufnahmen, Austausch- und Abspaltungsreaktionen gilt das «Prinzip der letzten Stelle» wenn mehr als eine Bindung gelöst wird. Bei Umlagerungen wird jedoch die bezeichnendste Bindungslösung angegeben[2]. So ist das Reaktionszeichen für die oben formulierte thio-Claisen Umlagerung CC∩SC und nicht CC∩CC.

Deoxygenierungen, Quaternierungen, Bildung von stabilen Radikalen und gewisse seltene Reaktionstypen bilden die am Schluss ange-

---

[1] Ausnahmen: Addition von metallorganischen Verbindungen, z.B.

$RCH{=}O + R'MgBr \rightarrow RCH(R'){-}OH$ CC⇓OC

$RCH{=}CHCOOR + LiCuR'_2 \rightarrow RCH(R'){-}CH_2COOR$ CC⇓CC

[2] Ähnlich wird die Bildung einer Peroxy-Funktion unter OO⇓CC und nicht unter OC⇓CC eingereiht.

ordneten Methoden in diesem Jahrbuch. Die entsprechenden Reaktionszeichen enthalten die Symbole El (Elektronenpaar), Het (heteropolare Bindung), Rad (Radikal), Res (Razemat-Trennung) und Oth (andere Reaktionstypen), z.B.

$R_2S{=}O \rightarrow R_2S{:}$ ElS ⇑ O

$R_3N + R'Cl \rightarrow R_3\overset{+}{N}R'\,Cl$ Het ⇓ N

Die Bildung des Reaktionszeichens wird durch folgende vereinfachende Annahmen erleichtert:

1. Die Bindungen für die Registrierung ergeben sich rein formal aus den Strukturformeln, ohne dass auf Reaktionsmechanismen Rücksicht genommen wird.
2. Doppel- und Dreifachbindungen werden 2 bzw. 3 Einfachbindungen gleichgesetzt.
3. Es werden in der Regel nur stabile organische Verbindungen berücksichtigt. Zwischenprodukte, wie z.B. Grignard-Verbindungen, Natrium-Malonester und anorganische Reaktionspartner, wie z.B. Salpetersäure, werden deshalb nicht zur Bildung des Reaktionszeichens herangezogen.

2. *Hilfsstoffe*

Eine weitere Unterteilung, die im Reaktionszeichen nicht mehr zum Ausdruck kommt, wird nach den für die Reaktion charakteristischen Hilfsstoffen vorgenommen. Die Reihenfolge der Hilfsstoffe richtet sich im wesentlichen nach dem periodischen System. Hilfsstoffe aus mehreren Bestandteilen werden nach demjenigen eingeteilt, der für die Reaktion verantwortlich ist, z.B. steht $KMnO_4$ bei Mn, NaClO bei Cl. Geht ein Bestandteil des Hilfsstoffs in das Reaktionsprodukt ein, dann ist der Rest als Träger dieses Bestandteils für die Einordnung massgebend; das ist z.B. bei einer Chlorierung mit PCl_5 Phosphor, bei einer Nitrosierung mit $NaNO_2$ Natrium.

High-Coverage Searches

A search through Synthetic Methods provides a selection of key references from the journal literature. For greater coverage, as for bibliographies, a supplementary search through the following publications is suggested.

Chemical Reactions Documentation Service[1]
designed for both current awareness and retrospective retrieval. Its monthly publication, the Journal of Synthetic Methods, covers the journal and patent literature, and provides 3,000 abstracts of recently published papers annually, together with 3,000 supplementary references.
One-line access is available to over 50,000 reactions, these including the data in all the abstracts in Synthetic Methods.

Science Citation Index[2]
for which Synthetic Methods serves as a source of starting references.

Chemical Abstract Service[3]
References may not be included in Synthetic Methods

1) to reactions which are routinely performed by well known procedures,
2) to subjects which can be easily located in handbooks and indexes of abstract journals, such as the ring system of heterocyclics or the metal in case of organometallic compounds, and
3) to inadequately described procedures, especially if yields are not indicated.

References to less accessible publications such as those in the Russian or Japanese language are usually only included if the method in question is not described elsewhere.

[1] Derwent Publications Ltd., 128 Theobalds Road, London WC1X 8RP, England.
[2] Institute for Scientific Information, Philadelphia, Pa., USA.
[3] Chemical Abstracts Service, Columbus, Ohio, USA.

Trends in Synthetic Organic Chemistry 1983

With research into leukotrienes very much in the ascendancy, a timely review by Samuelsson[1] puts into perspective these compounds considered to be mediators in allergic asthma. Interest in the linearly-fused tricyclopentane skeleton has also increased by virtue of its natural occurrence and likely physiological activity; Oppolzer's synthesis[2] by a reiterative intramolecular magnesium-ene reaction is a particular highlight, as also is the potential role of this same ring system as a building block for the preparation of complex polyhedra, such as dodecahedrane[3].

The spiropiperidine skeleton of the histrionicotoxins is obtainable both by intramolecular Michael amination[4] and via the popular acyl-immonium route[5]. The 1*H*-pyrrolo[1,2-*a*]indole system, important in mitomycin chemistry, has also been prepared by novel routes: one involving double ring closure of a dienazide[6], and another proceeding via a regiospecific Nenitzescu reaction with *p*-quinone monoketals[7]. Original syntheses of D-biotin[8] and codeine[9] have been reported, and novel N-functionalized monobactams may well prove of value in antibiotic therapy[10].

With a view to specifying unambiguously the steric course and product configuration in such reactions as the aldol condensation and Michael addition, a new notation based on the Cahn-Ingold-Prelog

[1] B. Samuelsson, Angew. Chem. Intern. Ed. *21,* 902–10 (1982).

[2] W. Oppolzer, K. Bättig, Tetrahedron Letters *23,* 4669–72 (1982); f. methods s. J. E. Pauw, A. C. Weedon, Tetrahedron Letters *23,* 5485–8 (1982); G. Mehta, K. S. Rao, ibid. *24,* 809–12 (1983).

[3] R. J. Ternansky, D. W. Balogh, L. A. Paquette, J. Am. Chem. Soc. *16,* 4503–4 (1982).

[4] S. A. Godleski, D. J. Heacock, J. Org. Chem. *47,* 4820–2 (1982); M. Glanzmann et al., Tetrahedron *38,* 2805–10 (1982).

[5] D. A. Evans, E. W. Thomas, R. E. Cherpeck, J. Am. Chem. Soc. *104,* 3695–700 (1982).

[6] Y. Naruta et al., Chem. Letters *1982,* 1859–62.

[7] R. M. Coates, P. A. MacManus, J. Org. Chem. *47,* 4822–4 (1982).

[8] E. G. Baggiolini et al., J. Am. Chem. Soc. *104,* 6460–2 (1982).

[9] W. H. Moos, R. D. Gless, H. Rapoport, J. Org. Chem. *48,* 227–38 (1983).

[10] D. M. Floyd et al., J. Org. Chem. *47,* 5160–7 (1982); J. Am. Chem. Soc. *104,* 6053–60 (1982).

system is now at hand[11]. The Baldwin nomenclature for classifying ring closures has also been extended with associated rules for intramolecular aldol condensations in polyketonic substrates[12].

In the aldol technology itself, there have been several notable developments: (Z)-enolates, obtainable with lithium bis[dimethyl(phenyl)-silyl]amide as base, direct the stereospecific synthesis[13] of the γ-alkoxy-β-hydroxyketo fragment, as occurring in the ansa chain of rifamycin S, while tin(II) enolates are key intermediate in asymmetric condensations[14]; a kinetic-controlled non-enolate route with α-mercuriketones indicates an unusual metal effect on stereospecificity[15], and a scheme based on Δ^2-oxazoline chemistry offers a useful alternative approach with complete regioselectivity[16].

A reaction-type related to the above – the addition of allylic organometallic compds. to aldehydes – has been neatly reviewed from diastereogenic considerations[17], and, in part, features in a general survey of stereo-regulated synthesis of acyclic systems with organometallics[18]. Of such reagents, allylic titanium complexes offer scope in ketone-versus-aldehyde chemoselectivity[19], while appropriately functionalized allylic tin reagents function as homoenolate equivalents[20]. In the same context of carbonylophilicity, organozirconium tributoxides have the same low basicity and high selectivity as titanium analogs, but are generally more reactive[21].

1,2,3-Trifunctionalized carbon chains are simply obtainable via ring closure of derivatized allyl alcohols and 2,3-oxidoalcohols. In the former method, the 3-iodo-2-aminoalcohol residue is secured in one step via the trichloroacetimidate[22], whereas in the latter case the 1,2,3-triol fragment may be prepared from the O-carbamyl derivative[23].

[11] D. Seebach, V. Prelog, Angew. Chem. Intern. Ed. *21,* 654–60 (1982).
[12] J. E. Baldwin, M. J. Lusch, Tetrahedron *38,* 2939–47 (1982).
[13] S. Masamune, J. W. Ellingboe, W. Choy, J. Am. Chem. Soc. *104,* 5526–8, 5528–31 (1982).
[14] N. Iwasawa, T. Mukaiyama, Chem. Letters *1982,* 1441–4; *1983,* 297–8.
[15] Y. Yamamoto, K. Maruyama, J. Am. Chem. Soc. *104,* 2323–5 (1982).
[16] D. P. Curran, J. Am. Chem. Soc. *104,* 4024–6 (1982); cf. Tetrahedron Letters *23,* 3123–6 (1982).
[17] R. W. Hoffmann, Angew. Chem. Intern. Ed. *21,* 555–642 (1982).
[18] Y. Yamamoto, Z. Maruyama, Heterocycles *18,* 357–86 (1982).
[19] M. T. Reetz, B. Wenderoth, Tetrahedron Letters *23,* 5259–62 (1982).
[20] A. J. Pratt, E. J. Thomas, J. Chem. Soc. Chem. Commun. *1982,* 1115–7; review of homoenolate anions and equivalents s. N. H. Werstiuk, Tetrahedron *39,* 205–68 (1983).
[21] Synth. Meth. *37,* 631; review s. B. Weidmann, D. Seebach, Angew. Chem. Intern. Ed. *22,* 31–45 (1983).
[22] G. Cardillo et al., J. Chem. Soc. Chem. Commun. *1982,* 1308–9.
[23] P. Ma et al., J. Org. Chem. *47,* 1378–80 (1982).

In the ever-expanding field of asymmetric synthesis, asymmetric induction on nucleophilic substitution is reported in C-α-alkylation with chiral sulfate esters[24] and in the replacement of sulfonyl groups[25]. Optically active hydrocarbons are available by cross-coupling with organozinc halides in the presence of a chiral ferrocenylphosphine-palladium complex[26], while asymmetric cyclopropanation is a welcome addition to pyrethroid chemistry[27]. The importance of amino-acids as chiral auxiliaries is highlighted in an excellent review[28].

In the synthesis of ethylene derivs., thiophosphinamides[29] are especially useful where Wittig reaction with the corresponding phosphonium ylids is sluggish, and a method based on α-carbalkoxyphosphonium salts[30] avoids the conventional basic treatment. Lithiomethylenetriphenylphosphorane, however, is more versatile than the non-metallated ylid, especially in its reactivity towards hindered ketones and epoxides[31]. A boron-variant of the Wittig synthesis proceeds by an unexpected deprotonation of tertiary dimesitylboranes, which also have a variety of other uses[32].

Important applications of [3.3]-sigmatropic shifts have been numerous. Carbanion-accelerated Claisen rearrangements[33] proceed under mild conditions, and an apparent homo-Claisen rearrangement utilizes established allylsilane chemistry[34]. Macrocyclics are available both via Claisen rearrangement[35] and reiterative oxy-Cope ring expansion[36], while the tandem Cope-Claisen sequence has been applied in construction of the estrogen skeleton[37]. Oxy-Cope rearrangements may also be performed by mercuration-demercuration[37a].

Cycloalkenes and γ-lactones are both obtainable via regiospecific ring closure with hydride-generated vinyl radicals[38]. A radical pathway is also presumed in a novel reductive cyclization of mercurial enones[39],

[24] P. Duhamel, J.-Y. Valnot, J. J. Eddine, Tetrahedron Letters *23,* 2863–6 (1982).
[25] M. Shibasaki, A. Nishida, S. Ikegami, J. Chem. Soc. Chem. Commun. *1982,* 1324–5.
[26] T. Hayashi et al., Bull. Chem. Soc. Japan *56,* 363–4 (1983).
[27] J. Mulzer, M. Kappert, Angew. Chem. Intern. Ed. *22,* 63–4 (1983).
[28] K. Drauz, A. Kleemann, J. Martens, Angew. Chem. Intern. Ed. *21,* 584–607 (1982).
[29] C. R. Johnson, R. C. Elliott, J. Am. Chem. Soc. *104,* 7041–4 (1982).
[30] J. L. Belletire, D. R. Walley, M. J. Bast, Synth. Commun. *12,* 469–75 (1982).
[31] E. J. Corey, J. Kang, J. Am. Chem. Soc. *104,* 4724–5 (1982).
[32] A. Pelter et al., Tetrahedron Letters *24,* 623–38 (1983).
[33] S. E. Denmark, M. A. Harmata, J. Am. Chem. Soc. *104,* 4972–4 (1982).
[34] S. R. Wilson, M. F. Price, J. Am. Chem. Soc. *104,* 1124–6 (1982).
[35] A. G. Cameron, D. W. Knight, Tetrahedron Letters *23,* 5455–8 (1982).
[36] D. L. J. Clive et al., J. Chem. Soc. Chem. Commun. *1982,* 828–30.
[37] F. E. Ziegler, H. Lim, J. Org. Chem. *47,* 5229–30 (1982).
[37a] N. Bluthe, M. Malacria, J. Gore, Tetrahedron Letters *23,* 4263–6 (1982).
[38] Synth. Meth. *37,* 955; Y. Ueno et al., ibid. *104,* 5564–6 (1982).
[39] S. Danishefsky, S. Chackalamannil, B.-J. Uang, J. Org. Chem. *47,* 2231–2 (1982).

although some limitations are evident. Rules which control the regiospecificity and reactivity of free radical substitution and addition reactions have been laid down and shown to be special cases of two laws which govern all radical reactions[40].

[4+2]-Cycloaddition is, as always, to the fore. A reiterative methodology has been devised for the synthesis of 6-membered linearly-condensed polycyclics[41], while CuCl is the catalyst of choice for reactions of dienes susceptible to polymerization[42]. An excellent review on heterodiene synthesis has been published[43].

Syntheses via allylpalladium complexes is a recurrent theme of considerable value. Macrocyclization is achievable at high concentration by using a polymerbased palladium(0) catalyst, the desired ring closure being facilitated by virtue of the low concentration of reacting molecules at the catalyst surface[44]. Trimethylenemethanepalladium complexes are viewed as key intermediates in a stereospecific bicyclo-[3.3.0]octane synthesis by intramolecular [3+2]-cycloaddition[45]. However, the regiospecificity associated with palladium-catalysed C-allylation with allyl esters may be conveniently reversed by using molybdenum hexacarbonyl[46].

In functional group chemistry, the importance of silyl enol ethers is highlighted in an exhaustive two-part review[47]. Simple replacement of hydroxyl by halogen can be achieved under mild conditions via 2-alkoxy-1,3,2-diazaphospholanes[48], whereas the reverse reaction proceeds[49] smoothly with $Hg(BF_4)_2$. Oxy-, aza-, and cyano-sulfenylation of simple ethylene derivs[50]., and 1,4-hydrocyanation[51] with tert-butyl isocyanide, are noteworthy, as also are developments in biscarbonylation[52] and applications of phosphinylamines[53] and methoxylamines[54] as electrophilic aminating agents.

[40] J. M. Tedder, Angew. Chem. Intern. Ed. *21,* 401–10 (1982).
[41] R. O. Angus, R. P. Johnson, J. Org. Chem. *48,* 273–6 (1983).
[42] T. J. Brocksom, M. G. Constantino, J. Org. Chem. *47,* 3450–3 (1982).
[43] S. M. Weinreb, R. R. Staib, Tetrahedron *38,* 3087–128 (1982).
[44] B. M. Trost, R. W. Warner, J. Am. Chem. Soc. *104,* 6112–4 (1982).
[45] B. M. Trost, D. M. T. Chan, J. Am. Chem. Soc. *104,* 3733–5 (1982).
[46] B. M. Trost, M. Lautens, J. Am. Chem. Soc. *104,* 5543–5 (1982).
[47] P. Brownbridge, Synthesis *1983,* 1–28, 85–104 (1983).
[48] S. Hanessian, Y. Leblanc, P. Lavallée, Tetrahedron Letters *23,* 4411–4 (1982).
[49] J. Barluenga et al., Synthesis *1983,* 53–5.
[50] Synth. Meth. *37,* 510.
[51] Y. Ito et al., J. Am. Chem. Soc. *104,* 6449–50 (1982).
[52] Synth. Meth. *37,* 827; F. Ozawa et al., Tetrahedron Letters *23,* 3383–6 (1982).
[53] G. Boche et al., Tetrahedron Letters *23,* 5399–402 (1982).
[54] P. Beak, B. J. Kokko, Tetrahedron Letters *24,* 561–4 (1983).

Higher-order organocuprates with stabilizing, solubilizing, and/or activating ligands, e.g. $Li_2[R_2Cu(CN)]$, have advantages over the conventional diorganocuprates[55], although experimentation with the latter is simplified by ultrasonication[56]. An improved preparation of allylmagnesium compounds is also welcome[57].

Among new auxiliaries, borane-diphenylamine[58] is a stable, crystalline hydroborating agent, and NB-enantride[59], a chiral lithium trialkylhydridoborate, is especially useful for the asymmetric reduction of acyclic ketones. A (μ_3-oxo)triiron complex[60] catalyses epoxidation with molecular oxygen, while pyridinium fluorochromate[61] is a less-acidic alternative to pyridinium chlorochromate. A new method for generation of diimide is also reported[62].

Amino groups may be protected as 2,5-dimethylpyrroles[63] which are resistant to strong nucleophiles, bases, and mineral acids. Protection of diols as O,O-silylene derivatives is easier with dialkylsilyl ditriflates than the corresponding dichlorosilanes[64].

A valuable review on DNA synthesis[65] concludes with the prediction that synthetic genes may soon be used to direct economical syntheses of biologically important proteins, and that "this branch of science may find its best application in solving basic problems such as health, food and energy so essential to mankind".

[55] B. H. Lipshutz, R. S. Wilhelm, J. Kozlowski, Tetrahedron Letters *23,* 3755–8 (1982).
[56] J. L. Luche et al., J. Org. Chem. *47,* 3805–6 (1982).
[57] W. Oppolzer et al., Tetrahedron Letters *23,* 3901–4 (1982).
[58] C. Camacho, G. Uribe, R. Contreras, Synthesis *1982,* 1027–30.
[59] M. M. Midland, A. Kazubski, J. Org. Chem. *47,* 2495–6 (1982).
[60] S. Ito, K. Inoue, M. Matsumoto, J. Am. Chem. Soc. *104,* 6450–2 (1982).
[61] M. N. Bhattacharjee et al., Synthesis *1982,* 588–90.
[62] Synth. Meth. *37,* 43.
[63] S. P. Breukelman, G. D. Meakins, M. D. Tirel, J. Chem. Soc. Chem. Commun. *1982,* 800–1.
[64] E. J. Corey, P. B. Hopkins, Tetrahedron Letters *23,* 4871–4 (1982).
[65] S. A. Narang, Tetrahedron *39,* 3–22 (1983).

Systematic Survey Vols 36 and 37
Systematische Übersicht Bd. 36 und 37

Abbreviations and Symbols
Abkürzungen und Symbole

Abbreviations – Abkürzungen

abs.	absolute	absolut
alc.	alcoholic	alkoholisch
aq.	aqueous	wäßrig
ar.	aromatic	aromatisch
atm.	atmosphere(s)	Atmosphäre(n)
cf.	compare	vergleiche
compd(s).	compound(s)	Verbindung(en)
concd.	concentrated	konzentriert
d.	density	Dichte
deriv(s).	derivative(s)	Derivat(e)
dil.	diluted	verdünnt
et al.	and others	und andere
E.	Example	Beispiel
e.g.	for example	zum Beispiel
F. e. s.	Further example(s) see	weitere(s) Beispiel(e) siehe
hr(s).	hour(s)	Stunde(n)
incl.	including	einschließlich
liq.	liquid	flüssig
min.	minutes	Minuten
M	molar	molar
prepn.	preparation	Darstellung
prim.	primary	primär
s.	see	siehe
s*37*	supplementary reference in Volume *37*	Ergänzungszitat in Band *37*
s. a.	see also	siehe auch
satd.	saturated	gesättigt
sec.	secondary	sekundär
soln.	solution	Lösung
startg. m. f.	starting material for (the preparation of...)	Ausgangsmaterial für (die Darstellung von...)
subst.	substituted	substituiert
sym.	symmetrical	symmetrisch
temp.	temperature(s)	Temperatur(en)
tert.	tertiary	tertiär
v.i.	via intermediates	über Zwischenprodukte
w.a.r.	without additional reagents	ohne Hilfsstoffe
Y*	Yield	Ausbeute

* Yields in parentheses refer to the immediately preceding step of a multi-step reaction.
Bei mehrstufigen Synthesen beziehen sich eingeklammerte Ausbeuteangaben auf die unmittelbar vorhergehende Stufe.

Symbols – Symbole	(on the right margin)	(am rechten Rand)
↯	Electrolysis	Elektrolyse
	Irradiation	Bestrahlung
○	Ring closure	Ringschluß
	Ring contraction	Ringverengerung
	Ring expansion	Ringerweiterung
	Ring opening	Ringöffnung
Ⓗ	Ring hydrogenation	Ringhydrierung
←	'see title on the left half of the page'	«siehe Titel auf der linken Hälfte der Seite»

Formation of H–O Bond

Uptake ⇓

Addition to Oxygen and Carbon HO ⇓ OC

Antimony trichloride *$SbCl_3$*

Glycols from oxido compds.  → C(OH)C(OH)

with periodic acid cf. *16,* 1; with $SbCl_3$ s. P.S. Kalsi et al., Indian J. Chem. *19B,* 459–62 (1980).

Rearrangement ∩

Oxygen/Carbon Type HO ∩ OC

Potassium carbonate *K_2CO_3*

2-Ethylenealcohols from oxido compds. 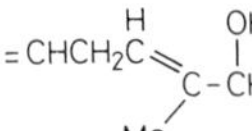 → C(OH)C=C

with Na_2CO_3 cf. *16,* 769; with K_2CO_3, simultaneous migration of double bonds, 2,4,6-trienols, s. E.J. Corey et al., Tetrahedron Letters *22,* 1077–80 (1981).

Titanium(IV) iso-propoxide *Ti(OPr-i)$_4$*

3-Ene-1,2-diols from 2,3-oxidoalcohols
Regio- and geo-specific ring opening

1.

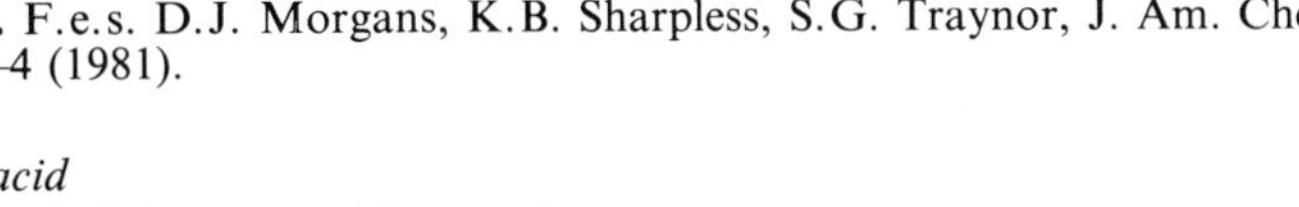

2,3-Epoxynerol treated with 1.2 equivalents Ti(IV) isopropoxide in methylene chloride at room temp. for 12 hrs. → product. Y 70–80%. – **Intramolecular assistance** by the adjacent **hydroxy group** (confirmed by absence of reaction with simple epoxides) introduces interesting new stereochemical aspects **in Lewis acid-catalysed reactions.** F.e.s. D.J. Morgans, K.B. Sharpless, S.G. Traynor, J. Am. Chem. Soc. *103,* 462–4 (1981).

Sulfuric acid *H_2SO_4*

2-Ethylenealcohols from oxido compds.
with *p*-toluenesulfonic acid cf. *16,* 2; with catalytic amounts of H_2SO_4 or methanesulfonic acid, limitations, s. A. van Zon, R. Huis, Rec. Trav. Chim. Pays-Bas *100,* 425–9 (1981).

Exchange ⇅

Remaining Elements ↑ HO⇅Rem

Potassium fluoride/crown ether ←

Protection of hydroxyl groups as *tert*-butyldimethylsilyl ethers OSi≤ → OH
Selective removal of the protective group
with tetra-*n*-butylammonium fluoride cf. *30,* 4; with KF in the presence of crown ethers, ribonucleotides, s. C. Chavis et al., J. Org. Chem. *47,* 202–6 (1982).

Tetra-n-*butylammonium fluoride* n-$Bu_4N^+F^-$

Protection of diols as Si,O-heterocyclics
Removal of the protective group

2. 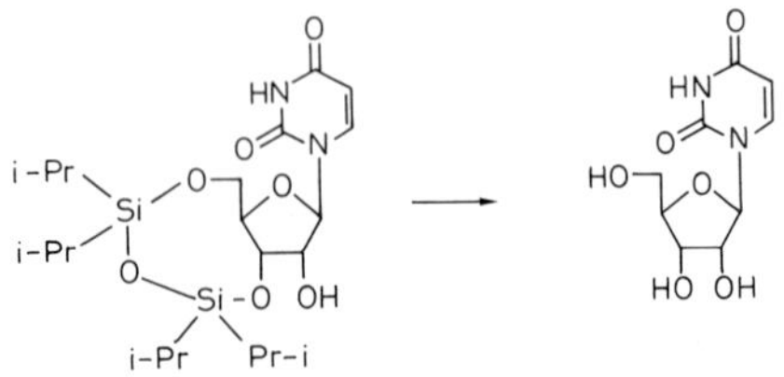

A mixture of startg. protected nucleoside and tetra-*n*-butylammonium fluoride in tetrahydrofuran allowed to react for 10 min. at room temp. → product. Y 100%. F.e. and preferential protection s. W.T. Markiewicz, J. Chem. Res. (M) *1979,* 181–97; with added pyridine hydrochloride s. C.A.A. van Boeckel, J.H. van Boom, Tetrahedron Letters *21,* 3705–8 (1980); protection of glycol groups via **isomerization** with mesitylenesulfonic acid s. ibid. *21,* 1571–74 (1980).

Carbon ↑ HO⇅C

Electrolysis ϟ

Protection of alcohols and amino groups COOAr → OH
as 4-ethoxy-1-naphthyl carbonates and carbamates, respectively
Anodic removal of the protective group

3. 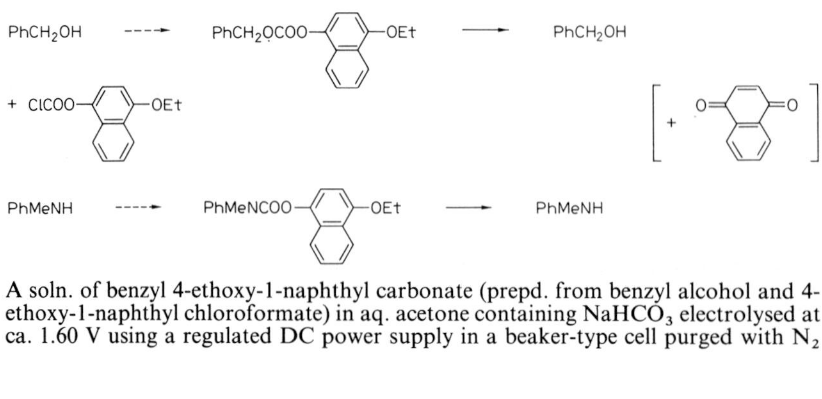

A soln. of benzyl 4-ethoxy-1-naphthyl carbonate (prepd. from benzyl alcohol and 4-ethoxy-1-naphthyl chloroformate) in aq. acetone containing $NaHCO_3$ electrolysed at ca. 1.60 V using a regulated DC power supply in a beaker-type cell purged with N_2

with a carbon felt anode and platinum wire cathode isolated in a fritted tube → benzyl alcohol. Y 86% by g.l.c. – Similarly, N-benzyl-N-methylcarbamate analog in the presence of $HClO_4$ → N-benzylmethylamine. Y 80% (by g.l.c. as the trifluoroacetamide). F.e.s. R.W. Johnson, E.R. Grover, L.J. MacPherson, Tetrahedron Letters *22,* 3719–20 (1981).

Lithium/ethylenediamine *$Li/NH_2CH_2CH_2NH_2$*

Hindered carboxylic acids from their esters COOR → COOH

with *n*-propyllithium cf. *36,* 6; with Li/ethylenediamine s. P. Sengupta, M. Sen, S. Das, Indian J. Chem. *19B,* 721–2 (1980).

Potassium hydroxide *KOH*

***a*-Hydroxycarboxylic acids from 1,3-dioxolan-4-ones** C

s. *37,* 791.

Potassium hydroxide/alumina *KOH/Al_2O_3*

Carboxylic acids from their esters

with "anhydrous hydroxide" cf. *15,* 6s*32;* with intimately ground KOH/Al_2O_3, "basic alumina", s. S.L. Regen, A.K. Mehrotra, Synth. Commun. *11,* 413–7 (1981).

Potassium tert-*butoxide* *KOBu*-t

O-Deacylation OAc → OH

s. *37,* 607.

Lithium tert-*butylmercaptide* *$LiSC(CH_3)_3$*

Preferential demethylation of phenolethers ArOMe → ArOH

with NaCN in dimethyl sulfoxide cf. *34,* 405; also with Li *tert*-butylmercaptide s. F.S. El-Feraly, Y.-M. Chan, J. Natur. Prod. (Lloydia) *44,* 557–61 (1981).

Sodium iodide s. under BF_3, BBr_3, and $MeSiCl_3$ *NaI*

Ammonia *NH_3*

Selective O-deacetylation of carbohydrate derivs. OAc → OH

with NH_3-satd. methanol cf. *28,* 483; with aq. NH_3 in pyridine, 3′-O-deacetylation of deoxyribonucleotides, s. S.C. Srivastava, Bio-Org. Chem. *10,* 161–8 (1981).

Triethylamine *Et_3N*

Protection of hydroxyl groups
as 2-(benzenesulfonyl)ethyl carbonates $OCOOCH_2CH_2SO_2Ph$ → OH
Selective removal of the protective group under mild conditions

4.

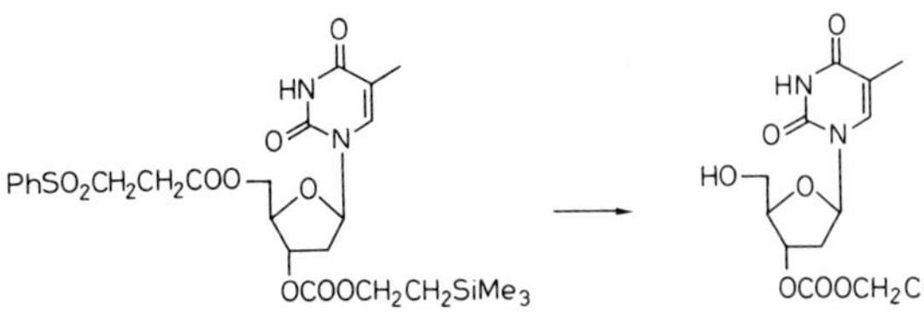

Startg. protected alcohol treated with 15 equivalents of triethylamine in pyridine, and the mixture allowed to stand at 20° for 20 hrs. → product. Y 99%. – Other base-labile protective groups remained intact. F.e. (98–85%) s. N. Balgobin, S. Josephson, J.B. Chattopadhyaya, Tetrahedron Letters *22,* 3667–70 (1981); protection **as 2-(methylthiomethoxy)ethyl carbonates** cf. S.S. Jones, C.B. Reese, S. Sibanda, ibid. *22,* 1933–6 (1981); as **2-(triphenylphosphonio)ethyl carbonates** cf. H. Kunz, H.-H. Bechtolsheimer, Synthesis *1982,* 303–4.

Triethylamine *Et_3N*

Protection of phosphoryl hydroxy groups as 2-(benzenesulfonyl)ethyl esters Removal of the protective group P–$OCH_2CH_2SO_2Ph$ → P–OH

5. 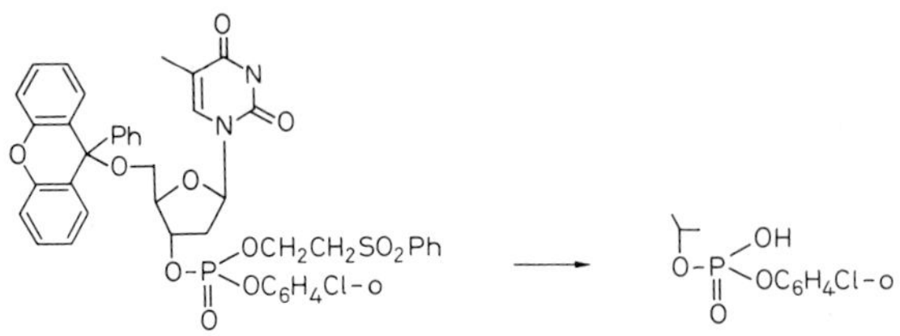

Startg. nucleotide in pyridine treated with 2 equivalents of triethylamine, and the mixture kept at 20° for 3 hrs. → product. Y ca. 100%. – The protecting group is stable under oligonucleotide coupling conditions. F.e.s. N. Balgobin, S. Josephson, J.B. Chattopadhyaya, Tetrahedron Letters *22,* 1915–8 (1981); **as 2-(diphenylmethylsilyl)ethyl esters** cf. S. Honda, T. Hata, ibid. *22,* 2093–6 (1981); **as *o*-bromophenyl esters** cf. Y. Stabinsky, R.T. Sakata, M.H. Caruthers, ibid. *23,* 275–8 (1982).

Ethylenediamine s. under Li *$NH_2CH_2CH_2NH_2$*

Cupric chloride *$CuCl_2$*

Preferential cleavage of O,O-isopropylidene derivs. (CO, CO)$C(CH_3)_2$ → (C–OH, C–OH)

with H_2SO_4 cf. *5,* 12; under mild, neutral conditions with $CuCl_2$ s. M. Iwata, H. Ohrui, Bull. Chem. Soc. Japan *54,* 2837–8 (1981).

Zinc/acetic acid *Zn/MeCOOH*

Protection of carboxyl groups as *p*-methoxyphenacyl esters Removal of the protective group COO–C–COR → COOH

s. *29,* 2s*33;* also photolytic deprotection s. F.S. Tjoeng, G.A. Heavner, Synthesis *1981,* 897–9.

Lithium tetrahydridoaluminate/titanocene dichloride *$LiAlH_4/(C_5H_5)_2TiCl_2$*

Cleavage of allyl ethers $OCH_2CH{=}CH_2$ → OH

s. *37,* 74

Boron fluoride/sodium iodide *BF_3/NaI*

Catechol monoesters from 2-alkoxy-1,3-benzodioxoles

6. 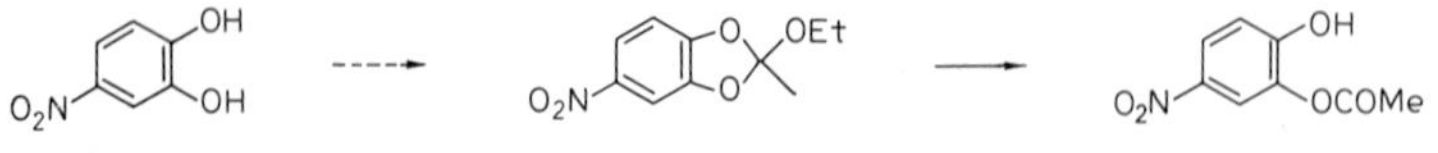

A soln. of 2-ethoxy-2-methyl-5-nitro-1,3-benzodioxole in methylene chloride at 0° treated with a soln. of NaI in acetone and BF_3-etherate under N_2, and the mixture kept at the same temp. for 5 min. → 4-nitrocatechol 2-monoacetate. Y 97%. – This reaction is useful as part of a 2-step **mono-O-acetylation of catechols.** F.e.s. F. Vellaccio et al., J. Org. Chem. *46,* 3087–91 (1981).

Boron bromide/sodium iodide/15-crown-5 polyether ←

Cleavage of ethers under mild conditions OMe → OH

of aryl methyl ethers *cf.* 19, 8; of aliphatic methyl ethers with BBr_3/NaH/15-crown-5 polyether s. H. Niwa, T. Hida, K. Yamada, Tetrahedron Letters *22,* 4239–40 (1981).

Enzymes ←

Carboxylic acids from their esters COOR → COOH

Enzymatic resolution

s. *28,* 13; asym. partial hydrolysis with pig liver esterase s. S. Iriuchijima, K. Hasegawa, G. Tsuchihashi, Agr. Biol. Chem. *46,* 1907–10 (1982); s.a. T. Izawa et al., Tennen Yuki Kagobutsu Toronkai Koen Yoshishu, 24th *1981,* 606 (Japan); Chem. Abstr. *96,* 181030 (1982).

15-Crown-5 polyether s. under BBr_3 ←

Ammonium formate s. under $(MeCOO)_2Pd$ $HCOONH_4$

Tributylhexadecylphosphonium bromide s. under H_2SO_4 $[C_{16}H_{33}PBu_3]^+Br^-$

Acetic acid s. under Zn *MeCOOH*

2,3-Dichloro-5,6-dicyanoquinone ←

Protection of hydroxyl groups as *p*-methoxybenzyl ethers $OCH_2Ar \rightarrow OH$

Selective removal of the protective group

7.

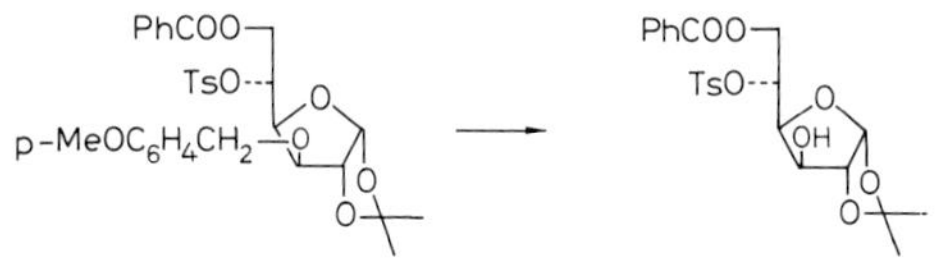

A soln. of startg. m. in 18:1 dichloromethane/water treated with 1.5 equivalents of 2,3-dichloro-5,6-dicyanoquinone at room temp. for 2 hrs. → product. Y 86%. – **Under** these **oxidative conditions** a wide range of protecting and functional groups remained unaffected, indicating more general applicability than the standard hydrogenolytic removal. F.e. and removal of 3,4-dimethoxybenzyl groups s. Y. Oikawa, T. Yoshioka, O. Yonemitsu, Tetrahedron Letters *23,* 885–8 (1982); **preferential protection** of nucleoside 2′-hydroxyl groups, and removal of the protective group with triphenylcarbonium fluoroborate, cf. H. Takaku, K. Kamaike, Chem. Letters *1982,* 189–92.

Trichloro(methyl)silane/sodium iodide $MeSiCl_3/NaI$

Cleavage of ethers OR → OH

8.

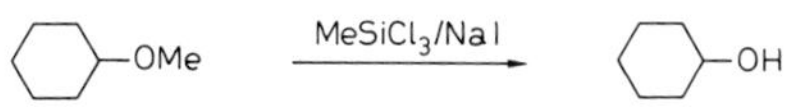

Trichloromethylsilane and cyclohexyl methyl ether added successively with continuous stirring to a soln. of NaI in acetonitrile under N_2, and stirring continued for 7 hrs. → cyclohexanol. Y 81%. – This new reagent system leaves alcohol and ester groups unaffected, and is thereby free of some of the difficulties encountered with trimethyliodosilane. F.e., incl. cleavage of benzyl, trityl, and tetrahydropyran-2-yl ethers, s. G.A. Olah et al., Angew. Chem. Intern. Ed. *20,* 690–1 (1981).

Titanocene dichloride s. under $LiAlH_4$ $(C_5H_5)_2TiCl_2$

Hydrazine H_2NNH_2

Protection of alcohol groups

as γ-ketocarboxylic acid esters $OCO(CH_2)_2COMe \rightarrow OH$

cf. *23,* 17s*32;* preferential protection, and selective removal of levulinoyl groups, s. Rec. Trav. Chim. Pays-Bas *100,* 65–72 (1981).

Diphosphorus tetraiodide or phosphorus triiodide P_2I_4, PI_3

Cleavage of acetals in non-aq. medium $C(OR)_2 \rightarrow CO$

9. $C_9H_{19}C(OMe)_2Me \rightarrow C_9H_{19}COMe$

A mixture of startg. m. and P_2I_4 in dichloromethane stirred at 20° for 12 min. → product. Y 91%. F.e., also with phosphorus triiodide, s. J.-N. Denis, A. Krief, Angew. Chem. Intern. Ed. *19,* 1006–7 (1980).

Methanesulfonic acid $MeSO_3H$

Removal of the Merrifield polymer support ←

with HF cf. *13,* 43s*24;* with $MeSO_3H$ in trifluoroacetic acid s. J.W. van Nispen et al., Rec. Trav. Chim. Pays-Bas *100,* 435–6 (1981).

Sulfuric acid/tributylhexadecylphosphonium bromide $H_2SO_4/[C_{16}H_{33}PBu_3]^+Br^-$

Carboxylic acids from carboxylic acid *tert*-butyl-esters COOBu-*t* → COOH

Selective and preferential hydrolysis in a 2-phase medium

10. $CH_3(CH_2)_6COOBu\text{-}t \rightarrow CH_3(CH_2)_6COOH$

A mixture of *tert*-butyl octanoate, aq. H_2SO_4, and a little tributylhexadecylphosphonium bromide stirred *at room temp.* for 1 min. → octanoic acid. Y 93% (94% conversion). – Alcohols, ethers, alkenes, and nitriles were unaffected, while phenyl and *prim*-alkyl carboxylates were hydrolysed only on heating at 110°. F.e. and with HBr s. D. Landini, F. Rolla, J. Org. Chem. *47,* 154–7 (1982).

Pyridine hydrochloride $C_5H_5N.HCl$

Carboxylic acids from their esters COOMe → COOH

s. *16,* 13s*32;* selective cleavage of *o*-subst. methyl esters s. I.G.C. Coutts, M. Edwards, D.J. Richards, Synthesis *1981,* 487–9.

Potassium pentacyanocobaltate(II) $K_3[Co(CN)_5]$

Carboxylic acids from benzyl esters $COOCH_2Ph \rightarrow COOH$

s. *37,*18

Palladium–carbon *Pd–C*

Protection of carboxyl as 2,4,6-trimethylbenzyl esters $COOCH_2C_6H_2(CH_3)_3 \rightarrow COOH$

Removal of protective groups from peptides

removal with trifluoroacetic acid s. *23,* 11; with both acid cleavage and hydrogenolysis with 10% palladium-on-charcoal s. D.E. Rivett, F.H.C. Stewart, Australian J. Chem. *33,* 2693–7 (1980).

Palladium(II) acetate/ammonium formate $(MeCOO)_2Pd/HCOONH_4$

Transfer hydrogenolysis $COOCH_2Ar \rightarrow COOH$

Removal of the polymer support in Merrifield peptide synthesis

s. *7,* 62s*36;* with Pd acetate and ammonium formate as H-donor cf. M.K. Anwer, A.F. Spatola, Tetrahedron Letters *22,* 4369–72 (1981).

Tetrakis(triphenylphosphine)palladium(0)/triphenylphosphine/potassium carboxylate $(Ph_3P)_4Pd/Ph_3P/RCOOK$

Carboxylic acids from allyl esters $COOCH_2CH{=}CH_2 \rightarrow COOH$

with $PdCl_2(Ph_3P)_2/HCOONH_4$ cf. *34,* 55; with tetrakis(triphenylphosphine)palladium(0) and triphenylphosphine in the presence of K 2-ethylhexanoate s. P.D. Jeffrey, S.W. McCombie, J. Org. Chem. *47,* 587–90 (1982).

Formation of H–N Bond

Uptake ⇓

Addition to Nitrogen and Sulfur HN⇓NS

Arylmercaptans *ArSH*

Diaminosulfanes from sulfinodiimines N=S=N → NH–S–NH

11. $PhSO_2N{=}S{=}NC_6H_4NO_2\text{-}p \xrightarrow{PhSH} PhSO_2NH\text{-}S\text{-}NHC_6H_4NO_2\text{-}p$

A soln. of thiophenol in benzene added slowly to a stirred soln. of the startg. sulfinodiimine in the same solvent, and stirring continued at room temp. until decolourized (2–4 hrs.) → product. Y 78%. F.e. s. E.S. Levchenko, N.P. Pelkis, Zh. Org. Khim. *18,* 453–4 (1982).

Exchange ⇂↾

Oxygen ↑ HN⇂↾O

Lithium/liq. ammonia/methanol *$Li/NH_3/CH_3OH$*

Amines from nitro compds. $NO_2 \rightarrow NH_2$

with Na in liq. NH_3 cf. *1,* 23s*28;* with Li in liq. NH_3/methanol s. A.P. Krapcho, T.A. Collins, Synth. Commun. *12,* 293–8 (1982).

Sodium formate/potassium dihydrogen phosphate *$HCOONa/KH_2PO_4$*

with Pd–C/triethylammonium formate cf. *33,* 76s*36;* non-catalytic reduction with formic acid salts in N-methyl-2-pyrrolidone s. J.H. Babler, S.J. Sarussi, Synth. Commun. *11,* 925–30 (1981).

Magnesium s. under $TiCl_4$ *Mg*

Calcium chloride s. under Fe *$CaCl_2$*

Sodium tetrahydridoborate s. under $NiCl_2$ *$NaBH_4$*

Sodium tetrahydridoborate/stannous chloride *$NaBH_4/SnCl_2$*

Selective reduction

12. $p\text{-}MeCOOC_6H_4\text{-}NO_2 \rightarrow p\text{-}MeCOOC_6H_4\text{-}NH_2$

A soln. of *p*-nitrophenyl acetate and 5 equivalents of $SnCl_2.2H_2O$ in 9:1 ethyl acetate/*tert*-butanol heated at 60° for 1 hr., 0.5 equivalents $NaBH_4$ added portionwise with stirring at 60°, and stirring continued for a further 2 hrs. at the same temp. → *p*-aminophenyl acetate. Y 96%. – Most other functional groups (e.g. nitriles, ketones, esters, oximes, and olefinic double bonds) were unaffected, F.e. s. T. Satoh et al., Chem. Pharm. Bull. *29,* 1443–5 (1981).

Methanol s. under Li — *CH_3OH*

Titanium tetrachloride/magnesium — *$TiCl_4/Mg$*
Hydrazines from N-nitrosamines — N–NO → $N–NH_2$

13. $Ph_2N–N{=}O \rightarrow Ph_2N–NH_2$

A soln. prepared from $TiCl_4$ and Mg powder in 4:1 methylene chloride-ether under inert gas added with stirring to a soln. of N-nitrosodiphenylamine in ether at room temp., after 30 min. aq. HCl added, and stirring continued for 1 hr. → N,N-diphenylhydrazine. Y 95%. – The system is very selective and leaves a range of functional groups intact. F.e.s. I.D. Entwistle, R.A.W. Johnstone, A.H. Wilby, Tetrahedron *38,* 419–23 (1982).

Hydrazine s. under Ni — *H_2NNH_2*

Iron/calcium chloride — *$Fe/CaCl_2$*
Amines from nitro compds. — $NO_2 \rightarrow NH_2$

14.

3-Aminopyridines. A mixture of 2-methoxy-5-nitropyridine and Fe dust in 85% ethanol refluxed for 2 hrs. in the presence of $CaCl_2$ → 3-amino-6-methoxypyridine. Y 97%. – This method is cheaper, more convenient, less hazardous, and easier in the work-up stage than methods described previously; intermediate nitroso and hydroxylamine derivs. are also obtainable when limited amounts of $CaCl_2$ are used. F.e.s. C.O. Okafor, M.L. Steenberg, J.P. Buckley, Chem. Pharm. Bull. *30,* 302–18 (1982).

Lithium cobalt(I) phthalocyanine ←
Selective reduction under mild conditions

15.

A soln. of lithium phthalocyaninecobalt(I) tetrahydrofuran complex in methanol treated under N_2 with *o*-nitrobenzaldehyde, and the mixture stirred at 20–25° for 90 hrs. → *o*-aminobenzaldehyde. Y 96%. – Halogeno-, cyano-, isocyano-, keto-, and aldehydo-groups, as well as carboxylic acid esters, acids, arenes, and carbon-carbon double bonds, were left intact. F.e., **also one-pot Friedländer quinoline ring synthesis,** s. H. Eckert, Angew. Chem. Intern. Ed. *20,* 208–10 (1981).

Bis(dimethylglyoximato)cobalt(II) s. under PtO_2 ←

Nickel/hydrazine — *Ni/H_2NNH_2*
s. *27,* 15s*30;* selective reduction s. F. Yuste, M. Saldaña, F. Walls, Tetrahedron Letters *23,* 147–8 (1982); solvent effect s. N.R. Ayyanger et al., Synthesis *1981,* 640–1.

Nickel chloride/sodium tetrahydridoborate *$NiCl_2/NaBH_4$*

Amines from nitro and nitroso compds. NO → NH_2
NO_2 → NH_2

with Pd–C/$NaBH_4$ cf. *17,* 28; with $NiCl_2$/$NaBH_4$, also from azoxy compds., azo compds., and hydroxylamines, s. A. Nose, T. Kudo, Chem. Pharm. Bull. *29,* 1159–61 (1981).

Palladium–graphite *Pd–C*

Amines from nitro compds. NO_2 → NH_2
with Pd–C cf. *2,* 49; with Pd–graphite s. D. Savoia et al., J. Chem. Soc. Chem. Commun. *1981,* 540–1.

Platinum oxide/bis(dimethylglyoximato)cobalt(II) ←
with Adams catalyst cf. *27,* 15; activation with e.g. bis(dimethylglyoximato)cobalt(II), 20 to 60 fold increase in rate of hydrogenation of aliphatic nitro compds., s. A.L. Fedorov et al., Zh. Org. Khim. *18,* 1335–6 (1982).

Nitrogen ↑ HN↓↑N

Reductions with
titanium tetrachloride/sodium tetrahydridoborate *$TiCl_4/NaBH_4$*

$>$N–NO → $>$NH

16. PhMeN–NO → PhMeNH

Sec. amines from N-nitrosamines. A soln. of startg. nitrosamine in dimethoxyethane added gradually to a stirred suspension of $TiCl_4$ and $NaBH_4$ in the same solvent at room temp., and stirring continued for 14 hrs. → N-methylaniline. Y 92%. F.e., also with $NiCl_2$, s. S. Kano et al. Synthesis *1980,* 741–2; reduction of carboxylic acids and amides, and S-deoxygenation of sulfoxides s. ibid. *1980,* 95–7; regiospecific hydration of ethylene derivs. s. J. Chem. Soc. Chem. Commun. *1980,* 414–5.

Halogen ↑ HN ↓↑ Hal

Triazenes *RN=N–NHR′*

Replacement of N-halogen by N-hydrogen N–Hal → NH
with Na metabisulfite cf. *27,* 548; with triazenes, sulfonamides, s. O.O. Orazi et al., Tetrahedron Letters *23,* 293–4 (1982).

Sulfur ↑ HN↓↑S

Trifluoroacetic acid *CF_3COOH*

Prim. amines from sulfinamides RNHS(O)R′ → RNH_2
Optically active compds.
s. *37,* 35

Thioanisole/sulfonic acids *$C_6H_5SCH_3/RSO_3H$*

4-Methoxy-2,3,6-trimethylbenzenesulfonyl as N-protective group NSO_2Ar → NH
for guanidines cf. *9,* 42s*36;* for amines, removal with thioanisole/$MeSO_3H$, s. J. Chem. Soc. Chem. Commun. *1982,* 445–6.

Remaining Elements ↑ HN⇅Rem

Hydrogen chloride *HCl*

Amines from N-bromophosphoromonoamidates $NPO(OR)_2 \rightarrow NH$
s. *37,* 447

Carbon ↑ HN⇅C

Without additional reagents *w.a.r.*

Protection of amino groups as halogenocarboxylic acid amides
Removal of the protective group $NCO(CH_2)_3Cl \rightarrow NH$
from chloracetamides cf. *24,* 34; from 4-chlorobutyramides s. Z. Mandić et al., Croat. Chem. Acta *54,* 127–39 (1981).

Electrolysis ϟ

Protection of amino groups as 4-ethoxy-1-naphthyl carbamates $NCOOAr \rightarrow NH$
Anodic removal of the protective group
s. *37,* 3

Sodium hydroxide s. under H_2O_2 *NaOH*

Potassium hydroxide *KOH*

Cyclic sec. amines from N,N-disubst. formamidines $NCH{=}NR \rightarrow NH$
s. *37,* 784

Triethylamine s. under Pd Et_3N

Sodium hydrogen carbonate $NaHCO_3$

N-Decarbovinyloxylation $NCOOCH{=}CH_2 \rightarrow NH$
s. *37,* 408

Cadmium *Cd*

Removal of 2,2,2-trichloroethoxycarbonyl protective groups $N\text{–}COOCH_2CCl_3 \rightarrow NH$
s. *24,* 9s*33;* selective removal with cadmium dust in 1:1 acetic acid/dimethylformamide cf. G. Hancock, I.J. Galpin, B.A. Morgan, Tetrahedron Letters *23,* 249–52 (1982).

Zinc bromide $ZnBr_2$

Selective N-deacylation $NAc \rightarrow NH$
with BF_3 cf. *30,* 120; with $ZnBr_2$, nucleosides, s. R. Kierzek et al., Tetrahedron Letters *22,* 3761–4 (1981).

Ammonium formate s. under $(MeCOO)_2Pd$ $HCOONH_4$

Enzyme ←

Protection of amino groups as acetamides
Removal of the protective group
via thioacylamines cf. *24,* 40; protection as phenylacetamides, enzymatic removal, s. F. Brtník, T. Barth, K. Jŏst, Collection Czech. Chem. Commun. *46,* 1983–9 (1981).

syn-2-Nitrobenzaldoxime/1,1,3,3-tetramethylguanidine ←

Protection of 4-pyrimidinyl hydroxyl groups in nucleosides N=COAr → NH–CO
Removal of O-aryl protective groups

17. 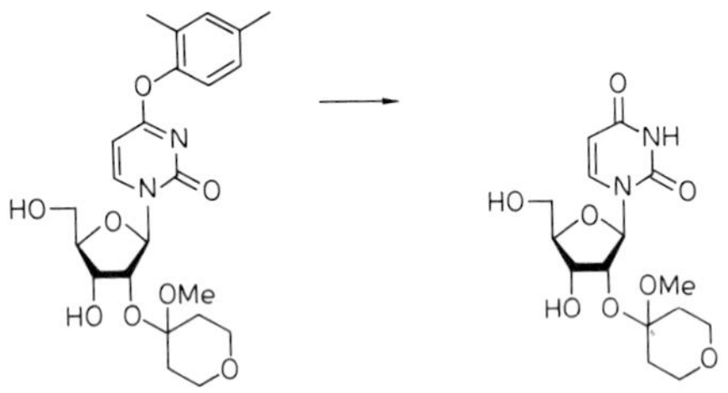

A 0.05 *M* soln. of startg. m. in anhydrous dioxane treated with 10 equivalents of *syn*-2-nitrobenzaldoxime and 9 equivalents of tetramethylguanidine at room temp. for 120 min. → 2′-O-(4-methoxytetrahydro-4-pyranyl)uridine. Y 83%. F.e. with O-phenyl- and O-(2-nitrophenyl)-derivs. s. S.S. Jones et al., Tetrahedron Letters *22,* 4755–8 (1981); **protection of 6-purinyl hydroxyl groups** as β-subst. ethyl derivs. cf. B.L. Gaffney, R.A. Jones, ibid. *23,* 2257–60 (1982); as dibutylphosphinothioyl derivs. cf. M. Sekine et al., J. Org. Chem. *47,* 571–3 (1982); Bull. Chem. Soc. Japan *54,* 3076–83 (1981).

Carboxylic acids s. under $(Ph_3P)_4Pd$ *RCOOH*

Hydrogen peroxide/sodium hydroxide $H_2O_2/NaOH$

N-Deformylation NCHO → NH
with aq. H_2O_2 cf. *16,* 27; in the presence of NaOH in aq. alcohol s. S. Kodato, S. Sugasawa, Proc. Jpn. Acad., Ser. B. *56,* 576 (1980) (Eng.); Chem. Abstr. *94,* 174542 (1980).

Hydrogen chloride *HCl*

s. *20,* 311; with alcohol as co-solvent, α-amino-α,β-ethylenecarboxylic acid esters, s. T. Moriya, K. Matsumoto, M. Miyoshi, Synthesis *1981,* 915–7.

Prim. amines from iminoesters RN=C(OR′)Ar → RNH_2
s. *37,* 792

Hydrogen bromide *HBr*

N-Deformylation NCHO → NH
s. *37,* 378

Potassium pentacyanocobaltate(II) $K_3[Co(CN)_5]$

Selective N-decarbobenzoxylation and O-debenzylation $NCOOCH_2Ph$ → NH
Homogeneous hydrogenolysis

18. 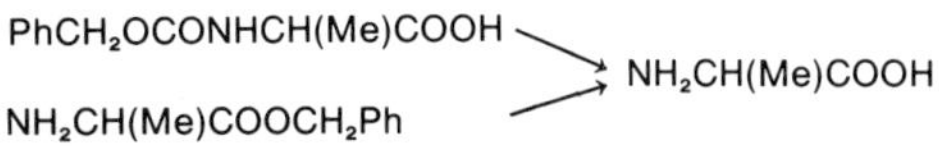

A soln. of benzyloxycarbonylalanine in methanol added to a little $K_3[Co(CN)_5]$ (prepd. *in situ* from $CoCl_2$ and KCN under H_2) in the same solvent, and hydrogenated for 3 hrs. at 20° → alanine. Y 94%. – Similarly, benzyl alaninate → alanine. Y 92%. – Benzyl ether groups remained intact. F.e.s. G. Losse, H.-U. Stiehl, Z. Chem. *21,* 188 (1981); N-decarbobenzoxylation with palladous acetate/ammonium formate cf. M.K. Anwer, A.F. Spatola, Tetrahedron Letters, *22,* 4369–72 (1981).

Cobalt(I) phthalocyanine ←

Protection of amino groups N–COO–CMe_2CCl_3 → NH
...as carbo-2,2,2-trichloro-*tert*-butoxyamines
Removal of the protective groups

s. *24*, 9s*33*; N^6-protected adenosine s. R.G.K. Schneiderwind, I. Ugi, Z. Naturforsch. *36B*, 1173-5 (1981).

Chlorotris(triphenylphosphine)rhodium(I) $(Ph_3P)_3RhCl$

...as diallylamines $N(CH_2CH{=}CH_2)_2$ → NH_2
Removal of the protective group

19.

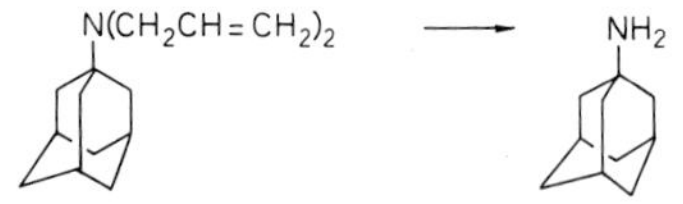

A soln. of 1-adamantyldiallylamine in 84:16 acetonitrile/water containing a little chlorotris(triphenylphosphine)rhodium(I) refluxed under N_2 for 2 hrs. with continuous removal of solvent → 1-adamantylamine. Y 65%. – The protecting group is resistant to acid, base, and nucleophiles. F.e.s. B.C. Laguzza, B. Ganem, Tetrahedron Letters *22*, 1483–6 (1981).

Palladium/triethylamine Pd/Et_3N

N-Decarbobenzoxylation $NCOOCH_2C_6H_5$ → NH

cf. *15*, 146s*31*; with triethylamine and N,N-dimethylacetamide as additives, sulfur-containing peptides, s. Org. Synth. *59*, 159–69 (1979).

Palladium(II) acetate/ammonium formate $(MeCOO)_2Pd/HCOONH_4$

s. *37*, 18

Tetrakis(triphenylphosphine)palladium(0)/ triphenylphosphine/carboxylic acid $(Ph_3P)_4Pd/Ph_3P/RCOOH$

N- and O-Decarballyloxylation $NCOOCH_2CH{=}CH_2$ → NH

with $Ni(CO)_4$ cf. *29*, 28; with $(Ph_3P)_4Pd/PPh_3$ in the presence of carboxylic acid, selectivity and limitations, s. P.D. Jeffrey, S.W. McCombie, J. Org. Chem. *47*, 587–90 (1982).

Via intermediates *v.i.*

N-Demethylation via urethans NMe → NH

with 2,2,2-trichloroethyl chloroformate s. *23*, 479s*27*; limitations with steroidal dimethylamines, cf. H. Kapnang and G. Charles, Tetrahedron Letters *21*, 2951–54 (1980).

Protection of amino groups NCOOR → NH
as carbo-(1,3-dithian-2-yl)methoxyamines
Removal of the protective group via S-oxidation

20.

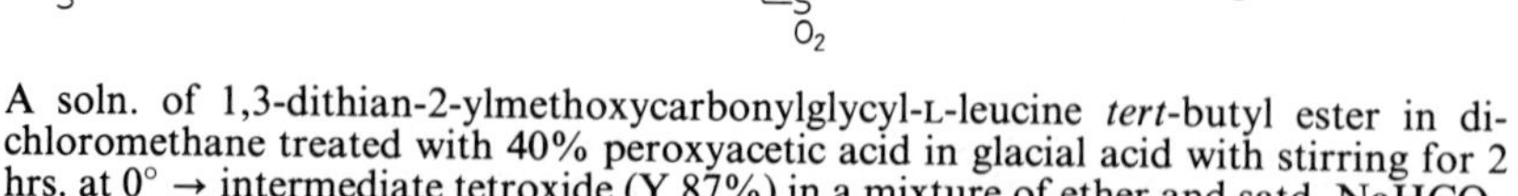

A soln. of 1,3-dithian-2-ylmethoxycarbonylglycyl-L-leucine *tert*-butyl ester in dichloromethane treated with 40% peroxyacetic acid in glacial acid with stirring for 2 hrs. at 0° → intermediate tetroxide (Y 87%) in a mixture of ether and satd. $NaHCO_3$ soln. stirred at room temp. for ca. 1 hr. → glycyl-L-leucine *tert*-butyl ester (Y 85%). – The protecting group is resistant to bases and trifluoroacetic acid. F.e.s. H. Kunz, R. Barthels, Chem. Ber. *115*, 833–45 (1982).

Elimination ⇑

Nitrogen ↑ HN⇑N

Isoamyl nitrite/lithium acetate *RONO/MeCOOLi*

Amines from hydrazines N–NH_2 → NH

21. PhMeN–NH_2 → PhMeNH

A soln. of N-methyl-N-phenylhydrazine and isoamyl nitrite in benzene containing acetic acid and Li acetate frozen in a dry ice/2-propanol bath at 0.5–1.0 mm., heated for 20 hrs. at 50°, and the mixture stirred with Na_2CO_3 for 30–60 min. → N-methylaniline. Y 90%. F.e.s. M. De Rosa, P. Haberfield, J. Org. Chem. *46,* 2639–43 (1981).

Formation of H–S Bond

Uptake ⇓

Addition to Nitrogen and Sulfur HS⇓NS

Thiophenol *PhSH*

***o*-Mercaptoamidines from 3-amino-1,2-benzisothiazoles** C

22.

A soln. of 3-amino-1,2-benzisothiazole in toluene treated with a soln. of thiophenol in the same solvent with stirring, and heated at 80° for 30 min. → 2-mercaptobenzamidine. Y 84%. F.e.s. H. Böshagen, W. Geiger, Ann. Chem. *1982,* 14–27.

Addition to Sulfur and Sulfur HS⇓SS

Sodium/liq. ammonia *Na/NH_3*

Mercaptosulfinic acids from thiolsultones C

23.

A soln. of dibenzo[*c,e*][1,2]dithiin S,S-dioxide in tetrahydrofuran added to liq. NH_3 cooled in a dry-ice/acetone bath, the mixture stirred, Na added in small pieces during 1 hr., and stirring continued for an additional 45 min. → disodium 2-mercapto-2′-biphenylsulfinate. Y 98%. F.e.s. L. Field, V. Eswarakrishnan, J. Org. Chem. *46,* 2025–9 (1981).

Exchange ⇅

Carbon ↑ HS⇅C

Sodium *Na*

1,3-Oxathiole ring opening ⊂
with NaOH cf. *9,* 55; *o*-mercaptophenols from 1,3-benzoxathioles with Na in pyridine s. A.M. Bernard et al., J. Heterocyc. Chem. *18,* 639–40 (1981).

Hydrazine H_2NNH_2

Thiazole ring opening ⊂
with NaOH cf. *2,* 608, *18,* 262; *o*-aminomercaptans from benzothiazoles with 85% hydrazine hydrate s. M.R. Chedekel, D.E. Sharp, G.A. Jeffery, Synth. Commun. *10,* 167–73 (1980).

Mercuric acetate/formic acid $(MeCOO)_2Hg/HCOOH$

Acetamidomethyl as S-protective group SR → SH
s. *21,* 60s27; protection procedure s. Org. Synth. *59,* 190–5 (1979).

Formation of H-Rem Bond

Exchange ⇅

Halogen ↑ HRem⇅Hal

Lithium tetrahydridoborate $LiBH_4$

Sec. boranes from halogenoboranes ⪖BHal → ⪖BH

24.

A soln. of $LiBH_4$ in tetrahydrofuran added to B-chloro-9-borabicyclo[3.3.1]nonane in the same solvent at 0°, and analysed by n.m.r. after 15 min. → 9-borabicyclo[3.3.1]nonane. Y 100%. F.e. and reducing agents s. H.C. Brown, S.U. Kulkarni, J. Organometal. Chem. *218,* 299–307 (1981).

*Tri-*n*-butyltin hydride* n-Bu_3SnH

Replacement of chlorine by hydrogen on phosphorus(III) ⪖PCl → ⪖PH
Cyclic hypophosphorus acid derivs.

25.

5,5-Dimethyl-2-chloro-1,3,2-dioxaphosphorinane treated with tri-*n*-butyltin hydride, and the mixture heated for 4 hrs. at 100° → product. Y 75%. F.e.s. E.E. Nifantiev et al., Tetrahedron *37,* 3183–94 (1981); Zh. Obstrch. Khim. *51,* 2379 (1981).

Formation of H–C Bond

Uptake ⇓

Addition to Oxygen and Carbon HC⇓OC

Lithium *Li*

Phenols from dibenzofurans ⊂
s. *9,* 59; regiospecificity s. T. Keumi et al., Nippon Kagaku Kaishi 1981 (*2*), 259 (Japan); Chem. Abstr. *95,* 42537 (1981).

Lithium/ethylamine *$Li/EtNH_2$*

Alcohols from cyclic ethers
with Na in liq. NH_3 cf. *27,* 37s*28;* regio- and stereo-specific ring opening with Li/ethylamine, 2- and 3-ethylenealcohols, s. T. Kobayashi, H. Tsuruta, Synthesis *1980,* 492–3.

Potassium–graphite *KC_8*

Graphite insertion compds. as reagents ←
Reviews
s. *28,* 29s*32;* graphite intercalation compds. as reagents in organic synthesis s.a. R. Setton, F. Beguin, S. Piroelle, Synthetic Metals *4,* 299 (1982); Chem. Abstr. *97,* 5404 (1982).

Potassium hydroxide s. under Zn *KOH*

n-*Butyllithium s. under 2,5-Dihydroanisole* n-*BuLi*

Sodium bromide s. under Ni *NaBr*

Ethylamine s. under Li *$EtNH_2$*

tert-*Butylamine s. under $Li[AlH(OR)_3]$* t-*$BuNH_2$*

1,4-Dihydropyridines/zinc iodide ←
Alcohols from oxo compds. CO → CH_2

26.

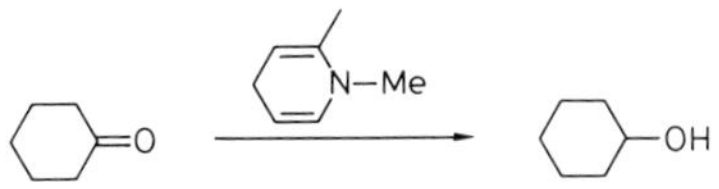

Sec. alcohols. 1,2-Dimethyl-1,4-dihydropyridine and cyclohexanone added to a stirred suspension of 2 equivalents ZnI_2 in *cyclohexane* at 20°, and worked up after 24 hrs. → cyclohexanol. Y 88%. F.e., reagents, and conditions s. A.J. de Koning et al., Rec. Trav. Chim. Pays-Bas *99,* 316–21 (1980); with *polymer-based 1,4-dihydropyridines* cf. R. Mathis et al., Tetrahedron Letters *22,* 59–60 (1981); with *chiral*

1,4-dihydropyridines, **asym. reduction,** s. M. Seki et al., J. Am. Chem. Soc. *103,* 4613–5 (1981); J. Chem. Soc. Chem. Commun. *1980,* 815–7; A. Ohno et al., Bull. Chem. Soc. Japan, *54,* 3478–81 (1981); P. Jouin, C.B. Troostwijk, R.M. Kellogg, J. Am. Chem. Soc. *103,* 2091–3 (1981); with $AlCl_3$ cf. Tetrahedron Letters *23,* 3185–8 (1982); **review of** developments in **dihydropyridine chemistry** from 1972–80 s. J. Kuthan, A. Kurfürst, Ind. Eng. Chem. Prod. Res. Dev. *21* 191–261 (1982).

Polymer-based 1,4-dihydropyridines ←
s. *37,* 26

Chiral 1,4-dihydropyridines ←
s. *37,* 26

Zinc/potassium hydroxide *Zn/KOH*
Arylcarboxylic acids from phthalides C
s. *37,* 925

Zinc iodide s. under 1,4-Dihydropyridines ZnI_2

Lithium tetrahydridoborate $LiBH_4$
1,3-Diols from β-hydroxyketones CO → CHOH
Stereospecific reduction
with tributylborane/$NaBH_4$ cf. *36,* 31; protected *cis*-1,3-diols with $LiBH_4$ s. M.J. Broadhurst, C.H. Hassall, G.J. Thomas, J. Chem. Soc. Chem. Commun. *1982,* 158–60.

Sodium tetrahydridoborate $NaBH_4$
Sec. alcohols from ketones
s. *29,* 933; 2-aminoalcohols, stereospecific reduction, s. B. Alcaide et al., Tetrahedron Letters *23,* 2033–6 (1982).

Sodium tetrahydridoborate–sand $NaBH_4–SiO_2$
Alcohols from oxo compds.
with $NaBH_4$ on alumina cf. *35,* 20; with $NaBH_4$ on 'Fontainebleau sand' s. G. Bram, E.D'Incan, A. Loupy, J. Chem. Soc. Chem. Commun. *1981,* 1066–7.

Sodium tetrahydridoborate/lanthanide chlorides ←
2-Ethylenealcohols from α,β-ethyleneketones C=C–CO → C=C–CHOH
cf. *26,* 809s*33;* f. lanthanide salts, optimization, s. J. Am. Chem. Soc. *103,* 5454–9 (1981).

Sodium tetrahydridoborate/palladium(II) acetate $NABH_4/(MeCOO)_2Pd$
Sec. alcohols from ketones CO → CHOH
Hindered compds.
s. *37,* 49

Zinc tetrahydridoborate $Zn(BH_4)_2$
2,3-Oxidoalcohols from α,β-oxidoketones
s. *18,* 66s*31;* stereospecific reduction s. T. Nakata, T. Tanaka, T. Oishi, Tetrahedron Letters *22,* 4723–6 (1981).

Lithium tetrahydridoaluminate/chiral aminodiols ←
Sec. alcohols from ketones
Asym. reduction
in the presence of (–)-N-methylephedrine cf. *33,* 43s*36;* with chiral aminodiols s. J.D. Morrison et al., Tetrahedron Letters *22,* 2619–22 (1981).

Lithium tetratritidoaluminate *$LiAlT_4$*

Oxido compds from α-diketones via glycols CO–CO → CHOH–CHOH

1,2-Ditritiooxido compds.

27. 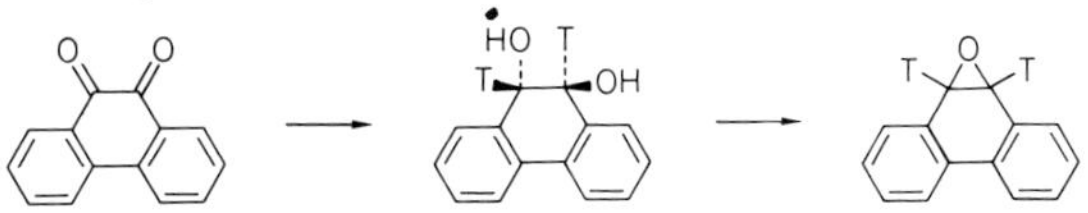

Phenanthrene-9,10-quinone contained in the thimble of a Soxhlet apparatus continuously extracted into a soln. of $LiAlH_4$ and $LiAl^3H_4$ in ether during 24 hrs. → *trans*-9,10-dihydroxy-9,10-dihydrophenanthrene-9,10-^{3}H (Y 61%) added with stirring to NaH in dimethylformamide at room temp., stirring continued for 45 min., N-tosylimidazole added, and stirred for a further hr. → phenanthrene-9,10-oxide-9,10-^{3}H (Y 85%). F.e.s. F. Oesch, A.J. Sparrow, K.L. Platt, J. Label. Compounds Radiopharm. *17,* 93–100 (1980).

*Lithium hydridotri-*sec*-butylborate* *Li[HB(*s*-Bu)$_3$]*

Sec. alcohols from ketones CO → CHOH

Stereospecific reduction

s. *25,* 32s*28;* glycosyl glycerides s. T. Mukaiyama, S. Tanaka, M. Asami, Chem. Letters *1982,* 433–6.

Sodium trihydridocyanoborate/boron fluoride *$Na[BH_3CN]BF_3$*

Alcohols from oxido compds. 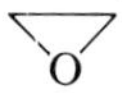 → CHC(OH)

Regiospecific ring opening

with $LiAlH_4/AlCl_3$ cf. *11,* 59; with $Na[BH_3CN]/BF_3$ s. R.O. Hutchins, I.M. Taffer, W. Burgoyne, J. Org. Chem. *46,* 5214–5 (1981).

Tetrabutylammonium tetrahydridoborate *$Bu_4N^+BH_4^-$*

Alcohols from oxo compds. CO → CHOH

with tetraethylammonium tetrahydridoborate cf. *18,* 58s*36;* with tetrabutylammonium tetrahydridoborate, comparisons, s. D.J. Raber, Tetrahedron Letters *22,* 5107–10 (1981).

Tetrabutylammonium dihydridodichloroborate *$Bu_4N^+BH_2Cl_2^-$*

Prim. alcohols from aldehydes CHO → CH_2OH

Retention of oxido groups

s. *37,* 52

B-3a-Pinanyl-9-borabicyclo[3.3.1]nonane ←

Alcohols from oxo compds. CO → CHOH

Asymmetric reduction

s. *33,* 44s*35;* s.a. H.C. Brown, G.G. Par, J. Org. Chem. *47,* 1606–8 (1982).

Sodium hydridotris(3,5-dimethylphenoxo)borate *Na[BH(OAr)$_3$]*

Prim. alcohols from aldehydes with retention of keto groups CHO → CH_2OH

s. *37,* 28

Sodium dihydridobis(2-methoxyethoxo)aluminate *$Na[AlH_2(OCH_2CH_2OCH_3)_2]$*

Alcohols from oxido compds. 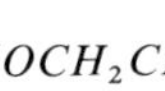 → CHC(OH)

s. *27,* 35; optically active 1,3-diols s. P. Ma et al., J. Org. Chem. *47,* 1378–80 (1982); J.M. Finan, Y. Kishi, Tetrahedron Letters *23,* 2719–22 (1982).

Lithium hydridotris(3-ethyl-3-pentoxo)aluminate *Li[AlH(OCEt₃)₃]*

Prim. alcohols from aldehydes with retention of keto groups CHO → CH_2OH

28. $$C_5H_{11}CHO \xrightarrow{Li[AlH(OCEt_3)_3]} C_5H_{11}CH_2OH$$

An equimolar mixture of hexanal and cyclohexanone treated with 1 equivalent Li tris[(3-ethyl-3-pentyl)oxy]aluminium hydride (prepd. from $LiAlH_4$ and 3-ethyl-3-pentanol) in tetrahydrofuran at –78°, and the mixture hydrolysed after 3 hrs. → 1-hexanol (Y ≳ 90%) and cyclohexanol (Y ca. 0.4%). – Unhindered and highly reactive ketones are effectively inert to this highly hindered hydridoaluminate. F.e.s. S. Krishnamurthy, J. Org. Chem. *46,* 4628–9 (1981); with sodium hydridotris(3,5-dimethylphenoxo)borate, prepd. *in situ* from $NaBH_4$ and 3,5-dimethylphenol, cf. S. Yamaguchi, K. Kabuto, F. Yasuhara, Chem. Letters *1981,* 461–2; f. method cf. J.H. Babler, S.J. Sarussi, J. Org. Chem. *46,* 3367–9 (1981).

*Lithium hydridotri-*tert*-butoxoaluminate* *Li[AlH(OCMe₃)₃]*

Sec. alcohols from ketones CO → CHOH

Stereospecific reduction

hydroxyketones cf. *32,* 26s*33;* optically active 3-hydroxythioesters, also diastereoisomer with potassium selectride (cf. *34,* 235; *12,* 64s*29*), s. K. Suzuki, A. Ikegawa, T. Mukaiyama, Chem. Letters *1982,* 899–902.

*Lithium hydridotri-*tert*-butoxoaluminate/* tert*-butylamine* *Li[AlH(OCMe₃)₃]/*t*-BuNH₂*

Sec. alcohols from ketones

Retention of aldehyde groups via protection as aldimines *in situ*

29. $$Me(CH_2)_4COMe \rightarrow Me(CH_2)_4CH(OH)Me$$

$$\left[Me(CH_2)_6CHO + H_2NBu\text{-}t \rightarrow Me(CH_2)_6CH{=}NBu\text{-}t \rightarrow Me(CH_2)_6CHO\right]$$

Dry *tert*-butylamine added under N_2 to a stirred mixture of octanal, 2-heptanone, and 4Å molecular sieves in dry tetrahydrofuran, left 30 min., Li tri-*tert*-butoxohydridoaluminate in the same solvent added, stirring continued for 30 min., treated sequentially at 0° with water and cold 2 *M* HCl, and worked up after a further 10 min. at room temp. → octanal (Y 98%) and heptan-2-ol (Y 100%). F.e. and limitations s. M.P. Paradisi, G.P. Zecchini, G. Ortar, Tetrahedron Letters *21,* 5085–8 (1980).

Reductions with *1-pyrrolylborane-tetrahydrofuran* ←

Alcohols from oxo compds. under mild conditions

30.

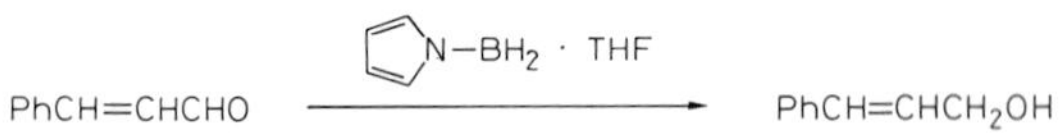

A soln. of 1-pyrrolylborane-tetrahydrofuran complex in tetrahydrofuran treated under N_2 with cinnamaldehyde at 0°, the mixture stirred at 0° for 3 hrs., and hydrolysed with 2 *M* HCl → 3-phenyl-2-propenol. Y 88%. F.e., **also ketones from acetylene derivs., and hydration of ethylene derivs.,** s. M. Añez et al., Synthesis *1980,* 214–6.

Reductions with *bis(trifluoroacetoxy)borane* *HB(OCOCF₃)₂*

31. $$THF \cdot BH_3 + CF_3COOH \downarrow$$

$$PhCHO \xrightarrow{[HB(OCOCF_3)_2]} PhCH_2OH$$

Benzaldehyde added under N_2 to a stirred soln. of bis(trifluoroacetoxy)borane (prepd. by mixing BH_3-tetrahydrofuran and trifluoroacetic acid) in tetrahydrofuran at 0–5°, and left to stand for 1 min. → benzyl alcohol. Y 90%. F. reactions, incl. indolines from indoles (s. Synth. Meth. *31,* 39s*34*), and selective reductions, s. B.E. Maryanoff, D.F. McComsey, S.O. Nortey, J. Org. Chem. *46,* 355–60 (1981).

Borane.phenylhydrazine *$PhNHNH_2.BH_3$*

Borane.2-aminopyridine ←
with borane.amine complexes s. *15,* 39s*36;* also with borane.phenylhydrazine, *erythro*-2-aminoalcohols, cf. T. Kametani et al., Yakugaku Zasshi *102,* 241–4 (1982) (Japan); Chem. Abstr. *97,* 5887 (1982); with borane.2-aminopyridine in aq. medium cf. Y. Okamoto. T. Osawa, T. Kinoshita, Synthesis *1982,* 462–3.

Sand s. under $NaBH_4$ *SiO_2*

Boron fluoride s. under $Na[BH_3CN]$ *BF_3*

Lanthanide chlorides s. under $NaBH_4$ ←

Samarium diiodide *SmI_2*
Lanthanide diiodides as coupling and reducing agents CHO → CH_2OH
cf. *35,* 443; s.a. Tetrahedron *37,* Suppl. No. 1, 175–80 (1981).

Isopropanol s. under $RhCl_3$ *i-PrOH*

2,5-Dihydroanisole/n-butyllithium ←
Reduction with metalated 1,4-cyclohexadienes under mild conditions CO → CHOH

32.

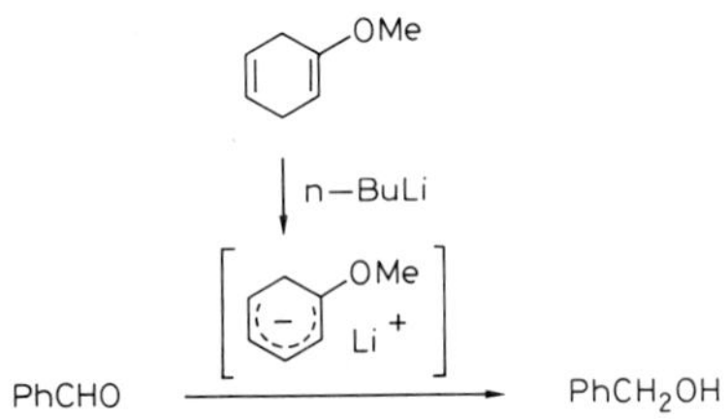

Alcohols from oxo compds. *n*-Butyllithium in petroleum ether added to a soln. of 2,5-dihydroanisole in tetrahydrofuran at –78°, the mixture allowed to warm to room temp., a soln. of benzaldehyde in tetrahydrofuran added at 0°, and the pale yellow mixture stirred at room temp. for 5 hrs. → benzyl alcohol. Y 100%. F.e., **also prim. alcohols from carboxylic acid esters,** and **the latter from 2 carboxylic acid chloride molecules,** s. M. Hiramatsu, T. Fujinami, S. Sakai, Chem. Letters *1982,* 7–8.

Chiral aminodiols s. under $LiAlH_4$ ←

Carboxylic acids s. under Ni *RCOOH*

Formic acid *HCOOH*
2-Aminoalcohols from oxazolidine ring C
with aq. acetic acid cf. *6,* 73; with formic acid s. S. Wawzonek, Org. Prep. Proc. Intern. *13,* 126–9 (1981).

Tartaric acid s. under Ni ←

Triorganotin hydrides *R_3SnH*
Sec. alcohols from ketones via alkoxystannanes CO → CHOH
s. *13,* 70; *19,* 132; α-hydroxy- from α-keto-carboxylic acid esters, also asym. reduction, s. A. Rahm, M. Pereyre, Bull. Soc. Chim. Belges *89,* 843–8 (1980).

Bis(dimethylglyoximato)cobalt(II)/quinine hydrochloride ←

Asymmetric hydrogenation CO → CHOH

s. *30*, 18; f.e. and chiral amines s. Bull. Chem. Soc. Japan *54*, 2124–35 (1981).

Nickel/zinc *Ni/Zn*

Stereospecific reduction

s. *11*, 63s*31*; cholestanols, stereospecificity with comparison of Urushibara catalysts and solvents, s. Can. J. Chem. *58*, 1061–8 (1980).

Nickel/tartaric acid/sodium bromide/carboxylic acids ←

Heterogeneous asymmetric hydrogenation

s. *36*, 36; with added carboxylic acid, especially pivalic acid, s. Chem. Letters *1982*, 315–6.

Potassium hydrido(phosphine)ruthenate complexes ←

Cationic rhodium(I) phosphine complex ←

Alcohols from oxo compds.

Homogeneous hydrogenation

from aldehydes with $RuCl_2(Ph_3P)_3$ cf. *23*, 51s*33*; also from ketones with potassium hydrido(phosphine)ruthenate complexes such as $K[(Ph_3P)_2Ph_2PC_6H_4RuH_2]$, and reduction of esters and nitriles, s. R.A. Grey, G.P. Pez, A. Wallo, J. Am. Chem. Soc. *103*, 7536–42 (1981); with cationic rhodium(I) phosphine and rhodium(I) hydride complexes, also by transfer hydrogenation with isopropanol (cf. Synth. Meth. *33*, 47s*36*), s. K. Tani et al., Chem. Letters *1982*, 261–4.

Rhodium trichloride/isopropanol *$RhCl_3$/i-PrOH*

Benzylalcohols from aromatic aldehydes $ArCHO \rightarrow ArCH_2OH$

s. *37*, 132

Palladium(II) acetate s. under $NaBH_4$ *$(MeCOO)_2Pd$*

Addition to Nitrogen and Carbon HC⇓NC

Sodium fluoride s. under i-Bu_2AlH *NaF*

Zinc *Zn*

N-Aminomethylation via cyanamides $N\text{–}CN \rightarrow NHCH_2NH_2$

33.

$$C_8H_{17}NH_2 \xrightarrow{BrCN} C_8H_{17}NHCN \rightarrow C_8H_{17}NHCH_2NH_2$$

Formaminals. Octyl cyanamide (prepared from octylamine and BrCN) in 15% aq. HCl heated with zinc dust until it went into soln., and heating continued for ca. 30 min. → N-aminomethyloctylamine. Y 90%. F.e.s. A. Malik, N. Afza, S. Siddiqui, Z. Naturforsch. B *37*, 512–8 (1982).

Borane-dimethyl sulfide *$BH_3.Me_2S$*

Prim. amines from nitriles $CN \rightarrow CH_2NH_2$

s. *37*, 46

Sodium tetrahydridoborate *$NaBH_4$*

Hydroxylamines from nitrones $C{=}NO \rightarrow CH{-}NOH$

with $LiAlH_4$ cf. *10*, 61; with $NaBH_4$, N-hydroxyazetidines, s. M.L.M. Pennings, D.N. Reinhoudt, Tetrahedron Letters *23*, 1003–6 (1982).

Tert. amines from azomethinium salts $C{=}NR_2^+ \rightarrow CHNR_2$

s. *16*, 52; 3-aminothioenolethers s. R.J. Sundberg, C.P. Walters, J.D. Bloom, J. Org. Chem. *46*, 3730–2 (1981).

Sodium tetrahydridoborate/chiral α-acylaminocarboxylic acids ←

Sec. amines from azomethines C=NR → CHNHR
Asymmetric reduction

34. 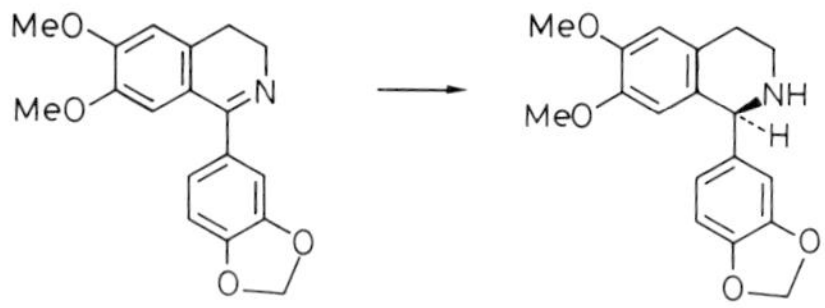

A soln. of startg. 3,4-dihydroisoquinoline deriv. in dichloromethane added to a stirred soln. of sodium triacoxyborohydride (prepared from N-benzyloxycarbonyl-L-proline and $NaBH_4$) in the same solvent, and the mixture stirred at room temp. for 22 hrs. → product. Y 90% (enantiomeric excess 86%). F.e. and reagents, s. K. Yamada, M. Takeda, T. Iwakuma, Tetrahedron Letters *22,* 3869–72 (1981).

Lithium tetrahydridoaluminate *$LiAlH_4$*

Optically active prim. amines from N-sulfinylimines C=NS(O)R → CHNHS(O)R
via sulfinamides by asymmetric reduction

35. 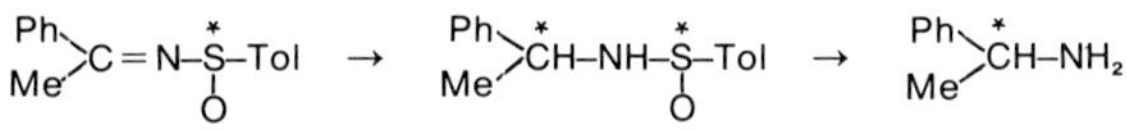

Startg. chiral N-sulfinylimine in ether added dropwise to a stirred soln. of $LiAlH_4$ in the same solvent at room temp., and stirring continued 10–15 hrs. under N_2 → intermediate (Y 80–90%) treated in methanol with a soln. of trifluoroacetic acid in the same solvent while stirring at 0°, and stirring continued for 1 hr. at the same temp. → (S)-1-phenylethylamine. Y 60–70%; 78% enantiomeric excess. F.e., also by reduction of rac. N.-sulfinylimines with chiral lithium trihydridoalkoxoaluminates, s. R. Annunziata, M. Cinquini, F. Cozzi, J. Chem. Soc. Perkin Trans. I *1982,* 339–43; **from N-phosphinylimines via phosphinic amides** cf. B. Krzyzanowska, W.J. Stec, Synthesis *1982,* 270–2.

Acylhydrazines from acylhydrazones C=N–NAc → CHNH–NAc
with $NABH_4$ cf. *25,* 37; with $LiAlH_4$ s. E.Y. Posyagina et al., Zh. Org. Khim. *10,* 2153–6 (1981).

*Lithium hydridotri-*sec*-butylborate* *Li[HB(s-Bu)₃]*

Sec. amines from azomethines C=NR → CHNHR
Stereospecific reduction
s. *35,* 22; f. methods s. J.E. Wrobel, B. Ganem, Tetrahedron Letters *22,* 3447–50 (1981).

Diisobutylaluminium hydride/sodium fluoride *i-Bu_2AlH/NaF*

Preferential N-alkylation of amines with oxo compds. C
via regiospecific reductive ring opening

36. 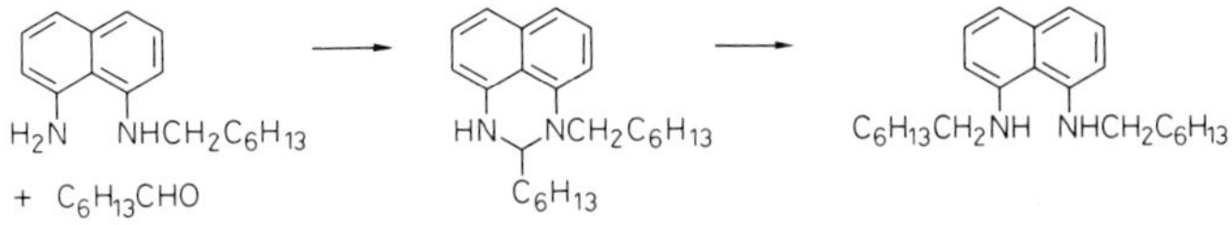

of cyclic aminals. Heptaldehyde added dropwise with stirring to a soln. of N-heptyl-1,8-diaminonaphthalene in benzene at 0°, and stirring continued at the same temp. for 30 min. → 1-heptyl-2-hexyl-2,3-dihydroperimidine, treated with a soln. of diisobutylaluminium hydride in hexane at 0°, stirred at 25° for 8 hrs., refluxed for 1 hr., re-cooled to 0°, dil. with benzene, NaF and water added sequentially, and vigorous stirring continued for 20 min. at 25° → N,N′-diheptyl-1,8-diaminonaphthalene.

Overall Y 88%. F.e., **also via cyclic amidines,** and prepn. of **polyazamacroheterocyclics from lactams via bicyclic amidines,** s. H. Yamamoto, K. Maruoka, J. Am. Chem. Soc. *103,* 4186–94 (1981).

Sodium trihydridocyanoborate/acetic acid *$Na[BH_3CN]/MeCOOH$*

Acoxylamines from acyloximes COO–N=C → COO–NHCH

with sodium amalgam cf. *27,* 42; with sodium trihydridocyanoborate/acetic acid, stereospecific reduction, s. D.D. Sternbach, W.C.L. Jamison, Tetrahedron Letters *22,* 3331–4 (1981).

Isopropanol s. under $(Ph_3P)_3RhCl$ *$(CH_3)_2CHOH$*

Chiral α-acylaminocarboxylic acids s. under $NaBH_4$ ←

Trichlorosilane *$HSiCl_3$*

Sec. amines from azomethines C=NR → CHNHR

37.

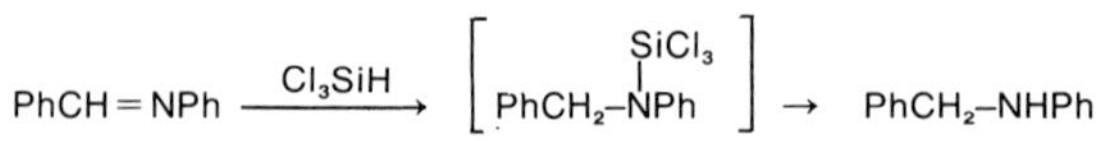

Trichlorosilane added dropwise over 10 min. to a soln. of N-(benzylidene)aniline in dry acetonitrile at 0°, refluxed for 4 hrs., re-cooled to 0°, 85% ethanolic KOH added dropwise over 2 hrs. (*caution:* highly exothermic), and the mixture stirred at room temp. for 12 hrs. → N-benzylaniline. Y 79% F.e.s. R.A. Benkeser, D.C. Snyder, J. Organometal. Chem. *225,* 107–15 (1982).

Nickel *Ni*

Hydrazines from hydrazones C=N–N → CHNH–N

with Pd–C cf. *17,* 75; using Raney-Ni in benzene, with simultaneous nitro group reduction, s. M.M. Kidwai, N.H. Khan, Sci. Res. News. *2(1),* 17 (1980); Chem. Abstr. *95,* 186743 (1981).

Chlorotris(triphenylphosphine)rhodium(I)/ isopropanol/sodium carbonate *$(Ph_3P)_3RhCl/(CH_3)_2CHOH/Na_2CO_3$*

Sec. amines from azomethines C=NR → CHNHR

with $(Ph_3P)_3RhCl$/organosilicon hydrides cf. *29,* 42; with $(Ph_3P)_3RhCl$/isopropanol/ Na_2CO_3, comparison of catalysts, s. R. Grigg, T.R.B. Mitchell, N. Tongpenyai, Synthesis *1981,* 442–4.

Palladium–carbon *Pd–C*

Carboxylic acid amides from 2-azetidiones C

with Ni cf. *17,* 74; with Pd–C, α-amino- and α-hydroxy-carboxylic acid amides, s. I. Ojima, S. Suga, R. Abe, Chem. Letters *1980,* 853–6; dipeptides s. Tetrahedron Letters *21,* 3907–10 (1980).

Addition to Carbon and Carbon HC⇓CC

Electrolysis/enzyme ←

Electroenzymatic hydrogenation C=C → CHCH

38.

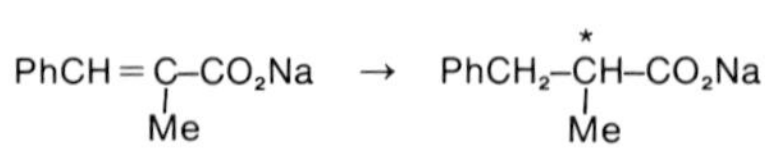

A soln. of sodium (E)-2-methylcinnamate in phosphate buffer (pH 6.4) containing EDTA and a little 1,1′-dimethyl-4,4′-bipyridinium dication (pre-reduced at –800 mV)

electrolysed under N_2 in the presence of the eneoate reductase from *Clostridium* sp. La 1 for 12–20 hrs. while maintaining the catholyte pH by addition of 1:1 H_3PO_4/H_2O → (R)-2-methyl-3-phenylpropionic acid. Y 95%. F.e., also microbiological reduction with bacterial cells, s. H. Simon et al., Angew. Chem. Intern. Ed. *20,* 861–8 (1981); enzyme-catalysed synthesis based on electrochemical generation of NAD(PH) from NAD(P) s. R. DiCosimo et al., J. Org. Chem. *46,* 4622–3 (1981).

Reductions with *sodium-potassiumalloy/18-crown-6 polyether* ←
Hydrogenation of carbon-carbon multiple bonds C≡C → CH=CH

39. $CH_3(CH_2)_2C\equiv C(CH_2)_2CH_3 \rightarrow CH_3(CH_2)_2CH=CH(CH_2)_2CH_3$

Ethylene from acetylene derivs. A little 18-crown-6 and *tert*-butyl alcohol in tetrahydrofuran stirred under N_2 at 0° while adding Na/K alloy in one portion by syringe, a soln. of 4-octyne in the same solvent added dropwise during 10 min., and stirring continued for a further 20 min. → 4-octene. Y 83%. F.e., **also** further reduction to **hydrocarbons,** and **benzene ring hydrogenation,** s. D.J. Mathre, W.C. Guida, Tetrahedron Letters *21,* 4773–6 (1980).

*Tetramminelithium(0)/*tert-*butanol* *$Li(NH_3)_4$/t-BuOH*
Reductions with lithium ammonia complex
s. *27,* 50; with tetramminelithium(0) complex/*tert*-butanol, *cis*-ethylene from acetylene derivs., s. R.E. Doolittle, Org. Prep. Proc. Intern. *13,* 179–84 (1981).

Triethylamine s. under i-Bu_2AlH *Et_3N*

1,4-Dihydropyridines ←
Ketones from α,β-ethyleneketones CO–C=C → CO–CHCH
s. *17,* 80; via *α,β*-ethyleneazomethinium salts, 3-ketosteroids, s. R.A. Gase, U.K. Pandit, Rec. Trav. Chim. Pays-Bas *99,* 334–40 (1981); optically active ketones with chiral 1,4-dihydropyridines s. N. Baba et al., Can. J. Chem. *58,* 387–92 (1980).

Sodium tetrahydridoborate (s.a. under $NbCl_5$) *$NaBH_4$*
Hydrogenation of carbon-carbon double bonds C=C → CHCH
s. *19,* 72; *γ*-lactones from *α,β*-ethylene-*γ*-lactones s. M. Sato, A. Kosasayama, F. Uchimaru, Chem. Pharm. Bull. *29,* 2885–92 (1981).

Tert. amines from enamines C=C–N⋜ → CHCH–N⋜
with $NaBH_4$/mercuric acetate cf. *28,* 38; with $NaBH_4$, *β*-aminocarboxylic acid esters, s. S. Carlsson and S.-O. Lawesson, Tetrahedron *38,* 413–7 (1982).

Sodium tetrahydridoborate/rhodium trichloride *$NaBH_4$/$RhCl_3$*
Benzene ring hydrogenation Ⓗ

40.

A soln. of N-acetyl-L-phenylalanine and $RhCl_3.3H_2O$ in ethanol stirred at 30° for 2 hrs., a soln. of $NaBH_4$ in the same solvent added dropwise during 30 min., and stirring continued at the same temp. for 1 hr. → N-acetyl-L-cyclohexylalanine. Y 94%. Carboxylic acids, esters, and amides are unaffected. F.e., **also piperidines from pyridines,** s. M. Nishiki et al., Tetrahedron Letters *23,* 193–6 (1982).

Lithium tetrahydridoaluminate *$LiAlH_4$*
trans-**Ethylene from acetylene derivs.** C≡C → CH=CH
α,β-ethylene-nitriles and -halides cf. *36,* 45; eneselenides s. J.V. Comassetto, J.T.B. Ferreira, N. Petragnani, J. Organometal. Chem. *216,* 287–94 (1981).

Diisobutylaluminium hydride *i-Bu_2AlH*

Cyclic ketones from α,β-ethyleneketones C=C–CO → CHCHCO

41. 

Isoflavanones. A soln. of startg. isoflavone in 15:7 toluene/tetrahydrofuran treated at –65° with a soln. of diisobutylaluminium hydride in toluene under N_2, the mixture kept at the same temp. for 1 hr., quenched with methanol, and allowed to warm to room temp. → product. Y 90%. F.e.s. S. Antus, A. Gottsegen, M. Nogradi, Synthesis *1981,* 574–6.

Sodium trihydridocyanoborate/trifluoroacetic acid *$Na[BH_3CN]/CF_3COOH$*

Cyclic N-sulfonylamines from enamines C=C → CHCH

s. *37,* 415

Diisobutylaluminium hydride/triethylamine *i-Bu_2AlH/Et_3N*

Mercaptals from ketene mercaptals C=C(SR)$_2$ → CHCH(SR)$_2$

with $LiAlH_4$ cf. *33,* 54; with diisobutylaluminium hydride/triethylamine, β-ketomercaptals, s. R.B. Gammill, D.M. Sobieray, P.M. Gold, J. Org. Chem. *46,* 3555–8 (1981).

Triethyldialuminium trichloride s. under Ni(acac)$_2$ *$Et_3Al_2Cl_3$*

tert-*Butanol s. under Li* t-*BuOH*

18-Crown-6 polyether s. under Na/K ←

Enzymes s. under Electrolysis ←

Ethyl acetate s. under NH_2OH *EtOAc*

Formic acid/triethylsilane/cation exchanger ←

Ionic hydrogenation C=C → CHCH

with trifluoroacetic acid/triethylsilane cf. *23,* 63; with formic acid/triethylsilane in the presence of cation exchanger KU-1 s. G.D. Kolomnikova, Izv. Akad. Nauk SSSR Ser. Khim. *1981,* 1162–4; with dichloro(ethyl)silane in the presence of $AlHal_3$, or with isopropylbenzenes as H-donor, cf. G.I. Bolestova et al., ibid. *1982,* 1322–6.

Trifluoroacetic acid s. under Na[BH$_3$CN] *CF_3COOH*

Lead s. under Pd–C *Pb*

Dicarbonylbiscyclopentadienyltitanium/water *$(C_5H_5)_2Ti(CO)_2/H_2O$*

Ethylene from acetylene derivs. C=C → CH=CH

with $(C_5H_5)_2TiCl_2$/bis(diisopropylamino)aluminium hydride cf. *34,* 43; with $(C_5H_5)_2Ti(CO)_2$/water, *cis*-ethylene derivs., s. B. Demerseman, P.H. Dixneuf, J. Chem. Soc. Chem. Commun. *1981,* 665–6.

*Tri-*n-*butyltin hydride/*
tetrakis(trimethylphosphine)palladium(0) n-*$Bu_3SnH/Pd(PPh_3)_4$*

Selective and preferential hydrogenation of carbon-carbon double bonds C=C → CHCH

42. PhCH=CHCHO → $PhCH_2CH_2CHO$

A stirred soln. of startg. α,β-ethylenealdehyde in tetrahydrofuran containing a little 3% tetrakis(triphenylphosphine)palladium(0) treated with 1.5 equivalents tributyltin hydride during 1.5 hrs., and worked up with H_2O as proton donor and methylene chloride → product. Y >99%. F.e., also with retention of isolated double bonds, s. E. Keinan, P.A. Gleize, Tetrahedron Letters *23,* 477–80 (1982); with acetic acid as H-donor, and $ZnCl_2$ as promotor, s. P. Four, F. Guibe, ibid. *23,* 1825–8 (1982).

Hydroxylamine/ethyl acetate *$NH_2OH/EtOAc$*

Hydrogenation of carbon-carbon double bonds with alternative diimide generation

43.

$$\begin{array}{c} NH_2OH \\ \downarrow MeCOOEt \end{array}$$

$$C_7H_{15}CH{=}CHCH_2OH \xrightarrow{[HN=NH]} C_7H_{15}CH_2CH_2CH_2OH$$

Powdered KOH added to a stirred soln. of hydroxylamine hydrochloride in dimethylformamide at 25–35° under N_2, the resulting mixture stirred for 10 min., filtered under N_2, the filtrate cooled in an ice-bath, treated with ethyl acetate, the soln. added dropwise with stirring during 1 hr. to 2-decen-1-ol at 90–100° under N_2, and heating continued for an additional 1 hr. → 1-decanol. Y 95%. F.e. and esters s. P.A. Wade, N.V. Amin, Synth. Commun. *12,* 287–91 (1982).

Triphenylphosphine s. under $Ni(acac)_2$ *PPh_3*

Sodium phosphinate s. under Pd–C *NaH_2PO_2*

Niobium pentachloride/sodium tetrahydridoborate *$NbCl_5/NaBH_4$*

(Z)-Ethylene from acetylene derivs. $C{\equiv}C \rightarrow CH{=}CH$
s. *37,* 742

Tetracarbonylhydridoferrate anion *$[FeH(CO)_4]^-$*

Sec. alcohols from *α,β*-ethyleneketones $C{=}C{-}CO \rightarrow CHCHCHOH$
with $NaBH_4$ cf. *24,* 269; with tetracarbonylhydridoferrate anion s. M. Yamashita et al., Bull. Chem. Soc. Japan *55,* 1329–30 (1982).

Nickel–graphite *Ni–C*

Reductions with graphite-supported Group VIII metals prepd. *in situ* $C{\equiv}C \rightarrow CH{=}CH$

44.

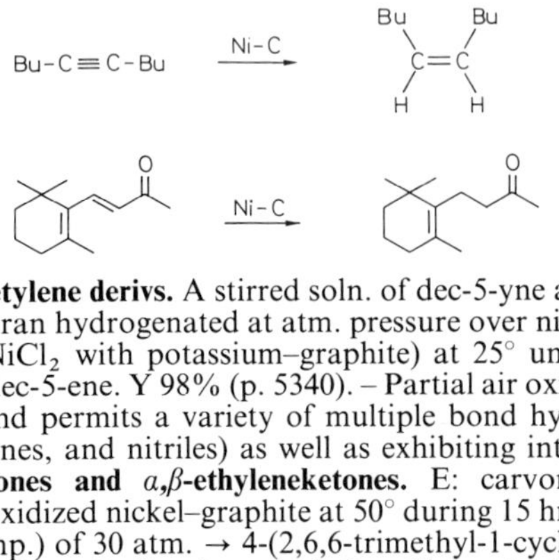

(Z)-Ethylene from acetylene derivs. A stirred soln. of dec-5-yne and a little *ethylenediamine* in tetrahydrofuran hydrogenated at atm. pressure over nickel–graphite (prepd. *in situ* by reducing $NiCl_2$ with potassium–graphite) at 25° until H_2 uptake ceased (30–140 min.) → *cis*-dec-5-ene. Y 98% (p. 5340). – Partial air oxidation of the catalyst reduces its activity and permits a variety of multiple bond hydrogenations (e.g. of ethylene derivs., ketones, and nitriles) as well as exhibiting interesting selectivity in **reduction of *β*-diketones and *α,β*-ethyleneketones.** E: carvone in methanol hydrogenated with air-oxidized nickel–graphite at 50° during 15 hrs. under an initial H_2 pressure (at room temp.) of 30 atm. → 4-(2,6,6-trimethyl-1-cyclohexen-1-yl)butan-2-one. Y 81% (p. 5344). F.e.s. D. Savoia et al., J. Org. Chem. *46,* 5340–3; 5344–8 (1981); *palladium–graphite* as catalyst cf. J. Chem. Soc. Chem. Commun. *1981,* 540–1; with *iron–graphite,* **ethylene derivs. from 1,2-dibromides,** and **ketones from *α*-bromoketones,** s. J. Org. Chem. *47,* 876–9 (1982).

Nickel/aluminium *Ni/Al*

Hydrogenation with Urushibara Ni-catalysts
ring hydrogenation of isocyclics cf. *15,* 63; ethylene from acetylene derivs., optimization of Urushibara Ni-catalysts, s. A.B. Fasman, N.A. Maksimova, B.Y. Pelmenshtein, Izv. Akad. Nauk Kaz. SSR Ser. Khim. *1981,* 31 (Russ); Chem. Abstr. *95,* 96940 (1981); Urushibara Ni/Zn catalyst s. ibid. *1981,* 26; Chem. Abstr. *95,* 96939.

Nickel acetylacetonate/triethyldialuminium trichloride/ triphenylphosphine *$Ni(acac)_2/Et_3Al_2Cl_3/PPh_3$*

Partial and preferential homogeneous hydrogenation C=C → CHCH
with $PdCl_2$ cf. *36,* 49; cycloalkenes with $Ni(acac)_2$/triethyldialuminium trichloride/ triphenylphosphine s. M. Sakai et al., Bull. Chem. Soc. Japan *55,* 343–4 (1982).

Rhodium–carbon *Rh–C*

β-Acoxycarboxylic from β-acoxy-α,β-ethylenecarboxylic acid esters C(OAc)=C–COOR → CH(OAc)CHCOOR

45.

Startg. m. added by syringe to tetrahydrofuran containing 5% Rh on carbon, and the mixture hydrogenated at room temp. under 1 atm. H_2 with vigorous stirring → product. Y 100%. F.e.s. J.D. Rozzell, Tetrahedron Letters *23,* 1767–70 (1982).

Rhodium trichloride s. under $NaBH_4$ *$RhCl_3$*

Chiral rhodium phosphine complexes ←

Asymmetric homogeneous hydrogenation C=C → CHCH
review s. *27,* 57s*36;* with chiral rhodium diphosphine complexes s. T. Yoshikuni, J.C. Bailar, Inorg. Chem. *21,* 2129–33 (1982); J. Padmakumari Amma, J.K. Stille, J. Org. Chem. *47,* 468–73 (1982); K. Tani et al., Chem. Letters *1982,* 265–8;
with chiral aminophosphine complexes s. K. Osakada et al., ibid. *1981,* 1691–4;
with chiral phosphinite complexes s. J. Bourson, L. Oliveros, J. Organometal. Chem. *229,* 77–84 (1982); J. Bakos, I. Tóth, L. Markó, J. Org. Chem. *46,* 5427–8 (1981);
with phosphinoferrocenyl complexes s. W.R. Cullen, J.D. Woollins, Can. J. Chem. *60,* 1793–9 (1982); f. catalysts s. J.W. Scott et al., ibid. *46,* 5086–93 (1981); J. Irrure, A. Bosch, J. Capdevila, Afinidad *38,* 201–3, 373 (1981); effect of basicity of added amine and conditions s. D. Sinou, Tetrahedron Letters *22,* 2987–90 (1981);
with *polymer-based phosphine complexes* s. G.L. Baker et al., J. Org. Chem. *46,* 2954–60 (1981);
with *water-soluble* transition metal diphosphine *complexes* s. R.G. Nuzzo et al., ibid. *46,* 2861–7 (1981);
prepn. of chiral dipeptides s. I. Ojima et al., ibid. *47,* 1329–34 (1982);
mechanism and stereoselectivity of asym. hydrogenation, review, s. J. Halpern, Science *217,* 401–7 (1982).

Palladium–carbon *Pd–C*

Hydrogenation of carbon-carbon double bonds C=C(COOH)NHCOR → CHCH(COOH)NHCOR
α-aminocarboxylic acid esters cf. *29,* 739; α-acylaminocarboxylic acids s. C. Cativiela, E. Melendez, Synthesis *1981,* 805-9.

γ-Iminolactones from β,γ-ethylene-γ-iminolactones C=C → CHCH
s. *37,* 666

Palladium–carbon/lead/sodium phosphinate *$Pd–C/Pb/NaH_2PO_2$*

***cis*-Ethylene from acetylene derivs.** C≡C → CH=CH
Transfer hydrogenation
with Pd–C/formic acid/triethylamine cf. *34,* 46; with Pb- and Mg-modified Pd–carbon und Na phosphinate as *H*-donor s. R.A.W. Johnstone, A.H. Wilby, Tetrahedron *37,* 3667–70 (1981); **review of catalytic transfer hydrogenation and hydrogenolysis** s. H. Haertner, Kontakte *1980,* 3–10 (Ger); Chem. Abstr. *93,* 219961 (1980).

Palladium–graphite *Pd–C*

Hydrogenation of carbon–carbon multiple bonds ←
s. *37,* 44

Tetrakis(triphenylphosphine)palladium(0) s. under n-Bu_3SnH *$Pd(PPh_3)_4$*

Exchange ⇂↾

Oxygen ↑ HC⇂↾O

n-*Butyllithium s. under 2,5-Dihydroanisole* n-*BuLi*

Potassium carboxylate s. under $(Ph_3P)_4Pd$ *RCOOK*

Caesium fluoride s. under $(EtO)_3SiH$ *CsF*

Triethylamine s. under Pd–C Et_3N

Zinc/formic acid *Zn/HCOOH*
Hydrocarbon from keto groups $CO \rightarrow CH_2$
s. *28,* 48; partial reduction s. Collection Czech. Chem. Commun. *46,* 1167–72 (1981).

Diborane/sodium salt B_2H_6/Na^+
Prim. alcohols from carboxylic acids $COOH \rightarrow CH_2OH$
s. *29,* 57; rapid reduction in tetrahydrofuran from Na carboxylates cf. N.M. Yoon, B.T. Cho, Tetrahedron Letters *23,* 2475–78 (1982).

Reductions with *borane-dimethyl sulfide* $BH_3.Me_2S$
Prim. alcohols from carboxylic acid esters $COOR \rightarrow CH_2OH$
Improved procedure

46. $PhCOOEt \rightarrow PhCH_2OH$

Ethyl benzoate added slowly at 25° to a *slight excess* (10%) of borane-dimethyl sulfide in *tetrahydrofuran,* the mixture refluxed for 4 hrs. with distillation of dimethyl sulfide through a Vigreux column maintained at ca. 40° by a heating jacket, and hydrolysed with aq. K_2CO_3 at room temp. → benzyl alcohol. Y 90%. F.e., incl. **hindered compds.,** and limitations s. H.C. Brown, Y.M. Choi, Synthesis *1981,* 439–40; **prim. amines from carboxylic acid amides** s. ibid. *1981,* 441–2, 996-7; **from nitriles** s. ibid. *1981,* 605–6.

Sodium tetrahydridoborate (s.a. under $TiCl_4$*)* $NaBH_4$
Alcohols from N-acoxydicarboxylic acid imides
N-Protected 2-aminoalcohols

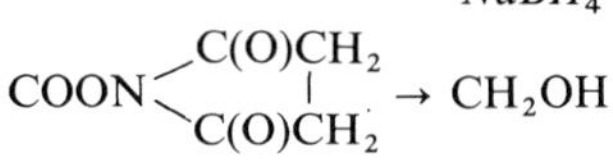

47.

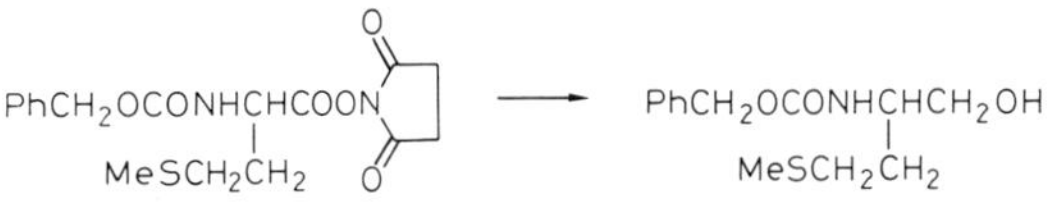

A soln. of startg. active ester in 1:1 tetrahydrofuran/methanol added dropwise during 15 min. to a soln. of $NaBH_4$ in 80% methanol at 10°, stirring continued for 15 min., and neutralized with 1 *M* HCl → product. Y 93%. F.e., also from N-hydroxy-5-norbornene-2,3-dicarboximide esters, s. M. Kubota, O. Nagase. H. Yajima, Chem. Pharm. Bull. *29, 1169–71 (1981).*

Sodium tetrahydridoborate/ tributylhexadecylphosphonium bromide $NaBH_4/[C_{16}H_{33}P(C_4H_9)_3]^+Br^-$
Hydrocarbons from sulfonic esters $OSO_2R \rightarrow H$
with $NaBH_4$ s. *25,* 62; with added tributylhexadecylphosphonium bromide as phase transfer catalyst s. F. Rolla, J. Org. Chem. *46,* 3909–11 (1981).

Reductions with *sodium tetrahydridoborate/methanesulfonic acid* *$NaBH_4/MeSO_3H$*

CON< → CH_2N<

48. PhNHCOMe → $PhNHCH_2Me$

Amines from carboxylic acid amides. A stirred mixture of startg. amide and $NaBH_4$ in *dimethyl sulfoxide* treated dropwise with a soln. of methanesulfonic acid in the same solvent during 30 min., and quenched with 10% NaOH → product. Y 77%. – The method is less hazardous than reduction with B_2H_6 in dimethyl sulfoxide or $NaBH_4$ in dimethylformamide. F.e. and limitations, **also cyclic amines from lactames,** and **prim. alcohols from carboxylic acids,** s. S.R. Wann, P.T. Thorsen, M. Kreevoy, J. Org. Chem. *46*, 2579–81 (1981).

Selective reductions with
sodium tetrahydridoborate/palladium(II) chloride *$NaBH_4/PdCl_2$*

CO → CH_2

49. PhCOPh → $PhCH_2Ph$

Hydrocarbons from ketones. $PdCl_2$ suspended a soln. of benzophenone in methanol, $NaBH_4$ added during 30 min. at 20°, and the mixture stirred for an additional 1 hr. → diphenylmethane. Y 83%. – The reduction proceeds quickly under mild conditions, at almost neutral pH, and has high selectivity. F.e. and selectivity, also from benzyl alcohols, replacement of aromatic halogen by hydrogen, and sec. alcohols from hindered ketones, s. T. Satoh et al., Chem. Letters *1981*, 1029–30.

Lithium tetrahydridoborate/B-methoxy-9-borabicyclo[3.3.1]nonane ←
Prim. alcohols from carboxylic acid esters COOR → CH_2OH
Selective reduction
with $LiBH_4$ cf. *11*, 510; with $LiBH_4$ and a little B-methoxy-9-borabicyclo-[3.3.1]nonane or methyl borate, also hindered compds., s. H.C. Brown, S. Narasimhan, J. Org. Chem. *47*, 1604–6 (1982).

Lithium tetrahydridoaluminate *$LiAlH_4$*
Cyclic ethers from lactones O–CO → O–CH_2

with $LiAlH_4/BF_3$ cf. *32*, 50; with $LiAlH_4$ in tetrahydrofuran, prepn. of crown ethers, s. D.J. Ager, I.O. Sutherland, J. Chem. Soc. Chem. Commun. *1982*, 248–9.

Lithium tetrahydridoaluminate/polypropylene glycol ←
Prim. alcohols from carboxylic acids COOH → CH_2OH
with $LiAlH_4$ cf. *18*, 947; from K carboxylates, phase transfer catalysis with polypropylene glycol, s. S. Szakács, S. Göbölös, J. Szammer, Monatsh. Chem. *112*, 883–6 (1981).

Lithium tetrahydridoaluminate/aluminium chloride *$LiAlH_4/AlCl_3$*
Benzofurans from aurones ←

50.

Startg. aurone added to a stirred soln. of $LiAlH_4$ in ether, stirred at room temp. for 60 min., treated with $AlCl_3$, and stirring continued for 105 min. → 2-(4-methoxybenzyl)benzofuran. Y 78%. F.e.s. K. Kurosawa, Y. Morita, Bull. Chem. Soc. Japan *54*, 635–6 (1981).

Lithium tetrahydridoaluminate/titanocene dichloride *$LiAlH_4/C_5H_5)_2TiCl_2$*
Hydrocarbon from carbonyl groups $CO \rightarrow CH_2$
s. *37,* 74

Lithium tetrahydridoaluminate/iodine *$LiAlH_4/I_2$*
2-Allenealcohols from 2-acetylene-1,4-diol monoethers $C{-}C{\equiv}C \rightarrow C{=}C{=}C$
s. *25,* 52; in the presence of solid I_2 cf. G.E. Keck, R.R. Webb, Tetrahedron Letters *23,* 3051–4 (1982).

B-3a-Pinanyl-9-borabicyclo[3.3.1]-nonane ←
γ-Lactones from γ-ketocarboxylic acid esters ○
Asymmetric reduction

51. 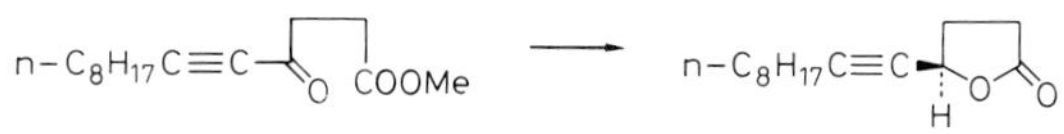

(+)-*a*-Pinene added to a soln. of 9-borabicyclo[3.3.1]nonane in tetrahydrofuran, the mixture stirred at reflux for 2–5 hrs., cooled to room temp., 1 equivalent methanol added followed by startg. ketoester, stirring continued for 7 days at room temp., a soln. of NaOH in methanol added, and refluxed for 15 min. → product. Y 70–75%, enantiomeric excess 85–90%. F.e.s. M.M. Midland, N.H. Nguyen, J. Org. Chem. *46,* 4107–8 (1981).

B-Methoxy-9-borabicyclo[3.3.1]nonane s. under $LiBH_4$ ←

*Potassium tri-*sec*-butylhydridoborate* *K[HB(*s*-Bu)$_3$]*
Lactones from dicarboxylic acid anhydrides ○
with $NaBH_4$ cf. *19,* 73; with K tri-*sec*-butylhydridoborate, reversal of regiospecificity, s. P. Morand, J. Salvator, M.M. Kayser, J. Chem. Soc. Chem. Commun. *1982,* 458–9; with Li tri-*sec*-butylhydridoborate cf. M.A. Makhlouf, B. Rickborn, J. Org. Chem. *46,* 4810–11 (1981).

Tetrabutylammonium dichlorodihydridoborate *$Bu_4N^+BH_2Cl_2^-$*
Selective reductions $C(OR)_2 \rightarrow CHOR$

52. 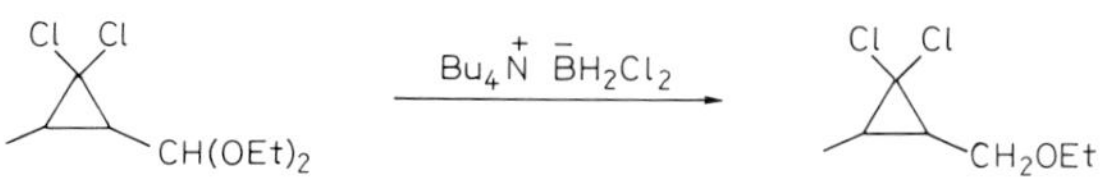

Ethers from acetals. A soln. of startg. m. in benzene added dropwise with stirring to a suspension of tetrabutylammonium dihydridodichloroborate in the same solvent at 0–5°, the mixture gradually heated to 65–70°, and left at this temp. for 6 hrs. → product. Y 88%. F.e., **also prim. alcohols from aldehydes** with retention of oxido groups, s. G.V. Kryshtal et al., Izv. Akad. Nauk SSSR Ser. Khim. *1981,* 1146–8.

Sodium cyanotrihydridoborate *$Na[BH_3CN]$*
Replacement of siloxy groups by hydrogen C
with reductive ring opening of cyclic peroxides
s. *37,* 119

Boron fluoride s. under Et_3SiH *BF_3*

Aluminium chloride s.a. under $LiAlH_4$ *$AlCl_3$*

Aluminium chloride/ethanethiol *$AlCl_3/EtSH$*

Defunctionalization of aryl derivs. ArOR → ArH
via aromatic thioethers under mild conditions
Arenes

53.

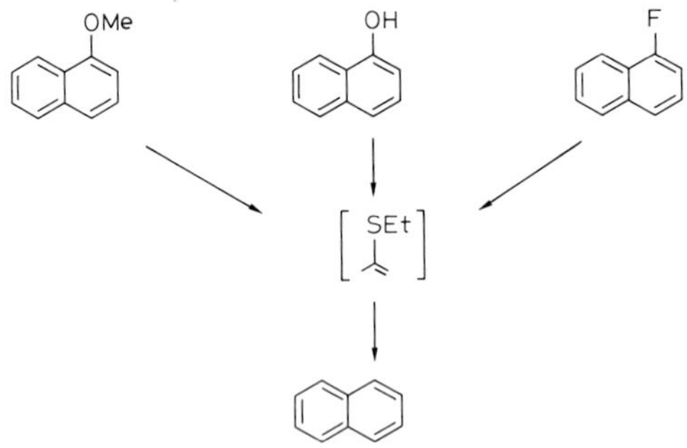

from phenolethers.	**from phenols.**	**from aryl halides.**
1-Methoxynaphthalene	1-Naphthol	1-Fluoronaphthalene

treated under N_2 with 2.5 equivalents of $AlCl_3$ and 5 equivalents of ethanethiol in methylene chloride at room temp. for

4.5 hrs. → product.	5 hrs. → product.	7 hrs. → product.
Y 86%.	Y 76%.	Y 87%.

F.e. (79–100%), also with isolation of intermediates, s. M. Node, E. Fujita et al., Tetrahedron Letters *23,* 689–92 (1982).

Polypropylene glycol s. under $LiAlH_4$ ←

*2,5-Dihydroanisole/*n*-butyllithium* ←

Prim. alcohols from carboxylic acid esters COOR → CH_2OH
s. *37,* 32

Formic acid. s. under Zn *HCOOH*

Ethanethiol s. under $AlCl_3$ *C_2H_5SH*

Triethylsilane/boron fluoride *Et_3SiH/BF_3*

Hydrocarbons from oxo compds. CO → CH_2
s. *29,* 63s*33;* s.a. Org. Synth. *60,* 108–12 (1981); indans from indan-1,3-diones s. O.K. Popova et al., Izv. Akad. Nauk SSSR Ser. Khim. *1981,* 2084–7; **review of indan-1,3-diones** s. E. Gudriniece, Latv. PSR Zinat. Akad. Vestis, Kim. Ser. *1980,* 645–62 (Russ.); Chem. Abstr. *94,* 139486 (1981).

Triethoxysilane/caesium fluoride *$(EtO)_3SiH/CsF$*

Reactions on alkali metal fluoride surfaces COOR → CH_2OH
s. *36,* 670; prim. alcohols from carboxylic acid esters s. Synthesis *1981,* 558–9.

Titanocene dichloride s. under $LiAlH_4$ *$(C_5H_5)_2TiCl_2$*

Titanium tetrachloride/sodium tetrahydridoborate *$TiCl_4/NaBH_4$*

Prim. alcohols from carboxylic acids COOH → CH_2OH
Prim. amines from amides CON$\lessgtr$ → $CH_2N\lessgtr$
s. *37,* 16

*Tri-*n*-butyltin hydride* n-Bu_3SnH

Replacement of acoxy groups by hydrogen OAc → H

s. *23,* 29; acetals from α-acoxyacetals s. H. Redlich, H.-J. Neumann, H. Paulsen, J. Chem. Res. (S) *1982,* 36.

Hydrocarbons from alcohols OH → OC(S)– → H

via xanthates cf. *31,* 49; via thionocarboxylic acid esters, 2′-deoxynucleosides, s. R.A. Lessor, N.J. Leonard, J. Org. Chem. *46,* 4300–1 (1981); via thionocarbonic acid esters cf. M.J. Robins, J.S. Wilson, J. Am. Chem. Soc. *103,* 932–3 (1981); via thionocarbamic acid esters with K/*tert*-butylamine/18-crown-6 cf. A.G.M. Barrett, P.A. Prokopiou, D.H.R. Barton, J. Chem. Soc. Perkin Trans. I *1981,* 1510–5.

Hydrocarbon groups and formic acid esters from selenocarbonic acid esters OCOSeR → H

54. 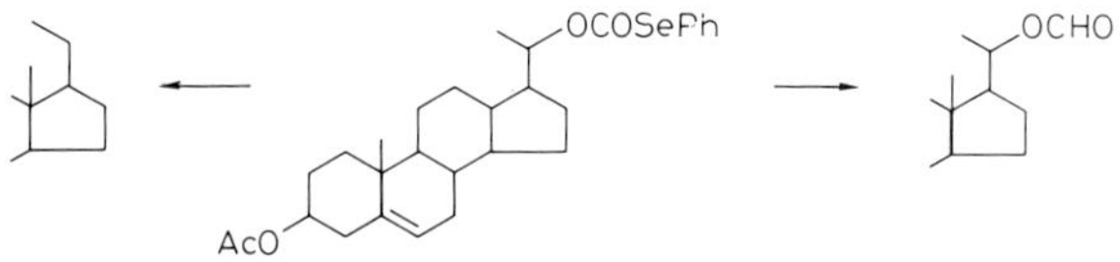

A soln. of 3β-acetoxy-20-benzeneselenocarbonyloxypregn-5-ene in

refluxing xylene treated with tri-*n*-butyltin hydride and a little azodiisobutyronitrile, and heating continued until t.l.c. indicated completed reaction (2–60 min.) → 3β-acetoxypregn-5-ene. Y 90%.

benzene containing tri-*n*-butyltin hydride irradiated through Pyrex with a low-pressure Hg-lamp until t.l.c. indicated completed reaction → 3β-acetoxypregn-5-en-20-yl formate. Y 87%.

F.e.s. J. Pfenninger, C. Heuberger, W. Graf, Helv. Chim. Acta *63,* 2328–37 (1980).

Deoxygenation of tert. alcohols via formiminoesters and thionoformic acid esters OC(S)H → H

55. $Me(CH_2)_{16}CMe_2OH + :C{=}NC_6H_4NMe_2\text{-}p \rightarrow Me(CH_2)_{16}CMe_2OCH{=}NC_6H_4NMe_2\text{-}p$

$\downarrow$

$Me(CH_2)_{16}CHMe_2 \leftarrow Me(CH_2)_{16}CMe_2OCH{=}S$

Startg. m. treated with a 10% excess of *p*-dimethylaminophenyl isocyanide in the presence of CuO at 85° → formimino ester (Y 90%), in dry pyridine treated with excess H_2S and several equivalents of 100% H_2SO_4 at −10° to 0° → thionoformic acid ester (Y 90%), in benzene treated with tri-*n*-butyltin hydride at reflux → product (Y 83%). – Related methods, e.g. Synth. Meth. *31,* 49, are only satisfactory for prim. and sec. alcohols. F.e.s. D.H.R. Barton et al., Tetrahedron Letters *23,* 2019–22 (1982).

*Tri-*n*-butyltin hydride/tetrakis(triphenylphosphine)palladium(0)* n-$Bu_3SnH/Pd(PPh_3)_4$

Ethylene derivs. from acoxy-2-ethylenes C=C–C(OAc) → C=C–CH
Selective and regiospecific reduction

56. PhCH=CHCH(OAc)CN → $PhCH{=}CHCH_2CN$

β,γ-Ethylenenitriles. Tri-*n*-butyltin hydride added to a mixture of the startg. m. and a little tetrakis(triphenylphosphine)palladium(0) in tetrahydrofuran → product. Y 99%. F.e., **also from 2-ethyleneamines,** and with retention of aldehyde, benzylic acoxy, and halogen s. E. Keinan, N. Greenspoon, Tetrahedron Letters *23,* 241–4 (1982).

Tributylhexadecylphosphonium bromide s. under $NaBH_4$ $[C_{16}H_{33}P(C_4H_9)_3]^+Br^-$

Methanesulfonic acid s. under $NaBH_4$ *$MeSO_3H$*

Iodine s. under $LiAlH_4$ *I_2*

Hydrogen iodide/hypophosphorous acid *HI/H_3PO_2*
Replacement of acoxy groups by hydrogen OAc → H
s. *37,* 925

Nickel–alumina *$Ni–Al_2O_3$*
Hydrocarbons from ketones CO → CH_2
with Raney-alloy (Ni/Al) cf. *19,* 56; with Ni on alumina, gas-phase hydrogenolysis of unfunctionalized ketones, s. W.F. Maier et al., Tetrahedron Letters *22,* 4227–30 (1981); **hydrogenolysis of alcohols** s. Z. Naturforsch. *37B,* 392–4 (1982).

Palladium–carbon/triethylamine *$Pd–C/Et_3N$*
Replacement of acoxy groups by hydrogen OAc → H
with Pd cf. *7,* 112; with Pd–C/triethylamine, 3-deoxyaldonolactones, s. K. Bock, I. Lundt, C. Pedersen, Acta. Chem. Scand. *35B,* 155–62 (1981).

Tetrakis(triphenylphosphine)palladium(0) s. under n-*Bu_3SiH* *$Pd(PPh_3)_4$*

Palladium(II) chloride s. under $NaBH_4$ *$PdCl_2$*

Via intermediates *v.i.*
Cyclic ethers from lactones via lactols O–CO → O–CH_2 (cyclic)
s. *37,* 76

Nitrogen ↑ HC⇃↾N

Electrolysis ϟ
Cathodic deamination NH_2 → H
s. *29,* 68; of pyrazolo[1,5-*a*]pyrimidine-7-amines s. C. Bellec et al., Can. J. Chem. *59,* 2826–32 (1981).

Sodium tetrahydridoborate *$NaBH_4$*
Alcohols from 3-acylthiazolidine-2-thiones ←

57.

A soln. of startg. 3-acylthiazolidine-2-thione in tetrahydrofuran treated with a soln. of $NaBH_4$ in aq. tetrahydrofuran, the mixture stirred at room temp. until the original colour of the soln. disappeared, and acidified with aq. HCl → product. Y 96%. F.e., also formation of aldehydes, s. Y. Nagao et al. J. Chem. Soc. Perkin Trans. I *1980,* 2470–3; **S-containing leaving groups,** especially **reactions of 3-acylthiazolidine-2-thiones, review,** s. Yakugaku Zasshi *102,* 401–27 (1982).

Sodium dihydridobis(2-methoxyethoxo)aluminate *$Na[AlH_2(OCH_2CH_2OMe)_2]$*
α,β-Oxidoaldehydes from glycidic acid amides CON< → CHO
s. *37,* 131

Tri-n-*butyltin hydride* n-Bu_3SnH

Replacement of aliphatic nitro groups by hydrogen $NO_2 \rightarrow H$

Selective reduction

58. $p\text{-}NCC_6H_4CMe_2NO_2 \rightarrow p\text{-}NCC_6H_4CHMe_2$

A mixture of the startg. nitro compd., tri-*n*-butyltin hydride, and a little azodiisobutyronitrile refluxed in benzene for 1.5 hrs. → product. Y 94%. – Many other functional groups, e.g. keto, ester, cyano, chloro, and organic sulfur groups were unaffected under these conditions. F.e. and limitations, **also replacement by deuterium,** s. N. Ono et al., Tetrahedron Letters *22,* 1705–8 (1981); s.a. D.D. Tanner et al., J. Am. Chem. Soc. *103,* 1557–9 (1981).

Tri-n-*butyltin hydride/tetrakis(triphenylphosphine)palladium(0)* n-$Bu_3SnH/Pd(PPh_3)_4$

Ethylene derivs. from 2-ethyleneamines C=C–C–N< → C=C–CH

s. *37,* 56

Nickel *Ni*

Prim. amines from hydrazones $C{=}N{-}NH_2 \rightarrow CH{-}NH_2$

s. *9,* 444; s.a. J. Wolinski, K. Biniecki, Acta. Pol. Pharm. *38,* 179 (1981); Chem. Abstr. *96,* 19934 (1982).

Titanium trichloride $TiCl_3$

Reductive elimination of N-heterocyclic diazo groups ←

s. *36,* 62; **review of heterocyclic diazo compds.** s. M.H. Elnagdi, M.E. Zayed, S. Abdou, Heterocycles *19,* 559–78 (1982).

Halogen ↑ HC↓↑Hal

Electrolysis ϟ

Replacement of bridgehead iodine by hydrogen I → H

59.

A soln. of 1-iodoadamantane in dimethylformamide containing tetra-*n*-butylammonium perchlorate electrolysed using a Hg pool cathode and Pt gauze anode at –2350 mV until the current reached background level → adamantane. Y 78%. F.e.s. R.S. Abeywickrema, E.W. Della, J. Org. Chem. *46,* 2352–5 (1981).

Potassium/dicyclohexyl-18-crown-6 polyether ←

Replacement of halogen by hydrogen F → H

of bromine cf. *23,* 92; of fluorine in the presence of dicyclohexyl-18-crown-6 polyether, s. T. Ohsawa et al., Tetrahedron Letters *22,* 2583–6 (1981).

Sodium hydroxide s. under Al and $RhCl_3$ *NaOH*

Sodium formate/tetrakis(triphenylphosphine)palladium(0) *$HCOONa/Pd(PPh_3)_4$*
Replacement of chlorine by hydrogen in N-heterocyclics Cl → H
with Pd–C/Pt–C cf. *5,* 75; with Na formate and a little tetrakis(triphenylphosphine)palladium(0), pyrazines, also with retention of cyclic N-oxide groups, s. Y. Akita, A. Ohta, Heterocycles *16,* 1325–8 (1981).

Triethylamine s. under $(EtO)_2P(O)H$ *Et_3N*

Ethyldiisopropylamine s. under Pd–C *$i\text{-}Pr_2NEt$*

Barium hydroxide s. under Pd–C *$Ba(OH)_2$*

Aluminium/mercuric chloride/sodium hydroxide *$Al/HgCl_2/NaOH$*
Bromides from 1,1-dibromides CBr_2 → CHBr
with *n*-BuLi/HBr cf. *26,* 70; with aluminium/mercuric chloride/NaOH s. A.I. Dyachenko, O.S. Korneva, O.M. Nefedov, Izv. Akad. Nauk SSSR Ser. Khim. *1980,* 2842; with Mg/alcohol cf. ibid. *1980,* 2432.

Sodium tetrahydridoborate (s.a. under $(C_5H_5)_2TiCl_2$) *$NaBH_4$*
Prim. alcohols from carboxylic acid halides COF → CH_2OH
from chlorides cf. *5,* 40; from fluorides s. Y.M. Vilenchik et al., Zh. Vses. Khim. Obshchest. *26,* 212–3 (1981).

Sodium tetrahydridoborate/ tributylhexadecyl phosphonium bromide *$NaBH_4/[C_{16}H_{33}P(C_4H_9)_3]^+Br^-$*
Hydrocarbons from halides Hal → H
s. *25,* 62; selective reductive, phase transfer catalysis in the presence of tributylhexadecylphosphonium bromide, s. F. Rolla, J. Org. Chem. *46,* 3909–11 (1981).

Sodium tetrahydridoborate/palladium(II) chloride *$NaBH_4/PdCl_2$*
Replacement of aromatic halogen by hydrogen ArHal → ArH
s. *37,* 49

Lithium tetrahydridoaluminate (s.a. under $TiCl_3$) *$LiAlH_4$*
Replacement of halogen by hydrogen under heterogeneous conditions Hal → H
Ultrasonication

60. $o\text{-}BrC_6H_4Me \rightarrow C_6H_5Me$

Arenes. A mixture of *o*-bromotoluene and $LiAlH_4$ in dimethoxyethane under N_2 ultrasonicated with an ultrasound laboratory cleaner for 5 hrs. at 35° → toluene. Y 98%. – High yields (70–99%) were obtained with both simple and deactivated ar. halides F.e.s. B.H. Han, P. Boudjouk, Tetrahedron Letters *23,* 1643–6 (1982).

Tetrahydridoboratobis(triphenylphosphine)copper(I) *$CuBH_4(PPh_3)_2$*
Aldehydes from carboxylic acid chlorides COCl → CHO
s. *11,* 135s*33;* comparison of phosphine complexes s. P.N. Davey, G.W.J. Fleet, P.C. Harding, J. Chem. Res. (S) *1981,* 336–7.

Alumina s. under R_3SnH *Al_2O_3*

Aluminium chloride/ethanethiol *$AlCl_3/EtSH$*
Replacement of aromatic halogen by hydrogen ArHal → ArH
s. *37,* 53

Dicyclohexyl-18-crown-6 polyether s. under K ←

Alcohols s. under $RhCl_3$ *ROH*

Isopropanol s. under $Fe(CO)_5$ *$(CH_3)_2CHOH$*

Ethanethiol s. under $AlCl_3$ *EtSH*

Trimethylsilyl iodide *Me_3SiI*

Ketones from α-halogenoketones COCHal → COCH
s. *26,* 71s*36;* s.a. T.-L. Ho, Synth. Commun. *11,* 101–3 (1981).

Titanocene dichloride/sodium tetrahydridoborate/air *$(C_5H_5)_2TiCl_2/NaBH_4/O_2$*
Preferential replacement of halogen by hydrogen Hal → H

61. $p\text{-}I\text{–}C_6H_4Cl \rightarrow C_6H_5Cl$

Dimethylformamide added to a mixture of 1-chloro-4-iodobenzene, $NaBH_4$, and a little titanocene dichloride, and stirred at 70° for 24 hrs. in *air* → chlorobenzene. Y 99%. – The reaction does not occur under inert atmospheres. F.e., also with retention of bromine, s. B. Meunier, J. Organometal. Chem. *204,* 345–6 (1981).

Titanium trichloride/lithium tetrahydridoaluminate *$TiCl_3/LiAlH_4$*
Replacement of halogen by hydrogen
with $TiCl_3$ cf. *29,* 79; with titanium slurries prepd. from $TiCl_3/LiAlH_4$, and f. reductions, s. R. Dams, M. Malinowski, H.-J. Geise, Rec. Trav. Chim. Pays-Bas *101,* 112–4 (1982).

Organotin hydrides–alumina *$R_3SnH\text{–}Al_2O_3$*

62.

A mixture of 1,1-dichloro-2,2-diphenylcyclopropane, Al_2O_3-supported organotin hydride, and cyclohexane heated at 40° for 2 hrs. → 1,1-diphenylcyclopropane. Y 100%. – Supporting organotin hydrides on a solid inorganic carrier improves stability and aids regeneration of the reagent. F.e., also with silica gel-supported organotin hydrides, s. H. Schumann, B. Pachaly, Angew. Chem. Intern. Ed. *20,* 1043–4 (1981).

Triphenylphosphine s. under $RhCl_3$ *Ph_3P*

Tributylhexadecylphosphonium bromide s. under $NaBH_4$ *$[C_{16}H_{33}P(C_4H_9)_3]^+Br^-$*

Diethylphosphonate/triethylamine *$(EtO)_2P(O)H/Et_3N$*
Partial replacement of bromine by hydrogen Br → H
with trimethyl phosphite cf. *32,* 64; with diethyl phosphite/triethylamine s. T. Hirao et al., J. Org. Chem. *46,* 3745–7 (1981).

Sodium diethyl phosphite s. under Te *$(EtO)_2PONa$*

Diphosphorus tetraiodide P_2I_4

Ketones from α-halogenoketones COCHal → COCH

63. 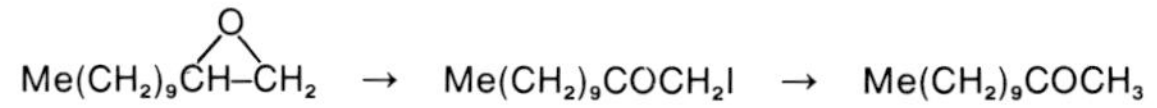

A mixture of startg. m. and P_2I_4 in dichloromethane stirred at 20° for 1 hr. → product. Y 91%. – This is part of a 2-step prepn. of **methyl ketones from oxido compds.,** contrasting with the direct acid-catalysed isomerization which usually proceeds with rearrangement or in the direction of the regioisomer. F.e., also with phosphorus triiodide, s. J.N. Denis, A. Krief, Tetrahedron Letters *22,* 1431–2 (1981).

Sodium dithionite $Na_2S_2O_4$

via pyridinium salts cf. *29,* 80; directly with $Na_2S_2O_4$ s. S.-K. Chung, Q.-Y. Hu, Synth. Commun. *12,* 261–6 (1982).

Tellurium/sodium diethyl phosphite $Te/(EtO)_2PONa$

64. 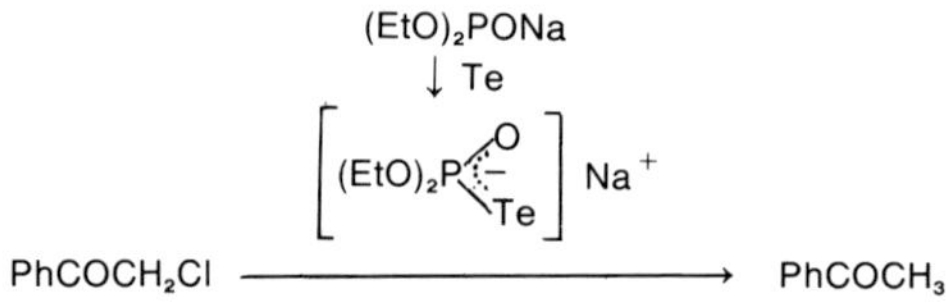

An ethanolic soln. of Na diethyl phosphite injected onto finely divided tellurium under argon, the mixture stirred to complete the formation of O,O-diethyl phosphorotelluroate, a soln. of 2-chloroacetophenone in ethanol introduced, and stirring continued at 25° for 1 hr. → acetophenone. Y 84%. F.e.s. D.L.J. Clive, P.L. Beaulieu, J. Org. Chem. *47,* 1124–6 (1982).

Iron–graphite Fe–C

Ketones from α-bromoketones COCBr → COCH

s. *37,* 956

Iron carbonyl/isopropyl alcohol $Fe(CO)_5/(CH_3)_2CHOH$

1,1-Di- from 1,1,1-tri-chlorides $CCl_3 \rightarrow CHCl_2$

s. *27,* 77; f. catalysts s. Izv. Akad. Nauk SSSR Ser. Khim. *1981,* 2090–4.

Rhodium trichloride/triphenylphosphine/alcohols/sodium hydroxide ←

Replacement of aromatic halogen by hydrogen ArHal → ArH

with NaOMe/$Pd(PPh_3)_4$ cf. *29,* 77s*33;* with NaOH/$RhCl_3/PPh_3$ in ethanol, also with Pd(II) acetate, s. T. Okamoto, S. Oka, Bull. Chem. Soc. Japan *54,* 1265–6 (1981).

Palladium–carbon/ethyldiisopropylamine $Pd\text{–}C/i\text{-}Pr_2NEt$

Aldehydes from carboxylic acid chlorides COCl → CHO

Modified Rosemund-Zetzsche reduction

s. *27,* 79; selectivity s. Rec. Trav. Chim. Pays-Bas *100,* 21–4 (1981).

Palladium–carbon/barium hydroxide $Pd\text{–}C/Ba(OH)_2$

Replacement of bromine by hydrogen Br → H

with Pd–carbon/Na acetate cf. *33,* 247; with $Ba(OH)_2$ as base and solubilizer, tetronic acid, s. D.G. Schmidt, H. Zimmer, Synth. Commun. *11,* 895–90 (1981).

Tetrakis(triphenylphosphine)palladium(0) s. under HCOONa $Pd(PPh_3)_4$

Palladium(II) chloride s. under $NaBH_4$ $PdCl_2$

Sulfur ↑ **HC↓↑S**

Isopropylmagnesium bromide/ dichlorobis(triphenylphosphine)nickel(II) *i-PrMgBr/$NiCl_2(PPh_3)_2$*

Ethylene derivs. from thioenolethers C=CSR → C=CH
Retention of configuration

65. 

A little dichlorobis(triphenylphosphine)nickel(II) added to a soln. of startg. (E)-thioenolether in dry tetrahydrofuran at room temp. under N_2, the mixture stirred for 5 min., a soln. of isopropylmagnesium bromide in ether added, and refluxed at 65° for 36 hrs. → (Z)-product. Y 76%. F.e.s. B.M. Trost, P.L. Ornstein, Tetrahedron Letters *22,* 3463–6 (1981).

Sodium tetrahydridoborate *$NaBH_4$*

Prim. alcohols from thiolic acid esters COSR → CH_2OH
Selective reduction

66. $CH_3(CH_2)_{14}COSPh \rightarrow CH_3(CH_2)_{14}CH_2OH$

$NaBH_4$ added to a soln. of phenyl hexadecanothioate in ethanol at 0°, the mixture stirred at room temp. under argon for 6 hrs. then cooled to 0° → 1-hexadecanol. Y 99%. – Nitrile, carboxylic acid and ester functions are unaffected under these conditions. F.e.s. H.-J. Liu, R.R. Bukownik, P.R. Pednekar, Synth. Commun. *11,* 599–603 (1981).

1-Benzyl-1,4-dihydronicotinamide ←

Nitro compds. from 1-nitrosulfones $C(NO_2)SO_2R \rightarrow CHNO_2$

67. $Me_2C(NO_2)SO_2Ph \rightarrow Me_2CHNO_2$

A mixture of startg. m. and 1-benzyl-1,4-dihydronicotinamide in dimethylformamide stirred under N_2 at room temp. for 6 hrs. → product. Y 95%. F.e.s. N. Ono et al., J. Chem. Soc. Chem. Commun. *1981,* 71–2.

Nickel *Ni*

Desulfuration of thiophene derivs C
long chain fatty acids cf. *11,* 139; macrocyclic ketones s. C. Galli et al., J. Org. Chem. *45,* 1906 (1980)

Replacement of arylthio groups by hydrogen SPh → H
s. *37,* 612

Remaining Elements ↑ **HC↓↑Rem**

Sodium hydroxide *NaOH*

α,β-Ethylene- from α,β-ethylene-α-silylnitriles C=CSi≡ → C=CH
s. *37,* 875

Sodium methoxide *NaOMe*

Replacement of vinylic silyl groups by hydrogen
with acetic acid cf. *34,* 66; with sodium methoxide, α,β-ethylenehalides from 1-halogenoenesilanes, s. H.-P. On, W. Lewis, G. Zweifel, Synthesis *1981,* 999–1000.

1-Benzyl-1,4-dihydronicotinamide/irradiation ←

Hydrocarbons from monoorganomercury compds. C–Hg– → C–H

s. *14,* 206s*37*

Trifluoroacetic acid CF_3COOH

Allenes from silyl-2-acetylenes C≡C–CSi⋞ → CH=C=C

68. $Me(CH_2)_5C{\equiv}CH$ -----→ $Me(CH_2)_5C{\equiv}CCH_2SiMe_3$ ——→ $Me(CH_2)_5CH{=}C{=}CH_2$

Startg. acetylenesilane in methylene chloride treated with trifluoroacetic acid at room temp. for ca. 1 hr. → 1,2-nonadiene. Y 52%. – This is part of a multistep **synthesis of terminal allenes from acetylene derivs. with addition of one C-atom.** F.e., **also halogenallenes and** α**-alleneketones** (using halogen and acyl chloride, respectively, as electrophile), s. T. Flood, P.E. Peterson, J. Org. Chem. *45,* 5006–7 (1980).

*Tri-*n*-butyltin hydride* n-Bu_3SnH

Hydrocarbon groups and aldehydes from selenolic acid esters COSeR → CHO

69.

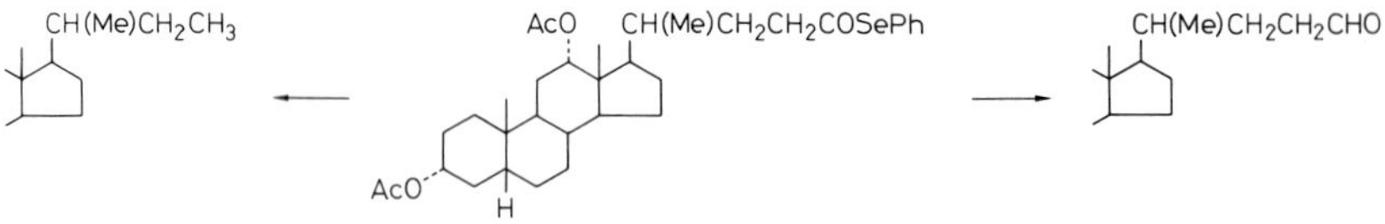

A refluxing soln. of phenyl 3α,12α-diacetoxy-5β-selenocholan-24-oate in

mesitylene | benzene

treated with tri-n-butyltin hydride and a little azodiisobutyronitrile, and heating continued until t.l.c. indicated completion of reaction (2–60 min.) →

3α,12α-diacetoxy-20α-ethyl-5β-pregnane. Y 84%. | 3α,12α-diacetoxy-5β-cholan-24-al. Y 92%.

F.e., also aldehydes photochemically, s. J. Pfenninger, C. Heuberger, W. Graf, Helv. Chim. Acta *63,* 2328–37 (1980).

Formic from selenolcarbonic acid esters ArSeCOOR → HCOOR

s. *37,* 54

Hydrocarbons from monoorganomercury compds. C–Hg– → C–H

with $NaBH_4$ cf. *14,* 206; with tri-n-butyltin hydride s. A.J. Bloodworth, J.L. Courtneidge, J. Chem. Soc. Chem. Commun. *1981,* 1117–8; with 1-benzyl-1,4-dihydronicotinamide under irradiation cf. H. Kurosawa, H. Okada, T. Hattori, Tetrahedron Letters *22,* 4495–8 (1981); **review** of demercuration with $NaBH_4$ s. E.A. Sullivan, Ventron Alembic *22,* 1–3 (1981); Chem. Abstr. *95,* 41561 (1981).

Carbon ↑ HC↓↑C

Potassium hydroxide *KOH*

Replacement of cyano groups by hydrogen CN → H

Alkali fusion

70.

A mixture of chlorpheniramine nitrile and 85% KOH pellets heated at 160° under N_2 with rapid stirring for 4 hrs. → chloropheniramine. Y 95%. F.e.s. C.E. Berkoff et al., Synth. Commun. *10,* 939–45 (1980).

Sodium tetrahydridoborate/hydrogen chloride *$NaBH_4/HCl$*

Sec. amines and ketones from oxo compds., via eneformamidines ←

Synthesis with addition of one C-atom

71.

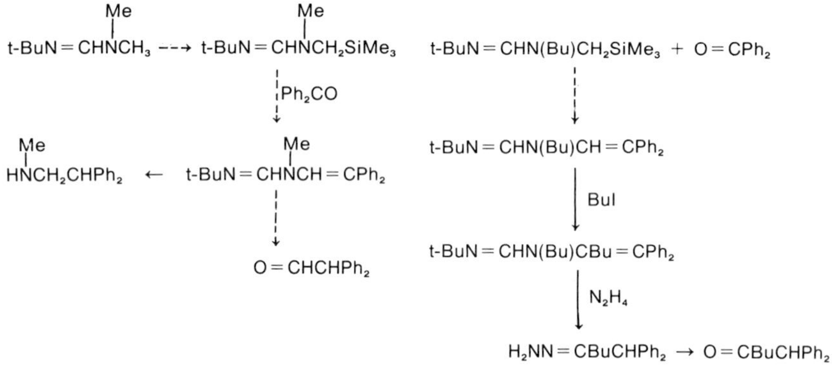

A soln. of startg. crude enamidine (prepd. from benzophenone and the corresponding formamidine via metalation and Peterson olefination)

in 80% ethanol treated with 10% HCl until pH 6 was attained, a soln. of $NaBH_4$ in the same solvent added dropwise between −5 and −15° with further HCl added to maintain acidity, the mixture stirred at 0° for 1 hr., made strongly alkaline (pH > 12) by treatment with NaOH pellets, the concd. ethereal extract dissolved in tetrahydrofuran, treated with 10% HCl, and stirred for 2 hrs. → product. Y 65% (based on startg. silylamidine).

in tetrahydrofuran at −75° treated with 2 equivalents *tert*-butyllithium, allowed to warm to −25°, stirred for 1.5 hrs., recooled to −78°, *n*-butyl iodide in the same solvent added, and allowed to warm to room temp. → intermediate enamidine, heated with hydrazine/acetic acid/ethanol/water for 6 hrs. → intermediate hydrazone, in tetrahydrofuran treated with aq. Cu(II) acetate, and refluxed for 30 min. → product. Y 61% (based on startg. silylamidine).

F.e., also reduction of the intermediate enamidines to **aldehydes,** s. A.I. Meyers, G.E. Jagdmann, J. Am. Chem. Soc. *104,* 877–9 (1982).

Lithium hydridotriethylborate *$Li[BHEt_3]$*

Prim. alcohols from trityl ketones $COCPh_3 \rightarrow CH_2OH$

Reductive C-cleavage

72. $Me(CH_2)_4C{\equiv}CCHMeCH_2COCPh_3 \rightarrow Me(CH_2)_4C{\equiv}CCHMeCH_2CH_2OH$

4-Acetylenealcohols. A soln. of 4-methyl-1,1,1-triphenyl-5-undecyn-2-one in tetrahydrofuran at −45° treated with 1 *M* Li hydridotriethylborate in the same solvent, the mixture warmed with stirring for 40 hrs., treated with 1 *M* HCl, and worked up with ether → 3-methyl-4-decyn-1-ol. Y 77%. F.e.s. R. Locher, D. Seebach, Angew. Chem. Intern. Ed. *20,* 569–70 (1981).

Ethanedithiol/boron fluoride *$HSCH_2CH_2SH/BF_3$*

C-Deacylation COR → H

with *p*-toluenesulfonic acid/ethylene glycol cf. *22,* 86s*32;* with ethanedithiol/boron fluoride-etherate, pyrrole ring, s. K.M. Smith, K.C. Langry, J. Chem. Soc. Chem. Commun. *1981,* 283–4.

Tri-n-*butyltin hydride* n-Bu_3SnH

Hydrocarbon groups from selenolic acid esters COSeR → H

Degradation with loss of one C-atom

s. *37,* 69

Phosphoric acid H_3PO_4

tert-**Butyl as C-blocking group** Ar-Bu-*t* → ArH

Removal of the blocking group

with $AlCl_3$ cf. *10,* 103; with phosphoric acid s. M. Tashiro, G. Fukata, T. Itoh, Synthesis *1981,* 489–90.

Rhodium–carbon *Rh–C*

Ketones from tetrahydro-4-pyrones C

73. 

A mixture of startg. tetrahydro-4-pyrone and H_2 passed over 5% Rh on C at 320–80° → product. Y 75%. F.e. and catalysts s. M.M. Vartanyan et al., Izv. Akad. Nauk SSSR Ser. Khim. *1981,* 912–3.

Elimination ⇑

Oxygen ↑ HC⇑O

Lithium tetrahydridoaluminate/titanocene dichloride $LiAlH_4/(C_5H_5)_2TiCl_2$

Replacement of allylic and benzylic hydroxyl groups by hydrogen OH → H

74. $Me_2C{=}CHCH_2CH_2CMe{=}CHCH_2OH \rightarrow Me_2C{=}CHCH_2CH_2CMe{=}CHCH_3$

$PhCH_2OH \rightarrow PhCH_3$

A soln. of 1 equivalent startg. allyl alcohol, 2 equivalents $LiAlH_4$, and a little titanocene dichloride in tetrahydrofuran stirred for 3 hrs. at –20° then at room temp. overnight → product. Y 85%. – Similarly, benzyl alcohol at reflux → toluene. Y 81%. F.e., also alcohols by **O-deallylation,** and hydrocarbon from carbonyl groups, s. F. Sato et al., Chem. Letters *1980,* 103–6.

Aluminium chloride/ethanethiol $AlCl_3/EtSH$

Arenes from phenols ArOH → ArH

s. *37,* 53

Acridanes ←

Reductions with hindered 1,4-dihydropyridines in acids ArC(OH) → ArCH

Alkylarenes from benzylalcohols

75. 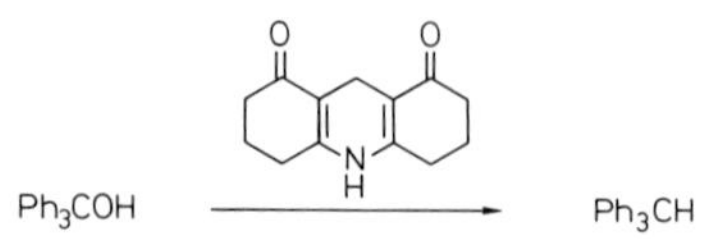

Trityl alcohol and acridane deriv. in dichloromethane allowed to react in the presence of trifluoroacetic acid (or boron fluoride-etherate) at room temp. for 15 min. →

triphenylmethane. Y 89%. – Unlike N-benzyl-1,4-dihydropyridine and Hantszch ester, the hindered acridine analogs are stable towards heat and strong acid. F.e., **also from styrenes,** s. S. Singh et al., J. Chem. Soc. Chem. Commun. *1982,* 453–4; f. reductions with 10-subst. decahydroacridines s. V.A. Kaminskii et al., Khim. Geterotsikl. Soedin., *1981,* 95–8.

Benzoyl chloride s. under $Fe(CO)_5$ C_6H_5COCl

Titanocene dichloride s. under $LiAlH_4$ $(C_5H_5)_2TiCl_2$

Triethylsilane/boron fluoride Et_3SiH/BF_3

Cyclic ethers from lactones via lactols ←

76.

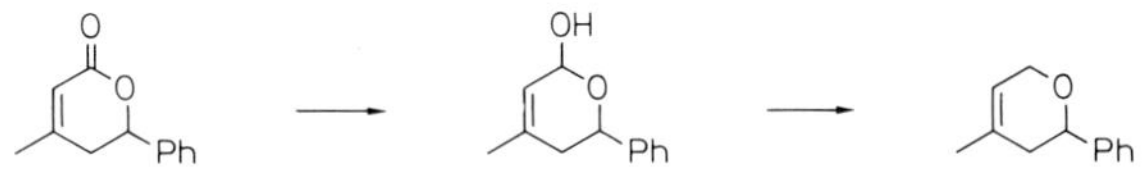

Startg. lactone treated with diisobutylaluminium hydride in toluene according to Synth. Meth. *24,* 802 → crude lactol and triethylsilane in dichloromethane cooled under N_2 to −78°, BF_3-etherate added dropwise, the mixture stirred for 1 hr., and allowed to warm to room temp. with vigorous stirring → product. Y 72%. – Benzyl ethers, hindered esters, unprotected alcohols, and carbon-carbon double bonds are unaffected. F.e.s. G.A. Kraus et al., J. Org. Chem. *46,* 2417–9 (1981).

Iron carbonyl/benzoyl chloride $Fe(CO)_5/C_6H_5COCl$

Hydrocarbons from benzyl alcohols ArC(OH) → ArCH

with $Fe(CO)_5$/K cf. *36,* 79; with $Fe(CO)_5$/benzoyl chloride, di- and triarylmethanes, s. T.-H. Luh, Synth. Commun. *11,* 829–33 (1981).

Via intermediates *v.i.*

Hydrocarbons from alcohols via mesylates OH → H

s. *11,* 149; electrochemical method, selective reduction, cf. T. Shono et al., Tetrahedron Letters *1979,* 2157–60.

Sulfur ↑ HC⇑S

Dibenzoyldiazene ←

Desulfuration of 2-thiouracils ←

77.

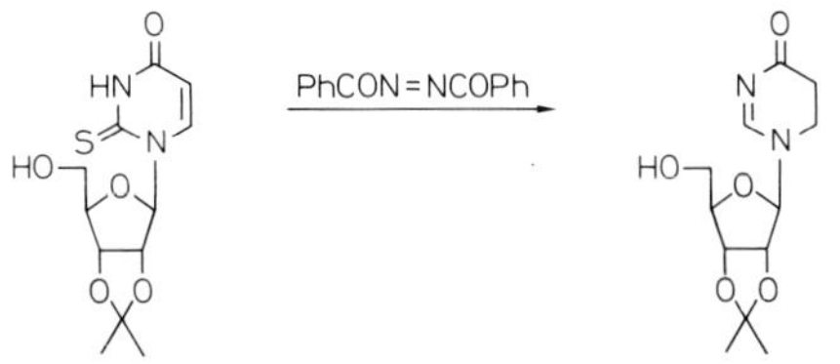

A soln. of 3 equivalents of dibenzoyldiazene in tetrahydrofuran added to an aq. soln. of 2′,3′-O-isopropylidene-2-thiouridine (pH 9) at room temp., stirred overnight, another 3 equivalents of dibenzoyldiazene added, and stirring continued a further 24 hrs. → product. Y 65%. F.e. and methods s. O. Mitsunobu et al., Tetrahedron Letters *23,* 517–20 (1982).

Carbon ↑ HC⇑C

Without additional reagents *w.a.r.*

Thermal decarboxylation COOH → H

in paraffin cf. *6,* 135; in 1,1,2,2-tetrachloroethane s. T. Fujii et al., Chem. Pharm. Bull. *29,* 3392–7 (1981).

Irradiation ⫻

Acylamines from α-acylaminocarboxylic acids AcNCCOOH → AcNCHR
Photodecarboxylation

78. $PhCONHCH_2COOH \rightarrow PhCONHCH_3$

Aroylamines. A soln. of N-benzoylglycine in acetone irradiated using a 400 W. high-pressure Hg lamp at room temp. for 3 hrs. under argon → N-methylbenzamide. Y 99%. F.e.s. H. Nakai et al., Synthesis *1982,* 141–3.

Piperidine/sulfuric acid $C_5H_{11}N/H_2SO_4$

Decarboxylation COOH → H

with pyridine cf. *26,* 86; with piperidine/sulfuric acid, chromones from chromone-3-carboxylic acids, s. C.K. Ghosh, S. Khan, Synthesis *1981,* 719–21.

Morpholine ←

Aldehydes from α-ketocarboxylic acids COCOOH → CHO

with N-ethylmorpholine cf. *21,* 120s*36;* with morpholine s. I.K. Stamos, Tetrahedron Letters *23,* 459–62 (1982).

Formation of O–N Bond

Uptake ⇓

Addition to Nitrogen ON⇓N

3-Bromo-5-hydroperoxy-4,4-dimethyl-3,5-diphenyl-Δ¹-pyrazoline ←

Amine oxides from tert. amines $R_3N \rightarrow R_3NO$

s. *37,* 120

Oxidations with *pentafluoroperoxybenzoic acid* $C_6F_5COO_2H$

79.

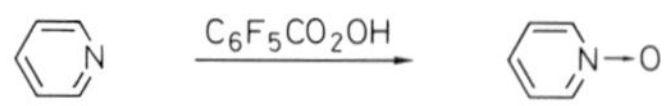

Cyclic N-Oxides. A soln. of pentafluoroperoxybenzoic acid in methylene chloride added slowly to a rapidly stirred suspension of pyridine and Na_2CO_3 in the same solvent, and the mixture stirred for 0.1 hr. at 25° → pyridine N-oxide. Y 94%. **Also oxido compds. from ethylene derivs.,** and **sulfoxides or sulfones from thioethers,** s. U.M. Dzhemilev et al., Izv. Akad. Nauk SSSR Ser. Khim. *1981,* 1320–2.

Nitric acid HNO_3

Aromatic nitro from nitroso compds. $NO \rightarrow NO_2$

without nitration s. *23,* 114; 5-nitro-8-hydroxyquinoline with 55% HNO_3 s. M.O. Lozinskii et al., Farm. Zh. (Kiev) *1981,* 65 (Ukrain); Chem. Abstr. *95,* 61950 (1981).

Exchange ⇂↾

Hydrogen ↑ ON⇂↾H

Sodium permanganate *$NaMnO_4$*

Nitro compds. from prim. amines $NH_2 \rightarrow NO_2$

s. *37,* 232

Elimination ⇑

Hydrogen ↑ ON⇑H

Phenyl iodosotrifluoroacetate *$PhI(OCOCF_3)_2$*

Furoxan ring from glyoximes ○

with NaOCl cf. *12,* 155; with iodosobenzene bis(trifluoroacetate) s. D.N. Nicolaides, J.K. Gallos, Synthesis *1981,* 638–9.

N-Chloroisatin ←

Nitroso compds. from hydroxylamines NHOH → N=O

s. *37,* 276

Oxygen ↑ ON⇑O

Sodium hydride *NaH*

1,2-Benzisoxazoles from *o*-hydroxyoxime acetates ○

s. *7,* 161; with NaH and f. methods cf. G.M. Shutske et al., J. Med. Chem. *25,* 36–44 (1982).

Formation of O–Hal Bond

Uptake ⇓

Addition to Halogen OHal⇓Hal

Pyridine *C_5H_5N*

Iodosoacetates from iodides $I \rightarrow I(OAc)_2$

s. *27,* 543; **review of iodosocarboxylates** s. A. Varvoglis, Chem. Soc. Rev. *10,* 377–407 (1981).

Potassium bromate *$KBrO_3$*

Iodoxy compds. from iodides $I \rightarrow IO_2$

with peroxyacetic acid cf. *19,* 125; with $KBrO_3$ s. A. Banerjee et al., J. Indian Chem. Soc. *58,* 605–7 (1981).

Formation of O–S Bond

Uptake ⇓

Addition to Hydrogen and Sulfur OS⇓HS

Tetraphenylporphyrin/irradiation ←

Sulfonic acids from mercaptans SH → SO_3H
with crystal violet as sensitizer cf. *31,* 82; with tetraphenylporphyrin s. S.J. Jongsma, J. Cornelisse, Tetrahedron Letters *22,* 2919–22 (1981).

Addition to Sulfur OS⇓S

Irradiation ⫻

Sulfines from thioketones C=S → C=S=O
s. *21,* 130s*36;* **review of photooxidation of organosulfur compds.** s. W. Ando, Sulfur Reports *1,* 147–207 (1981).

Sodium hydroxide s. under NaOCl *NaOH*

*Tetra-*n*-butylammonium bromide s. under NaOCl* n-$Bu_4N^+Br^-$

3-Bromo-5-hydroperoxy-4,4-dimethyl-3,5-diphenyl-Δ¹-pyrazoline ←

Sulfoxides from thioethers >S → >SO
s. *37,* 120

Peroxyacetic acid (s.a. under $(acac)_2MoO_2$) $MeCOO_2H$

Cyclic 1,1-disulfones from mercaptals
s. *37,* 20

1,1-Disulfones from mercaptals $C(SR)_2 \rightarrow C(SO_2R)_2$
with peroxypropionic acid cf. *9,* 342; with peroxyacetic acid s. E.H. Mørkved, M.W. Cronyn, J. Pharm. Sci. *71,* 59–63 (1982).

Pentafluoroperoxybenzoic acid $C_6F_5COO_2H$

Sulfoxides and sulfones from thioethers >S → >SO → $>SO_2$
s. *37,* 79

Bis(p-methoxyphenyl) selenoxide Ar_2SeO

Sulfoxides from thioethers >S → >SO
s. *37,* 702

Iodylbenzene $PhIO_2$

S-Oxidation of thioethers
s. *37,* 266

Iodosobenzene/meso-*tetraphenylporphyrinatoiron(III)chloride* ←
Sulfoxides from thioethers

80. $Ph_2S \longrightarrow Ph_2SO \dashrightarrow Ph_2SO_2$

A mixture of diphenyl sulfide, iodosobenzene, and a little *meso*-tetraphenylporphyrinatoiron(III) chloride in methylene chloride stirred at −18° for 50 min. under argon → diphenyl sulfoxide. Y 87%. – The method is useful for oxidizing substrates such as nitroaryl derivs., which are resistant to the conventional peracids. F.e. and catalyst, also oxidation to **sulfones,** s. W. Ando, R. Tajima, T. Takata, Tetrahedron Letters *23,* 1685–8 (1982); with H_2O_2 as re-oxidant in the presence of imidazole cf. S. Oae, Y. Watanabe, K. Fujimori, ibid. *23,* 1189–92 (1982).

Titanium trichloride/hydrogen peroxide $TiCl_3/H_2O_2$
... under mild conditions

81. $Ph_2S \rightarrow Ph_2SO$

A soln. of H_2O_2 in methanol added dropwise to a stirred soln. of diphenyl sulfide and $TiCl_3$ in 5:1 methanol/water at room temp., and stirring continued for 5 min. → diphenyl sulfoxide. Y 100%. – The product is uncontaminated with sulfone. F.e.s. Y. Watanabe, T. Numata, S. Oae, Synthesis *1981,* 204–6.

Peroxymonophosphoric acid H_3PO_5
Sulfones from sulfoxides $\gtrless SO \rightarrow \gtrless SO_2$
s. *37,* 102

Hydrogen peroxide (s.a. under $TiCl_3$, Na_2WO_4*)* H_2O_2
Sulfoxides from thioethers $\gtrless S \rightarrow \gtrless SO$
s. *5,* 101; improved procedure s. J. Drabowicz, M. Mikolajczyk, Synth. Commun. *11,* 1025–30 (1981).

Hydrogen peroxide/meso-*tetraphenylporphyrinatoiron(III)chloride* ←
s. *37,* 80

Potassium peroxymonosulfate $KHSO_5$
Selective oxidation of thioethers to sulfones under mild conditions $\gtrless S \rightarrow \gtrless SO_2$

82. $PhSMe \rightarrow [PhS(O)Me] \rightarrow PhS(O)_2Me$

A soln. of thioanisole in methanol at 0° treated with 3 equivalents $KHSO_5$ in water, and the slurry stirred for 4 hrs. at room temp. → methyl phenyl sulfone. Y 100%. – The reaction is highly selective leaving hydroxyl, keto, and methylene groups intact. F.e., also isolation of the intermediate **sulfoxides,** s. B.M. Trost, D.P. Curran, Tetrahedron Letters *22,* 1287–90 (1981).

Molybdenyl acetylacetonate/peroxyacetic acid $(acac)_2MoO_2/MeCOO_2H$
Sulfoxides and sulfones from thioethers $\gtrless S \rightarrow \gtrless SO \rightarrow \gtrless SO_2$
with $MoCl_5/H_2O_2$ or $Mo(CO)_6/H_2O_2$ cf. *17,* 153s*28;* with molybdenyl acetylacetonate/peroxyacetic acid, selective oxidation, s. J. Rouchaud, C. Moons, J. Meyer, Bull. Soc. Chim. France II *1980,* 441–3.

Sodium tungstate/hydrogen peroxide Na_2WO_4/H_2O_2
Sulfones from thioethers $\gtrless S \rightarrow \gtrless SO_2$
with tungstic acid/H_2O_2 cf. *18,* 142; with Na_2WO_4/H_2O_2, 3-sulfonylpyridines, s. C.S. Giam, K. Kikukawa, D.A. Trujillo, Org. Prep. Proc. Intern. **13,** 137–40 (1981).

Sodium hypochlorite/sodium hydroxide *NaOCl/NaOH*

Sulfoximines from sulfilimines ⪰S=NH → ⪰S(O)=NH

with *m*-chloroperoxybenzoic acid/K_2CO_3 cf. *26*, 94s*36;* with NaOCl/NaOH, N-unsubst. sulfilimines, s. N. Furukawa et al., Synthesis *1982*, 77–8.

Sodium hypochlorite/tetra-n-*butylammonium bromide* *NaOCl/$Bu_4N^+Br^-$*

N-Sulfonylsulfoximines
from N-sulfonylsulfilimines ⪰S=N−SO_2 → ⪰S(O)=N−SO_2

with $KMnO_4$/pyridine cf. *26*, 94; with NaOH and tetra-*n*-butylammonium bromide as phase transfer catalyst s. N. Furukawa et al., Tetrahedron Letters *22*, 3989–92 (1981).

Sodium permanganate *$NaMnO_4$*

Sulfones from thioethers ⪰S → ⪰SO_2

s. *37*, 232

Benzyltriethylammonium permanganate *$[PhCH_2NEt_3]^+MnO_4^-$*

Sulfones from sulfoxides ⪰SO → ⪰SO_2

with $KMnO_4$ cf. *25*, 82; with benzyltriethylammonium permanganate, also from thioethers, s. D. Scholz, Monatsh. Chem. *112*, 241–3 (1981).

Polymer-based periodate and iodate ←

Sulfoxides from thioethers ⪰S → ⪰SO

s. *37*, 231

meso-*Tetraphenylporphyrinatoiron(III) chloride s. under C_6H_5IO* ←

Rearrangement ⋂

Oxygen/Nitrogen type OS⋂ON

Polyphosphoric acid/bis(4-methoxyphenyl) sulfide *$(HPO_3)_n/Ar_2S$*

o-**Sulfinylcarboxylic acid amides from** *o*-**(arylthio)hydroxamic acids** ←

83. o-PhS–C_6H_4–CONHOH → o-PhS(O)–C_6H_4–$CONH_2$

A mixture of finely ground 2-(phenylthio)benzohydroxamic acid, a little bis(4-methoxyphenyl) sulfide as oxygen transfer agent, and polyphosphoric acid heated at 125° for 2 hrs. → 2-(phenylsulfinyl)benzamide. Y 89%. F.e.s. G.P. Dhareshwar, P.N. Chhaya, B.D. Hosangadi, Indian J. Chem. *19B*, 831–5 (1980).

Exchange ⇅

Oxygen ↑ OS⇅O

Without additional reagents *w.a.r.*

Ditriflates from diols OH → OSO_2CF_3

s. *37*, 108

Potassium hydroxide *KOH*

Sulfamic acids from *o*-hydroxyaryl sulfamidates ←

s. *37,* 277

Phenyl phosphorodichloridate/pyridine *$PhOP(O)Cl_2/C_5H_5N$*

Sulfinic acid derivs. from sulfinic acids OH → OS(O)R

esters with dicyclohexylcarbodiimide s. *27,* 108; also amides and thiosulfinic acid esters with phenyl phosphorodichloridate and pyridine s. M. Furukawa et al., Synthesis *1980,* 937–9.

Nitrogen ↑ OS↓↑N

Sodium methoxide *NaOMe*

O-Tosylation of carbohydrates OH → OTs

with N-*p*-toluenesulfonylimidazole cf. *29,* 101; with N-*p*-toluenesulfinylimidazole via *p*-toluenesulfinic acid esters s. H. Redlich, W.-U. Meyer, Ann. Chem. *1981,* 1354–60.

Halogen ↑ OS↓↑Hal

Sodium hydride *NaH*

O-Nonafluorobutanesulfonation OH → OSO_2R

s. *16,* 151s*28;* **review of perfluoroalkanesulfonic esters,** prepn. and applications, s. Synthesis *1982,* 85–126.

Sulfur ↑ OS↓↑S

Hexamethyldisiloxane/dimethyl sulfoxide *$Me_3SiOSiMe_3/Me_2SO$*

Sulfinyl esters S(O)Hal → S(O)OR

from sulfinyl chlorides and chlorosulfite esters

84. PhS(O)Cl + MeOS(O)Cl → PhS(O)OMe

A few drops of dimethyl sulfoxide added to a mixture of phenylsulfinyl chloride, methyl chlorosulfite, and hexamethyldisiloxane (as Cl^- acceptor), at room temp. for 8 hrs. → product. Y 86%. F.e.s. J. Drabowicz, Chem. Letters *1981,* 1753–4.

Remaining Elements ↑ OS↓↑Rem

Triethylamine *Et_3N*

Preferential protection of carbohydrate hydroxyl via cyclic dialkoxystannanes

s. *32,* 85; protection as (2-methoxyethoxy)methyl ethers s. M.A. Nished, M.S. Chowdhary, L. Anderson, Carbohyd. Res. *102,* 99–110 (1982).

Carbon ↑ OS↓↑C

Without additional reagents *w.a.r.*

Iodosohydroxy tosylates from iodosoacetates $I(OAc)_2$ → I(OH)OTs

s. *33,* 93; **review of iodosocarboxylates** s. A. Varvoglis, Chem. Soc. Rev. *10,* 377–407 (1981).

Formation of O–Rem Bond

Uptake ⇓

Addition to Hydrogen and Oxygen ORem ⇓ HO

Via intermediates *v.i.*

O-Phosphorylation via S,S-diaryl dithiophosphates OH → $OP(O)(OH)_2$

85.

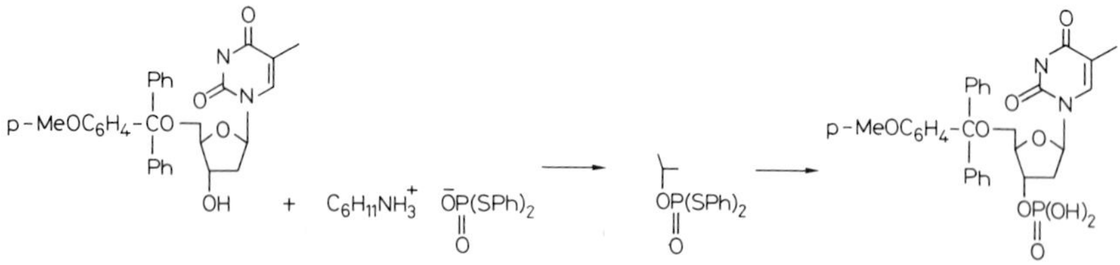

2,4,6-Triisopropylbenzenesulfonyl chloride added to a soln. of cyclohexylammonium S,S-diphenyl phosphorodithioate in pyridine, the mixture kept at room temp. for 30 min., treated with startg. nucleoside, and stirred for 20 hrs. → intermediate dithiophosphate (Y 95%), in pyridine treated with water and Ag acetate, the mixture stirred vigorously at room temp. for 18 hrs., and H_2S introduced at 0° until a clear supernatant was obtained → product (Y 96%). – The S,S-diaryl dithiophosphate group is stable under a variety of conditions, and useful in the protection of nucleoside 3'- or 5'-hydroxyl groups during oligonucleotide synthesis. F.e., also partial removal of arylthio groups, s. M. Sekine, K. Hamaoki, T. Hata, Bull. Chem. Soc. Japan *54,* 3815–27 (1981).

Addition to Sulfur and Remaining Elements ORem ⇓ SRem

m-*Chloroperoxybenzoic acid* m-$ClC_6H_4COO_2H$

Phosphoric sulfonic anhydrides from thiophosphoric esters ⪾P(O)SR → ⪾$P(O)OSO_2R$

s. *37,* 93

Addition to Remaining Elements ORem ⇓ Rem

m-*Chloroperoxybenzoic acid* m-$ClC_6H_4COO_2H$

Phosphoric from phosphorous esters $(RO)_3P$ → $(RO)_3PO$

with Cl_2 cf. *22,* 104s*31;* with *m*-chloroperoxybenzoic acid s. K.K. Ogilvie, M.J. Nemer, Tetrahedron Letters *22,* 2531–2 (1981); under O_2 in the presence of azodiisobutyronitrile, retention of configuration, cf. T.M. Gajda, A.E. Sopchik, W.G. Bentrude, ibid. *22,* 4167–70 (1981).

*Bis(*p*-methoxyphenyl)selenoxide* Ar_2SeO

Phosphine oxides from phosphines R_3P → R_3PO

s. *37,* 702

Sodium periodate — *$NaIO_4$*

Tellurones from telluroxides — $\geq TeO \rightarrow \geq TeO_2$

86. $(p\text{-}MeOC_6H_4)_2TeO \rightarrow (p\text{-}MeOC_6H_4)_2TeO_2$

Aq. $NaIO_4$ added dropwise to a stirred soln. of bis(*p*-methoxyphenyl) telluroxide in methanol at room temp., and the mixture stirred for 3 hrs. → bis(*p*-methoxyphenyl) tellurone. Y 82%. – The tellurone is a relatively mild oxidizing agent, useful, e.g., in preparing **oxo compds. from** the corresponding benzylic **alcohols.** F. reactions s. L. Engman, M.P. Cava, J. Chem. Soc. Chem. Commun. *1982,* 164–5.

Rearrangement ∩

Hydrogen/Oxygen Type ORem ∩ HO

Mesitylenesulfonic acid — *$ArSO_3H$*

Isomerization of Si,O-heterocyclics ←

s. *37,* 2

Oxygen/Sulfur Type ORem ∩ OS

Iodine — *I_2*

Thiolic acid esters from sulfoxides — $\geq P\text{–}C\text{–}S(O)R \rightarrow \geq P(O)C(SR)$

via α-sulfinylphosphines and α-organothiophosphine oxides

87.

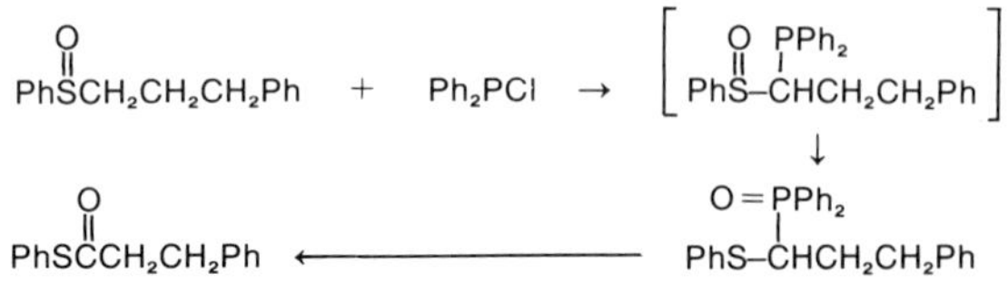

A soln. of the startg. sulfoxide in dry tetrahydrofuran treated dropwise with methyllithium at –78°, stirred for 30 min., the mixture added during 1–2 min. by cannula to a soln. of chlorodiphenylphosphine in the same solvent at –78° under N_2, warmed to 0° after 5 min., iodine added, and quenched with $Na_2S_2O_3$, after a further 5 min. → intermediate (Y 81%) and methyllithium in tetrahydrofuran oxygenated by bubbling O_2 into the soln. *(caution)* at –78° for 15 min. followed at –44° for 20 min. → product (Y 67%). F.e.s. E. Vedejs et al., J. Org. Chem. *46,* 5253–4 (1981).

Sulfur/Remaining Elements Type ORem ∩ SRem

Sodium hydride — *NaH*

2-(Phosphoryloxy)thioenol derivs. from α-(phosphorylthio)ketones ←

S→O-Phosphoryl group migration

88.

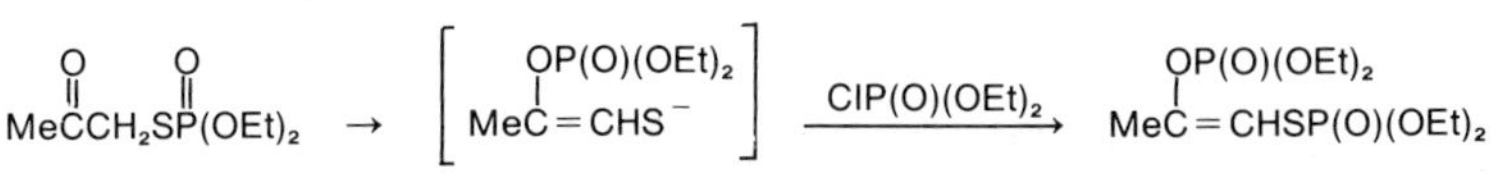

A soln. of startg. β-ketophosphorothiolate in dry ether added with stirring to a suspension of NaH in the same solvent at 20–50°, when evolution of H_2 had ceased the mixture stirred for a further 15 min., startg. halide added, and hydrolysed after a further 6 hrs. → product. Y 95%. F.e. and electrophiles s. G. Sturtz, M. Baboulene, Tetrahedron *37,* 3067–70 (1981).

Exchange ↓↑

Hydrogen ↑ ORem↓↑H

*Sodium hydroxide/tetra-*n*-butylammonium hydrogen sulfate* $NaOH$/n-$Bu_4N^+HSO_4^-$

Enol phosphates from ketones CHCO → C=C–OPO(OR)$_2$
with Li diisopropylamide cf. *33,* 749; with NaOH/tetra-*n*-butylammonium hydrogen sulfate, geospecific O-phosphorylation of *β*-dicarbonyl compds., s. R.A. Jones, S.S. Badesha, Synth. Commun. *11,* 557–60 (1981).

1,8-Diazabicyclo[5.4.0]undec-7-ene ←

Enoxysilanes from oxo compds. CHCO → C=C–OSi$\equiv$
with triethylamine cf. *29,* 829; with 1,8-diazabicyclo[5.4.0]undec-7-ene, also silylacetylenes from terminal acetylene derivs., s. Y. Taniguchi, J. Inanaga, M. Yamaguchi, Bull. Chem. Soc. Japan *54,* 3229–30 (1981).

Zinc chloride $ZnCl_2$

O-Silylation with organosilicon hydrides SiH → SiOR
with Na heptanoate cf. *11,* 170; with ZnCl$_2$, siloxy-2-acetylenes, s. S.F. Karaev, S.O. Guseinov, Zh. Obshch. Khim. *51,* 1107–10 (1981).

Diphenyl disulfide *PhSSPh*

Spirophosphoranes from cyclic phosphorous esters ○
with diethyl azodicarboxylate cf. *32,* 90; with diphenyl disulfide s. Y. Kimura, M. Miyamoto, T. Saegusa, J. Org. Chem. *47,* 916–9 (1982).

*Tetra-*n*-butylammonium hydrogen sulfate s. under NaOH* n-$Bu_4N^+HSO_4$

Chloroplatinic acid H_2PtCl_6

Borinic acid silyl esters from borinic acids and organosilicon hydrides R$_2$BOSi$\equiv$

89. $$Et_3SiH + HOBPh_2 \rightarrow Et_3SiOBPh_2$$

A little H$_2$PtCl$_6$ added to a mixture of diphenylborinic acid and triethylsilane in ether at room temp., and worked up when evolution of H$_2$ ceased → triethylsiloxydiphenylborane. Y 80%. F.e. and reactions s. E.V. Chistova et al., Zh. Obshch. Khim. *51,* 1078–85 (1981).

Oxygen ↑ ORem↓↑O

Without additional reagents *w.a.r.*

O-Silyl O-alkyl keteneacetals as silylating agents OH → OSi$\equiv$
s. *35,* 62; O-, N- and S-*tert*-butyldimethylsilylation s. Synthesis *1981,* 451–2.

Triethylamine NEt_3

Silylation with trimethylsilyl triflate CHCO → C=C–OSi$\equiv$
s. *31,* 93; *33,* 100; 1,3-disiloxy-1,3-dienes from *β*-diketones s. K. Krägeloh, G. Simchen, Synthesis *1981,* 30–2; prepn. of trimethylsilyl triflate *in situ* from allyltrimethylsilane and CF$_3$SO$_3$H cf. G.A. Olah et al., J. Org. Chem. *46,* 5212–4 (1981); enoxysilane from ketones or *α*-silylketones s. H. Emde et al., Ann. Chem. *1981,* 1643–57; N,N-disilylenamines s. H. Ahlbrecht, E.-O. Düber, Synthesis *1982,* 273–5.

Pyridine *C_5H_5N*

S,S-Diaryl dithiophosphates
from S,S-diaryl dithiophosphoric acids $(ArS)_2P(O)OH \rightarrow (ArS)_2P(O)OR$
s. *37,* 85

2,6-Lutidine ←

O-Silylation with trialkylsilyl triflates $OH \rightarrow OSiR_3$

90. $t\text{-}BuOH + t\text{-}BuMe_2SiOSO_2CF_3 \rightarrow t\text{-}BuOSiMe_2Bu\text{-}t$

1 Equivalent of *tert*-butyl alcohol treated with 1.5 equivalents of *tert*-butyldimethylsilyl triflate and 2 equivalents of 2,6-lutidine in methylene chloride at 25° for 10 min. → product. Y 90%. – The method is especially useful for the silylation **of hindered tert. alcohols or unreactive sec. alcohols;** as a complementary reagent, the highly congested triisopropylsilyl triflate is valuable for the **preferential silylation** of prim. and sec. alcohols. F.e., **also enoxysilanes from oxo compds.,** s. E.J. Corey et al., Tetrahedron Letters *22,* 3455–8 (1981).

Benzotriazolyloxytris(dimethylamino)phosphonium hexafluorophosphate ←

Oligonucleotide synthesis ←
Triester method
s. *17,* 169s*36;* coupling with benzotriazolyloxytris(dimethylamino)phosphonium hexafluorophosphate, optimization, s. D. Molko, A. Guy, R. Téoule, Nucleosides and Nucleotides *1,* 65–80 (1982).

Arenesulfonylazolides ←

s. *17,* 169s*36;* coupling with N-*p*-nitrobenzenesulfonyl-4-nitroimidazolide s. C.A. Leach, F. Waldmeier, C. Tamm, Helv. Chim. Acta *64,* 2515–23 (1981); with N-2,4,6-triisopropylbenzenesulfonyl-5-(2-pyridyl)tetrazolide s. E. Ohtsuka, Z. Tozuka, M. Ikehara, Tetrahedron Letters *22,* 4483–6 (1981); phosphorylation with *p*-chlorophenyl 2-cyanoethyl phosphorochloridate s. Helv. Chim. Acta *64,* 2142–7 (1981); with *p*-chlorophenyl 2-cyanoethyl cyclohexylammonium phosphate s. N.T. Thuong, M. Chassignol, C. Barbier, Tetrahedron Letters *22,* 851–4 (1981); with 4-chlorophenyl 5-chloro-8-quinolyl hydrogen phosphate s. H. Takaku, K. Kamaike, K. Kasuga, Chem. Letters *1982,* 197–200; with bis(benzotriazol-1-yl) phosphates s. G. van der Marel et al., Tetrahedron Letters *22,* 3887–90 (1981); with *p*-chlorophenyl N-(*p*-methoxyphenyl)chlorophosphoramidate s. E. Ohtsuka, S. Shibahara, M. Ikehara, Chem. Pharm. Bull. *29,* 2440–8 (1981);

review of oligonucleotide synthesis s. H. Paulsen, Angew. Chem. Intern. Ed. *21,* 155–73 (1982).

Arenesulfonyl chlorides *$ArSO_2Cl$*

s. *17,* 169s*36;* with added N-methylimidazole s. V.A. Efimov, S.V. Reverdatto, O.G. Chakhmakhcheva, Tetrahedron Letters *23,* 961–4 (1982); A.L. Kayushin, Y. A. Berlin, M.N. Kolosov, Bioorg. Khim. *8,* 660–6 (1982).

Arenesulfonyl chlorides/pyridine *$ArSO_2Cl/C_5H_5N$*

Silica-supported oligonucleotide synthesis ←
s. *17,* 169s*36;* continuous flow method on a polydimethylacrylamide-kieselguhr support s. M.J. Gait et al., J. Chem. Soc. Chem. Commun. *1982,* 37–40; semi-automated method s. E. Ohtsuka, H. Takashima, M. Ikehara, Tetrahedron Letters *23,* 3081–4 (1982); **from phosphoromonoamidites,** also liquid phase method, cf. S.L. Beaucage, M.H. Caruthers, ibid. *22,* 1859–62 (1981).

p-Toluenesulfonic acid *TsOH*

O-Silylation with α,β-ethylene-β-siloxyketones

91. 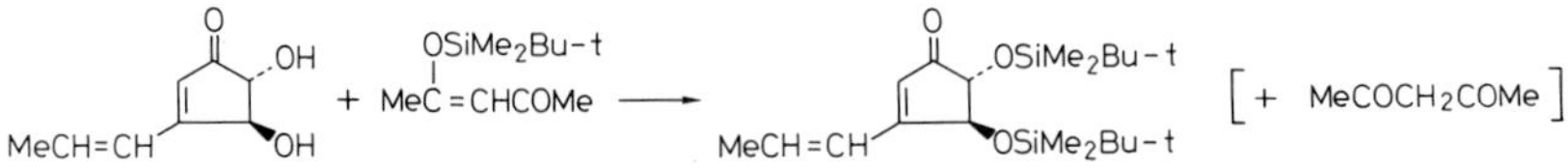

of alcohols. Pentane-2,4-dione *tert*-butyldimethylsilyl enol ether added to a soln. of terrein and a little *p*-toluenesulfonic acid in dimethylformamide, and the mixture stirred overnight at room temp. → terrein bis(*tert*-butyldimethylsilyl ether). Y 92%. – The method is particularly useful for sensitive and hindered compds. F.e., **also disiloxanes from silanols,** s. T. Veysoglu, L.A. Mitscher, Tetrahedron Letters *22,* 1299–302 (1981); **O-trimethylsilylation** under neutral conditions cf. ibid *22,* 1303–6 (1981).

Sulfuric acid *H_2SO_4*

O-Silylation with disiloxanes

with pyridinium tosylate cf. *34,* 91; with H_2SO_4, carboxylic acid silyl esters, s. H. Matsumoto et al., Chem. Letters *1980,* 1475–8.

Nitrogen ↑ ORem↓↑N

Without additional reagents *w.a.r.*

Alcoholysis of phosphorus azolides ←

s. *33,* 102; phosphorous acid esters with phosphorimidazolidites, s. D.A. Predvoditelev et al., Zh. Org. Khim. *18,* 1326–7 (1982).

O-Silylation with N,N′-bis(trimethylsilyl)urea OR → OSi≡

92. $Me(CH_2)_8OH + Me_3SiNHCONHSiMe_3 \rightarrow Me(CH_2)_8OSiMe_3$

of alcohols. A well-stirred soln. of *n*-nonanol in dichloromethane treated with N,N′-bis(trimethylsilyl)urea, and the mixture heated at 40° for 2 hrs. → product. Y 93%. F.e., **also O-silylation of carboxylic acids,** s. W. Verboom, G.W. Visser, D.N. Reinhoudt, Synthesis *1981,* 807–9; O-silylation of carboxylic acids **with 3-trimethylsilyl-2-oxazolidone** cf. C. Palomo, ibid. *1981,* 809–11; of β-ketocarboxylic acid esters and β-diketones s. ibid. *1982,* 280–1.

Phosphoric sulfonic anhydrides $(RO)_2P(O)O\text{–}SO_2R'$

93. 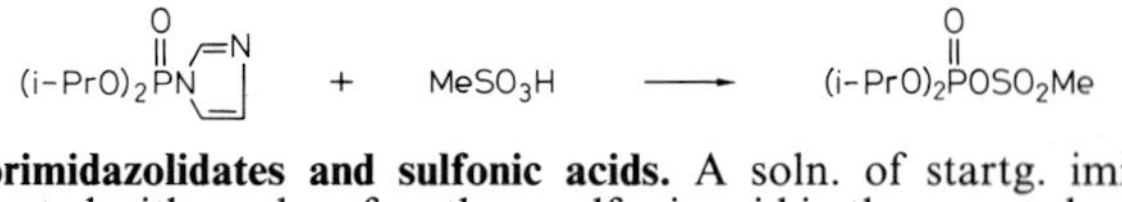

from phosphorimidazolidates and sulfonic acids. A soln. of startg. imidazolide in acetonitrile treated with a soln. of methanesulfonic acid in the same solvent at 0°, and the mixture stirred at room temp. for 2 hrs. → methanesulfonic phosphoric anhydride diisopropyl ester. Y 100% by n.m.r. F.e. and methods, **also phosphinic sulfonic anhydrides.** s. W. Dabkowski et al., Chem. Ber. *115,* 1636–43 (1982); **from thiophosphoric esters** cf. Y. Segall, J.E. Casida, A.C.S. Symp. Series 171 (Phosphorus Chem.) 337 (1981); Chem. Abstr. *96,* 69111 (1982).

Irradiation ///

Phosphonic amide esters from phosphinic azides ←

94. 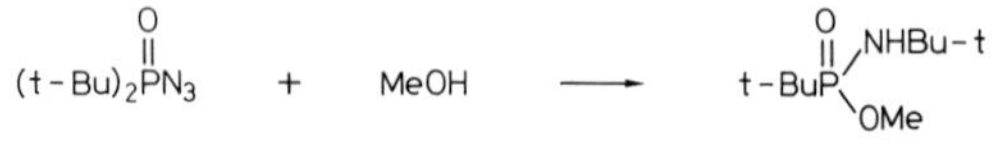

A stirred soln. of di-*tert*-butylphosphinic azide in methanol irradiated using a 125 W. medium-pressure Hg lamp for 19 hrs. → methyl N,P-di-*tert*-butylphosphonamidate.

Y 71%. F.e. and limitations, **also phosphonic diamides,** s. M.J.P. Harger, M.A. Stephen, J. Chem. Soc. Perkin Trans. I *1981,* 736–40; from diarylphosphinic azides s. Tetrahedron *38,* 1511–5 (1982).

Sulfuric acid H_2SO_4

Optically active mixed phosphonic esters from phosphonic amide esters P–N⋷ → P–OR

with proline derivs. cf. *34,* 92; with α-methylbenzylamine derivs. s. T. Otsuki, Y. Okamoto, H. Sakurai, Synthesis *1981,* 811–3.

Halogen ↑ ORem ↓↑Hal

Without additional reagents *w.a.r.*

Enoxysilanes from α-halogenoketones CHalCO → C=C–OSi⋷

with Zn cf. *32,* 96; with bis(triethylsilyl)mercury s. D.V. Gendin, O.A. Kruglaya, N.S. Vyazankin, Zh. Obshch. Khim. *51,* 1614–21 (1981); 2-bromoenoxysilanes from *α,α*-dibromoketones s. ibid. *50,* 1659–60 (1980).

Sodium hydride *NaH*

2-(Phosphoryloxy)thioenol derivs. from α-(phosphorylthio)ketones ←

s. *37,* 88

Potassium carbonate/Aliquat 336 ←

O-Silylation of alcohol groups OH → OSi⋷

with imidazole cf. *30,* 61; with K_2CO_3/Aliquat 336, phase transfer catalysis, s. M. Lissel, J. Weiffen, Synth. Commun. *11,* 545–9 (1981); with $AgNO_3$/pyridine, preferential O-silylation of nucleoside hydroxyl groups, cf. G.H. Hakimelahi, Z.A. Proba, K.K. Ogilvie, Tetrahedron Letters *22,* 4775–8, 5243–6 (1981); ibid. *23,* 1997–2000 (1982); Can. J. Chem. *60,* 1106–13 (1982).

Sodium iodide/triethylamine NaI/Et_3N

O-Silylation with (2,4,6-tri-*tert*-butylphenoxy)dimethylsilyl chloride $OSiMe_2OAr$

95.

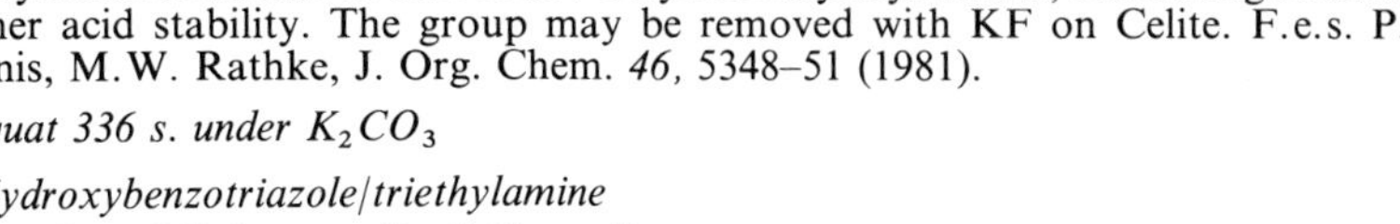

Enoxysilanes. A mixture of acetophenone, NaI, triethylamine, and (2,4,6-tri-*tert*-butylphenoxy)dimethylsilyl chloride in acetonitrile refluxed for 3 hrs. → product. Y 86%. – The silylating agent is inexpensive, and the hydrolytic stability of the obtained enoxysilanes is similar to that of *tert*-butyldimethylsilyl derivs., but having somewhat higher acid stability. The group may be removed with KF on Celite. F.e.s. P.A. Manis, M.W. Rathke, J. Org. Chem. *46,* 5348–51 (1981).

Aliquat 336 s. under K_2CO_3 ←

1-Hydroxybenzotriazole/triethylamine ←

Protection of diols as cyclic dialkoxysilanes ←

96.

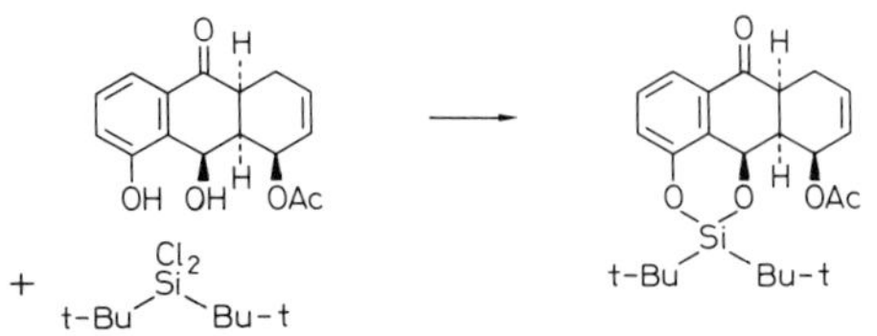

Triethylamine and di-*tert*-butyldichlorosilane added sequentially to a soln. of startg. diol and a little 1-hydroxybenzotriazole in acetonitrile, and heated in an oil bath at

65° for 30 min. → product. Y 84%. – O,O-Di-*tert*-butylsilylene derivs. are soluble in tetrahydrofuran below 25° (unlike O,O-isopropylidene derivs.), resistant to Lewis acids and chromatography, and do not complicate n.m.r. spectra. – The protecting group may be efficiently removed, with retention of ester and *tert*-butyldimethylsilyl ether groups, on treatment with pyridinium fluoride in tetrahydrofuran at 25°. F.e.s. B.M. Trost, C.G. Caldwell, Tetrahedron Letters *22,* 4999–5002 (1981).

Sulfur ↑ ORem⇃↾S

Without additional reagents *w.a.r.*

Dithiophosphonic O-esters from perthiophosphonic anhydrides C

s. *18,* 633; s.a. J. Wu, X. Sun, Huaxue Xuebao *39,* 568 (1981) (Ch); Chem. Abstr. *96,* 199785 (1982).

Caesium fluoride *CsF*

Phosphorus(V) from thiophosphorus(V) esters P–SR → P–OR′

97. $Ph_2P(O)SBu + HOBu \rightarrow Ph_2P(O)OBu$

A soln. of *n*-butyl diphenylthiophosphinate and butanol in chloroform containing a suspension of CsF stirred at room temp. for 3 days → *n*-butyl diphenylphosphinate. Y 80%. Also phosphonic and phosphoric esters s. L. Horner, R. Gehring, H. Lindel, Phosphorus-Sulfur *11,* 349–56 (1981).

Silver acetate *MeCOOAg*

Phosphoric monoesters from S,S-diaryl dithiophosphates P–SAr → P–OH

s. *37,* 85

Mercuric alkoxide $Hg(OR)_2$

Alkoxysilanes from disilthianes $R_3SiSSi\equiv \rightarrow R_3SiOR'$

98.

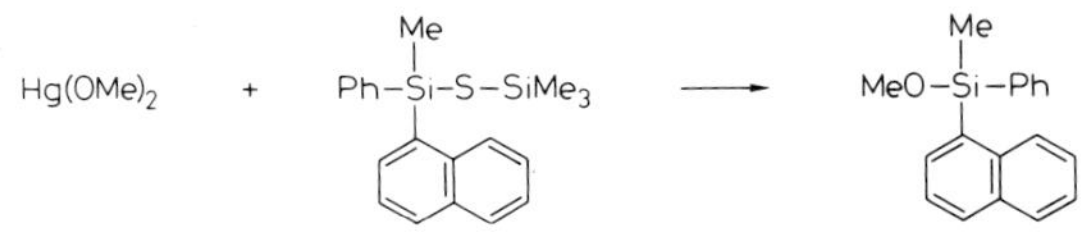

A soln. of startg. disilthiane in heptane added dropwise with stirring and gentle warming to Hg(II) methoxide in heptane, and heated for 15 min. at 90–95° → product. Y 86%. – Reaction proceeds largely with inversion of configuration. F.e., **also enoxy- and acoxy-silanes, and chlorosilanes** with retention of configuration, s. V.A. Chauzov, N.S. Kitaeva, Y.I. Baukov, Zh. Obshch. Khim. *51,* 1790–4 (1981).

Remaining Elements ↑ ORem⇃↾Rem

Without additional reagents *w.a.r.*

Enoxygermanes $C{=}C(OGe\equiv)$

99.

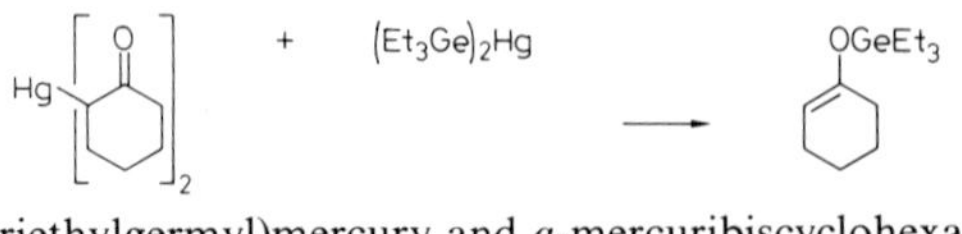

A mixture of bis(triethylgermyl)mercury and *α*-mercuribiscyclohexanone in tetrahydrofuran heated at 65° for 6 hrs. → triethyl(1-cyclohex-1-enyloxy)germane. Y 72%. F.e.s. L.I. Belousova et al. Izv. Akad. Nauk SSSR Ser. Khim. *1980,* 2126–9; **from** ***α*-bromoketones** cf. Zh. Obshch. Khim. *49,* 951 (1979); *51,* 1614–21 (1981).

Enoxysilanes from
α-silylketones and trialkylsilyl triflates COCSi⋜ → C(OSi⋜)=C
s. *37,* 90

Cyclic enediol phosphates from 1,2-di(siloxy)-1-ethylene derivs. ○

100.

A soln. of methyl difluorophosphate in toluene added dropwise to a soln. of startg. m. in the same solvent at 85°, the mixture stirred at 80° for 30 min., cooled to 60°, and distilled at 0.2 torr → product. Y 90%. F.e.s. R. Schwartz, I. Ugi, Angew. Chem. Intern. Ed. *20,* 789–91 (1981).

Formation of O–C Bond

Uptake ⇓

Addition to Hydrogen and Carbon OC ⇓ HC

Copper/cuprous chloride *Cu/CuCl*
Catechols from phenols ArH → ArOH

101.

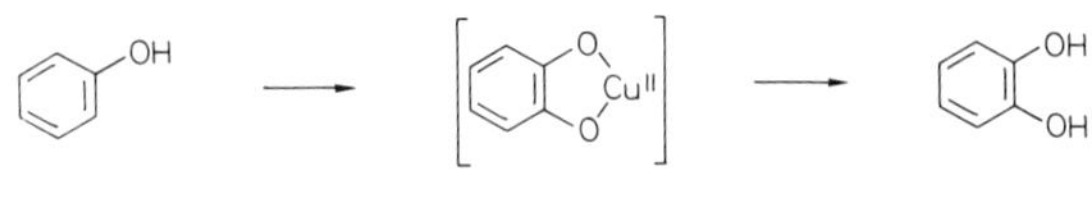

A soln. of phenol in acetonitrile oxygenated in the presence of Cu and a catalytic amount of CuCl, and the mixture acidified → product. Y 70–90%. F.e.s. P. Capdevielle, M. Maumy, Tetrahedron Letters *23,* 1577–80 (1982).

Silica gel s. under SeO_2 SiO_2

m-*Chloroperoxybenzoic acid* m-$ClC_6H_4COO_2H$
4-Hydroxy-Δ^2-isoxazol-5-ones from Δ^2-isoxazol-5-ones H → OH
s. *37,* 210

Phenyl iodosoacetate $PhI(OAc)_2$
α-Hydroxycarboxylic acid esters from carboxylic acid esters
s. *37,* 139

Oxidations with *peroxymonophosphoric acid* H_3PO_5
Carboxylic acids from aldehydes CHO → COOH
Arylcarboxylic acids

102. PhCHO → PhCOOH

A soln. of H_3PO_5, benzaldehyde, and a little ethylenediaminetetraacetic acid in 40% ethanol kept at 25° and pH 7.3 untiliodometric titration indicated the reaction to be completed → benzoic acid. Y 98%. F.e., **also sulfones from sulfoxides,** s. Y. Ogata, Y. Sawaki, Y. Tsukamoto, Bull. Chem. Soc. Japan *54,* 2061–4 (1981); **phenols from arenes** s. Tetrahedron *37,* 1485–6 (1981).

Hydrogen peroxide/antimony pentafluoride/hydrogen fluoride $H_2O_2/SbF_5/HF$
Resorcinols from phenols in superacidic medium ArH → ArOH

103.

H_2O_2 added to a soln. of *p*-cresol in 2:5 SbF_5/HF at –40°, and the mixture stirred for 30 min. → product. Y 78%. F.e.s. J.-P. Gesson, J.-C. Jacquesy, M.-P. Jouannetaud, J. Chem. Soc. Chem. Commun. *1980,* 1128–9; **review of protonation in superacidic medium** s. D. Fărcasiu, Accounts Chem. Res. *15,* 46–51 (1982).

*Selenium dioxide–silica gel/*tert-*butyl hydroperoxide* SeO_2–SiO_2/t-*BuOOH*
Allylic oxidation of methyl groups $C=CCH_3 \rightarrow C=CCH_2OH$
Cyclic 2-ethylene-*prim*-alcohols

104.

5% SeO_2 supported on silica gel in dichloromethane or hexane stirred with excess of 70% *tert*-butyl hydroperoxide for 15 min. at room temp. under argon, a soln. of 6,10-diacetoxybicyclohumulene in the same solvent added dropwise, and the mixture allowed to stand at room temp. for 3 hrs. → 6,10-diacetoxybicyclohumulen-12-ol. Y 80%. – No epoxidation was detected under these conditions. F.e.s. B. R. Chhabra et al., Chem. Letters *1981,* 1703–6; **with E→Z-rearrangement** s. M. Haruna, K. Ito, J. Chem. Soc. Chem. Commun. *1981,* 483–4.

Oxidation with *chromium(V) complexes* $[bipyH_2][CrOCl_5]$
Carboxylic acids from aldehydes CHO → COOH

105.

$$p\text{-}NO_2C_6H_4CHO \xrightarrow{[BipyH_2][CrOCl_5]} p\text{-}NO_2C_6H_4COOH$$

The chromium(V) complex, prepd. from 2,2′-bipyridyl and CrO_3 in concd. HCl, added at room temp. under N_2 to a soln. of *p*-nitrobenzaldehyde in dichloromethane, and the mixture stirred for 30 min. → *p*-nitrobenzoic acid. Y 95%. F.e. and complexes s. T.K. Chakraborty, S. Chandrasekaran, Synth. Commun. *10,* 951–6 (1980); **oxo compds. from alcohols** s. Tetrahedron Letters *21,* 1583–6 (1980).

tert-*Butyl hypochlorite s. under* SeO_2 t-*BuOCl*

Hydrogen fluoride s. under H_2O_2 *HF*

Manganese(II) acetate $(MeCOO)_2Mn$

s. *8,* 157; f. catalysts, isobutyric acid, s. M. Kaledowska, L. Nowakowski, Z. Nedzynska, Zesz. Nauk.-Wyzsza Szk. Pedagog. im. Powstancow Slask. Opolu, (Ser.): Chem. *1981,* 4, 95 (Pol.); Chem. Abstr. *95,* 186591 (1981).

Via intermediates *v.i.*
***p*-Quinols from phenols via *p*-cyclopentenylphenols** ArH → ArOH

106.

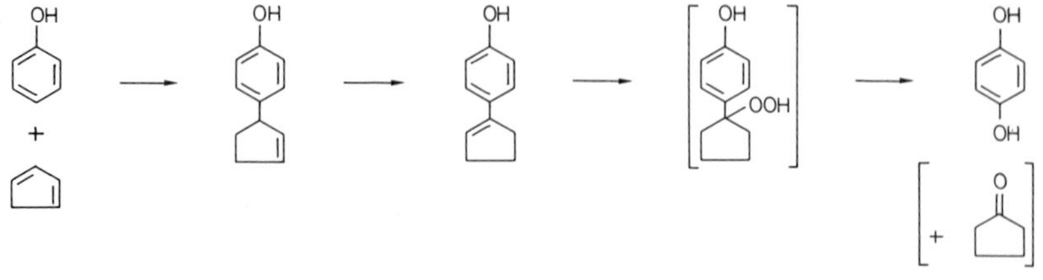

Phenol treated with cyclopentadiene in the presence of 85% phosphoric acid at room temp. (cf. Synth. Meth. *13,* 690) → 4-(cyclopenten-3-yl)phenol (Y >80%) in dry benzene treated with a little bis(benzonitrile)dichloropalladium(II), and refluxed for 2 hrs. → 4-(cyclopenten-1-yl)phenol (Y 91%) in acetonitrile treated sequentially with 30% H_2O_2 and concd. HCl, and stirred at 50° for 3 hrs. → *p*-quinol (Y 92%). F.e.s. D.V. Rao, F.A. Stuber, Tetrahedron Letters *22,* 2337–40 (1981).

Addition to Oxygen and Nitrogen OC ⇓ ON

Sodium tetraphenylborate *$NaBPh_4$*

N-Acetoxypyridinium salts from pyridine N-oxides ⋛NO → ⋛$\overset{+}{N}$–OAc
perchlorates cf. *21,* 158; tetraphenylborates with $NaBPh_4$ s. E.V. Titov, K.Y. Chotii, V.I. Rybachenko, Zh. Obshch. Khim. *51,* 682–8 (1981).

Addition to Oxygen and Carbon OC ⇓ OC

Without additional reagents *w.a.r.*

Regio- and stereo-specific ring opening of oxido compds. with carboxylic and sulfonic acids  → C(OH)C(O–)

107. 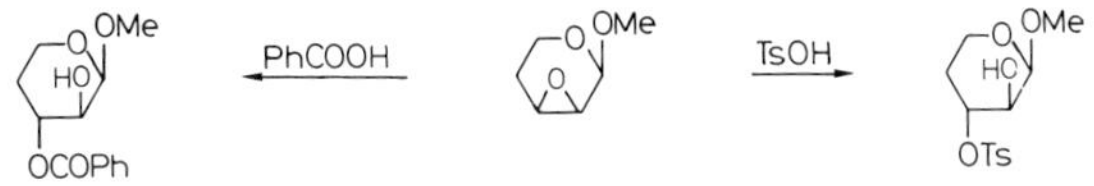

Glycol monoesters	**Glycol monotosylates**
Startg. epoxide in hexamethylphosphoramide treated with	
3 equivalents of benzoic acid at 130° for 5 hrs. → monobenzoate. Y. 70%.	1 equivalent of *p*-toluenesulfonic acid at 80° for 1.5 hrs. → monotosylate. Y 89%.

F.e.s. D. Picq, D. Anker, H. Pacheco, Tetrahedron Letters *22,* 4517–20 (1981); glycol monoesters with neutral alumina in ether cf. G.H. Posner, M. Hulce, R.K. Rose, Synth. Commun. *11,* 737–41 (1981).

Ditriflates C

108. 

from cyclic ethers. A soln. of triflic anhydride in methylene chloride added to a soln. of tetrahydropyran in the same solvent at –50°, and allowed to warm to 20° → 1,5-pentanediol triflate. Y 94%, F.e., **also from diols,** s. E. Lindner, G. von Au, H.-J. Eberle, Chem. Ber. *114,* 810–3 (1981).

Acylaminoalcohols from cyclic iminoesters
2-acylaminoalcohols with triethylamine cf. *28,* 85; 3-acylaminoalcohols without base s. T.M. Pavel, Zh. Org. Khim. *18,* 178–81 (1982); O.N. Chernysh, S.I. Yakimovich, ibid. *18,* 181–6 (1982).

Sodium hydroxide *NaOH*

α-Acylamino-α,β-ethylenecarboxylic acids from 4-alkylidene-5-oxazolones
s. *16,* 931; geospecific ring opening s. C. Catwiela, E. Meléndez, Synthesis *1980,* 901–2.

2,6-Dimethylpyridine s. under 2,6-Dimethylpyridinium tosylate ←

Alumina Al_2O_3

Phthalide from 3,4-dihydroisocoumarin ring with oxirane ring opening ←

109.

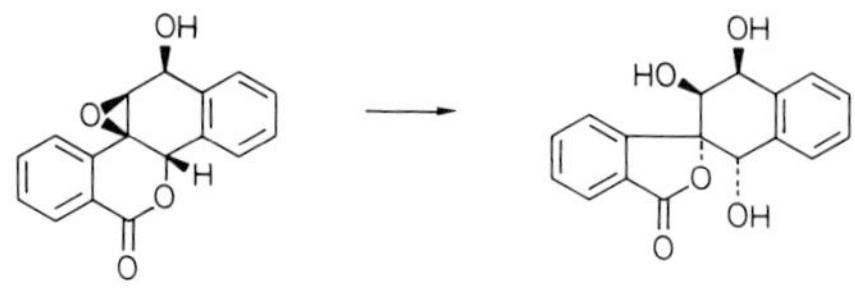

Neutral Al_2O_3 (grade III) added to a soln. of 10b,11-epoxy-*cis*-12-hydroxy-*cis*-4b,10b,11,12-tetrahydronaphtho[1,2-*c*]isocoumarin in 1:1 chloroform/methanol, and the mixture allowed to stand at room temp. for 75 hrs. → 1,2,3,4-tetrahydro-1, *trans*-3,*trans*-4-trihydroxynaphthalene-2-spiro-3′-*cis*-3′-O-phthalide. Y 75%. F.e.s. Y. Harigaya et al., Chem. Pharm. Bull. *29,* 2557–64 (1981).

Glycol monoesters from oxido compds. $\triangledown$(O) → C(OH)C(OR)

s. *37,* 107

Boron fluoride BF_3

Spiroorthocarboxylic acid esters from lactones and oxido compds. ←

110.

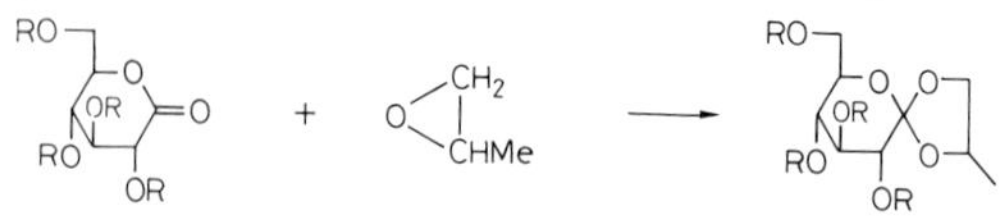

R = PhCH$_2$

Carbohydrate derivs. A stirred soln. of tetra-O-benzyl-D-glucono-1,5-lactone and BF_3-etherate in dichloromethane treated dropwise with 1,2-epoxypropane during 1 hr. at 28–30°, and stirring continued for 1 hr. → 1,2-O-(2,3,4,6-tetra-O-benzyl-D-glucopyranosylidene)-1,2-propanediol. Y 65%. F.e., **also from glycols,** s. M. Tamaru, S. Horito, J. Yoshimura, Bull. Chem. Soc. Japan *53,* 3687–90 (1980).

Cation exchanger ←

Dicarboxylic acid esters from anhydrides C

Phthalic acid esters

s. *16,* 184; with Wofatit KPS s. M. Kaledkowska, L. Nowakowski, A. Stepien, Zesz. Nauk.-Wyzsza Szk. Pedagog. im. Powstancow Slask. Opolu, (Ser.): Chem. *1981,* 5, 101–9 (Pol); Chem. Abstr. *96,* 199234 (1982).

m-*Chloroperoxybenzoic acid* m-$ClC_6H_4COO_2H$

Oxidative furan ring opening

with HNO_3 cf. *5,* 122; *28,* 90; with *m*-chloroperoxybenzoic acid, macrocyclic polyketones, s. P.D. Williams, E. LeGoff, J. Org. Chem. *46,* 4143–7 (1981); macrocyclic lactones by Rose Bengal sensitized photooxidation cf. B.L. Feringa, Tetrahedron Letters *22,* 1443–6 (1981).

2,6-Dimethylpyridinium tosylate/2,6-dimethylpyridine ←

Glycosaminides from carbohydrate Δ^2-oxazoline derivs. C

s. *14,* 174; with 2,6-dimethylpyridinium tosylate and a little 2,6-dimethylpyridine cf. R. Gigg, R. Conant, Carbohyd. Res. *100,* C1–C4 (1982).

Phosphorus trichloride PCl_3

Acylals from aldehydes CHO → CH(OAc)$_2$

with $HClO_4$ cf. *33,* 669; with PCl_3 s. J.K. Michie, J.A. Miller, Synthesis *1981,* 824.

Hydrogen chloride HCl

α-Acylaminocarboxylic acids from Δ^2-5-oxazolones C

s. *37,* 215

Dihydrotetrakis(triphenylphosphine)ruthenium $RuH_2(PPh_3)_4$

Carboxylic acid esters from two aldehyde molecules 2 CHO → $COOCH_2$

Tishckenko reaction

s. *4,* 154s*34;* f.e. and catalysts s. T. Ito et al., Bull. Chem. Soc. Japan *55,* 504–12 (1982).

Addition to Nitrogen and Carbon OC ⇓ NC

Without additional reagents *w.a.r.*

N-Acylthionocarbamic acid esters from acylisothiocyanates CON=C=S → CONHC(S)OR

111. 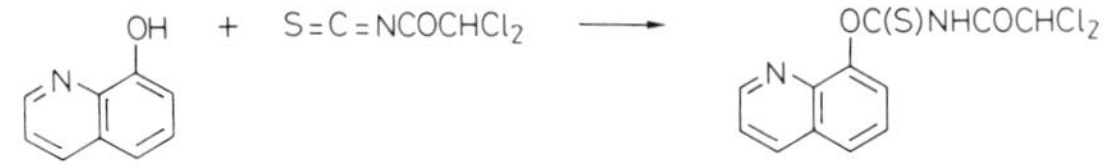

A soln. of startg. acylisothiocyanate and 8-hydroxyquinoline in ether refluxed for 3 hrs. → product. Y 80–90%. F.e.s. A.A. Hijazi, H.M. Al-Hazimi, S.S. Al-Showiman, J. Chem. Soc. Pak. *2,* 29 (1980); Chem. Abstr. *94,* 30535 (1981).

Sodium alkoxides *NaOR*

Iminoesters from Δ^1-azirines C

112. 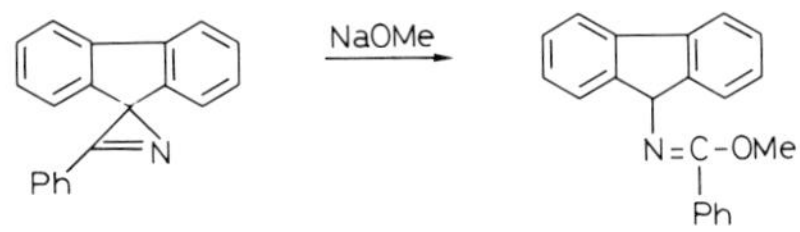

Fluorene-9′-spiro-2-(3-phenyl)-2*H*-azirine added to a soln. of Na in dry methanol, and the mixture refluxed for 1 hr. → methyl N-(9-fluorenyl)benzenecarboximidate. Y 85%. F.e.s. A.H. Schulthess, H.-J. Hansen, Helv. Chim. Acta *64,* 1322–36 (1981).

Cupric oxide *CuO*

Formiminoesters from isonitriles C=N → HC(=N)OR

s. *37,* 55

m-*Chloroperoxybenzoic acid* m-$ClC_6H_4COO_2H$

Oxaziridines from imines O

2-Sulfonyloxaziridines

s. *29,* 132s*36;* improved procedure, also optically active compds. in the presence of (–)-benzylquinidinium or (+)-benzylcinchoninium chloride, s. J. Org. Chem. *47,* 1774–5 (1982).

Addition to Carbon and Carbon OC ⇓ CC

Irradiation (s.a. under Tetraphenylporphyrin, Sensitox, O_2, and V_2O_5) ⫻

1-Alkoxy-β,γ-ethylenecarboxylic acid amides from 5,6-dihydro-2-pyridone ring C

113. 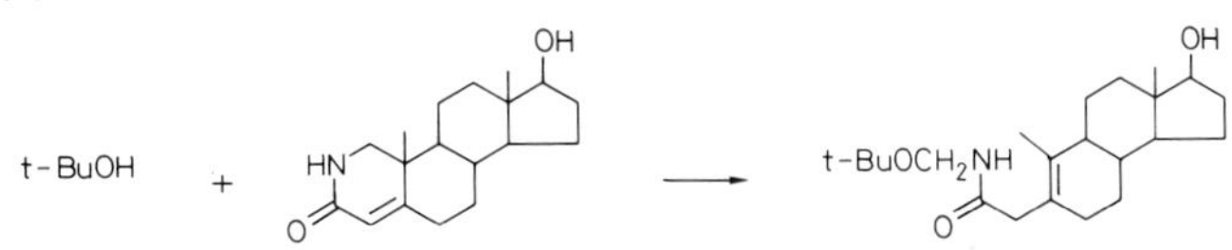

A soln. of 17β-hydroxy-2-aza-4-androsten-3-one in *tert*-butyl alcohol irradiated with a low-pressure Hg lamp for 3 hrs. → 1-*tert*-butoxy-17β-hydroxy-2-aza-1,10-*seco*-androst-5(10)-en-3-one. Y 75%. F.e.s. A. Cánovas et al., Helv. Chim. Acta *63,* 2380–9 (1980).

Electrolysis ϟ

α-(Organothio)aldehydes from thioenolethers C=CH(SR) → C(SR)CHO

114. $PhCH{=}CHSC_6H_4Me\text{-}p \rightarrow PhCH(CHO)SC_6H_4Me\text{-}p$

Startg. thioenolether in aq. acetonitrile containing 0.2 *M* $NaClO_4$ electrolysed at 1.2 V in a divided cell using a platinum plate anode until 1.95 F/mol of electricity had passed at 25° → product. Y 93% by g.l.c. F.e.s. A. Matsumoto, K. Suda, C. Yijima, J. Chem. Soc. Chem. Commun. *1981,* 263.

Sodium hydroxide s. under H_2O_2 *NaOH*

Potassium carbonate s. under H_2O_2 K_2CO_3

Sodium hydrogen carbonate s. under m-$ClC_6H_4COO_2H$ $NaHCO_3$

Potassium fluoride s. under m-$ClC_6H_4COO_2H$ *KF*

Chiral amines s. under Al(i-Bu)$_3$ ←

Diethylamine/hydrogen chloride Et_2NH/HCl

Ketones from allenes C=C=C → CHCOCH

with catecholatoborane/hydrogen peroxide cf. *35,* 27; with diethylamine/hydrogen chloride, β-ketophosphonic esters, s. H.-J. Altenbach, R. Korff, Tetrahedron Letters *22,* 5175–8 (1981).

Pyridine s. under NaOCl and CuCl C_5H_5N

Trimethylamine oxide s. under $BH_3.Me_2S$ Me_3NO

Silver carbonate Ag_2CO_3

Enolesters from acetylene derivs. C≡C → C(OAc)=CH

115. $AcOCH_2C{\equiv}CCH_2OAc + AcOH \rightarrow AcOCH_2C(OAc){=}CHCH_2C{=}CHCH_2OAc$

Ag_2CO_3 added in small portions during 30 min. at 80° under N_2 to a stirred soln. of 1,4-diacetoxy-2-butyne in acetic acid, and the mixture stirred at 110° for 6 hrs. → 1,2,4-triacetoxy-2-butene. Y 80%. F.e. and limitations s. Y. Ishino et al., Chem. Letters *1981,* 641–4; the role of **silver salts in organic processes, review,** s. J.R. Long, Aldrichimica Acta *14,* 63–70 (1981).

Cuprous chloride s. under $PdCl_2$ *CuCl*

Cuprous chloride/pyridine $CuCl/C_5H_5N$

***o*-Formamidoketones from 3-subst. indoles** ⊂

by ozonolysis cf. *2,* 168; by oxygenation with CuCl/pyridine, or with Cu(II) salts in the absence of O_2, s. J. Tsuji et al., Bull. Chem. Soc. Japan *54,* 2369–73 (1981).

Mercuric acetate/sodium tetrahydridoborate $(MeCOO)_2Hg/NaBH_4$

Mild directed hydration of carbon-carbon double bonds via hydroxymercuration C=C → CHC(OH)

s. *22,* 143; hydration of ethylenealcohols, acoxy- and alkoxy-ethylenes s. J. Org. Chem. *46,* 531–8 (1981); of ethylenehalides s. ibid. *46,* 930–9 (1981); f. Hg(II) salts s. ibid. *46,* 3810–2 (1981).

Diborane/hydrogen peroxide *B_2H_6/H_2O_2*

Diols from dienes
Stereospecific hydration

116. 

1,4-Diols. 1 *M* BH_3-tetrahydrofuran added dropwise over 1.5 hrs, to a soln. of bi-1-cyclohexen-1-yl in tetrahydrofuran at –25 to –21° under N_2, stirred at –25 to –15° for 1 hr. then at –15 to –10° for 5.5 hrs., stored at 5° for 12 hrs., stirred again at room temp. for 12 hrs., quenched by dropwise addition of ethanol, aq. NaOH and 30% H_2O_2 added at such a rate that the temp. did not exceed 50°, stirring continued at 46–52° for 2 hrs., cooled to room temp., and satd. with aq. K_2CO_3 → *d,l*-trans,trans-(1,1′-bicyclohexyl)-2,2′-diol. Y 98%. F.e.s. T.W. Bell, J. Am. Chem. Soc. *103,* 1163–71 (1981); with thexylborane, **remote functionalization via intramolecular hydroboration,** cf. W.C. Still, K.P. Darst, ibid. *102,* 7385–7 (1980).

Borane-dimethyl sulfide/trimethylamine oxide *$BH_3.Me_2S/Me_3NO$*

Reactions via ene-1-boranyl-1-silanes C≡CSi⋲ → CH_2COSi⋲

s. *33,* 124; acylsilanes via silylacetylenes with $BH_3.Me_2S$ and trimethylamine oxide s. Synthesis *1981,* 288–9.

Borane-dimethyl sulfide/molybdenum peroxide complex ←

Alcohols from ethylene derivs. C=C → CHC(OH)

Stereospecific hydration

117. 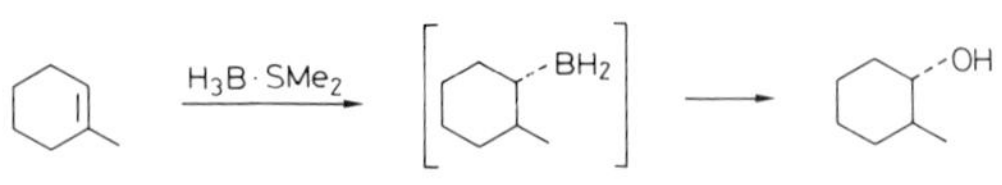

$H_3B.SMe_2$ added to a soln. of 1-methylcyclohexene in tetrahydrofuran, stirred at 0° for 1 hr., warmed to room temp. during 2 hrs., methanol added, the soln. stirred a further 30 min., cooled to 0°, MoO_5-pyridine-hexamethylphosphoramide complex added over 5 min., the mixture stirred for 30 min. at 0° then at room temp. for 1 hr., and treated with 3*M* NaOH → *trans*-2-methylcyclohexanol. Y 94%. – The molybdenum peroxide complex is a mild alternative to trialkylamine N-oxides for use under anhydrous conditions. M.M. Midland, S.B. Preston, J. Org. Chem. *45,* 4514–5 (1980).

Sodium tetrahydridoborate s. under $(MeCOO)_2Hg$ and $TiCl_4$ *$NaBH_4$*

Chiral sec. boranes *R_2BH*

Asymmetric synthesis

with bis(isopinocamphenyl)borane cf. *16,* 196; with isopinocamphenylborane s. H.C. Brown, P.K. Jadhar, J. Org. Chem. *46,* 5047–8 (1981); with di(longifolyl)borane cf. ibid. *46,* 2988–90 (1981).

9-Borabicyclo[3.3.1]nonane ←

Hydration of carbon-carbon double bonds

s. *16,* 194s*31;* of functionalized alkenes s. H.C. Brown, J.C. Chen, J. Org. Chem. *46,* 3978–88 (1981); prepn. of highly pure 9-borabicyclo[3.3.1]nonane dimer s. ibid. *46,* 4599–600 (1981).

Triisobutylaluminium/nickel(II) complex/chiral amine ←

Optically active alcohols from ethylene derivs.
vis asym. hydroalumination

118. 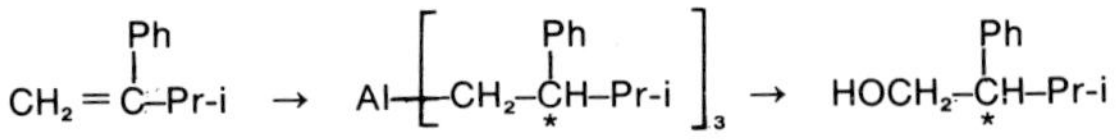

(–)-N,N-Dimethylmenthylamine added under argon to triisobutylaluminium at –20°, treated with a little bis(N-methylsalicylideneamine)nickel(II) at 0°, stirred for 2 min., 2-phenyl-3-methyl-1-butene injected by syringe, stirred at room temp. for 40 hrs., heptane added, dry air bubbled through the soln. for 8 hrs., and hydrolysed with dil. H_2SO_4 → (–)-(R)-2-phenyl-3-methyl-1-butanol. Y 61%. F.e., **also optically active hydrocarbons,** s. G. Giacomelli, L. Bertero, L. Lardicci, Tetrahedron Letters *22,* 883–6 (1981).

Borane-triethylamine/hydrogen peroxide $H_3B.NEt_3/H_2O_2$

Alcohols from ethylene derivs.
with H_3BSMe_2 cf. *21,* 174s*29;* with borane-triethylamine, 3-aminoalcohols from 2-ethyleneamines, s. M. Baboulène et al., Bull. Soc. Chim. France II *1980,* 565–70; with crystalline borane-triphenylphosphine and -triphenyl phosphite, liberating borane on addition of sulfur, cf. A. Pelter, R. Rosser, S. Mills, J. Chem. Soc. Chem. Commun. *1981,* 1014–5.

Borane-triphenylphosphine/sulfur/hydrogen peroxide $H_3B.PPh_3/S_8/H_2O_2$

s. *21,* 174s*37*

1-Pyrrolylborane-tetrahydrofuran ←

Alcohols from ethylene derivs. C=C → CH(OH)
Ketones from acetylene derivs. C≡C → CH_2CO
s. *37,* 30

Thallium(III) acetate $Tl(OAc)_3$

Methyl ketones from terminal acetylene derivs. C≡CH → $COCH_3$

with $TlCl_3$ cf. *23,* 150; with Tl(III) acetate, also enolesters via organothallium(III) complexes, s. S. Uemura et al., J. Chem. Soc. Perkin Trans. I *1981,* 991–4.

Acetone s. under $KHSO_5$ Me_2CO

Tetraphenylporphyrin/irradiation ←

***o*-β-Hydroxyphenols from α-aryloxo compds.** O
via 2-arylenoxysilanes and 3-siloxy-3,8a-dihydro-1,2-benzodioxins

119. 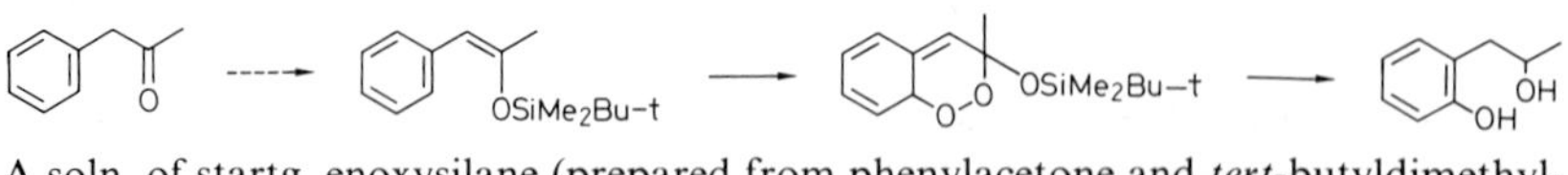

A soln. of startg. enoxysilane (prepared from phenylacetone and *tert*-butyldimethylsilyl chloride) in trichlorofluoromethane irradiated with a tungsten/bromine lamp through a 5% aq. $K_2Cr_2O_7$ filter soln. at 0° in the presence of tetraphenylporphyrin → intermediate *endo*-peroxide (Y ca 100%), slowly added in soln. to excess of methanolic $Na[BH_3CN]$ at 0°, and worked up with water → product (Y 66%). – This is part of a multistep synthesis of ***o*-alkylphenols.** F.e.s. I. Saito et al., Tetrahedron Letters *23,* 1717–20 (1982).

Sensitox/irradiation ←

Triacylamines from oxazoles C
s. *37,* 251

Rose Bengal/irradiation ←

Reviews of singlet oxygen ←

s. *30,* 86; *26,* 125; s.a. A.A. Gorman, M.A.J. Rodgers, Chem. Soc. Rev. *10,* 205–31 1981; H.H. Wasserman, J.I. Ives, Tetrahedron *37,* 1825–52 (1981).

Benzonitrile s. under H_2O_2 *PhCN*

tert-*Butyl hydroperoxide s. under $(acac)_2VO$ and $(C_5H_5)_2Zr(Cl)H$* t-*BuOOH*

Oxidations with
3-bromo-5-hydroperoxy-4,4-dimethyl-3,5-diphenyl-Δ^1-pyrazoline ←

Oxido compds. from ethylene derivs. C=C →

120. 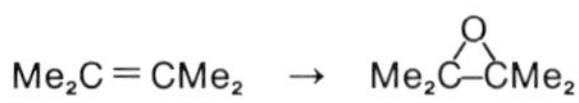

An excess of 2,3-dimethyl-2-butene in deuteriochloroform added at 34° to a soln. of 3-bromo-5-hydroperoxy-4,4-dimethyl-3,5-diphenyl-Δ^1-pyrazoline in the same solvent until n.m.r. indicated completion of reaction → product. Y 84%. F.e.s. A.L. Baumstark, R.S. Pilcher, J. Org. Chem. *47,* 1141–3 (1982); **sulfoxides from thioether, and amine oxides from tert. amines** s. Tetrahedron Letters *22,* 4591–4 (1981).

Peroxytrifluoroacetic acid/disodium hydrogen phosphate *CF_3COO_2H/Na_2HPO_4*

Enollactones from cyclic α,β-ethyleneketones
Baeyer-Villiger oxidation

121. 

90% H_2O_2 added to a soln. of trifluoroacetic anhydride in methylene chloride at 0°, the mixture stirred for 1 hr., transferred dropwise via a cold syringe to a soln. of 1-bromocyclohexen-3-one in the same solvent, and the mixture stirred at 0° over anhydrous powdered Na_2HPO_4 for 4 hrs. → 4-bromo-2-oxacyclohept-3-en-1-one. Y 68%. F.e.s. G.A. Krafft, J.A. Katzenellenbogen, J. Am. Chem. Soc. *103,* 5459–66 (1981).

m-*Chloroperoxybenzoic acid* m-*$ClC_6H_4COO_2H$*

Alcohols from ketals via orthocarbonic acid esters $(OR)_2 \rightarrow OH$
Double Baeyer-Villiger oxidation

122. $(C_7H_{15})_2C(OEt)_2 \rightarrow [\ (C_7H_{15}O)_2C(OEt)_2 \rightarrow (C_7H_{15}O)_2C{=}O\ ] \rightarrow C_7H_{15}OH$

A well-stirred suspension of 3–4 molar equivalents *m*-chloroperoxybenzoic acid in methylene chloride treated with startg. ketal while maintaining the temp. at 15–30° with ice-cooling, the crude intermediate (containing mainly orthoester and diethyl carbonate) worked up after 5 hrs., hydrolysed with 0.16 *M* HCl in methylene chloride, and reduced with $LiAlH_4$ → heptanol. Y 75%. F.e. and isolation of intermediate, **also diols from cyclic ketals,** s. W.F. Bailey, M.-J. Shih, J. Am. Chem. Soc. *104,* 1769–71 (1982).

m-*Chloroperoxybenzoic acid/ sodium hydrogen carbonate* m-*$ClC_6H_4COO_2H/NaHCO_3$*

ε-Alkylidene-ε-lactones from 4,5,6,7-tetrahydrobenzofurans

123. 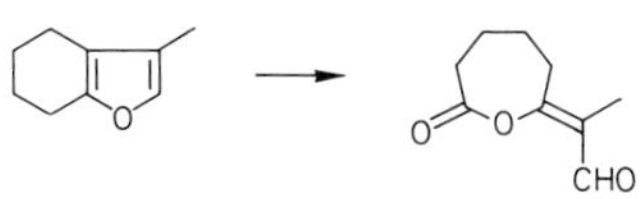

A mixture of *m*-chloroperoxybenzoic acid and $NaHCO_3$ in dichloromethane at 0° treated with a soln. of 3-methyl-4,5,6,7-tetrahydrobenzofuran in the same solvent, the mixture allowed to stand at the same temp. for 15 min., further *m*-chloroperoxybenzoic acid in dichloromethane added, and stirred for 15 min. → product. Y 81%. F.e.s. S.B. Gingerich et al., J. Org. Chem. *46,* 2589–91 (1981).

Oxidations with
m-*chloroperoxybenzoic acid/potassium fluoride* m-$ClC_6H_4COO_2H/KF$
Oxido compds. from ethylene derivs. C=C → 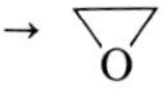

124.

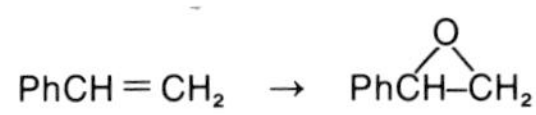

Styrene added to 3 equivalents of 1:1 *m*-chloroperoxybenzoic acid/KF complex in dichloromethane, the mixture stirred at room temp. overnight, filtered, 3 equivalents of freshly activated KF added to the filtrate to ensure complete elimination of acid, filtered again, and the solvent removed → styrene oxide. Y >95% by g.l.c. and n.m.r. – The simplified work-up avoids the usual basic treatment, thereby providing a useful method for the prepn. of sensitive compds. F.e., also **Baeyer-Villiger oxidation** of ar. aldehydes, s. F. Camps et al., Tetrahedron Letters *22,* 3895–6 (1981).

m-*Chloroperoxybenzoic acid/*
disodium hydrogen phosphate m-$ClC_6H_4COO_2H/Na_2HPO_4$
Oxaspiropentanes from alkylidenecyclopropanes
s. *37,* 695

p-*Nitroperoxybenzoic acid* p-$O_2NC_6H_4COO_2H$
Cyclopent-2-enones from 1,2,4-trienes ○
with mercuric acetate/perchloric acid cf. *33,* 127; 4-vinylcyclopent-2-enones with *p*-nitroperoxybenzoic acid s. J. Chem. Res. (S) *1981,* 244–5.

Pentafluoroperoxybenzoic acid $C_6F_5COO_2H$
Oxido compds. from ethylene derivs. C=C → 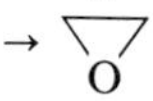
s. *37,* 79

O-Alkylmonoperoxycarbonic acids $ROCOO_2H$
with O-phenylmonoperoxycarbonic acid cf. *31,* 121; with O-alkylmonoperoxycarbonic acids, prepd. *in situ* from H_2O_2 and N-carbalkoxyimidazoles, s. Y. Tsunokawa, S. Iwasaki, S. Okuda, Tetrahedron Letters *23,* 2113–6 (1982).

Selenoxides s. under OsO_4 R_2SeO

Iodosobenzene/dichlorotris(triphenylphosphine)-
ruthenium(II) $C_6H_5IO/RuCl_2(PPh_3)_3$
α-**Diketones from acetylene derivs.** C≡C → COCO

125. Ph–C≡C–Ph → PhCOCOPh

A little dichlorotris(triphenylphosphine)ruthenium(II) in methylene chloride added to a suspension of iodosobenzene in the same solvent, the mixture treated immediately with a soln. of diphenylacetylene in the same solvent, and stirred for 15 min. → benzil. Y 86%. F.e., **also carboxylic acids from terminal acetylene derivs.,** s. P. Müller, J. Godoy, Helv. Chim. Acta *64,* 2531–3 (1981); *α*-**ketocarboxylic acid amides and esters** from ynamimes and alkoxyacetylenes, respectively s. Tetrahedron Letters *23,* 3661–4 (1982).

N-Bromosuccinimide $CH_2CO\backslash$ / $CH_2CO/$ NBr

1,4-Dioxaspiro[4.5]deca-6,8-diene-2,10-diones from catechols via 1,4-benzodioxin-2-ones ○

126. 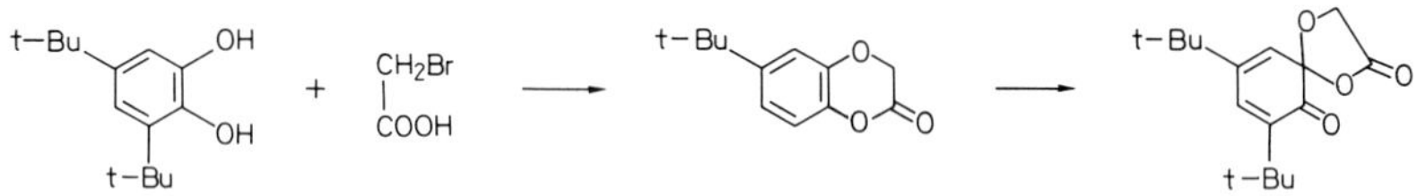

A mixture of 3,5-di-*tert*-butylcatechol, bromoacetic acid, and NaOH in ethanol refluxed under N_2 for 5 hrs. → 6,8-di-*tert*-butyl-1,4-benzodioxan-2-one (Y 83%), in 1:1 acetonitrile/water treated dropwise while stirring with 0.88 *M* NaOH followed by N-bromosuccinimide, and worked up after 5 min. → product (Y 94%). F.e.s. T.H. Chang et al., J. Chinese Chem. Soc. *27,* 97–105 (1980) (Eng.).

Silica gel SiO_2

Acetylation with ketene OH → $OCOCH_3$

review s. *6,* 366s*28;* acetylation of phenols, incl. hindered compds., with silica gel s. T. Chihara, S. Teratani, H. Ogawa, J. Chem. Soc. Chem. Commun. *1981,* 1120.

Chlorohydridobis(cyclopentadienyl)zirconium/ tert-butyl hydroperoxide $(C_5H_5)_2Zr(Cl)H$/t-*BuOOH*

Alcohols from ethylene derivs. via hydrozirconation C=C → CHC(OH)

s. *30,* 603; prim. alcohols on oxidation with *tert*-butyl hydroperoxide s. T. Gibson, Tetrahedron Letters, *23,* 157–8 (1982).

Titanium tetrachloride/sodium tetrahydridoborate $TiCl_4/NaBH_4$

***anti*-Markownikov addition**

s. *37,* 16

Disodium hydrogen phosphate Na_2HPO_4

s. under n-$ClC_6H_4COO_2H$ *and* CF_3COO_2H

Vanadium pentoxide/irradiation V_2O_5////

2,5-Dialkoxy-2,5-dihydrofurans from furans ←

by electrolysis cf. *11,* 196; s.a. V.P. Peresada et al., Khim.-Farmatsevt. Zh. *1982,* 537; by irradiation in the presence of a little V_2O_5 s. B.L. Feringa, R.J. Butselaar, Tetrahedron Letters *23,* 1941–2 (1982).

Vanadyl acetylacetonate/tert-butyl hydroperoxide $(acac)_2VO$/t-*BuOOH*

Oxido compds. from ethylene derivs. C=C → 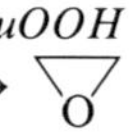

s. *24,* 149s*36;* epoxidation of monoterpenes, comparison of catalysts, s. D.V. Banthorpe, S.E. Barrow, Chem. Ind. *1981,* 502–3; role of trimethylsilyl group in stereospecificity s. H. Tomioka et al., Tetrahedron Letters *23,* 3387–90 (1982); with heterogeneous Mo-catalysts, e.g. MoO_2, cf. V.S. Markevich, V.N. Ul'yanova, V.A. Loginova, Neftepererab. Neftekhim. (Moscow) *1979* (II), 47; Chem. Abstr. *92,* 215157 (1980); **3,4-oxidoalcohols,** stereospecifity, s. E.D. Mihelich, K. Daniels, D.J. Eickhoff, J. Am. Chem. Soc. *103,* 7690–92 (1981).

Oxygen/irradiation O_2////

Reactions with polymer-based sensitizers ←

s. *24,* 143s*36;* with aerosil-bound sensitizers cf. L. Horner, J. Klaus, Ann. Chem. *1981,* 792–810.

Hydrogen peroxide (s.a. under H_3BNEt_3 H_3BPPh_3) — H_2O_2

Hydrogen peroxide/sodium hydroxide — $H_2O_2/NaOH$

Oxidative ring opening of *o*-quinones

dicarboxylic acid esters with Pb tetraacetate cf. *27,* 146s*32;* dicarboxylic acids with H_2O_2/NaOH s. A.N. Grinev et al., Khim.-Farmatsevt. Zh. *16,* No. 4, 49–54 (1982).

Hydrogen peroxide/poly[(S)-alanine]/sodium hydroxide ←

Asym. induction with synthetic enzymes — C=C →

Oxido compds. from ethylene derivs.

3-Phase medium

127.

Optically active α,β-oxidoketones. A soln. of chalcone in toluene treated with a little poly[(S)-alanine], the mixture stirred for 30 min., solns of NaOH and 30% H_2O_2 slowly added dropwise, and stirred at room temp. for 24 hrs. → (–)-1,3-diphenyl-2(R), 3(S)-epoxypropan-1-one. Y 85% (optical Y 93%). S. Juliá, J. Masana, J.C. Vega, Angew. Chem. Intern. Ed. *19,* 929–30 (1980).

Hydrogen peroxide/benzonitrile/potassium carbonate — $H_2O_2/C_6H_5CN/K_2CO_3$

Oxido compds. from ethylene derivs.

with $KHCO_3$ as catalyst cf. *18,* 191; in the presence of K_2CO_3, higher yields and shorter reaction times, s. C.H. Gagnieu, A.V. Grouiller, J. Chem. Soc. Perkin Trans. I *1982,* 1009–11.

Sulfur s. under $H_3B.PPh_3$ — S_8

Poly[vinyl(pyridinium tosylate)] ←

Protection of hydroxyl groups as tetrahydro-2-pyranyl ethers ←

with $POCl_3$ cf. *20,* 62; with poly(4- and 2-vinyl-pyridinium tosylate) s. F.M. Menger, C.H. Chu, J. Org. Chem. *46,* 5044–5 (1981).

Bis(trimethylsilyl) peroxide/trimethylsilyl triflate — $(Me_3Si)_2O_2/Me_3SiOSO_2CF_3$

Baeyer-Villiger oxidation

s. *36,* 129; in the presence of trimethylsilyl triflate cf. M. Suzuki, H. Takada, R. Noyori, J. Org. Chem. *47,* 902–4 (1982).

Potassium peroxymonosulfate/acetone — $KHSO_5/Me_2CO$

Oxido compds. from ethylene derivs. — C=C →

Stereospecific epoxidation under mild conditions

128.

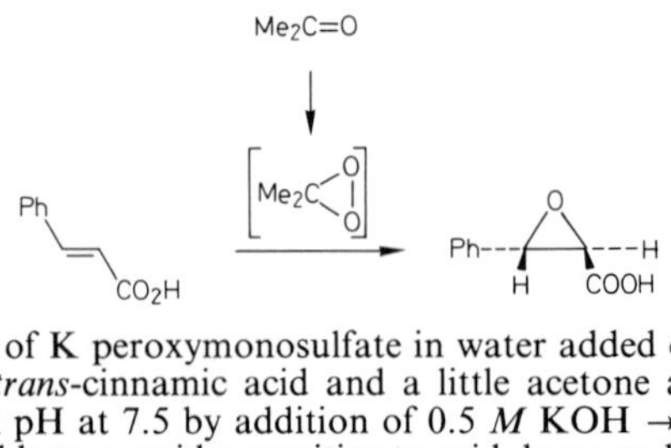

A freshly prepd. soln. of K peroxymonosulfate in water added dropwise during 2–4 hrs. to a mixture of *trans*-cinnamic acid and a little acetone at 2° in water while maintaining the initial pH at 7.5 by addition of 0.5 *M* KOH → product. Y 95%. – The method is applicable to epoxides sensitive to acid, base, and temperatures above 50°. F.e., regiospecificity, and with phase transfer catalysis, s. R. Curci et al., J. Org. Chem. *45,* 4758–60 (1980).

Sulfuric acid H_2SO_4

Homoconjugate ring opening of cyclopropylketones 

s. *31,* 125; of tricyclo[3.2.0.0^{2,7}]heptan-6-ones, nucleophiles, s. J.C. Gilbert, T. Luo, U. Patel, J. Org. Chem. *46,* 4042–6 (1981).

Chromium trioxide CrO_3

Ketocarboxylic acids by oxidative ring opening

s. *4,* 173; *o*-ketoacetic acids from indenes, and f. methods s. G. Brancaccio et al., J. Med. Chem. *24,* 998–1000 (1981).

*Molybdenum dioxide/*tert-*butyl hydroperoxide* MoO_2/t-$BuOOH$

Oxido compds. from ethylene derivs. C=C → 

s. *24,* 149s*37*

Molybdenum peroxide complex (s.a. under $H_3B.SMe_2$*)* ←

2,3-Oxidoalcohols from 2-ethylenealcohols C(OH)C=C → C(OH)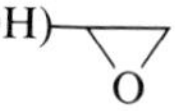

s. *37,* 228

Sodium hypochlorite/pyridine $NaOCl/C_5H_5N$

Oxido compds. from ethylene derivs. C=C → 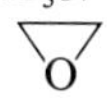

s. *18,* 193; α,β-oxido-γ-lactones with NaOCl in pyridine s. A.A. Jakubowski et al., J. Org. Chem. *47,* 1221–8 (1982).

Hydrogen chloride s. under Et_2NH HCl

Nitro-cobalt(III)complexes/bis(benzonitrile)dichloropalladium(II) ←

Oxo compds. from ethylene derivs. C=CH → CH–CO

s. *37,* 235

Nickel(II) complex s. under $Al(i\text{-}Bu)_3$ ←

Dichlorotris(triphenylphosphine)ruthenium(II) s. under C_6H_5IO $RuCl_2(Ph_3P)_3$

Palladium(II) chloride/cuprous chloride $PdCl_2/CuCl$

Ketones from ethylene derivs.
Wacker oxidation

methyl ketones cf. *19,* 200; β- and γ-alkoxyketones s. J. Tsuji, H. Nagashima, K. Hori, Tetrahedron Letters *23,* 2679–82 (1982); γ-ketocarboxylic acid esters and γ-diketones s. Chem. Letters *1982,* 859–60; preferential oxidation with bis(acetonitrile)chloronitropalladium(II) cf. A. Heumann, F. Chauvet, B. Waegell, Tetrahedron Letters *23,* 2767–8 (1982); with $PdSO_4$ and heteropolyacids cf. H. Ogawa et al., J. Chem. Soc. Chem. Commun. *1981,* 1274–5.

Bis(benzonitrile)dichloropalladium(II)
s. under nitro-cobalt(III) complexes $PdCl_2(PhCN)_2$

Osmium tetroxide/selenoxides OsO_4/R_2SeO

Glycols from ethylene derivs. C=C → C(OH)C(OH)

129. $MeCH{=}CH_2 \rightarrow MeCH(OH)CH_2OH$

A soln. of diphenyl selenoxide in a 1:1 mixture of acetone and 0.05 *M* K_2CO_3 (contained in a Fischer-Porter glass pressure reactor) treated with a little OsO_4 in 4% K_2CO_3 via syringe, propylene introduced, the mixture stirred overnight at room temp., and worked up with 10% $NaHSO_3$ → propylene glycol. Y 95%. F.e., also with methyl phenyl selenide in the presence of photochemically generated singlet oxygen, s. A.G. Abatjoglou, D.R. Bryant, Tetrahedron Letters *22,* 2051–4 (1981).

Via intermediates *v.i.*

3,4-Oxidoalcohols from 3-ethylenealcohols

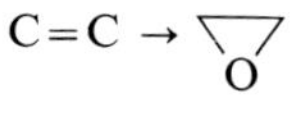

s. *37,* 450

Rearrangement ∩

Hydrogen/Oxygen Type OC∩HO

Ammonium hydroxide *NH_4OH*

Rearrangement of hydroxyurethans ←

of acyclic hydroxyurethans s. *14,* 382; of cyclic hydroxyurethans with concd. aq. NH_3 s. J. Tadanier et al., Carbohyd. Res. *98,* 11–23 (1981).

p-*Toluenesulfonic acid* *TsOH*

6*H*-1,2-Oxazine-6-carboxylic acid esters from 5-alkoxy-2-furyloximes

130.

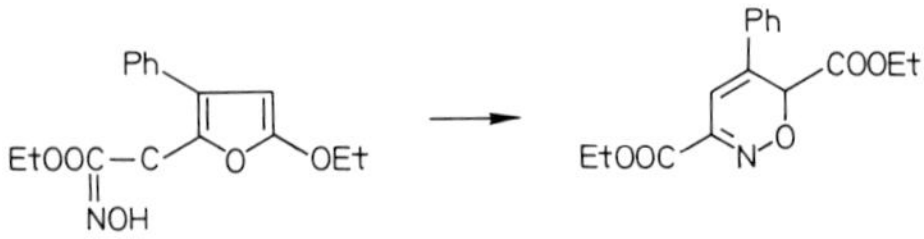

Startg. oxime and a little *p*-toluenesulfonic acid monohydrate in benzene refluxed for 5 min. → product. Y 80%. F.e.s. Y. Nakada et al., Tetrahedron Letters *22,* 473–6 (1981).

Via intermediates *v.i.*

Optically active *α,β*-oxidoaldehydes from *α,β*-ethylenecarboxylic acids via asym. bromolactonization

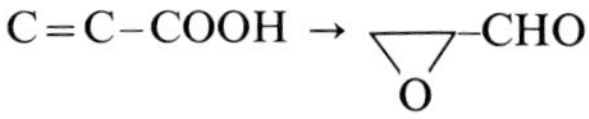

131.

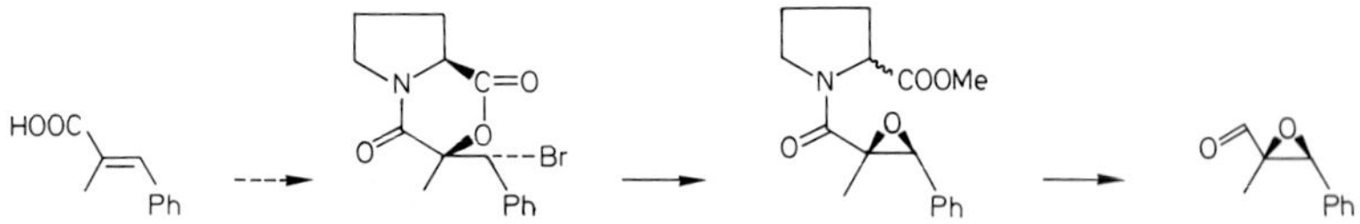

A soln. of Na methoxide in methanol added during 5 min. to a stirred soln. of startg. bromolactone in the same solvent at –78° under argon, and stirring continued at the same temp. for 4 hrs. → methyl N-(2-methyl-3-phenyl-2(R),3(S)-epoxypropionyl)prolinate (Y 90%) in ether treated dropwise with a soln. of Na dihydridobis(2-methoxyethoxo)aluminate in the same solvent during 5 min. under argon at 0°, the mixture stirred at the same temp. for 1 hr., satd. NH_4Cl added, and stirring continued for 30 min. → (+)-2-methyl-3-phenyl-2(R), 3(S)-epoxypropanal (Y 89%). F.e.s. M. Hayashi, S. Terashima, K. Koga, Tetrahedron *37,* 2797–803 (1981).

Hydrogen/Carbon Type OC∩HC

Irradiation s. under $Fe(CO)_5$

Sodium carbonate s. under $RhCl_3$ *Na_2CO_3*

Potassium carbonate K_2CO_3

Anomerization of carbohydrate derivs. under non-acidic conditions ←
with *tert*-butylmagnesium bromide cf. *34,* 129; with K_2CO_3 s. H.J. Koeners et al., Rec. Trav. Chim. Pays-Bas *99,* 355–62 (1980); with $TiCl_4$ cf. S. Koto et al., Bull. Chem. Soc. Japan *55,* 1092–6 (1982).

Glyoxalase ←

α-Hydroxycarboxylic acid esters from α-ketoaldehydes COCHO → CH(OH)COOR
with N,N-diethylcysteamine cf. *11,* 215; with enzymes, optically active compds., s. M.A.K. Patterson, R.P. Szajewski, G.M. Whitesides, J. Org. Chem. *46,* 4682–5 (1981).

Triphenylphosphine s. under $RhCl_3$ Ph_3P

Molybdenyl acetylacetonate $(acac)_2MoO_2$

One-step epimerization of carbohydrate 2-hydroxyl groups ←
with $Ba(OH)_2$ cf. *2,* 837; with molybdenyl acetylacetonate and f. catalysts s. Y. Abe, T. Takizawa, T. Kunieda, Chem. Pharm. Bull. *28,* 1324–6 (1980).

Iron carbonyl/irradiation $Fe(CO)_5$/hν

Oxo from oxido compds. epoxide (O) → CHCO

with $Co_2(CO)_8$ cf. *18,* 202; photochemical method with $Fe(CO)_5$, from phenyl-subst. epoxides, s. T. Kobayashi, M. Nitta, Chem. Letters *1982,* 325–8.

Rhodium trichloride/triphenylphosphine/sodium carbonate $RhCl_3/Ph_3P/Na_2CO_3$

Lactones from dialdehydes ○

Intramolecular Tishchenko reaction

132. 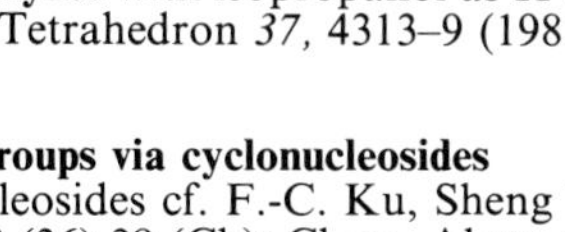

A mixture of $RhCl_3.3H_2O$, triphenylphosphine, and Na_2CO_3 in ethanol refluxed with stirring under N_2 for 15 min., startg. dialdehyde added to the orange-yellow precipitate, and refluxing continued for 2 hrs. → lactone. Y 93%. F.e. and catalyst, **also benzyl alcohols from ar. aldehydes** with isopropanol as H-donor, s. R. Grigg, T.R.B. Mitchell, S. Sutthivaiyakit, Tetrahedron *37,* 4313–9 (1981).

Via intermediates *v.i.*

Epimerization of hydroxyl groups via cyclonucleosides ←
s. *19,* 206; via 2,2′-cyclonucleosides cf. F.-C. Ku, Sheng Wu Hua Hsueh Yu Sheng Wu Wu Li Chin Chan *1980* (36) 39 (Ch); Chem. Abstr. *95,* 98190 (1981).

1,2-Carbonyl transposition COCH → CHCO
s. *37,* 142

Epimerization of hydroxyl ←
s. *37,* 148

Oxygen/Oxygen Type OC∩OO

Dichlorotris(triphenylphosphine)ruthenium(II) $RuCl_2(PPh_3)_3$

1,3-Dioxido compds. from 3,6-dihydro-1,2-dioxin ring ←
s. *36,* 245; 3,5,7-trioxatricyclo[4.1.0.0^{2,4}]heptanes s. M.L. Graziano, M.R. Iesce, R. Scarpati, J. Chem. Soc. Chem. Commun. *1981,* 720; with dichlorotris(triphenylphosphine)ruthenium(II) cf. M. Suzuki, R. Noyori, N. Hamanaka, J. Am. Chem. Soc. *104,* 2024–5 (1982); with $Rh_2(CO)_4Cl_2$ cf. J.-P. Hagenbuch et al., Helv. Chim. Acta *65,* 887–95 (1982).

Oxygen/Nitrogen Type **OC∩ON**

Trimethylsilyl polyphosphate ←
Beckmann rearrangement C(=NOH)R → CONHR

133. 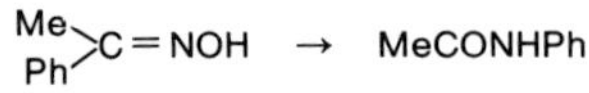

Acetophenone oxime added to a soln. of trimethylsilyl polyphosphate (prepd. by heating a mixture of P_2O_5 and hexamethyldisiloxane) in benzene, and worked up after 6 hrs. at room temp. → product. Y 96%. – Similar yields were obtained with ethyl polyphosphate, but the latter is more difficult to prepare. F.e.s. T. Imamoto, H. Yokoyama, M. Yokoyama, Tetrahedron Letters *22,* 1803–4 (1981).

Antimony pentachloride $SbCl_5$
***o*-Wallach rearrangement** ←
o-hydroxyazo compds. cf. *30,* 96s*31;* 2-pyridones from pyridine N-oxides, also *o*-aminophenols from ar. amine N-oxides, cf. J. Yamamoto et al., Tetrahedron *37,* 1871–3 (1981).

Hydrogen chloride *HCl*
Beckmann rearrangement
Ring expansion
with H_2SO_4 cf. *4,* 182; with dry HCl in acetonitrile at 70°, laurolactam, s. J. Stresinka, A. Chovanova, L. Lukacovic, Chem. Prum. *31,* 250 (1981) (Slo); Chem. Abstr. *95,* 42311 (1981).

Nitrogen/Carbon Type **OC∩NC**

Hydrogen chloride *HCl*
N-Acyl to O-acyl group migration ←
s. *37,* 607

Sulfur/Carbon Type **OC∩SC**

Camphorsulfonic acid ←
Macrocyclic 2-mercaptolactones from 2′-hydroxythiolactones

134. 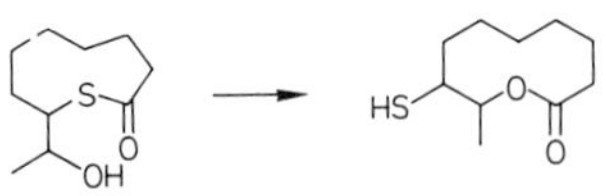

Startg. m. in methylene chloride treated with a little camphorsulfonic acid → product. Y 91%. F.e. and prepn. of startg. m. s. E. Vedejs, D.W. Powell, J. Am. Chem. Soc. *104,* 2046–8 (1982).

Carbon/Carbon Type **OC∩CC**

Sodium hydride *NaH*
Lactones by retro-Dieckmann cleavage
s. *31,* 144; macrocyclic lactones s. V. Bhat, R.C. Cookson, J. Chem. Soc. Chem. Commun. *1981,* 1123–4.

Potassium tert-*butoxide* t-*BuOK*

Cyclopropanol ring opening
with NaH cf. *32,* 663; with K *tert*-butoxide, as part of a ring expansion of ketones, s. Y. Hoyano, V. Patel, J.B. Stothers, Can. J. Chem. *58,* 2730–2 (1980).

2,5-Dihydrofurans from 2-allenealcohols ○
s. *25,* 123; with dicarbonyl(cyclopentadienyl)(isobutylene)iron cation cf. D.F. Marten, J. Org. Chem. *46,* 5422–5 (1981).

Silver nitrate *$AgNO_3$*

O-Heterocyclics from allenealcohols
2,5-dihydrofurans cf. *36,* 148; 2-vinyltetrahydrofurans from 4-allenealcohols s. P. Audin et al., Bull. Soc. Chim. France II *1981,* 313–8.

Mercuric sulfate/sulfuric acid *$HgSO_4/H_2SO_4$*

Furans from acetyleneoxido compds. ○
s. *24,* 168; steroid derivs. s. H. Berbalk, K. Eichinger, R. Schuster, Synthesis *1981,* 613–6.

Bis[1,2-(diphenylphosphino)ethane]palladium(0) ←

α,β-Ethylene- from β-methylene-γ-lactones ←
Rearrangement

135.

3-Sprio-1′-cyclohexane-4-methylenedihydro-2(5*H*)-furanone treated under 60 atm. CO_2 with an equimolar amount of bis[(1,2-diphenylphosphino)ethane]palladium(0) in benzene at 125° for 21 hrs. → 4-methyl-5-spiro-1′-cyclohexane-2(5*H*)-furanone. Y 76%. F.e.s. Y. Inoue et al., Chem. Letters *1980,* 1521–2.

Bis(acetonitrile)dichloropalladium(II) *$PdCl_2(MeCN)_2$*

Isomerization of acoxy-2-ethylenes C(OAc)C=C → C=C–C(OAc)
s. *35,* 87; 5-acoxy-1,3-dienes s. B.T. Golding, C. Pierpoint, R. Aneja, J. Chem. Soc. Chem. Commun. *1981,* 1030–1; **1,3-chirality transfer** s. P.A. Grieco et al., J. Am. Chem. Soc. *102,* 7587–8 (1980); steroid side chains s. S. Takatsuto, M. Ishiguro, N. Ikekawa, J. Chem. Soc. Chem. Commun. *1982,* 258–60; with $PdCl_2$/Na[BH_3CN], carbohydrate derivs., s. L.V. Dunkerton, K.T. Brady, F. Mohamed, Tetrahedron Letters *23,* 599–602 (1982).

Via intermediates *v.i.*

Isomerization of α,β-ethyleneketones ←
via 1-alkylthio-1,3-dienes cf. *36,* 151; via α′-arylthio-α,β-ethylenetosylhydrazones and arylthio-2-ethylenes s. T. Mimura, T. Nakai, Chem. Letters *1981,* 1579–82.

Exchange ↓↑

Hydrogen ↑ OC↓↑H

Electrolysis ϟ

α-Alkoxyurethans from urethans H → OR
s. *37,* 876

Electrolysis/sodium cyanide *↯/NaCN*

Carboxylic acid esters from aldehydes CHO → COOR

136. PhCHO + HOMe → PhCOOMe

Arylcarboxylic acid esters. A soln. of benzaldehyde in methanol containing NaCN electrolysed at ca 17° in a divided cell using a Pt anode at a potential of 2.30 V with a constant current of 0.8 A until 5.6 F/mol electricity passed (9.4 hrs.) → methyl benzoate. Y 80%. – Aliphatic aldehydes gave low yields and *α,β*-unsatd. aldehydes were unreactive. F.e.s. T. Chiba et al., Bull. Chem. Soc. Japan *55,* 335–6 (1982).

Electrolysis/palladium–carbon *↯/Pd–C*

Preferential replacement of hydrogen by acoxy groups H → OAr

Aromatic monoacetoxylation

137.

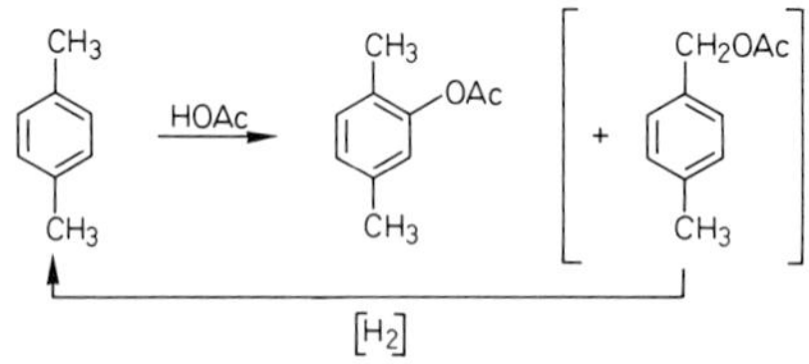

A stirred slurry of *p*-xylene, K acetate, and a little 10% Pd-C in acetic acid electrolysed at 18° in an undivided cell using a graphite cylinder as cathode and a bent Pt foil as anode until 2 F/mol had passed → 2,5-dimethylphenyl acetate. Y 75%. – Concurrent side-chain acetoxylation is avoided by catalytic hydrogenolysis of the by-product with H_2 generated at the cathode. F.e.s. L. Eberson, E. Oberrauch, Acta Chem. Scand. Ser. B *35,* 193–6 (1981).

Lithium alkoxides s. under t-*BuOCl* *LiOR*

Sodium alkoxides s. under PhI(OAc)$_2$ *NaOR*

Sodium peroxide Na_2O_2

2-Ene-1,4-diones from ethyleneketones ←

from *β,γ*-ethyleneketones with $Cu(NO_3)_2$ cf. *20,* 121; from *α,β*-ethyleneketones with Na_2O_2, steroid derivs., s. H.L. Holland, U. Daum, E. Riemland, Tetrahedron Letters *22,* 5127–8 (1981).

Sodium cyanide s. under ↯ *NaCN*

Sodium polysulfide Na_2S_x

Aromatic aminoaldehydes from nitrohydrocarbons ←

p-aminobenzaldehyde s. *1,* 162; in the presence of dimethylformamide, industrial prepn., s. K. Kato et al., Nippon Kagaku Kaishi *1981,* 255 (Japan); Chem. Abstr. *95,* 6705 (1981).

Sodium acetate s. under $PdCl_2$ *MeCOONa*

Triethylamine s. under SO_2Cl_2 Et_3N

Pyridine C_5H_5N

Ring closures

with N-(dichloromethylene)dimethylammonium chloride

Cyclic carbonic acid esters from diols

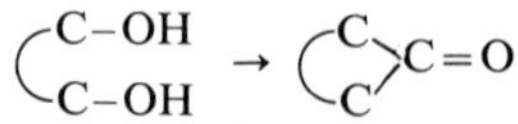

s. *36,* 153; carbohydrate derivs. with pyridine, limitations. s. Australian J. Chem. *35,* 581–9 (1982).

Cupric oxide *CuO*

Carboxylic acid esters $2\ RCH_2OH \rightarrow RCOOCH_2R$
from two prim. alcohol molecules
with *tert*-butyl hypochlorite cf. *9,* 363; with CuO s. B. Berthon et al., Tetrahedron Letters *22,* 4073–6 (1981).

Cupric sulfate/persulfate $CuSO_4/S_2O_8^{2-}$

Keto from methylene groups $CH_2 \rightarrow CO$
with $AgNO_3$ cf. *27,* 177; with $CuSO_4$, aryloxo compds., s. P.T. Perumal, M.V. Bhatt, Indian J. Chem. *20B,* 153–5 (1981).

Cupric chloride s. under $PdCl_2$ $CuCl_2$

Mercuric tetrafluoroborate $Hg(BF_4)_2$

trans-**Alkoxy-2-ethylenes and alcohols** CH–C=C → C=C–C(OR)
from ethylene derivs with double bond migration

138. $PhCH_2CH{=}CH_2 + HOMe \rightarrow PhCH{=}CHCH_2OMe$

A soln. of allylbenzene and methanol in tetrahydrofuran treated with $Hg(BF_4)_2$, the mixture refluxed with stirring for 2 hrs., the ppt. removed, and the filtrate hydrolysed with 3 *M* aq. KOH → product. Y 71%. F.e.s. J. Barluenga, L. Alonso-Cires, G. Asensio, Tetrahedron Letters *22,* 2239–40 (1981).

Silica gel s. under SO_2Cl_2 SiO_2

Thallium(III) nitrate $Tl(NO_3)_3$

Cyclic carboxylic acid esters from ketones ○
Stereospecific ring contraction
s. *33,* 128; cyclic *β,γ*-ethylenecarboxylic acid esters s. E. Mincione, P. Barraco, M.L. Forcellese, Gazz. Chim. Ital. *110,* 515–7 (1980).

Acetone s. under $RuH_2(PPh_3)_4$ Me_2CO

N,N′-Carbonyldiimidazole ←

1,3,4-Dioxazol-2-ones from hydroxamic acids ○
with phosgene cf. *35,* 88; with N,N′-carbonyldiimidazole s. D. Geffken, Ann. Chem. *1982,* 211–8.

m-*Chloroperoxybenzoic acid* m-$ClC_6H_4COO_2H$

2-Azetidinones from 2-carboxyazetidines ←
s. *33,* 164; f. method cf. H.H. Wasserman et al., J. Org. Chem. *46,* 2991–9 (1981).

Phenyl iodosoacetate/sodium alkoxides $PhI(OAc)_2/NaOR$

α-**Alkoxycarboxylic acid esters from carboxylic acid esters** H → OR

139. $PhCH_2COOMe + HOMe \rightarrow PhCH(OMe)COOMe$

A mixture of startg. ester, phenyl iodosoacetate, and 3 equivalents Na methoxide in methanol stirred at room temp. for 3 days → product. Y 70%. F.e., **also *α*-hydroxycarboxylic acid esters,** s. R.M. Moriarty, H. Hu, Tetrahedron Letters *22,* 2747–50 (1981).

Phenyl iodosotrifluoroacetate *PhI(OCOCF$_3$)$_2$*

1,3-Diol trifluoroacetates from cyclopropanes C

140. 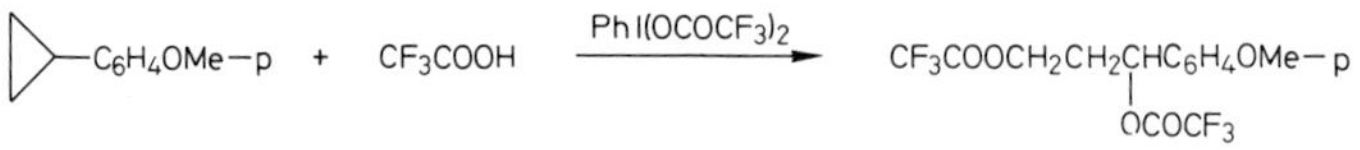

A soln. of phenyl iodosotrifluoroacetate in trifluoroacetic acid added gradually with stirring to a soln. of startg. cyclopropane in chloroform at –10°, and stirring continued for a further 1 hr. → 1,3-bis(trifluoroacetoxy)-1-(*p*-methoxyphenyl)propane. Y 98%. F.e.s. Y.S. Shabarov, E.V. Pisanova, L.G. Saginova, Zh. Org. Khim. *17,* 1337–9 (1981).

Lead dioxide *PbO$_2$*

β-(1,4,2-Dioxazol-5-yl)-α,β-ethyleneketones from furans and hydroxamic acids ←
Acylnitroso compds. as intermediates

141. 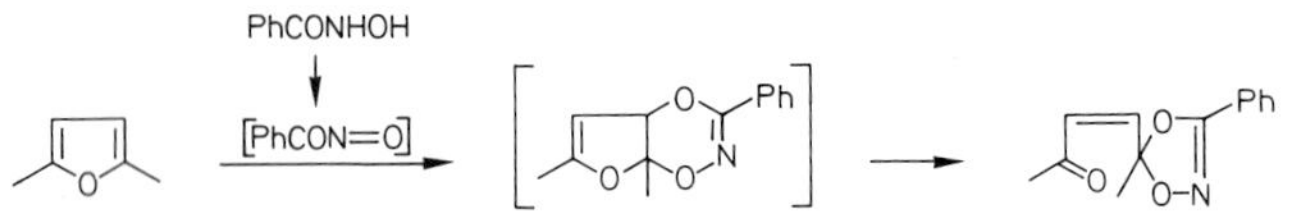

A mixture of benzohydroxamic acid and 2,5-dimethylfuran in ethyl acetate treated with PbO_2, and stirred at room temp. for 1 day → (Z)-4-(3-phenyl-5-methyl-1,4,2-dioxazol-5-yl)but-3-en-2-one. Y 83%. F.e., also with Ag_2O as oxidant, and isolation of the intermediate, s. D. Mackay, L.H. Dao, J.M. Dust, J. Chem. Soc. Perkin Trans. I *1980,* 2408–16.

Lead tetraacetate *(MeCOO)$_4$Pb*

Urethans from formamides NHCHO → NHCOOR
with diethyl azodicarboxylate cf. *31,* 437; with Pb tetraacetate, N-arylurethans, s. R. Leardini, G. Zanardi, Synthesis *1982,* 225–6.

Aliphatic 1-acetoxyazo compds. from hydrazones C=NNHR → C(OAc)(N=NR)
s. *16,* 219; f.e. and reactions s. D.W.K. Yeung, J. Warkentin, Can. J. Chem. *58,* 2386–93 (1980).

Bismuth(III) nitrate s. under Pd–C *Bi(NO$_3$)$_3$*

Sulfuryl chloride/triethylamine/silica gel *SO$_2$Cl$_2$/Et$_3$N/SiO$_2$*

Cyclic α-diketone monothioketals from monothioketals ←

142. 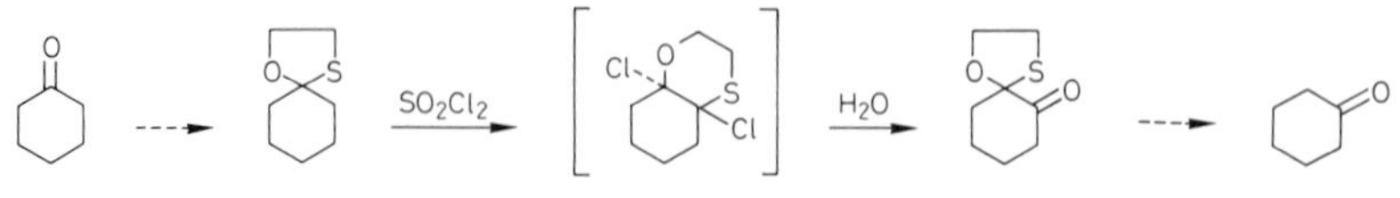

A soln. of SO_2Cl_2 in carbon tetrachloride added dropwise to a stirred soln. of startg. thioketal in the same solvent at 0° during 30–60 min., treated with triethylamine in the same solvent, allowed to warm to room temp., concd. in vacuo, silica gel and water added, and the mixture stirred until hydrolysis complete → 6-oxospiro[cyclohexane-1,2′-1′,3′-oxathiolan]. Y 94%. – This is part of a novel 4-step **1,2-carbonyl transposition.** F.e., **also cyclic α-diketone monomercaptals from mercaptals,** s. P.C. Bulman-Page et al., J. Chem. Soc. Perkin Trans. I *1981,* 457–61.

Potassium nitrosodisulfonate *$ON(SO_3K)_2$*

Teuber oxidation ←

Quinones from phenols

s. *8,* 206; 2-phase system, also from anilines, s. D. Scholz, Sci. Pharm. (Austria) *50* (1), 78 (1982); Chem. Abstr. *97,* 5906 (1982).

Persulfate s. under $CuSO_4$ *$S_2O_8^{2-}$*

tert-*Butyl hypochlorite/lithium alkoxide* t-*BuOCl/LiOR*

Replacement of hydrogen by alkoxy groups H → OR

s. *29,* 177s*32;* 1-alkoxyacylamines from acylamines s. P. Feyen, W. Schröck, Angew. Chem. Intern. Ed. *20,* 808–9 (1981).

Sodium permanganate *$NaMnO_4$*

Carboxylic acids from prim. alcohols $CH_2OH \rightarrow COOH$

s. *37,* 232

Potassium hexacyanoferrate(III) *$K_3[Fe(CN)_6]$*

2-Pyridones from pyridinium salts ←

with NaOH cf. *31,* 162; with K hexacyanoferrate(III) s. T. Fujii, T. Hiraga, M. Ohta, Chem. Pharm. Bull. *29,* 2503–8 (1981).

Ruthenium dioxide *RuO_4*

Oxidations of N-heterocyclics ←

s. *24,* 507; f.e.s. G. Bettoni et al., Tetrahedron *37,* 4159–64 (1981).

Dihydridotetrakis(triphenylphosphine)ruthenium/acetone *$RuH_2(Ph_3P)_4/Me_2CO$*

Carboxylic acid esters from two prim. alcohol molecules $2\ CH_2OH \rightarrow COOCH_2$

s. *37,* 236

Palladium–carbon s.a. under ϟ *Pd–C*

Palladium–carbon/bismuth(III) nitrate *$Pd–C/Bi(NO_3)_3$*

Carboxylic acids from prim. alcohols $CH_2OH \rightarrow COOH$

Aryloxyacetic acids

143. $$PhOCH_2CH_2OH \rightarrow PhOCH_2COOH$$

A mixture of 10% Pd–C, a little $Bi(NO_3)_3.5H_2O$, 1.3 *M* aq. NaOH, and 2-phenoxyethanol in a cylindrical glass vessel heated at 90° with continuous stirring under a stream of O_2 until 1 mol. absorbed (ca. 1 hr.) → phenoxyacetic acid. Y 100%. F.e.s. H. Fiege, K. Wedemeyer, Angew. Chem. Intern. Ed. *20,* 783–4 (1981).

Palladium(II) chloride/cupric chloride/sodium acetate *$PdCl_2/CuCl_2/MeCOONa$*

Regiospecific oxidative substitution of β,γ-ethylenesulfones ←

144. $$n\text{-}C_9H_{19}CH{=}CHCH_2SO_2C_6H_4Me\text{-}p \xrightarrow{HOAc} n\text{-}C_9H_{19}\underset{\displaystyle OAc}{CH}CH{=}CHSO_2C_6H_4Me\text{-}p$$

γ-Acoxy-α,β-ethylenesulfones. A suspension of $CuCl_2$, Na acetate, and NaCl in acetic acid containing a little $PdCl_2$ stirred at 90° for 2 hrs., startg. *p*-tolyl sulfone added, and the mixture heated at 80° for 60 hrs. → product. Y 86%. F.e., **also γ-chloro-α,β-ethylenesulfones,** s. K. Ogura, N. Shibuya, H. Iida, Tetrahedron Letters *22,* 1519–22 (1981).

Oxygen ↑ **OC↓↑O**

Without additional reagents *w.a.r.*

O-β,β,β-Trifluoroethylation $OH \rightarrow OCH_2CF_3$

145. $PhCH_2OH + Ph_3P(OCH_2CF_3)_2 \rightarrow PhCH_2OCH_2CF_3$

2,2,2-Trifluoroethyl ethers. A soln. of bis(2,2,2-trifluoroethoxy)triphenylphosphorane in 1:1 dichloromethane/ether treated with benzyl alcohol, and the mixture stirred at room temp. for 1 hr. → 2,2,2-trifluoroethyl benzyl ether, Y 84%. F.e., **also carboxylic acid 2,2,2-trifluoroethyl esters,** s. T. Kubota et al., J. Org. Chem. *45,* 5052–7 (1980).

Reactions of trichloroacetiminoesters ←
mixed phosphoric esters s. *14,* 605; glycosyl phosphates s. R. R. Schmidt, M. Stumpp, J. Michel, Tetrahedron Letters *23,* 405–8 (1982); C-aryl glycosides from activated arenes s. ibid. *23,* 409–12 (1982).

Glycol esters from cyclic orthocarboxylic acid esters C
with BF_3 and Ag carboxylates cf. *31,* 107; without reagent, carbohydrate derivs., s. J. Leroux, A.S. Perkin, Carbohyd. Res. *94,* 108–14 (1981).

Electrolysis ϟ

4,5-Dihydroxy-Δ^2-isoxazolines from α,β-ethylene-β-nitroketones O

146.

A soln. of startg. nitroketone in 0.05 *M* H_2SO_4 containing 50% acetonitrile (pH 1.12) electrolysed at a potential of –0.4 V until ca. 2 F/mol consumed → 4,5-dihydroxy-3,4-diphenyl-5-methyl-2-isoxazoline. Y 91%. F.e.s. C. Bellec et al., Can. J. Chem. *59,* 527–31 (1981).

Sodium hydroxide *NaOH*

Replacement of ammonium groups by nucleophiles $NR_3^+ \rightarrow OR$
by fluorine cf. *28,* 507; by alkoxy groups with NaOH s. B.L. Gaffney, R.A. Jones, Tetrahedron Letters *23,* 2253–6 (1982); enolethers from quaternary enammonium salts s. M. Dandárová, J. Kováč, D. Végh, Collection Czech. Chem. Commun. *45,* 3557–63 (1980).

Potassium hydroxide *KOH*

Alkoxy-2-ethylenes from 1,2-disulfonates Carbohydrate derivs. $C(OSO_2R)C(OSO_2R)CH \rightarrow C{=}C{-}C(OR)$

147.

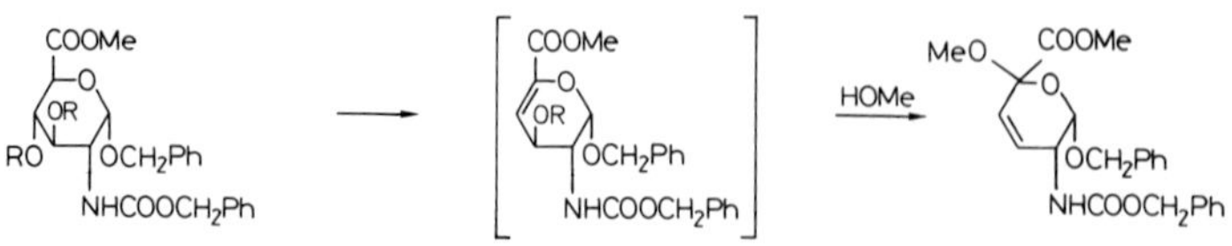

R = SO_2Me

A soln. of the startg. m. in methanol treated with a soln. KOH (or 1,8-diazabicyclo[5.4.0]undec-7-ene) in the same solvent at room temp., and reaction monitored by t.l.c. until complete → product. Y ca. 100%. F.e.s. J. Kiss, W. Arnold, Helv. Chim. Acta *64,* 1566–9 (1981).

Caesium carboxylates *RCOOCs*

Acoxy compds. from sulfonyl esters with inversion of configuration $OSO_2R \rightarrow AOc$

148. 

(S)-2-Octyl methanesulfonate treated with Cs propionate in dimethylformamide at 40° for 24 hrs. → (R)-2-octyl propionate. Y 100%. – This is part of multistep **epimerization of hydroxyl groups.** F. e. s. W. H. Kruizinga, B. Strijtveen, R. M. Kellogg, J. Org. Chem. *46,* 4321–3 (1981).

Sodium fluoride s. under $(CF_3CO)_2O$ *NaF*

Potassium fluoride or caesium fluoride *KF or CsF*

O-Acetylation OH → OAc

with KF and $(CF_3CO)_2O$ cf. *33,* 169; without $(CF_3CO)_2O$, also with CsF, of alcohols and phenols, s. E. Lee, Y. Soo, W.P. Hong, Bull. Korean Chem. Soc. *1,* 144 (1980) (Eng); Chem. Abstr. *95,* 6730 (1981).

Triethylamine s. under SO_2ClF, Ph_3P, *and N-(Diphenoxyphosphoroyl)-*Δ^4*-2-oxazolone* Et_3N

Silver trifluoromethanesulfonate/ p-nitrobenzenesulfonyl chloride $CF_3SO_3Ag/ArSO_2Cl$

Stereospecific glycosylation OH → OAr

s. *18,* 237s*36;* aryl *α*-glycosides s. Bull. Chem. Soc. Japan *54,* 1895–6 (1981); with added N,N-dimethylacetamide and triethylamine s. Chem. Letters *1982,* 1039–40.

Silica gel s. under $AlCl_3$ SiO_2

Boron fluoride BF_3

Preferential acylation of alcohol groups with retention of phenol groups OH → OAc

with pyridine cf. *28,* 86; with BF_3-etherate s. Y. Nagao et al., Chem. Pharm. Bull. *29,* 3202–7 (1981).

Carboxylic acid esters from carboxylic alkoxyformic anhydrides COO–O–CHO → COOR

s. *16,* 341; *α*-ketocarboxylic acid esters, direct method from the acid without isolation of anhydride, cf. J.M. Domagala, Tetrahedron Letters *21,* 4997–5000 (1980).

Aluminium chloride/silica gel $AlCl_3/SiO_2$

Carbohydrate O,O-isopropylidene derivs.

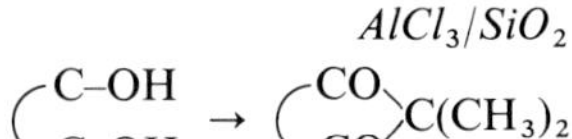

149. 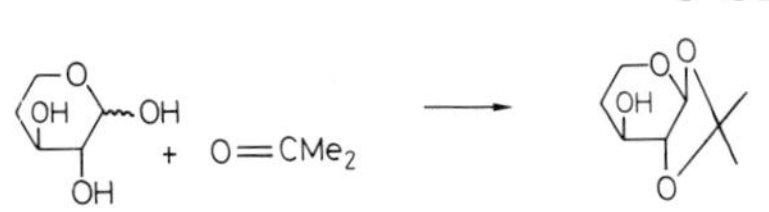

4-Deoxy-DL-*threo*-pentopyranose, dispersed in silica gel, introduced at the top of column filled with a suspension of silica gel in acetone, eluted with 4% $AlCl_3$ in the same solvent, and the eluent collected in aq. K_2CO_3 of pH 8.5 at 0° → 4-deoxy-1,2-O-isopropylidene-*β*-DL-*threo*-pentopyranose. Y 70%. F. methods s. C. Gagnieu, A. Grouiller, Carbohyd. Res. *84,* 61–9 (1980).

*Tetra-*n*-butylammonium bromide s. under Me_3SiBr* — n-$Bu_4N^+Br^-$

Tetraethylammonium perchlorate s. under $MeSO_3H$ — $Et_4N^+ClO_4^-$

Diethyl azodicarboxylate s. under Ph_3P — *EtOOC–N=N–COOEt*

2-Halogeno-1,3,5-trinitrobenzenes ←
Carboxylic acid esters from carboxylic acids COOH → COOR

150. PhCOOH + HOBu-s → PhCOOBu-s

2-Chloro-1,3,5-trinitrobenzene added to a stirred mixture of benzoic acid and *sec*-butanol in pyridine, and stirring continued for 3 hrs. → *sec*-butyl benzoate. Y 93%. – The reaction failed with phenols. F.e.s. S. Takimoto et al., Bull. Chem. Soc. Japan *54,* 1470–3 (1981); with 2-fluoro-1,3,5-trinitrobenzene in the presence of 4-dimethylaminopyridine cf. S. Kim, S. Yang, Synth. Commun. *11,* 121–4 (1981).

Trifluoroacetic anhydride/sodium fluoride — $(CF_3CO)_2O/NaF$
s. *6,* 211; fatty acid methyl and aryl esters, also with added NaF as base, s. J.S. Grossert, W.M.N. Ratnayake, T. Swee, Can. J. Chem. *59,* 2617–20 (1981).

Cation exchanger ←
with KU-2, methacrylates s. *11,* 232s*22; n*-dodecyl methacrylate, comparison with other catalysts, s. N.M. Kukui et al., Khim. Tekhnol. Topl. Masel *1981* (5), 51; Chem. Abstr. *95,* 149912 (1981).

Carbon tetrachloride s. under Ph_3P — CCl_4

*Trimethylsilyl bromide/cobalt bromide/ tetra-*n*-butylammonium bromide* — $Me_3SiBr/CoBr_2/$n-$Bu_4N^+Br^-$
α-Glycosylation OH → OR
s. *28,* 145s*37*

Zirconium(IV) hydroxide/sulfuric acid — $Zr(OH)_4/H_2SO_4$
Carboxylic acid esters from carboxylic acids COOH → COOR
Superacidic catalysis
with Nafion-H cf. *35,* 105; with $Zr(OH)_4$ and H_2SO_4, acetic acid esters, s. M. Hino, K. Arata, Chem. Letters *1981,* 1671–2.

Triphenylphosphine/diethyl azodicarboxylate — *Ph_3P/EtOOC–N=N–COOEt*
Cyclic hydroxamic acid esters
from hydroxamic acids and alcohols N–OH → N–OR

151.

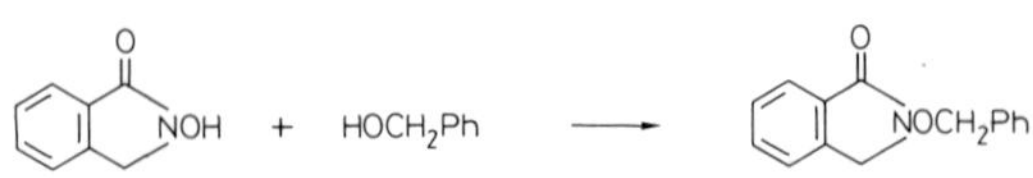

A mixture of startg. N-hydroxylactam, benzyl alcohol, and triphenylphosphine in tetrahydrofuran treated with diethyl azodicarboxylate, and left at room temp. for 24 hrs. → product. Y 94%. F.e., **also cyclic acyl hydroxamates from carboxylic acids,** s. E. Grochowski, T. Bolesławska, Pol. J. Chem. *55,* 615–21 (1981).

Triphenylphosphine/carbon tetrachloride/triethylamine — $Ph_3P/CCl_4/Et_3N$
Carboxylic acid aryl esters from carboxylic acids and phenols COOH → COOAr
with benzotriazolyl-1-oxytris(dimethylamino)phosphonium hexafluorophosphate cf. *33,* 174; with triphenylphosphine/CCl_4 in the presence of triethylamine s. S. Hashimoto, I. Furukawa, Bull. Chem. Soc. Japan *54,* 2227–8 (1981); without CCl_4 at high temp. cf. ibid. *54,* 2839–40 (1981).

N-(Diphenoxyphosphoryl)-Δ^4-2-oxazolones/triethylamine ←
Carboxylic acid esters from carboxylic acids COOH → COOR
s. *37,* 333

Thionyl chloride *$SOCl_2$*
Aminocarboxylic acid esters from acids
methyl esters s. *9,* 344; *n*-propyl esters s. S.P. Campana Filho, G. Goissis, J. Chromatogr. *236,* 197 (1982).

Sulfuryl chlorofluoride/triethylamine *SO_2ClF/NEt_3*
Carboxylic acid esters from carboxylic acids
with SO_2Cl_2 cf. *16,* 240; with SO_2ClF/triethylamine s. G.A. Olah, S.C. Narang, A. Garcia-Luna, Synthesis *1981,* 790–1.

p-*Nitrobenzenesulfonyl chloride s. under CF_3SO_3Ag* *$ArSO_2Cl$*

Methanesulfonic acid/cobalt bromide/
tetraethylammonium perchlorate *$MeSO_3H/CoBr_2/Et_4N^+ClO_4^-$*
α-Glycosylation OH → OR
with methanesulfonic acid s. *28,* 145; with added $CoBr_2$ and tetraethylammonium perchlorate s. S. Koto, N. Morishima, S. Zen, Bull. Chem. Soc. Japan *55,* 1543–7 (1982); with trimethylsilyl bromide/$CoBr_2$/tetra-*n*-butylammonium bromide cf. Chem. Letters *1981,* 427–8.

Trifluoromethanesulfonic acid *CF_3SO_3H*
O-Benzylation OH → OCH_2Ph

152.

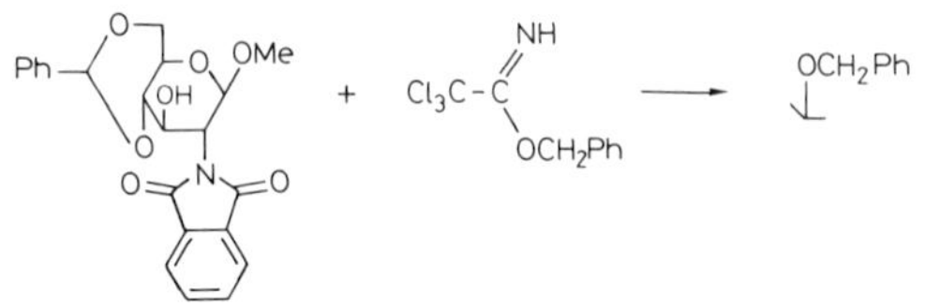

of alcohols. A stirred soln. of startg. m. and benzyl trichloroacetimidate in 2:1 cyclohexane-dichloromethane treated with a little triflic acid, and the mixture allowed to stand at room temp. for 18 hrs. → product. Y 98%. – The reaction conditions leave acid- and alkali-labile groups intact. F.e.s. T. Iversen, D.R. Bundle, J. Chem. Soc. Chem. Commun. *1981,* 1240–1; O-benzylation **of phenols** with benzyl tosylate, preferential conversion, s. P.M. Dewick, Synth. Commun. *11,* 853–7 (1981).

p-*Toluenesulfonic acid* *TsOH*
Glycosidation with simultaneous acetalation ←
s. *36,* 171; benzylidene acetals s. Synthesis *1982,* 29–32
1,3-Dioxolan-4-ones from α-hydroxycarboxylic acids and aldehydes ○
s. *37,* 791

Pyridinium tosylate *$C_5H_5NH^+OTs^-$*
Protection of carbohydrates as O,O-anisylidene derivs. $\begin{matrix}\text{C–OH}\\\text{C–OH}\end{matrix} \rightarrow \begin{matrix}\text{CO}\\\text{CO}\end{matrix}\!>\!\text{CAr}$

as O,O-(*p*-methoxybenzylidene) derivs. with HCl cf. *21,* 217; as O,O-1-(*p*-methoxyphenyl)ethylidene derivs. with pyridinium tosylate s. B.H. Lipshutz, M.C. Morey, J. Org. Chem. *46,* 2419–23 (1981); as O,O-(*o*-methoxybenzylidene) derivs. cf. V. Box, R. Hollingsworth, E. Roberts, Heterocycles *14,* 1713–6 (1980).

Sulfuric acid (s.a. under $Zr(OH)_4$) H_2SO_4

Carboxylic acid esters from carboxylic acids COOH → COOR
Ethylenecarboxylic acid esters
s. *12,* 209; 2-ethylhexyl acrylate, optimum conditions, s. J. Wasilewski et al., Chem. Stosow. *24,* 629 (1980) (Pol); Chem. Abstr. *95,* 149899 (1981).

Ferric chloride $FeCl_3$

Carbohydrate O,O-isopropylidene derivs.

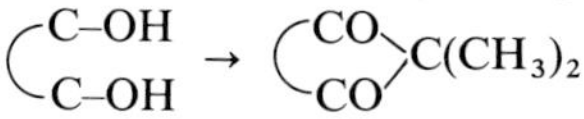

s. *17,* 236; with preceding cleavage of methyl glycosides s. F. Dasgupta, P.P. Singh, H.C. Srivastava, Indian J. Chem. *19B,* 1056–9 (1980).

Cobalt bromide s. under Me_3SiBr and $MeSO_3H$ $CoBr_2$

Via intermediates *v.i.*

Iminoesters from carboxylic acid amides
via iminoester fluoroborates $CONH_2$ → C(=NH)OR
satd. iminoesters s. *22,* 188; *α,β*-ethyleneiminoesters s. K. Sato et al., Synthesis *1982,* 137–8.

Sulfonyl esters from sulfonic acids and alcohols SO_3H → SO_3R
via sulfonyl chlorides cf. *4,* 213; via ammoniosulfonyl esters, also by transesterification, s. J.F. King, M. Aslam, Tetrahedron Letters *22,* 3573–6 (1981).

Nitrogen ↑ OC↓↑N

Without additional reagents *w.a.r.*

O-Acylation with 1-acylimidazoles OH → OAc
with N-bromosuccinimide cf. *32,* 161; with 1-acylimidazoles prepd. *in situ* by various methods, acylation of sensitive alcohols containing a silicon-oxygen bond, s. J.-P. Pometan et al., Eur. J. Med. Chem. *16,* 425–9 (1981).

Electrolysis ⚡

Ketones from ketoximes $R_2C{=}NOH$ → R_2CO

153.

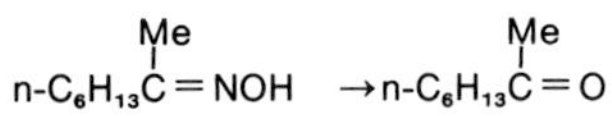

A well-stirred soln. of startg. oxime in 20% aq. acetonitrile containing tetraethylammonium fluoroborate electrolysed at 10° between graphite electrodes with a current density of 0.2 A/15 cm^2 electrode area until 2 F/mol had passed → 2-octanone. Y 90%. F.e.s. M.R. Van De Mark, A.M. Scornavacca, Org. Prep. Proc. Intern. *13,* 395–9 (1981).

Potassium hydroxide *KOH*

Carboxylic acid derivs. from carboxylic acid amides CON< → COOR
via N-nitrosocarboxylic acid amides
carboxylic acids. cf. *16,* 247; carboxylic acid benzhydryl esters, retention of 2-azetidinone ring, s. H.P. Isenring, W. Hofheinz, Synthesis *1981,* 385–7.

Sodium methoxide s. under $PhI(OAc)_2$ *NaOMe*

Lithium diisopropylamide s. under Molybdenum peroxide complex *i-Pr_2NLi*

Sodium carbonate *Na_2CO_3*

1,3-Dialkoxy compds. from N-cyclopropyl-N-nitrosoureas

154. 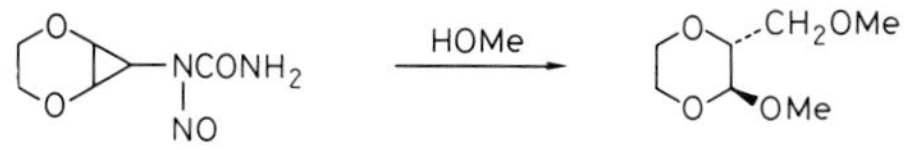

A soln. of N-(2,5-dioxabicyclo[4.1.0]hept-*exo*-7-yl)-N-nitrosourea in methanol added to a suspension of Na_2CO_3 in the same solvent, and the mixture stirred at room temp. for 12 hrs. → *trans*-2-methoxy-3-(methoxymethyl)-1,4-dioxane. Y 82%. F.e., **also alkoxy-2-ethylenes,** s. H. Jendralla, W. Pflaumbaum, Chem. Ber. *115,* 229–39 (1982); ring expansion to **1,3-dioxocines** s. ibid. *115,* 201–9; 220–8 (1982).

Potassium carbonate *K_2CO_3*

Aldehydes from sulfinamides $CH_2NS(O)R \rightarrow CHO$

s. *37,* 184

Sodium nitrite/sodium lauryl sulfate/sulfuric acid ←

Carboxylic acid amides from amidoximes C(=NOH)N< → CON<

155. 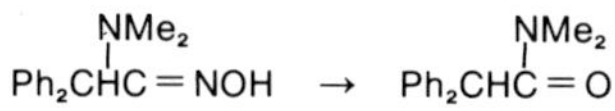

Startg. m. added to an aq. soln. of $NaNO_2$ and a little Na lauryl sulfate, the vigorously stirred suspension treated slowly with aq. H_2SO_4, and stirring continued for 4 hrs. → N,N-dimethyldiphenylacetamide. Y 97%. F.e.s. Z. Eckstein et al., Synth. Commun. *12,* 201–8 (1982).

Sodium salt *Na^+*

Nitronic acid esters
from nitronates and diazonium salts C=NOONa → C=NOOAr

156. 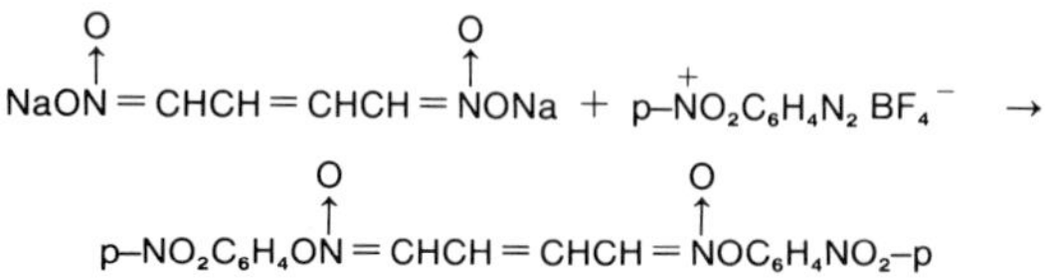

A soln. of *p*-nitrobenzenediazonium fluoroborate in water added dropwise with stirring at 0° to a soln. of 1,4-dinitro-2-butene disodium salt, and stirred a further 15 min. → di(*p*-nitrophenyl) 2-butene-1,4-bisnitronate. Y 93%. F.e.s. E.S. Lipina et al., Zh. Org. Khim. *16,* 2404–13 (1980).

Cupric hexafluoroacetylacetonate *$[(CF_3CO)_2CH]_2Cu$*

1,3-Dioxoles from α-diazoketones and aldehydes

157. 

A soln. of *p*-nitrobenzenediazonium fluoroborate in water added dropwise with stirring at 0° to a soln. of 1,4-dinitro-2-butene disodium salt, and stirred a further 15 min. → di(*p*-nitrophenyl) 2-butene-1,4-bisnitronate. Y 93%. F.e.s. E.S. Lipina et al., Zh. Org. Khim. *16,* 2404–13 (1980).

Silver cyanide *AgCN*

Aldehydes from pyridinium salts ←

with Na 1-oxido-4,6-diphenyl-2-pyridone cf. *36,* 184; with AgCN s. P.M. Fresneda et al., Synthesis *1981,* 711.

Zinc s. under Me_3SiCl *Zn*

Boron fluoride BF_3

Ketones from hydrazones $R_2C{=}N{-}NMe_2 \rightarrow R_2C{=}O$

from tosylhydrazones cf. *32,* 159; from N,N-dimethylhydrazones, retention of enol-acetate groups, s. R.E. Gawley, E.J. Termine, Synth. Commun. *12,* 15–8 (1982).

Ammonium ceric nitrate $(NH_4)_2Ce(NO_3)_6$

Protection of carboxyl groups as N-acyl-5,6-dihydrophenanthridines ←

Selective removal of the protective group

158.

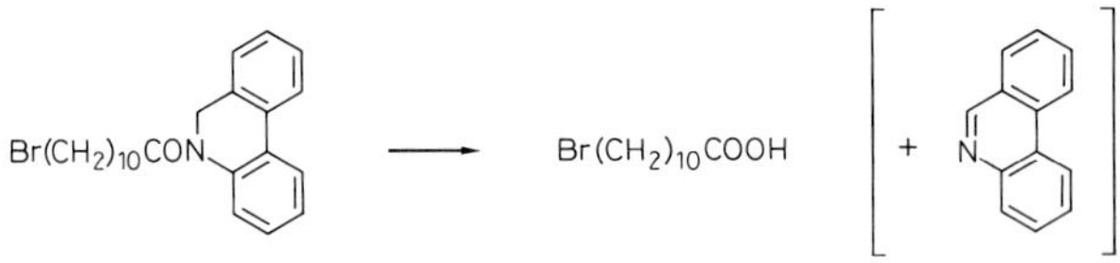

A soln. of the startg. amide in aq. acetonitrile treated with ca. 3 equivalents of ammonium ceric nitrate, and stirred for 15 min. at room temp. → 11-bromoundecanoic acid. Y 95%. Halide, carbolkoxy, and keto groups were unaffected, and the protecting group was resistant to alkali and Grignard conditions. F.e. and protection procedure s. T. Uchimaru, K. Narasaka, T. Mukaiyama, Chem. Letters *1981,* 1551–4.

Benzyltributylammonium chloride s. under $KMnO_4$ $[PhCH_2NBu_3]^+Cl^-$

Amberlyst 15 ←

Carboxylic acid esters
from N-unsubst. carboxylic acid amides or hydrazides
under mild conditions

$CONH_2$, $CONHNH_2$ → COOR

159.

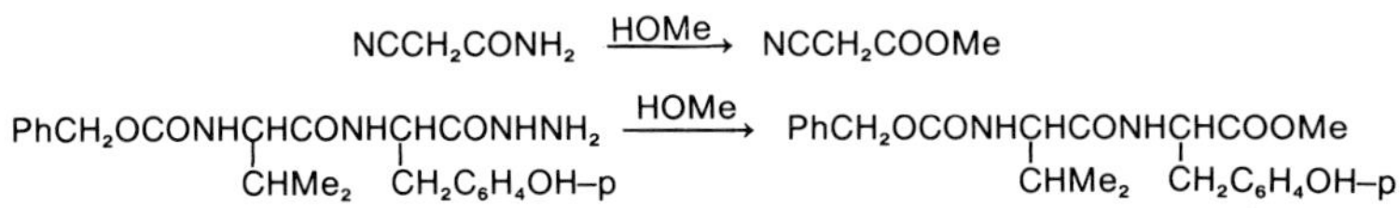

A mixture of cyanoacetamide and a 15-fold excess of Amberlyst 15 resin in the minimum of methanol to cover the resin stirred gently in a tightly capped flask at 60° for 18 hrs. → methyl cyanoacetate. Y 85%. – N-Subst. amide bonds were unaffected, thereby offering useful applications in removing terminal amide or hydrazide residues in peptides. E: DL-carbobenzoxyvalyltyrosine hydrazide → methyl DL-carbobenzoxyvalyltyrosinate. Y 62%. F.e. and **carboxylic acids** in water s. W.J. Greenlee, E.D. Thorsett, J. Org. Chem. *46,* 5351–3 (1981).

Benzoyl tert-*butyl nitroxide* *PhCON(O·)Bu*-t

3-Indolenones from 3 subst. indoles ←

ipso-**Substitution**

160.

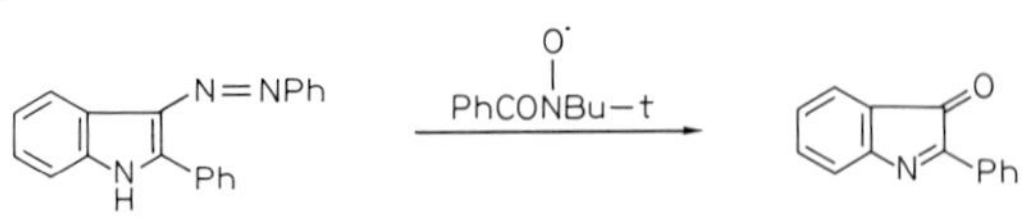

A mixture of 2.2 equivalents benzoyl *tert*-butyl nitroxide and 1 equivalent startg. indole in benzene allowed to react at 25° for 24 hrs. → 2-phenyl-3*H*-indol-3-one. Y 80%. F.e.s. C. Berti et al., J. Chem. Soc. Chem. Commun. *1981,* 694–5; **review of** *ipso*-**substitution** s. M. Tiecco, Pure Appl. Chem. *53,* 239–58 (1981).

Acetic acid s. under H_2SO_4 *MeCOOH*

Oxalic acid $(COOH)_2$

γ-Lactones from γ-iminolactones ←

s. *37,* 666

Phenyl iodosoacetate/sodium alkoxide *$PhI(OAc)_2/NaOR$*

Acetals from sym. azines C=NN=C → $C(OR)_2$

161. PhCH=N–N=CHPh + NaOMe → 2 $PhCH(OMe)_2$

Startg. azine, followed after 5 min. by 2 equivalents of phenyl iodosoacetate, added to 10 equivalents of Na methoxide in methanol, and stirred at room temp. for 2 days → product. Y 88%. F.e.s. R.M. Moriarty, H. Hu, Tetrahedron Letters *23,* 1537–40 (1982).

N-Bromosuccinimide *CH_2CO / CH_2CO >NBr*

Carboxylic acids from carboxylic acid hydrazides $CONHNH_2$ → COOH

s. *20,* 141; s.a. J. Meienhofer et al., Int. J. Pept. Protein Res. *18(1),* 6 (1981) (Eng); Chem. Abstr. *95,* 187647 (1981).

Trimethylsilyl chloride/zinc *Me_3SiCl/Zn*

Ketones from 1-nitroethylene derivs. $C=C(NO_2)$ → CHCO

with Pd–C cf. *12,* 266; with Me_3SiCl/Zn s. M. Husain, N.H. Khan, Synth. Commun. *11,* 185–207 (1981).

p-*Toluenesulfonic acid* *TsOH*

Carboxylic acid esters from ethylene-N-sulfonylcarboxylic acid hydrazides ←

162. PhCH=CHCONHNHTs + PrOH → $PhCH_2CH_2COOPr$

A soln. of startg. m. in *n*-propanol containing a little *p*-toluenesulfonic acid heated at 97° for 48 hrs. → product. Y 86%. F.e.s. F. Bellesia et al., Gazz. Chim. Ital. *111,* 511–3 (1981).

Sulfuric acid s.a. under $NaNO_2$ and Ni *H_2SO_4*

Sulfuric acid/acetic acid *$H_2SO_4/MeCOOH$*

Carboxylic acids from acid amides CON< → COOH

s. *37,* 658

Pyridinium dichromate *$(C_5H_5NH)_2Cr_2O_7$*

Cleavage of oximes C=NOH → CO

with pyridinium chlorochromate cf. *34,* 172; with pyridinium dichromate s. S. Satish, N. Kalyanam, Chem. Ind. *1981,* 809–10.

Molybdenum peroxide complex/lithium diisopropylamide ←

Ketones from sec nitro compds. $CHNO_2$ → CO

Carboxylic acids from prim. nitro compds. CH_2NO_2 → COOH

163.

A soln. of nitrocyclohexane in tetrahydrofuran added dropwise over 5 min. to a soln. of diisopropylamine and *n*-butyllithium/hexane in the same solvent at –78°, molybdenum peroxide-pyridine-hexamethylphosphoramide complex added quickly, the mixture allowed to warm to room temp. over 3 hrs., and quenched with satd. aq. Na_2SO_3 → cyclohexanone. Y 86%. – Similarly, *α*-nitrotoluene → benzoic acid. Y 74%. F.e.s. M.R. Galobardes, H.W. Pinnick, Tetrahedron Letters *22,* 5235–8 (1981).

Hydrogen chloride *HCl*

Aldehydes from 2-vinyl-1,3,2-benzodisilazoles C

s. *37,* 165

Potassium permanganate/benzyltributylammonium chloride *$KMnO_4/[PhCH_2NBu_3]^+Cl^-$*

Carboxylic acids from prim. nitro compds. $CH_2NO_2 \rightarrow COOH$
with $KMnO_4$ cf. *18,* 256; with $KMnO_4$ and benzyltributylammonium chloride as phase transfer catalyst, carbohydrate derivs., s. Y.A. Zhdanov, Y.E. Alekseev, S.S. Doroshenko, Dokl. Akad. Nauk SSSR *265,* 1147–9 (1982) [Chem.] (Russ); Chem. Abstr. *96,* 218162 (1982).

Nickel/sulfuric acid *Ni/H_2SO_4*

Aldehydes from nitriles $CN \rightarrow CHO$
s. *23,* 223; α-acoxy-aldehydes from -nitriles s. P. Tinapp, Arch. Pharm. *314,* 512–4 (1981).

Via intermediates *v.i.*

Aldehydes from prim. amines via 2-pyridylaldimines $CH_2NH_2 \rightarrow CHO$

164.

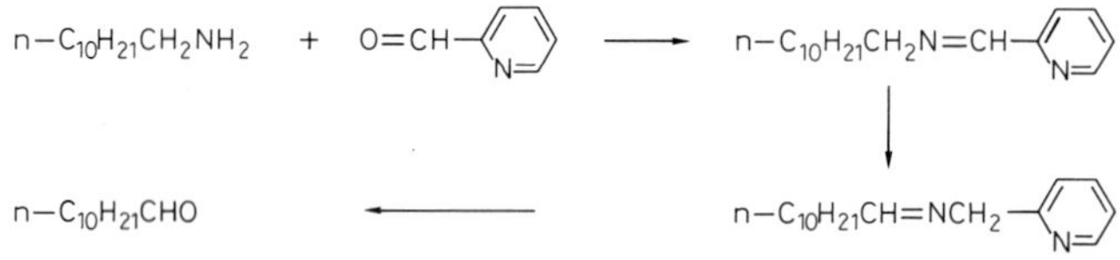

A soln. of 2-pyridinecarboxaldehyde and *n*-undecylamine in ether stirred at room temp. for 40 min. → N-undecyl-2-picolinimine (Y 98%) in tetrahydrofuran added dropwise to a soln. of Li diisopropylamide in 5:3 tetrahydrofuran/hexane, the mixture stirred at –70° for 1 hr., quenched by quickly pouring into a well stirred soln. of oxalic acid in 1:4:4 water/methanol/tetrahydrofuran, and stirring continued at room temp. for 2 hrs. → undecanal (Y 94%). F.e. and ketones in low yield s. J.H. Babler, B.J. Invergo, J. Org. Chem. *46,* 1937–8 (1981).

Aldehydes from nitriles via 2-vinyl-1,3,2-benzodisilazoles $CN \rightarrow CHO$

165.

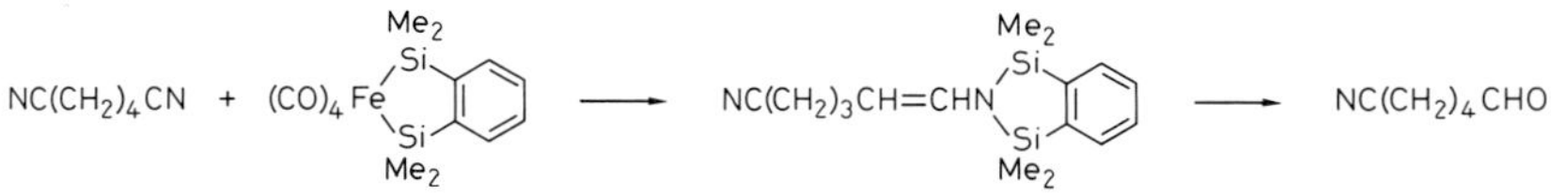

An equimolar soln. of startg. nitrile and disilylated iron carbonyl complex (prepared from 1,2-bis(dimethylsilyl)benzene and $Fe(CO)_5$) in hexane irradiated at room temp. with a 450 W. high-pressure Hg-lamp in a quartz reaction vessel for 34 hrs. → intermediate (Y 70%) hydrolysed with dil. HCl → aldehyde (Y 79%). F.e. and selectivity, also application to the **protection of nitriles,** s. R.J.P. Corriu, J.J.E. Moreau, M. Pataud-Sat, J. Org. Chem. *46,* 3372–4 (1981).

Aldehydes from amines via sulfinamides $CH_2NHR \rightarrow CHO$
s. *37,* 184

Halogen ↑ OC↓↑Hal

Without additional reagents *w.a.r.*

Reactions with peroxydisulfuryl difluoride
Fluorosulfonic esters from iodides $I \rightarrow OSO_2F$

166. $$CF_3I + (FSO_2O)_2 \rightarrow [CF_3I(OSO_2F)_2] \rightarrow CF_3OSO_2F$$

Peroxydisulfuryl difluoride added dropwise with stirring to a soln. of trifluoroiodomethane in trifluorotrichloroethane at –60 to –50°, allowed to warm to 0° during 2 hrs., and left overnight → trifluoromethyl fluorosulfonate. Y 85%. F.e.s. A.V. Fokin et al., Izv. Akad. Nauk SSSR Ser. Khim. *1980,* 2641–3; fluorosulfonyl **aryl esters from arenes** s. ibid. *1981,* 2638–40.

β-(Organothio)-γ-lactones O
from β,γ-ethylenecarboxylic acid derivs. and sulfenyl chlorides
via γ-chloro-β-(organothio)carboxylic acid derivs.

167.
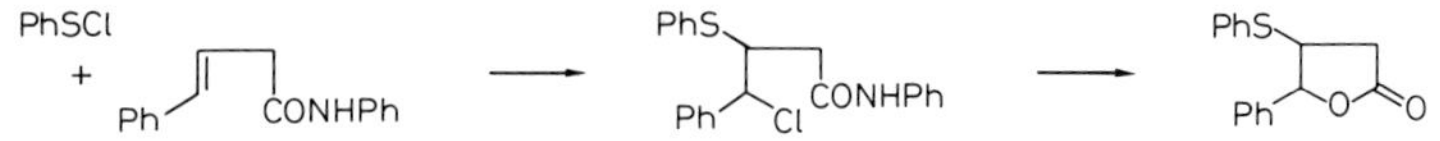

Benzenesulfenyl chloride in carbon tetrachloride added dropwise at −30° to a soln. of startg. amide in the same solvent, and the mixture left for 10–15 hrs. at 0° → intermediate (Y 90%), heated under reflux in nitromethane with *exclusion of moisture* → product (Y 98%). F.e.s. N.K. Potsyute, L.P. Rasteikene, I.L. Knunyants, Izv. Akad. Nauk SSSR Ser. Khim. *1980,* 2363–9.

Ring closures of 2-halogenurethans
2-oxazolidones cf. *9,* 45; 1,3-dioxolan-2-ones s. K. Ponsold, P. Grosse, J. Prakt. Chem. *323,* 801–18 (1981).

Sodium hydride/sodium iodide *NaH/NaI*
Protection of phenol groups as phenylthiomethyl ethers ArOH → $ArOCH_2SPh$

168. $p\text{-}BrC_6H_4OH + ClCH_2SPh \rightarrow p\text{-}BrC_6H_4OCH_2SPh$

NaH added to a soln. of *p*-bromophenol in hexamethylphosphoramide under N_2, the mixture stirred at room temp. for 30 min., treated with chloromethyl phenyl sulfide and NaI, and stirring continued for 18 hrs. → *p*-bromophenyl phenylthiomethyl ether. Y 87%. – The protective group is essentially inert to conditions required for hydrolysis of the related methylthiomethyl ethers, but may be removed efficiently by treatment at reflux with a soln. of $HgCl_2$ in aq. acetonitrile. F.e.s. R.A. Holton, R.V. Nelson, Syn. Commun. *10,* 911–4 (1980).

Sodium hydroxide (s.a. under Montmorillonite) *NaOH*
α-Hydroperoxyoximes from 2-nitrosohalide dimers ←

169.
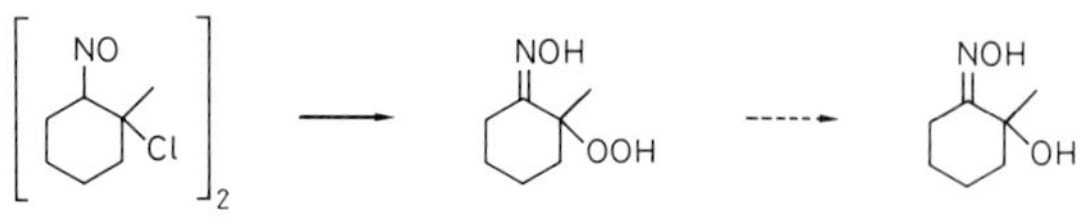

Startg. m. treated with a 3-fold excess of H_2O_2 in the presence of an equimolecular amount of 5% NaOH → product. Y 71%. F.e. and reduction to **α-hydroxyoximes** with KI/acetic acid, also **α-peroxyoximes** with *t*-butyl hydroperoxide, s. M. Schulz, J. Römbach, Z. Chem. *21,* 404–5 (1981).

γ-(Acyloximino)oxo compds.
from cyclic 1-hydroxynitrones and carboxylic acid chlorides C

170.
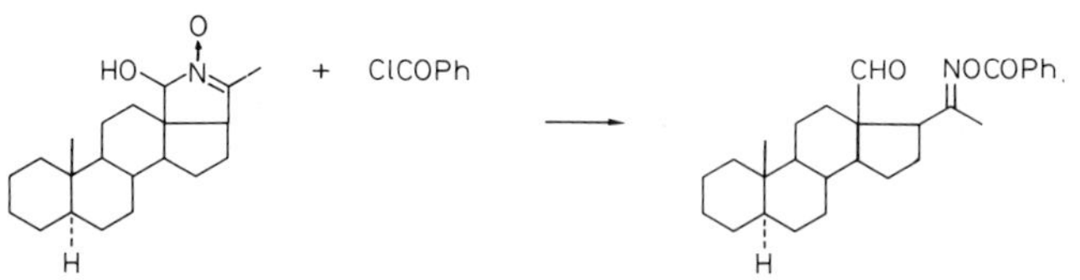

Benzoyl chloride added to a stirred mixture of startg. hydroxynitrone in benzene aq. NaOH, and worked up after 18 hrs. → 20-O-benzoyloximino-5α-pregnan-18-al. Y 100%. F.e.s. J.P. Alazard, H. Dadoun, X. Lusinchi, Tetrahedron *37,* Suppl. No. 1, 41–53 (1981).

Sodium hydroxide *NaOH*

O,O-Methylenedioxy compds. from diols ○

with NaH cf. *30,* 128; with NaOH, carbohydrate derivs., s. A. Lipták, V.A. Oláh, J. Kerékgyárto, Synthesis *1982,* 421–3.

1,4-Benzodioxin-2-ones from catechols and α-halogenocarboxylic acids

s. *37,* 126

2-Imidazolones from 2-chloroimidazolium salts ←

from chlorides cf. *18,* 391; from tetrafluoroborates s. H. Wamhoff et al., Angew. Chem. Intern. Ed. *20,* 612 (1981).

*Sodium hydroxide/tetra-*n-*butylammonium chloride* $NaOH/n\text{-}Bu_4N^+Cl^-$

O-Acylation OH → OAc

Phase transfer catalysis

of phenols by solid-liq. phase transfer catalysis cf. *35,* 119; of carbohydrates with tetra-*n*-butylammonium chloride as liq.-liq. phase transfer catalyst s. W. Szeja, Pol. J. Chem. *54,* 1301–4 (1980); preferential O-acylation of prim. hydroxyl groups s. ibid. *54,* 1323–5 (1980); of hindered phenols with acetic anhydride cf. K.-T. Liu, M.-Y. Kuo, J. Chinese Chem. Soc. *28,* 209–11 (1981).

*Sodium hydroxide/tetra-*n-*butylammonium hydrogen sulfate* $NaOH/n\text{-}Bu_4N^+HSO_4^-$

Preferential protection of hydroxyl groups as *o*-methoxyphenoxymethyl ethers $OH \rightarrow OCH_2OAr$

Phase transfer catalysis

171. $o\text{-}MeOC_6H_4OCH_2Cl + HO(CH_2)_7Me \rightarrow o\text{-}MeOC_6H_4OCH_2O(CH_2)_7Me$

A mixture of *n*-octanol and 2-methyldodecan-2-ol in benzene/aq. NaOH treated with guaiacylmethyl chloride in the presence of a little tetra-*n*-butylammonium hydrogen sulfate at 20° for 45 min. → guaiacylmethyl *n*-octyl ether. Y 95% (unreacted tert. alcohol Y 95%). – The protective group is resistant to strong bases, oxidants, and reductants, but only moderately stable in acidic media; removal of the group is effected by treatment with $ZnBr_2$ in methylene chloride. F.e., also preferential protection of sec. hydroxyl groups, s. B. Loubinoux, G. Coudert, G. Guillaumet, Tetrahedron Letters *22,* 1973–6 (1981).

Sodium hydroxide/benzyltriethylammonium bromide $NaOH/PhCH_2NEt_3^+Br^-$

Aryl glycosides Br → OAr

with $CdCO_3$ cf. *27,* 218; aryl β-glycosides with NaOH and benzyltriethylammonium bromide as phase transfer catalyst, s. D. Dess et al., Synthesis *1981,* 883–5.

Potassium hydroxide/polyethylene glycol ←

Williamson ether synthesis Hal → OR

with KH/cryptate(2.2.2) cf. *7,* 281s*36;* with KOH/polyethylene glycol s. S. Slaoui et al., Tetrahedron Letters *23,* 1681–4 (1982).

Potassium hydroxide/dimethyl sulfoxide KOH/Me_2SO

O-Alkylation of carbohydrates OH → OR

with NaH/Me_2SO, O-methylation, cf. *29,* 410; with KOH/Me_2SO s. P. Fugedi, P. Nanasi, J. Carbohyd. Nucleosides Nucleotides *8,* 547–55 (1981).

Sodium alkoxides *NaOR*

Carboxylic acid esters from 1-chloro-1,1-difluorides $CClF_2 \rightarrow COOR$

s. *37,* 668

Sodium methoxide *NaOMe*

Favorskii rearrangement of cyclic *α,α,α′*-**trihalogenoketones**

Regiospecific ring contraction by one C-atom

172. 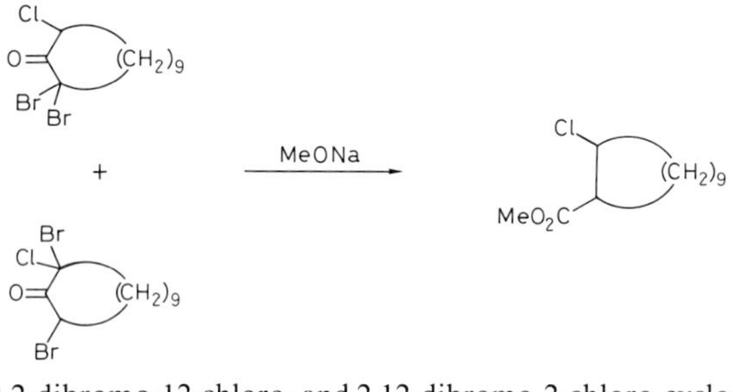

A mixture of 2,2-dibromo-12-chloro- and 2,12-dibromo-2-chloro-cyclododecanone in benzene treated with Na methoxide in methanol at room temp. with rapid stirring for 1 hr. → methyl 2-chloro-1-cycloundecene-1-carboxylate. Y 97%. – The method is of value in the synthesis of **unsym. macrocyclic ketones,** the ring contraction being a useful alternative to the low-yielding benzilic acid rearrangement. A. Abad et al., Tetrahedron Letters *22,* 1733–6 (1981).

Ring contraction of cyclic ketones

s. *24,* 210; cyclic carboxylic acid amides s. C. Bischoff, E. Schröder, J. Prakt. Chem. *323,* 616–20 (1981).

Sodium methoxide/dimethyl sulfoxide *NaOMe/Me₂SO*

α-**Alkoxy-***α,β*-**ethyleneketones from** *α*-**hydroxyketones** CHCH(OH) → C=C(OR)

173. 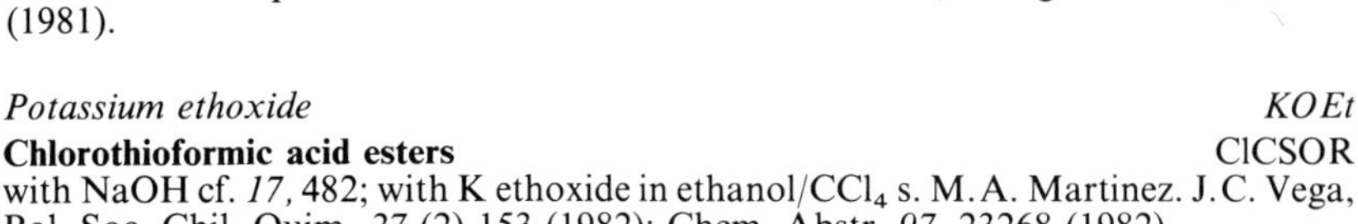

Startg. *α*-hydroxyketone treated with an excess of Na methoxide in 10:1 dimethyl sulfoxide/methanol during 35 min. at 55°, the mixture cooled to 10°, and methyl iodide added → product. Y 93%. F.e.s. P.A. Grieco et al., J. Org. Chem. *46,* 1022–4 (1981).

Potassium ethoxide *KOEt*

Chlorothioformic acid esters ClCSOR

with NaOH cf. *17,* 482; with K ethoxide in ethanol/CCl_4 s. M.A. Martinez. J.C. Vega, Bol. Soc. Chil. Quim. *27* (2) 153 (1982); Chem. Abstr. *97,* 23268 (1982).

Sodium aroxides *NaOAr*

Reactions using solid-phase cosolvents Br → OR

174. $Me(CH_2)_6Br + NaOPh \rightarrow Me(CH_2)_6OPh$

Inexpensive and easily recoverable poly(acrylamide) gels function effectively as cosolvents in reactions, such as **nucleophilic substitution,** normally requiring polar aprotic solvents. E: A soln. of 1-bromoheptane in dioxane containing poly(N,N-dimethylacrylamide), cross-linked with 1 mol. % N,N′-dimethyl-N,N′-ethylenebis(acrylamide), treated at 75° with Na phenoxide for 5 hrs. in a selaed tube → n-heptyl phenyl ether. Y 75%. – In the absence of the co-solvent only 2% of the ether is formed in 17 hrs. F.e. in both 2- and 3-phase media s. S.L. Regen, A. Mehrotra, A. Singh, J. Org. Chem. *46,* 2182–4 (1981).

Potassium tert-*butoxide* t-*BuOK*

Carboxylic acid amides from α-chlorazomethines via regiospecific iminocyclopropane ring opening Favorskii-type rearrangement C(Cl)C(=N)R → C(R)CONH

175.

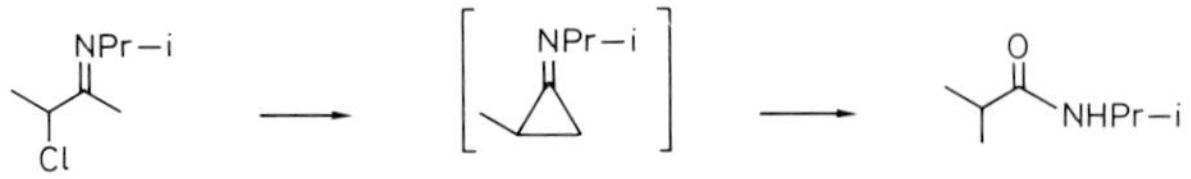

Startg. α-chlorazomethine treated with a soln. of K *tert*-butoxide in tetrahydrofuran, and the mixture allowed to react at room temp. for 2 hrs. → product. Y 62–71%. F.e.s. N. De Kimpe et al., Tedrahedron Letters *22,* 1837–40 (1981).

n-*Butyllithium s. under* $B(OBu)_3$ n-*BuLi*

Potassium carbonate s. under $KHCO_3$ K_2CO_3

Anion exchanger-supported carbonate ←

Prim. alcohols from halides Hal → OH

176.

$$Me(CH_2)_7Br \xrightarrow{CO_3^{=}} Me(CH_2)_7OH$$

A mixture of Amberlyst A26 (carbonate form) and 1-bromooctane in benzene refluxed for 4 hrs. with vigorous stirring → 1-octanol. Y 90%. F.e., also 1,2,3-triols from 2,3-oxidoalcohols, s. G. Gardillo et al., Synthesis *1981,* 793–4.

Potassium hydrogen carbonate/potassium carbonate/Aliquat 336 ←

Sym. carbonic acid esters from halides 2 RHal → ROCOOR

Solid-liq. phase transfer catalysis

177.

$$2\ m\text{-}ClC_6H_4CH_2Br \xrightarrow{K_2CO_3} (m\text{-}ClC_6H_4CH_2O)_2C{=}O$$

A little Aliquat 336 and 3-chlorobenzyl bromide in toluene added to a mixture of $KHCO_3$ and K_2CO_3, and stirred at 100° for 8–15 hrs. → carbonic acid bis(3-chlorobenzyl) ester. Y 86%. F.e., **also mixed carbonic acid esters,** s. M. Lissel, E.V. Dehmlow, Chem. Ber. *114,* 1210–5 (1981); with organotin compds. as catalyst cf. T. Fujinami, S. Sato, S. Sakai, Chem. Letters *1981,* 749–52.

Potassium acetate/18-crown-6 polyether ←

Acoxy-2-ethylenes from halogenocyclopropanes C

Geospecific ring opening

178.

A mixture of 2-phenylcyclopropyl bromide, K acetate, and 18-crown-6 in dimethyl sulfoxide stirred under N_2 at 85° for 18 hrs. → (E)-product. Y 66%. F.e.s. D.H. Buchanan, G. McComas, Tetrahedron Letters *21,* 4317–20 (1980).

Caesium fluoride *CsF*

Acyl glycosides OH → OCOR

from acyl chlorides in the presence of pyridine cf. *24,* 213; from acyl fluorides in the presence of CsF, stereospecificity, s. S. Shoda, T. Mukaiyama, Chem. Letters *1982,* 861–2.

Sodium iodide s. under NaH *NaI*

Sodium salt *Na^+*

N-Glycosyloxy-N-heterocyclics Br → ON<

from N-hydroxy-N-heterocyclics and glycosyl halides

179.

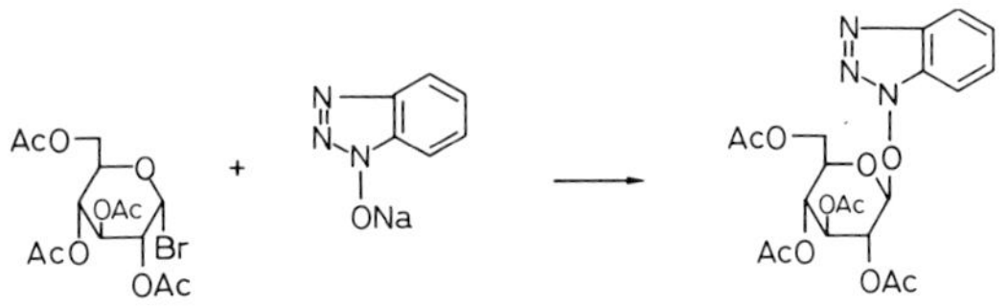

Startg. N-hydroxy deriv. and Na ethoxide in ethanol stirred at room temp. until formation of Na salt complete, solvent replaced by acetonitrile and *α*-bromoacetoglucose, refluxed for a few hrs., and left overnight → product. Y 78%. F.e.s. E. Grochowski, E. Falent-Kwastowa, Pol. J. Chem. *54,* 2229–32 (1980).

Triethylamine s. under Silver imidazolide *Et_3N*

Triethylenediamine ←

O-2,4-Dinitrophenylation OH → OAr

with NEt_3 cf. *12,* 285; with triethylenediamine, 2,4-dinitrophenyl glycosides, s. H.J. Koeners et al., Rec. Trav. Chim. Pays-Bas *99,* 355–62 (1980).

4-Dimethylaminopyridine ←

1,3-Dioxolane-2-thiones from glycols ○

s. *37,* 937

Pentaalkylguanidines **as hindered bases** *$R_2NC(=NR)NR_2$*

Carboxylic acid esters from carboxylic acids COOH → COOR

180.

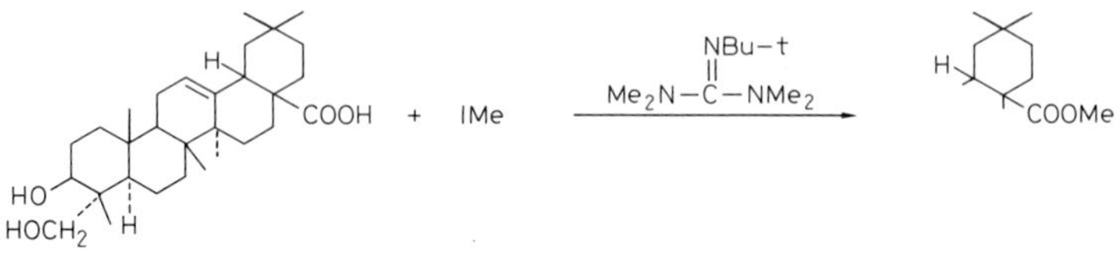

A soln. of hederagenin in dimethylformamide treated rapidly at room temp. with methyl iodide in the presence of N[1]-*tert*-butyl-N[2],N[2],N[3],N[3]-tetramethylguanidine → methyl ester. Y 88%. F.e. and prepn. of the bases s. D.H.R. Barton, J.D. Elliott, S.D. Géro, J. Chem. Soc. Chem. Commun. *1981,* 1136–7.

Silver imidazolide/mercuric chloride/triethylamine ←

Cyclic orthocarboxylic acid esters of carbohydrates ○

with Ag salicylate cf. *27,* 217; with Ag imidazolide/$HgCl_2$/NEt_3 s. B.-M. Dahlin et al., Acta. Chem. Scand. *35B,* 669–70 (1981).

Zinc oxide *ZnO*

Solvolysis of tert. halides ←

s. *36,* 198; also of allylic and benzylic halides s. Tetrahedron *38,* 1843–6 (1982).

Mercuric chloride s. under Silver imidazolide *$HgCl_2$*

*Tributyl borate/*n*-butyllithium/hydrogen peroxide* *$B(OBu)_3$/*n*-BuLi/H_2O_2*

Phenols from bromides ArBr → ArOH

with Mg/trimethyl borate cf. *25,* 163a; with *n*-butyllithium/tributyl borate/H_2O_2 s. P. Jacob, A.T. Shulgin, Synth. Commun. *11,* 969–77 (1981).

Boron oxide s. under SO_3 B_2O_3

Silver silicate/aluminium oxide ←

Acylglycosides from acyl halogenosugars OH → OR

with Ag salicylate cf. *26,* 203s*28;* with Ag silicate/Al_2O_3 s. H. Paulsen, O. Lockhoff, Chem. Ber. *114,* 3102–14 (1981); with Ag imidazolide/$ZnCl_2$/tetra-*n*-butylammonium chloride, *α*-glycosides from *α*-glycosyl chlorides, incl. reaction of hindered alcohols, cf. P.J. Garegg, C. Ortega, B. Samuelsson, Acta Chem. Scand. *35B,* 631–3 (1981).

Quaternary ammonium-exchanged montmorillonite/sodium hydroxide ←

Symmetrical formals from alcohols 2 ROH → $ROCH_2OR$

3-Phase catalysis

181. $CH_2Cl_2 + 2\ HOBu \rightarrow CH_2(OBu)_2$

Tixogel VP (a quaternary ammonium-exchanged montmorillonite) added to a vigorously stirred soln. of *n*-butanol in *dichloromethane,* after 5 min. 50% aq. NaOH added in one portion, and the mixture gently refluxed with stirring for 3 days → product. Y 80%. – The catalyst is inexpensive and fulfils the requirements of high hydrophilicity and low nucleophilicity. F.e., also using dibromomethane s. A. Cornélis, P. Laszlo, Synthesis *1982,* 162–3.

Silver fluoroborate $AgBF_4$

Carboxylic acids from *α*-halogenoketones ←

with NaOH cf. *10,* 189; with $AgBF_4$, arylacetic acids, s. C. Giordano et al., Tetrahedron Letters *23,* 1385–6 (1982).

Cerium(IV) salts Ce^{4+}

Quinones from halogenophenols ←

with CrO_3 cf. *11,* 294; with Ce(IV) or Mn(III) salts in aq. acetonitrile s. M.B. Gopinathan, M.V. Bhatt, Indian J. Chem. *20B,* 71–2 (1981).

Polyethylene glycol s. under KOH ←

18-Crown-6-polyether s. under MeCOOK ←

Aldehydes s. under $FeCl_3$ *RCHO*

*Tetra-*n*-butylammonium chloride s. under NaOH* n-$Bu_4N^+Cl^-$

*Tetra-*n*-butylammonium hydrogen sulfate s. under NaOH* n-$Bu_4N^+HSO_4^-$

Benzyltriethylammonium bromide s. under NaOH $[PhCH_2NEt_3]^+Br^-$

Aliquat 336 s. under $KHCO_3$ ←

6-Nitroquinoline ←

Selective and preferential O-tritylation OH → $OCPh_3$

with pyridine cf. *8,* 275; of ribonucleosides s. G.H. Hakimelahi, Z.A. Proba, K.K. Ogilvie, Can. J. Chem. *60,* 1106–13 (1982); with 6-nitroquinoline s. J. Okupniak, R.W. Adamiak, M. Wiewiórowski, Pol. J. Chem. *55,* 679–82 (1981).

m-*Chloroperoxybenzoic acid* m-$ClC_6H_4COO_2H$

Oxidative substitution of iodine ←

alcohols cf. *34,* 955; ethers, incl. bridgehead substitution s. R.I. Davidson, P. J. Kropp, J. Org. Chem. *47,* 1904–9 (1982).

Dimethyl sulfoxide s. under KOH and NaOMe Me_2SO

Stannous chloride/silver perchlorate *$SnCl_2/AgClO_4$*

α-Glycosides from β-glycosyl fluorides F → OR

182.

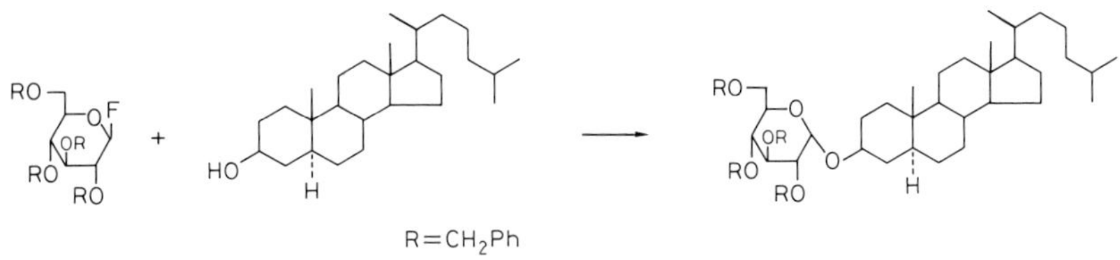

An ethereal soln. of 3β-cholestanol and 2,3,4,6-tetra-O-benzyl-β-D-glucopyranosyl fluoride added to a mixture of $SnCl_2$, $AgClO_4$, and 4Å molecular sieves at –15°, and stirred at this temp. for 24 hrs. → 3β-cholestanyl 2,3,4,6-tetra-O-benzyl-α-D-glucopyranoside. Y 88%. F.e.s. T. Mukaiyama, Y. Murai, S. Shoda, Chem. Letters *1981*, 431–2.

Stannic chloride *$SnCl_4$*

Urethans from carbamyl chlorides OH → OCON<

with pyridine cf. *11*, 284; with $SnCl_4$, phenyl N,N-diethylcarbamate, s. I. Isakova et al., Zh. Org. Khim. *18*, 1287–91 (1982).

Ozone *O_3*

Dicarboxylic acid esters from cyclic 1,2-ethylene-1,2-dichlorides C

s. *37*, 217

Hydrogen peroxide s. under $B(OBu)_3$ *H_2O_2*

Sulfur trioxide/boron oxide *SO_3/B_2O_3*

Carboxylic acids from 1,1,1-trifluorides CF_3 → COOR

with H_2SO_4 cf. *6*, 258; with SO_3 in the presence of B_2O_3, 4,4′-oxybis(tetrafluorophenylcarboxylic acid), s. G.P. Tataurov et al., Khim. Prom-st., Ser.; Reakt Osobo Chist. Veshchestva *1980* (6) 44, Chem. Abstr. *95*, 42577 (1981).

Sulfuric acid *H_2SO_4*

Carboxylic acid amides from iminohalides C(I)=N → CONH

without reagents cf. *17*, 259; with H_2SO_4, perfluoro-carboxylic acid amides from -iminoiodides, and f. reactions s. M. Tordeux, C. Wakselman, Tetrahedron *37*, 315–8 (1981).

Silver perchlorate s. under $SnCl_2$ *$AgClO_4$*

Ferric chloride/aldehydes *$FeCl_3/RCHO$*

Interchange of halogen and oxygen
Oxo compds. and dihalogenomethyl compds.
from 1,1-dihalides and aldehydes $CHal_2$ → CO

183. $Ph_2CBr_2 + PhCHO \rightarrow Ph_2C{=}O + PhCHBr_2$

Dibromodiphenylmethane added to a vigorously stirred mixture of benzaldehyde and a little $FeCl_3$ at 80° over ca. 30 min., and stirring continued at this temp. for a further 30 min. → diphenyl ketone (Y 85%) and benzal bromide (Y 82%). F.e.s. G.S. Litvinenko, G.F. Dvorko, N.M. Maksimenko, Zh. Obshch. Khim. *51*, 2574–7 (1981).

Via intermediates *v.i.*

Aldehydes via sulfinamides

CH_2Hal ⇉ CHO
CH_2N<

184. 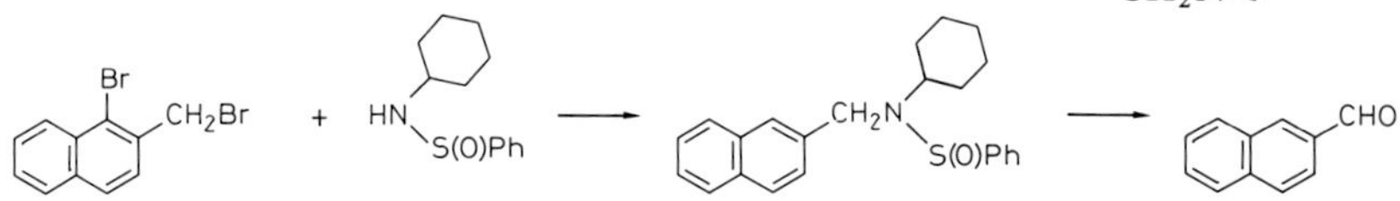

from halides. A soln. of N-cyclohexylbenzenesulfinamide in 1,2-dimethoxyethane added dropwise to a suspension of NaH in the same solvent, stirred for 1 hr. at room temp. and at 50–70° for 3 hrs., cooled to room temp., a soln. of 1-bromo-2-(bromomethyl)naphthalene in the same solvent added, and the mixture stirred at 80° overnight → 1-bromo-2-naphthaldehyde. Y 68%. F.e. and with isolation of intermediate, **also from amines** and benzenesulfinyl chloride with cleavage of sulfinamide by K_2CO_3, s. B.M. Trost, G. Liu, J. Org. Chem. *46,* 4617–20 (1981).

Sulfur ↑ **OC⇂↾S**

Sodium hydride/dibenzo[18]crown-6 polyether ←

Stereospecific glycosidation with triflates OH → OR

185. 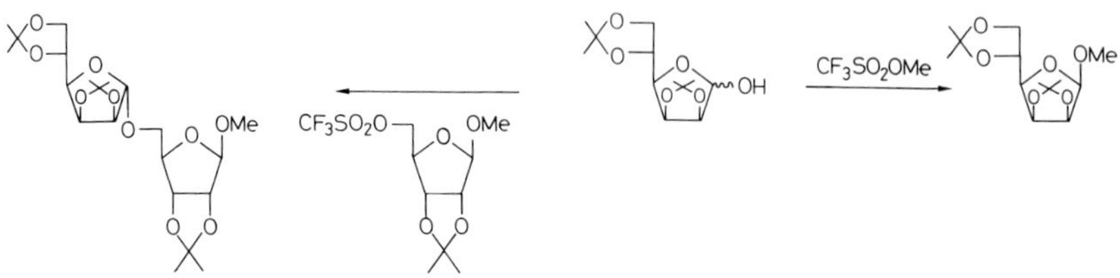

α-Glycosides. | **β-Glycosides.**

A soln. of 2,3:5,6-di-O-isopropylidene-D-mannose in tetrahydrofuran

stirred with dibenzol[18]crown-6 at room temp., NaH added, the mixture treated after 15 min. with methyl 2,3-O-isopropylidene-5-O-(trifluoromethanesulfonyl)-β-D-ribofuranoside, and worked up after 8 hrs. → α-glycoside. Y 72%.	treated with NaH at room temp., after 15 min. methyl triflate added via pipette, and worked up after 12 hrs. → β-glycoside. Y 61%.

F.e.s. R.R. Schmidt, U. Moering, M. Reichrath, Chem. Ber. *115,* 39–49 (1982).

Sodium hydroxide s. under $KMnO_4$ *NaOH*

Sodium alkoxides *NaOR*

3-α-Subst. indoles via 3-alkylideneindolenines ←

186. 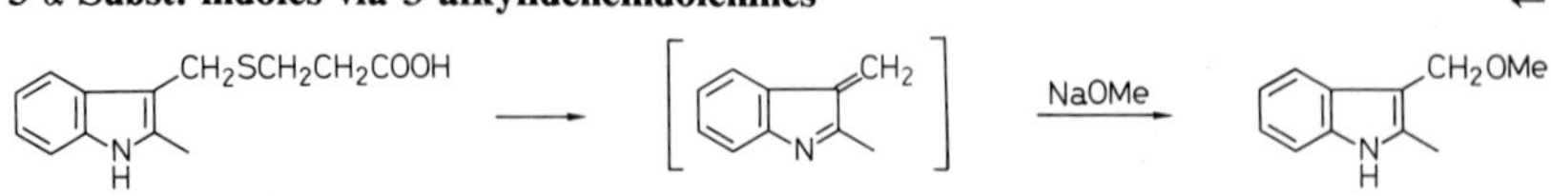

A mixture of 3-[(2-methylindol-3-yl)methylthio]propionic acid and Na methoxide in 4:5 methanol/pyridine refluxed for 50 min. → 3-methoxymethyl-2-methylindole. Y 80%. F.e. and nucleophiles, **also di-3-indolylmethanes,** s. R. Bennett, A. Maggiolo, T. Shah, J. Heterocyc. Chem. *18,* 391–4 (1981); **from 3-α-aminoindoles** cf. M. Somei, Y. Karasawa, C. Kaneko, Heterocycles *16,* 941–9 (1981).

Potassium superoxide/18-crown-6 polyether ←

Uracils from thiouracils CS → CO

with chloroacetic acid/HCl cf. *19,* 283; with KO_2/18-crown-6 polyether, from thio- and dithio-uracils, s. E. Katori et al., Chem. Pharm. Bull. *29,* 3075–7 (1981).

Sodium nitrite/hydrogen chloride *$NaNO_2/HCl$*

Replacement of carbonyl oxygen by sulfur
Carboxylic acid amides from thioamides

with $AgNO_3$ cf. *16,* 285; with $NaNO_2$/HCl s. K.A. Jorgensen, A.-B.A.G. Ghattas, S.-O. Lawesson, Tetrahedron *38,* 1163–8 (1982).

Triethylamine *Et_3N*

Acylation with 2-acylthiobenzoxazoles OH → OAc
Carboxylic acid esters and amides

187.

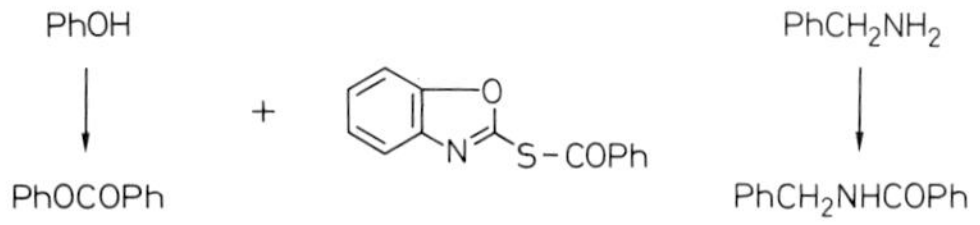

Esters. Phenol and triethylamine | **Amides.** Benzylamine
added to a soln. of S-(2-benzoxazolyl) thiobenzoate in tetrahydrofuran at 15°, and stirred for
24 hrs. → product. Y 93%. | 10 min. → product. Y 81%.
– The acyl transfer reagents are crystalline solids with excellent hydrolytic stability. F.e., **also selective N-acylation,** s. M. Ueda, K. Seki, Y. Imai, Synthesis *1981,* 991–3.

Pyridine s. under Pyridinium tribromide *C_5H_5N*

Lithium dialkylcuprates *$Li[CuR_2]$*

Carboxylic acid esters from thiolic acid 2-pyridyl esters COSR → COOR′

188.

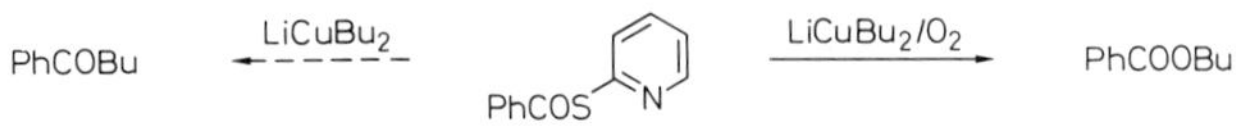

A soln. of Li di-*n*-butylcuprate in ether stirred under O_2 for 5 min. at –78°, added with stirring to a soln. of 2-pyridyl thiobenzoate in the same solvent at the same temp. under O_2, and the mixture allowed to warm to room temp. during 3 hrs. → n-butyl benzoate. Y 99%. F.e., also synthesis of **ketones** under N_2, s. S. Kim, J.I. Lee, J. Chem. Soc. Chem. Commun. *1981,* 1231–2.

Mercuric oxide/fluoroboric acid *HgO/HBF_4*

Cleavage of cyclic monothioacetals under mild conditions 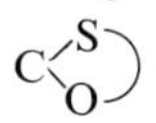 → CO

189.

Aq. 35% HBF_4 added at room temp. to a suspension of HgO in tetrahydrofuran, a soln. of 2-(*p*-tolyl)-1,3-benzoxathiole in the same solvent added in one portion with vigorous stirring, and worked up after ca. 5 min. → *p*-tolualdehyde. Y 92%. F.e., **also cleavage of cyclic mercaptals** (cf. Synth. Meth. *26,* 211), **and thioorthocarboxylic acid esters,** s. I. Degani, R. Fochi, V. Regondi, Synthesis *1981,* 51–3.

Mercuric chloride/mercuric oxide *$HgCl_2/HgO$*

Acetals from mercaptals $C(S–)_2 \rightarrow C(O–)_2$

with methyl fluorosulfonate cf. *31,* 216; with $HgCl_2$/HgO s. F.J. Lopez Aparicio, F.Z. Benitez, F.S. Gonzalez, Carbohyd. Res. *90,* 315–8 (1981); method s. An. Quim. *70,* 1055–61 (1974).

Mercuric perchlorate $Hg(ClO_4)_2$

Carboxylic acid esters from 2-alkylthio-1,3-benzodithioles C

s. *37,* 836

Trimethyloxonium fluoroborate $Me_3O^+BF_4^-$

Ketones from cyclic mercaptals

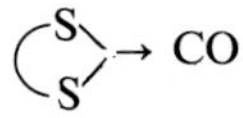

s. *37,* 190

Fluoroboric acid s. under HgO HBF_4

Ammonium ceric nitrate $(NH_4)_2Ce(NO_3)_6$

Oxo compds. from 1,3-dithianes

with N-bromosuccinimide cf. *27,* 230; with ammonium ceric nitrate s. H.-J. Cristau et al., Synth. Commun. *11,* 423–8 (1981).

18-Crown-6 polyether s. under NaH and KO_2 ←

*Tetra-*n-*butylammonium bromide s. under Pyridinium tribromide* n-$Bu_4N^+Br^-$

Dimethyl sulfoxide s. under tert-*Butyl halides* Me_2SO

*Iodoxybenzene/*p-*toluenesulfonic acid* $PhIO_2/TsOH$

Oxo compds. from mercaptals $C(S-)_2 \rightarrow CO$

s. *37,* 266

tert-*Butyl halides/dimethyl sulfoxide* t-$BuHal/Me_2SO$

Cleavage of cyclic mercaptals under mild, neutral conditions

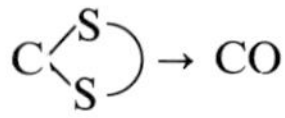

190.

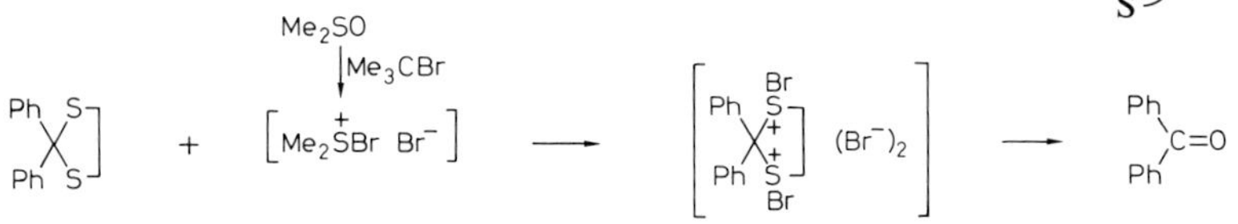

Startg. mercaptal added to a stirred mixture of *tert*-butyl bromide and dimethyl sulfoxide, and heated at 70–75° under N_2 for 4 hrs. → benzophenone. Y 95%. F.e., also with *tert*-butyl iodide, s. G.A. Olah, A.K. Mehrotra, S.C. Narang, Synthesis *1982,* 151–2; with trimethyloxonium fluoroborate cf. I. Stahl, ibid. *1981,* 135–7.

Titanium tetrachloride $TiCl_4$

Ketones from thioenolethers $C{=}CSPh \rightarrow CHCO$

s. *37,* 901

Sulfuryl chloride SO_2Cl_2

Carboxylic acid esters from halides $CH(SR)SO_2R \rightarrow COOR$

Synthesis with addition of one C-atom

191.

$$n\text{-}C_{12}H_{25}Br + MeSCH_2SO_2Me \rightarrow n\text{-}C_{12}H_{25}CH(SMe)SO_2Me$$
$$\downarrow$$
$$n\text{-}C_{12}M_{25}COOMe \xleftarrow{HOMe} n\text{-}C_{12}H_{25}CCl(SMe)SO_2Me$$

via 1-sulfonylthioethers. A mixture of 50% aq. NaOH and a little trioctylmethylammonium chloride added to a soln. of dodecyl bromide and formaldehyde dimethyl dithioacetal S,S-dioxide in toluene, and stirred at 60° for 30 hrs. → tridecanal dimethyl dithioacetal S,S-dioxide (Y 83%) and SO_2Cl_2 in chloroform refluxed for 1 hr., and the intermediate chloro deriv. refluxed in methanol for 24 hrs. → product (Y 87%). F.e., **also via 1-sulfinylsulfones,** and regiospecific synthesis of β,γ-ethylenecarboxylic acid esters from β,γ-ethylenehalides, s. K. Ogura, J. Watanabe, H. Iida, Tetrahedron Letters *22,* 4499–502 (1981).

p-Toluenesulfonic acid s. under $PhIO_2$ *TsOH*

Pyridinium tribromide/tetra-n-butylammonium bromide/pyridine $[C_5H_5NH]^+Br_3^-/(n\text{-}Bu)_4N^+Br^-/C_5H_5N$

Oxo compds. from mercaptals $C(S\text{-})_2 \rightarrow CO$

192. 

Tetra-*n*-butylammonium bromide added with vigorous stirring to a mixture of 2-phenyl-1,3-dithiane, pyridine, and pyridinium tribromide in dichloromethane/water at 0°, and stirring continued at room temp. for 2 hrs. → benzaldehyde. Y 87%. – This method is especially useful for the hydrolysis of α-thioketalized aldehydes and carboxylic acid esters, whereas cleavage of α-thioketalized acids was best achieved in aq. acetonitrile. F.e.s. G.S. Bates, J. O'Doherty, J. Org. Chem. *46,* 1745–7 (1981).

Hydrogen chloride s. under $NaNO_2$ *HCl*

Potassium permanganate/sodium hydroxide *$KMnO_4/NaOH$*

Carboxylic acids from α-nitrosulfones $CH(NO_2)SO_2R \rightarrow COOH$

193. n-BuCH$(NO_2)SO_2$Ph → n-BuCOOH

An aq. soln. of $KMnO_4$ added dropwise over 30 min. to a soln. of 1-nitro-1-(phenylsulfonyl)pentane in 5% NaOH, and stirred a further 30 min. → valeric acid. Y 89%. **F.e.** and **reactions** s. P.A. Wade et al., J. Org. Chem. *46,* 765–70 (1981).

Nickel peroxide *NiO_2*

Ureas from thioureas NCSN< → NCON<

with H_2O_2/NaOH cf. *17,* 288; with NiO_2 s. H. Ogura, S. Mineo, K. Nakagawa, Chem. Pharm. Bull. *29,* 1518–24 (1981).

Remaining Elements ↑ OC⇃↾Rem

Without additional reagents *w.a.r.*

Carboxylic acid methyl esters from carboxylic acids COOH → COOMe

194. 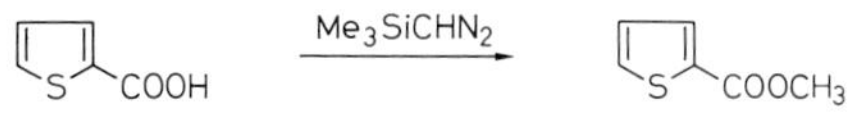

A stirred soln. of thiophene-2-carboxylic acid in 2:7 methanol/benzene treated with a soln. of trimethylsilyldiazomethane in benzene, and the mixture stirred at room temp. for 30 min. → thiophene-2-carboxylic acid methyl ester. Y 100%. – The methylating agent is safer than diazomethane. F.e.s. N. Hashimoto, T. Aoyama, T. Shioiri, Chem. Pharm. Bull. *29,* 1475–8 (1981); α-aminocarboxylic acid methyl esters s. ibid. *30,* 1921–4 (1982).

Partial O-phenylation of diols under mild conditions OH → OPh

195. 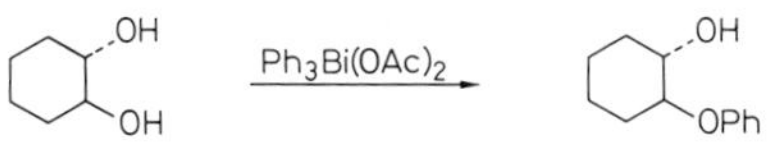

Triphenylbismuth(v) diacetate added to a soln. of *trans*-cyclohexane-1,2-diol in dichloromethane, and the mixture refluxed for 4 hrs. → *trans*-2-phenoxycyclohexanol. Y 88%. F.e. and limitations s. S. David, A. Thieffry, Tetrahedron Letters *22,* 5063–6 (1981).

Sodium alkoxides *NaOR*

5-α-Alkoxy- from 5-α-phosphonio-imidazoles ←

s. *37*, 388

Methyllithium *MeLi*

Thiolic acid esters from α-organothiophosphine oxides CH(SR)P(O)< → COSR

s. *37*, 87

Sodium hydrogen carbonate s. under m-*$ClC_6H_4COO_2H$* *$NaHCO_3$*

Potassium fluoride s. under $AgBF_4$ *KF*

Triethylamine s. under $CF_3SO_2OSiMe_3$ *Et_3N*

1,8-Diazabicyclo[5.4.0]undec-7-ene ←

O-Acylation with acylphosphonic esters OH → OCOPh

196. $PhCH_2CH_2OH + PhC(O)P(O)(OEt)_2 \rightarrow PhCH_2CH_2OCOPh$

O-Benzoylation. A mixture of 2 equivalents diethyl benzoylphosphonate and 1 equivalent phenethyl alcohol in dichloromethane treated with 1 equivalent 1,8-diazabicyclo[5.4.0]undec-7-ene at room temp. for 10 min. → phenethyl benzoate. Y 99%. F.e., **also preferential O-benzoylation** of prim. hydroxyl with retention of sec. hydroxyl groups, s. M. Sekine, A. Kume, T. Hata, Tetrahedron Letters *22*, 3617–20 (1981).

N-Methylmorpholine N-oxide s. under OsO_4 ←

Silver(I) carboxylates/iodine *$RCOOAg/I_2$*

α-Acoxyoxo compds. from enoxysilanes C(OSi≡)=C → CO–C(OAc)

with Pb tetraacetate cf. *31*, 218; with I_2/Ag(I) carboxylates, cyclic α-acoxyketones, s. J. Org. Chem. *46*, 2717–21 (1981).

Mercuric oxide s. under $Pd(PPh_3)_4$ *HgO*

Silver fluoroborate/potassium fluoride *$AgBF_4/KF$*

Ketones from silanes via α-bromosilanes C(Br)Si≡ → CO

Benzocyclobutenedione monoketals

197.

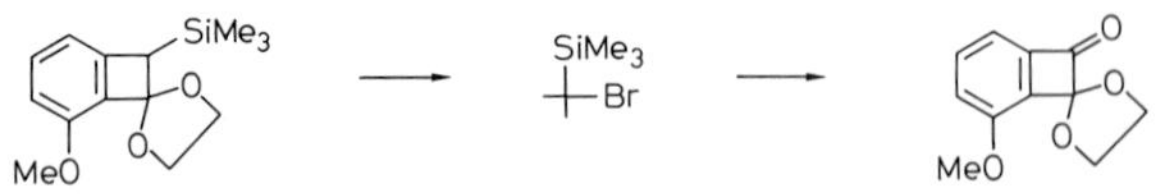

A mixture of N-bromosuccinimide and the startg. m. in carbon tetrachloride irradiated with a 200 W unfrosted incandescent lamp at 60° for 20 min. under N_2 → intermediate (Y 96%), added to a soln. of $AgBF_4$ in dimethyl sulfoxide at 60°, stirred at this temp. for 16 hrs., cooled to room temp., treated with $KF.2H_2O$, and stirring continued for a further 1 hr. → product (Y 85%). F.e.s. J.S. Swenton et al., J. Org. Chem. *46*, 4825–36 (1981).

Triethylammonium fluoride s. under m-*$ClC_6H_4COO_2H$* *$[Et_3NH]^+F^-$*

tert-*Butyl hydroperoxide s. under $(Ac_2CH)_2MoO_2$* t-*BuOOH* *m-Chloroperoxybenzoic acid* $m\text{-}ClC_6H_4COO_2H$

Ketones from enesilanes C=C–Si≡ → CHCO

s. *30,* 143; 1,5-diketones via bicyclic ketals s. L.A. Paquette, Y.-K- Han, J. Am. Chem. Soc. *103,* 1831–5 (1981).

m-Chloroperoxybenzoic acid/triethylammonium fluoride $m\text{-}ClC_6H_4COO_2H/[Et_3NH]^+F^-$

α-Hydroxycarboxylic acid derivs. from O-silyl keteneacetals C=C(OSi≡)(OR) → C(OH)COOR

acids from ketene disilyl acetals cf. *31,* 217; α-hydroxycarboxylic acid esters from O-silyl O-alkyl keteneacetals with added triethylammonium fluoride s. G.M. Rubottom, R. Marrero, Synth. Commun. *11,* 505–11 (1981).

m-Chloroperoxybenzoic acid/sodium hydrogen carbonate/sulfuric acid $m\text{-}ClC_6H_4COO_2H/NaHCO_3/H_2SO_4$

2-Methylene-1,4-diols from oxido compds. via 2-methylene-4-hydroxysilanes C(Si≡)C=C → C=C–C(OH)

198.

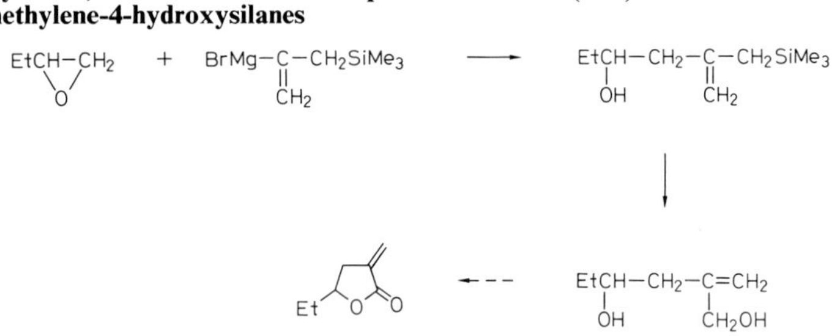

Startg. epoxide in tetrahydrofuran treated with 1-trimethylsilylmethylvinylmagnesium bromide in the presence of CuI at –30° for 1 hr. then at 0° for 2 hrs. → intermediate (Y 83%) in methylene chloride treated with *m*-chloroperoxybenzoic acid and $NaHCO_3$ at 0° for 1–1.5 hrs., followed by 3:1 tetrahydrofuran/0.15 *M* aq. H_2SO_4 → product (Y 80%). F.e., method, and oxidation to **α-methylene-γ-lactones** with MnO_2, s. H. Nishiyama et al., Tetrahedron Letters *23,* 1267–70 (1982).

Iodosobenzene C_6H_5IO

Protection of amino groups as rhenium(I) complexes ←
Removal of the protective group

199.

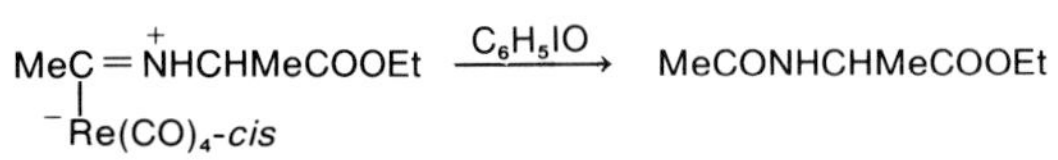

A soln. of startg. rhenium carbonyl complex in methylene chloride treated portionwise with iodosobenzene at 25° until the vigorous evolution of CO subsided → ethyl N-acetyl-L-alaninate. Y 67%. – The protecting group is stable in ethanolic KOH and ethereal HCl, but readily removed in the above oxidative media or by acidic solvolysis. F.e. and use in **peptide synthesis** s. A.J. Baskar, C.M. Lukehart, K. Srinivasan, J. Am. Chem. Soc. *103,* 1467–72 (1981).

Trimethylsilyl triflate $CF_3SO_2OSiMe_3$

O-Tritylation $OCPh_3$

200. $CH_2{=}CMeCOOSiMe_3 + Ph_3COSiMe_3 \rightarrow CH_2{=}CMeCOOCPh_3$

A mixture of trimethyltrityloxysilane and trimethylsilyl methacrylate in dry methylene chloride treated with a little trimethylsilyl triflate in the same solvent at 0° under argon, and stirred for 6 hrs. at this temp. → trityl methacrylate. Y 73%. F.e. and trityl ethers s. S. Murata, R. Noyori, Tetrahedron Letters *22,* 2107–8 (1981).

Trimethylsilyl triflate/triethylamine $CF_3SO_2OSiMe_3/Et_3N$

Glycosyl acetals from silyl glycosides $C(OR)_2 \rightarrow C(OR)(OR')$

201.

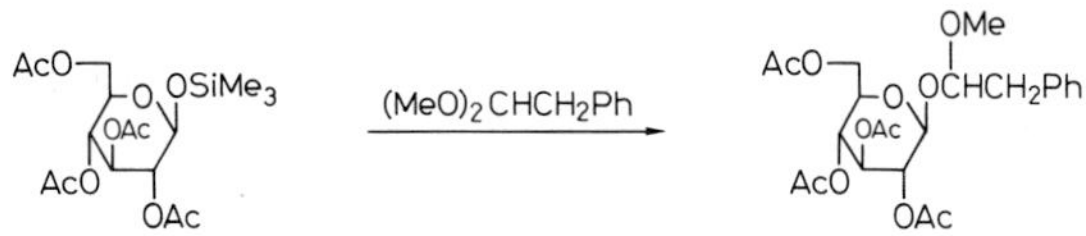

A little trimethylsilyl triflate in methylene chloride added to a soln. of 2,3,4,6-tetra-O-acetyl-1-O-trimethylsilyl-β-D-glucopyranose and startg. acetal in the same solvent under inert gas at –70°, the mixture stirred at this temp. for ca. 20 hrs., and triethylamine added → product. Y 84%. F.e.s. L.-F. Tietze, R. Fischer, Angew. Chem. Intern. Ed. *20,* 969–70 (1981).

Sulfuric acid s. under m-$ClC_6H_4COO_2H$ H_2SO_4

*Molybdenyl acetylacetonate/*tert-*butyl hydroperoxide* $(acac)_2MoO_2$/t-*BuOOH*

Oxidative cleavage of enoxysilanes ⊂

202.

Dicarboxylic acids. A soln. of startg. enoxysilane and a little molybdenyl acetylacetonate in benzene treated dropwise with a soln. of *tert*-butyl hydroperoxide in the same solvent under N_2, stirred at 60° for 48 hrs., 3:1:1 trifluoroacetic acid/tetrahydrofuran/water added, stirring continued at the same temp., and the crystals collected after 24 hrs. → product. Y 86%. F.e. and ketocarboxylic acids, also ketones or carboxylic acids from acyclic enoxysilanes, s. K. Kaneda et al., Tetrahedron Letters *22,* 2595–8 (1981).

Iodine s. under (RCOO)Ag I_2

Osmium tetroxide/N-methylmorpholine N-oxide ←

α-Hydroxyketones from enoxysilanes C=C–Si⋶ → CHCO

with *m*-chloroperoxybenzoic acid cf. *30,* 144; with OsO_4/N-methylmorpholine N-oxide s. J.P. McCormick, W. Tomasik, M.W. Johnson, Tetrahedron Letters *22,* 607–10 (1981).

Tetrakis(triphenylphosphine)palladium(0)/mercuric oxide $Pd(PPh_3)_4/HgO$

α-Alkoxyketones
from enoxysilanes and alkyl hypochlorites C(OSi⋶)=C → COC(OR)
Palladium-catalysed polarity reversal

203.

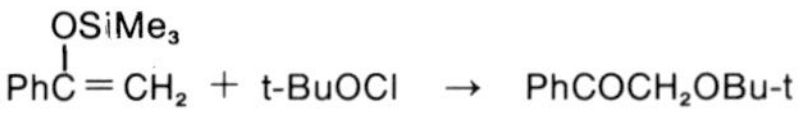

Solns. of 1-phenyl-1-trimethylsilyloxyethene and *tert*-butyl hypochlorite in toluene added sequentially and dropwise to a little tetrakis(triphenylphosphine)palladium(0) and HgO suspended in the same solvent at –78°, and the mixture stirred under argon for 1 hr. at the same temp. in the dark → *α-tert*-butoxyacetophenone. Y 64%. F.e., **also α-ketoacetals from 2-siloxyenolethers,** s. T. Nakatsuka, T. Mukaiyama, Chem. Letters *1982,* 369–72.

Carbon ↑ OC⇃↾C

Without additional reagents *w.a.r.*

O-Acylation with acylcyanides OH → OAc

with NEt_3 cf. *31,* 223; without reagents s. J. Castells et al., Tetrahedron *38,* 337–46 (1982); of hydroxylamines, prepn. of ar. acoxylamines, s. S. Prabhakar, A.M. Lobo, M.M. Marques, Tetrahedron Letters *23,* 1391–4 (1982).

O-Trichloroacetylation under neutral conditions OH → $OCOCCl_3$

204. $CCl_3COCCl_3 + HOMe \rightarrow CCl_3COOMe$

of alcohols. Methanol and 0.5 equivalents dimethylformamide stirred for 10 min. at 0° under N_2, hexachloroacetone added dropwise, and the sealed reaction vessel heated at 55° for 1.5 hrs. → methyl trichloroacetate. Y 93%. F.e.s. R.S. Freedlander et al., J. Org. Chem. *46,* 3519–21 (1981).

Irradiation s. under Diazo compds. and TiO_2

Electrolysis

Carboxylic acid esters from *p*-quinol monoesters and alcohols COOR → COOR
Anodic transesterification

205. $Me(CH_2)_4COO\text{–}C_6H_4\text{–}OH\text{–}p + HO(CH_2)_5Me \rightarrow Me(CH_2)_4COO(CH_2)_5Me$

A soln. of startg. *p*-quinol monoester and a 5–10 fold excess of hexanol in acetonitrile or methylene chloride electrolysed under N_2 in a beakertype cell with a carbon felt anode and platinum wire cathode isolated in a fritted tube at a controlled potential using a regulated DC power supply with a current density of 5–10 mA/cm^2 until 2.1 equivalents of charge passed at –30° → product. Y 95% by g.l.c. F.e.s. R.W. Johnson et al., Tetrahedron Letters *22,* 3715–8 (1981).

Oxocarboxylic acid esters

206.
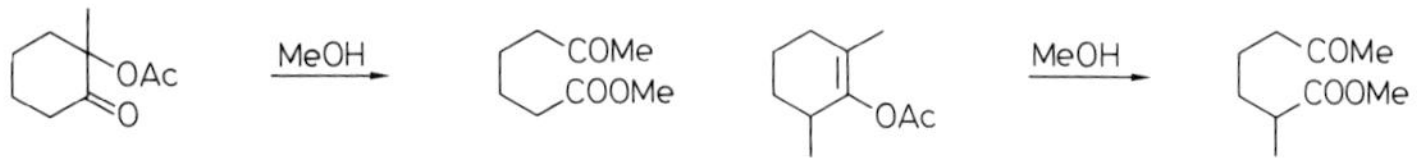

from cyclic *α*-acoxyketones	**from cyclic enolesters**
A soln. of $LiClO_4$ and startg. *α*-acoxyketone in methanol at 30–34°	A soln. of $LiClO_4$ and startg. enolester in 10:1 methanol/acetic acid at 2–3°
electrolysed at the Pt anode of a divided cell under a constant voltage of 20 V until 5.5 F/mol passed → product. Y 82%.	electrolysed at the Pt anode of a divided cell under a constant voltage of 20 V until 6.7 F/mol passed → product. Y 72%.

F.e.s. S. Torii, T. Inokuchi, R. Oi, J. Org. Chem. *47,* 47–52 (1982).

Sodium hydroxide *NaOH*

Carboxylic acids from trichloromethyl ketones $COCCl_3$ → COOH
with $NaHCO_3$ cf. *7,* 320; with NaOH, indole-3-carboxylic acids, s. V.E. Golubev, V.A. Shumakov, Deposited Doc. *1981,* VINITI, 913–81 4 pp. (Russ); Chem. Abstr. *97,* 23574 (1982).

Sodium methoxide s. under H_2O_2 *NaOMe*

Sodium ethoxide *NaOEt*

N-Acylcarboxylic acid amides
from *α*-acylaminocarboxylic acid esters CH(COOR)NHCOR → CONHCOR
Oxidative decarbalkoxylation

207. $PhCH(COOEt)NHCOC_6H_4NO_2\text{–}p \rightarrow PhCONHCOC_6H_4NO_2\text{–}p$

An O_2-stream bubbled through a stirred soln. of startg. N-acylphenylglycinate and 3 equivalents Na ethoxide in dry hexamethylphosphoramide at room temp. for 5 min. → product. Y 92%. F.e. and methods s. C. Yijima, F. Hino, K. Suda, Synthesis *1981,* 610–1.

Organolithium compd. s. under OsO_4 *RLi*

Potassium superoxide/18-crown-6 polyether ←

6-Hydroxy-5,6-dihydro-2-pyrones from 2,5-cyclohexadien-2-ol-1-ones O

208. 

Startg. m. treated with KO_2 in benzene containing 18-crown-6 polyether at room temp. for 16 hrs. under dry air → product. Y 90%. F.e.s. A.A. Frimer, P. Gilinsky-Sharon, G. Aljadeff, Tetrahedron Letters *23,* 1301–4 (1982).

Potassium sulfate s. under $KHSO_5$ K_2SO_4

Potassium hydrogen sulfate s. under $KHSO_5$ $KHSO_4$

Potassium fluoride/18-crown-6 polyether ←

Transesterification of phosphonic esters P–OR → P–OR′

with Na alkoxides cf. *24,* 236; with KF/18-crown-6 polyether s. J. Szewczyk, B. Lejczak, P. Kafarski, Synthesis *1982,* 409–12, 412–4.

Sodium iodide *NaI*

Acoxy compds from ethers OR → OAc

s. *37,* 479

Trimethylamine/silica gel Et_3N/SiO_2

α,β-Ethyleneketones from 5,6-dihydro-4*H*-1,2-oxazines via 5,6-dihydro-4*H*-1,2-oxazinium salts C

209. 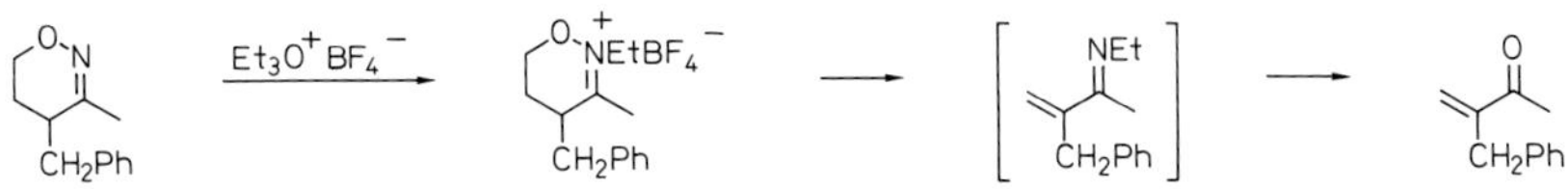

α-Methyleneketones. A mixture of startg. 1,2-oxazine (prepd. via metallation of 3-methyl-4*H*-5,6-dihydro-1,2-oxazine with Li dimethylamide at low temp.) and triethyloxonium fluoroborate in methylene chloride allowed to react at room temp. for 30 min. → intermediate (Y 98%) in chloroform treated with 1 equivalent 2 *M* trimethylamine in carbon tetrachloride at –65°, after 1 min. the trimethylammonium fluoroborate salt filtered off, the filtrate allowed to warm to room temp., and hydrolysed by passing the soln. through SiO_2 (containing 10% H_2O) at 0° eluting with 1:1 chloroform/hexane → product (Y >80%). F.e.s. R. Lidor, S. Shatzmiller, J. Am. Chem. Soc. *103,* 5916–7 (1981).

1,5-Diazabicyclo[*4.3.0*]*non-5-ene* ←

α-Ketocarboxylic acids from Δ^2-isoxazol-5-ones via 4-hydroxy-Δ^2-isoxazol-5-ones

210. 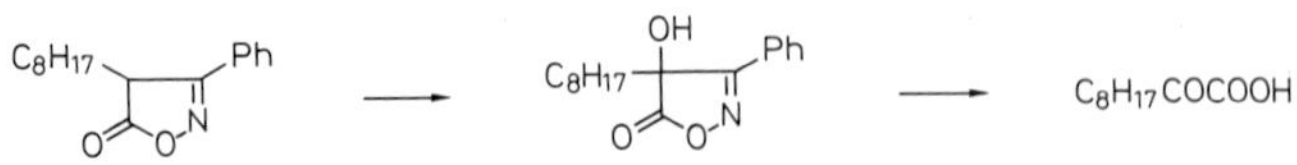

1.3 Equivalents of *m*-chloroperoxybenzoic acid in dichloromethane added to a soln. of startg. m. in the same solvent, and the mixture stirred at room temp. for 12 hrs. → 4-hydroxy-Δ^2-isoxazol-5-one deriv. (Y 89%), in benzene treated with a 100% molar excess of 1,5-diazabicyclo[4.3.0]non-5-ene, and the mixture refluxed for 3 hrs. → 2-oxodecanoic acid (Y 98%). F.e. and methods s. C. Baldoli et al., Gazz. Chim. Ital. *111,* 347–50 (1981); stereospecific **4-hydroxylation of Δ^2-isoxazoles** cf. W. Schwab, V. Jäger, Angew. Chem. Intern. Ed. *20,* 603–5 (1981).

4-(Dimethylamino)pyridine ←

Carbonic acid esters by transesterification 2 ArOH → ArOCOOAr
mixed carbonic acid esters cf. *26,* 219; carbonic acid aryl esters from 2 phenol molecules and bis(*o*-nitrophenyl) carbonate in the presence of 4-(dimethylamino)pyridine s. D.J. Brunelle, Tetrahedron Letters *23,* 1739–42 (1982).

N-Methylmorpholin N-oxide s. under OsO_4 ←

Calcium carbonate s. under $(MeCOO)_4Ph$ *$CaCO_3$*

Lithium tetrahydridoaluminate *$LiAlH_4$*
Transesterification COOR → COOR′
s. *37,* 214

Alumina *Al_2O_3*
Preferential O-acylation of prim. hydroxyl groups OH → OAc
Transesterification COOR → COOR′

211. $PhCHOH(CH_2)_4OH + EtOAc \rightarrow PhCHOH(CH_2)_4OAc$

A soln. of startg. diol in ethyl acetate stirred over Woelm W-200-N neutral alumina at 75–80° for 1 hr. → product. Y 92% based on recovered startg. m. F.e. and with retention of phenol groups, also N-acylation of prim. ar. amines, s. G.H. Posner, M. Oda, Tetrahedron Letters *22,* 5003–6 (1981); transesterification in the presence of base- and acid-sensitive groups s. Synthesis *1981,* 789–90.

Montmorillonite ←
Transacetalation OCH_2OEt
Ethoxymethyl ethers
s. *18,* 297s*37*

Boron fluoride *BF_3*
Glycosylation with iminoesters $O{-}C(=N)CCl_3 \rightarrow OR'$
in the presence of *p*-toluenesulfonic acid cf. *33,* 241; with trichloroacetimidates in the presence of BF_3-etherate s. R.R. Schmidt, G. Grundler, Synthesis *1981,* 885.
Glycosides from acyl glycosides OAc → OR
with $SnCl_4$ cf. *34,* 209; with BF_3, *β,β,β*-trichloroethyl glycosides, s. G. Magnusson et al., Acta Chem. Scand. *35B,* 213–6 (1981).

18-Crown-6 polyether s. under KO_2 and KF ←

Diazo compds./methylene blue/irradiation ←
Carboxylic from *α*-ketocarboxylic acids COCOOH → COOH
Photooxidative decarboxylation

212. PhCOCOOH → PhCOOH ----→ PhCOOMe
+ PhCOC(N_2)Ph   + PhCOCOPh

A soln. of phenylglyoxylic acid in 10:3 acetonitrile/pyridine added to a soln. of benzoyl(phenyl)diazomethane in acetonitrile containing methylene blue, and irradiated with water-cooling using a 500 W halogen lamp while introducing O_2 for 80 min. → benzoic acid, isolated as the methyl ester (Y 88%), and benzil (90%). F.e.s. W. Ando, H. Miyazaki, T. Akasaka, Tetrahedron Letters *23,* 1197–200 (1982).

Methylene blue s. under Diazo compds. ←

Acetonitrile s. under RuO_4 CH_3CN

Silica gel. s. under Me_3N SiO_2

Titanium dioxide/irradiation $TiO_2/h\nu$

Photocatalytic oxidation with semiconductors
Ketones from ethylene derivs. C=C → CO
C-Cleavage

213. $Ph_2C{=}CH_2 \rightarrow Ph_2C{=}O$

A stirred suspension of TiO_2 and startg. alkene in 20:1 acetonitrile/trifluoroethanol irradiated in a Rayonet reactor at 350 ± 30 nm under a slow stream of air for 6 hrs. → benzophenone. Y 84%. F.e., also with ZnO or CdS, s. M.A. Fox, C.-C. Chen, J. Am. Chem. Soc. *103,* 6757–9 (1981); photocatalytic oxidation of lactams and acylamines with TiO_2 s. J.W. Pavlik, S. Tantayanon, ibid. *103,* 6755–7 (1981).

Titanium(IV) alkoxides $Ti(OR)_4$

Transesterification under mild, neutral conditions COOR → COOR′

214. $t\text{-}BuMe_2SiOCHMeCH_2COOEt + HOPr\text{-}i \xrightarrow{Ti(OPr\text{-}i)_4} t\text{-}BuMe_2SiOCHMeCH_2COOPr\text{-}i$

Tetraisopropyl titanate added to a soln. of ethyl 3-(*tert*-butyldimethylsiloxy)butanoate in isopropanol, the mixture refluxed for 6 hrs., cooled to ca. 45°, then quenched with 1 *M* HCl → isopropyl 3-(*tert*-butyldimethylsiloxy)butanoate. Y 83%. – Under the reaction conditions employed, a variety of functional groups and chiral centers, both in the ester and alcohol components, remained unaffected. F.e. and neutral work up procedures s. D. Seebach et al., Synthesis *1982,* 138–41; with Li tetraalkoxyaluminates, prepd. from $LiAlH_4$ and alcohols in ether, cf. S. Kumar et al., Indian J. Chem. *19B,* 856–8 (1980); with bis(trimethylsilyl) sulfate cf. Y. Morizawa et al., Synthesis *1981,* 899–901; with trimethylsilyl iodide/I_2 cf. G.A. Olah et al., ibid. *1981,* 142–3.

Trimethylsilyl iodide/iodine Me_3SiI/I_2
s. *37,* 214

Lead tetraacetate $(MeCOO)_4Pb$

Replacement of carboxyl by acoxy groups COOH → OAc
by electrolysis cf. *17,* 292s*31;* with Pb tetraacetate, 4-acoxy-2-azetidinones, s. M. Brener, K.-H. Ongania, Monatsh. Chem. *112,* 867–9 (1981).

Sym. ketones from halides via Δ^2-5-oxazolones and α-acylaminocarboxylic acids $R_2C(NHAc)COOH \rightarrow R_2CO$

215.

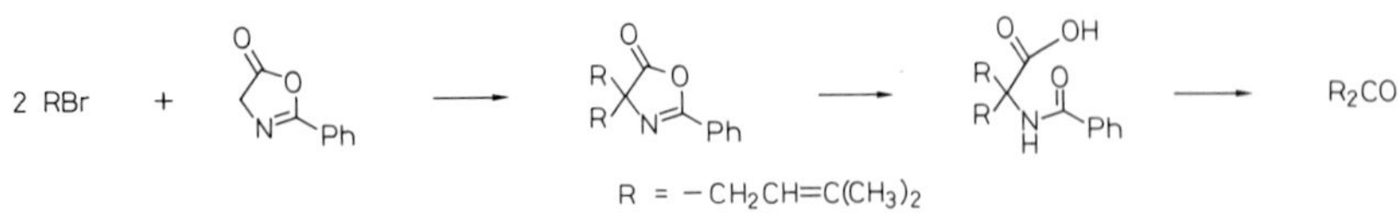

Ethyldiisopropylamine added dropwise to 2-phenyl-2-oxazolin-5-one and startg. bromide in dry dimethylformamide containing NaI, and stirred for 6 hrs. → 4,4-dialkyl deriv. (Y 95%) in aq. acetone stirred overnight with 2 *M* HCl → 2-(benzoylamino)-2-(3,3-dimethylallyl)-5-methyl-4-hexenoic acid (Y 86%) in tetrahydrofuran containing hexamethylphosphoramide treated with Pb tetraacetate for 2 hrs. at 5–8°, warmed to 25° with stirring, further oxidant added to complete reaction, treated with 1:1 water/petroleum ether followed by oxalic acid, and stirring continued for 2 hrs. → 2,8-dimethyl-2,7-nonadien-5-one (Y 80%). F.e.s. R. Lohmar, W. Steglich, Chem. Ber. *113,* 3706–15 (1980).

Lead tetraacetate/calcium carbonate — *$(MeCOO)_4Pb/CaCO_3$*

α-Acoxyketones from 2,3-oxidoalcohols 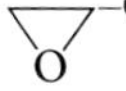–C(OH) → COC(OAc)

C-Cleavage

216. 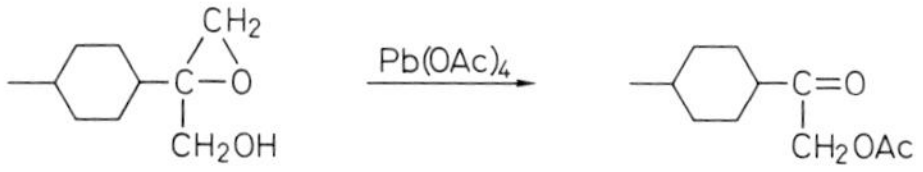

Pb tetraacetate and $CaCO_3$ added to a soln. of startg. epoxyalcohol in benzene, and refluxed under N_2 for 2 hrs. → product. Y 87%. F.e.s. T.J. Brocksom, J.T.B. Ferreira, A.L. Braga, J. Chem. Res. (S) *1981,* 334–5.

Ozone — *O_3*

Ozonolysis ←

of α,β-ethylenehalides s. *32,* 180; ozonolysis of ethylene derivs., review of recent advances, s. V.N. Odinokov, G.A. Tolstikov, Russ. Chem. Rev. *50,* 636–56 (1981).

Carboxylic acid esters from 1,2-ethylene-1,2-dichlorides CCl=CCl → COOR

Ozonolysis

217. 

A soln. of *trans*-3,4-dichloro-3-hexene in methanol treated with ozone at –40° during 3 hrs., the mixture purged with N_2, and stirred at room temp. for 40 hrs. → methyl propionate. Y 92% by n.m.r. F.e., **also dicarboxylic acid esters** from cyclic 1,2-ethylene-1,2-dichlorides, s. T.-S. Huh, J. Neumeister, K. Griesbaum, Can. J. Chem. *59,* 3188–91 (1981).

Hydrogen peroxide/sodium methoxide — *$H_2O_2/NaOMe$*

Oxidative ring opening of isatins ⊂

o-aminocarboxylic acids cf. *8,* 303; esters with H_2O_2/NaOMe s. G. Reissenweber, D. Mangold, Angew. Chem. Intern. Ed. *20,* 882–3 (1981).

Hydrogen peroxide/hydrogen chloride — *H_2O_2/HCl*

Phenols from 1-arylcyclopentenes ←

s. *37,* 106

Bis(trimethylsilyl) sulfate — *$(Me_3SiO)_2SO_2$*

Transesterification COOR → COOR′

s. *37,* 214

Nafion-H ←

Transacetalation OH → OCH_2OR

with TsOH cf. *18,* 297; with Nafion H, methoxymethyl ethers from alcohols, s. G.A. Olah et al., Synthesis *1981,* 471–2; ethoxymethyl ethers with Montmorillonite cf. U.-A. Schaper, ibid. *1981,* 794–6.

p-*Toluenesulfonic acid* — *TsOH*

Disproportionation of acetals ←

with sulfanilic acid cf. *23,* 258; with *p*-toluenesulfonic acid, formals, s. B.C. Barot, H.W. Pinnick, J. Org. Chem. *46,* 2981–3 (1981).

Carboxylic acid esters from α,α-dialkoxynitriles $C(OR)_2CN$ → COOR

s. *37,* 794

Potassium peroxomonosulfate/potassium sulfate/ potassium hydrogen sulfate *$KHSO_5/K_2SO_4/KHSO_4$*

α,β-Ethylene-γ-hydroxyketones from (1,3-dien)olethers C(OH)−C=C−CO

with *m*-chloroperoxybenzoic acid as oxidant cf. *31,* 226; with $KHSO_5$ in the presence of K_2SO_4 and $KHSO_4$ s. S.N. Suryawanshi, P.L. Fuchs, Tetrahedron Letters *22,* 4201–4 (1981).

Iodine s. under Me_3SiI *I_2*

Sodium periodate s. under RuO_4 *$NaIO_4$*

Triphenylcarbonium perchlorate *$Ph_3C^+ClO_4^-$*

Disaccharides from trityl ethers ←

and carbohydrate 1,2-O,O-(1-cyanoethylidene) derivs. cf. *36,* 231; s.a. Bioorg. Khim. *8,* 1133–42 (1982); carbohydrate monothioorthocarboxylic acid esters s. Carbohyd. Res. *99,* 189–93 (1982).

Benzylhexadecyldimethylammonium periodate s. under OsO_4 *$R_4N^+IO_4^-$*

Hydrogen chloride (s.a. under H_2O_2) *HCl*

Phthalides via *o*-(Δ^2-oxazolin-2-yl)benzyl bromides ○

218.

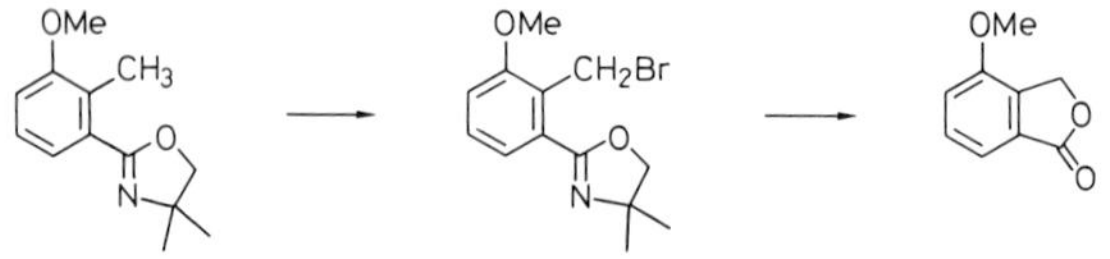

A soln. of 2-(2-methyl-3-methoxyphenyl)-4,4-dimethyl-2-oxazoline in carbon tetrachloride treated with N-bromosuccinimide and benzoyl peroxide, the mixture heated at reflux while irradiating with a 200 W light bulb for 1 hr., the solvent removed, the residue dissolved in 6 *M* HCl, and refluxed overnight → product. Y 96%. F.e., **also via *o*-(Δ^2-oxazolin-2-yl)benzyl alcohols,** s. A.I. Meyers, W.B. Avila, J. Org. Chem. *46,* 3881–6 (1981); direct method **from aldehydes** cf. M.S. Newman et al., Bull. Soc. Chim. Belges *88,* 871–3 (1979); s.a. J. Org. Chem. *45,* 2301–4 (1980).

Carboxylic acids from α-amino-α,β-ethylenenitriles C=C(N⋚)CN → COOH

s. *37,* 848

Cobalt(II) Schiff-base complex/oxygen *$[Co(salpr)]/O_2$*

***p*-Quinones from *p*-quinone methids** ←

219.

A mixture of startg. m. and the cobalt(II) Schiff-base complex [Co(Salpr)] in dichloromethane allowed to react in a stream of O_2 for 1 hr. at 20° → 2,6-di-*tert*-butyl-*p*-benzoquinone. Y 100%. F.e. and complexes s. A. Nishinaga et al., Tetrahedron Letters *21,* 4849–52 (1980).

Ruthenium tetroxide/sodium periodate/acetonitrile *$RuO_4/NaIO_4/CH_3CN$*

Improved oxidations with ruthenium tetroxide

Carboxylic acids from ethylene derivs. CH=CH → COOH

220. BuCH=CHBu → 2 BuCOOH

Oxidations catalysed by RuO_4 are significantly more rapid and milder in the presence of *acetonitrile* as cosolvent. E: A mixture of (E)-5-decene and 4·1 equivalents of

$NaIO_4$ in 3:2:2 water/acetonitrile/carbon tetrachloride treated with a little $RuCl_3 \cdot (H_2O)_n$, and the 2 phase mixture stirred vigorously for 2 hrs. at room temp. → pentanoic acid. Y 88%. – In 1:1 water/carbon tetrachloride, no reaction was detected. F.e. and reactions s. P.H.J. Carlsen et al., J. Org. Chem. *46,* 3936–8 (1981).

Osmium tetroxide/N–methylmorpholine N-oxide ←

α-Hydroxyaldehydes from enolethers C=CH(OR) → C(OH)CHO
Branched carbohydrates

221.
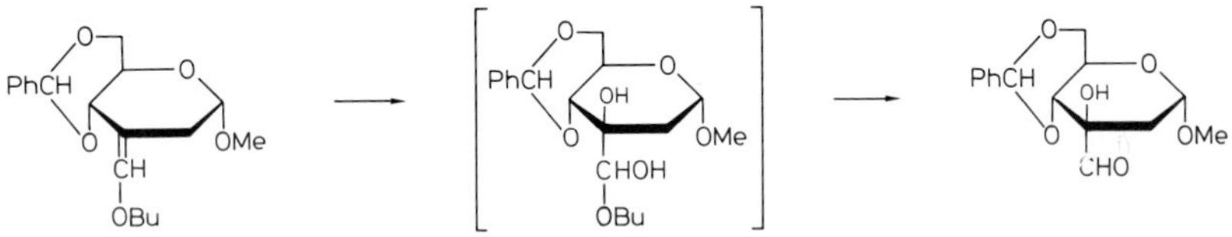

A mixture of the startg. methyl 4,6-O-benzylidene-2-deoxy-α-D-*erythro*-hexopyranosid-3-ulose enolether, N-methylmorpholine N-oxide, and a little OsO_4 in aq. acetone heated at 45° for 18 hrs. → methyl 4,6-O-benzylidene-2-deoxy-3-C-formyl-α-D-arabinohexopyranoside. Y 100%. F.e.s. I. Dyong, J. Weigand, W. Meyer, Tetrahedron Letters *22,* 2969–70 (1981).

Osmium tetroxide/benzylhexadecyldimethylammonium periodate/alkyllithium $OsO_4/R_4N^+IO_4^-/RLi$

Alcohols from ethylene derivs. via oxo compds. C=C → C(OH)R
One-pot conversion
s. *35,* 148; synthesis of alcohols using carbanions in place of hydride ion, and with benzylhexadecyldimethylammonium periodate as re-oxidant in place of $NaIO_4$, s. K. Inomata, Y. Nakayama, H. Kotake, Bull. Chem. Soc. Japan *53,* 565–6 (1980).

Elimination ⇑

Hydrogen ↑ OC⇑H

Electrolysis/poly-4-vinylpyridine hydrobromide ←

Catalysed electrochemical oxidation with polymer-based reagents
Oxo compds. from alcohols CHOH → CO

222.

A slowly stirred soln. of octan-2-ol in acetonitrile electrolysed at 50° between Pt electrodes in a divided cell containing poly-4-vinylpyridine hydrobromide beads at an applied potential of 10–30 V (60 mA) until 2.4 mF/mol passed (ca. 21.5 hrs.) → octan-2-one. Y 81%. – In this method, where work-up is uncomplicated by separation and recovery of electrolytes, the polymeric oxidant generated *in situ* functions catalytically, the reduced form being continuously re-oxidized as the current passes. F.e.s. J. Yoshida, R. Nakai, N. Kawabata, J. Org. Chem. *45,* 5269–73 (1980).

Potassium carbonate s. under $PdCl_2$ K_2CO_3

Pyridine s. under CuCl *C_5H_5N*

Trimethylamine N-oxide-antimony pentachloride *$(Me_3NO)SbCl_5$*

Copper-magnesium oxide *Cu-MgO*

Ketones from sec. alcohols in the vapor phase

s. *31,* 235; with a Ca-modified catalyst, mathematical model, cyclohexanone, cf. Vestsi Akad. Navuk BSSR, Ser. Khim. Navuk *1981* (6), 112 (Russ); Chem. Abstr. *96,* 103707 (1982).

Cupric sulfate s. under $KMnO_4$ *$CuSO_4$*

Cuprous chloride/pyridine *$CuCl/C_5H_5N$*

***o*-Quinones from *o*-quinols** ←

with MnO_2 cf. *18,* 306s*31;* with CuCl/pyridine s. T.R. Demmin, M.D. Swerdloff, M.M. Rogić, J. Am. Chem. Soc. *103,* 5795–804 (1981).

Cupric chloride s. under $S_2O_8^{2-}$ *$CuCl_2$*

Magnesium oxide s. under Cu *MgO*

Silica gel s. under IO_4^- *SiO_2*

Molecular sieve s. under 2,3-Dichloro-5,6-dicyanoquinone ←

Thallium(III) nitrate *$Tl(NO_3)_3$*

Quinone monoketals ←

with $FeCl_3/K_2CO_3$ cf. *35,* 159; with $Tl(NO_3)_3$, optically active compds., s. R.N. Warrener, P.S. Gee, R.A. Russell, J. Chem. Soc. Chem. Commun. *1981,* 1100–1.

Ammonium ceric nitrate/sodium hypobromite *$(NH_4)_2Ce(NO_3)_6/NaOBr$*

Ketones from sec. alcohols CHOH → CO

Preferential oxidation

223.

A suspension of NaOBr in 3:7 water/acetonitrile treated with the startg. diol and a little ammonium ceric nitrate, and the mixture heated at 80° for 30 min. → product. Y 89%. – Prim. alcohols were recovered unchanged, and carbon-carbon bond cleavage, which often occurs when stoichiometric amounts of Ce(IV) salts are used, was not observed. F.e. and limitations s. H. Tomioka, K. Oshima, H. Nozaki, Tetrahedron Letters *23,* 539–42 (1982).

Acetone s. under $RuH_2(PPh_3)_4$ *Me_2CO*

Diethyl azodicarboxylate/ triphenylphosphine/ethyl nitroacetate *$EtOOC{-}N{=}N{-}COOEt/Ph_3P/NO_2CH_2COOEt$*

Oxo compds. from alcohols via nitronic acid esters

224. $n\text{-}C_{13}H_{25}CH_2OH + O_2NCH_2CO_2Et \rightarrow [\, n\text{-}C_{13}H_{25}CH_2O\text{-}N(O){=}CHCO_2Et \,] \rightarrow n\text{-}C_{13}H_{25}CHO$

A soln. of diethyl azodicarboxylate in tetrahydrofuran added dropwise to a soln. of tetradecanol, triphenylphosphine, and ethyl nitroacetate in the same solvent at room temp., stirred for 2 hrs., and refluxed for a further 2 hrs. → tetradecanal. Y 85%. F.e.s. O. Mitsunobu, N. Yoshida, Tetrahedron Letters *22,* 2295–6 (1981).

Liver alcohol dehydrogenase ←

Asymmetric oxidation of diols to lactones O

225. 

γ-Lactones. Startg. diol treated with horse liver alcohol dehydrogenase for 3 days at 20° in a buffered soln. (pH 9) containing nicotinamide adenine dinucleotide coenzyme → (1S, 2R)-lactone. Y 72%; enantiomeric excess 100%. F.e., incl. **δ-lactones,** s. I.J. Jakovac et al., J. Chem. Soc. Chem. Commun. *1980,* 515–6; without expensive coenzyme cf. H. Ohta, H. Tetsukawa, N. Noto, J. Org. Chem. *47,* 2400–4 (1982). by asym. transfer dehydrogenation cf. Y. Ishii et al., Chem. Letters 1982, 1179–82.

Amberlite IRA 900 hypochlorite s. under Ca(OCl)$_2$ ←

Acetic acid s. under NaOCl *MeCOOH*

Dimethyl sulfoxide/oxalyl chloride *$Me_2SO/ClCOCOCl$*

Oxo compds. from alcohols CHOH → CO

s. *21,* 285s*33; β*-keto- from *β*-hydroxy-carbonyl compds. s. A.B. Smith, P.A. Levenberg, Synthesis *1981,* 567–71.

Dimethyl sulfoxide/molybdenum peroxide *Me_2SO/MoO_4*

with acetic anhydride cf. *21,* 285; with a little yellow molybdenum peroxide and other catalysts s. Y. Masuyama, A. Tsuhako, Y. Kurusu, Tetrahedron Letters *22,* 3973–6 (1981); benzylic oxidation of tricarbonylchromium complexes (with Ac_2O) s. S.G. Levine, B. Gopalakrishnan, ibid. *23,* 1239–40 (1982).

*Bis(*p*-methoxyphenyl) selenoxide* *(p-$MeOC_6H_4)_2SeO$*

Quinones from quinols ←

s. *37,* 492

*Bis(*p*-methoxyphenyl) tellurone* *(p-$MeOC_6H_4)_2TeO_2$*

Oxo compds. from alcohols CHOH → CO

s. *37,* 86

Iodoxybenzene *$PhIO_2$*

Aromatic aldehydes and *o*-quinones from benzyl alcohols and phenols, respectively ArCHOH → ArCO

s. *37,* 266

2,3-Dichloro-5,6-dicyanoquinone/molecular sieve ←

O,O-(*p*-Methoxybenzylidene) derivs. by oxidative ring closure 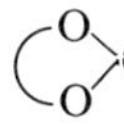

226. 

A soln. of startg. m. in anhydrous dichloromethane treated with 1.5 equivalents of 2,3-dichloro-5,6-dicyanoquinone in the presence of powdered 3 Å molecular sieves at room temp. for 1.5 hrs. → product. Y 80%. – **Removal of the protective group** can be achieved conventionally with 80% acetic acid, or the product may be oxidized further to the corresponding alkali-labile diol mono(*p*-methoxybenzoate). F.e. and protection as 3,4-dimethoxybenzylidene derivs. s. Y. Oikawa, T. Yoshioka, O. Yonemitsu, Tetrahedron Letters *23,* 889–92 (1982).

N-Chlorosuccinimide $\begin{matrix}CH_2CO\\CH_2CO\end{matrix}\!>\!NCl$

1-(Arylthio)lactones from lactones via (arylthio)carboxylic acids ○

227. 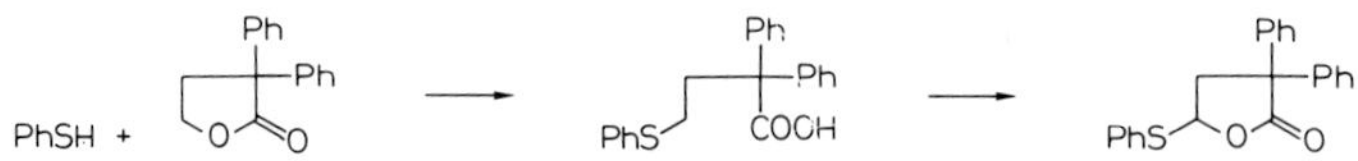

A mixture of startg. γ-lactone and benzenethiol treated with BBr_3 (cf. Synth. Meth. *35,* 372) → intermediate (Y 69%), in toluene treated with 1 equivalent N-chlorosuccinimide at room temp. under N_2 for 1 hr. → product (Y 85%). F.e., and conversion to **enollactones** by conventional oxidative elimination, s. M. Ochiai et al., Chem. Letters *1981,* 283–4; γ-**(arylthio)-**γ**-lactones** from γ-sulfinylcarboxylic acids cf. M. Watanabe et al., Bull. Chem. Soc. Japan *54,* 817–21 (1981).

N-Chloroisatin ←

Oxo compds. from alcohols CHOH → CO

s. *37,* 276

Oxalyl chloride s. under Me_2SO *ClCOCOCl*

Triphenylphosphine s. under EtOOC–N=N–COOEt Ph_3P

Oxidations with *organobismuth(V) compds.* ←

s. *35,* 154, 155; selectivity and f. oxidations s. Tetrahedron *37,* Suppl. No. 1, 73–9 (1981); **cleavage of glycols** with triphenylbismuth/N-bromosuccinimide in the presence of K_2CO_3 cf. J. Chem. Soc. Chem. Commun. 1981, 1232–3.

Antimony pentachloride s. $(Me_3NO)SbCl_5$ $SbCl_5$

Sodium persulfate/cupric chloride $S_2O_8{}^{2-}/CuCl_2$

Lactones from carboxylic acids ○

with $(NH_4)_2S_2O_8$ cf. *31,* 238; γ-lactones with Na persulfate, s. G.I. Nikishin, E.I. Troyansky, I.V. Svitanko, Izv. Akad. Nauk SSSR Ser. Khim. *1981,* 1436.

Chromium(V) complexes ←

Oxo compds. from alcohols CHOH → CO

s. *37,* 105.

Pyridinium dichromate $(C_5H_5NH)_2Cr_2O_7$

Aldehydes from prim. alcohols CH_2OH → CHO

s. *25,* 185s*34;* protected α-aminoaldehydes s. C.F. Stanfield, J.E. Parker, P. Kanellis, J. Org. Chem. *46,* 4797–8 (1981).

Pyridinium chlorochromate $C_5H_5NH^+CrO_3Cl^-$

Lactols from diols ○

s. *24,* 175s*36;* **review of pyridinium chlorochromate** s. G. Piancatelli, A. Scettri, M. D'Auria, Synthesis *1982,* 245–58.

Molybdenum peroxide s. under Me_2SO MoO_4

Oxoperoxomolybdenum complex ←

Oxo compds. from alcohols CHOH → CO

228. 

A soln. of oxoperoxobis(N-phenylbenzohydroxamato)molybdenum(VI) and cyclododecanol in 1,2-dichloroethane refluxed for 4 hrs. → cyclododecanone. Y 90%. F.e. and prepn. of complex, **also 2,3-oxidoalcohols from 2-ethylenealcohols,** s. H. Tomioka, et al., Tetrahedron Letters *21,* 4843–6 (1980).

Oxidations with *bromine/nickel(II) benzoate* $Br_2/(PhCOO)_2Ni$

229. $Me(CH_2)_5CH(OH)Me \rightarrow Me(CH_2)_5COMe$

Br_2 in acetonitrile added dropwise to a stirred mixture of octan-2-ol and anhydrous Ni(II) benzoate in the same solvent at 20° over 45 min., and worked up after 4 hrs. → octan-2-one. Y 97%. F.e., **including α-diketones from glycols,** s. M.P. Doyle, R.L. Dow, Synth. Commun. *10,* 881–8 (1980); **γ-lactones from 1,4-diols,** regiospecific ring closure, s.a. J. Org. Chem. *46,* 4806–8 (1981).

Sodium hypochlorite/acetic acid *NaOCl/MeCOOH*

Ketones from sec. alcohols

s. *37,* 230

Sodium hypobromite s. under $(NH_4)_2Ce(NO_3)_6$ *NaOBr*

Calcium hypochlorite/Amberlite IRA 900 hypochlorite ←

Solid–solid–liq. 3-phase catalysis

230. 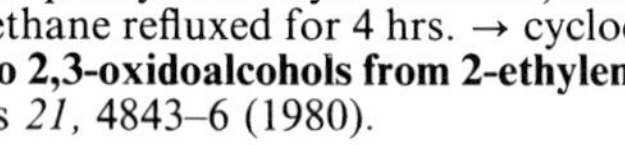

Preferential oxidation. Cycloheptanol and *n*-heptanol added to a well stirred suspension of $Ca(OCl)_2$ and a little Amberlite IRA 900 (hypochlorite form) in carbon tetrachloride keeping the temp. below 40°, and the mixture stirred at room temp. for 3 hrs. → cycloheptanone (Y 85%) and *n*-heptanol (Y 98%). – Work up is simple and the catalyst can be used repeatedly without any loss of activity. F.e. and prepn. of catalyst s. M. Schneider, J.-V. Weber, P. Faller, J. Org. Chem. *47,* 364–5 (1982); with $Ca(OCl)_2$ in acetonitrile/acetic acid cf. S.O. Nwaukwa, P.M. Keehn, Tetrahedron Letters *23,* 35–8 (1982); preferential oxidation with aq. NaOCl in acetic acid s. R.V. Stevens, K.T. Chapman, H.N. Weller, J. Org. Chem. *45,* 2030 (1980).

Periodate–silica gel $IO_4^- - SiO_2$

Quinones from quinols ←

s. *37,* 231

Oxidations with *polymer-based periodate and iodate* ←

231. 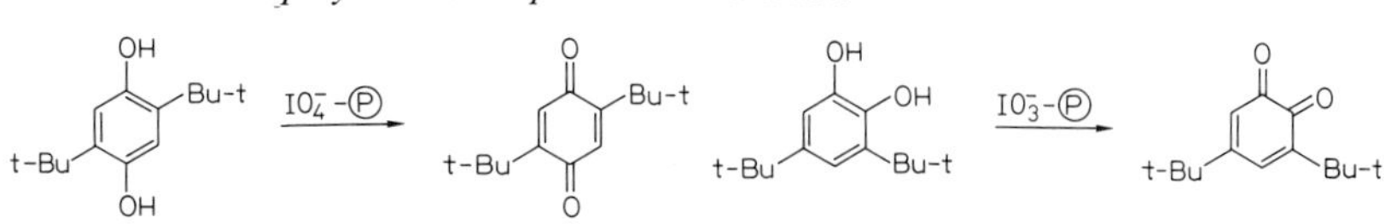

Polymer-supported quaternary ammonium periodates and iodates can be used to oxidize a variety of functions in both protic and aprotic solvents. E: **Quinones from quinols.** A soln. of 2,5-di-*tert*-butylquinol in acetone kept in contact with a column of Amberlyst A26 (periodate form) for 1 hr. at 20° → 2,5-di-*tert*-butyl-1,4-

benzoquinone. Y 97%. – Similarly, 3,5-di-*tert*-butylcatechol with Amberlyst A26 (iodate form) → 3,5-di-*tert*-butyl-*o*-benzoquinone. Y 100%. F.e., **also cleavage of glycols,** prepn. of **sulfoxides from thioethers,** and f. oxidations s. C.R. Harrison, P. Hodge, J. Chem. Soc. Perkin Trans. I *1982,* 509–11; oxidations with periodate-silica gel, incl. prepn. of **azo from hydrazo compds.,** cf. ibid. *1981,* 2970–3.

Poly-4-vinylpyridine hydrobromide s. under ↯ ←

Selective oxidations with *sodium permanganate monohydrate* *$NaMnO_4.H_2O$*

Ketones from sec. alcohols CHOH → CO

232. $Me(CH_2)_5CH(OH)Me \rightarrow Me(CH_2)_5COMe$

2-Octanol in hexane stirred with solid $NaMnO_4$ monohydrate at reflux for 2.5 hrs. → 2-octanone. Y 95%. – Alkenes, alkynes, epoxides, and amides were unaffected under these conditions, and allyl alcohols reacted slowly. F.e., also carboxylic acids from prim. alcohols, sulfones from thioethers, and nitro compds. from prim. amines, s. F.M. Menger, C. Lee, Tetrahedron Letters *22,* 1655–6 (1981).

Potassium permanganate/cupric sulfate *$KMnO_4/CuSO_4$*

s. *35,* 362s*35; α,β*-ethyleneketones, also with $KMnO_4$ on bentonite, s. N.A. Noureldin, D.G. Lee, Tetrahedron Letters *22,* 4889–90 (1981).

Oxidations with *bis(pyridine)silver(I) permanganate* *$[Ag(C_5H_5N)_2]MnO_4$*

Oxo compds. from alcohols

233. $PhCH_2OH \xrightarrow{[Ag(C_5H_5N)_2]MnO_4} PhCHO$

A stirred soln. of benzyl alcohol in benzene treated portionswise with bispyridinesilver(I) permanganate (prepared by adding pyridine to aq. $KMnO_4$ followed by $AgNO_3$), and the mixture refluxed for 30 min. → benzaldehyde. Y 100%. – This mild and selective oxidant, readily soluble in polar solvents, may be stored for several days in the dark, and is an useful alternative to MnO_2. F.e., **also sym. azo compds. from amines,** s. H. Firouzabadi, B. Vessal, M. Naderi, Tetrahedron Letters *23,* 1847–50 (1982).

Ferric nitrate–clay ←

... under mild conditions

234. $PhCH_2OH \rightarrow PhCHO$

$Fe(NO_3)_3$ impregnated on K 10 bentonite clay added in small portions with vigorous stirring to a soln. of benzyl alcohol in pentane (or hexane), and stirring continued under reduced pressure for 90 min. after which evolution of NO_2 ceased → benzaldehyde. Y 81%. F.e., limitations, and prepn. of reagent s. Cornelis A, Laszlo P., Synthesis *1980,* 849–50.

Oxidations with *cobalt(III) nitro-complexes* ←

235. $PhCH_2OH \rightarrow PhCHO$

From alcohols. A soln. of benzyl alcohol in 1,2-dichloroethane added under argon to a stoichiometric amount of nitro(pyridine)tetraphenylporphyrinatocobalt(III), treated with *BF_3-etherate,* and the stoppered mixture heated at 60° for 45 min. → benzaldehyde. Y 99%. F.e., also with molecular O_2 catalysed by cobalt nitro-complexes, s. B.S. Tovrog et al., J. Am. Chem. Soc. *103,* 3522–6 (1981); **from ethylene derivs.** in the presence of bis(benzonitrile)dichloropalladium(II) s. R.M. Ranshel, J.L. Villafranca ibid. *102,* 6616–8 (1980).

Nickel(II) benzoate s. under Br_2 *$(PhCOO)_2Ni$*

Dichlorotris(triphenylphosphine)ruthenium(II)/oxygen $RuCl_2(PPh_3)_3/O_2$

Oxo compds. from alcohols

with N-methylmorpholine-N-oxide as re-oxidant cf. *15,* 261s*31;* under O_2, α,β-ethyleneoxo compds., s. M. Matsumoto, S. Ito, J. Chem. Soc. Chem. Commun. *1981,* 907–8; preferential oxidation of prim. alcohols without re-oxidant s. H. Tomioka et al., Tetrahedron Letters *22,* 1605–8 (1981).

Dihydridotetrakis(triphenylphosphine)ruthenium/acetone $RuH_2(PPh_3)_4/Me_2CO$

Lactones from diols ○

Regiospecific ring closure by homogeneous dehydrogenation

236.

A mixture of startg. diol, 3 equivalents of acetone (as H-acceptor), and a little dihydridotetrakis(triphenylphosphine)ruthenium in dry toluene heated at 180° under argon for 3 hrs. → product. Y 71%. F.e., **also carboxylic acid esters from 2 prim. alcohol molecules,** s. S. Murahashi et al., Tetrahedron Letters *22,* 5327–30 (1981).

Palladium(II) chloride/potassium carbonate $PdCl_2/K_2CO_3$

Ketones from sec. alcohols CHOH → CO

with $PdCl_2$/Na acetate cf. *33,* 252; with K_2CO_3 as base in carbon tetrachloride, also carboxylic acid esters from prim. alcohols, s. H. Nagashima, J. Tsuji, Chem. Letters *1981,* 1171–2.

Oxygen ↑ OC⇑O

Without additional reagents *w.a.r.*

1,2,5-Oxadiazin-6-one ring from *o*-nitrosourethans ○

237.

A suspension of 1-(2-chlorophenyl)-3-methyl-4-nitroso-5-ethoxycarbonylaminopyrazole in petroleum ether boiled for 5 hrs. in a flask fitted with a Vigreux column with periodic addition of further solvent to maintain the volume → product. Y 92%. F.e.s. P. Giori et al., Farmaco, Ed. Sci. *36,* 1019–28 (1981).

Sodium hydride *NaH*

(s.a. under Bis[1,2-bis(diphenylphosphino)ethane]palladium(0))

Δ^4-2-Oxazolones from α-carbalkoxyaminoketones

238.

with N-alkylation. A soln. of the startg. ketone in dimethylformamide added under N_2 with stirring to a suspension of 2 equivalents NaH in the same solvent, heated at 45–50° for 30 min., methyl iodide added cautiously with cooling, and the mixture kept at room temp. for 30 min. → product. Y >80%. F.e., **also C-α-alkylation of α-acylaminoketones,** and hydrolysis to **alkylated α-aminoketones,** s. R.E. Bowman, J. Chem. Soc. Perkin Trans. I *1980,* 2126–33.

Sodium hydride/N-tosylimidazole ←

Oxido compds. from glycols CHOHCHOH → 
1,2-Ditritiooxido compds.
s. *37*, 27

Sodium methoxide *NaOMe*

Prepn. of chiral crown polyethers ←
s. *14*, 326s*36*; **review of chiral crown heterocyclics** s. S.T. Jolley, J.S. Bradshaw, R.M. Izart, J. Heterocyc. Chem. *19*, 3–19 (1982).

n-*Butyllithium s. under TsCl* n-*BuLi*

Potassium hydrogen carbonate s. under Polymer-based phosphonium salts $KHCO_3$

Caesium carbonate Cs_2CO_3

Macrocyclic lactones from mesyloxycarboxylic acids ○
s. *37*, 241

Triethylamine s. under Ph_3P Et_3N

Diethyl azodicarboxylate/triphenylphosphine $EtOOCN{=}NCOOEt/Ph_3P$

Cyclic ethers from diols
oxido compds. s. *34*, 88s*36*; also from 1,4- and 1,5-diols s. C.N. Barry, S.A. Evans, J. Org. Chem. *46*, 3361–4 (1981).

Lactones from hydroxycarboxylic acids
Macrocyclic lactones
with PPh_3/2,2′-bipyridyl disulfide cf. *30*, 168; with PPh_3/diethyl azodicarboxylate, s. R.J. Bass et al., J. Chem. Soc. Perkin Trans. I *1981*, 124–31.

*1-Phenyl-Δ^2-tetrazoline-5-thione/*tert*-butyl isocyanide* ←

Lactones from hydroxycarboxylic acids ○

239. 

Macrocyclic lactones. A suspension of 1-phenyl-Δ^2-tetrazoline-5-thione in dry toluene treated with *tert*-butyl isocyanide, stirred at room temp. for 5 min. until homogeneous, added to ω-hydroxypendadecanoic acid, the mixture dil. with toluene, and refluxed for 30 min. → pentadecanolide. Y 80%. – The method is free from work-up complications unlike many of the conventional macrolide ring closures. F.e.s. U. Schmidt, M. Dietsche, Angew. Chem. Intern. Ed. *20*, 771–2 (1981).

Hexachloroethane s. under Ph_3P ←

Acetyl chloride *MeCOCl*

1,1′-Spirodilactones from ketodicarboxylic acids

240. 

Startg. ketodicarboxylic acid dissolved in neat acetyl chloride at room temp., and worked up when evolution of gas ceased → spiro[5.5]divalerolactone. Y 66%. F.e.s. T.R. Hoye, D.R. Peck, P.K. Trumper, J. Am. Chem. Soc. *103*, 5618–20 (1981).

Polymer-based phosphonium salt/potassium hydrogen carbonate ←

Macrocyclic lactones from mesyloxycarboxylic acids
3-Phase catalysis

241.

12-Methanesulfonyloxydodecanoic acid added in one portion to a mixture of toluene, a little 1% cross-linked polymer-based phosphonium mesylate, and aq. $KHCO_3$, and the resulting 3-phase mixture heated for 20 hrs. at 90° → product. – The method avoids the use of conventional high-dilution techniques. F.e.s. S.L. Regen, Y. Kimura, J. Am. Chem. Soc. *104,* 2064-5 (1982); with caesium carbonate in dimethylformamide cf. M. Barbier, J. Chem. Soc. Chem. Commun. *1982,* 668–9.

Triphenylphosphine s.a. under EtOOC–N=N–COOEt Ph_3P

Triphenylphosphine/hexachloroethane/triethylamine $Ph_3P/C_2Cl_6/Et_3N$

1,3-Oxazin-6-one ring from ***o*****-acylaminocarboxylic acid esters** ○

242.

Triethylamine added to a soln. of startg. N-acylaminoester, triphenylphosphine, and hexachloroethane in methylene chloride under argon, and stirred 2 hrs. under reflux → 2-(4-methoxyphenyl)-5,6,7,8-tetrahydro-4*H*-[1]benzothieno[2,3-*d*][1,3]oxazin-4-one. Y 98%. F.e.s. D. Achakzi et al., Chem. Ber. *114,* 3188–94 (1981).

N-Tosylimidazole s. under NaH ←

p-*Toluenesulfonyl chloride*/n-*butyllithium* *TsCl*/n-*BuLi*

Oxetanes from 1,3-diols

with $NaOH/H_2SO_4$ cf. *15,* 262; with tosyl chloride/*n*-butyllithium s. P. Picard et al., Synthesis *1981,* 550–1.

p-*Toluenesulfonic acid* *TsOH*

Furan ring from ***β,γ*****-oxidoketones** ○

with BF_3 cf. *20,* 197; with TsOH s. R.A. Cormier, M.D. Francis, Synth. Commun. *11,* 365–9 (1981).

N-Cyclohexyl-N'-2-(N-methylmorpholinio)ethylcarbodiimide p-*toluenesulfonate*

Δ^2-5-Oxazolones from ***α*****-acylaminocarboxylic acids**

with ethylchloroformate/triethylamine cf. *25,* 189; with N-cyclohexyl-N'-2-(N-methylmorpholinio)ethylcarbodiimide *p*-toluenesulfonate s. C.F. Hoyng, M.G. McKenna, D.L. Walters, Synthesis *1982,* 191–3.

Dimethyl sulfate Me_2SO_4

Polysubst. tetrahydrofurans from 1,4-diols ○
Intramolecular assistance by arylthio groups

243.

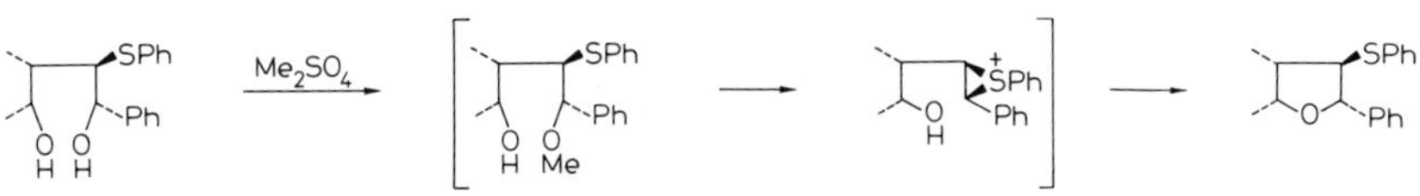

Startg. diol treated with dimethyl sulfate in methylene chloride at 0° for 10 min. under

N_2 → product. Y 85%. – Participation of the phenylthio group after initial alkylation ensures net **retention of configuration** at the 4 contiguous carbon atoms. F.e. and methylating agents, also from 1,4-diol monoethers, s. D.R. Williams, J.G. Phillips, B.A. Barner, J. Am. Chem. Soc. *103,* 7398–9 (1981).

Sulfuric acid *H_2SO_4*

5-Hydroxybenzofurans from *p*-quinones via regiospecific [2+2]-cycloaddition of enolesters ←

244.

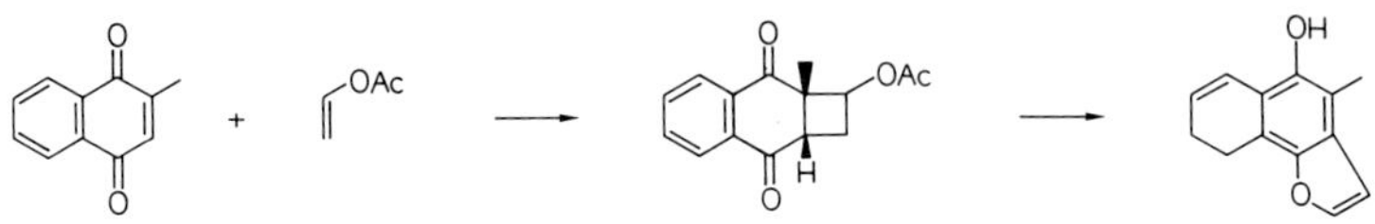

A soln. of 2-methyl-1,4-naphthoquinone and vinyl acetate in benzene irradiated through Pyrex with a 450 W Hanovia high-pressure Hg-lamp for 1 hr. at room temp. → adduct (Y 98%), in tetrahydrofuran treated with 5% aq. H_2SO_4, and refluxed under N_2 for 16 hrs. → product (Y 92%). F.e.s. H.-J. Liu, W.H. Chan, Can. J. Chem. *58,* 2196–8 (1980).

Hydrogen chloride *HCl*

Chromans from *o*-γ-hydroxyphenols ○

with P_2O_5 cf. *12,* 341; with HCl, retention of configuration and f. methods, s. N. Cohen, R.J. Lopresti, G. Neukom, J. Org. Chem. *46,* 2445–50 (1981).

Furans from 4-alkoxy-2,3-dienols

s. *37,* 610

5,6-Dihydro-2-pyrones

245.

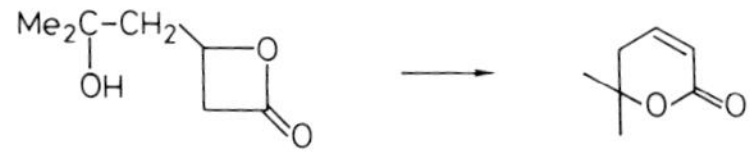

from δ-hydroxy-β-lactones. A soln. of 4-(2-hydroxy-2-methylpropyl)oxetan-2-one in ether satd. with HCl with ice-cooling, and the mixture allowed to stand at room temp. for 2 days. → 5,6-dihydro-6,6-dimethyl-2*H*-pyran-2-one. Y 65%. F.e.s. T. Kato et al., Chem. Pharm. Bull. *29,* 1624–8 (1981); **from δ,ε-ethylene-β-hydroxycarboxylic acids** cf. T. Fujita et al., J. Appl. Chem. Biotechnol. *32,* 421–6 (1982).

Bis[1,2-bis(diphenylphosphino)ethane]palladium(0)/sodium hydride ←

Spirocyclics from cyclic acoxy-2-ethylenes

246.

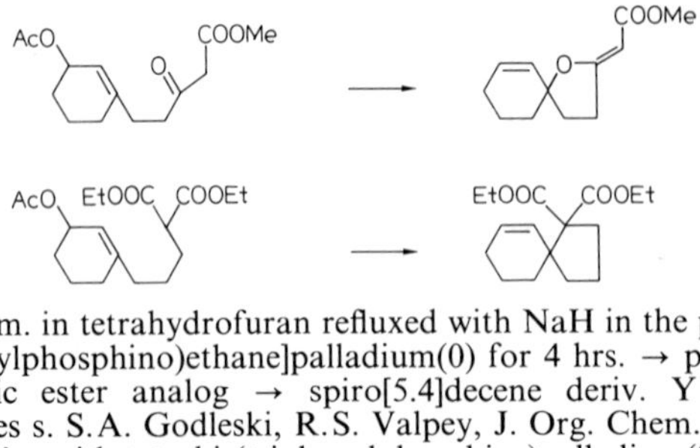

A soln. of startg. m. in tetrahydrofuran refluxed with NaH in the presence of a little bis[1,2-bis(diphenylphosphino)ethane]palladium(0) for 4 hrs. → product. Y 63%. – Similarly, malonic ester analog → spiro[5.4]decene deriv. Y 66%. F.e. incl. spiro[5.5]undecenes s. S.A. Godleski, R.S. Valpey, J. Org. Chem. *47,* 381–3 (1982); **spiro-N-heterocyclics** with tetrakis(triphenylphosphine)palladium(0)/triethylamine s. Tetrahedron Letters *22,* 2247–50 (1981).

Nitrogen ↑ OC ⇑ N

Without additional reagents *w.a.r.*

1,3-O,N- and 1,3-S,N-heterocyclics from diamines and carboxylic acid chlorides via N-ω-acylaminopyridinium salts ○

247.

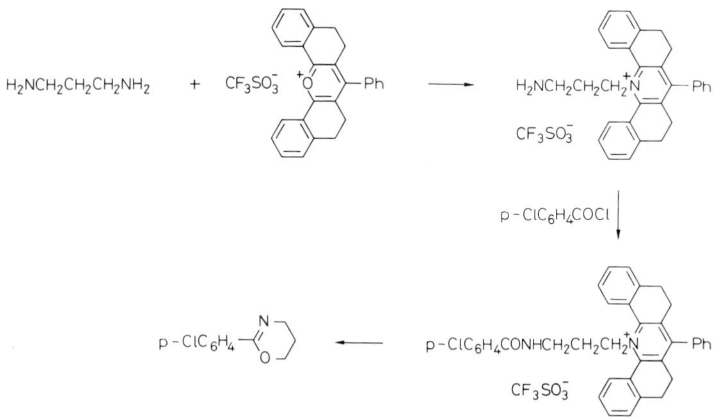

5,6-Dihydro-4*H*-1,3-oxazines. A mixture of 1,3-diaminopropane and triethylamine added to a suspension of 7-phenyl-5,6,8,9-tetrahydrodibenzo[*c,h*]xanthylium trifluoromethanesulfonate in methylene chloride, stirred for 1 hr. at 20°, 10 drops of acetic acid added, and stirring continued for 3 hrs. → intermediate N-(3-aminopropyl)pyridinium salt (Y 92%) in methylene chloride treated dropwise with triethylamine followed by *p*-chlorobenzoyl chloride, and stirred for 3 hrs. at 20° → intermediate N-(3-acylaminopropyl)pyridinium salt (Y 77%) in dioxane refluxed for 4 hrs. → product (Y 92%). F.e., incl. Δ^2-oxazolines, S-analogs, and 2-amino-derivs. from isocyanates, s. A.R. Katritzky et al., Tetrahedron *37*, 2383–90 (1981).

1,2,4-Oxadiazoles from tetrazol-1-yloximes

248.

A soln. of startg. tetrazole in *p*-xylene refluxed for 15–30 min. → product. Y 100%. F.e., also pyrolysis without solvent, s. J. Plenkiewicz, T. Zdrojewski, Bull. Soc. Chim. Belges *90*, 193–4 (1981).

Diisobutylaluminium hydride/hydrogen chloride *i-Bu₂AlH/HCl*

Furans from α,β-ethylene-γ-hydroxynitriles

249.

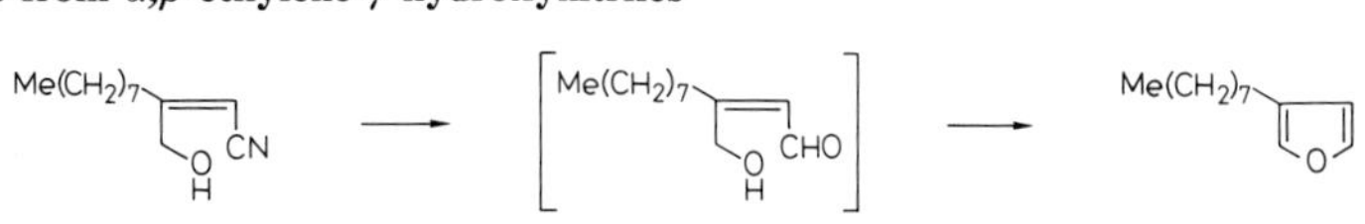

A soln. of 3-hydroxymethyl-2-undecenenitrile in ether treated with a soln. of diisobutylaluminium hydride in hexane at –10° for 1 hr., and the crude aldehyde stirred with 1 *M* HCl in ether at room temp. for 2 hrs. → 3-octylfuran. Y 74%. F.e.s. T. Mandai et al., Tetrahedron Letters *22*, 2187–90 (1981).

Phosphate buffer ←

2-Oxazolidones from 2-hydroxy-N′-nitrosoureas

250.

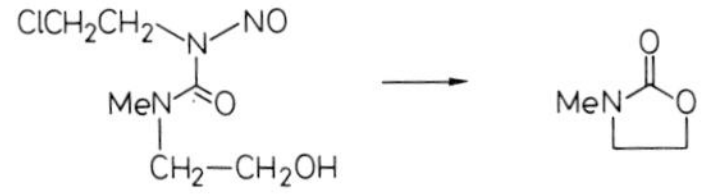

A soln. of 1-(2-chloroethyl)-3-(2-hydroxyethyl)-3-methyl-1-nitrosourea in 1 *M* phosphate buffer (pH 7.4) stirred for 30 min. at 5° → 3-methyloxazolidin-2-one. Y 92%. F.e. and preparation of startg. m. s. K. Tsujihara et al., Chem. Pharm. Bull. *29,* 2509–15 (1981).

Hypophosphorous acid s. under $(MeCOO)_2Cr$ H_3PO_2

p-*Toluenesulfonic acid* *TsOH*

Macrocyclic lactones from protected halogenhydrins via 2-ω-hydroxyoxazoles and hydroxytriacylamines Oxazoles as protected carboxylic acids

251.

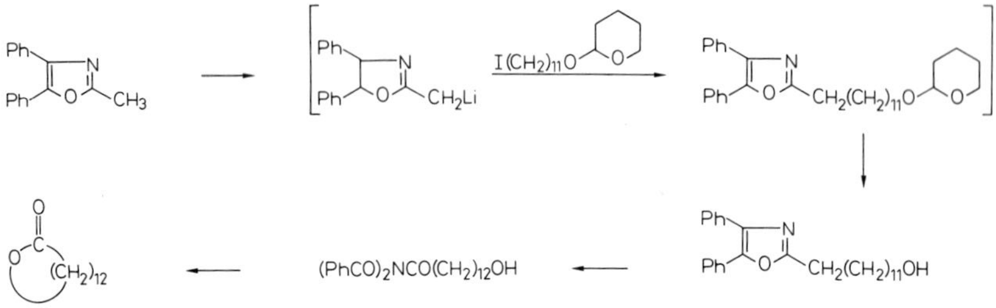

A soln. of 2-methyl-4,5-diphenyloxazole in tetrahydrofuran added to a soln. of Li diisopropylamide in the same solvent at –78° over 30 min., stirred for an additional 1 hr., treated with 2-(11-iodoundecyloxy)tetrahydropyran at –78°, worked up after 10 min., and the residue hydrolysed in tetrahydrofuran with 2 *M* HCl at 60° for 3 hrs. → hydroxyoxazole (Y 57%), in methylene chloride oxygenated in the presence of a little Sensitox while irradiating the mixture with a 650 W tungsten-halogen lamp for 30 min., solvent replaced by benzene, and added over 24 hrs. to refluxing benzene containing a little *p*-touenesulfonic acid → 13-tridecanolide (Y 75%). F.e.s. H.H. Wasserman, R.J. Gambale, M.J. Pulwer, Tetrahedron Letters *22,* 1737–40 (1981); *a*-lateral metallation of oxazoles s.a. J. Org. Chem. *46,* 1410 (1981).

Chromium(II) acetate/hypophosphorous acid $(MeCOO)_4Cr_2/H_3PO_2$

Ketones from Δ^3-5-oxazolones C

with $Ca(OH)_2$ cf. *33,* 262; with Cr(II) acetate/H_3PO_2 s. U. Niewöhner, W. Steglich, Angew. Chem. Intern. Ed. *20,* 395–6 (1981).

Hydrogen chloride s. under i-Bu_2AlH *HCl*

Halogen ↑ OC ⇑ Hal

Sodium hydride *NaH*

Δ^2-Oxazolines from 2-halogenacylamines O

with NaOH cf. *10,* 241; with NaH s. F. Franco, J.M. Muchowski, J. Heterocyc. Chem. *17,* 1613 (1980).

*Sodium hydroxide/tetra-*n*-butylammonium hydrogen sulfate* *NaOH/*n*-$Bu_4N^+HSO_4^-$*

Cyclic ethers from *tert*-aminoalcohols via bromhydrins ○

252.

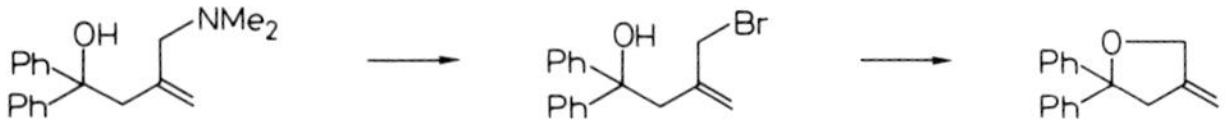

3-Methylenetetrahydrofurans. Cyanogen bromide in methylene chloride added to a soln. of 4-(dimethylamino)-3-methylene-1,1-diphenylbutanol in the same solvent at 0°, and stirred for 15 min. at this temp. then for 16 hrs. at room temp. → intermediate (Y 85%), in methylene chloride stirred vigorously at room temp. with 50% NaOH and a little tetrabutylammonium hydrogen sulfate for 16 hrs. → 4-methylene-2,2-diphenyltetrahydrofuran (Y 89%). F.e.s. J.J. Fitt, H.W. Gschwend, J. Org. Chem. *46,* 3349–52 (1981).

Potassium hydroxide *KOH*

Oxido compds. from chlorhydrins CClC(OH) → epoxide

s. *6,* 306; optically active compds. s. B. Koppenhoefer, R. Weber, V. Schurig, Synthesis *1982,* 316-8.

Potassium hydroxide/methanol *KOH/MeOH*

3-Oxetanone ring

s. *33,* 265; f. method cf. M. Pons, S.S. Simons, J. Org. Chem. *46,* 3262–4 (1981).

Silver(I) trifluoroacetate *CF_3COOAg*

Regiospecific bromocyclopropane ring opening with simultaneous ring closure ←

253.

to 1-vinyllactones. A soln. of startg. cyclopropane in 2,2,2-trifluoroethanol treated with Ag trifluoroacetate at 25° for 19 hrs. → product. Y 85%. F.e., **also ring closure to cyclic 2-ethyleneethers,** s. R.L. Danheiser et al., Tetrahedron Letters *22,* 4205–8 (1981).

Methanol s. under KOH *MeOH*

*Tetra-*n*-butylammonium hydrogen sulfate s. under NaOH* n-$Bu_4N^+HSO_4^-$

Sulfur ↑ OC⇑S

Sodium hydroxide *NaOH*

Intramolecular nucleophilic 3-substitution of indoles ○

N-Sulfonyl as leaving group

254.

A soln. of startg. ketone in methanol and aq. 2 *M* NaOH heated on a steam bath for 5 min. → 6*H*-pyrido[4′,3′:5,6]oxepino[3,2-*b*]indol-5(12*H*)-one. Y 83%. F.e.s. M.G. Beal et al., J. Chem. Soc. Perkin Trans. I *1982,* 435–9.

Potassium tert-*butoxide* t-*BuOK*

Crown polyether template ring closure
Macrocyclic lactones from hydroxythiolic acid esters

255.

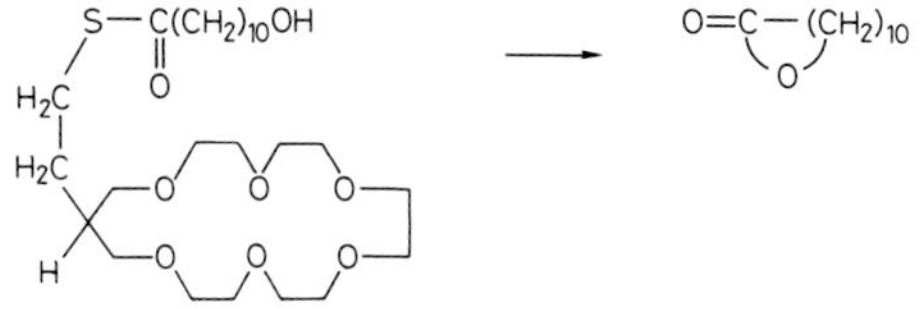

A soln. of startg. m. in benzene and K *tert*-butoxide in tetrahydrofuran added simultaneously by syringe over 3.2 hrs. to a high-dilution vessel containing activated 4Å molecular sieves in freshly distilled, stirred tetrahydrofuran cooled with tap water → 11-undecanolide. Y 77%. – Medium dilution method gave a lower yield (63%). F.e.s. W.H. Rastetter, D.P. Phillion, J. Org. Chem. *46,* 3209–14 (1981).

Remaining Elements ↑ OC⥣Rem

Without additional reagents *w.a.r.*

1,4-Cyclohexadione ring from 1,2-di(siloxy)cyclobut-1-enes C
via 1,4-di(siloxy)bicyclo[2.2.0]hexane ring
Stereospecific conversion

256.

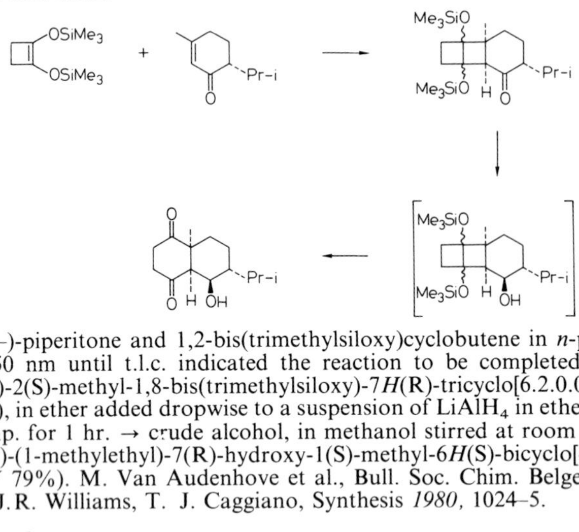

A soln. of (–)-piperitone and 1,2-bis(trimethylsiloxy)cyclobutene in *n*-pentane irradiated at 350 nm until t.l.c. indicated the reaction to be completed → 5(R)-(1-methylethyl)-2(S)-methyl-1,8-bis(trimethylsiloxy)-7*H*(R)-tricyclo[6.2.0.0^{2,7}]decan-6-one (Y 70%), in ether added dropwise to a suspension of $LiAlH_4$ in ether, and stirred at room temp. for 1 hr. → crude alcohol, in methanol stirred at room temp. for 24 hrs. → 8(R)-(1-methylethyl)-7(R)-hydroxy-1(S)-methyl-6*H*(S)-bicyclo[4.4.0]decane-2,5-dione (Y 79%). M. Van Audenhove et al., Bull. Soc. Chim. Belges *90,* 255–62 (1981); s.a. J.R. Williams, T. J. Caggiano, Synthesis *1980,* 1024–5.

Sodium alkoxides *NaOR*

Ketoenolethers from cyclic α,β-ethylene-γ-hydroxyselenones
Grob-type fragmentation

257.

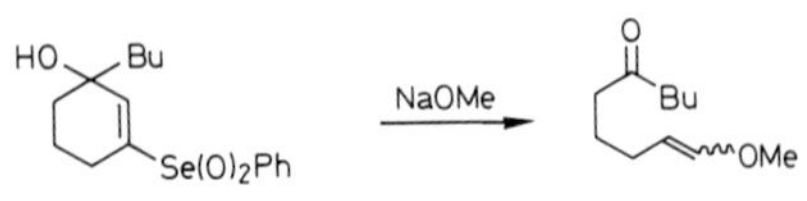

1-Butyl-3-(phenylselenonyl)-2-cyclohexen-1-ol treated with Na methoxide in methanol overnight at room temp., and heated at 45° for 3 hrs. → 10-methoxy-9-decen-5-one. Y 86%. F.e., **also acetyleneketones** with NaH, s. M. Shimizu, R. Ando, I. Kuwajima, J. Org. Chem. *46,* 5246–8 (1981).

Ammonium ceric nitrate/sodium bromate *$(NH_4)_2Ce(NO_3)_6/NaBrO_3$*

Oxo compds. from alkoxysilanes CH(OSi⋮) → CO

s. *37,* 264

Benzenesulfenyl chloride/silver nitrate *PhSCl/AgNO$_3$*

α,β-Ethylenealdehydes
from 3-hydroxyenesilanes C(OH)C=CHSi⋮ → C=C–CHO

258.

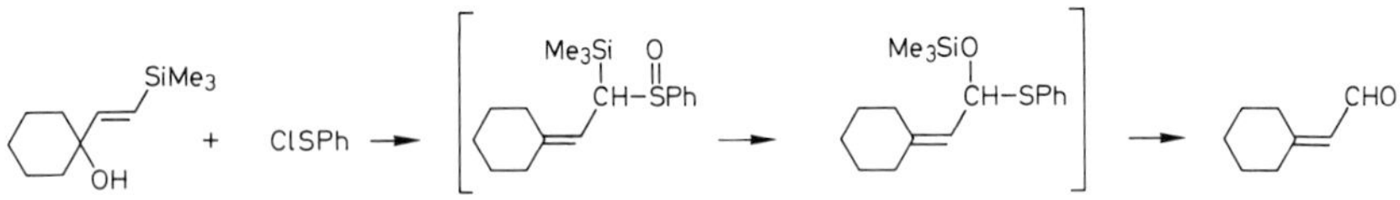

Benzenesulfenyl chloride added dropwise to a stirred soln. of 1-(2-trimethylsilylvinyl)-1-hydroxycyclohexane and *triethylamine* in dry ether at –25°, poured into water, extracted with ether, and the oil obtained treated with $AgNO_3$ in aq. acetonitrile → product. – The intermediate *β,γ*-**ethylene-*α*-silylsulfoxides** are isolable at low temp. F.e.s. I. Cutting, P.J. Parsons, Tetrahedron Letters *22,* 2021–2 (1981).

*Tetra-*n-*butylammonium fluoride* n-$Bu_4N^+F^-$

Cyclic ethers from siloxymesylates
2,3-Dihydro-4*H*-1,4-benzoxazine ring ○

259.

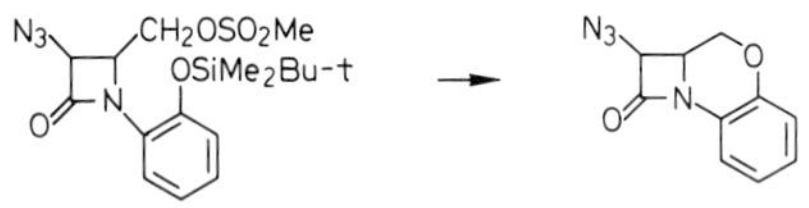

A soln. of startg. mesylate in tetrahydrofuran at 0° treated dropwise with a soln. of tetra-*n*-butylammonium fluoride in the same solvent during 15 min. → product. Y 100%. F.e.s. G. Just, Y.S. Tsantrizos, A. Ugolini, Can. J. Chem. *59,* 2981–7 (1981).

Carbon ↑ OC⇑C

Without additional reagents *w.a.r.*

Regiospecific synthesis of *α*-aminoketones CH(OH)C–CN → COCH
from amines and 2 different aldehyde molecules
via *α*-aminonitriles and *α*-amino-*β*-hydroxynitriles

260.

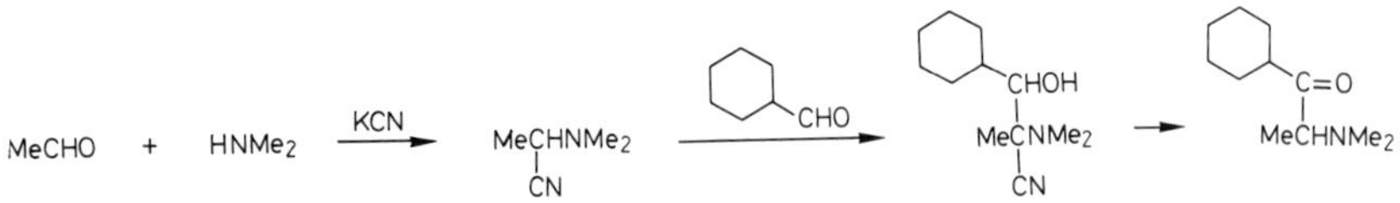

An aq. soln. of dimethylamine hydrochloride treated with aq. KCN at 0° followed by 1.1 equivalents of acetaldehyde at 25° during 2 hrs. → *α*-aminonitrile, in tetrahydrofuran treated with Li diisopropylamide at –78° during 0.5–5 hrs., cyclohexanecarboxaldehyde added, quenched at the same temp. with satd. NH_4Cl soln., extracted with cold ether, and the adduct distilled with elimination of HCN → product. Overall Y 72%. F.e.s. D. Enders, H. Lotter, Tetrahedron Letters *23,* 639–42 (1982).

Without additional reagents *w.a.r.*

Oxo compds. from 1,3-dioxolan-4-ones C

Flash vacuum pyrolysis

261. 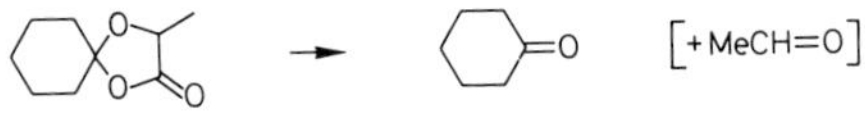

2,2-Pentamethylene-5-methyl-1,3-dioxolan-4-one flash vacuum pyrolysed at 800° and 0.5 torr → cyclohexanone. Y 93%. F.e.s. T.B. Cameron, F.M. El-Kabbani, H.W. Pinnick, J. Am. Chem. Soc. *103,* 5414–7 (1981).

Aldehydes from glycidic acid *tert*-butyl esters 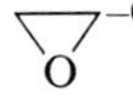 –COOBu-*t* → CHO

s. *26,* 859; also with K *tert*-butoxide/acetic acid s. R.K. Bansal, K. Sethi, S.K. Jain, Indian J. Chem. *20B,* 121–4 (1981).

Potassium hydroxide *KOH*

3,4-Oxidoalcohols from cyclic 2-bromocarbonic acid esters ←
s. *37,* 450

Potassium carbonate K_2CO_3

Oxazolo[4,5-*b*]quinolin-2(4*H*)-ones from 4a,5-epoxy-5-deazaflavins

262. 

Startg. 4a,5-epoxy-3-methyl-5-deazaflavin in 10% aq. K_2CO_3 stirred at room temp. for 5 hrs. → product. Y 93%. F.e.s. F. Yoneda, Y. Sakuma, Tetrahedron Letters *22,* 3977–80 (1981).

Sodium bromide s. under Me_3SiCl *NaBr*

Silver(II) oxide *AgO*

Quinones from quinolethers ←
s. *23,* 310s*31;* isochroman-5,8-diones s. J.I. Retamal et al., Synth. Commun. *12,* 279–85 (1982).

Mercuric acetate $(MeCOO)_2Hg$

Carbonic acid esters from orthoformic acid esters $CH(OR)_3 \rightarrow CO(OR)_2$

263. $(EtO)_3CH \rightarrow (EtO)_2C=O$

A mixture of triethoxymethane and a little Hg acetate heated in a stainless-steel autoclave for 2 hrs. at 190° → diethyl carbonate. Y 72%. F.e.s. S.M. Kalashnikov et al., Zh. Prikl. Khim. *55,* 484 (1982).

Triphenylcarbonium fluoroborate $Ph_3C^+BF_4^-$

Lactones from benzyloxycarboxylic acid esters

264. 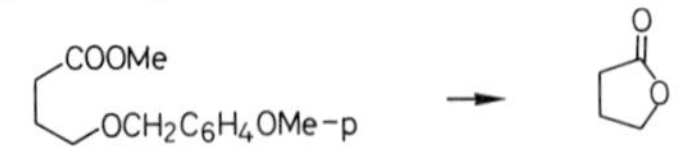

A soln. of startg. ester in acetonitrile treated with triphenylcarbonium fluoroborate at room temp. for several hrs. until g.c. indicated the reaction to be completed → product. Y 91%. F.e. and method s. T.R. Hoye, M.J. Kurth, V. Lo, Tetrahedron Letters *22,* 815–8 (1981).

Ammonium ceric nitrate $(NH_4)_2Ce(NO_3)_6$

Ketones from tert. alcohols ⋝C(OH)R → ⋝CO
with acetyl hypoiodite cf. *29,* 268s*31;* oxidative C-debenzylation with ammonium ceric nitrate s. K. Shankaran, D.G. Talekar, A.S. Rao, Synth. Commun. *12,* 141–9 (1982).

Ammonium ceric nitrate/sodium bromate $(NH_4)_2Ce(NO_3)_6/NaBrO_3$

Oxo compds. from ethers under mild conditions CHOR → CO

265. 

A stirred mixture of cyclohexyl methyl ether, $NaBrO_3$, and a little ammonium ceric nitrate in 7:3 water/acetonitrile heated under N_2 at ca. 80° for 12 hrs. → cyclohexanone. Y 81%. F.e., **also from alkoxysilanes,** s. G.A. Olah, B.G.B. Gupta, A.P. Fung, Synthesis *1980,* 897–8.

Iodoxybenzene/trichloroacetic acid $PhIO_2/CCl_3COOH$

Oxidations with iodoxyarenes C(OH)C(OH) → 2 CO

266. 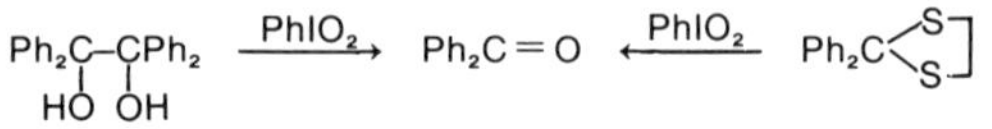

Cleavage of glycols.
A soln. of benzpinacol in dichloromethane treated with iodoxybenzene in the presence of a little trichloroacetic acid at room temp. for 5 hrs. → benzophenone. Y. 97%.

Oxo compds. from mercaptals.
A soln. of benzophenone-1,3-dithiolane in benzene treated with iodoxybenzene in the presence of a little *p*-toluenesulfonic acid at room temp. for 2 hrs. → benzophenone. Y 99%.

F.e. and catalysts, also controlled **S-oxidation of thioethers,** and prepn. of **ar. aldehydes and** *o***-quinones from benzyl alcohols and phenols, respectively,** s. D.H.R. Barton et al., Tetrahedron Letters *23,* 957–60 (1982).

N-Bromosuccinimide 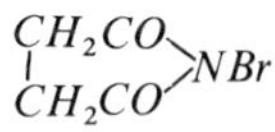

Aldehydes from ethers CH_2OR → CHO

with $RhCl_3$/triphenylphosphine cf. *32,* 237; with N-bromosuccinimide, *β*-chloro-*α,β*-ethylenealdehydes, s. K. Schulze et al., Synthesis *1981,* 483–4.

N-Iodosuccinimide 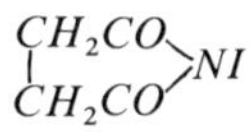

Cleavage of glycols C(OH)C(OH) → 2 CO

267. 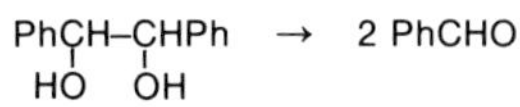

A soln. of 1,2-diphenyl-1,2-ethanediol in tetrahydrofuran treated with N-iodosuccinimide, and the mixture stirred for 15 min. in the dark → benzaldehyde. Y 85–90%. F.e.s. T.R. Beebe, P. Hii, P. Reinking, J. Org. Chem. *46,* 1927–9 (1981).

Trimethylsilyl chloride/sodium bromide $Me_3SiCl/NaBr$

Reactions with trimethylsilyl bromide prepd. *in situ* $HC(OR)_3$ → HCOOR

268. 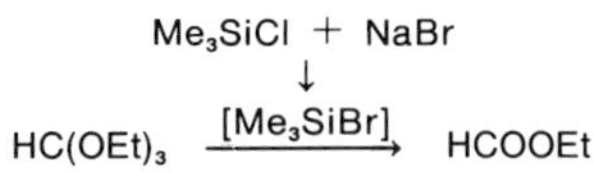

Formic from orthoformic acid esters. Chlorotrimethylsilane added with stirring to a suspension of NaBr in triethyl orthoformate, warmed at 30° for 8 min., and refluxed for 2 hrs. → ethyl formate. Y 92%. Also thioethers from sulfoxides (cf. Synth. Meth. *34,* 999), bromocarboxylic acid and phosphorous silyl esters from lactones and trialkyl phosphites, respectively, and *α***-bromoketones from ketones,** s. A.H. Schmidt, M. Russ, Chem. Ber. *114,* 1099–110 (1981).

Lead tetraacetate *$(MeCOO)_4Pb$*

Oxidative cleavage of glycol O,O-stannylene derivs. in aprotic medium

269.

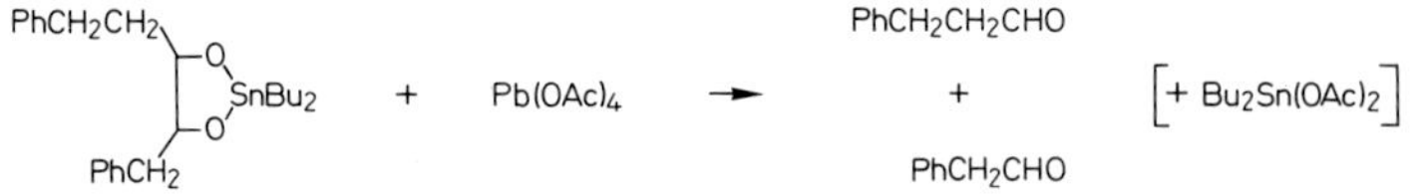

Pb tetraacetate added to a soln. of startg. stannylene deriv. in methylene chloride under N_2, and worked up after 5 min. at room temp. → phenylacetaldehyde (85%) and 3-phenylpropionaldehyde (91%). In contrast with the conventional glycol cleavage, formation of acetic acid is avoided. F.e. and oxidants s. S. David, A. Thiéffry, Tetrahedron Letters *22,* 2885–8 (1981).

Triphenylphosphine s. under $(MeCOO)_2Pd$ *Ph_3P*

Pyridinium tosylate *$[C_5H_5NH]^+TsO^-$*

Ketones from 1,3-dioxolanes

s. *36,* 169; removal of 4-chloromethyl-1,3-dioxolane protecting groups s. V.M. Thuy, H. Petit, P. Maitte, Bull. Soc. Chim. Belges *89,* 759 (1980).

Bis(trimethylsilyl) sulfate $(Me_3SiO)_2SO_2$

γ-Lactones from cyclopropanecarboxylic acid esters

270.

A mixture of diethyl 2-vinylcyclopropane-1,1-dicarboxylate and 2 equivalents of bis(trimethylsilyl) sulfate in 1,2-dichloroethane refluxed for 1 hr. → product. Y 98%. F.e.s. Y. Morizawa, T. Hiyama, H. Nozaki, Tetrahedron Letters *22,* 2297–300 (1981); with silica gel cf. M.E. Alonso et al., J. Heterocyc. Chem. *19,* 369–71 (1982); **via γ-iodocarboxylic acids** cf. S.P. Brown, B.S. Bal, H.W. Pinnick, Tetrahedron Letters *22,* 4891–4 (1981).

Hydrogen peroxide/molybdenum trioxide *H_2O_2/MoO_3*

Lactones from lactolides O-CH(OR) → O-CO

with *m*-chloroperoxybenzoic acid/BF_3 cf. *33,* 283; with H_2O_2/MoO_3, 5,6-dihydro-2-pyrones, s. M. Chmielewski, J. Jurczak, J. Org. Chem. *46,* 2230–3 (1981).

Sodium bromate s. under $(NH_4)_2Ce(NO_3)_6$ *$NaBrO_3$*

Polymer-based periodate and iodate ←

Cleavage of glycols C(OH)C(OH) → 2 CO

s. *37,* 231

Palladium(II) acetate/triphenylphosphine *$(MeCOO)_2Pd/Ph_3P$*

Allyl ethers from carbonates under mild, non-basic conditions $OCOOCH_2CH=CH_2 \rightarrow OCH_2CH=CH_2$

271.

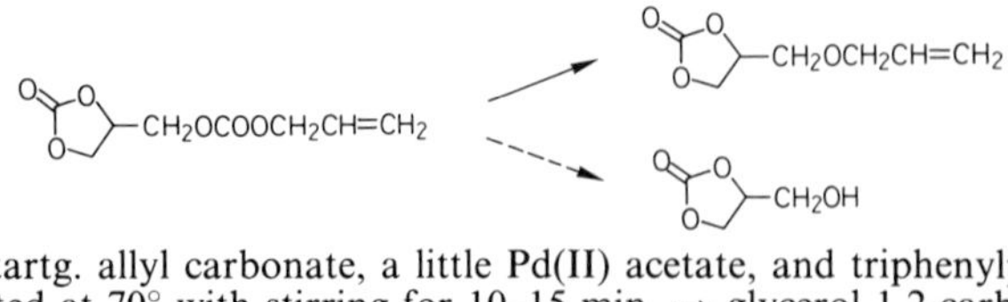

A soln. of startg. allyl carbonate, a little Pd(II) acetate, and triphenylphosphine in benzene heated at 70° with stirring for 10–15 min. → glycerol 1,2-carbonate 3-allyl ether. Y 82%. F.e., also hydrogenolysis to the parent alcohols with tri-*n*-butyltin hydride and a little tetrakis(triphenylphosphine)palladium (cf. Synth. Meth. *29,* 28s*37*), retention of 1,3-dioxolan-2-one groups, s. F. Guibe, Y. Saint M'Leux, Tetrahedron Letters *22,* 3591–4 (1981).

Formation of N–N Bond

Uptake ⇓

Addition to Nitrogen NN⇓N

tert-*Butyl hypochlorite* t-*BuOCl*

Cyanonitrene as intermediate
N-Cyanaminimides from tert. amines $R_3N \rightarrow R_3N{-}NCN$

272.

$$\begin{array}{c} \text{NaNHCN} \\ \downarrow \\ Et_3N \xrightarrow{[:NCN]} Et_3\overset{+}{N}{-}\overset{-}{N}CN \end{array}$$

A soln. of triethylamine in methanol added to a stirred soln. of *tert*-butyl hypochlorite in the same solvent at <–50°, stirring continued for 1 hr., treated slowly with a freshly prepared soln. of sodium cyanamide in methanol at the same temp., allowed to warm to room temp. with stirring for at least 3 hrs., solvent replaced by methylene chloride, and stirred for 2–4 hrs. at room temp. → product. Y 60–100%. – This method of generating cyanonitrene is more economical than the conventional azide route. F.e. and reactions s. M.G.K. Hutchins, D. Swern, Tetrahedron Letters *22,* 4599–602 (1981).

Rearrangement ∩

Oxygen/Nitrogen Type NN∩ON

Without additional reagents *w.a.r.*

3-Acylaminopyrazole ring from 3-*o*-amino-1,2,4-oxadiazoles ○

s. *36,* 262; f.e.s. D. Korbonits, I. Kanzel-Szvoboda, K. Horváth, J. Chem. Soc. Perkin Trans. I *1982,* 759–66.

Exchange ↓↑

Hydrogen ↑ NN↓↑H

Without additional reagents *w.a.r.*

Triazene 1-oxides and azoxy compds.
from nitroso compds. and hydrazines $\gtrdot N{-}NH_2 \rightarrow \gtrdot N{-}N{=}N(O)R$

273. $Ph_2N{-}NH_2 + 3\ PhNO \rightarrow Ph_2N{-}N{=}N(O)Ph + PhN{=}N(O)Ph$

A soln. of 1,1-diphenylhydrazine in 96% ethanol added over 15 min. to a stirred soln. of 3 equivalents nitrosobenzene in the same solvent at –5°, stirring continued at this temp. for a further 30 min., and allowed to warm slowly to 5° → (Z)-1,3,3-triphenyltriazene-1-oxide (Y 79%) and (Z)-1,2-diphenyldiazene oxide (Y 85%). F.e., incl. 2-phthalimidodiazene 1-oxides, and reactions s. L. Hoesch, B. Köppel, Helv. Chim. Acta *64,* 864–89; 890–904 (1981); **prepn. of azoxy compds., review,** s. V.N. Yandovskii, B.V. Gidaspov, I.V. Tselinskii, Russ. Chem. Rev. *50,* 164–79 (1981).

Phenyllithium/cuprous chloride *PhLi/CuCl*

Sym. hydrazines from sec. amines 2 >NH → >N–N<

with $KMnO_4$ cf. *16,* 354; with phenyllithium/cuprous chloride s. J. Jackisch, J. Legler, T. Kauffmann, Chem. Ber. *115,* 659–68 (1982).

Sodium nitrite *$NaNO_2$*

Tetrazole ring from *o*-hydrazino-N-heterocyclics ○

s. *16,* 351; s.a. S.B. Mahajan, S.S. Sangapure, Y.S. Agasimundin, Indian J. Chem. *19B,* 596–8 (1980); in the presence of polyphosphoric acid s. K.C. Joshi, P. Chand, J. Heterocyc. Chem. *17,* 1783–4 (1980).

Cuprous chloride s. under PhLi *CuCl*

Bis(pyridine)silver(I) permanganate *$(C_5H_5N)_2Ag^+MnO_4^-$*

Sym. azo compds. from amines $2\ NH_2 \rightarrow N{=}N$

s. *37,* 233

Oxygen ↑ NN⇅O

Zinc *Zn*

Sym. azo from nitro compds. $2\ NO_2 \rightarrow N{=}N$

s. *3,* 251; optically active compds. s. G. Heppke, G. Scherowsky et al., Chem. Ber. *114,* 2501–18 (1981).

Hydrazines *R_2NNH_2*

Azoxy from nitroso compds. 2 RN=O → RN(O)=NR

s. *37,* 273

Halogen ↑ NN⇅Hal

Potassium alkoxides *KOR*

Phosphonohydrazidates from N-chlorophosphonic diamides RP(=O)(OR)N–

274.

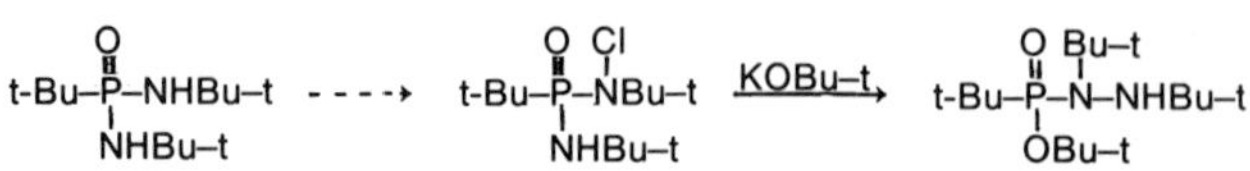

A suspension of K *tert*-butoxide in tetrahydrofuran treated with a soln. of startg. m. in the same solvent at 0°, and the mixture stirred at 0–25° for 15 hrs. → P,1,2-tri-*tert*-butylhydrazidophosphonic acid *tert*-butyl ester. Y 91%. F.e. in high yield, **also from phosphonic diamides** without isolation of the N-chloro deriv., s. H. Quast, M. Heuschmann, M.O. Abdel-Rahman, Ann. Chem. *1981,* 943-6.

Carbon ↑ NN⇂↾C

Nitryl fluoroborate NO_2BF_4

Deacylative N-nitration NAc → N–NO_2

N-Sulfonyl-N-nitramines from N-acylsulfonamides

275. 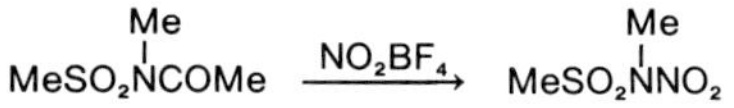

The startg. N-acylsulfonamide added in small portions to a vigorously stirred suspension of nitryl fluoroborate in acetonitrile at –25°, allowed to warm to ca. 20°, and stirring continued for 30 min. → product. Y 83%. F.e., **also N-nitro- from N-acylurethans,** s. O.A. Lukyanov et al., Izv. Akad. Nauk SSSR Ser. Khim. *1981,* 2138–41.

Elimination ⇑

Hydrogen ↑ NN⇑H

Electrolysis ϟ

1,2,4-Triazoline-3,5-diones from 1,2,4-triazole-3,5-diones NHNH → N=N

with N-bromosuccinimide cf. *19,* 356s*31;* electrochemical prepn. *in situ* s. H. Wamhoff, G. Kunz, Angew. Chem. Intern. Ed. *20,* 797 (1981).

Sodium hydrogen carbonate s. under m-$ClC_6H_4COO_2H$ $NaHCO_3$

Cupric sulfate/pyridine $CuSO_4/C_5H_5N$

Oxidative 1,2,3-triazole ring closure ○

s. *6,* 325; 1,2,3-triazole-1-oxides s. M. Bergtrup, J. Holm, J. Chem. Soc. Perkin Trans. I *1981,* 503–13.

Silica gel s. under IO_4^- SiO_2

m-*Chloroperoxybenzoic acid/ sodium hydrogen carbonate* m-$ClC_6H_4COO_2H/NaHCO_3$

Diazo compds. from hydrazones C=N–NH_2 → C=N_2

with HgO cf. *3,* 255; with *m*-chloroperoxybenzoic acid/$NaHCO_3$, 2-phase medium, s. K.L. Handoo, S.K. Handoo, Indian J. Chem. *20B,* 699–700 (1981).

Oxidation with *N-chloroisatin* ←

Azo from hydrazo compds. NHNH → N=N

276. 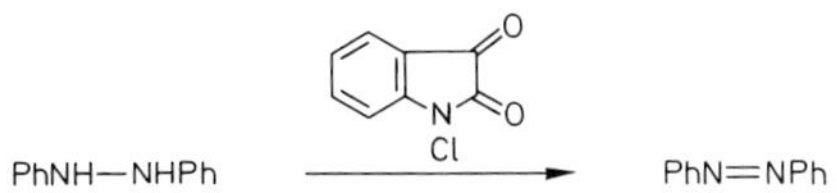

N-Chloroisatin and hydrazobenzene mixed in benzene at room temp. for 30 min. → azobenzene. Y 92%. Also with 2,2-diphenyl-2,3-dihydro-3-oxo-1,5,7-trichloroindole as oxidant (Y 84%). F.e., **also nitroso compds from hydroxylamines, and oxo compds. from alcohols** with irradiation, s. C. Berti, L. Greci, Synth. Commun. *11,* 681–6 (1981).

Lead tetraacetate $(MeCOO)_4Pb$

Oxidative 1,2,3-triazole ring closure ○

s. *34,* 255; *v*-triazolo-[1,5-*a*]pyrazines and -[1,5-*b*]pyridazines s. Bull. Soc. Chim. Belges *91,* 153–61 (1982).

Bromine *Br_2*

Tetrazolium salts from formazans
with N-bromosuccinimide cf. *10,* 255; with Br_2 s. V.P. Shchipanov, Khim. Geterotsikl. Soedin. *1981,* 987–90.

Periodate–silica gel *IO_4^{-}–SiO_2*

Azo from hydrazo compds. NHNH → N=N
s. *37,* 231

Oxygen ↑ NN⇑O

Triethylamine *Et_3N*

Δ^2-1,3,4-Oxadiazol-5-ones from acylureas ○
s. *30,* 193; f. method cf. H. Singh, L.D.S. Yadav, B.K. Bhattacharya, J. Indian Chem. Soc. *57,* 1006–10 (1980).

Sulfur ↑ NN⇑S

Potassium hydroxide *KOH*

Diazo compds. from sulfonylhydrazones $C=NNHSO_2Ar \rightarrow C=N_2$
from tosylhydrazones with NaOH cf. *31,* 307s*34;* from 2,4,6-triisopropylbenzenesulfonylhydrazones with KOH s. C.C. Dudman, C.B. Reese, Synthesis *1982,* 419–21.

Carbon ↑ NN⇑C

Potassium hydroxide *KOH*

Cyclic azo compds. from 1,2,4-triazolidine-3,5-dione ring ⊂
s. *36,* 294; f. method cf. W. Adam, L.A. Arias, O. De Lucchi, Synthesis *1981,* 543–5.

Formation of N-Hal Bond

Exchange ⇂↑

Hydrogen ↑ NHal⇂↑H

tert-*Butyl hypochlorite/chiral trifluoromethylcarbinols* ←

N-Chlorination NH → NCl
s. *25,* 209; **asym. N-chlorination** of aziridines in the presence of chiral trifluoromethylcarbinols s. A. Forni, I. Moretti et al., J. Chem. Soc. Chem. Commun. *1981,* 588–90.

Formation of N–S Bond

Uptake ⇓

Addition to Oxygen and Sulfur NS⇓OS

Triethylamine *Et_3N*

Sulfamic acid derivs. from amines via *o*-hydroxysulfamate esters ←

277.

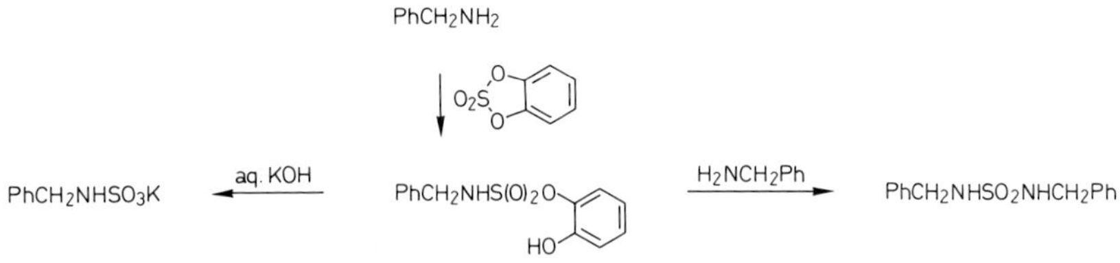

Sulfamic acids. | **Sulfamides.**

A soln. of benzylamine and triethylamine in dimethylformamide treated dropwise with a soln. of catechol sulfate in methylene chloride while stirring vigorously at 0° under argon, the mixture kept at the same temp. for 2.5 hrs., and poured into 1% HCl → intermediate (Y 98%)

treated sequentially with 1 *M* KOH and water, and refluxed under argon for 30 min. → product. Y 99% (p. 5371). | and benzylamine in dioxane refluxed under argon until t.l.c. indicated completion of reaction → product. Y 91% (p. 5373).

F.e.s. G.E. DuBois et al., J. Org. Chem. *45,* 5371–3, 5373–5 (1980).

Rearrangement ∩

Oxygen/Nitrogen Type NS∩ON

Without additional reagents *w.a.r.*

N-(2-Pyridylthio)lactams by electrocyclic rearrangement ←

278.

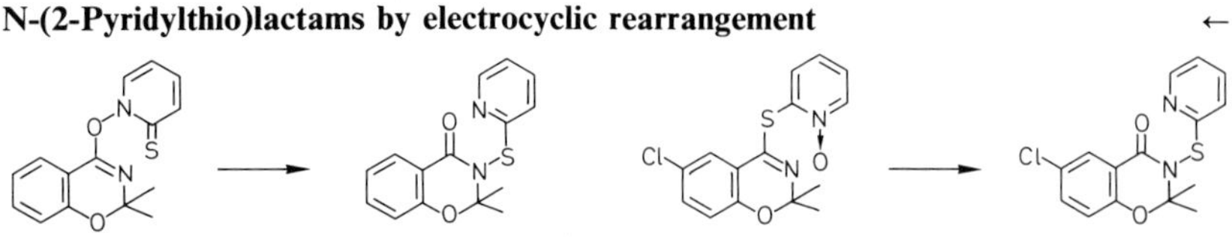

A soln. of

startg. thiopyridone | startg. N-oxide

in chloroform refluxed for 1–5 hrs.

→ product. Y 96%. | → product. Y 96%.

F.e. and rearrangements s. R. Tachikawa, K. Wachi et al., Chem. Pharm. Bull. *29,* 3529–35 (1981).

Oxygen/Sulfur Type NS⥠OS

n-*Butyllithium* n-*BuLi*

***o*-(Sulfonylamino)phenols from *o*-(sulfonyloxy)amines** ←
O→N-Sulfonyl group migration

279.

A soln. of 2-aminophenyl 4-toluenesulfonate in ether treated at 0° with *n*-butyllithium, the mixture stirred at room temp. for 1 hr., and hydrolysed with water → N-(2-hydroxyphenyl)-4-toluenesulfonamide. Y 95%. F.e. and pyridine analog s. K.K. Andersen et al., J. Org. Chem. *47*, 1884–9 (1982).

Sulfur/Carbon Type NS⥠SC

Without additional reagents *w.a.r.*

N-(2-Pyridylthio)lactams by electrocyclic rearrangement ←
s. *37*, 278

Exchange ⇅

Hydrogen ↑ NS⇅H

Without additional reagents *w.a.r.*

(N-sulfinyl)bis(isopropyl)immonium hexachloroantimonates from sec. amines $R_2NH \rightarrow R_2\overset{+}{N}{=}S{=}O$

280.

$$\text{i-Pr}_2\text{NH} + \text{SOCl}_2 \xrightarrow{\text{SbCl}_5} \text{i-Pr}_2\overset{+}{\text{N}}{=}\text{S}{=}\text{O}\ \text{SbCl}_6^-$$

A soln. of diisopropylamine in dichloromethane added dropwise to a stirred soln. of $SOCl_2$in the same solvent at –70° under N_2, precipitated amine hydrochloride removed by filtration at room temp., filtrate cooled to –30°, a soln. of $SbCl_5$ in dichloromethane added dropwise, and stirring continued for 15 min. at room temp. → N-(2-propyl)-N-sulfinyl-2-propanaminium hexachloroantimonate. Y 71%. F.e.s. M.A. Pérez, G. Kresze, Synthesis *1981*, 707–8; thionylimmonium **fluoroborates from thionylimines** cf. Angew. Chem. Intern. Ed. *17*, 63 (1978).

N-Chlorosuccinimide/triethylamine $\begin{matrix} CH_2CO \\ | \\ CH_2CO \end{matrix}\!\!>\!NCl/Et_3N$

Sulfonic acid derivs. from sulfinic acids ←
amides with Cl_2 cf. *4*, 298; also sulfonyl esters and chlorides with N-chlorosuccinimide/triethylamine s. M. Furukawa et al., Chem. Pharm. Bull. *29*, 623–8 (1981); amides with NaOCl cf. F.E. Scully, K. Bowdring, J. Org. Chem. *46*, 5077–81 (1981).

Bromine/triethylamine Br_2Et_3N

S-Tosylimination $C{=}S \rightarrow C{=}S{=}NTs$
with chloramine-T cf. *32*, 250; with sulfonamides and Br_2/triethylamine s. A. Koide et al., Synthesis *1981*, 486–7.

Sodium hypochlorite *NaOCl*

Sulfonamides from sulfinic acids and amines S(O)OH → $SO_2N\lt$

s. *4*, 298s*37*

Oxygen ↑ NS⇃↾O

Without additional reagents *w.a.r.*

N-Carbalkoxysulfilimines from thioethers >S → >S=NCOOR

281.

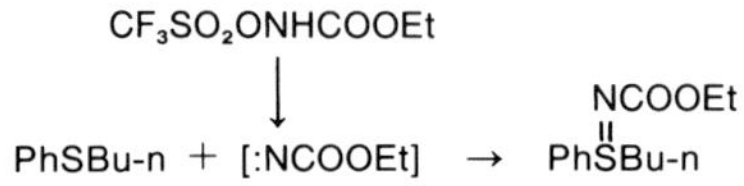

Ethyl N-[(trifluoromethanesulfonyl)oxyl]carbamate added to a soln. of phenyl *n*-butyl sulfide in methylene chloride, the mixture stirred at room temp. for 4–6 hrs., and washed with satd. $NaHCO_3$ soln. → S-*n*-butyl-S-phenyl-N-(ethoxycarbonyl)sulfilimine. Y 97%. – With allylic and propargylic thioethers, the sulfilimine initially formed undergoes [2.3]-sigmatropic rearrangement to **N-allyl- and N-allenylsulfenamides,** respectively. F.e. and reactions s. Y. Tamura et al., J. Org. Chem. *46*, 1732–4 (1981).

Sulfamides from ***o*****-hydroxysulfamate esters** ←

s. *37*, 277

Nitrogen ↑ NS⇃↾N

Without additional reagents *w.a.r.*

Sulfenamides from amines and 3-(organothio)hydantoins NH → NSR

282.

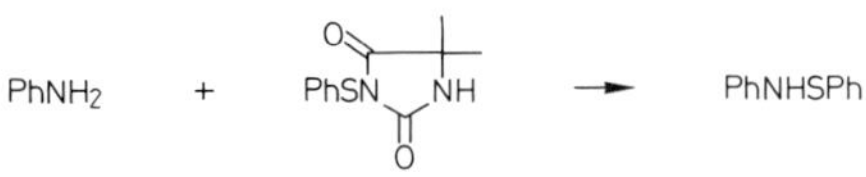

A soln. of 5,5-dimethyl-3-(phenylthio)hydantoin and aniline in ethanol stirred vigorously with boiling for 6 hrs., cooled to room temp., and poured into water → N-phenylbenzenesulfenamide. Y 79%. – Sulfenylated hydantoins are excellent sulfur transfer agents as they are inexpensive, stable, convenient to handle, and provide improved yields with simple work-up conditions. F.e., **also sym. diaminosulfanes from amines** with N,N′-thiodicaprolactam and N,N′-thiobis(5,5-dimethylhydantoin), s. G. Sosnovsky, J.A. Krogh, Ann. Chem. *1982*, 121–36.

Halogen ↑ NS⇃↾Hal

Triethylamine *Et_3N*

S-Imination with N-chlorides >S → >S=N–

N-acyliminosulfonyl chlorides cf. *22*, 301; N-acyliminosulfinyl chlorides s. Zh. Org. Khim. *18*, 162–8 (1982); N-sulfonyliminosulfinamides s. I.V. Koval, T.G. Oleinik, M.M. Kremlev, ibid. *17*, 2174–6 (1981); N-acyl-N′-sulfonylsulfinamidines s. ibid. *17*, 533–9; 565–9 (1981).

Remaining Elements ↑ NS⇂↿Rem

n-*Butyllithium* n-*BuLi*

Thionylimines from aminosilanes NHSi⋚ → N=S=O

283.

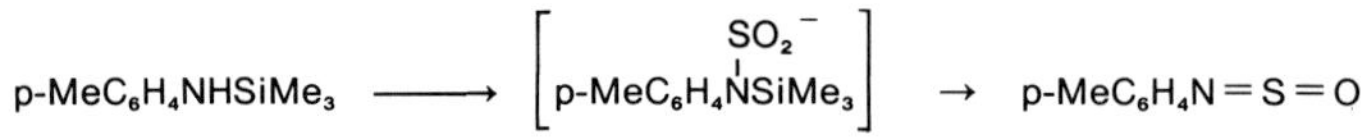

A soln. of *n*-butyllithium in hexane slowly added to a stirred soln. of startg. N-trimethylsilylamine in tetrahydrofuran at 0° under N_2, the mixture stirred at room temp. for 1 hr., the intermediate anion added to an excess of SO_2 in the same solvent at –78°, and stirring continued at room temp. for 1 hr. → product. Y 90%. F.e. and method s. P.A.T.W. Porskamp, B. Zwanenburg, Synthesis *1981,* 368–9.

Elimination ⥮

Hydrogen ↑ NS⥮H

Bromine Br_2

Isothiazoles by oxidative ring closure ○

s. *18,* 354; 2,5-diaminoisothiazolium salts s. K. Grohe, H. Heitzer, Ann. Chem. *1982,* 884–93; 5-alkylthioisothiazoles with I_2 cf. F.C. James, H.D. Krebs, Australian J. Chem. *35,* 393–403 (1982).

5-Imino-Δ³-isothiazolines
from β-amino-α,β-ethylenecarboxylic acid thioamides
s. *37,* 433

Nitrogen ↑ NS⥮N

Without additional reagents *w.a.r.*

5-Amino-1,2,4-thiadiazoles ○
from imidoylhydrazinium betaines and isothiocyanates
via-N-thiocarbamylamidrazone betaines

284.

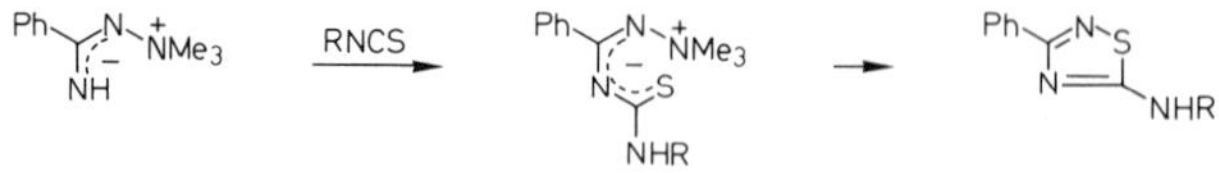

R= 2–naphthyl

A soln of startg. isothiocyanate in benzene treated with (trimethylammonio)phenylamidinate (prepd. according to Synth. Meth. *33,* 332), and the mixture warmed on a steam bath until the solid dissolved → intermediate (Y 100%) heated to its decomposition temp. (173–4°) → 5-(2-naphthylamino)-3-phenyl-1,2,4-thiadiazole (Y 72%). F.e., also one-pot procedure in lower yield, s. R.F. Smith, T.P. Feltz, J. Heterocyc. Chem. *18,* 201–3 (1981).

Without additional reagents *w.a.r.*

1,4,2-Benzodithiazines from 2-azido-1,3-benzodithioles ◯

285. 

A soln. of 1,3-benzodithiol-2-yl azide in toluene refluxed for 2 hrs. → 1,4,2-benzodithiazine. Y 90%. F.e.s. J. Nakayama et al., J. Chem. Soc. Perkin Trans. I *1981,* 618–20.

Carbon ↑ NS⇑C

Without additional reagents *w.a.r.*

Benz[*d*]isothiazoles from O-acyl-*o*-alkylthiooximes ◯

286. 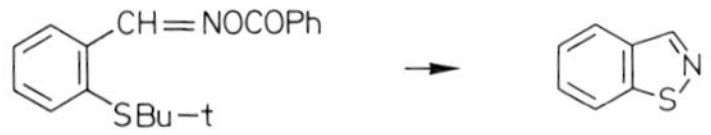

A soln. of (E)-O-benzoyl-2-(*tert*-butylthio)benzaldoxime in methanol refluxed for 60 min. → 1,2-benzisothiazole. Y 74%. F.e.s. A.J. Lawson, Phosphorus-Sulfur *12,* 357–67 (1982).

Thionyl chloride $SOCl_2$

Benz[*d*]isothiazol-3-ones from *o*-sulfinylcarboxylic acid amides

287. 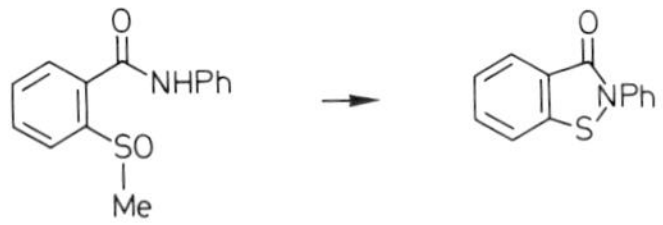

$SOCl_2$ added to a stirred suspension of startg. m. in dichloromethane, and the mixture refluxed for 10 hrs. → product. Y 98%. F.e.s. Y. Uchida, S. Kozuka, J. Chem. Soc. Chem. Commun. *1981,* 510–1, f. methods s. Bull. Chem. Soc. Japan *54,* 1781–6 (1981).

Formation of N–Rem Bond

Exchange ⇂↿

Halogen ↑ NRem⇂↿Hal

Triethylamine Et_3N

Protection of prim. amino groups as cyclic disilazanes ◯

288. 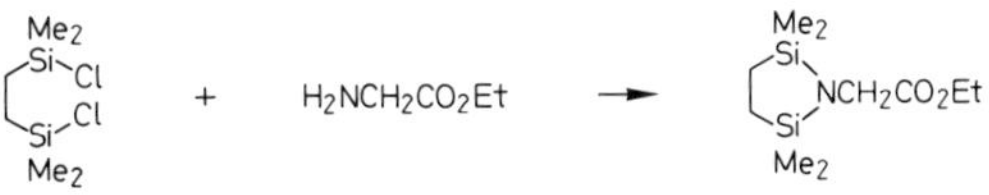

A soln. of 1,2-bis(chlorodimethylsilyl)ethane in dichloromethane added via syringe to a stirred soln. of ethyl glycinate and triethylamine in the same solvent, the mixture stirred at room temp. under argon for 2 hrs., and worked up with aq. NaH_2PO_4 → product. Y 92%. – The protective group is stable under very basic conditions (organolithium reagents and amide bases), but easily removed by mild acids. F.e.s. S. Djuric, J. Venit, P. Magnus, Tetrahedron Letters *22*, 1787–90 (1981).

Sulfur ↑ NRem ⇃↾ S

Without additional reagents *w.a.r.*

Thiophosphonic diamides
from dithiaphosphetane disulfides and amines RP(=S)(N<)$_2$

289. 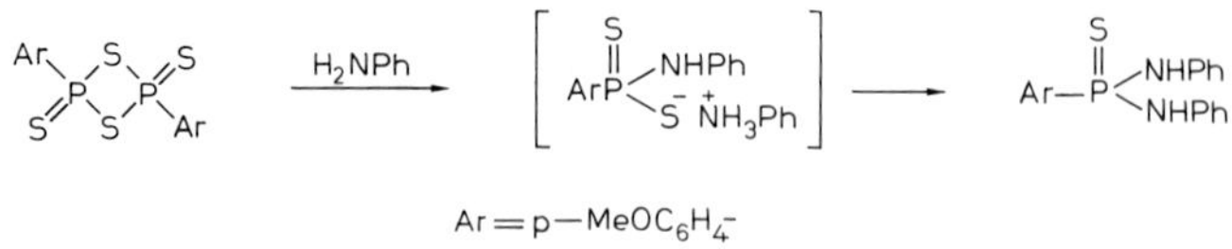

A soln. of aniline and 2,4-bis(*p*-methoxyphenyl)-1,3,2,4-dithiadiphosphetane-2,4-disulfide in xylene refluxed for 6 hrs. → product. Y 73%. F.e., also isolation of intermediate, s. K. Clausen, A.A. El-Barbary, S.-O. Lawesson, Tetrahedron *37*, 1019–25 (1981).

Formation of N–C Bond

Uptake ⇓

Addition to Oxygen and Carbon NC ⇓ OC

Without additional reagents *w.a.r.*

2-Aminoalcohols from oxido compds.  → C(OH)–C–N<

s. *1*, 277/8; N-bridged di- and tri-saccharides s. U. Spohr, W. Meyer zu Reckendorf, Ann. Chem. *1981*, 2139–63.

1-Amino-6-hydroxy-5,6-dihydrouracils from 1,3-oxazine-2,4-diones ○

290. 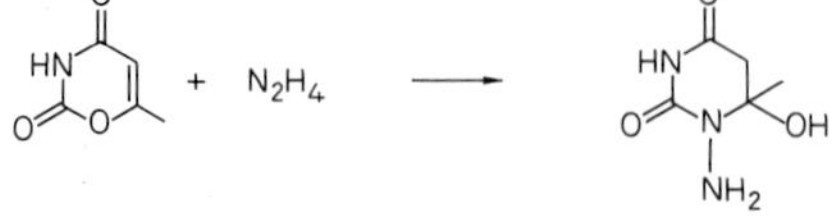

Hydrazine hydrate added to a stirred suspension of finely powdered 6-methyl-1,3-oxazine-2,4-dione in ethanol, and stirring continued for 10 hrs. → 1-amino-6-hydroxy-6-methyl-5,6-dihydrouracil. Y 80%. F.e.s. M. Yogo, K. Hirota, S. Senda, J. Chem. Soc. Perkin Trans. I *1982*, 473–5.

2-Amino-1,3-oxazin-4-ones
s. *37*, 307

Without additional reagents *w.a.r.*

Review of cycloadditions with acylisocyanates s. *28,* 266s*29;* **review of addition of** aliphatic and aromatic **acylisocyanates to unsatd. compds.** s. B.A. Arbuzov, N.N. Zobova, Synthesis *1982,* 433–50.

Ring opening of 4*H*-3,1-benzoxazin-4-ones with N-nucleophiles C
with amines cf. *29,* 391; with hydrazines s. A.N. Osman, S. Botros, Pharmazie *37,* 17–8 (1982); with NH_3, *o*-ureidocarboxylic acid amides, s. I. Bitter, L. Szöcs, L. Töke, Acta Chim. Acad. Sci. Hung. *107,* 57–66 (1981).

Hydroxycarboxylic acid amides from lactones
s. *2,* 320; coumarin and thiocoumarin ring opening s. I.E. El-Kholy, M.M. Mishrikey, H.M. Feid-Allah, J. Heterocyc. Chem. *18,* 105–10 (1981).

Sodium hydroxide *NaOH*

2-Aminoalcohols from oxido compds. 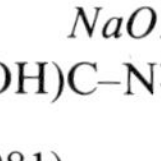 → C(OH)C–N<

s. *13,* 361; with ar. amines s. R. Krug, W. Jugelt, Z. Chem. *21,* 406–7 (1981).

Borate buffer ←

Aspartine-containing peptides ←
s. *22,* 313; f. method cf. F.J. Vinick. S. Jung, Tetrahedron Letters *23,* 1315–8 (1982).

Addition to Nitrogen and Remaining Elements NC⇓NRem

Without additional reagents *w.a.r.*

Sec. amines from phosphine imines and amines $N{=}PR_3 \rightarrow NHR'$
s. *26,* 291; **review of phosphine imines** s. E.W. Abel, S.A. Mucklijohn, Phosphorus-Sulfur *9,* 235–66 (1981).

Addition to Nitrogen and Carbon NC⇓NC

Without additional reagents *w.a.r.*

Quaternary N-aminomethylammonium salts
from tert. amines and methyleneammonium salts $R_3N \rightarrow R_3\overset{+}{N}CH_2N$<

291. 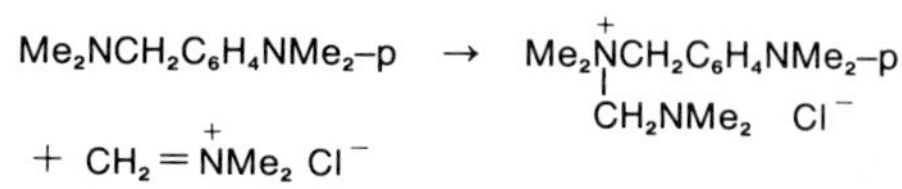

Startg. 4-(aminomethyl)-N,N-dimethylaniline added with stirring to a suspension of startg. methyleneammonium salt in acetonitrile at room temp., and stirring continued for 1 hr. → N-(4-dimethylaminobenzyl)-N-(dimethylaminomethyl)dimethylammonium chloride. Y 80%. F.e.s. H. Böhme, U. Bomke, J.-P. Denis, Ann. Chem. *315,* 40–7 (1982).

N-Hydroxyisoureas
from hydroxylamines and cyanic acid esters $ROC{\equiv}N \rightarrow ROC({=}NH)N{-}OH$

292. 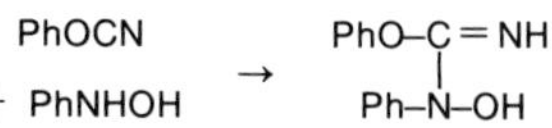

An ice-cooled soln. of N-phenylhydroxylamine in ether treated dropwise with a soln. of phenyl cyanate in the same solvent → 1-hydroxy-1,3-diphenylisourea. Y 96%. F.e., also conversion to **3-imino-1,2,4-oxadiazolidin-5-ones** with isocyanates (p. 10), s. M. Neitzel, G. Zinner, Arch. Pharm. *314,* 2–9; 10–8 (1981).

Without additional reagents *w.a.r.*

Unsym. ureas from amines and isocyanates N=C=O → NHCON<

s. *17*, 403; selective N-carbamylation, retention of hydroxyl groups, s. R.S. Bezwada, S.S. Stivala, J. Chem. Eng. Data *25*, 292–3 (1980).

Guanidines from carbodiimides −N=C=N− → (−NH)(−NH)C=N

s. *11*, 536; acylguanidines s. M. Richter, K. Strauss, M. Augustin, Z. Chem. *22*, 103–4 (1982).

Thioureas from isothiocyanates and amines N=C=S → NHCSN<

s. *18*, 555; in the molten state cf. X. Ribas, M. Ventura, An. Secc. Cienc. Col. Univ. Gerona (Univ. Auton. Barcelona) *1980*, 209–16; Chem. Abstr. *97*, 23416 (1982).

Isodithiobiurets from isothioureas and isothiocyanates C=NH → C=NCSNHR

s. *35*, 209; s.a. I.D. Bedekar, A.R. Joshi, M.G. Paranjpe, J. Indian Chem. Soc. *18*, 1014–5 (1981).

Dipolar cycloaddition with cyanamides ○

s. *10*, 270; with N,N-dicyanamides s. P.H. Benders, D.N. Reinhoudt, D.M.W. van den Ham, Rec. Trav. Chim. Pays-Bas *100*, 330–3 (1981).

Ring closures with diazonium N-betaines

s. *36*, 288; 1,2,4-triazine ring from 2-ketophosphonium ylids s. J. Heterocyc. Chem. *18*, 675–7 (1981).

Sodium methoxide *NaOMe*

Amidoximes from nitriles CN → C(=NOH)NH_2

with K_2CO_3 cf. *4*, 311; with Na methoxide, *α*-(arylthio)amidoximes, s. I. Červená et al., Collection Czech. Chem. Commun. *46*, 1188–98 (1981).

5-Amino-*v*-triazolo[1,5-*a*]pyrimidine ring from *o*-azidonitriles and nitriles ○

Double ring closure

293.

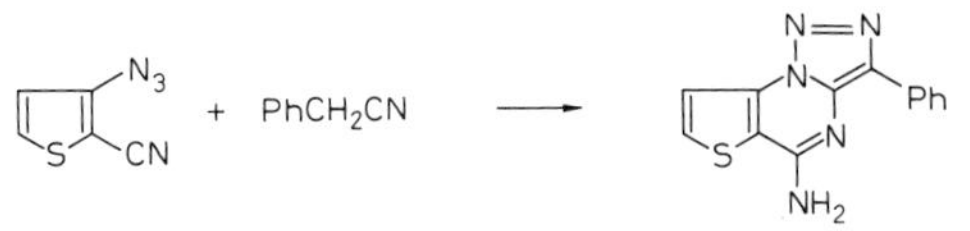

A soln. of Na in methanol added in one portion to a soln. of 3-azido-2-cyanothiophene and phenylacetonitrile in the same solvent, stirred at room temp. overnight, and cooled in a refrigerator → 3-phenyl-5-amino-1,2,3-triazolo[1,5-*a*]thieno[3,2-*d*]pyrimidine. Y 82%. F.e. and **1,2,3-triazole ring opening** (p. 1771) s. C. Westerlund, J. Heterocyc. Chem. *17*, 1765–9; 1771–5 (1980).

Triethylamine s. under Mesitylenesulfonyloxylamine *Et_3N*

Triethylaluminium *Et_3Al*

Amidazines from amidrazones and nitriles 2 RC(NH_2)=N−NH_2 → RC(NH_2)=N−N=C(NH_2)R

294.

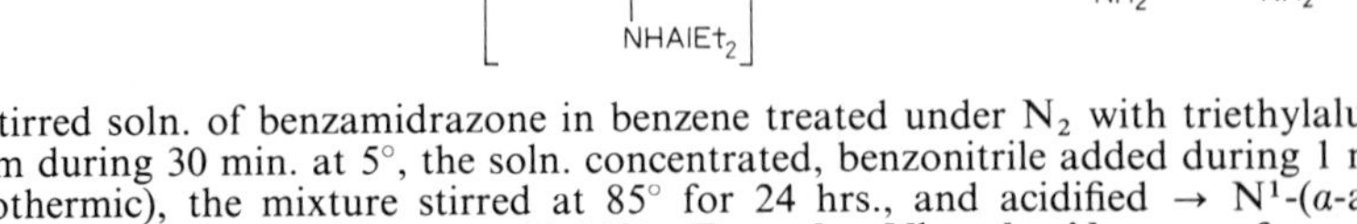

A stirred soln. of benzamidrazone in benzene treated under N_2 with triethylaluminium during 30 min. at 5°, the soln. concentrated, benzonitrile added during 1 min. (exothermic), the mixture stirred at 85° for 24 hrs., and acidified → N[1]-(*α*-aminobenzylidene)benzamidrazone. Y 95%. F.e., **also N[1]-acylamidrazones from carboxylic acid hydrazides,** s. T. Kauffmann, L. Bán, D. Kuhlmann, Chem. Ber. *114*, 3684–90 (1981).

Mesitylenesulfonyloxylamine/triethylamine *$ArSO_2ONH_2/Et_3N$*

1,1-Aminoazo compds. from hydrazones C=N–NH → $C(NH_2)N=N$

295. 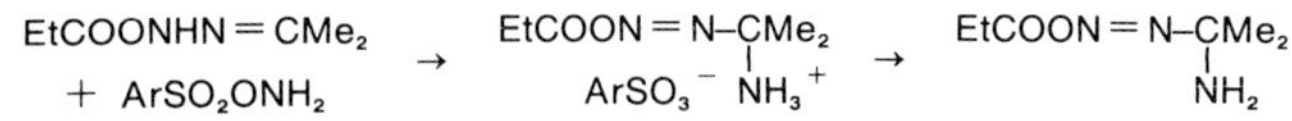

A soln. of startg. hydrazone in ether treated dropwise under N_2 with mesitylenesulfonyloxylamine in the same solvent at 25°, and stirred for 1 hr. → intermediate salt, in dichloromethane treated dropwise with triethylamine with stirring, and the mixture worked up after a further hr. → product. Y 70%. F.e. and ring closure to **1,2,4-triazoles** s. P. Métra, J. Hamelin, Can. J. Chem. *60,* 285–90 (1982).

Bis(dibenzylideneacetone)palladium(0) *$\{(PhCH=CH)_2CO\}_2Pd$*

Enisocyanates from Δ^1-azirines C

Carbonylation

s. *36,* 291; with bis(dibenzylideneacetone)palladium(0) as catalyst s. Organometallics *1,* 70–4 (1982).

Addition to Sulfur and Carbon NC⇓SC

Without additional reagents *w.a.r.*

Ring closure with sulfonyl isothiocyanates ○

s. *30,* 221; **review of sulfonyl isocyanates and sulfonyl isothiocyanates** s. J.W. McFarland, Sulfur Reports *1,* 215–62 (1981).

Sodium hydroxide/benzyltriethylammonium chloride *$NaOH/[PhCH_2NEt_3]^+Cl^-$*

Dithiocarbamic acid esters from amines NH → NCSSR

cf. *19,* 380; phase transfer catalysis with $[PhCH_2NEt_3]^+Cl^-$ s. X. Huang, Z. Chen. M. Tang, Hangzhou Daxue Xuebao, Ziran Kexueban *9,* 81–4 (1982) (Chin); Chem. Abstr. *97,* 23417 (1982); cyclic dithiourethans from dihalides s. W. Hanefeld, Arch. Pharm. *310,* 409–17 (1977); with NaOH in dimethyl sulfoxide cf. J. Garin et al., Synthesis *1981,* 961.

Addition to Carbon NC⇓CC

Without additional reagents *w.a.r.*

Enamines from acetylene derivs. C≡C → CH=CN<

β-Amino-α,β-ethyleneketones

s. *7,* 217s*36;* s.a. E.V. Vasileva et al., Zh. Org. Khim. *17,* 1378–81 (1981); A.S. Zanina et al., ibid. *18,* 772 (1982).

1,2,3-Triazoles from acetylene derivs. and azides ○

s. *8,* 404; 1-glycosyl-1,2,3-triazoles s. F. Chrétien, B. Gross, J. Heterocyc, Chem. *19,* 263–7 (1982).

4-Amino-2-imino-2,5-dihydrofurans from γ-hydroxy-α,β-acetylenenitriles

296. 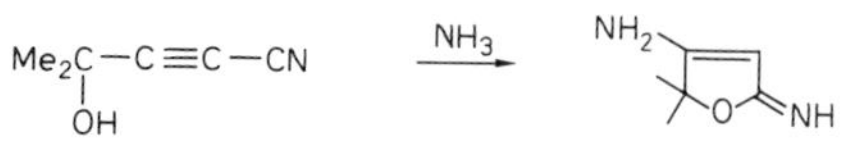

Startg. nitrile treated with liq. NH_3 containing ca. 0.5–3% water → 2-imino-4-amino-5,5-dimethyl-2,5-dihydrofuran. Y 96%. F.e.s. Y.M. Skvortsov et al., Zh. Org. Khim. *17,* 884–5 (1981).

Lithium diisopropylamide s. under $PdCl_2(MeCN)_2$ *$i\text{-}Pr_2NLi$*

Sodium nitrite/acetic acid *$NaNO_2/MeCOOH$*

β-Nitro- from α,β-ethylene-ketones $COC=C \rightarrow COCH\text{–}CNO_2$

297. $MeCOCH=CH_2 + NaNO_2 \rightarrow MeCOCH_2CH_2NO_2$

Acetic acid added to a stirred mixture of $NaNO_2$ and but-3-en-2-one in tetrahydrofuran during 10 min. at 20–25°, and stirring continued for 18 hrs. → 4-nitrobutan-2-one. Y 82%. F.e.s. T. Miyakoshi, S. Saito, J. Kumanotani, Chem. Letters *1981,* 1677–8.

Ammonium hydroxide s. under NaOCl *NH_4OH*

Mercuric nitrate/sodium tetrahydridoborate *$Hg(NO_3)_2/NaBH_4$*

Amidomercuration-demercuration
Acylamines from ethylene derivs. $C=C \rightarrow CH\text{–}C\text{–}NAc$
Markownikoff addition

298.

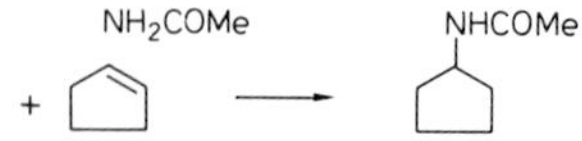

Anhydrous $Hg(NO_3)_2$ added to a stirred soln. of startg. olefin and acetamide in dichloromethane, the mixture heated at reflux for 6–24 hrs., cooled to 0°, then treated with *n*-butylamine, 10% NaOH, and a soln. of $NaBH_4$ in 10% NaOH → product. Y 97%. F.e., **also ureas and urethans,** s. J. Barluenga et al., J. Chem. Soc. Chem. Commun. *1981,* 670–1; **sulfonylamines** and intramolecular process s. ibid. *1981,* 1178–9.

Amines from ethylene derivs. $C=C \rightarrow CH\text{–}CN\lessdot$
by 2-aminomercuration-demercuration
s. *31,* 313; with $NaBH_4$ in place of Li powder, 2- and 3-aminosilanes, s. Synthesis *1982,* 414–7.

Boron fluoride *BF_3*

Δ^2-Imidazolines from aziridines and nitriles
s. *30,* 217; s.a. I.G. Tishchenko, O.N. Bubel, V.A. Konovalov, Khim. Geterotsikl. Soedin. *1981,* 952–7.

Acetic acid s. under $NaNO_2$ *MeCOOH*

Chloramine-T s. under OsO_4 *TsN(Cl)Na*

Lead tetraacetate s. under $PdCl_2(C_6H_5CN)_2$ *$(MeCOO)_4Pb$*

Hydrazoic acid *HN_3*

Lactams from cyclic ketones ←
Schmidt reaction
s. *4,* 326s*26;* 1,2,3,4-tetrahydro-2*H*-1,4-benzodiazepin-2-ones s. V.K. Venugopal et al., Indian J. Chem. *20B,* 156–7 (1981); regiospecific lactam formation by **nitrogen insertion reactions of bridged bicyclic ketones** s. G.R. Krow, Tetrahedron *37,* 1283–307 (1981).

Sodium hypochlorite/ammonium hydroxide *$NaOCl/NH_4OH$*

Prim. amines from ethylene derivs. $C=C \rightarrow CH\text{–}CNH_2$
with chloramine cf. *20,* 248; with $NaOCl/NH_4OH$ s. G.W. Kabalka et al., J. Org. Chem. *46,* 4296–8 (1981); **prim. amines-^{15}N** s. J. Chem. Soc. Chem. Commun. *1982,* 62.

Hydrogen chloride *HCl*

Addition of amines to ynamines C≡CN< → CH=C(N<)$_2$

299. MeCH=CH–C≡C–NPr$_2$ + HNEtPh → MeCH=CH–CH=C(NPr$_2$)(NEtPh)

Ketene aminals. Startg. amine and a little ethereal HCl added to startg. 1-dialkylamino-3-penten-1-yne, and heated with stirring at 70–80° for 3 hrs. → product. F.e.s. S.E. Tolchinsky et al., Zh. Org. Khim. *15,* 1824–31 (1979); **amidines** cf. ibid. *17,* 1807–11 (1981).

Bis(acetonitrile)dichloropalladium(II)/ lithium diisopropylamide *$PdCl_2(MeCN)_2$/i-Pr_2NLi*

N-Alkylation with ethylene derivs. NH → NR

300.

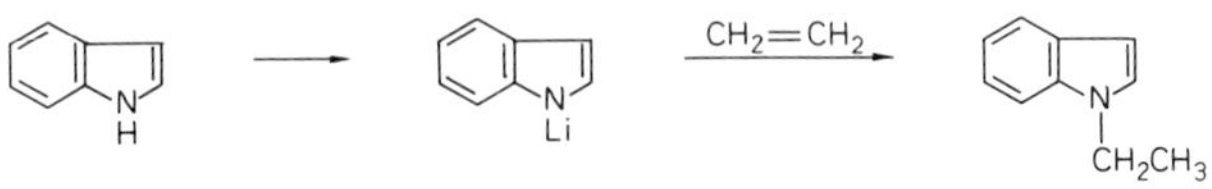

of indoles. A soln. of Li diisopropylamide in tetrahydrofuran added to a soln. of indole in the same solvent at –78°, allowed to warm to room temp. hexamethylphosphoramide added to aid solubility, recooled to –78°, the soln. added dropwise via a precooled syringe to a soln. of ethylene-dichlorobis(acetonitrile)palladium complex and *triethylamine* in 20:1 tetrahydrofuran/hexamethylphosphoramide at –60°, stirred under argon for 45 min. at –55°, the inert atmosphere replaced by H_2, and allowed to warm slowly to room temp. overnight → N-ethylindole. Y 62%. F.e.s. L.S. Hegedus, P.M. Winton, S. Varaprath, J. Org. Chem. *46,* 2215–21 (1981).

Bis(benzonitrile)dichloropalladium(II)/ lead tetraacetate *$PdCl_2(PhCN)_2/(MeCOO)_4Pb$*

Oxyamination of ethylene derivs. C=C →C(OAc)–C(N<)
s. *36,* 303; of aryloxy-2-ethylenes s. J.-E. Bäckvall, S.E. Byström, J. Org. Chem. *47,* 1126–8 (1982).

Osmium tetroxide/chloramine-T *OsO_4/TsN(Cl)Na*

2-Tosylaminoalcohols by Sharpless oxyamination C=C → C(OH)–C–NHTs
s. *31,* 312; carbohydrate derivs. s. A. Banaszek, Pol. J. Chem. *55,* 583 (1981).

Rearrangement ∩

Hydrogen/Nitrogen Type NC∩HN

Without additional reagents *w.a.r.*

Thioureas from N-subst. ammonium thiocyanates N–CSNH$_2$
s. *10,* 281; mathematical optimization, simplex method, s. G. Minkova, V. Baeva, Khim. Ind. (Sofia) *1980,* 402 (Bulg.); Chem. Abstr. *95,* 24485 (1981).

Sodium hydride *NaH*

Isomerization of oxamides ←

301.

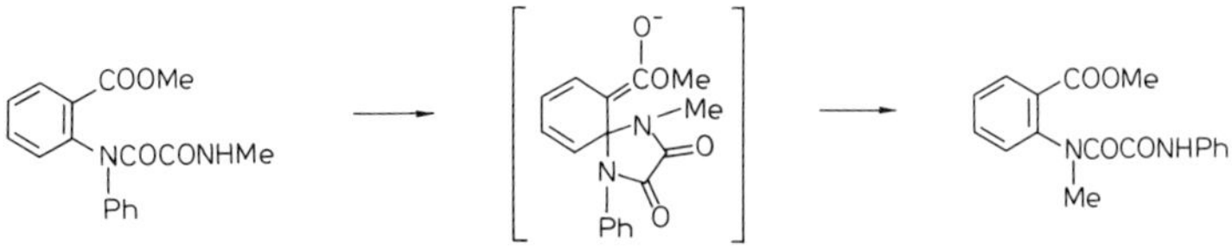

A mixture of NaH in dimethylformamide treated with 2-[(2-methylamino)-1,2-dioxoethyl)phenylamino]benzoic acid methyl ester, and poured into water after 1 hr.

→ 2-[(1,2-dioxo-2-(methylamino)ethyl)phenylamino]benzoic acid methyl ester. Y 90%. The reaction proceeds via intramolecular Michael-retro-Michael conversion with the benzoate residue functioning as the receptor of the terminal nitrogen. F.e.s. N.P. Peet, S. Sunder, R.J. Barbuch, J. Heterocyc. Chem. *17,* 1513–8 (1980).

Hydrogen chloride *HCl*

Pyrazolo[1,5-*a*]pyrimidines ○

s. *30,* 224; f. method cf. B.B. Gavrilenko, Zh. Org. Khim. *18,* 1079–84 (1982).

Hydrogen/Carbon Type NC∩HC

Without additional reagents *w.a.r.*

Ring closures with N′-aryl-N-iminoamidines ○

1,4-dihydro-1,2,4-benzotriazines with HBr cf. *24,* 342; 4,5-dihydro-1*H*-1,3,4-benzotriazepines without reagent s. R. Fusco, F. Sannicolo, Tetrahedron Letters *23,* 1829–30 (1982).

Lithium diisopropylamide *i-Pr$_2$NLi*

Isomerization of azomethines C=N–CH → CH–N=C

s. *37,* 164

Oxygen/Nitrogen Type NC∩ON

Sodium methoxide *NaOMe*

Δ^2-1,2,4-Triazolin-5-ones from 3-amino-1,2,4-oxadiazoles ←

302.

25% Na methoxide in methanol added to a soln. of 3-(3-chlorophenyl)amino-5-methoxy-1,2,4-oxadiazole in the same solvent, and the mixture refluxed for 3 hrs. → product. Y 82%. F.e.s. S.S. Jones, D.B. Staiger, D.F. Chodosh, J. Org. Chem. *47,* 1969–71 (1982).

Sodium alkoxide/irradiation *NaOEt/hν*

Cyclic hydroxamic acids from nitro compds. via nitronates ○

303.

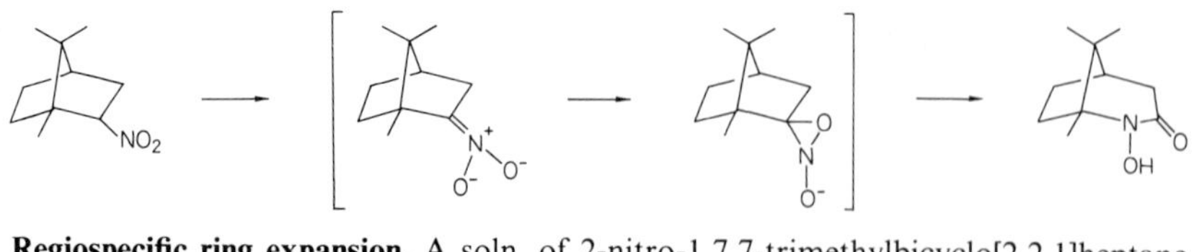

Regiospecific ring expansion. A soln. of 2-nitro-1,7,7-trimethylbicyclo[2.2.1]heptane in ethanolic Na ethoxide irradiated with a low-pressure Hg-lamp for 4.5 hrs. → 2-hydroxy-1,8,8-trimethyl-2-azabicyclo[3.2.1]octan-3-one. Y 76%. F.e., also formation of acyclic hydroxamic acids (cf. Synth. Meth. *36,* 143), s. K. Yamada et al., J. Am. Chem. Soc. *103,* 7003–5 (1981).

Oxygen/Carbon Type NC∩OC

Without additional reagents *w.a.r.*

2-Ethyleneacylamines from 2-ethyleneiminoesters by [3.3]-sigmatropic rearrangement ←

s. *29,* 325; carbohydrate derivs. s. I. Dyong, J. Weigand, H. Merten, Tetrahedron Letters *22,* 2965–8 (1981).

Sodium hydroxide *NaOH*

2,4-Pyrrolidion-5-ylacetic acids from 5-oxa-1-azaspiro[3.4]oct-7-ene-2,6-diones ←

304.

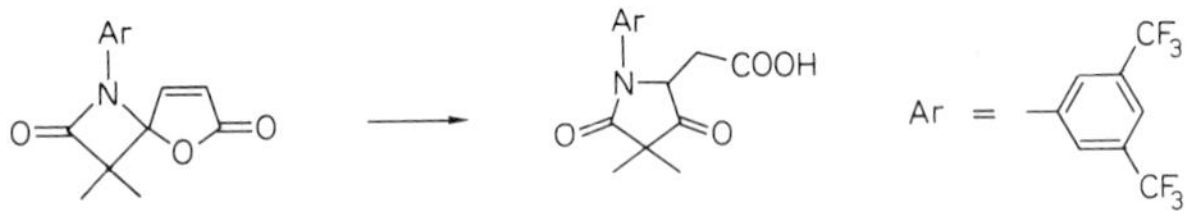

A stirred suspension of startg. spirolactam in methanol treated with 1 *M* NaOH at room temp., and stirring continued until a homogeneous soln. resulted (1–12 hrs.) → product. Y 94%. F.e. and reactions s. M. Roth, Helv. Chim. Acta *64,* 1930–6 (1981).

Potassium tert-*butoxide* t-*BuOK*

4-α-Hydroxy-2-oxazolidones from 2,3-oxidourethans ○

305.

A soln. of startg. epoxide in tetrahydrofuran treated with K *tert*-butoxide at –10° to 0° → product. Y 81%. F.e., **also 4-α-hydroxy-1,3-dioxolan-2-ones from 2,3-oxidocarbonic acid esters** with $AlCl_3$, s. N. Minami, S.S. Ko, Y. Kishi, J. Am. Chem. Soc. *104,* 1109–11 (1982); with NaH cf. W.R. Roush, R.J. Brown, J. Org. Chem. *47,* 1371–3 (1982); 4-α-hydroxy-1,3-dioxolan-2-ones from 2,3-oxidourethans with 5% $HClO_4$ s. T. Katsuki et al., ibid. *47,* 1373–8 (1982).

Boron fluoride BF_3

Cyclic formamides from hydroxyisonitriles

306.

N-Formylindolines. A soln. of *o*-(2-hydroxy-2,4-dimethyl-3-pentenyl)phenyl isocyanide in dichloromethane treated at –78° with BF_3-etherate, the mixture stirred for 10 min. at the same temp., warmed to 0°, and stirred for a further 1 hr. → N-formyl-2-methyl-2-(2-methyl-1-propenyl)indoline. Y 80%. F.e. and Lewis acid catalysts s. Y. Ito, K. Kobayashi, T. Saegusa, Chem. Letters *1980,* 1563–6.

Acetic acid *MeCOOH*

Uracils via 2-amino-1,3-oxazin-4-ones

307.

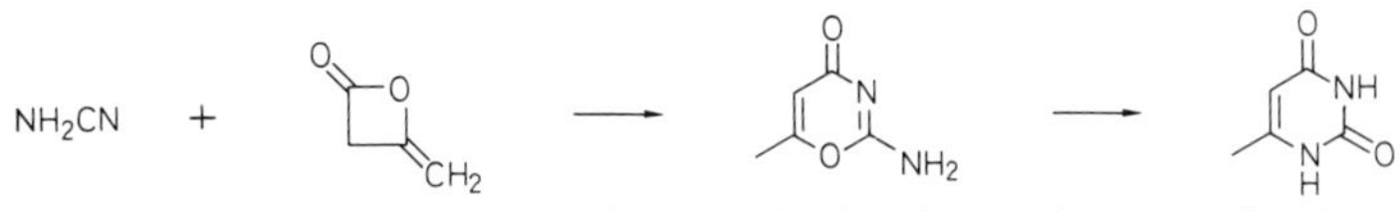

Diketene added dropwise in 4 portions at 24 hr. intervals to a stirred soln. of cyanamide in water at room temp. → 2-amino-6-methyl-4*H*-1,3-oxazin-4-one (Y ca. 58%) refluxed in acetic acid for 24 hrs. → 6-methyluracil (Y 76%). F.e.s. T. Kato, Chem. Pharm. Bull. *29,* 862–6 (1981).

Polyphosphoric acid *$(HPO_3)_n$*

2,4-Quinazolinediones from *o*-ureidocarboxylic acids via 2-imino-1,2-dihydro-4*H*-3,1-benzoxazin-4-ones

s. *21,* 378; from 2-amino-3,1-benzoxazin-4-ones via *o*-ureidocarboxylic acid amides cf. I. Bitter, L. Szöcs, L. Töke, Acta. Chim. Acad. Sci. Hung. *107,* 57–66 (1981).

Sulfuric acid *H_2SO_4*

2-Pyridones from 2-pyrone imines ←

s. *31,* 305; 3,4-dihydro-2-pyridones s. C. Seoane et al., J. Heterocyc. Chem. *18,* 309–14 (1981).

Nitrogen/Nitrogen Type NC∩NN

Without additional reagents *w.a.r.*

Tetrazole ring from azidonitriles ○

Intramolecular 1,3-dipolar cycloaddition

s. *14,* 402; tetrazolo[1,5-*a*]quinolines, also 2-cyanoindoles from the geoisomers, s. L. Garanti, G. Zecchi, J. Org. Chem. *45,* 4767–9 (1980).

Nitrogen/Carbon Type NC∩NC

Anion exchanger ←

Dimroth rearrangement ←

with NaOH cf. *30,* 228; with Dowex 1 (OH^-) resin s. Y. Yamazaki, H. Maeda, Agr. Biol. Chem. *45,* 2631–2 (1981).

Sulfur/Carbon Type NC∩SC

n-*Butyllithium* n-*BuLi*

5-Mercaptoimidazoles from 5-aminothiazoles ←

308.

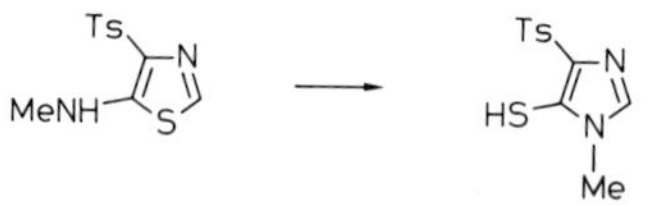

A soln. of *n*-butyllithium in hexane added to 5-(methylamino)-4-tosylthiazole (prepn. s. 642) in tetrahydrofuran at –70°, cooling bath removed, warmed to 20° during 10 min., then stirred for a further 15 min. → 1-methyl-4-tosylimidazole-5-thiol. Y 89%. F.e.s. S.P.J.M. van Nispen et al., Rec. Trav. Chim. Pays-Bas *101,* 28–34 (1982).

Bis(benzonitrile)dichloropalladium(II) *$PdCl_2(PhCN)_2$*

N-Alkylcarboxylic acid amides from S-allylthioiminoesters ←

s. *37,* 703

Carbon/Carbon Type **NC∩CC**

Irradiation ⚡

Intramolecular[2+2]-cycloaddition of ethyleneazo compds. ○
Diazacage compds.

309. 

A soln. of startg. m. in acetonitrile-d_3 contained in an n.m.r. tube UV-irradiated for 5.5 hrs. → product. Y 90%. L.A. Paquette et al., J. Org. Chem. *45,* 4922–6 (1980).

Electrolysis/potassium bromide *↯/KBr*

Electrochemical Hofmann degradation $CONH_2 \rightarrow NHCOOR$
Urethans from carboxylic acid amides

310. $Me(CH_2)_7CONH_2 \rightarrow [\,Me(CH_2)_7N{=}C{=}O\,] \xrightarrow{MeOH} Me(CH_2)_7NHCOOMe$

A stirred, water-cooled soln. of KBr and startg. amide in methanol electrolysed between platinum anode and cathode with a constant current of 0.1–0.4 A until 9 F/mol of electricity passed → product. Y 73% by g.l.c. – The **mild, neutral conditions** are an obvious advantage over the classical method (cf. Synth. Meth. *11,* 402). F.e., also with isolation of the intermediate **isocyanates,** s. T. Shono et al., Chem. Letters *1982,* 565–8.

Diethylamine s. under $PdCl_2(PhCN)_2$ *Et_2NH*

Mercuric acetate/sodium tetrahydridoborate *$(MeCOO)_2Hg/NaBH_4$*

N-Heterocyclics from ethyleneamines
via intramolecular aminomercuration-demercuration ○
with $HgCl_2$ cf. *27,* 345; with Hg(II) acetate, piperidine ring, s. Y. Saito et al., Tetrahedron Letters *21,* 75–8 (1980); by transannular ring closure, *cis*-hexahydro-5(1*H*)-indolizinones, s. S.R. Wilson, R.A. Sawicki, J. Heterocyc. Chem. *19,* 81–3 (1982); intramolecular acylaminomercuration-demercuration, N-acylpyrrolidines, s. K.E. Harding, S.R. Burks, J. Org. Chem. *46,* 3920–2 (1981).

Palladium(II) acetate/pyridine *$(MeCOO)_2Pd/C_5H_5N$*

2-Oxazolidones via Hofmann-type degradation ○
s. *28,* 309; from *β*.hydroxycarboxylic acids, f. method, cf. J.M. Caroon, A.F. Kluge et al., J. Med. Chem. *24,* 1320–8 (1981).

Bis(benzonitrile)dichloropalladium(II)/diethylamine/trifluoromethanesulfonic acid *$PdCl_2(PhCN)_2/Et_2NH/CF_3SO_3H$*

Cyclic amines from ethyleneamines
Regiospecific ring closure

311. 

Piperidines. A soln. of startg. aminoalkene and trifluoromethanesulfonic acid in tetrahydrofuran added to 1 equivalent bis(benzonitrile)dichloropalladium(II) in the same solvent at 0°, stirred 5 min., treated with 2 equivalents diethylamine at –65° during 30 min., stirred for 60 min., flushed with H_2, pressurized at ca. 1.8 atm. for ca. 30 min. while warming to 0°, and stirred with 5 *M* KOH → product. Y 76%. F.e. incl. **pyrrolidines** s. B. Pugin, L.M. Venanzi, J. Organometal. Chem. *214,* 125–33 (1981).

Exchange ⇃↾

Hydrogen ↑ NC⇃↾H

Electrolysis ϟ

Anodic generation of nitrilimines ○

s-Triazolo[4,3-a]pyridinium salts from pyridines and hydrazones

312. 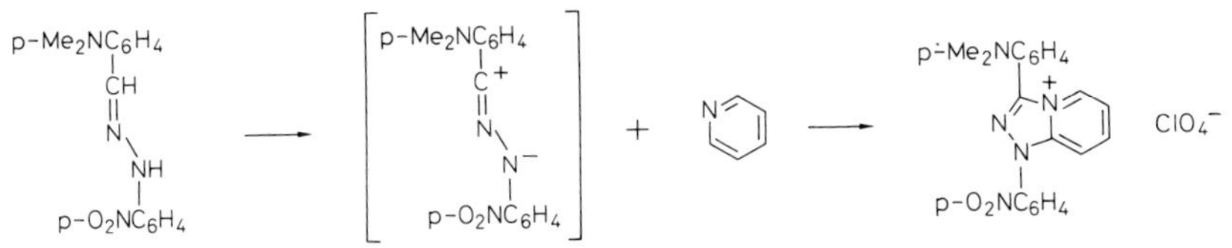

A soln. of startg. hydrazone and pyridine in acetonitrile containing tetraethylammonium perchlorate electrolysed at 1.05 V in the anode compartment of a divided cell fitted with a Pt-gauze anode until the current fell from 200 to 20 mA → 1-*p*-nitrophenyl-3-(*p*-dimethylaminophenyl)-*s*-triazolo[4,3-*a*]pyridinium perchlorate. Y 80%. F.e.s. I. Tabaković, S. Crljenak, Heterocycles *16*, 699–702 (1981).

Sodium hydroxide s. under TsN_3 *NaOH*

Potassium hydroxide *KOH*

3,6-Dihydro-1,3,4-oxadiazine-2-thiones from α-hydroxyhydrazones

313. 

A soln. of the startg. m. in dry dimethylformamide treated with KOH and dry CS_2, heated with stirring at 60–70° for 30 min., and stirring continued at room temp. for 6 hrs. → product. Y 88%. F.e.s. A.I. Dychenko, P.S. Pelkis, Ukr. Khim. Zh. *47*, 634–7 (1981).

Sodium selenite Na_2SeO_3

Hydrazones from diazonium salts $CH_2 \rightarrow C{=}NNHAr$

with Na acetate cf. *9*, 423; with sodium selenite s. E.P. Nesynov, T.F. Aldokhina, Zh. Obshch. Khim. *49*, 1087–93 (1979).

Silver nitrate/boron fluoride $AgNO_3/BF_3$

Aromatic nitration $H \rightarrow NO_2$

s. *23*, 380; with added $AgNO_3$ cf. G.A. Olah et al., J. Org. Chem. *46*, 3533–7 (1981).

2-Chloropyridinium iodide/sodium trithiocarbonate ←

Sym. thioureas from prim. amines $2\ RNH_2 \rightarrow RNHCSNHR$

314. $2\ PhCH_2NH_2 + (NaS)_2CS \rightarrow (PhCH_2NH)_2CS$

A soln. of 2-chloropyridinium iodide and sodium trithiocarbonate in dichloromethane stirred at room temp. for 0.5 hrs., benzylamine added, and stirring continued at reflux for 3 hrs. → N,N′-dibenzylthiourea. Y 100%. – No product was obtained from sec. amines or weakly basic prim. amines. F.e., **also cyclic thioureas,** s. Y. Takikawa et al., Chem. Letters *1982*, 641–2.

Trifluoroacetic anhydride s. under NH_4NO_3 $(CF_3CO)_2O$

Sodium trithiocarbonate s. under 2-Chloropyridinium iodide Na_2CS_3

Oxalyl chloride $(COCl)_2$

Thiolcarbalkoxyisocyanates from thiolcarbamic acid esters $RSCONH_2 \rightarrow RSCON{=}C{=}O$

315. $MeSCONH_2 + (COCl)_2 \rightarrow MeSCON{=}C{=}O$

S-Methyl thiocarbamate in methylene chloride added with stirring and cooling to a soln. of oxalyl chloride in the same solvent at 0–5°, and refluxed for 2.5 hrs. → S-methylthiocarbonyl isocyanate. Y 68%. F.e.s. Y.A. Strepikheev et al., Zh. Vses. Khim. Obshchest. *26,* 352 (1981).

2-Azido-3-ethylbenzothiazolium fluoroborate ←

Azo group transfer by azidinium salts ←
s. *33,* 335; 1-phosphoniodiazo compds. s. M. Regitz et al., Ann. Chem. *1981,* 1865–73.

Tosyl azide TsN_3

Diazo group transfer $CH_2 \rightarrow C{=}N_2$
s. *20,* 271; α.diazophosphonic esters, prepn. and reactions, s. P.A. Bartlett et al., J. Org. Chem. *47,* 1284–91 (1982); with NaOH and Aliquat 336 as phase transfer catalyst, di-*tert*-butyl diazomalonate, s. H.J. Ledon, Org. Synth. *59,* 66 (1979).

Tosyl azide/sodium hydroxide/Aliquat 336 ←
s. *20,* 271s*37*

Ammonium nitrate/trifluoroacetic anhydride $NH_4NO_3/(CF_3CO)_2O$

Aromatic nitration $H \rightarrow NO_2$

316. $Ph\text{–}H \xrightarrow{NH_4NO_3} Ph\text{–}NO_2$

Chloroform added to a mixture of benzene and NH_4NO_3 in trifluoroacetic anhydride, and stirred at room temp. for 2 hrs. → nitrobenzene. Y 95–99%. F.e., metal nitrates, and limitations, s. J.V. Crivello, J. Org. Chem. *46,* 3056–60 (1981); **review** of recent preparative aspects s. G.A. Olak et al., Proc. Natl. Acad. Sci. USA *79,* 4487–94 (1982).

tert-*Butyl thionitrite/hydrogen chloride* t-*BuS–N=O/HCl*

Hydroximinochlorides from prim. chlorides $CH_2Cl \rightarrow C(=NOH)Cl$

317. $t\text{-}BuCOCH_2Cl + t\text{-}BuS\text{–}N{=}O \rightarrow t\text{-}BuCOC(Cl){=}NOH$

α-**Ketohydroximinochlorides.** A soln. of *tert*-butyl thionitrite in ether added over 20 min. to a stirred soln. of *tert*-butyl chloromethyl ketone in ether containing HCl gas at 25°, and stirring continued at this temp. for a further 3 hrs. → product. Y 85%. – Bromoketones gave low yields. F.e.s. Y.H. Kim, S.H. Jung, Tetrahedron Letters *22,* 2371–2 (1981).

Sulfuric acid H_2SO_4

Nitration $H \rightarrow NO_2$
s. *2,* 346; adiabatic process, nitrobenzene, s. A.A. Guenkel, H.C. Prime, J.M. Rae, Chem. Eng. (N.Y.) *88* (16), 50 (1981); Chem. Abstr. *95,* 203469 (1981).

Periodate s. under OsO_4 IO_4^-

Aliquat 336 s. under TsN_3 ←

Hydrogen chloride s. under tert-*Butyl thionitrite* *HCl*

Osmium tetroxide/periodate OsO_4/IO_4^-

Isoquinolines from indenes ◌

by ozonolysis cf. *36,* 327; with OsO_4/periodate in phosphate buffer (pH 8) s. D.S. Dime, S. McLean, J. Org. Chem. *46,* 5000–3 (1981).

Oxygen ↑ NC↓↑O

Without additional reagents *w.a.r.*

N-Alkylation with amide acetals NH → NR

s. *27,* 373; preferential N-alkylation of cytosine s. D.L. Helfer, R.S. Hosmane, N.J. Leonard, J. Org. Chem. *46,* 4803–4 (1981); Synthesis *1981,* 118–9.

N-Aminomethylation NH → $NCH_2N\lt$

of phthalimides cf. *11,* 423; of cyanamides s. G. Zinner, J. Wettke, Arch. Pharm. *314,* 889–91 (1981).

Enamines from enolethers C=C–OR → C=C–N<

with HCl cf. *19,* 480; without reagent, *β*-amino- from *β*-aryloxy-*α,β*-ethylenesulfones, retention of configuration, s. M.A. Vasileva et al., Zh. Org. Khim. *17,* 1405–8 (1981).

Urethans from mixed carbonic acid esters and amines ROCOOPh → ROCON<

s. *27,* 404; from alkyl phenyl carbonates s. V.V. Mikheev, N.V. Svetlakov, R.M. Garipov, Zh. Org. Khim. *18,* 246–51 (1982).

Hydrazonoesters from orthocarboxylic acid esters $C(OR)_3$ → C(OR)=N–N<

s. *34,* 352; N-sulfonylhydrazonoesters without reagent s. P.G. Williard, L.A. Grab, J.P. Springer, Tetrahedron Letters *22,* 2731–4 (1981).

Sulfonylformamidrazones SO_2N–N–CH=NR

s. *34,* 332s*36;* f. method cf. Tetrahedron *38,* 369–72 (1982).

2-Hydroxyaminimides
from carboxylic acid esters, hydrazines, and oxido compds. ←

318.

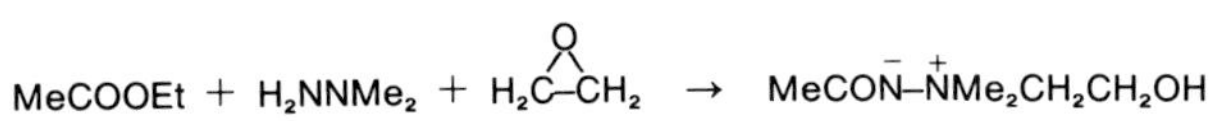

1,1-Dimethylhydrazine added rapidly at room temp. with stirring to ethyl acetate under N_2, ethylene oxide passed through the mixture, then stirring continued for 4.75 hrs. → product. Y 94%. F.e.s. R.J. Small, Org. Prep. Proc. Intern. *13,* 55–8 (1981); s.a. M. Tichniouin et al., Eur. J. Med. Chem. *17,* 265–70 (1982).

Aromatization by ring closure of *α,β*-ethylene-*δ*-ketoaldehydes ○

319.

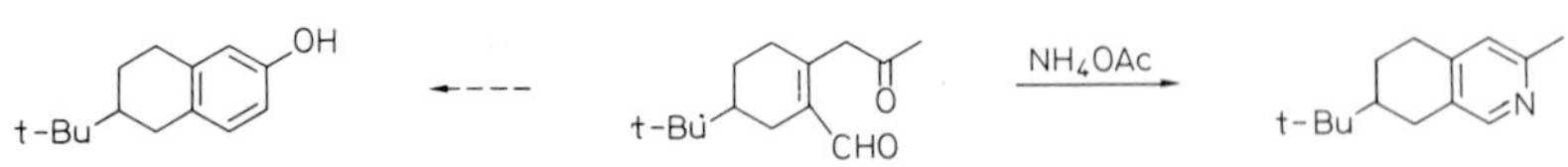

Pyridine ring. A soln. of crude startg. m. (prepd. by Wacker oxidation of the corresponding dienal) in ethanol treated with ca. 5 equivalents of ammonium acetate for 0.5 hrs. → product. Y 91% (based on startg. dienal). F.e.s. M.A. Tius, A. Thurkauf, J.W. Truesdell, Tetrahedron Letters *23,* 2819–22 (1982); **phenol ring** with methanolic KOH cf. ibid. *23,* 2823–4 (1982).

4(3H)-Pyrimidinones from amidines

and *β*-ketocarboxylic acid esters cf. *25,* 257; and *α,β*-acetylenecarboxylic acid esters s. K.A. Gupta, A.K. Saxena, P.C. Jain, Synthesis *1981,* 905–7.

Cyclic aminals from diamines and oxo compds.

s. *37,* 36

Without additional reagents *w.a.r.*

Ring closures with 2,3-dihydroxy-1,4-dioxane as glyoxal equivalent

320. 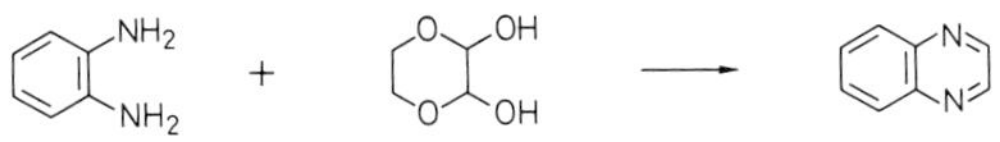

Pyridazine ring from *o*-diamines. A soln. of 2,3-dihydroxy-1,4-dioxane in ethanol treated with *o*-phenylenediamine, and stirred at room temp. for 30 min. → quinoxaline. Y 95%. – The reagent is stable, unlike anhydrous glyoxal which has a tendency to polymerize. F.e. and reactions s. M.C. Venuti, Synthesis *1982,* 61–3.

Isocarbostyrils from isocoumarins
s. *24,* 386; from 3-alkoxy-3,4-dihydroisocoumarins cf. M. Somei et al., Chem. Pharm. Bull. *29,* 249–53 (1981).

Imidazoles
from *α*-diketones s. *15,* 340; from *α*-diketone monoximes, 2-arylimidazoles, cf. A.C. Veronese et al., J. Heterocyc. Chem. *17,* 1723–5 (1980).

N-Glycosyl-N-heterocyclics from aldose hydrazones

321. 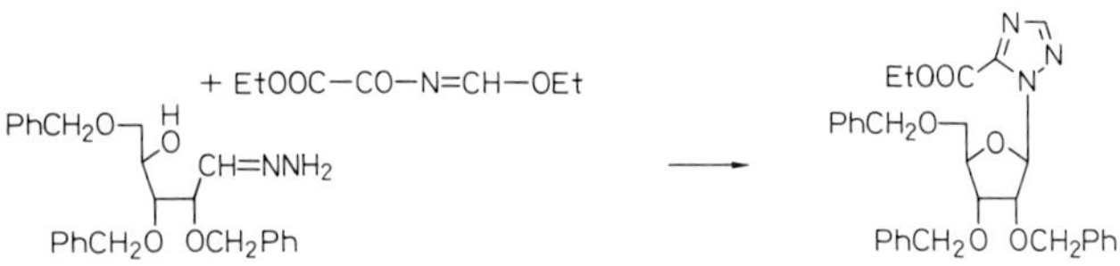

1,2,4-Triazole nucleosides. A soln. of startg. ribose hydrazone in methylene chloride treated at 0° with a soln. of [N-(ethoxymethylene)ethanedioic acid amide ethyl ester in the same solvent, the mixture allowed to react at room temp. for 30 min., the solvent replaced by xylene, and the soln. heated at 130° for 30 min → 1-(2,3,5-tri-O-benzyl-*β*-D-ribofuranosyl)-1*H*-1,2,4-triazol-5-carboxylic acid ethyl ester. Y 75%. F.e. and methods s. R.R. Schmidt, D. Heermann, Chem. Ber. *114,* 2825–33 (1981); **also pyrazole nucleosides** s. Ann. Chem. *1981,* 2309–17.

Polyazamacroheterocyclics from diamines and dialdehydes
s. *29,* 354; s.a. P.G. Owston et al., J. Chem. Soc. Chem. Commun. *1980,* 1218–20.

Phthalazones from phthalides ◌
s. *12,* 427s*29;* from 3-alkylidenephthalides, retention of carboxyl, cf. A. Mroczkiewicz, Acta Pol. Pharm. *38,* 263 (1981) (Pol.); Chem. Abstr. *96,* 181225 (1981); 3,4-dihydro-1(2*H*)-phthalazones s. M. Kuzuya et al., Chem. Pharm. Bull. *28* 3561–9 (1980).

Sodium hydride *NaH*

Double ring closures with cyanogen bromide ○
5*H*-Benzoxazolo[3,2-*a*]quinazolin-5-ones

322. 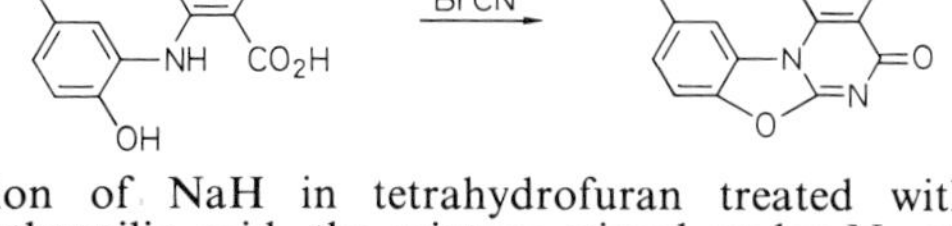

A cold suspension of NaH in tetrahydrofuran treated with N-(5-chloro-2-hydroxyphenyl)anthranilic acid, the mixture stirred under N_2 at 0° for 10 min., cyanogen bromide in the same solvent added in one portion to the dark green soln., stirring continued at the same temp. for 1 hr. and at room temp. for 2 hrs., the mixture allowed to stand overnight, and refluxed for 30 min. → 10-chloro-5*H*-benzoxazolo[3,2-*a*]quinazolin-5-one. Y 100%. F.e. and method s. D.H. Kim, J. Hetereocyc. Chem. *18,* 287–91 (1981); **3,4-dihydro-1*H*-pyrimido[1,2-*a*]benzimidazol-2-ones** cf. E. Szarvasi et al., Eur. J. Med. Chem. *16,* 327–32 (1981).

Potassium hydroxide *KOH*

3,4-Dihydro-2(1*H*)-pyrimidinone ring from *α,β*-ethyleneketones
s. *25,* 276; 3,4-dihydro-2(1*H*)-pyrimidinethiones using thiourea and KOH cf. M.I. Ali, A.E.G. Hammam, J. Chem. Eng. Data *26,* 342–3 (1981).

Sodium methoxide/acetyl chloride *NaOMe/MeCOCl*

1,3-Dipolar cycloaddition via nitrile oxides
1,2,4-Oxadiazoles from nitro compds.
with phenyl isocyanate/NEt_3 cf. *21,* 436; with Na methoxide in methanol in the presence of acetyl chloride, also Δ^2-1,2,4-oxadiazolines, isoxazoles, and Δ^2-isoxazolines, s. K. Harada, E. Kaji, S. Zen, Chem. Pharm. Bull. *28,* 3296–303 (1980).

Sodium ethoxide *NaOEt*

4-Amino-Δ^4-pyrrol-2-ones from 2-azetidinone-4-carboxaldehydes ◌

323. 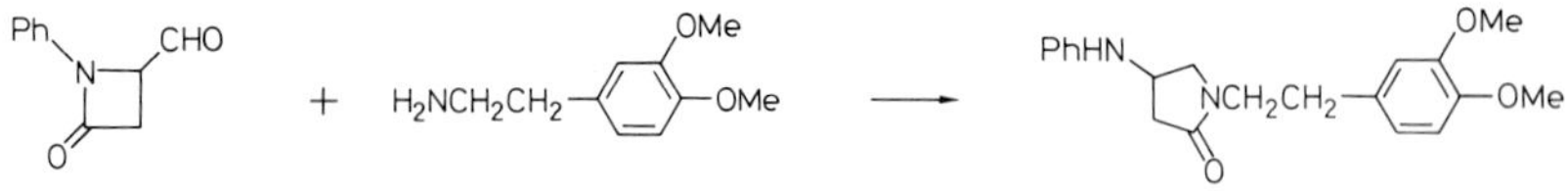

A mixture of 4-formyl-1-phenyl-2-azetidinone and startg. amine in ethanol refluxed for 2 hrs., Na ethoxide added, and refluxing continued for a further 1.5 hrs. → 4-anilino-2,3-dihydro-1-(3,4-dimethoxyphenethyl)-2-oxopyrrole. Y 75%. F.e., **also 3-*α*-amino-4-hydroxy-2-pyrrolidones from 3-oxiranyl-2-azetidinones,** s. S. Kano, S. Shibuya, T. Ebata, J. Heterocyc. Chem. *18,* 1239–40 (1981).

Potassium carbonate *K_2CO_3*

2-Aminopyrimidines from *β*-diketones ○
with guanidine carbonate cf. *1,* 344; with guanidines in the presence of K_2CO_3 s. A. Kreutzberger, B. Richter, Arch. Pharm. *315,* 2–7 (1982); N-condensed pyrimidine ring s. ibid. *315,* 47–52 (1982).

Sodium azide *NaN_3*

Azides from alcohols via alkoxyphosphonium salts OH → N_3
s. *33,* 360; glycosyl azides s. Tetrahedron *38,* 103–12 (1982); inversion of configuration in carbohydrate derivs., s. H. Kunz, P. Schmidt, Ann. Chem. *1982,* 1245–60.

2-Ethyleneazides from 2-ethylenemesylates C=C–C–OMs → N_3C–C=C
via [3.3]-sigmatropic rearrangement
Carbohydrate derivs.

324. 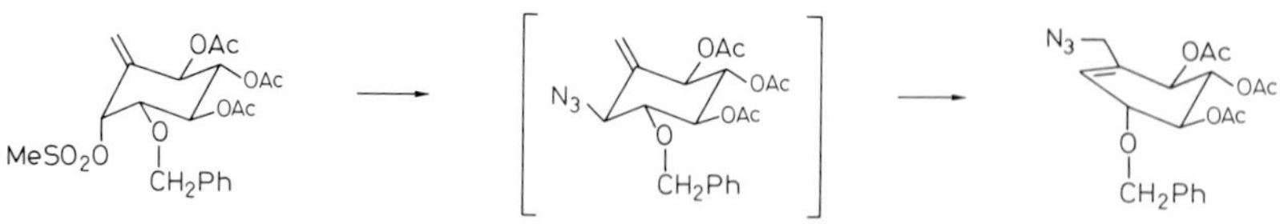

A soln. of startg. mesylate in dimethylformamide treated with NaN_3, and the mixture stirred at 70° for 1 hr. → 1L-(1,3/2,4)-2,3,4-tri-O-acetyl-5-C-azidomethyl-1-O-benzoyltetrahydroxycyclohex-5-ene, Y 59%. F.e.s. H. Paulsen, F.R. Heiker, Carbohyd. Res. *102,* 83–98 (1982).

Sodium azide/trifluoroacetic acid *NaN_3/CF_3COOH*

Azides from alcohols OH → N_3
in the presence of H_2SO_4 cf. *33,* 379; with CF_3COOH, 2-azidoadamantane deriv., s. A. Kalir, D. Balderman, Org. Synth. *60,* 104–8 (1981).

Ammonium hydroxide *NH_4OH*

Pyridines from pyrylium salts ○

with NH_4-carbonate cf. *8,* 468; with NH_4OH, 2,6-di-*tert*-butyl-4-methylpyridine, s. A.G. Anderson, P.J. Stang, Org. Synth. *60,* 34–40 (1981).

Sodium salt *Na^+*

Pyrazoles from β-dioxo compds.

from β-diketones s. *7,* 437; from dialdehydes, 4-aminopyrazoles, s. C. Reichardt, K. Schagerer, Ann. Chem. *1982,* 530–5.

N,N-Dimethylcyclohexylamine s. under P_2O_5 ←

Benzylamine s. under Pd–C *$C_6H_5CH_2NH_2$*

Triethylamine (s.a. under $(PhO)_2P(O)N_3$, $PhOP(O)(Cl)NHPh$, $(EtO)_2P(O)CN$, N-(Diphenoxyphosphoryl)-Δ^4-2-oxazolone, Ph_3P) *Et_3N*

N-Acylation with acyloximes NH → NAc
Peptide synthesis

with acetic acid cf. *20,* 293s*26;* with triethylamine s. T. Kubiak, S. Drabarek, Pol. J. Chem. *55,* 73–8 (1981); using 2-pyridyl methyl ketoxime derivs. cf. T. Pluciński, G. Kupryszewski, ibid. *55,* 573–82 (1981).

N-Acylation with N-hydroxysuccinimide esters

s. *32,* 317; prepn. of silica gel-supported α-carbalkoxyaminocarboxylic acid amides s. N. Watanabe, Chem. Letters *1981,* 1331–2.

Peptide synthesis under mild conditions COOH → CON<
Protection of amino groups as cobalt(III) complexes
with simultaneous activation of carbalkoxy groups
Cathodic removal of the protective group

325. 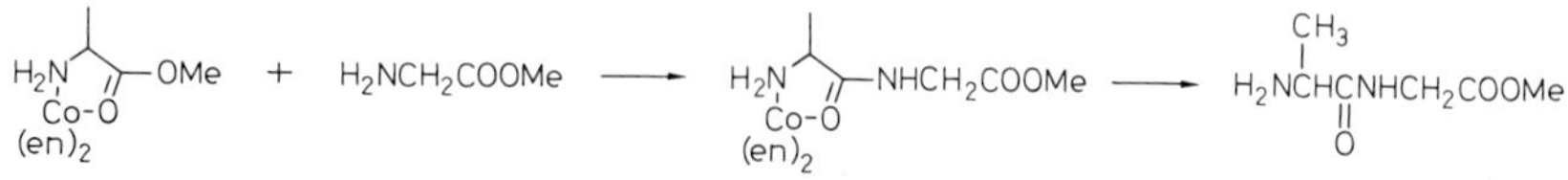

A soln. of glycine methyl ester hydrochloride and triethylamine in dimethyl sulfoxide added to a soln. of bis(ethylenediamine)alaninatocobalt(III) in the same solvent, and quenched after 5 min. with acetic acid → intermediate dipeptide complex (Y 89%) in water containing pyridinium acetate electrolysed at –1.0 V for 2 hrs. at pH <5, the soln. concentrated, H_2S bubbled through the mixture for 20 min., CoS filtered off, and the filtrate worked up → dipeptide (Y 96%). F.e.s. C.R. Clark et al., J. Am. Chem. Soc. *103,* 7023–5, 7025–6 (1981).

Protection of guanidino groups in arginine-containing peptides

s. *21,* 426; **review** s. Z. Procházka, K. Jŏst, K. Bláha, Chem. Listy *75,* 699 (1981).

Piperidine *$C_5H_{11}N$*

Benzoxazoles from o-aminophenols and aldehydes ○

s. *21,* 437; with piperidine as catalyst cf. A.S. Hammam et al., J. Appl. Chem. Biotechnol. *32,* 485–8 (1982).

Pyridine (s.a. under $SOCl_2$ and MnO_2) *C_5H_5N*

β-Ketocarboxylic acid amides COOR → CON<
from β-ketocarboxylic acid esters and amines

s. *12,* 453; s.a. Z. Ejmocki, J. Korczak, Z. Eckstein, Przem. Chem. *59,* 597 (1980) (Pol.); Chem. Abstr. *94,* 191862 (1981).

2-Aminopyridine ←

1-Subst. 2(1*H*)-pyridones from 2-pyrones ○

1-amino-2(1*H*)-pyridones with N_2H_4 cf. *34,* 333; 1-hydroxy-2(1*H*)-pyridones using $NH_2OH.HCl$ and 2-aminopyridine as base s. G. Lohaus, W. Dittmar, Arzneimittel-Forsch. *31,* 1311–6 (1981).

4-Dimethylaminopyridine s. under RN=C=NR ←

Ethylmagnesium bromide *EtMgBr*

Oxo compd. N-derivs. from oxo compds. CO → C=N–N<

diphenyl ketimines cf. *24,* 401s*36;* hydrazones s. Bull. Chem. Soc. Japan *54,* 2337–40 (1981).

Trimethylaluminium Me_3Al

1,3-Heterocyclics from carboxylic acid esters ○

326.

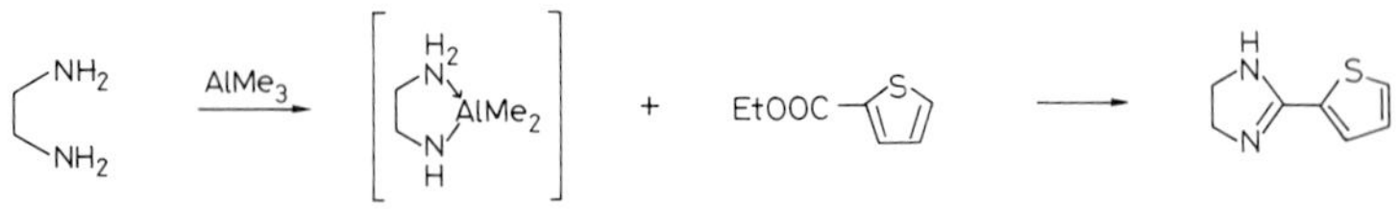

Δ²-Imidazolines. Ethylenediamine added dropwise to a stirred soln. of trimethylaluminium in toluene maintaining the temp. below 10°, when evolution of methane ceased ethyl thiophene-2-carboxylate added gradually at room temp., refluxed for 3 hrs. under argon, cooled, treated dropwise with water/methanol/methylene chloride, and refluxed on a steam bath for 15 min. → 2-(2-thienyl)-2-imidazoline. Y 92%. F.e. and limitations, **also benzimidazoles and benzothiazoles,** s. G. Neef, U. Eder, G. Sauer, J. Org. Chem. *46,* 2824–6 (1981).

Sodium trihydridocyanoborate $Na[BH_3CN]$

Amines from oxo compds. CO → CHN<

s. *19,* 448s*29;* benzylaminoalkylphosphonates s. G. Fabre, N. Collignon, P. Savignac, Can. J. Chem. *59,* 2864–9 (1981); *β*-(dialkylamino)phosphonates s. Phosphorus-Sulfur *10,* 81–6 (1981); sec. methylamines s. L.L. Martin et al., J. Med. Chem. *25,* 346–51 (1982).

Alumina Al_2O_3

Benzimidazoles from *o*-nitramines ○

s. *26,* 369; f. method, also 4(3*H*)-quinazolones from *o*-nitrocarboxylic acid amides, s. C.V.C. Rao, K.K. Reddy, N.V.S. Rao, Indian J. Chem. *19B,* 655–8 (1980).

Molecular sieve ←

2-Aminoenazo compds. $C(NH_2)=C–N=NR$

327.

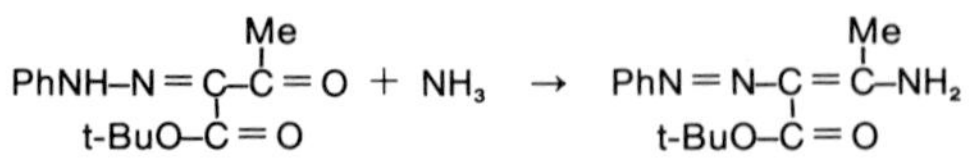

from *α*-diketone monohydrazones. A soln. of startg. ketohydrazone in ether repeatedly satd. with NH_3 with stirring during 13 days in the presence of freshly regenerated molecular sieve 3 Å → *tert*-butyl 3-amino-2-phenylazo-2-butenoate, Y 75%. F.e., **also from enamines and diazonium salts,** s. J. Gasteiger, U. Strauss, Chem. Ber. *114,* 2336–47 (1981).

Boron fluoride BF_3

Enamines from ketones CHCO → C=C–N<

with $MgSO_4$ cf. *19*, 446s*32;* with BF_3, β-amino-α,β-ethyleneketones from β-diketones with low-boiling amines, s. M. Azzaro, S. Geribaldi, B. Videau, Synthesis *1981*, 880–1.

N-Glycosyl-N-heterocyclics from acyl glycosides OAc → N<

with $ClCH_2COOH$ cf. *21*, 444; with BF_3 s. C. Chavis et al., J. Carbohydr. Nucleosides Nucleotides *8*, 507–18 (1981).

N-Sulfonylimines from sulfonamides and aldehydes CHO → CH=NSO_2R

328. $p\text{-}NO_2C_6H_4CHO + H_2NSO_2C_6H_4Me\text{-}p \rightarrow p\text{-}NO_2C_6H_4CH{=}NSO_2C_6H_4Me\text{-}p$

A refluxing soln. of *p*-nitrobenzaldehyde and p-toluenesulfonamide in benzene treated with BF_3-etherate, and heating continued for 16 hrs. with continuous entrainment of water → N-(*p*-nitrobenzylidene)-*p*-toluenesulfonamide. Y 66%. F.e.s. W.R. McKay, G.R. Proctor, J. Chem. Soc. Perkin Trans. I *1981*, 2435–42; reactions of product s. ibid. *1981*, 2443–50.

Fluoroborates ←

4*H*-1,2-Diazepines from pyrylium salts

s. *26*, 370; *review of 1,2-diazepines* s. V. Snieckus, J. Streith, Acc. Chem. Res. *14*, 348–55 (1981).

*Tetra-*n*-butylammonium iodide s. under N,N'-Carbonyldiimidazole* n-$Bu_4N^+I^-$

Urea NH_2CONH_2

Dicarboxylic acid imides from anhydrides

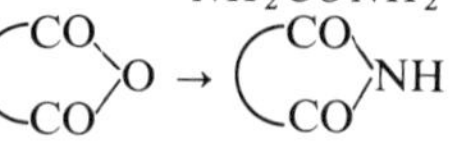

s. *19*, 460; also from diacids s. G.C. Crockett, T.H. Koch et al., Synth. Commun. *11*, 447–54 (1981).

Dicyclohexylcarbodiimide (s.a. under N-Hydroxytartrimide) *RN=C=NR*

Peptides from *o*-nitrophenyl esters COOAr → CON<

s. *14*, 450s*28;* with 4-picolyl ester groups as solubilizing attachment s. F.H.C. Stewart, Australian J. Chem. *34*, 2013–7 (1981).

Dicyclohexylcarbodiimide/4-dimethylaminopyridine ←

Peptide synthesis COOH → CON<

with Et_3N as additive cf. *19*, 454; with 4-dimethylaminopyridine s. S.S. Wang, R.B. Merrifield et al., Int. J. Pept. Protein Res. *18*, 459–67 (1981); Pept. Synth. Struct., Funct., Proc. Am. Pept. Symp., 7th *1981*, 197–9; Chem. Abstr. *97*, 39360 (1982).

Basic carbodiimide/1-hydroxybenzotriazole ←

Peptide synthesis

s. *22*, 394s*33;* in 1,2-dichloroethane without isolation of the intermediate peptide fragments, 'hold in solution' method, s. S. Nozaki, I. Muramatsu, Bull. Chem. Soc. Japan *55*, 2165–8 (1982).

N,N'-Carbonyldiimidazole ←

Nitriles from aldoximes CH=NOH → CN

s. *28*, 440; optically active compds. s. C. Botteghi, G. Chelucci, M. Marchetti, Synth. Commun. *12*, 25–33 (1982).

*N,N′-Carbonyldiimidazole/tetra-*n*-butylammonium iodide* ←

Δ^2-Oxazolines from aziridines and carboxylic acids via N-acylaziridines ○

329. 

N,N′-Carbonyldiimidazole added in several portions over 5 min. to a soln. of acetic acid in tetrahydrofuran, stirred at room temp. for 30 min., 2(S)-benzylaziridine added in one portion, stirring continued for a further hr., treated with a little tetra-*n*-butylammonium iodide, and refluxed for 2.5 hrs. → (–)-2-methyl-4(S)-benzyl-2-oxazoline. Y 100%. F.e., also isolation of the intermediate, s. G.S. Bates, M.A. Varelas, Can. J. Chem. *58,* 2562–6 (1980).

(+)-N-Hydroxytartrimide/dicyclohexylcarbodiimide ←

Asymmetric peptide synthesis COOH → CON⋜

330. 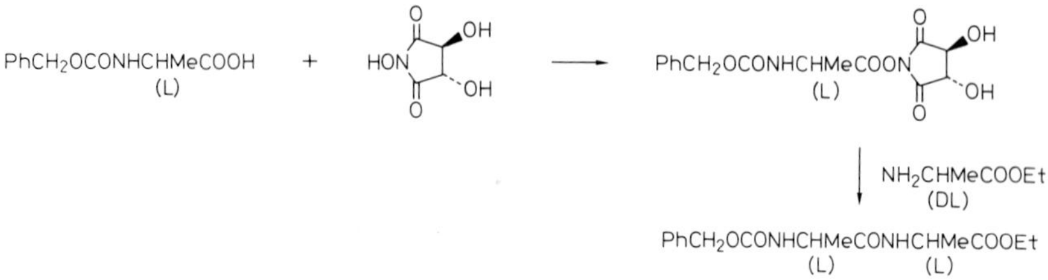

A soln. of (+)-N-hydroxytartrimide in tetrahydrofuran treated with N-carbobenzoxy-L-alanine at 0° for 24 hrs. in the presence of dicyclohexylcarbodiimide, and stirred with racemic ethyl alaninate at the same temp. for a further 24 hrs. → ethyl N-carbobenzoxy-L-alanyl-L-alanine. Y 93%; optical Y 100%. F.e.s. T. Teramoto, M. Deguchi, T. Kurosaki, Tetrahedron Letters *22,* 1109–12 (1981).

1-Hydroxybenzotriazole ←

N-Nitrosoureas from N-nitrosourethans and amines N(=O)N–COOR → N(=O)N–CON⋜

s. *36,* 348; s.a. J. Med. Chem. *25,* 178–82 (1982).

Proteolytic enzymes ←

Peptide synthesis COOH → CON⋜

s. *35,* 250; coupling of silica-supported amides s. A. Könnecke, S. Dettlaff, H.-D. Jakubke, Monatsh. Chem. *113,* 331–7 (1982); **review of protease-catalysed peptide synthesis** s. H.-D. Jakubke, P. Kuhl, Pharmazie *37,* 89–106 (1982).

Anion exchanger ←

5-Aminoisoxazoles from α-cyanoketones ○

with Na acetate cf. *24,* 391; with Amberlite IRA-400 s. S. Dokic, L. Servis, B. Suskovic, Hem. Ind. *34,* 325 (1980) (Serbo-Croatian); Chem. Abstr. *94,* 192200 (1981).

Acetic acid (s.a. under HCl) *MeCOOH*

Unsym. azines from hydrazones ⋝C=NN=C⋜

s. *12,* 469; 1,2-bis(aldazines) from glyoxal s. R.R. Gallucci, J. Chem. Eng. Data *27,* 217–9 (1982).

Imidazole ring from *o*-quinones ○

s. *23,* 424; s.a. M.A. Constenla, M.A. Calvo, Ing. Cienc. Quim. *5*(1), 31 (1981) (Span.); Chem. Abstr. *96,* 142760 (1982).

Trifluoroacetic acid s. under NaN_3 CF_3COOH

N,N'-Diacylsulfinodiimines $RCON{=}S{=}NCOR$

1,1-Di(acylamines) from aldehydes CHO → $CH(NAc)_2$
with HCl cf. *25*, 287; with N,N'-diacylsulfinodiimines s. N.Y. Derkach, G.G. Barashenkov, E.I. Slyusarenko, Zh. Org. Khim. *18*, 70–8 (1982).

*N-Phenylselenophthalimide/tri-*n*-butylphosphine* ←

Carboxylic acid amides from carboxylic acids and amines COOH → CON<

331.

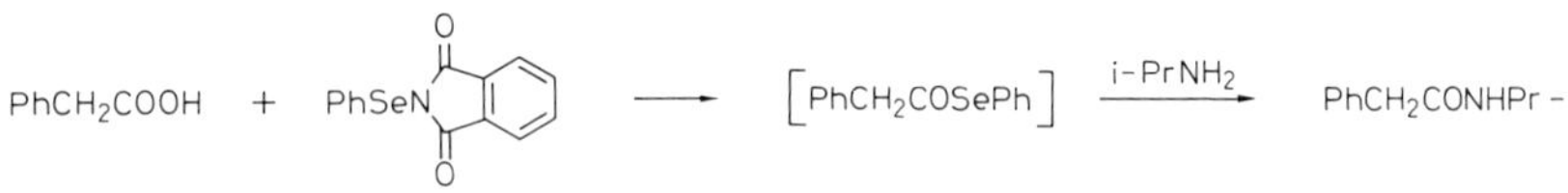

A soln. of phenylacetic acid and N-phenylselenophthalimide in tetrahydrofuran treated at room temp. with tri-*n*-butylphosphine followed by the addition of isopropylamine, and worked up after 2–3 hrs. → N-isopropylphenylacetamide. Y 95%. – F.e., also isolation of the intermediate selenolic acid esters (cf. Synth. Meth. *33*, 605), and prepn. of selenides from alcohols (cf. Synth. Meth. *32*, 599), s. P.A. Grieco et al., J. Org. Chem. *46*, 1215–7 (1981).

Carbon tetrachloride s. under Ph_3P CCl_4

Acetyl chloride s. under NaOMe *MeCOCl*

Titanium tetrachloride $TiCl_4$

Azomethines from amines and oxo compds. CO + H_2N → C=N
s. *23*, 428; α-halogenazomethines s. N. De Kimpe et al., Synthesis *1982*, 43–6.

*Tri-*n*-butylphosphine s. under N-Phenylselenophthalimide* n-Bu_3P

Triphenylphosphine/carbon tetrachloride/triethylamine $Ph_3P/CCl_4/Et_3N$

1,3-Heterocyclics from carboxylic acids ○

Redoxidative ring closure

332.

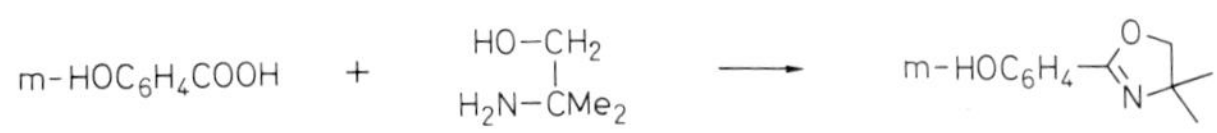

Δ^2-Oxazolines from 2-aminoalcohols. A soln. of 3 equivalents triphenylphosphine in 1:1 pyridine/acetonitrile added within 3–4 hrs. to a mixture of *m*-hydroxybenzoic acid, startg. aminoalcohol, 3 equivalents triethylamine, and carbon tetrachloride in the same solvent system → product. Y 67%. F.e. and bases, **also Δ^2-thiazolines** from 2-aminomercaptans, **Δ^2-imidazolines** from 1,2-diamines, **and 4*H*-5,6-dihydro-1,3-oxazines** from 3-aminoalcohols, s. H. Vorbrüggen, K. Krolikiewicz, Tetrahedron Letters *22*, 4471–4 (1981).

Hexamethylphosphoramide s. under $SOCl_2$ $(Me_2N)_3PO$

Nitroaryl phosphinates $R_2P(O)(OAr)$

Peptide synthesis COOH → CON<
s. *28*, 353s*37*

Phenyl N-phenylphosphoramidochloridate/triethylamine $PhOP(O)(Cl)NHPh/Et_3N$

Activation of carboxylic acids ←
with phenyl N-phenylphosphoramidochloridate
s. *37*, 333

N-Acylphosphoromonoamidites *$(RO)_2PNR'Ac$*
Peptide synthesis COOH → CON<
s. *25,* 278s*37*

Diethyl phosphorocyanidate/triethylamine *$(EtO)_2P(O)CN/Et_3N$*
Liquid phase peptide synthesis with soluble supports
cf. *28,* 353s*33;* containing a *p*-alkoxybenzyl alcohol group s. Z. Physiol. Chem. *362,* 1385–91 (1981).

Diphenyl phosphorazidate/triethylamine *$(PhO)_2PON_3/Et_3N$*
Isocyanates from carboxylic acids COOH → N=C=O
with NaN_3 cf. *13,* 430; with diphenyl phosphorazidate/triethylamine (cf. Synth. Meth. *31,* 310) s. P.L. Gendler, H. Rapoport, J. Med. Chem. *24,* 33–8 (1981).

N-(Diphenoxyphosphoryl)-Δ⁴-2-oxazolone/triethylamine ←
Activation of carboxylic acids as N-acyl-Δ⁴-2-oxazolones COOH → CON<

333.
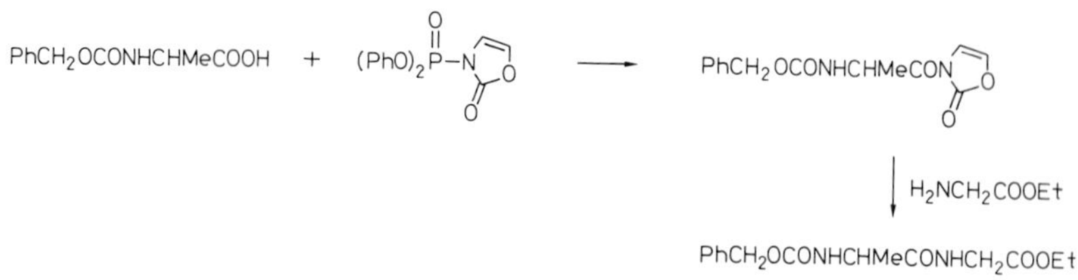

Peptide synthesis. A soln. of startg. protected aminoacid in acetonitrile treated with N-(diphenoxyphosphoryl)-Δ⁴-2-oxazolone and triethylamine at room temp. → intermediate oxazolone (Y 95%), in acetonitrile treated at room temp. with glycine ethyl ester → product. Y 98%. F.e., **also selective N-acylation,** and prepn. of **carboxylic and thiolic acid esters from alcohols and mercaptans,** respectively, s. T. Kunieda et al., Tetrahedron Letters *22,* 1257–8 (1981); with *phenyl N-phenylphosphoramidochloridate* cf. R. Mestres, C. Palomo, Synthesis *1982,* 288–91; peptide synthesis with *norborn-5-ene-2,3-dicarboximido diphenyl phosphate* s. Y. Kiso et al., J. Chem. Soc. Chem. Commun. *1980,* 1029–30; thiolic acid esters s.a. Chem. Letters *1981,* 1427–8.

2,2′-Spirobi(1,3,2-benzodioxaphosphole) ←
Peptide synthesis
with triphenyl phosphite cf. *25,* 278; with 2,2′-spirobi(1,3,2-benzodioxaphosphole) s. J.I.G. Cadogan, I. Gosney et al., J. Chem. Soc. Chem. Commun. *1982,* 298–9; f. phosphorus(III) reagents s. W. Kawanobe et al., Pept. Chem. *1981* (Pub. *1982*) 19th, 1–6; Chem. Abstr. *97,* 24205 (1982); with N-acylphosphoromonoamidites s. Chem. Letters *1982,* 825–8.

Phosphorus pentoxide/N,N-dimethylcyclohexylamine ←
Replacement of oxygen by amino groups $OH \rightarrow NH_2$

334.
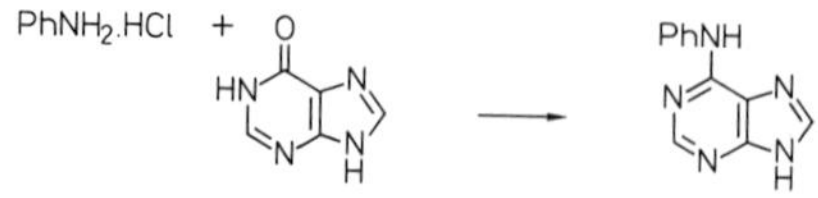

Adenines. A mixture of P_2O_5, N,N-dimethylcyclohexylamine, and aniline hydrochloride heated on an oil bath at 150°, hypoxanthine added when homogeneous, and stirred at the same temp. for 24 hrs. → N⁶-phenyladenine. Y 83%. F.e., also with sec. amines, s. N.S. Girgis, E.B. Pedersen, Synthesis *1982,* 480–2; **2- and 4-aminoquinolines** s. Chem. Scripta *18,* 240–1 (1981); **4-aminoquinazolines** s. Acta Chem. Scand. *34B,* 637–42 (1980).

4(3*H*)-Pyrimidinone ring
from *o*-acylaminocarboxylic acid esters and prim. amines O

335. 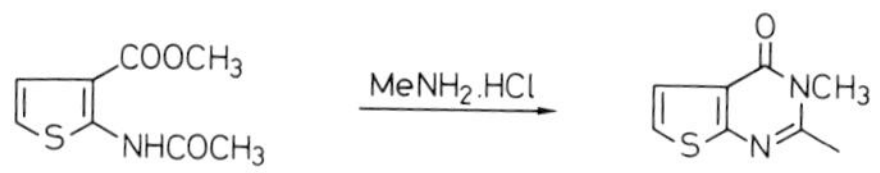

Thieno[2,3-*d*]pyrimidin-4(3*H*)-ones. A mixture of methyl 2-acetylamino-3-thiophenecarboxylate, methylamine hydrochloride, P_2O_5, and N,N-dimethylcyclohexylamine heated with stirring at 180° for 45 min., cooled to 100°, 2 *M* NaOH added to pH 8–9, then stirring continued for a further 1 hr. → 2,3-dimethylthieno[2,3-*d*]pyrimidin-4(3*H*)-one. Y 66%. F.e.s. K.E. Nielsen, E.B. Pedersen, Chem. Scripta *18,* 135–8 (1981); **hypoxanthines** s. Tetrahedron *38,* 1435–41 (1982); **pyrido[2,3-*d*]pyrimidin-4(3*H*)-ones** s. Ann. Chem. *1982,* 1012–5; **4(3*H*)-quinazolones** s. Acta Chem. Scand. *34B,* 637–42 (1980); **4-aminoquinazolines from *o*-acylaminonitriles** cf. Chem. Scripta *18,* 242–4 (1981).

Norborn-5-ene-2,3-dicarboximido diphenyl phosphate ←
Peptide synthesis COOH → CON<
s. *37,* 333

*Bis(*o*-nitrophenyl) phenylphosphonate* *RP(O)(OR)*$_2$
with diethyl phosphoryl cyanide cf. *28,* 353; with bis(*o*-nitrophenyl) phenylphosphonate s. Y. Watanabe et al., Chem. Letters *1981,* 65–8, 285–8; with nitroaryl phosphinates cf. ibid. *1981,* 1367–70.

Tris(dimethylamino)arsine $(R_3N)_3As$
Amination of oxo compds. ←
s. *22,* 399; **review of aminoarsines as reagents** s. F. Kober, Synthesis *1982,* 173–84.

1-Sulfonyloxybenzotriazoles ←
Peptide synthesis COOH → CON<
s. *35,* 254; with 1-*β*-naphthalenesulfonyloxybenzotriazole s. S.D. Sharma, K.B. Mathur, Indian J. Chem. *20B,* 227–9 (1981).

Thionyl chloride/pyridine $SOCl_2/C_5H_5N$
N-Formylation NH → NCHO
with formic acid s. *17,* 433; with formic anhydride prepared *in situ* from HCOOH, $SOCl_2$, and pyridine cf. L. Grehn, U. Ragnarsson, J. Org. Chem. *46,* 3492–7 (1981).

Thionyl chloride/hexamethylphosphoramide $SOCl_2/(Me_2N)_3PO$
Carboxylic acid amides from carboxylic acids and amines COOH → CON<
with RSO_2Cl/pyridine cf. *11,* 235; with $SOCl_2$/hexamethylphosphoramide s. K. Hirai et al., J. Med. Chem. *24,* 20–7 (1981).

Chlorosulfonyl isocyanate $ClSO_2NCO$
Carboxylic acid derivs. from carboxylic acids

336. $PhCOOH + ClSO_2N{=}C{=}O \xrightarrow{Et_3N} [PhCOOCONSO_2Cl]^-[Et_3NH]^+ \xrightarrow{PhNH_2} PhCONHPh$

Carboxylic acid amides. A soln. of chlorosulfonyl isocyanate in dry dichloromethane added to a stirred soln. of benzoic acid and dry *triethylamine* in the same solvent at 0°, the mixture stirred at 0–10° for 3 hrs., dry aniline added, asnd stirring continued at room temp. for 5 hrs. → benzanilide. Y 90%. F.e., **also carboxylic acid esters and anhydrides,** s. K.S. Keshavamurthy, Y.D. Vankar, D.N. Dhar, Synthesis *1982,* 506–8.

Chlorosulfonyl isocyanate *$ClSO_2NCO$*

Nitriles from carboxylic acids COOH → CN

s. *23,* 521; optically active compds. s. C. Botteghi, G. Chelucci, M. Marchetti, Synth. Commun. *12,* 25–33 (1982).

p-*Toluenesulfonic acid* *TsOH*

Enacylamines from acetals and carboxylic acid amides $CHC(OR)_2$ → C=C–NAc

s. *23,* 430; *β*-acylamino-*α,β*-ethyleneketones s. P.S. Khokhlov et al., Zh. Org. Khim. *18,* 1010–1 (1982).

Sulfuric acid *H_2SO_4*

Acylureas from carboxylic acids COOH → $CONHCONH_2$

with $SOCl_2$ cf. *22,* 409; with fuming H_2SO_4 s. N.A. Kravchenya, L.G. Timoshinina, Vesti Akad. Navuk BSSR, Ser. Khim. Navuk *1982,* 109–11; Chem. Abstr. *97,* 23276 (1982).

Hydrochlorides ←

Double ring closures with *α*-thiocyanoketones ○

s. *19,* 479; 4*H*-thiazolo[3,2-*a*]thieno[3,2-*e*]pyrimidin-4-ones s. H.K. Gakhar et al., J. Indian Chem. Soc. *58,* 1017–8 (1981); 5*H*-1,3,4-thiadiazolo[2,3-*b*]quinazolin-5-ones s. A.S. Shawali et al., J. Heterocyc. Chem. *19,* 73–5 (1982).

Benzylamine hydrochloride *$PhCH_2NH_2.HCl$*

Replacement of hydroxyl by prim. amino groups OH → NH_2

with phenyl phosphorodiamidate cf. *29,* 372s*34;* with benzylamine hydrochloride, limitations, s. W. Stadlbauer, T. Kappe, Synthesis *1981,* 833–4.

Hydrogen chloride *HCl*

3-Alkoxypyridine N-oxides
from enolethers and *β*-halogeno-*α*-isonitrosoketones ○
via 3-acyl-6-alkoxy-5,6-dihydro-4*H*-1,2-oxazines

337.

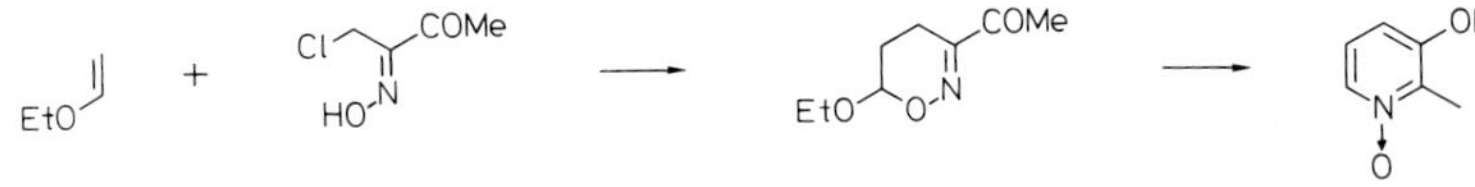

A mixture of 1-chlorobutane-2,3-dione 2-oxime and a 10 fold excess of startg. enolether treated with Na_2CO_3 → 3-acetyl-6-ethoxy-5,6-dihydro-4*H*-1,2-oxazine (Y 100%), in aq. methanol satd. with HCl allowed to react at room temp. for 3 hrs., and worked up with base → 3 methoxy-2-methylpyridine 1-oxide (Y 67%). F.e.s. T.L. Gilchrist, G.M. Iskander, A.K. Yagoub, J. Chem. Soc. Chem. Commun. *1981,* 696–8.

Hydrogen chloride/acetic acid *HCl/MeCOOH*

Hydantoins from *α*-ketoaldehydes and ureas

338.

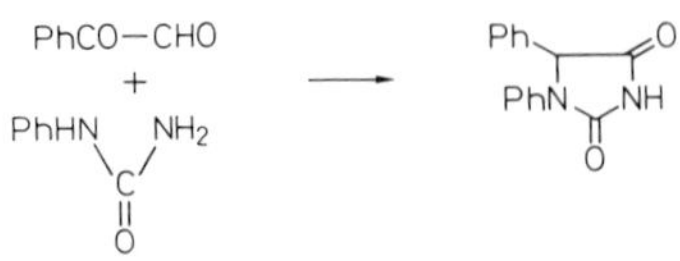

A mixture of phenylurea and phenylglyoxal in ethanol containing concd. HCl and glacial acetic acid as catalyst refluxed for 4 hrs. → 1,5-diphenylhydantoin. Y 72%. F.e.s. K.C. Joshi, V.N. Pathak, M.K. Goyal, J. Heterocyc. Chem. *18,* 1651–3 (1981).

Hydrogen bromide/acetic acid *HBr/MeCOOH*

Solid phase peptide synthesis COOH → CON<

s. *19,* 33; **review of solid-phase peptide synthesis** s. Peptides (N.Y.) *2,* 1 (1980); Chem. Abstr. *93,* 114938 (1980); with a dialkoxybenzyl alcohol resin cf. R.C. Sheppard, B.J. Williams, J. Chem. Soc. Chem. Commun. *1982,* 587–89; with α-(4-bromomethyl-3-nitrobenzamido)benzylcopoly(styrene-divinylbenzene) resin cf. ibid. *38,* 1193–208 (1982); with a hydroxymethylphenoxymethyl cf. ibid. *38,* 1183–92 (1982); with subst. benzhydrylamine resins, classification of acid lability, s. M. Christensen, O. Schou, V.S. Pedersen, Acta Chem. Scand. *35B,* 573–81 (1981); continuous flow solid phase synthesis with gel polymers of low pressure cf. E. Atherton et al., J. Chem. Soc. Chem. Commun. *1981,* 1151–2.

Manganese dioxide/pyridine MnO_2/C_5H_5N

Azomethines from alcohols $CH_2OH \rightarrow CH{=}N$

s. *24,* 421; pyridylaldimines with added pyridine to aid solubility s. M. Iwata, Bull. Chem. Soc. Japan *54,* 2835–6 (1981).

Dichlorotris(triphenylphosphine)ruthenium(II) $RuCl_2(Ph_3P)_3$

N-Alkylation of anilines $ArNH_2 \rightarrow ArNHR$

s. *37,* 339

Rhodium–carbon/ferric chloride $Rh\text{–}C/FeCl_3$

Isocyanates from nitro compds. $NO_2 \rightarrow N{=}C{=}O$

s. *22,* 420; **carbonylation of nitro compds., review,** s. V.I. Manov-Yuvenskii, B.F. Nefedov, Russ. Chem. Rev. *50,* 470–80 (1981).

Hydridotetrakis(triphenylphosphine)rhodium $RhH(PPh_3)_4$

N-Monoalkylation of amines with alcohols NH + HOR → NR

Homogeneous catalysis

339. $C_6H_{11}NH_2 + MeOH \rightarrow C_6H_{11}NHMe$

Sec. from prim. amines. A soln. of cyclohexylamine in methanol refluxed for 10 hrs. in the presence of a little hydridotetrakis(triphenylphosphine)rhodium → N-methylcyclohexylamine. Y 99%. F.e., **also tert. from sec. amines,** and cyclic amines from aminoalcohols, s. R. Grigg et al., J. Chem. Soc. Chem. Commun. *1981,* 611–2; N-alkylation of aniline with dichlorotris(triphenylphosphine)ruthenium(II) cf. Y. Watanabe, Y. Tsuji, Y. Ohsugi, Tetrahedron Letters *22,* 2667–70 (1981); with dihydridotetrakis(triphenylphosphine)ruthenium cf. S. Murahashi, K. Kondo, T. Hakata, ibid. *23,* 229–32 (1982).

Palladium–carbon *Pd–C*

Sec. from prim. amines and oxo compds. CO → CHNHR

Preferential N-alkylation

terminally N-alkylated peptides cf. *7,* 459; terminally N-alkylated spermidine and spermine hydrochlorides s. F.L. Weitl, K.N. Raymond, J. Org. Chem. *46,* 5234–7 (1981).

Palladium–carbon/benzylamine $Pd\text{–}C/C_6H_5CH_2NH_2$

Prim. amines from oxo compds. via benzylamines $CO \rightarrow CHNH_2$

s. *18,* 465; aminoalkylphosphonic acids s. G. Fabre, N. Collignon, P. Savignac, Can. J. Chem. *59,* 2864–9 (1981).

Via intermediates *v.i.*

Tetrazoles from carboxylic acid amides via iminochlorides ○

s. *6,* 441; tetrazole-5-carboxylates s. D.H. Klaubert et al., J. Med. Chem. *24,* 748–52 (1981).

Via intermediates *v.i.*

Nitriles from aldehydes via aldoximes CHO → CN

s. *30,* 277; with nitroethane/pyridine hydrochloride, ar. nitriles, cf. D. Dauzonne, P. Demerseman, R. Royer, Synthesis *1981,* 739–40.

Azomethines from ketals $C(OR)_2$ → C=NR

s. *4,* 392; via 1-alkoxyhalides, *α*-ketoazomethines, cf. N. De Kimpe et al., Bull. Soc. Chim. Belges *91,* 99–100 (1982).

Nitrogen ↑ NC↓↑N

Without additional reagents *w.a.r.*

Acylation with poly(3-acyl-Δ^4-2-oxazolones) NH → NAc

340.

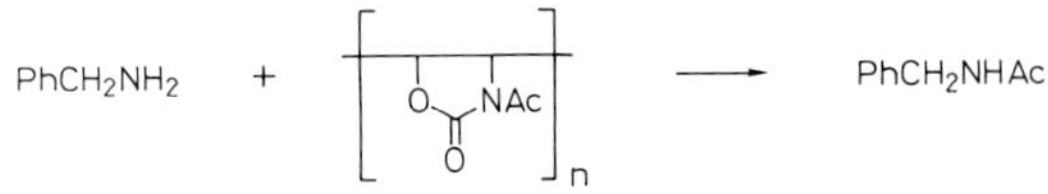

N-Acylation. A soln. of benzylamine in tetrahydrofuran treated with 2 equivalents poly(3-acetyl-2-oxazolone) at room temp. for 6 hrs. → product. Y 93%. – Work-up is simple and the deacylated polymer can be regenerated by conventional reagents (e.g. acetic anhydridepyridine) for repeated use without significant loss of activity. F.e., also **selective and preferential N-acylation,** and **O-acylation** in the presence of tertiary amines, Lewis acids, or F^- as catalyst, s. T. Kunieda et al., Tetrahedron Letters *23,* 1159–60 (1982).

Asym. N-acylation with 3-acylthiazolidine-2-thiones

341.

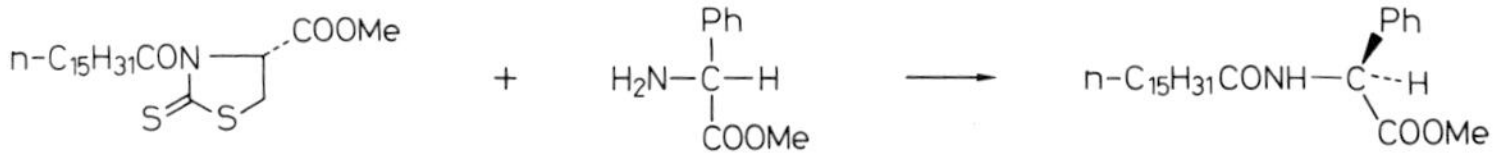

A soln. of (±)-phenylglycine methyl ester in methylene chloride added with stirring to a soln. of 3-hexadecanoyl-4(R)-methoxycarbonylthiazolidine-2-thione in the same solvent under N_2, stirring continued at room temp. for 12 hrs., and quenched with 10% HCl → (S)-N-hexadecanoylphenylglycine methyl ester. Y 93%; enantiomeric excess 64%. F.e.s. Y. Nagao et al., Tetrahedron Letters *23,* 201–4 (1982); application to the **determination of abs. configuration of chiral amines** s. ibid. *23,* 205–8 (1982); determination by circular dichroism of N-salicylidene derivs. cf. H.E. Smith et al., J. Org. Chem. *47,* 2525–31 (1982).

Formylation with N,N-diformylacetamide NH → NCHO

342.

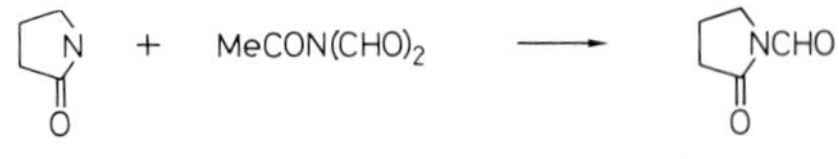

N-Formylation. A soln. of 2-pyrrolidone and N,N-diformylacetamide in benzene refluxed under N_2 until reaction complete by t.l.c. → 1-formyl-2-pyrrolidone. Y 75%. F.e., **also preferential N-formylation and O-formylation,** s. J.C. Gramain, R. Rémuson, Synthesis *1982,* 264–6.

Acylimines from *α*-diazoketones and imines C=NH → C=NAc
Ketenes as intermediates

343.

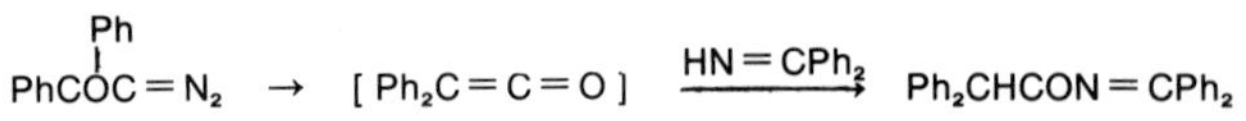

A soln. of 2-diazo-1,2-diphenylethanone and diphenylmethanimine in dry benzene heated under reflux for 6 hrs., and kept overnight at room temp. → N-

(diphenylmethylene)diphenylacetamide. Y 90%. F.e. and reactions, also α-alkoxynitriles in low yield, s. K.N. Mehrotra, O. Prasad, Bull. Chem. Soc. Japan *54,* 604–6 (1981).

N-Carbalkoxylation with N-carbalkoxypyridinium salts NH → NCOOR
s. *25,* 293s*26;* 4-carbalkoxyaminopyrimidines s. G. Auzou, R. Rips, J. Heterocyc. Chem. *18,* 835–6 (1981).

Transamination of aminomethylene compds. $C{=}CHNR_2 \rightarrow C{=}CHNR'_2$
s. *22,* 425; β-subst. N-benzylenamines s. V.G. Granik et al., Zh. Org. Khim. *17,* 2421–8 (1981).

N-Cyanamidines from amidines $C({=}NH)NH_2 \rightarrow C({=}NH)NHCN$

344. 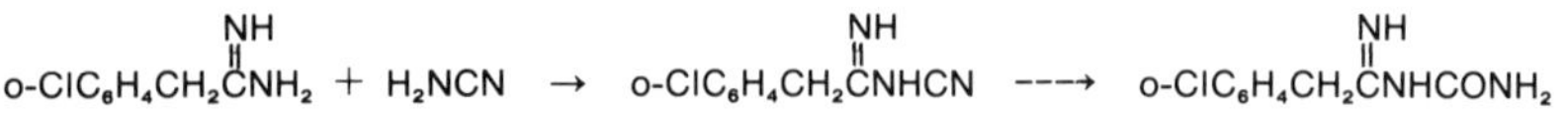

2-(2-Chlorophenyl)acetamidine in aq. ethanol treated with solid cyanamide, and the soln. degassed with a slow stream of N_2 → N-cyano-2-(2-chlorophenyl)acetamidine. Y 83%. F.e. and conversion to **N-carbamylamidines** s. J.B. Bream, C.W. Picard, T.G. White, Eur. J. Med. Chem. *16,* 175–9 (1981).

N-Sulfonylaziridine ring from ethylene derivs. and sulfonic acid azides 

s. *22,* 123; 2,7-diazabicyclo[4.1.0]hept-3-enes s. B.K. Warren, E.E. Knaus, J. Med. Chem. *24,* 462–4 (1981).

4(3*H*)-Pyrimidine ring from *o*-aminocarboxylic acid amides ○
s. *23,* 409; with formamide, 1*H*-pyrazolo[4,3-*d*]pyrimidin-7(6*H*)-ones, s. P.G. Baraldi et al., Synthesis *1981,* 727–9.

4(3*H*)-Quinazolones from *o*-aminocarboxylic acid amides and nitrilium salts
s. *37,* 749

Tricyclic orthocarboxylic acid amides
s. *36,* 367; s.a. Tetrahedron Letters *22,* 4365–8 (1981).

Δ^2-Oxazolines from 2-aminoalcohols
and iminoesters cf. *8,* 491; and iminoester fluoroborates, optically active compds., s. A.I. Meyers, M.A. Hanagan, A.L. Mazzu, Heterocycles *15,* 361–7 (1981).

Phthalimides from prim. amines by exchange of nitrogen
s. *16,* 471; from N-hydroxyphthalimides without reagent cf. H.A. Baddawy et al., Indian J. Chem. *19B,* 112–4 (1980).

4-Aminopyrimidines from 2-cyanoeneformamidines

345. 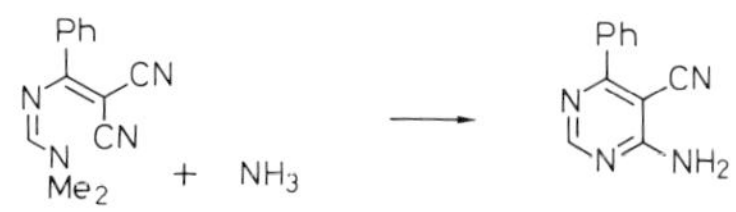

A suspension of 2-[(N,N-dimethylaminomethylene)amino]-1,1-dicyano-2-phenylethene in concd. aq. ammonia heated under reflux for 30 min. → 4-amino-5-cyano-6-phenylpyrimidine. Y 97%. F.e.s. M. Mittelbach, H. Junek, J. Heterocyc. Chem. *17,* 1385–7 (1980).

1,2,4-Triazine ring from enazides and tetrazoles

346. 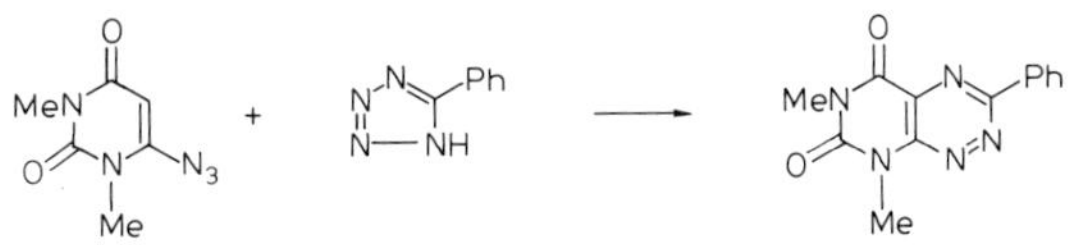

Pyrimido[5,4-*e*]-*as*-triazine-5,7(6*H*,8*H*)-diones. A mixture of 6-azido-1,3-dimethyluracil and 1 equivalent 5-phenyltetrazole in dimethylformamide refluxed for 30 min.

→ 3-phenylfervenulin. Y 75%. F.e. and isolation of intermediate, **also alloxazines from benzotriazoles,** s. K. Hirota et al., Heterocycles *15,* 285–8 (1981); **from carboxylic acid hydrazides** (cf. Synth. Meth. *33,* 387) s. J. Am. Chem. Soc. *100,* 7661–4 (1978); f. method cf. Heterocycles *16,* 559–61 (1981).

Electrolysis ϟ

Anodic acylation with hydroxamic acids NH → NAc

347. $PhCONHOH + NH_2Me \rightarrow PhCONHMe$

N-Acylation. A soln. of benzohydroxamic acid in acetonitrile containing 4 equivalents of methylamine and 0.1 *M* $NaClO_4$ electrolysed using a glassy-carbon anode at 0.40 V with passage of 2.2 coulombs/mole of the hydroxamic acid → product. Y 59%. F.e., **also selective N-acylation and O-acylation,** and prepn. of **ureas from amines by anodic carbamylation with N-hydroxyureas,** s. M. Masui, S. Ozaki, Tetrahedron Letters *23,* 2867–70 (1982).

Sodium methoxide *NaOMe*

Pyrimidines from vinylogous amidinium salts ○

2 alkylthiopyrimidines with pyridine cf. *33,* 392; 2-subst. derivs. from amidines with Na methoxide s. J. Prakt. Chem. *323,* 199–206 (1981).

Sodium acetate (s.a. under N_2O_4) *MeCOONa*

1,4-Dihydro-1,2,4,5-tetrazines ○

from two pyridinium and diazonium salt molecules

s. *36,* 370; from 3-pyridyldiazonium salts s. Ind. J. Chem. *20B,* 503–5 (1981).

Sodium azide *NaN_3*

Replacement of nitro by azido groups $NO_2 \rightarrow N_3$

s. *21,* 522; 1-azidoanthraquinones s. L.M. Gornostaev, V.T. Sakilidi, Zh. Org. Khim. *17,* 2217–21 (1981).

2*H*-1,2,3-Triazoles from 1-nitroethylene derivs. ○

348.

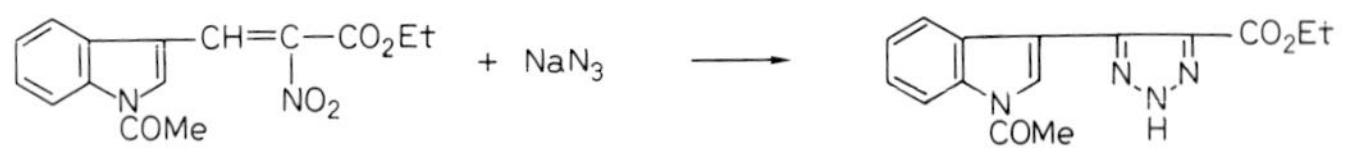

Startg. nitroalkene added to a soln. of NaN_3 in dry dimethylformamide, the mixture warmed 5 min. at 50–60°, and left at room temp. → product. Y 96%. F.e.s. V.S. Velezheva, Y.V. Erofeev, N.N. Sovorov, Zh. Org. Khim. *16,* 2157–63 (1980).

Triethylamine *Et_3N*

Pyrazine ring from *o*-diamines and 1-sulfonyl-Δ^4-2-imidazolones ○

349.

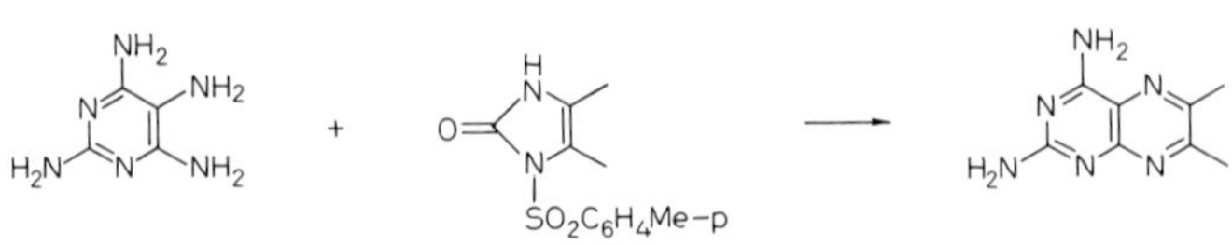

Pteridines. A mixture of startg. ms. and triethylamine in water-dimethylformamide stirred for 11 days at ca. 20° → product. Y 77%. F.e.s. S.I. Zavyalov, A.G. Zavozin, Izv. Akad. Nauk SSSR Ser. Khim. *1981,* 1666–8.

Pyridine *C_5H_5N*

Protection of prim. amino groups as N-alkyl-3,5-dinitro-4-pyridones

350.

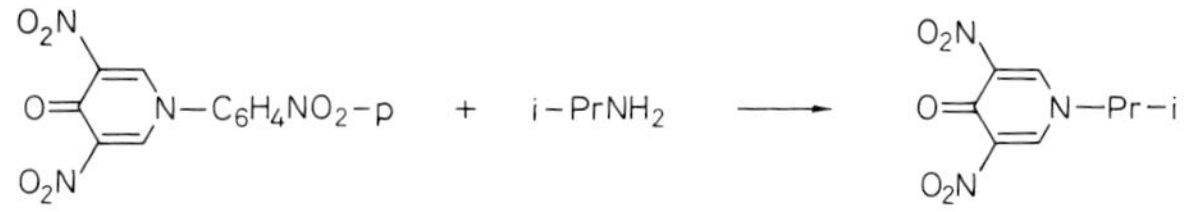

A soln. of 3,5-dinitro-1-(*p*-nitrophenyl)-4-pyridone in pyridine treated with a soln. of isopropylamine in the same solvent at room temp. → 1-isopropyl-3,5-dinitro-4-pyridone. Y 100%. – The startg. amine can be recovered in good yield by the addition of a second prim. amine (e.g. aniline or propylamine) under mild conditions. F.e.s. E. Matsumura et al., Tetrahedron Letters *22,* 757–8 (1981).

N,N'-Carbonyldiimidazole ←

1,2,4-Triazin-3(2*H*)-one ring from *o*-aminoazo compds.

351.

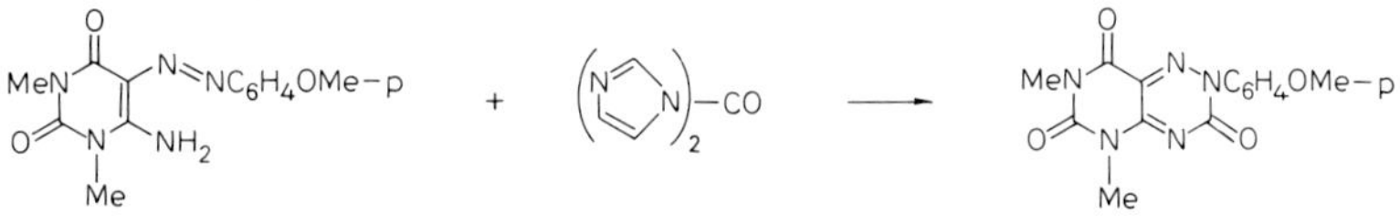

A mixture of startg. uracil and N,N'-carbonyldiimidazole fused with stirring at 125–140° for 2 hrs., cooled, and poured into ether → product. Y 95%. F.e., also with urea in lower yield, s. F. Yoneda, M. Higuchi, J. Heterocyc. Chem. *17,* 1365–8 (1980).

Acetic acid *MeCOOH*

Hydrazones from eneformamidines C=CN–CH=N → CHC=N–NH$_2$

s. *37,* 71

1*H*-1,5-Benzodiazepine ring from *o*-diamines ○

and vinylogous amidium salts cf. *34,* 381; and *o*-aminoaldehydes in the presence of acetic acid s. G. Roma, A. Ermili et al., J. Heterocyc. Chem. *18,* 1619–23 (1981).

Nitrogen dioxide/sodium acetate *N_2O_4/MeCOONa*

Transamination of N-subst. carboxylic acid amides via N-nitrosocarboxylic acid amides CONHR → CONHR'

352.

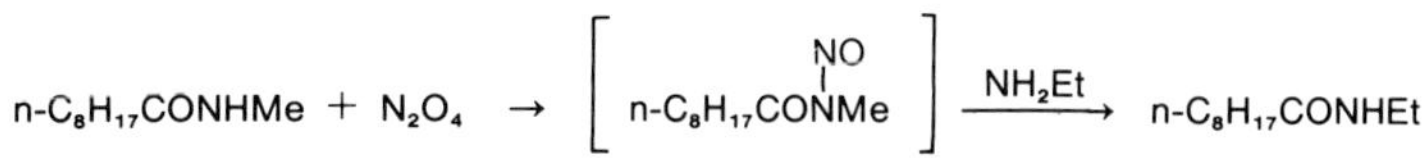

Anhydrous Na acetate added to a stirred soln. of startg. amide in dichloromethane cooled to –15°, a stream of N_2O_4 bubbled through the suspension until t.l.c. indicated complete N-nitrosation (< 1 hr.), ethylamine added, and maintained at room temp. for 5 hrs. → product. Y 98%. F.e.s. J. Garcia, J. Vilarrasa, Tetrahedron Letters *23,* 1127–8 (1982).

Hydroxylamine-O-sulfonic acid *H_2NOSO_3H*

1,2,4-Triazoles from formamidines ○

s. *36,* 376; N-condensed 1,2,4-triazoles with hydroxylamine-O-sulfonic acid cf. J. Org. Chem. *46,* 3123–4 (1981); **review of hydroxylamine-O-sulfonic acid** s. R.G. Wallace, Aldrichimica Acta *13,* 3–11 (1980).

Hydrogen chloride *HCl*

Pyrimidine ring from ***o*****-aminocarbonyl compds. and nitriles**

353. 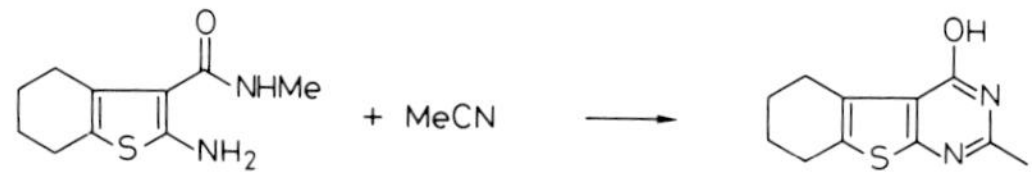

4-Hydroxypyrimidine ring. A stream of HCl passed through a mixture of 2-amino-3-(N-methylcarboxamido)-4,5,6,7-tetrahydrobenzo[*b*]thiophene in acetonitrile for 4–5 hrs., poured into ice water, and made alkaline with 10% NH_4OH → 2-methyl-4-hydroxy-5,6,7,8-tetrahydrobenzo[*b*]thieno[2,3-*d*]pyrimidine. Y 80%. F.e., also from *o*-aminocarboxylic acid esters, and prepn. of 4-aryl- and **4-amino-pyrimidine ring** from *o*-amino-ketones and -nitriles, respectively, s. K.G. Dave et al., J. Heterocyc. Chem. *17,* 1497–500 (1980).

Rhenium(VII) sulfide/hydrogen chloride *Re_2S_7/HCl*

Enacylamines from azides $CHCHN_3$ → C=CNHAc

α**-Acylamino-*****α,β*****-ethylenecarboxylic acid esters**

354. n-Pr–CH_2–CH(N_3)COOMe + Ac_2O → n-Pr–CH=C(NHAc)COOMe

A soln. of startg. 2-azidocarboxylic acid ester in 2:3 acetic anhydride/acetic acid containing HCl warmed in the presence of a little Re_2S_7 → product. Y 91%. F.e.s. F. Effenberger, T. Beisswenger, Angew. Chem. Intern. Ed. *21,* 203 (1982).

Iron pentacarbonyl/acetic acid *$Fe(CO)_5$/MeCOOH*

2-Sulfonylguanidines from carbodiimides N=C=N → NHC(=NTs)NH

s. *33,* 394; recent advances in **carbodiimide chemistry, review,** s. A. Williams, I.T. Ibrahim, Chem. Rev. *81,* 589–636 (1981).

Nickel *Ni*

One-step purine synthesis ○

s. *28,* 382; f. method cf. F. Yoneda, R. Koga, M. Higuchi, Chem. Letters *1982,* 365–8.

Dichlorotris(triphenylphosphine)ruthenium(II) *$RuCl_2(PPh_3)_3$*

Sym. tert. and sec. from prim. amines RNH_2 ↗ RNHR ↘ R_3N

s. *37,* 430

Palladium catalysts ←

Tert. amines from nitriles 3 RCN → $(RCH_2)_3N$

with Pt–C cf. *23,* 58; with Pd catalysts s. J. Koubek, J. Pasak, J. Horyna, Chem. Prum. *31,* 349 (1981) (Czech); Chem. Abstr. *95,* 186562 (1981).

Bis(benzonitrile)dichloropalladium(II) *$PdCl_2(PhCN)_2$*

N-Carbalkoxyiminoesters ←

from alkoxyethylenes and azidoformic acid esters

355. CH_2=CMeCH_2OEt + N_3COOEt → CH_3CHMeC(OEt)=NCOOEt

A mixture of startg. m. and ethyl azidoformate containing a little bis(benzonitrile)dichloropalladium(II) heated in a sealed tube at 80° for 24 hrs. → product. Y 95%. F.e., incl. conversion from alkoxy-3- and alkoxy-4-ethylenes, as well as enolethers, s. T. Migita et al., Chem. Letters *1982,* 1015–6.

Halogen ↑ NC⇃↾Hal

Without additional reagents *w.a.r.*

Tert. from sec. amines and chlorides $R_2NH \rightarrow R_2NR'$
with dimethylamine in dioxane cf. *16,* 503; in aq. isopropanol, N,N-dimethylbenzylamine, s. G. L. Kamalov et al., Khim. Prom-st., Ser. Reackt. Osobo Chist. Veshchestva *1980,* 22; Chem. Abstr. *95,* 42520 (1981).

Reactions of *α,α′*-dibromoketones with prim. amines ←

356. 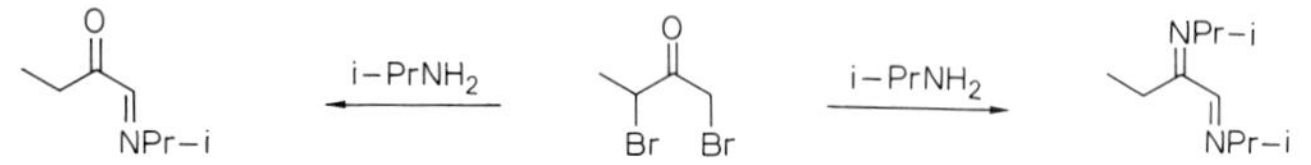

α-Ketoimines	**1,2-di(azomethines).**

5 Equivalents of isopropylamine added to a 10% soln. of 1,3-dibromo-2-butanone in

ether at room temp., and the mixture stirred for 20 min. → product. Y ca. 85% by g.l.c.	pentane at room temp., and the mixture stirred for 17 hrs. → product. Y ca. 100% by g.l.c.

F.e.s. N. de Kimpe et al., Tetrahedron Letters *23,* 789–92 (1982).

6-α-Anilino- from 5-bromo-uracils ←

357. 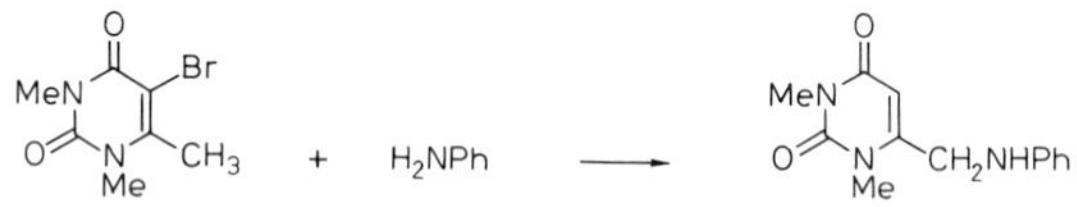

A mixture of startg. 5-bromouracil and aniline in dimethylformamide refluxed for 5 hrs. → product. Y 85%. F.e. and nucleophiles s. K. Hirota et al., J. Chem. Soc. Perkin Trans. I *1981,* 2943–7.

Imidazole ring from α-halogenoketones ○
s. *13,* 535; 1-acylaminoimidazoles s. G. Cerioni et al., J. Heterocyc. Chem. *18,* 1379–82 (1981).

5-Imino-Δ³-1,2,4-thiadiazolines from 5-amino-1,2,3,4-thiatriazoles and iminochlorides

358. 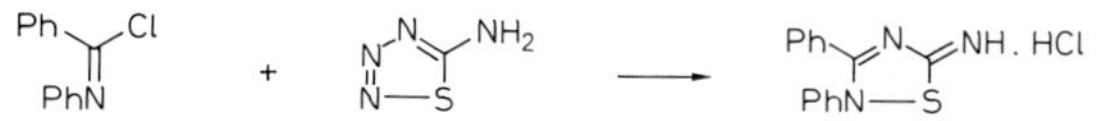

A mixture of 5-amino-1,2,3,4-thiatriazole and startg. imidoyl chloride in acetonitrile allowed to react at room temp. for 2 hrs. → 5-imino-1,2,4-thiadiazolinium chloride. Y 87%. F.e. and ring closures s. G. L'abbé, G. Vermeulen, Bull. Soc. Chim. Belges *90,* 89–98 (1981).

5*H*-Quinazolino[1,2-*a*]quinazoline-5,8(6*H*)-diones
Double ring closure

359. 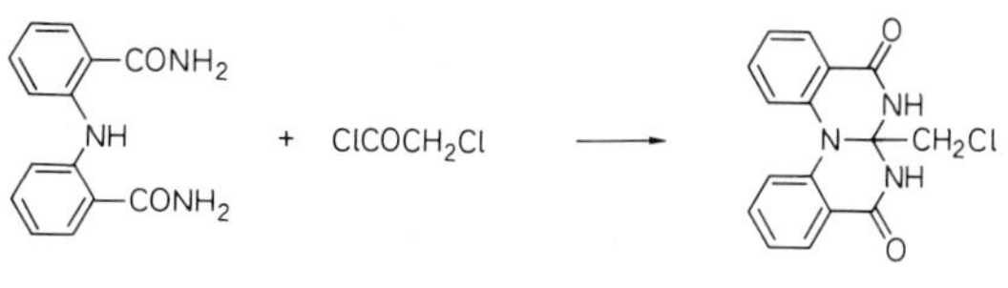

Chloroacetyl chloride added to a suspension of 2,2′-dicarbamoyldiphenylamine in chloroform at room temp., and the mixture refluxed for 8 hrs. → 6a-(chloromethyl)-

6a,7-dihydro-5*H*-quinazolino[1,2-*a*]quinazoline-5,8(6*H*)-dione. Y 83%. F.e. and **4(1*H*)-quinazolones** s. K. Ozaki, Y. Yamada, T. Oine, J. Org. Chem. *46,* 1571–5 (1981).

Without additional reagents (cont.) *w.a.r.*

6-Membered heterocyclics from 1,2-oxido-1-halides

360. 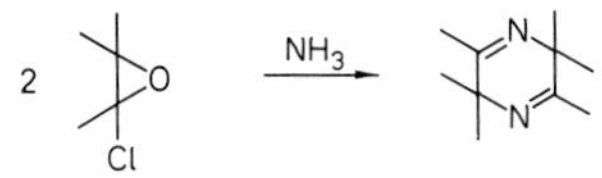

2,5-Dihydropyrazines. A stream of NH_3 gas passed through a soln. of the startg. m. in methylene chloride at −35°, and the stream continued while the mixture, containing precipitated NH_4Cl, allowed to warm slowly to room temp. → 2,5-dihydro-2,2,3,5,5,6-hexamethylpyrazine. Y 80%. F.e., **also 2-hydroxy-2,3-dihydro-1,4-oxathiins and 2-hydroxy-2,3-dihydro-1,4-benzoxazines,** s. C. Herzig, J. Gasteiger, Chem. Ber. *114,* 2348–54 (1981).

Reactions with sulfonyliminocarbonylchlorides

s. *22,* 448; 2-sulfonylamino-1,3-oxazin-6-one ring s. Synthesis *1981,* 965.

8,2′-Cyclonucleosides with two-atom bridges

Hydrazo-8,2′-cyclonucleosides

s. *33,* 408; also 8,2′-aminooxycyclonucleosides with NH_2OH s. T. Sasaki et al., J. Org. Chem. *46,* 5176–81 (1981).

Carboxylic acids and derivs. from 1,1-dihalogenocyclopropanes C

carboxylic acids cf. *24,* 225; carboxylic acid amides, γ-keto-derivs., s. I.G. Tischenko, O.G. Kulinkovich, N.V. Masalov, Synthesis *1982,* 268–70.

Sodium *Na*

Urethans from azines and halogenoformic acid esters C=N−N=C → CHNHCOOR

361. $Ph_2C{=}N{-}N{=}CPh_2 + ClCOOEt \rightarrow Ph_2CHNHCOOEt$

A soln. of startg. azine in dry dioxane heated to reflux with small pieces of Na under N_2, after ca. 15–20 min. the violet to dark brown mixture stirred for 16–28 hrs. at reflux, cooled to 60–70°, ethyl chloroformate added, and the mixture allowed to stand at room temp. for 6 hrs. → product. Y 90%. F.e.s. K.N. Mehrotra, S.C. Joshi, Indian J. Chem. *19B,* 794–5 (1980).

1-Sulfonylaminothioethers from thioethers H → $NHSO_2R$

362. 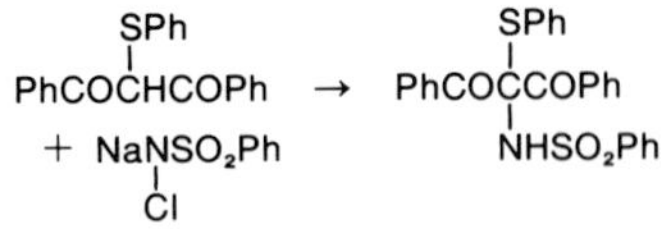

Startg. N-sodio compd. added to a soln. of startg. thioether in acetone, and stirred at room temp. ca. 4 hrs. → product. Y 86%. F.e.s. I.V. Koval, V.V. Andrushchenko, M.M. Kremlev, Zh. Org. Khim. *18,* 841–5 (1982).

Sodium hydride *NaH*

N-Alkylation NH → NR

s. *37,* 238

... of sulfinamides

s. *37,* 184

Mixed *o*-di(acylamines) from N-arylhydroxamic acids and iminochlorides via hetero-Cope rearrangement ←

363. 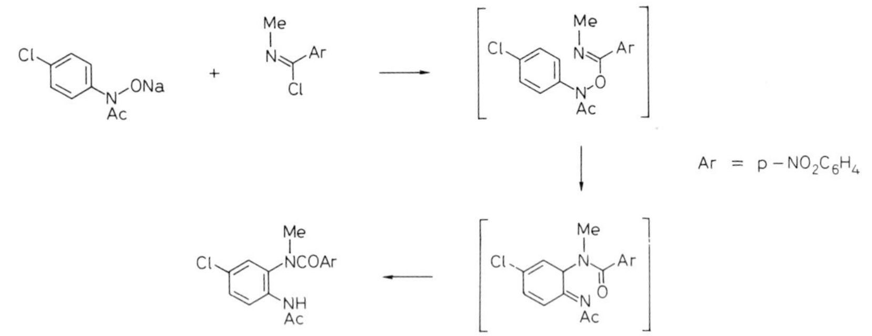

Startg. hydroxamic acid treated with a suspension of NaH in tetrahydrofuran, startg. iminochloride added, and the mixture stirred at room temp. for 2–6 hrs. → product. Y 80%. F.e. and limitations s. R. Höfelmeier, S. Blechert, Angew. Chem. Intern. Ed. *21,* 370 (1982).

*Sodium hydroxide/potassium carbonate/ tetra-*n*-butylammonium hydrogen sulfate* $NaOH/K_2CO_3$/n-$Bu_4N^+HSO_4^-$

Hydrazines from bromides via N-alkylation of phosphinic hydrazides $P(O)NHNH_2 \rightarrow P(O)NRNH_2$

s. *35,* 274; 1,2-dialkylhydrazines by stepwise alkylation via N′-acylphosphinic hydrazides s. Synthesis *1981,* 537–9.

Sodium hydroxide/anion exchanger ←

2,5-Piperazinediones from two *α*-halogenocarboxylic acid amide molecules ○

with PhLi cf. *28,* 392; 3-phase catalysis with ion exchange resin Duolite A-109 (Cl-form) and NaOH s. T. Okawara, T. Matsuda, M. Furukawa, Chem. Pharm. Bull. *30,* 1225–33 (1982).

*Sodium hydroxide/tetra-*n*-butylammonium bromide* $NaOH$/n-$Bu_4N^+Br^-$

Regiospecific N-alkylation of adenines NH → NR

s. *33,* 403s*36;* with NaH in dimethylformamide cf. M. Rasmussen, J.M. Hope, Australian J. Chem. *35,* 525–34 (1982); 9-alkylation by phase transfer catalysis with NaOH/tetra-*n*-butylammonium bromide s. M. Hedayatullah, J. Heterocyc. Chem. *19,* 249–51 (1982).

Sodium hydroxide/benzyltriethylammonium chloride $NaOH/[PhCH_2NEt_3]^+Cl^-$

Nucleosides from N-heterocyclics

with $Hg(CN)_2$ cf. *24,* 473; with NaOH and benzyltriethylammonium chloride as phase transfer catalyst s. F. Seela, H.-D. Winkeler, J. Org. Chem. *47,* 226–30 (1982); with tetrabutylammonium hydrogen sulfate s. Ann. Chem. *1982,* 499–506.

N-Alkylation by phase transfer catalysis

of 2-benzimidazolones s. *21,* 503s*36;* of acridones and carbazoles s. H. Nishi, H. Kohno, T. Kano, Bull. Chem. Soc. Japan *54,* 1897–8 (1981).

2-Azetidinones from *β*-bromocarboxylic acid chlorides 

in pyridine cf. *19,* 520; with NaOH and benzyltriethylammonium chloride as phase transfer catalyst s. T. Okawara et al., Chem. Letters *1981,* 185–8; s.a. Chem. Pharm. Bull. *30,* 1225–33 (1982).

Potassium hydroxide *KOH*

Cyclic quaternary ammonium salts from sec. amines and dihalides ○
s. *8,* 505; with KOH s. M.M. Movsumzade et al., Azerb. Khim. Zh. *1981,* 56–60.

Potassium hydroxide/acetone *KOH/Me_2CO*

N-Alkylation NH → NR
of carboxylic acid amides cf. *9,* 512; of alkoxylamines s. V.V. Dovlatyan, V.A. Pivazyan, K.A. Eliazyan, Arm. Khim. Zh. *1980,* 865–7; of imidazole ring s. Y. Kikugawa, Synthesis *1981,* 124–5.

Potassium amide *KNH_2*

Replacement of bromine by amino groups with position shift Br → NH_2
s. *19,* 510; anilines s. B. Roth et al., J. Med. Chem. *24,* 933–41 (1981).

Potassium carbonate s. under NaOH *K_2CO_3*

*Potassium thiocyanate/tetra-*n*-butylammonium bromide* *$KSCN$/n-$Bu_4N^+Br^-$*

Acylisothiocyanates from carboxylic acid chlorides COCl → CON=C=S
with $Pb(SCN)_2$ cf. *6,* 480; with KSCN and tetra-*n*-butylammonium bromide as phase transfer catalyst s. W.P. Reeves et al., Synth. Commun. *11,* 781–5 (1981).

Sodium azide *NaN_3*

Enazides from 1,2-dibromides CBrCHBr → C=C−N_3
s. *35,* 278; α-azido-α,β-ethylenecarboxylic acid esters s. M. Kakimoto, M. Kai, K. Kondo, Chem. Letters *1982,* 525–6.

Sodium azide/trimethylamine *NaN_3/Me_3N*

Replacement of chlorine by azido groups Cl → N_3
s. *37,* 478

Potassium fluoride–alumina *$KF–Al_2O_3$*

N-Alkylation under mild conditions NH → NR

364.

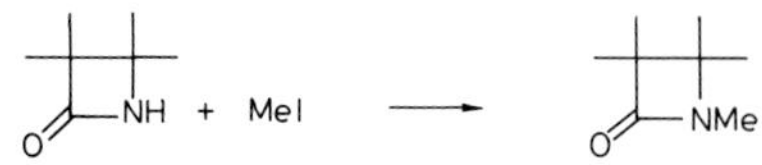

of lactams. Startg. lactam treated with methyl iodide in 1,2-dimethoxyethane at room temp. in the presence of 2.5 equivalents KF-alumina for 3 hrs. → product. Y 100%. F.e., **also of carboxylic acid amides and uracils,** s. J. Yamawaki, T. Ando, T. Hanafusa, Chem. Letters *1981,* 1143–6.

Ammonia s. under Me_3SiCl *NH_3*

Trimethylamine s. under NaN_3 *Me_3N*

Triethylamine (s.a. under R_2SiCl_2) *Et_3N*

N-Alkylideneureas from 1-bromoisocyanates C(Hal)N=C=O → C=NCON<

365.

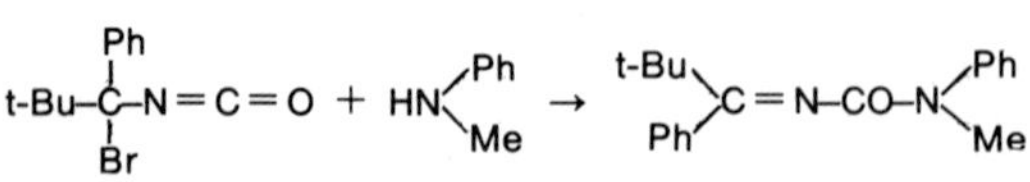

A soln. of N-methylaniline and triethylamine in tetrahydrofuran added with stirring to a soln. of (1-bromo-2,2-dimethyl-1-phenylpropyl) isocyanate in the same solvent at 0°, warmed to 22°, ether added, and stirring continued for 15 min. → 3-(2,2-dimethyl-1-phenylpropylidene)-1-methyl-1-phenylurea. Y 99%. F.e.s. R. Reck, J.C. Jochims, Chem. Ber. *115,* 860–70 (1982).

3,5-Pyrazolidiones from α-dicarbonyl compds. ○
from α-dicarboxylic acid chlorides s. *17,* 330; from malonic acid esters cf. E. Bellora et al., Farmaco, Ed. Sci. *36,* 432–44 (1981).

Ring closures with N-acylchloroformamidines

366.

3-Amino-1,2,4-triazoles. A soln. of N-(4-nitrobenzoyl)-1-pyrrolidinecarbimidoyl chloride in 1,2-dimethoxyethane treated with triethylamine and a soln. of startg. hydrazine in the same solvent at room temp., and the mixture refluxed for 1 hr. → 1-(4-methylphenyl)-5-(4-nitrophenyl)-3-(1-pyrrolidinyl)-1*H*-1,2,4-triazole. Y 83%. F.e. and ring closures s. W. Ried, H.-E. Erle, Chem. Ber. *115,* 475–82 (1982); **4-amino-1,3,5-oxadiazine-2-thiones** s. Ann. Chem. *1982,* 201–10.

2,5-Diamino-1,3,4-oxadiazoles from semicarbazides and iminocarbonylhalides

367.

A stirred soln. of 4-phenylsemicarbazide and triethylamine in tetrahydrofuran treated with N-phenylcarbimide dichloride at room temp., and the mixture allowed to stand for several hrs. → 2,5-diphenylamino-1,3,4-oxadiazole. Y 88%. F.e., also 1,2,4-triazolidine-3,5-diones in low yield, s. G. Zinner, M. Heitmann, Arch. Pharm. *314,* 193–6 (1981).

Triethylamine/hydrogen chloride Et_3N/HCl

N-Amino-N-heterocyclics from dihalides
with diethylaniline/HCl from benzyl carbazate cf. *15,* 384; with triethylamine/HCl from *tert*-butyl carbazate, N-aminoisoindolines, s. G. Cignarella et al., J. Med. Chem. *24,* 1003–6 (1981).

tert-*Butylamine* t-$BuNH_2$

O,N-Heterocyclics from N-alkoxyphthalimides

368.

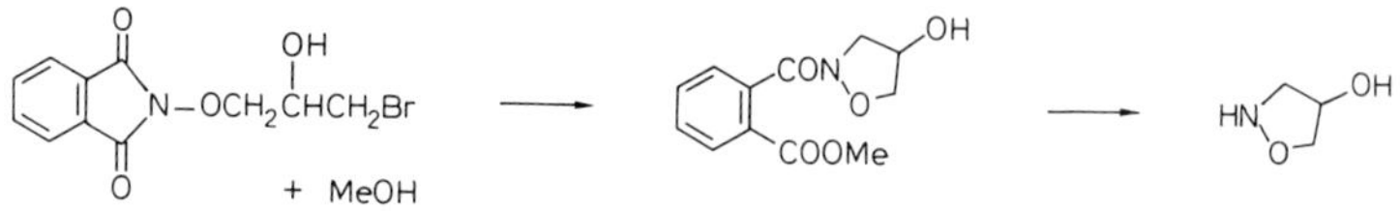

Isoxazolidines. A soln. of N-(3-bromo-2-hydroxypropyloxy)phthalimide and *tert*-butylamine in methanol stirred at room temp. for 1 hr. → 4-hydroxy-2-(2-methoxycarbonylbenzoyl)tetrahydroisoxazole (Y ca. 100%), heated in refluxing 4 *M* HCl for 4 hrs. → 4-hydroxytetrahydroisoxazole (Y ca. 100%). F.e.s. N. Amlaiky, G. Leclerc, Synthesis *1982,* 426–8; f. reactions, incl. prepn. of **5,6-dihydro-1,4,2-dioxazines,** s. J. Org. Chem. *47,* 517–23 (1982).

Ethyldiisopropylamine s. under Me_3SiCl i-Pr_2NEt

Copper *Cu*

N-Arylation NH → NAr
s. *13,* 496; of carboxylic acid amides and lactams, also with Cu_2O, s. T. Yamamoto, Y. Kurata, Chem. Ind. *1981,* 737–8.

Silver nitrate/trimethyl phosphite *$AgNO_3/(MeO)_3P$*

Replacement of bromine by nitro groups Br → NO_2

369. 

α-**Nitronitriles.** A soln. of $AgNO_3$-trimethyl phosphite in acetonitrile added dropwise during 10 min. to a stirred soln. of startg. m. in the same solvent under N_2, and stirring continued for 30 min. → product. Y 80%. F.e. and with $AgNO_3$-triphenylphosphine and -triisopropyl phosphite s. R. Ketari, A. Foucand, J. Org. Chem. *46,* 4498–501 (1981).

Magnesium *Mg*

1-Arylthio-1-triazenes RNHN=N–C–SAr

s. *37,* 377

Borane-tetrahydrofuran/
sodium dihydrobis(2-methoxyethoxo)aluminate *$H_3B.THF/$ $Na[AlH_2(OCH_2CH_2OMe)_2]$*

Reductive N-carbalkoxylation of N-heterocyclics ←
1-Carbalkoxy-1,2-dihydroquinolines

370. 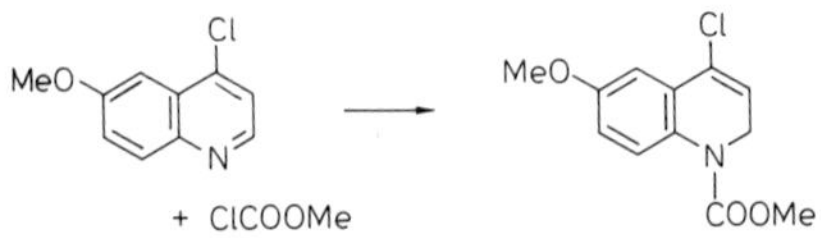

A soln. of BH_3-tetrahydrofuran complex in tetrahydrofuran added at –78° to a soln. of 4-chloro-6-methoxyquinoline in the same solvent, after 30 min. a soln. of Na dihydrobis(2-methoxyethoxy)aluminium in benzene/tetrahydrofuran added, the mixture stirred for 30 min. at –78°, methyl chloroformate added in one portion, and stirring continued for 18 hrs. at room temp. → methyl 1,2-dihydro-4-chloro-6-methoxyquinoline-1-carboxylate. Y 88%. F.e., also with diisobutylaluminium hydride, s. D.E. Minter, P.L. Stotter, J. Org. Chem. *46,* 3965–70 (1981).

Sodium trihydridocyanoborate *$Na[BH_3CN]$*

Transannular ring opening of N-condensed N-heterocyclics C
with chloroformic acid esters

371. 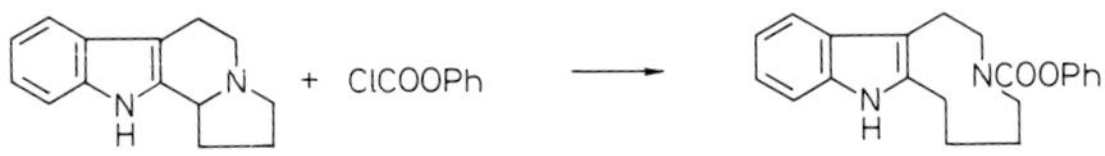

Startg. hexahydroindolo[3,2-*g*]indolizine treated with phenyl chloroformate at –70° for 1 hr., a soln. of $Na[BH_3CN]$ in tetrahydrofuran added, and the mixture allowed to warm to room temp. overnight → product. Y 83%. F.e.s. M.G. Calverley, J. Chem. Soc. Chem. Commun. *1981,* 1209–10.

Sodium dihydrobis(2-methoxyethoxo)aluminate *$Na[AlH_2(OCH_2CH_2OMe)_2]$*
s. under $H_3B.THF$

Alumina s. under KF *Al_2O_3*

18-Crown-6 polyether ←

N-Alkylation of phthalimides 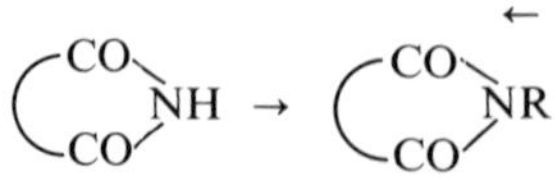

s. *27,* 480s*31;* with a catalytic amount of 18-crown-6 polyether cf. K. Soai, A. Ookawa, K. Kato, Bull. Chem. Soc. Japan *55,* 1671–2 (1982).

Acetone s. under KOH *Me_2CO*

1-Cyanopyrrolidine ←

1,3-Benzoxazine-2,4-diones from *o*-hydroxycarboxylic acid chlorides ○

372.

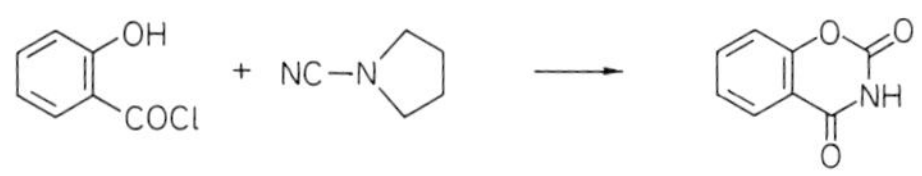

2-Hydroxybenzoyl chloride treated with 1-cyanopyrrolidine, cooled in an ice-bath, and allowed to warm slowly to room temp. with evolution of HCl → 2*H*-1,3-benzoxazine-2,4(3*H*)-dione. Y 79%. F.e.s. W. Ried, H.-E. Erle, Chem. Ber. *115,* 1662–4 (1982).

Anion exchanger s. under NaOH ←

Trimethylsilyl chloride/ammonia *Me_3SiCl/NH_3*

Protection *in situ* of hydroxyl groups as silyl ethers NH → NAc

Selective N-acylation

373.

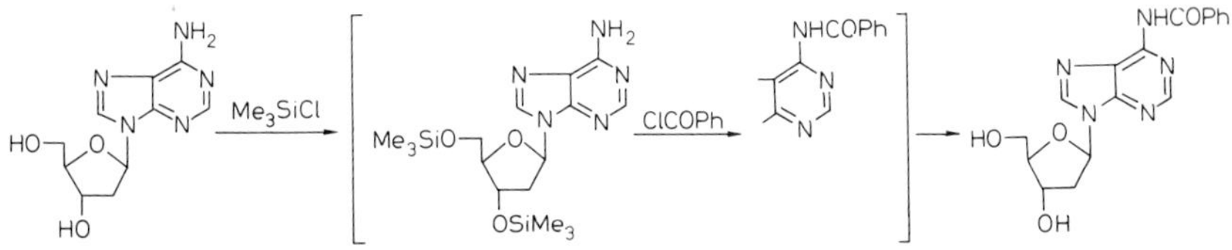

of deoxynucleosides. A suspension of 2-deoxyadenosine in dry pyridine treated with trimethylsilyl chloride, stirred for 15 min., benzoyl chloride added, stirring continued at room temp. for 2 hrs., cooled to 0°, treated sequentially with water followed after 5 min. by 29% aq. ammonia, and stirring continued at room temp. for 30 min. → 6-N-benzoyl-2′-deoxyadenosine. Y 94%. F.e., also with preferential protection of the 5′-hydroxyl group, s. G.S. Ti, B.L. Gaffney, R.A. Jones, J. Am. Chem. Soc. *104,* 1316–9 (1982).

Trimethylsilyl chloride/ethyldiisopropylamine *Me_3SiCl/i-Pr_2NEt*

Protection *in situ* of sulfhydryl groups as silyl sulfides

Selective N-acylation

374. $Me_3SiCl + HSCH_2CH_2NH_2 \rightarrow [Me_3SiSCH_2CH_2NH_2] \xrightarrow{HalCOCH_2Ph} HSCH_2CH_2NHCOCH_2Ph$

Acetonitrile and ethyldiisopropylamine added to 2-aminoethanethiol hydrochloride with ice-cooling under N_2, after 5 min. chlorotrimethylsilane added in one portion, the mixture kept at 0° for 10 min., treated sequentially with a soln. of startg. acyl halide in acetonitrile and more ethyldiisopropylamine at the same temp., and the mixture stirred for 30 min. at 0° and for 0.5–2 hrs. at room temp. → product. Y 95%. F.e.s. I. Shinkai et al., Synthesis *1980,* 924–6.

Dichlorosilanes/triethylamine *R_2SiCl_2/Et_3N*

Protection *in situ* of *β*-hydroxycarboxylic acids as O,O-silylene derivs. NH → $NCPh_3$

Preferential N-tritylation

375.

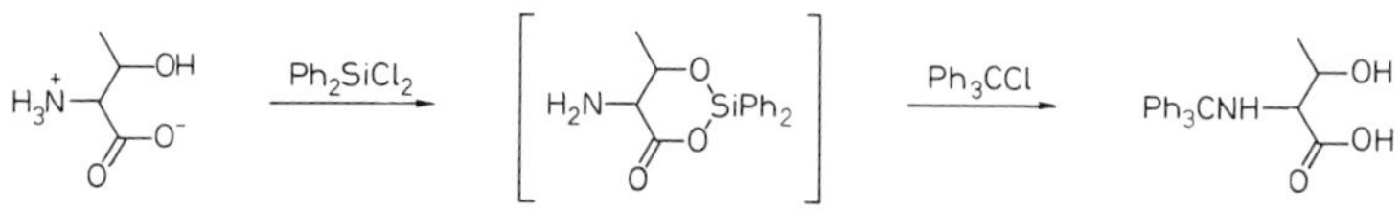

A mixture of threonine and dichlorodiphenylsilane in methylene chloride refluxed for 2 hrs. with stirring, allowed to cool to room temp., triethylamine added followed by a soln. of triphenylmethyl chloride in methylene chloride, the suspension stirred for 5 hrs. at room temp., excess methanol and triethylamine added, and the volatiles removed by rotary evaporator → N-tritylthreonine. Y 92%. F.e.s. K. Barlos, D. Papaioannou, D. Theodoropoulos, J. Org. Chem. *47,* 1324–6 (1982).

Trichloro(phenyl)silane *$PhSiCl_3$*

Acylisocyanates from carboxylic acid chlorides COCl → CONCO

with $SnCl_4$ and trimethylsilyl isocyanate cf. *32,* 387; with trichloro(phenyl)silane and N,O-bis(trimethylsilyl)urethan s. Zh. Obshch. Khim. *51,* 239–40 (1981).

Trimethyl phosphite s. under $AgNO_3$ *$(MeO)_3P$*

Sulfamide *$SO(NH_2)_2$*

Nitriles from carboxylic acid chlorides COCl → CN

376. $$p\text{-}MeOC_6H_4COCl \xrightarrow{NH_2SO_2NH_2} p\text{-}MeOC_6H_4CN$$

A soln. of the startg. m. and sulfamide in sulfolane heated at 120° for 3 hrs. → product. Y 97%. F.e. of both alkyl and aryl nitriles s. A. Hulkenberg, J.J. Troost, Tetrahedron Letters *23,* 1505–8 (1982).

*Tetra-*n-*butylammonium bromide s. under NaOH and KSCN* n-$Bu_4N^+Br^-$

*Tetra-*n-*butylammonium hydrogen sulfate s. under NaOH* n-$Bu_4N^+HSO_4^-$

Benzyltriethylammonium chloride s. under NaOH *$[PhCH_2NEt_3]^+Cl^-$*

Hydrogen chloride s. under Et_3N *HCl*

Via intermediates *v.i.*

Partial N-acylation of amines NH → NAc

s. *37,* 398

Anilines from aromatic bromides via 1-arylthio-1-triazenes Br → NH_2

377. $$PhMgBr + PhSCH_2N_3 \rightarrow PhNHN{=}NCH_2SPh \rightarrow PhNH_2$$

Azidomethyl phenyl sulfide in tetrahydrofuran at –78° treated with a soln. of phenylmagnesium bromide in the same solvent, the mixture kept at the same temp. for 1–2 hrs., warmed to –20°, and quenched with satd, aq. NH_4Cl → intermediate triazene (Y 73%) treated with aq. formic acid, and the mixture stirred for 2–24 hrs. → aniline (Y 90%). F.e. incl. hindered compds., also via organolithium compds. in the presence of $MgBr_2$, and without isolation of the intermediate, s. B.M. Trost, W.H. Pearson, J. Am. Chem. Soc. *103,* 2483–5 (1981).

***a-sec*-Amino from *a*-bromo-ketones and formiminoesters via *a*-formamidoketones** Br → NHR

378.

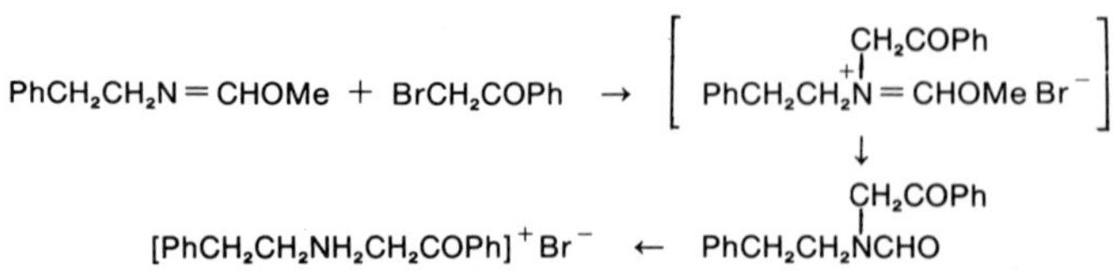

A soln. of phenacyl bromide in toluene heated at 90–100° for a few min., the startg. iminoester added in one portion, and worked up after 5 hrs. → intermediate (Y 86%) in 1.045 *M* methanolic HBr refluxed for 12–48 hrs. → N-(2-phenylethyl)phenacylamine hydrobromide (Y 98%). F.e.s. A. Guzmán, J.M. Muchowski, N.T. Naal, J. Org. Chem. *46,* 1224–7 (1981).

Amines from chlorides via azides Cl → N_3 → NH_2
s. *13,* 501; 4-aminopyrimidines s. W.K. Hagmann et al., J. Org. Chem. *46,* 1413–23 (1981).

1,3-O,N- and 1,3-S,N-Heterocyclics ○
from diamines and carboxylic acid chlorides
s. *37,* 247

Sulfur ↑ NC↓↑S

Without additional reagents *w.a.r.*

Carboxylic acid amides from amines NH_2 → $CONH_2$
s. *37,* 187

Thioacylation with bis(thioacyl) sulfides NH→ NC(S)R

379. $(p\text{-}MeC_6H_4CS)_2S + HNEt_2 \rightarrow p\text{-}MeC_6H_4CSNEt_2\ [+\ p\text{-}MeC_6H_4CS_2^-\ \overset{+}{N}H_2Et_2]$

Carboxylic acid thioamides. 2.4 Equivalents diethylamine added dropwise to a soln. of bis(4-methylthiobenzoyl) sulfide (prepd. according to Synth. Meth. *33,* 561) in *n*-hexane at 0°, the mixture stirred for 1 hr., the precipitated diethylammonium dithiocarboxylate removed, and the filtrate worked up → product. Y 92%. F.e., **also thiono- and dithio-carboxylic acid esters,** s. S. Kato et al., Ann. Chem. *1982,* 1229–44.

Δ^2-Imidazolines from 1,2-diamines ○

380.

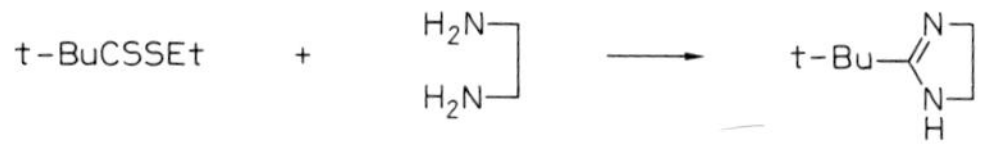

and dithiocarboxylic acid esters. A soln. of the startg. dithioester in benzene added to 1,2-diaminoethane with stirring at ca. 0°, and stirring continued for a few min. → product. Y 63%. F.e., **also from nitriles, carboxylic acid thioamides, and N-thioacylurethans,** s. G. Levesque, J.-C. Gressier, M. Proust, Synthesis *1981,* 963–5.

Ring closures with dithiocarbamic acid esters
2-Thiouracil ring from *o*-aminocarboxylic acids

381.

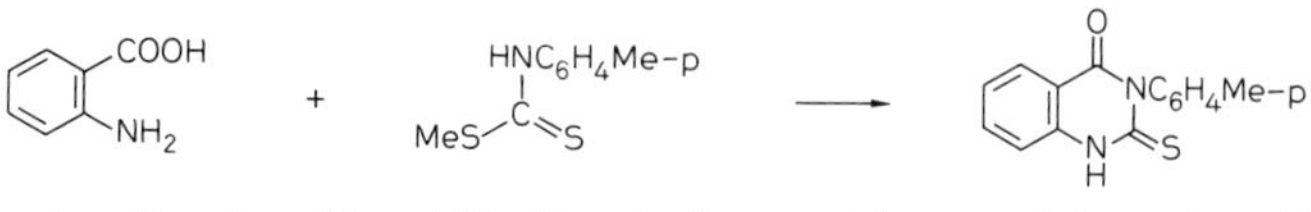

A stirred soln. of anthranilic acid in dimethylformamide treated dropwise with a soln. of methyl N-(*p*-methylphenyl)dithiocarbamate in the same solvent at room temp., and the mixture refluxed for 4–6 hrs. → product. Y 83%. F.e.s. J. Mayoral, E. Melendez et al., Synthesis *1981,* 962; **2-aminobenzimidazoles and 2-iminoimidazolidines** from *o*- and 1,2-diamines, respectively, s. ibid. *1982,* 482–4.

Ring closures with N-carbalkoxyiminodithiocarbonic acid esters

382.

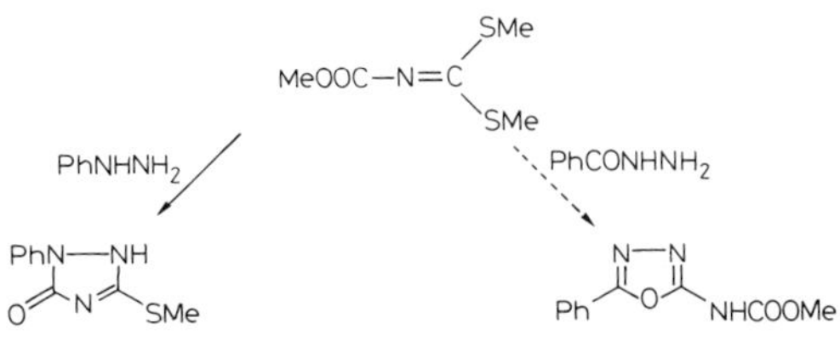

3-Alkylthio-Δ^3-1,2,4-triazolin-5-ones. A mixture of dimethyl N-carbomethoxyiminodithiocarbonate and phenylhydrazine in ethanol heated until no further methanethiol was evolved → 1-phenyl-3-methylthio-5-oxo-1,2,4-triazoline. Y 80%. F.e., **also 2-carbalkoxyamino-1,3,4-oxadiazoles from carboxylic acid hydrazides,** s. R. Evers, E. Fischer, M. Pulkenat, Z. Chem. *20,* 371–2 (1980); **5-subst. 2-carbalkoxyamino-1,3,4-thiadiazoles** s. ibid. *20,* 413–4 (1980).

Ring closures with N-dithiocarbalkoxyisothiouronium salts

383.

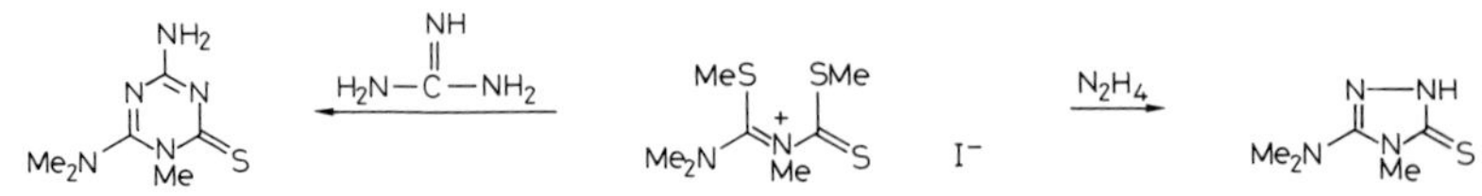

4,5-Diamino-1,3,5-triazine-2(1*H*)-thiones.	**3-Amino-Δ^2-1,2,4-triazoline-5-thiones.**
Guanidine hydrochloride and KOH	Hydrazine hydrate

added to a soln. of [dimethylamino(methylthio)methylene](dithiomethoxycarbonyl)-methylimmonium iodide in ethanol, and the mixture refluxed for

2 hrs. → product. Y 78%.	4 hrs. → product. Y 74%.

F.e. and reactions s. M. Yokoyama, K. Arai, T. Imamoto, J. Chem. Soc. Perkin Trans. I *1982,* 1059–62.

Lactams from sulfonylcarboxylic acid esters and amines
5,6-Dihydro-2(1*H*)-pyridone ring

384.

A soln. of the startg. m. in acetonitrile treated with dry methylamine for 20 min. at room temp., and stirred for a further 2 hrs. → product. Y 85%. F.e.s. E.A. Glazer, J.E. Presslitz, J. Med. Chem. *25,* 868–70 (1982).

3,4-Diamino- from 3-alkylthio-1,2,4-triazin-5(4*H*)-ones

385.

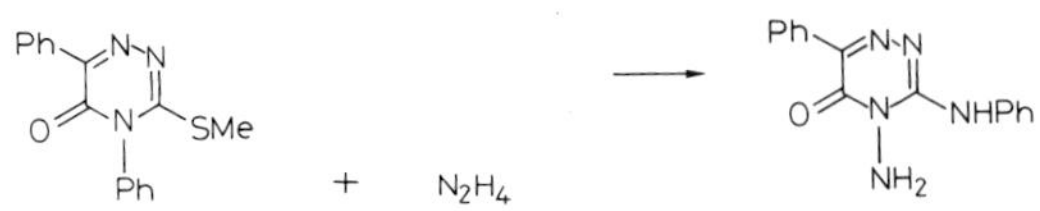

Hydrazine hydrate added to a soln. of startg. m. in ethanol, and the mixture stirred for 3 hrs. at reflux → product. Y ca. 80%. F.e.s. Y.A. Ibrahim, M.M. Eid, S.A.L. Abdel-Hady, J. Heterocyc. Chem. *17,* 1733–5 (1980).

Sodium hydroxide *NaOH*

2-Aminopyrimidine 1-oxides from 2(1*H*)-pyrimidinethiones

386.

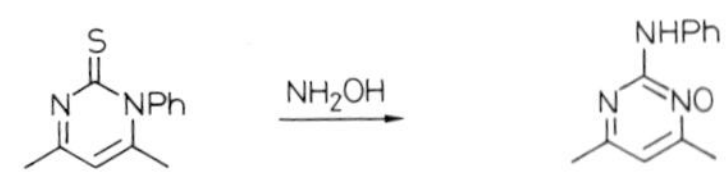

A mixture of startg. m., hydroxylamine hydrochloride, and NaOH in ethanol refluxed for 2 hrs. → 2-anilino-4,6-dimethylpyrimidine 1-oxide. Y 80%. F.e., **also isoxazoles from 2(1*H*)-pyrimidinones,** s. C. Kashima et al., Chem. Pharm. Bull. *29,* 2516–19 (1981).

*Sodium hydroxide/tetra-*n*-butylammonium bromide* *NaOH/*n*-$Bu_4N^+Br^-$*

N-Alkylation with dialkyl sulfates NH → NR

s. *6,* 410; solid-liq. phase transfer catalysis with NaOH/tetra-*n*-butylammonium bromide, N-alkylation of *o*-aminobenzophenones, s. G. Mouzin, H. Cousse, J.-M. Autin, Synthesis *1981,* 448–9.

Potassium hydroxide *KOH*

Ring closures with N-dithiocarbalkoxyisothiouronium salts ○

s. *37,* 383

Sodium acetate *MeCOONa*

1,2,4-Oxadiazoles from N-acylcarboxylic acid thioamides

387.

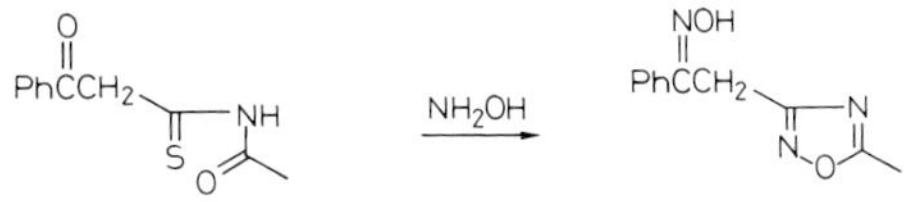

Startg. m. added to a mixture of hydroxylamine hydrochloride and Na acetate in ethanol, and stirred for 2 hrs. at 50° → product. Y 85%. F.e.s. G. Ronsisvalle, F. Guerrera, M.A. Siracusa, Tetrahedron *37,* 1415–20 (1981).

1,8-Bis(dimethylamino)naphthalene ←

388.

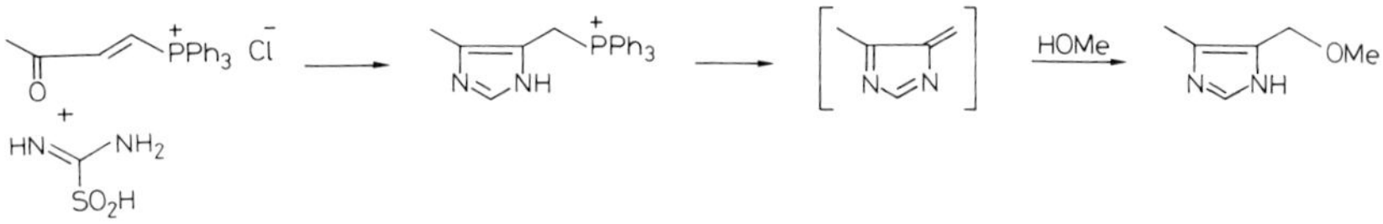

5-α-Alkoxyimidazoles. 1,8-Bis(dimethylamino)naphthalene added to a stirred soln. of startg. vinylphosphonium chloride and formamidine sulfinic acid in dimethyl sulfoxide, and the mixture heated to 80° → (Y 89%), in methanol treated while stirring with 25%-Na-methoxide in methanol at ambient temp., and the mixture refluxed for 20 min. → product (Y 80%). F.e., also **5-α-alkylthioimidazoles,** s. R.L. Webb, J.L. Lewis, J. Heterocyc. Chem. *18,* 1301–3 (1981).

5-α-Subst. imidazoles from α,β-ethylene-β-phosphonioketones
via 5-α-phosphonioimidazoles
Triphenylphosphine as leaving group

Boric acid *$B(OH)_3$*

Selective N-alkylation of nucleosides NH → NR

with sulfoxonium hydroxides cf. *34,* 418; with dimethyl sulfate or trimethyl phosphate in the presence of $B(OH)_3$ s. Bull. Chem. Soc. Japan *54,* 1569–70 (1981).

Boron fluoride *BF_3*

1,2,4-Triazole ring from carboxylic acid hydrazides ○

and thiolactams cf. *28,* 404; and N-carbalkoxyisothioureas, 3-carbalkoxyamino-1,2,4-triazoles with BF_3, s. R.V. Venkataratnam, K. Mohan, Chem. Ind. *1981,* 396.

Diethoxy disulfide *EtOSSOEt*

3,5-Diamino-1,2,4-thiadiazoles from two thiourea molecules

s. *37,* 432

Tributylphosphine *Bu_3P*

Thioacylureas from 1,2,4-dithiazol-3-ones via thioacylisocyanates

389.

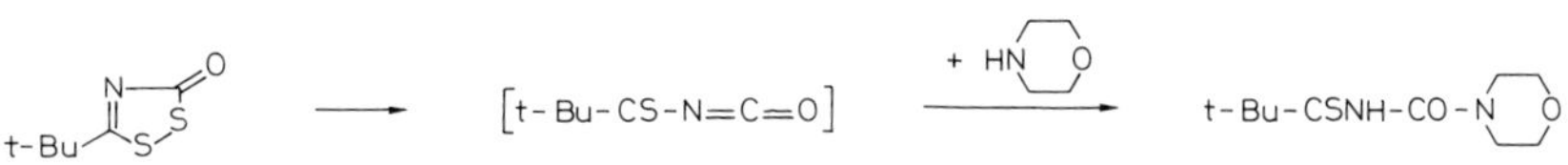

A mixture of tributylphosphine and morpholine in methylene chloride added to a soln. of 5-*tert*-butyl-1,2,4-dithiazol-3-one in the same solvent with stirring and ice-cooling → N-(morpholinocarbonyl)pivalic acid thioamide. Y 87%. F.e., also prepn. of thioacylisocyanates **from 4,5-thiazolinediones,** and f. reactions s. J. Goerdeler, K. Nandi, Chem. Ber. *114,* 549–63; 808–9 (1981).

Triphenylphosphine *Ph_3P*

Carboxylic acid amides
from carboxylic acid anhydrides and sulfenamides ⪰NSAr → ⪰NCOR

390.

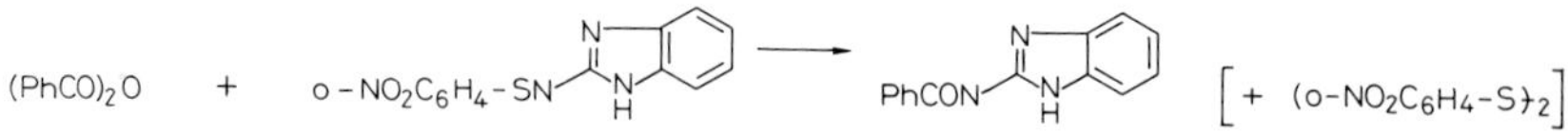

A mixture of startg. sulfenamide, benzoic anhydride, and triphenylphosphine as mild base in dioxane stirred at 18–20° for 2–4 hrs. → product. Y 96%. F.e. and reagents, **also from carboxylic acid halides,** s. E.A. Parfenov, V.A. Fomin, Zh. Obshch. Khim. *51,* 1144–53 (1981); **from thiolic acid esters** cf. ibid. *51,* 1129–36 (1981).

Sulfur *S*

Sym. thioureas from amines $2\ RNH_2 \rightarrow RNHCSNHR$
s. *6,* 489; f.e.s. V.S. Shaulov, E.V. Glebova, T.P. Vishnyakova, Izv. Vyssh. Ushebn: Zaved., Khim. Khim. Tekhnol. *25,* 119 (1982); Chem. Abstr. *96,* 217414 (1982).

*Tetra-*n-*butylammonium bromide s. under NaOH* n-$Bu_4N^+Br^-$

Via intermediates *v.i.*

Carboxylic acid thioamides
from dithiocarboxylic acids and amines
via organomercury dithiocarboxylates CSSH → C(S)N⪯

391.

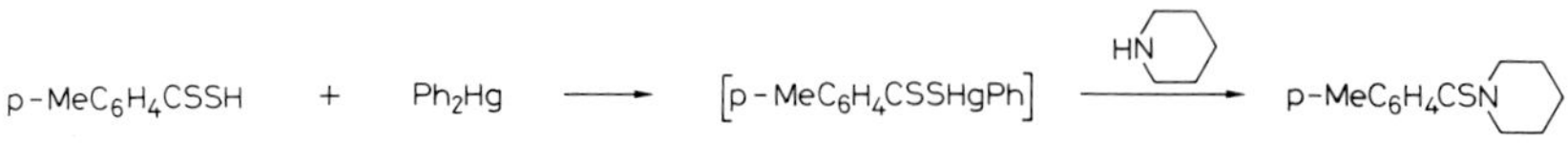

A soln. of 4-methyldithiobenzoic acid in ether treated with a soln. of diphenylmercury in dichloromethane, and the mixture stirred at room temp. for 1 hr. → phenylmercury 4-methyldithiobenzoate (Y 80%) in dichloromethane treated with piperidine, and stirring continued at room temp. for 1 hr. → N-pentamethylene-4-methylthiobenzamide (Y 92%). F.e.s. S. Kato et al., Z. Naturforsch. *36B,* 783–90 (1981).

Remaining Elements ↑ NC⇃↾Rem

Without additional reagents *w.a.r.*

Isocyanates from phosphine imines RN=P⪷ → RN=C=O

392. $PhNH_2 + PPh_3 \dashrightarrow PhN{=}PPh_3 \xrightarrow{CO_2} PhN{=}C{=}O\ [+\ Ph_3P{=}O]$

A stream of dry CO_2 passed through a soln. of startg. phosphine imine in benzene, and refluxed for 2 hrs. → product. Y 97%. F.e., **also isothiocyanates** with CS_2, s. P. Molina, M. Alajarin, A. Arques, Synthesis *1982,* 596–7; **carbodiimides** with 0.5 equivalents CS_2 cf. Synth. Commun. *12,* 573–7 (1982).

N-Sulfonylimines from bis(N-sulfonyl)seleninodiimides $CO \rightarrow C{=}NSO_2R$

393. $PhCHO + PhSO_2N{=}Se{=}NSO_2Ph \rightarrow PhCH{=}NSO_2Ph$

A soln. of benzaldehyde in methylene chloride added to a soln. of the startg. diimide in the same solvent, and the mixture stirred at 30–40° for 2 hrs. → product. Y 89%. F.e. and reactions, **also 1,1-diacylamines via N-acylimines,** s. N.Y. Derkach, G.G. Barashenkov, E.I. Slyusarenko, Zh. Org. Khim. *18,* 70–8 (1982).

Sodium methoxide *NaOMe*

Sec. from prim. amines via aminosilanes NHSi≡ → NHR

394. $PhNH_2 \dashrightarrow PhNHSiMe_3 \xrightarrow{BrCH_2CH=CH_2} PhNHCH_2CH=CH_2$

A mixture of anilinotrimethylsilane (prepd. by conventional silylation of aniline), Na methoxide, and allyl bromide in 5:1 tetrahydrofuran/diglyme stirred for 18 hrs. at 40° → N-allylaniline. Y 72%. F.e.s. W. Ando, H. Tsumaki, Chem. Letters *1981,* 693–4.

Sodium hydrogen carbonate/ benzyltriethylammonium chloride *$NaHCO_3/[PhCH_2NEt_3]^+Cl^-$*

Urethans from trialkylboranes
Phase transfer catalysis $R_3B \rightarrow RNHCOOR'$

395. $p\text{-}NO_2C_6H_4SO_2ONHCOOEt \rightarrow [\,:N\text{–}COOEt\,] \xrightarrow{Bu_3B} BuNHCOOEt$

A mixture of ethyl *p*-nitrobenzenesulfonyloxycarbamate and a little benzyltriethylammonium chloride in dichloromethane treated with a soln. of tributylborane in the same solvent, the mixture stirred vigorously at room temp., aq. $NaHCO_3$ added slowly dropwise, and stirring continued for 2 hrs. → ethyl N-butylcarbamate. Y 95%. F.e.s. I. Akimoto, A. Suzuki, Syn. Commun. *11,* 475–80 (1981); **nitrenes from N,O-disilylhydroxylamines** cf. Y.H. Chang, F.-T. Chiu, G. Zon, J. Org. Chem. *46,* 342–54 (1981).

Sodium iodide s. under Me_3SiCl *NaI*

Triethylamine *Et_3N*

N-Acyl- from
N-phosphoryl-phthalimides and carboxylic acids

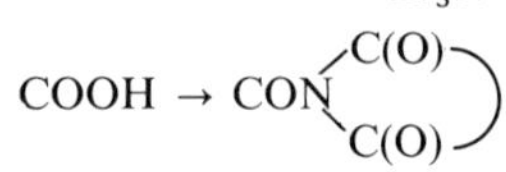

396.

A soln. of acetic acid and triethylamine in toluene treated with diphenyl phosphorophthalimide, and the mixture refluxed for 2 hrs. → product. Y 85%. F.e.s. C. Palomo, Synthesis *1981,* 993–4.

Magnesium *Mg*

Azo compds. from diazonium fluoroborates and Grignard compds. N=N
via organozinc chlorides cf. *13,* 513; direct method s. M.E. Garst, D. Lukton, Synth. Commun. *10,* 155–60 (1980).

Mercuric oxide/mercuric bromide *$HgO/HgBr_2$*

Cyclic N-glycosyl- from N,N′-disilyl-ureas ←

397.

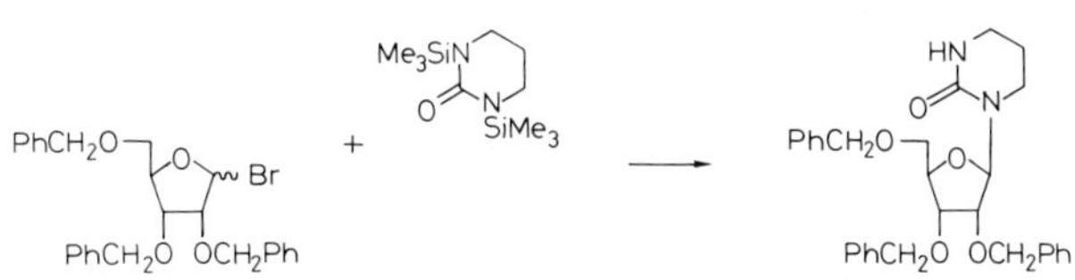

Startg. cyclic urea (prepd. by silylation of the unsubst. urea with excess of N,N-bis(trimethylsilyl)trifluoroacetamide in acetonitrile at room temp.) in benzene added rapidly to a refluxing mixture of HgO and $HgBr_2$ in the same solvent, after 10 min. a soln. of 2,3,5-tri-O-benzoyl-D-ribosyl bromide in benzene rapidly added, and refluxing continued for 18 hrs. → product. Y 76%. F.e.s. P.S. Liu et al., J. Med. Chem. *24,* 662–6 (1981).

Mercuric bromide s. under HgO *$HgBr_2$*

Trimethylsilyl chloride/sodium iodide *Me_3SiCl/NaI*

N-Glycosyl-N-heterocyclics from acyl glycosides OAc → N⊃

with $SnCl_4$ cf. *26,* 446; with trimethylsilyl chloride/sodium iodide s. A. Matsuda, Y. Kurasawa, K.A. Watanabe, Synthesis *1981,* 748; acyclic nucleoside analogs s. M.J. Robins, P.W. Hatfield, Can. J. Chem. *60,* 547–53 (1982).

Benzyltriethylammonium chloride s. under $NaHCO_3$ *$[PhCH_2NEt_3]^+Cl^-$*

Hydrogen fluoride *HF*

Partial N-acylation of amines via 2,1,3-Si,N,N-heterocyclics C

398.

A dil. soln. of benzoyl chloride in tetrahydrofuran added dropwise via syringe pump to a stirred soln. of cyclic silazane (prepared from the corresponding diamine and bis(diethylamino)dimethylsilane) in the same solvent at room temp., stirring continued for an additional 30 min., the mixture cooled with ice, treated with 40% aq. HF, and stirred at room temp. for a further 30 min. → product. Y 75%. F.e., incl. 1,3-diamine derivs., s. E. Schwartz, A. Shanzer, Tetrahedron Letters *23,* 979–82 (1982).

Carbon ↑ NC⇂↾C

Without additional reagents *w.a.r.*

Aromatic amines from halides and β-aminonitriles ArHal → ArN<

399.

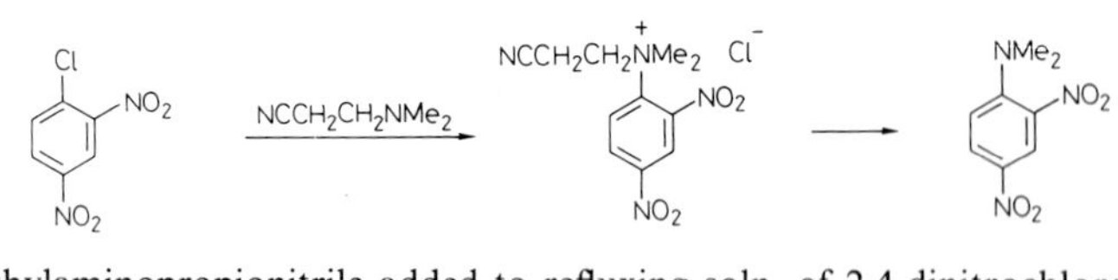

3-Dimethylaminopropionitrile added to refluxing soln. of 2,4-dinitrochlorobenzene in methanol, and left for 2 hrs. → 2,4-dinitro-N,N-dimethylaniline. Y 97%. F.e.s. L.L. Plakidin, V.N. Vostrova, Zh. Org. Khim. *18,* 342–52 (1982).

Replacement of cyano by amino groups CN → N<

s. *21,* 512; preferential replacement, 4-amino-3-cyanocinnolines, s. D.E. Ames, D. Bull, Tetrahedron *37,* 2489–91 (1981).

Carboxylic acid amides from acylcyanides COCN → CON<

s. *21,* 515; **chemistry of acylcyanides, review,** s. S. Hünig, R. Schaller, Angew. Chem. Intern. Ed. *21,* 36–49 (1982).

Formamides from formiminoesters and halides Br → NRCHO

α-Formamidoketones

s. *37,* 378

N-Acetoacetylation of carboxylic acid amides with 2,2,6-trimethyl-1,3-dioxin-4-one NH → $NCOCH_2COMe$

400.

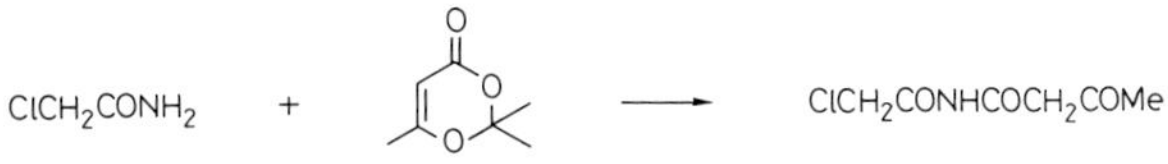

A mixture of chloracetamide and 2,2,6-trimethyl-1,3-dioxin-4-one (diketene-acetone adduct) in xylene refluxed for 30 min. → product. Y 90%. F.e.s. M. Sato, N. Kanuma, T. Kato, Chem. Pharm. Bull. *30,* 1315–21 (1982).

Sym. azines from glycidic acid esters ←

401.

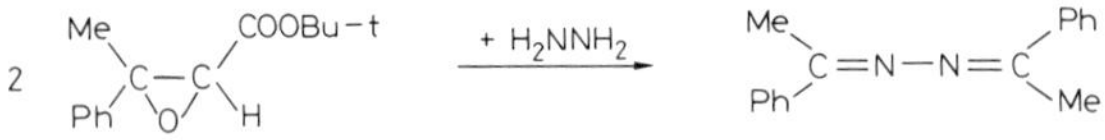

A mixture of *tert*-butyl β-methyl-β-phenylglycidate and excess hydrazine hydrate refluxed for 2 hrs. → acetophenone azine. Y 69%. F.e.s. K. Sethi, M.Z. Kirmani, Indian J. Chem. *20B,* 421–2 (1981).

Trichloromethyl as leaving group
s. *36,* 411; *22,* 493*s25;* f. nucleophiles s. J.T. Shaw et al., J. Heterocyc. Chem. *19,* 357–61 (1982).

Ring closures with N-(dichloromethylene)dimethylammonium chloride O
Quinazoline-2,4-diones from *o*-aminocarboxylic acid esters

402.

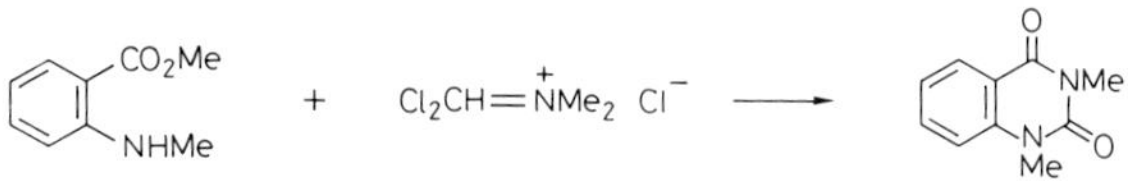

A mixture of (dichloromethylene)dimethylammonium chloride and methyl N-methylanthranilate in dichloroethane refluxed for 2 hrs. → 1,3-dimethyl-(1*H,* 3*H*)-quinazoline-2,4-dione. Y 76%. F.e., **also 2-amino-3,1-benzoxazin-4-ones,** s. I. Bitter, L. Szöcs, L. Töke, Acta Chim. Acad. Sci. Hung. *107,* 171–9 (1981); **3-amino-1,2,4-trazoles** s. J.S. Davidson, K. Karunaratne, Chem. Ind. *1981,* 648–9.

4-Imino-2-imidazolidones from α,β-diiminocarboxylic acid amides and isocyanates

403.

A soln. of the startg. m. in dry benzene treated with phenyl isocyanate, and the mixture refluxed for 3–5 hrs. → product. Y 71%. F.e., **also from α,β-diiminocarboxylic acid thioamides,** s. J. Moskal, J. Bronowski, A. Rogowski, Monatsh. Chem. *112,* 1405–16 (1981).

2-*o*-Hydroxyquinazolines from 4-chloro-2*H*-1,3-benzoxazines O C

404.

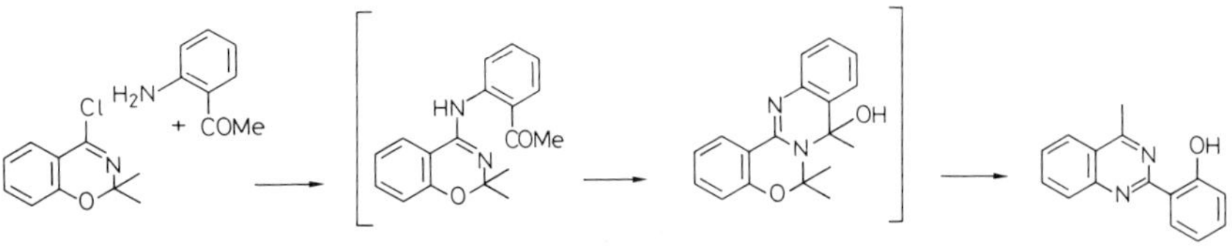

A mixture of 4-chloro-2,2-dimethyl-2*H*-1,3-benzoxazine and 2-aminoacetophenone in chloroform refluxed for 3 hrs. with stirring → 2-(2-hydroxyphenyl)-4-methylquinazoline. Y 92%. F.e. and isolation of the intermediate 4-amino-2*H*-1,3-benzoxazines, **also 2-*o*-hydroxy-4*H*-3,1-benzoxazines from *o*-aminobenzylalcohols** on refluxing in acetonitrile, s. R. Tachikawa, K. Wachi, A. Terada, Chem. Pharm. Bull. *30,* 559–63 (1982); **2-*o*-hydroxybenz-oxazoles, -othiazoles, and -imidazoles** s. ibid. Heterocycles *16,* 1165–8 (1981).

Sodium hydroxide/acetic acid *NaOH/MeCOOH*

Hydantoins from 5,6-*α*-diacoxyuracils

405. 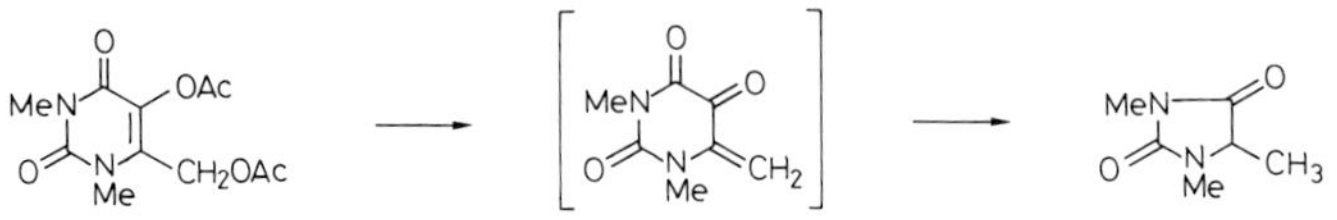

5-Acetoxy-6-(acetoxymethyl)-1,3-dimethyluracil treated with 0.022 *M* NaOH for 10 min. at room temp., the soln. acidified to pH 4 with acetic acid, and the mixture stored for 30 min. at room temp. → DL-1,3,5-trimethylhydantoin. Y 78%. F.e.s. B.A. Otter, I.M. Sasson, R.P. Gagnier, J. Org. Chem. *47,* 508–13 (1982).

Sodium ethoxide *NaOEt*

2-Alkylthiopyrimidines from pyrylium salts and isothioureas

406. 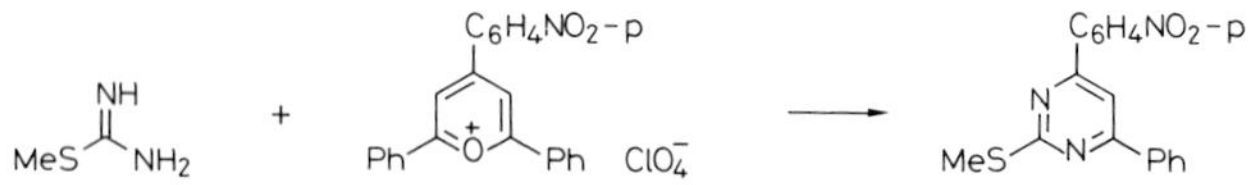

A soln. of Na ethoxide , followed after 15 min. by startg. pyrylium perchlorate, added to a suspension of startg. isothiourea in ethanol, stirred for 4 hrs., and left 20 hrs. in a refrigerator → product. Y 73%. F.e.s. E.A. Zvezdina, M.P. Zhdanova, G.N. Dorofeenko, Khim. Geterotsikl. Soedin. *1979,* 324–8.

Lithium diisopropylamide/hydrogen chloride *i-Pr$_2$NLi/HCl*

Carboxylic acid amides from aldehydes
via *α*-siloxynitriles CH(OSi⋲)CN → CON⋜
Electrophilic amination with phosphinyloxylamines

407. 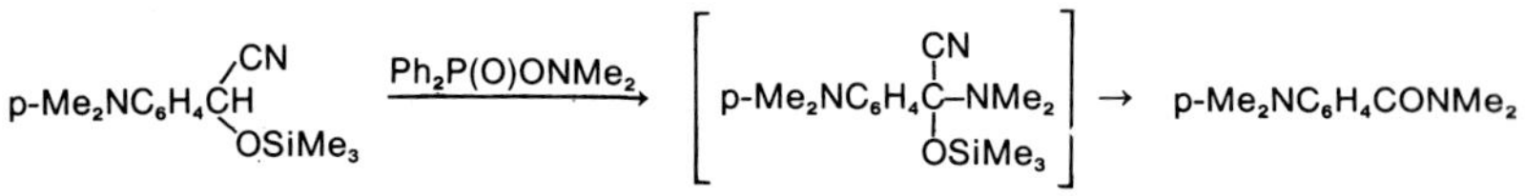

Startg. *α*-siloxynitrile metalated according to Synth. Meth. *29,* 808 with lithium diisopropylamide, treated with N,N-dimethyl-O-(diphenylphosphinyl)hydroxylamine at –78° to 20° for 5 hrs., and hydrolysed with 2 *M* HCl before work-up → product. Y 98%. F.e.s. G. Boche, F. Bosold, M. Niessner, Tetrahedron Letters *23,* 3255–6 (1982).

Sodium hydrogen carbonate *NaHCO$_3$*

Δ^4-2-Imidazolone from imidazole ring via Bamberger ring opening

408. 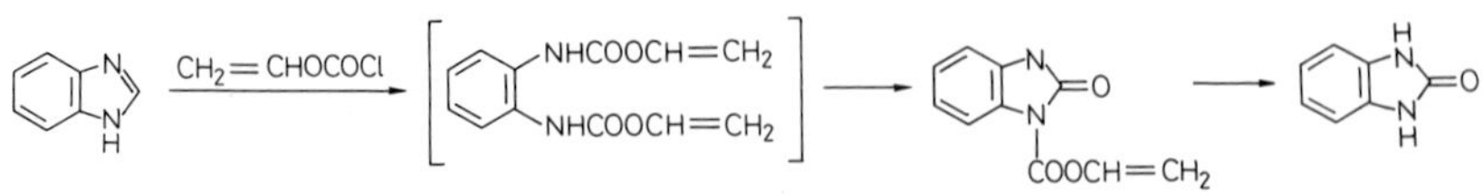

A stirred soln. of benzimidazole and $NaHCO_3$ in 50% aq. tetrahydrofuran at 0° treated with vinyl chloroformate, kept at the same temp. for 1 hr., and refluxed for 1 hr. → 1-vinyloxycarbonylbenzimidazol-2-one (Y 100%) in 50% aq. ethanol treated with 2 equivalents $NaHCO_3$, and the mixture refluxed for 1 hr. → benzimidazol-2-one (Y 100%). F.e.s. R.F. Pratt, K.K. Kraus, Tetrahedron Letters *22,* 2431–4 (1981).

Sodium acetate *MeCOONa*

2-Aminooxazoles from α-halogenoketones
s. *22,* 495; **review of 2-aminooxazoles and derivs.** s. L.S. Peshakova, V.B. Kalcheva, D.A. Simov, Khim. Geterotsikl. Soedin. *1981,* 1011–24.

Potassium cyanide *KCN*

4-Aminoquinazoline ring from *o*-(alkoxalylamino)nitriles and nitriles

409. 

2-Ethoxalylaminonaphthalene-1-carbonitrile heated with ethyl cyanoacetate and KCN in dimethylformamide at 70° with stirring for 24 hrs. → ethyl 1-amino-3-benzo[*f*]quinazolineacetate. Y 63%. F.e.s. Y. Tomioka, M. Yamazaki, Heterocycles *16,* 2115–9 (1981).

Triethylamine (s.a. under Polymer-based sulfonyl azides) *Et_3N*

N-Carbamylcarbodiimides from amines $NH_2 \rightarrow N{=}C{=}NCONH_2$

410. 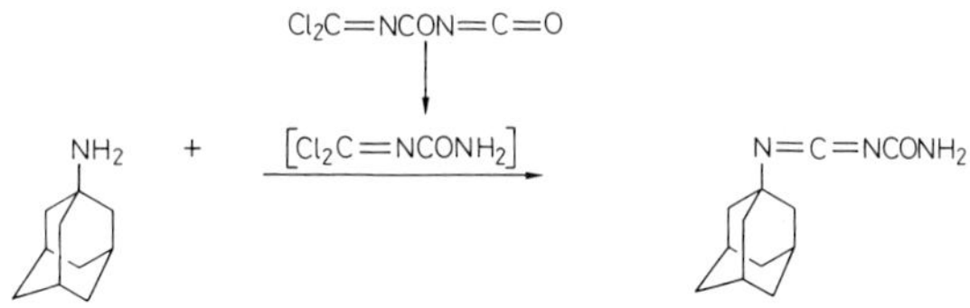

A little water in ether added to a soln. of *dichloromethylenecarbamoyl isocyanate* in ether at 22°, stirred for 1 hr., a soln. of 1-aminoadamantane and triethylamine in ether added at 0°, and stirring continued at 22° for a further 1 hr. → N-(1-adamantyl)-N′-carbamylcarbodiimide. Y 80%. F.e.s. R. Bunnenberg, J.C. Jochims, Chem. Ber. *114,* 2064–74 (1981).

***α-tert*-Amino- from α-bromo-carboxylic acid amides and tert. enamines** $Br \rightarrow NR_2$

411. 

A stirred soln. of 1-(1-pyrrolidinyl)cyclohexene and triethylamine in benzene or toluene treated with 2-bromoacetanilide, and stirring continued at 25° for 1–2 hrs. → 2-(1-pyrrolidinyl)acetanilide. Y 85%. F.e.s. A.A. El-Barbary, S. Carlsson, S.-O. Lawesson, Tetrahedron *38,* 405–12 (1982).

4-(Dimethylamino)pyridine ←

***o*-Aminocarboxylic acid amides from 3,1-benzoxazine-2,4-diones** C
s. *33,* 441; with 4-(dimethylamino)pyridine, also *o*-aminocarboxylic acid esters, s. M.C. Venuti, Synthesis *1982,* 266–8; esters from lactam acetals cf. J. Singh, N. Anand et al., Indian J. Chem. *20B,* 596–7 (1981).

Acetic acid s. under NaOH *CH_3COOH*

3-Methyl-1-phenyl-3-phospholene 1-oxide ←

Carbodiimides from 2 isocyanate molecules $2\ RN{=}C{=}O \rightarrow RN{=}C{=}NR$
s. *17,* 526; N,N′-divinylcarbodiimides with 3-methyl-1-phenyl-3-phospholene 1-oxide s. V.N. Fetyukhin, V.V. Kobrzhitskii, L.I. Samarai, Zh. Org. Khim. *17,* 2452–3 (1981).

Polymer-based sulfonyl azides/triethylamine ←

Diazo compds. from aldehydes $CH{-}CHO \rightarrow C{=}N_2$
with tosyl azide cf. *20,* 271; with polymer-based sulfonyl azides s. H. Dürr et al., Z. Naturforsch. *36B* 1149–52 (1981).

Hydrogen chloride s. under Pr_2NLi *HCl*

Dichlorotris(triphenylphosphine)ruthenium(II) $RuCl_2(PPh_3)_3$

Tert. methylamines from two tert. dimethylamine molecules $2\ RNMe_2 \rightarrow R_2NMe$

412. $2\ PhCH_2NMe_2 \rightarrow (PhCH_2)_2NMe\ [+\ Me_3N]$

Startg. m. heated without solvent at 180° for 1.5 hrs. in the presence of a little dichlorotris(triphenylphosphine)ruthenium(II) → product. Y 90%. F.e., **also from two sec. methylamine molecules,** s. A. Arcelli, Bui-The-Kai, G. Porzi, J. Organometal. Chem. *231,* C31–4 (1982).

Elimination ⇑

Hydrogen ↑ NC⇑H

Electrolysis/nickel hydroxide *ϟ/Ni(OH)$_2$*

Nitriles from prim. amines $CH_2NH_2 \rightarrow CN$

413. $Me(CH_2)_8CH_2NH_2 \rightarrow Me(CH_2)_8CN$

A soln. of 1-decylamine in 1:1 *tert*-butanol/aq. KOH electrolysed at 40° for 4 hrs. using a *nickel/nickel hydroxide anode* with a current of 2 A and voltage of 2.0 V → decanenitrile. Y 86%. F.e.s. U. Feldhues, H.J. Schäfer, Synthesis *1982,* 145–6.

Sodium hydroxide *NaOH*

1,2,3-Triazoles from 3-triazenopyridinium salts C O

414.

Aq. NaOH added dropwise to a suspension of 1-methyl-3-(3-phenyl-1-triazenyl)pyridinium iodide in 1:1 water/dichloromethane at 10–12°, and the mixture stirred at room temp. for 24 hrs. → 1-phenyl-4-(3-methyliminopropenyl)-1,2,3-triazole. Y 80%. F.e.s. S. Tanaka, K. Wachi, A. Terada, Chem. Pharm. Bull. *28,* 2083–8 (1980); **pyrazoles** cf. ibid. *28,* 1265–9 (1980).

Benzoquinone/bis(acetonitrile)dichloropalladium(II) ←

Cyclic sec. amines from ethylenesulfonylamines via cyclic N-sulfonylenamines and sulfonylamines Regiospecific ring closure ○

415.

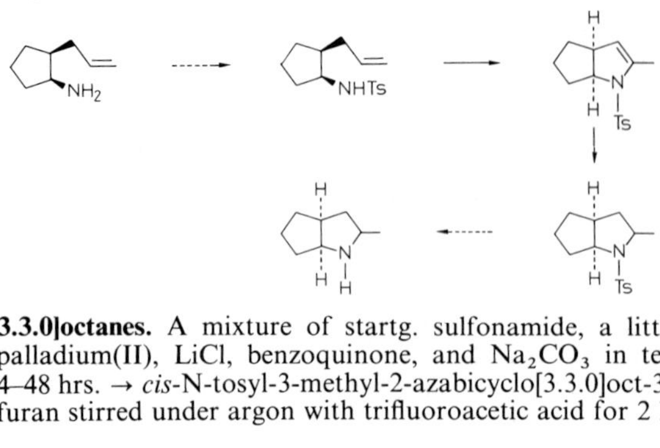

2-Azabicyclo[3.3.0]octanes. A mixture of startg. sulfonamide, a little bis(acetonitrile)dichloropalladium(II), LiCl, benzoquinone, and Na_2CO_3 in tetrahydrofuran refluxed for 24–48 hrs. → *cis*-N-tosyl-3-methyl-2-azabicyclo[3.3.0]oct-3-ene (Y 85%), in tetrahydrofuran stirred under argon with trifluoroacetic acid for 2 hrs., a soln. of

Na trihydridocyanoborate in the same solvent added by syringe, and stirred overnight → *cis*-N-tosyl-3-methyl-2-azabicyclo[3.3.0]octane (Y 78%). F.e., reductive methods, and limitations, also photochemical detosylation s. L.S. Hegedus, J.M. McKearin, J. Am. Chem. Soc. *104,* 2444–51 (1982).

Trifluoroacetic anhydride s. under Ph_2SeO *$(CF_3CO)_2O$*

Diphenyl selenoxide/trifluoroacetic anhydride *$Ph_2SeO/(CF_3CO)_2O$*

Oxidations with *diphenylselenium bis(trifluoroacetate)*

3,4-Dihydro- from 1,2,3,4-tetrahydro-isoquinolines CHNH → C=N

416.

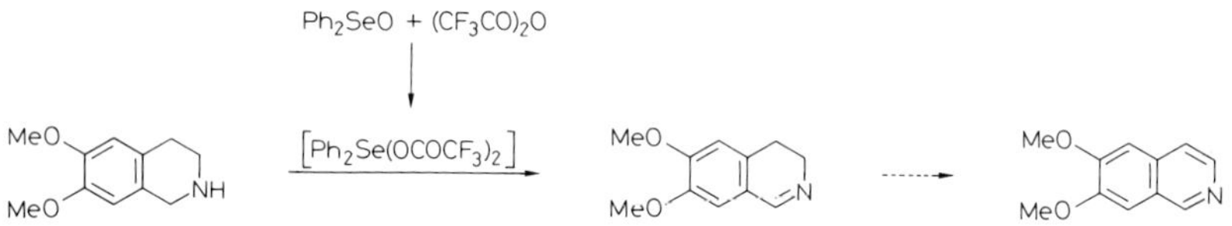

Trifluoroacetic anhydride in dimethoxyethane added by syringe to an equimolar amount of diphenyl selenoxide in the same solvent, stirred at 25° for 15 min., added dropwise during 30 min. to a soln. of startg. sec. amine in dimethoxyethane, and worked up after 3–12 hrs. → product. Y 80%. F.e., **also isoquinolines** with excess of oxidant, and prepn. of **cyclic α-aminonitriles from tert. amines,** s. J.P. Marino, R.D. Larsen, J. Am. Chem. Soc. *103,* 4642–3 (1981).

Lead tetraacetate *$(MeCOO)_4Pb$*

1,2,3-Benzotriazines from 1-aminoindazoles ○

s. *28,* 431; s.a. B.J. Clark, R. Grayshan, J. Chem. Res. (S) *1981,* 324–5.

Iodic acid *HIO_3*

Aminochromes

s. *8,* 545; **review of selective oxidations with iodic acid** s. U. Avico et al., Cron. Chim. *59,* 15–18, 32 (1979); Chem. Abstr. *95,* 23602 (1981).

Hydrogen chloride *HCl*

Pyrazoles from α,β-ethylenehydrazones or pyrazolines

s. *23,* 494; f. methods cf. H.M. Feid-Allah, Pharmazie *36,* 754–6 (1981).

Manganese dioxide *MnO_2*

Quinazolines from 3,4-dihydroquinazolines CHNH → C=N

s. *37,* 424

Nickel hydroxide s. under ϟ *$Ni(OH)_2$*

Bis(acetonitrile)dichloropalladium(II) s. under Benzoquinone *$PdCl_2(MeCN)_2$*

Dichlorobis(triphenylphosphine)palladium(II) *$PdCl_2(PPh_3)_2$*

Intramolecular amination of olefins ○

with $PdCl_2(MeCN)_2$ cf. *32,* 417; with $PdCl_2(PPh_3)_2$, **4(1*H*)-quinolones,** s. A. Kasahara et al., Chem. Ind. *1981,* 121.

Oxygen ↑ NC⇑O

Without additional reagents *w.a.r.*

1,2,4-Triazoles by cyclodehydration ○

s. *20,* 379; **N-condensed 1,2,4-triazoles,** thermal method in *o*-dichlorobenzene, cf. D.R. Shridhar et al., Indian J. Chem. *20B,* 132–4 (1981).

Without additional reagents (cont.) *w.a.r.*

Ring closures with N-oxides to benzo[1,2-*b*]indolizines
with ethyldiisopropylamine hydrobromide cf. *29,* 457; flash vacuum pyrolysis s. A. Ohsawa, T. Kawaguchi, H. Igeta, Chem. Letters *1981,* 1737–8.

Irradiation

Abnormal Beckmann rearrangement ←
ethylenenitriles s. *17,* 549; photochemical method, alkoxynitriles, cf. M. Bös, W. Fleischhacker, Ann. Chem. *1982,* 112–7.

Electrolysis

Reductive ring closure of nitroketones ○
with Ni/N_2H_4 cf. *17,* 553; by electrolysis, 4-hydroxy-4*H*-1,4-benzothiazine 1,1-dioxides, s. C.-P. Maschmeier, H. Tanneberg, H. Matschiner, Z. Chem. *21,* 219 (1981).

Sodium hydroxide *NaOH*

1,2,4-Triazine-3,5(2*H*,4*H*)-diones from α-ketocarboxylic acid semicarbazones
with Na ethoxide cf. *21,* 533; with NaOH in ethylene glycol s. A. Nováček et al., Collection Czech. Chem. Commun. *46,* 2203–6 (1981); nucleoside analogs s. H. Hřebabecký, ibid. *47,* 994–9 (1982).

Potassium hydroxide *KOH*

3-Mercapto-1,2,4-triazoles from acylthiosemicarbazides
with Na ethoxide cf. *6,* 510; with KOH, 3-mercapto-4-allyl-5-(4-methoxybenzyl)-1,2,4-triazole, s. T.R. Ovsepyan, A.K. Avetisyan, Sint. Geterotsikl. Soedin. *1981,* 12, 35 (Russ.); Chem. Abstr. *95,* 150519 (1981).

Uracil-6-carboxylic acids
from 3-aminomaleimide-1-carboxylic acid esters

417.

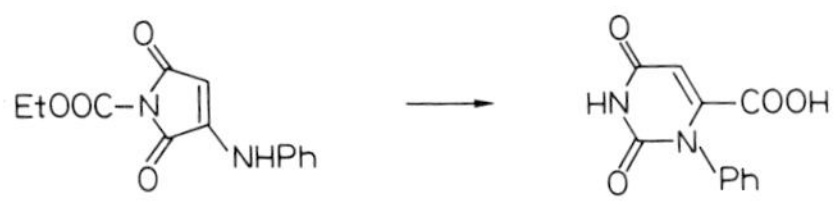

A suspension of 1-ethoxycarbonyl-3-phenylaminomaleimide in aq. KOH kept at 65° for 2 hrs. → 1-phenyluracil-6-carboxylic acid. Y 90%. F.e.s. J. Seres et al., Tetrahedron *37,* 1565–9 (1981).

Sodium methoxide *NaOMe*

2-Amino-Δ^1-azirines from O-sulfonylamidoximes ○

418.

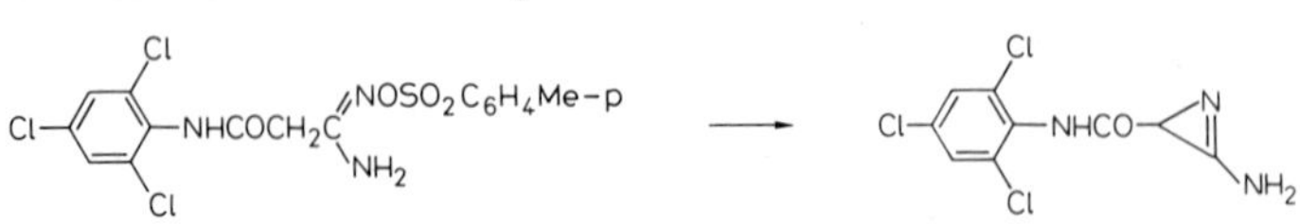

A slurry of startg. tosylate in methanol treated dropwise at 25° with a soln. of Na methoxide in methanol, and the reaction continued for 20 min. → product. Y 88%. F.e.s. J.A. Hyatt, J. Org. Chem. *46,* 3953–5 (1981).

Pyrido[1,2-*a*]-1,3,5-triazines
s. *28,* 435; from 2-pyridylisothiocyanates cf. M. Marchalín, J. Světlík, A. Martvoň, Collection Czech. Chem. Commun. *46,* 2557–63 (1981).

Potassium hydrogen carbonate s. under $MeSO_2Cl$ $KHCO_3$

Triethylamine s. under $(Ph_3P)_4Pd$ Et_3N

Pyridine C_5H_5N

Cyclic peptides by the azide route

s. *13,* 530s*29;* with 1-hydroxybenzotriazole as additive cf. R.F. Nutt, D.F. Veber, R. Saperstein, J. Am. Chem. Soc. *102,* 6539–45 (1980).

Tris(dimethylamino)borane $(Me_2N)_3B$

Macrocyclic lactams from aminocarboxylic acid esters by boron-template ring closure
Polyazamacrocyclics

419.

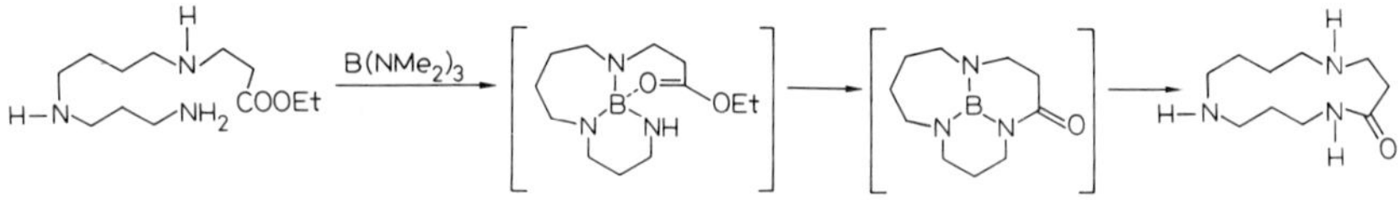

A soln. of tris(dimethylamino)borane in xylene added to a stirred soln. of ethyl 4,9,13-triazatridecanoate in the same solvent, stirring continued at 25° for 0.5 hrs., and refluxed for 15 hrs. → 2,6,11-triazacyclotridecanone. Y 77%. F.e.s. H. Yamamoto, K. Maruoka, J. Am. Chem. Soc. *103,* 6133–6 (1981); **lactams by template ring closure, review,** s. H.H. Wasserman, J.S. Wu, Heterocycles *17,* 581–605 (1982).

Alumina Al_2O_3

Pyrroles from dihydro-1,2-oxazines

s. *20,* 381; via 5-hydroxy-Δ^1-pyrrolines cf. H. Saiki, T. Mukai, Chem. Letters *1981,* 1561–4.

*Tetra-*n*-butylammonium hydrogen sulfate s. under* $MeSO_2Cl$ n-$Bu_4N^+HSO_4^-$

Dimethylchloroformiminium chloride $[Me_2N{=}CHCl]^+Cl^-$

Nitriles from aldoximes CH=NOH → CN

420. $Me_2C{=}C{=}CHCMe_2CH{=}NOH \rightarrow Me_2C{=}C{=}CHCMe_2CN$

A soln. of startg. aldoxime in acetonitrile allowed to react with a soln. of dimethylchloroformiminium chloride in the same solvent at –10° → product. Y 95%. F.e.s. J.-P. Dulcere, Tetrahedron Letters *22,* 1599–600 (1981).

Chlorosilanes ←

Isocyanates from urethans NHCOOR → N=C=O

with $COCl_2$ cf. *23,* 500; with chlorosilanes, e.g. $PhSiCl_3$, s. V.F. Mironov, V.P. Kozyukov, G.I. Orlov, Zh. Obshch. Khim. *51,* 1814–8 (1981).

Reactions with *pyrocatechyl phosphorotrichloridate* ←

s. *18,* 543; **review** of chlorine exchange reactions with pyrocatechyl phosphorotrichloridate s. J. Gloede, Z. Chem. *22,* 126–34 (1982).

*Methanesulfonyl chloride/potassium hydrogen carbonate/ tetra-*n*-butylammonium hydrogen sulfate* $MeSO_2Cl/KHCO_3/$ n-$Bu_4N^+HSO_4^-$

Lactam ring closure

2-Azetidinone ring

with $SOCl_2$ cf. *16,* 541; with methanesulfonyl chloride and $KHCO_3$ in the presence of tetra-*n*-butylammonium hydrogen sulfate s. Y. Watanabe, T. Mukaiyama, Chem. Letters *1981,* 443–4.

Tetrakis(triphenylphosphine)palladium(0)/triethylamine $(Ph_3P)_4Pd/Et_3N$

Spiro-N-heterocyclics from cyclic acoxy-2-ethylenes

s. *37,* 246

Palladium(II) chloride *$PdCl_2$*

Pyrroles from α,β-acetyleneketones via 2-hydroxy-3-acetyleneamines

421.

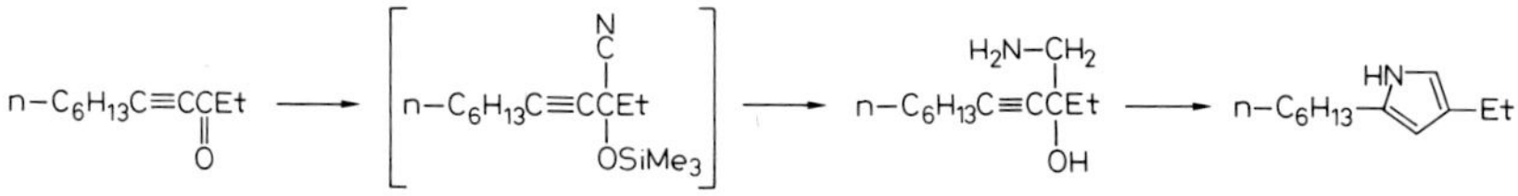

A mixture of startg. ynone and trimethylsilyl cyanide treated with a soln. of diethylaluminium chloride in benzene, the mixture stirred at room temp. for 1 hr., excess trimethylsilyl cyanide removed, and the residue treated with a soln. of $LiAlH_4$ in ether at room temp. for 1 hr. → 3-aminomethyl-4-undecyn-3-ol (Y 80%) and a little $PdCl_2$ in acetonitrile refluxed for 3 hrs. → 4-ethyl-2-hexylpyrrole (Y 84%). F.e. and with BF_3 s. K. Utimoto, H. Miwa, H. Nozaki, Tetrahedron Letters *22*, 4277–8 (1981).

Nitrogen ↑ NC⇑N

Without additional reagents *w.a.r.*

Indoles from β-styrylazides ○

s. *24*, 528; 7-subst. derivs. s. P.T. Kim et al., J. Heterocyc. Chem. *18*, 1373–7 (1981); isoquinolines in the presence of iodine cf. D.M.B. Hickey, C.J. Moody, C.W. Rees, J. Chem. Soc. Chem. Commun. *1982*, 3–4.

5-Cyanouracils

with NaOH cf. *28*, 454; thermal method s. O.S. Wolfbeis, Ann. Chem. *1982*, 182–5.

Benzotriazoles from 1-(o-nitroaryl)tetrazoles ⊂ ○

422.

A soln. of the startg. tetrazole in nitrobenzene refluxed for 0.5 hrs. → product. Y 81%. F.e.s. N.A. Evans, Australian J. Chem. *34*, 691–5 (1981).

Phthalazine-1,4-diones from N-phthalimidoazimines ○

423.

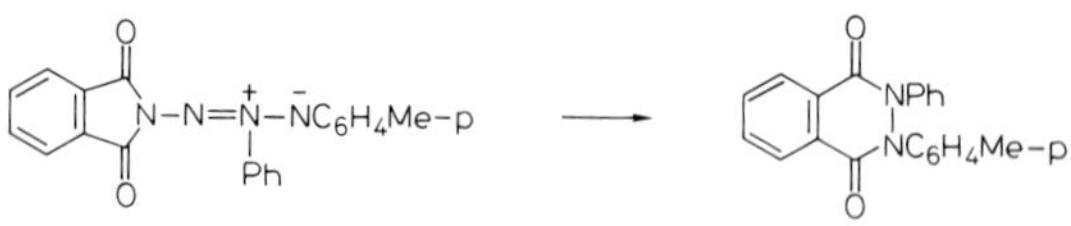

Startg. N-ylid in chloroform refluxed for 30 hrs. → N-phenyl-N,N′-phthaloyl-N′-*p*-tolylhydrazine. Y 79–85%. F.e.s. L. Hoesch, Helv. Chim. Acta *64*, 38–48 (1981).

1-Cyanoimidazoles by thermolysis

s. *28*, 450; from 2-azidopyrazines s. A. Ohta et al., Heterocycles *14*, 1963–6 (1980).

Boron fluoride/acetic acid *$BF_3/MeCOOH$*

1,2,4-Benzotriazines from formazans ○

Bamberger synthesis

with H_2SO_4 cf. *11*, 568; with BF_3/acetic acid s. R.H. Atallah, M.Z. Nazer, Tetrahedron *38*, 1793–6 (1982).

Fluoroboric acid *HBF_4*

3,4-Dihydroquinazolines from azidoazomethines

424. 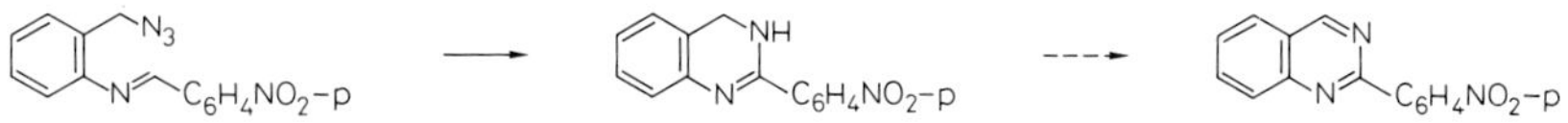

A soln. of 2-azidomethyl-N-(*p*-nitrobenzylidene)aniline in acetonitrile treated with fluoroboric acid-ether complex for 5 min. at room temp., and the resultant fluoroborate heated at 70° for 30 min. → product. Y 60%. F.e. and solvents, also oxidation to **quinazolines** with MnO_2, s. R. Kreher, U. Bergmann, Heterocycles *16,* 1693–6 (1981).

Acetic anhydride s. under $(MeCOO)_4Pb$ *$(MeCO)_2O$*

Acetic acid s. under BF_3 *CH_3COOH*

Benzoyl chloride *PhCOCl*

Azomethinium salts from monoacyl-1,1-diamines AcN–C–N< → C=$\overset{+}{N}$<

425. 

Methyleneammonium salts. A mixture of N-[(cyclohexylmethylamino)methyl]-N-methylformamide and benzoyl chloride in benzene allowed to stand at 20° for 16 hrs. → N-cyclohexyl-N-methylmethaniminium chloride. Y 74%. F.e.s. H. Böhme, E. Raude, Chem. Ber. *114,* 3421–9 (1981).

Lead tetraacetate/acetic anhydride *$(MeCOO)_4Pb/(MeCO)_2O$*

Pyrazoles from 1,3-di(hydrazones) ○

426. 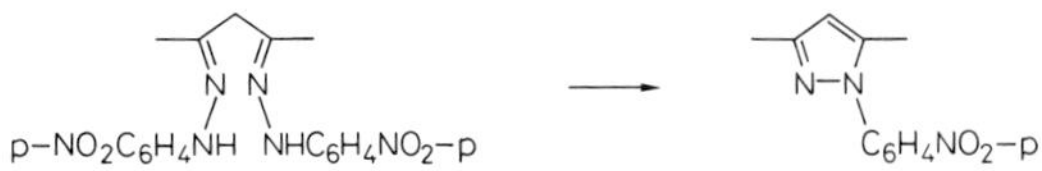

A soln. of pentane-2,4-dione bis(4-nitrophenylhydrazone) in acetic acid containing a little acetic anhydride treated with Pb tetraacetate, and stirred at 25–30° for 4 hrs. → 1-(4-nitrophenyl)-3,5-dimethylpyrazole. Y 95%. F.e.s. R.N. Butler, J.P. James, J. Chem. Soc. Perkin Trans. I *1982,* 553–5.

Triphenylphosphine *Ph_3P*

2-Alkoxy-1,3-benzoxazin-4-ones from *o*-azidocarbonylcarbonic acid esters via intramolecular Wittig-type synthesis with phosphine imines

427. 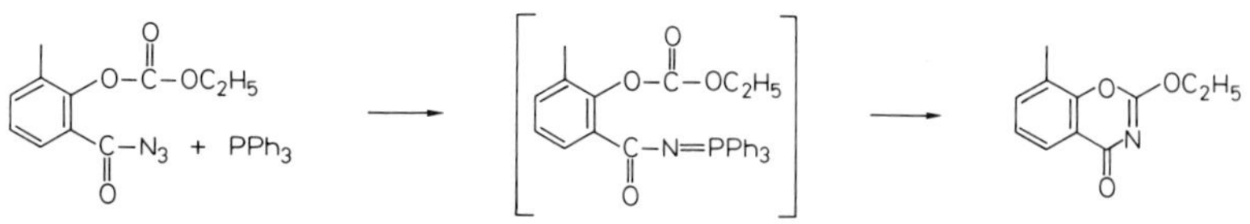

A stirred soln. of startg. azide in toluene at –20° treated with a soln. of triphenylphosphine in light petroleum, and the mixture refluxed for 10 min. → product. Y 72%. F.e. and isolation of intermediate s. H.B. Stegmann, F. Stöcker, G. Wax, Synthesis *1981,* 816–7.

Antimony pentachloride *$SbCl_5$*

Di(alkylidene)ammonium salts from 2,4-diaza-1,4-dienes $C=\overset{+}{N}=C$

428. 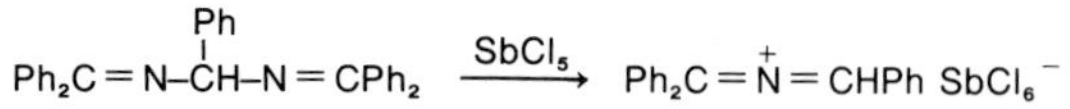

A stirred soln. of startg. m. in chloroform treated with acetyl chloride at room temp., the mixture cooled to 0°, a soln. of $SbCl_5$ in the same solvent added dropwise, and stirred at room temp. in the dark for 3 hrs. under N_2 → product. Y 76%. F.e.s. E.-U. Würthwein, Angew. Chem. Intern. Ed. *20,* 99–100 (1981).

Sulfuric acid *H_2SO_4*

Ring closure with diazo compds. ○

with formic acid cf. *29,* 465; with concd. H_2SO_4, 3-azetidinones, s. A.M. Sipyagin, V.G. Kartsev, Zh. Org. Khim. *16,* 2447–8 (1980).

Indoles from 1-azidobenzocyclobutenes

429. 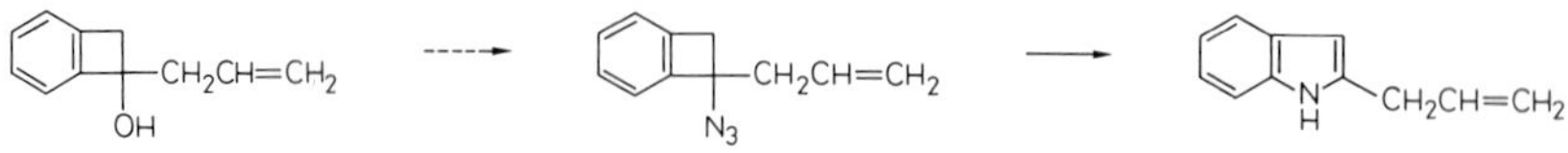

A soln. of startg. azide in chloroform treated with concd. H_2SO_4 at 0° → product. Y 95%. F.e., **also** directly **from benzocyclobuten-1-ols via Schmidt reaction,** s. G. Adam, J. Abdrieux, M. Plat, Tetrahedron Letters *22,* 3181–4 (1981).

Hydrogen chloride *HCl*

Δ^2-Pyrazol-5-ones from hydrazines ○

s. *18,* 550; with *β*-ketohydroxamic acids cf. K. Tabei, E. Kawashima, T. Kato, Chem. Pharm. Bull. *29,* 244–9 (1981).

Benzimidazol-2-ones

from *o*-aminoureas s. *36,* 443; **from *o*-aminourethans** cf. M. Bianchi et al., Eur. J. Med. Chem. *16,* 321–6 (1981).

Dichlorotris(triphenylphosphine)ruthenium(II) *$RuCl_2(PPh_3)_3$*

Cyclic amines from diamines

430. 

A mixture of diamine and a little dichlorotris(triphenylphosphine)ruthenium(II) in diphenyl ether heated at 180° for 5 hrs. in a sealed glass tube under N_2 → product. Y 90%. F.e., **also sym. tert. from prim. amines,** s. B.-T. Khai, C. Concilio, G. Porzi, J. Org. Chem. *46,* 1759–60 (1981); **sym. sec. from prim. amines** s. J. Organometal. Chem. *208,* 269 (1981).

Halogen ↑ NC ⇑ Hal

Without additional reagents *w.a.r.*

Nitriles from N,N-dichloramines CH_2NCl_2 → CN

with CsF cf. *24,* 535; **by pyrolysis** during g.l.c., also N-chlorimines and ethylene derivs., s. J.T. Roberts, B.R. Rittberg, P. Kovacic, J. Org. Chem. *46,* 4111–5 (1981); ibid. *45,* 5239–43 (1980).

Sodium hydroxide/anion exchanger ←

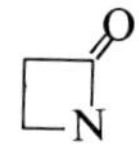

2-Azetidinones from β-bromocarboxylic acid amides
Phase transfer catalysis
s. *12,* 565s*36;* 3-phase catalysis with Duolite A-109 (Cl^-), a strongly basic polystyrene-based quaternary ion exchange resin, s. T. Okawara et al., Chem. Letters *1981,* 185–8; s.a. Chem. Pharm. Bull. *30,* 1225–33 (1982); with 50% NaOH/benzene and added α-aminocarboxylic acids as catalyst cf. ibid. *30,* 1574–8 (1982).

1,8-Diazabicyclo[5.4.0]undec-7-ene ←

N-Sulfinylimines C=NS(O)R

431. 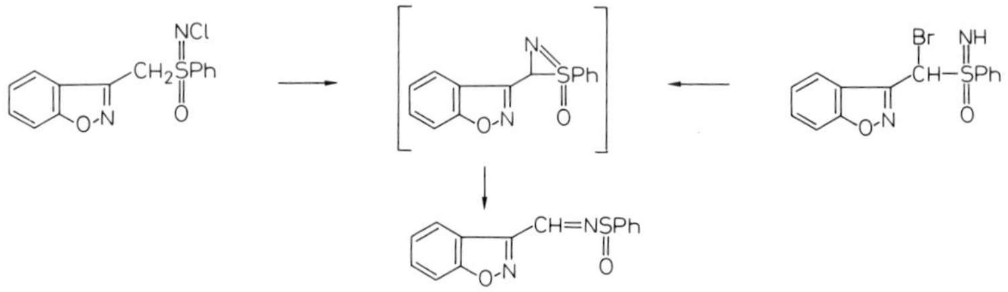

from N-halogenosulfoximines. A soln. of startg. N-chlorosulfoximine in dichloromethane treated with a slight excess of 1,8-diazabicyclo[5.4.0]undec-7-ene at room temp. for 5 min. → 3-phenylsulfinyliminomethyl-1,2-benzisoxazole. Y 83%.

from α-halogenosulfoximines. A soln. of startg. α-bromosulfoximine (prepn. s. 451) in chloroform treated with a slight excess of 1,8-diazabicyclo[5.4.0]undec-7-ene at room temp. for 2 hrs. → 3-phenylsulfinyliminomethyl-1,2-benzisoxazole. Y 86%.

F.e., also with K_2CO_3, s. T. Yoshida et al., J. Chem. Soc. Chem. Commun. *1982,* 106–7.

Anion exchanger s. under NaOH ←

Sulfur ↑ NC⇑S

Sodium hydroxide *NaOH*

1,2,4-Benzotriazines ○

s. *4,* 465; f. method cf. Bull. Chem. Soc. Japan *55,* 859–64 (1982).

Mercuric oxide *HgO*

Carbodiimides from thioureas NHCSNH → N=C=N
1-Tosylcarbodiimides
s. *37,* 521

Diethoxy disulfide/molecular sieve ←

Reactions of thioureas with diethoxy disulfide ←

432. 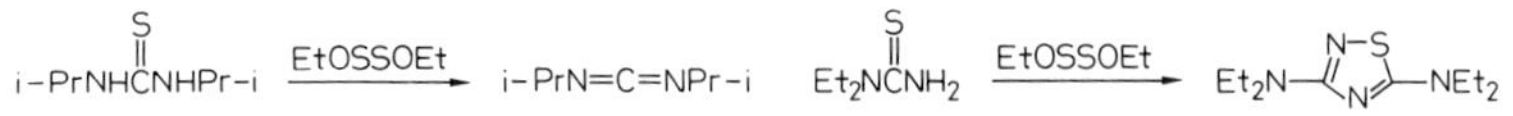

Carbodiimides.
Molecular sieves added to a mixture of startg. thiourea and diethoxy disulfide in dichloromethane, and refluxed for 4 hrs. → di-isopropylcarbodiimide. Y 66%.

3,5-Diamino-1,2,4-thiadiazoles.
A suspension of startg. thiourea and diethoxy disulfide in carbon tetrachloride refluxed for 7 hrs. → 3,5-bis(diethylamino)-1,2,4-thiadiazole. Y 75%.

F.e., **also amidinothioureas,** s. H. Kagami, S. Motoki et al., Bull. Chem. Soc. Japan *53,* 3658–60 (1980).

*Tri-*n-*butylphosphine* n-Bu_3P

1,2-Dihydropyrimidines from β-amino-α,β-ethylenecarboxylic acid thioamides via 5-imino-Δ^3-isothiazolines
α-Iminoketenimines as intermediates

433. 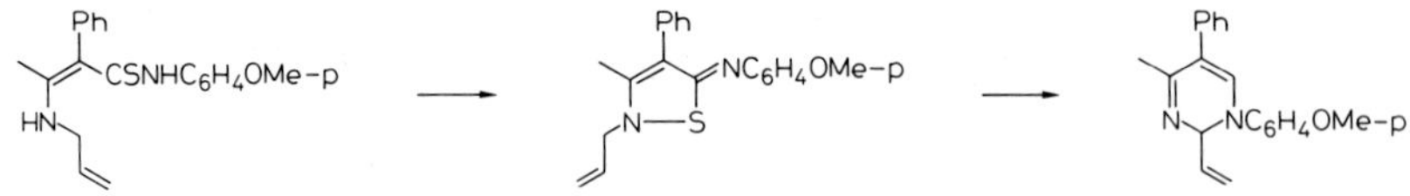

A soln. of Br_2 in acetic acid added dropwise with stirring to a soln. of 3-allylamino-N-(4-methoxyphenyl)-2-phenylthiocrotonamide and pyridine in ethyl acetate at 0–10° until a permanent coloration resulted → 2-allyl-N-(4-methoxyphenyl)-3-methyl-4-phenyl-3-isothiazoline-5-imine (Y 85%) in methylene chloride treated with tri-butylphosphine under argon, and worked up after ca. 12 hrs. → 1,2-dihydro-1-(4-methoxyphenyl)-4-methyl-5-phenyl-2-vinylpyrimidine (Y 85%). F.e.s. J. Goerdeler, C. Lindner, F. Zander, Chem. Ber. *114,* 536–48 (1981).

Remaining Elements ↑ NC⇑Rem

Without additional reagents *w.a.r.*

4-Quinazolones from (α-azoalkylidene)phosphoranes

434. 

A soln. of the startg. m. in toluene refluxed under N_2 for 20 min. → product. Y 86%. F.e.s. A. Alemagna, S. Maiorana et al., J. Chem. Soc. Chem. Commun. *1981,* 894–5.

Carbon ↑ NC⇑C

Without additional reagents *w.a.r.*

Two cyanoketene molecules from 2,5-diazido-*p*-quinones
s. *26,* 496; **synthesis and cycloadditions of cyanoketenes, review,** s. H.W. Moore, M.D. Gheorghiu, Chem. Soc. Rev. *10,* 289–328 (1981).

Imidazole ←

2-Oxazolidones from 5-β-hydroxy-1,3,4-dioxazol-2-ones

435. 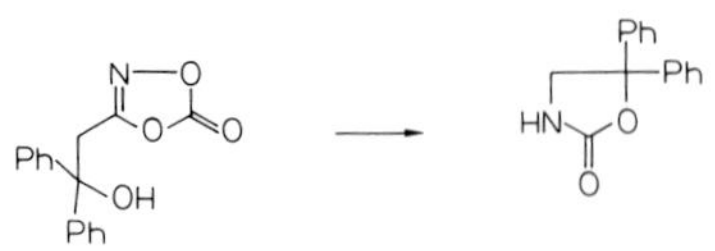

A mixture of startg. m. and imidazole in dichloromethane allowed to stand at room temp. for 12 hrs. → 5,5-diphenyloxazolidin-2-one. Y 72%. F.e., also prepn. of startg. m. from hydroxamic acids by carbonylation with N,N′-carbonyldiimidazole (cf. Synth. Meth. *35,* 88), s. D. Geffken, Ann. Chem. *1982,* 219–25; s.a. ibid. *1982,* 211–8.

Dicyclohexylcarbodiimide *RN=C=NR*

Azomethines from α-aminonitriles C(CN)NH → C=N
with KOH/methanol cf. *23,* 524; with dicyclohexylcarbodiimide s. K. Findeisen, H. Heitzer, K. Dehnicke, Synthesis *1981,* 702–4.

Acetic anhydride s. under $(MeCOO)_4Pb$ $(MeCO)_2O$

Phenyl iodosotrifluoroacetate $PhI(OCOCF_3)_2$

Amines from carboxylic acid amides $CONH_2 \rightarrow NH_2$
Hofmann degradation

s. *25,* 372s*34;* optically active monoacyl-1,1-diamines s. P. Pallai, M. Goodman, J. Chem. Soc. Chem. Commun. *1982,* 280–1.

Ethyl chloroformate *ClCOOEt*

3-Subst. 1,2,4-triazoles from N-1,2,4-triazol-4-ylamidines ←

s. *26,* 497; 1-acyl derivs. with acetic anhydride cf. P.C. Wade et al., J. Med. Chem. *25,* 331–3 (1982).

Lead tetraacetate/acetic anhydride $(MeCOO)_4Pb/(MeCO)_2O$

N-Halogenimines from *a-prim*-aminocarboxylic acid hydrohalides
Oxidative decarboxylation

436.

A suspension of 2′-aminodispiro[cyclohexane-1,1′-[1*H*]cyclopenta[*b*]quinoxaline-3′(2′*H*)-cyclohexane]-2′-carboxylic acid hydrochloride in acetic acid containing a little *pyridine* and acetic anhydride treated with Pb tetraacetate, and the mixture stirred at room temp. for 5 days → dispiro[cyclohexane-1,1′-[1*H*]cyclopenta[*b*]quinoxaline-3′(2′*H*)-cyclohexane]-2′-chlorimine. Y 79%. F.e., also N-bromimines with N-bromosuccinimide, s. S. Mohr, Chem. Ber. *114,* 2146–57 (1981).

Bromine Br_2

Nitriles from *a*-aminocarboxylic acids $CH(NH_2)COOH \rightarrow CN$
Degradation with loss of one C-atom

with NaOCl cf. *13,* 548; with Br_2 s. S.E. Poritere, R.A. Paegle, M.Y. Lidak, Khim. Geterotsikl. Soedin. *1982,* 539–41.

Pyridine hydrochloride C_5H_5NHCl

Benzoxazoles via Beckmann rearrangement ○

of *o*-alkoxyoximes cf. *26,* 505; of *o*-hydroxyoximes with $POCl_3$/dimethylacetamide s. S. Fujita, K. Koyama, Y. Inagaki, Synthesis *1982,* 68–9.

Palladium–carbon *Pd–C*

Lactams from carbalkoxyaminocarboxylic acids
via N-hydroxysuccinimide esters

437.

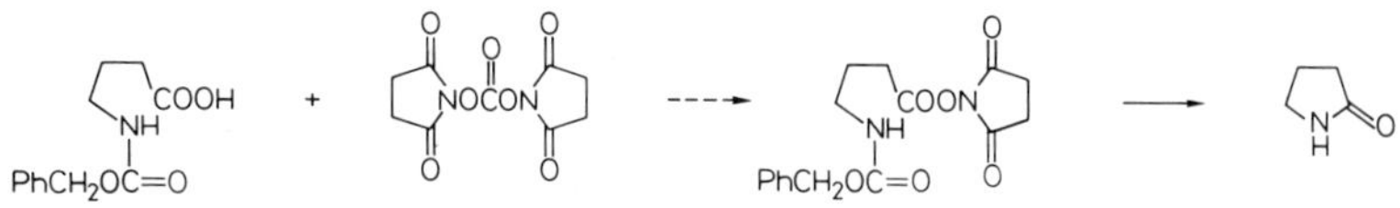

Startg. Z-protected aminoacid treated with N,N′-disuccinimidyl carbonate according to Synth. Meth. *36,* 85 → intermediate ester (Y 100%), in ethyl acetate under high dilution conditions hydrogenated with 10% Pd on carbon → product (Y 86%). Also 6-, 7-, and 8-membered lactams s. H. Ogura, K. Takeda, Heterocycles *15,* 467–8 (1981).

Formation of Hal–N Bond

Exchange ↓↑

Hydrogen ↑ HalN↓↑H

Xenon difluoride *XeF_2*

N-Fluorodicarboxylic from dicarboxylic acid imides NH → NF

s. *24,* 574s*37*

Formation of Hal–S Bond

Exchange ↓↑

Oxygen ↑ HalS↓↑O

Phosphoryl chloride/N,N-dimethylacetamide *$POCl_3/MeCONMe_2$*

Sulfonyl chlorides from sulfonic acids $SO_3H \rightarrow SO_2Cl$

s. *9,* 586; with added N,N-dimethylacetamide from the Na salt cf. S. Fujita, Synthesis *1982,* 423–4.

Carbon ↑ HalS↓↑C

Acetic acid *CH_3COOH*

Sulfonamides from thiocarbamic acid esters via sulfonyl chlorides $SC(O)NR_2 \rightarrow SO_2Cl$

438.

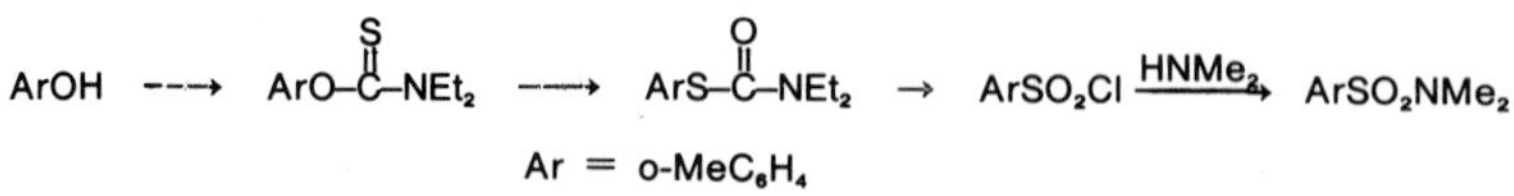

A soln. of startg. thiocarbamic acid ester, prepared from *o*-cresol via Newman-Kwart rearrangement (Synth. Meth. *22,* 625), in 80% acetic acid allowed to react with 3–4 equivalents Cl_2 at 0° → intermediate sulfonylchloride (Y 95%), in benzene containing excess 40% aq. dimethylamine refluxed for serveral hrs. → product (Y 94%). – This is part of a multistep preparation of sulfonic acid amides **from hindered phenols.** F.e.s. A. Wagenaar, J.B.F.N. Engberts, Rec. Trav. Chim. Pays-Bas *101,* 91–4 (1982).

Formation of Hal–Rem Bond

Uptake ⇓

Addition to Remaining Elements HalRem⇓Rem

Sulfuryl chloride *SO_2Cl_2*

Tellurium dichlorides from tellurides $R_2Te \rightarrow R_2TeCl_2$

s. *37*, 487

Exchange ⇂↾

Halogen ↑ HalRem⇂↾Hal

Arsenic trifluoride *AsF_3*

Replacement of P-chlorine by P-fluorine P–Cl → P–F

with SbF_3 cf. *19*, 589; with AsF_3, phosphonic difluorides, s. W. Althoff, M. Fild, R. Schmutzler, Chem. Ber. *114*, 1082–90 (1981); with NaF in the presence of crown ether, phosphorodifluoridates, s. F. Effenberger, G. König, H. Klenk, Synthesis *1981*, 70–2.

Sulfur ↑ HalRem⇂↾S

Phosphorus trihalide *$PHal_3$*

Replacement of P-thioalkyl groups by P-halogen P–SR → P-Hal

by fluorine with 2,4-dinitrofluorobenzene cf. *22*, 542; dithiophosphorohalidites with PCl_3 or PBr_3 s. O.G. Sinyashin et al., Zh. Obshch. Khim. *51*, 2410–3 (1981).

Formation of Hal–C Bond

Uptake ⇓

Addition to Oxygen and Carbon HalC ⇓ OC

Sodium iodide *NaI*

2-Acoxyiodides
from carboxylic acid chlorides and oxido compds.

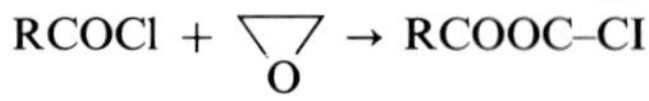

439.

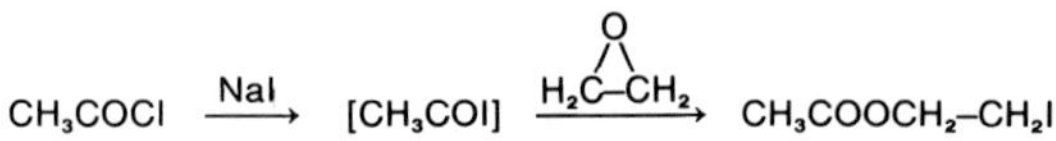

A stirred soln. of acetyl chloride in acetonitrile treated rapidly with a soln. of dry NaI in acetonitrile, cooled to 0°, ethylene oxide distilled into the mixture, and stirring continued for 2 hrs. → 2-iodoethyl acetate. Y 81%. F.e.s. K. Belsner, H.M.R. Hoffmann, Synthesis *1982*, 239–40.

Triethylamine *Et_3N*

2-Siloxyhalides from oxido compds.

s. *37*, 461

Trimethylsilyl chloride *Me_3SiCl*

1,2-Chlorohydrins from oxido compds.

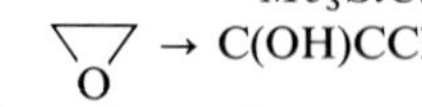

with $FeCl_3$ cf. *4*, 475s*32;* with Me_3SiCl s. M. Husain, N.H. Khan, Synth. Commun. *11*, 185–207 (1981).

Triphenylphosphine *Ph_3P*

2-Siloxyhalides from oxido compds.

s. *37*, 461

Thionyl chloride *$SOCl_2$*

Acoxychlorides from cyclic ethers and carboxylic acids C

with HCl cf. *29*, 487; with $SOCl_2$ s. A.I. Yurtanov, I.V. Martynov, Izv. Akad. Nauk SSSR Ser. Khim. *1981*, 1149–50.

Hydrogen chloride *HCl*

Halogenative ring opening of Δ^2-oxazolines

s. *8*, 347; s.a. G.S. Poindexter, Synthesis *1981*, 541–3.

Addition to Nitrogen and Carbon **HalC⇓NC**

Without additional reagents *w.a.r.*

Acylchloroformamidines
from cyanamides and carboxylic acid chlorides RCON=C(Cl)–N<

440.

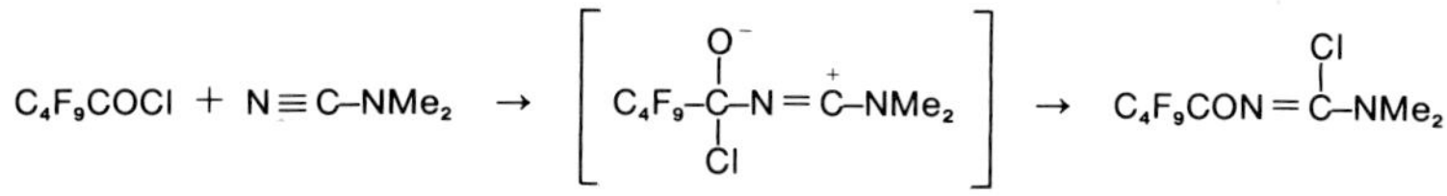

A mixture of dimethylcyanamide and perfluoropentanoyl chloride allowed to stand for 30 min. at 15° → N,N-dimethyl-N′-perfluoropentanoylchloroformamidine. Y 72%. F.e.s. V.N. Lebedev, V.V. Yurechko, T.P. Vishnyakova, Zh., Vses. Khim. Obshchest. *26,* 465–6 (1981).

Pyridine C_5H_5N

2-Fluoramines from aziridines  → CFC–N<

s. *34,* 474; f. methods s. G.M. Alvernhe et al., J. Org. Chem. *46,* 4938–48 (1981).

Addition to Carbon **HalC⇓CC**

Without additional reagents *w.a.r.*

2-Chlorofluorides from ethylene derivs. C=C → CFCCl

441.

1.1 Equivalents ClF gas passed into a soln. of startg. alkene in chloroform at 25–27° in the dark → product. Y 98%. F.e., method, and regiospecificity s. L.S. Boguslavskaya et al., Zh. Org. Khim. *16,* 2525–33 (1980); **replacement of bromine and chlorine by fluorine** with ClF s. ibid. *18,* 938–45, 946–53 (1982).

2-Iodofluorosulfonyl esters from ethylene derivs. C=C → ICOSO$_2$F

442.

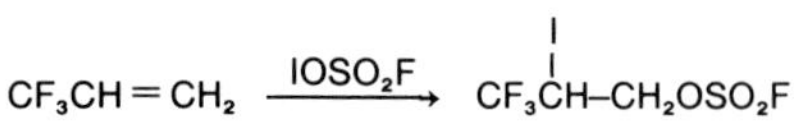

A soln. of iodine fluorosulfate in Freon 113 (CCl_2FCClF_2) at 20–30° treated with excess 1,1,1-trifluoropropene, and allowed to stand for 16 hrs. → 2-iodo-3,3,3-trifluoropropyl fluorosulfate. Y 75%. F.e.s. A.V. Fokin et al., Izv. Akad. Nauk SSSR Ser. Khim. *1981,* 2376–9.

Sodium iodide s. under BH_3 *NaI*

Mercuric acetate/potassium tetrahydridoborate $(MeCOO)_2Hg/KBH_4$

1-Alkoxy-2-ethylene-2-iodides
from allenes and alcohols C=C=C → C(OR)–CI=C
Regiospecific conversion

443.

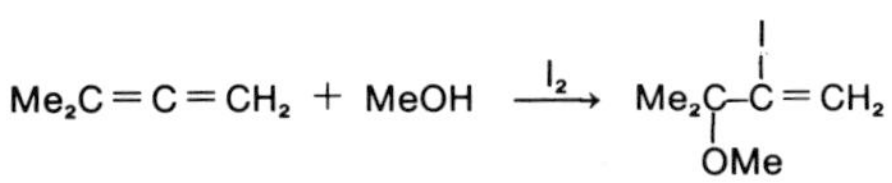

Hg(II) acetate added to a soln. of 1,1-dimethylallene in tetrahydrofuran/methanol, the mixture kept at room temp. for 30 min., I_2 added, stirring continued for 15–30

min., the mixture poured into a soln. of KBH_4 in aq. KOH, and stirred vigorously for 15 min. → product. Y 76% (95% purity). – The regioisomer was obtained by using 1 equivalent of I_2 in the absence of Hg(II). F.e. and **2-ethylene-2,1-iodohydrins** s. C. Georgoulis, W. Šmadja, J.M. Valery, Synthesis *1981*, 572–4.

Borane-tetrahydrofuran/chloramine-T/sodium iodide — *$BH_3.THF/TsN(Cl)Na/NaI$*

Hydrohalogenation via trialkylboranes — C=C → CHCHal

444.

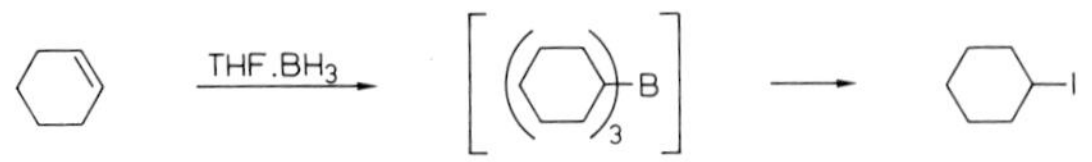

Iodides. A stirred soln. of cyclohexene in tetrahydrofuran treated with BH_3–tetrahydrofuran at 0°, stirred at 25° for 1 hr. under N_2, methanolic Na acetate, aq. NaI, and methanolic chloramine T added sequentially, and stirring continued for 1 min. → iodocyclohexane. Y 99%. – F.e.s. G.W. Kabalka, E.E. Gooch, J. Org. Chem. *46*, 2582–4 (1981); **labelled iodides** s. Intern. J. Appl. Radiation Isotopes *33*, 223–4 (1982); **bromides** with BrCl or Br_2 (cf. Synth. Meth. *24*, 560s*36*) s. J. Org. Chem. *46*, 3113–5 (1981); carbohydrate derivs. s. L.D. Hall, J.-R. Neeser, Can. J. Chem. *60*, 2082–6 (1982).

Diborane/iodine monochloride-^{125}I — *$B_2H_6/^{125}ICl$*

***anti*-Markownikoff hydroiodination**

s. *24*, 560s*36*; with ^{125}ICl, **labelled iodides,** s. G.W. Kabalka, E.E. Gooch, C. Otto, J. Radioanal. Chem. *65*, 115 (1981); Chem. Abstr. *95*, 203260 (1981); with dicyclohexylborane/ICl cf. Synth. Commun. *11*, 521–5 (1981).

Potassium tetrahydridoborate s. under $(MeCOO)_2Hg$ — *KBH_4*

Amberlyst A 26 iodide — ←

1,2-Dichlorides from ethylene derivs. — C=C → CClCCl

s. *37*, 453

N-Bromosuccinimide/dimethyl sulfoxide — *CH_2–CO / CH_2–CO >NBr/Me_2SO*

1,2-Bromohydrins from ethylene derivs. — C=C → C(OH)CBr

s. *24*, 562; s.a. Org. Synth. *59*, 16–20 (1979).

Chloramine-T s. under BH_3·THF — *TsN(Cl)Na*

Trimethylsilyl iodide — *Me_3SiI*

γ-Iodoketones from cyclopropylketones — C

s. *35*, 349; f.e.s. J. Org. Chem. *46*, 2412–4 (1981).

Titanium tetrachloride — *$TiCl_4$*

3-Halogeno- from cyclopropyl-silanes

445.

A stirred soln. of *exo*-7-(trimethylsilyl)bicyclo[4.1.0]heptane in dry methylene chloride treated dropwise at –78° with $TiCl_4$, stirred for 5 hrs., water added, and stirred at room temp. for a further 15 min. → *trans*-1-[(trimethylsilyl)methyl]-2-chlorocyclohexane. Y 84%. F.e. and regiospecificity s. R.G. Daniels, L.A. Paquette, J. Org. Chem. *46*, 2901–10 (1981).

Sulfuryl chloride *SO_2Cl_2*

Chlorination of carbon-carbon multiple bonds C≡C → CCl=CCl

of ethylene derivs. *3,* 429; 1,2-dichloro ethylenes from acetylene derivs. s. S. Uemura, C. Masaki et al., Bull. Chem. Soc. Japan *54,* 2843–4 (1981).

Iodine monochloride-^{125}I s. under B_2H_6 *ICl*

Perchloric acid/N-bromoacetamide *$HClO_4$/MeCONHBr*

Bromohydrins from ethylene derivs. C=C → C(OH)CBr

s. *11,* 597; regio- and stereo-chemistry of addition to *β*-methylstyrenes s. B.A. Marples, C.G. Saint, Synth. Commun. *12,* 545–55 (1982).

*3-Bromoimidazo[1,2-*b*]pyridazinium tribromide* ←

1,2-Dibromides from ethylene derivs. C=C → CBrCBr

s. *37,* 464

Via intermediates *v.i.*

***α,β*-Ethylenehalides from acetylene derivs.** C≡C → CH=CHal
via *α,β*-ethyleneboronic acids

s. *29,* 497; *α,β*-ethyleneiodides with ICl s. G.W. Kabalka, E. Gooch, H.C. Hsu, Synth. Commun. *11,* 247–51 (1981); labelled iodides from boronic acids with $Na^{125}I$/chloramine-T cf. J. Labelled Compounds Radiopharm. *19,* 795–9 (1982).

***cis*-1,2-Chloroiodoethylenes from acetylene derivs.** C≡C → CCl=CI
via regio- and stereo-specific addition of tellurium tetrachloride

446. 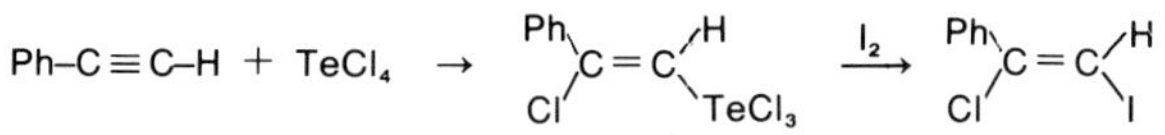

A soln. of phenylacetylene in carbon tetrachloride treated at 76° with an equimolar amount of $TeCl_4$ for 1 hr. → intermediate (Y 86%), in acetonitrile treated with I_2, and the mixture refluxed for 2 hrs. → (Z)-product (Y 94%). F.e. and bromo-analogs s. S. Uemura, H. Miyoshi, M. Okano, Chem. Letters *1979,* 1357–8.

2-Bromoamines from ethylene derivs. C=C → BrC–CN<
via *anti*-Markownikoff addition of N-bromophosphoromonoamidates

447. 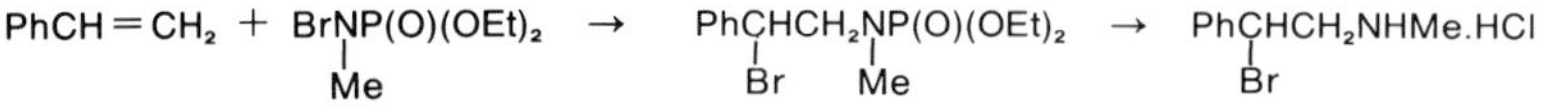

Styrene added to a soln. of diethyl N-bromo-N-methylphosphoramidate in dichloromethane, and the mixture refluxed for 3 hrs. → diethyl N-methyl-N-(2-bromo-2-phenylethyl)phosphoramidate (Y 76%), in tetrahydrofuran treated with dry HCl at 0–5° during 1 hr., and allowed to stand at room temp. for 24 hrs. → product. Y 68%. F.e.s. K. Osowska, A. Zwierzak, Synthesis *1979,* 577–9; addition of **N,N-dibromophosphoromonoamidates** s. Tetrahedron *37,* 2675–81 (1981).

Rearrangement ∩

Oxygen/Halogen Type HalC∩OHal

Sodium hydrogen carbonate s. under $FeSO_4$ *$NaHCO_3$*

Ferrous sulfate/sodium hydrogen carbonate *$FeSO_4/NaHCO_3$*

1,4-Chlorhydrins from alkyl hypochlorites ←

photochemically cf. *17,* 593; with $FeSO_4/NaHCO_3$ s. Ž. Čeković, G. Djokić, Tetrahedron *37,* 4263–8 (1981).

Nitrogen/Carbon Type HalC∩NC

Without additional reagents *w.a.r.*

2-Oxazolidones from N-carbalkoxyaziridines ←

s. *24,* 568; 3-acyl-2-oxazolidones from carboxylic acid halides cf. S.V. Baranov, V.B. Mochalin, Zh. Vses. Khim. Obshchest. *26,* 715 (1981).

Exchange ⇅

Hydrogen ↑ HalC⇅H

Without additional reagents *w.a.r.*

α′-Bromination of α-cyanoketones H → Br

448.

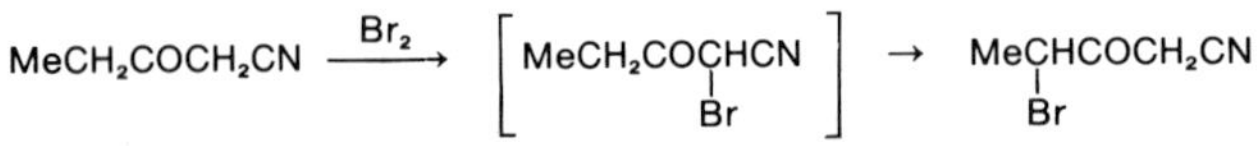

A stirred soln. of startg. α-cyanoketone in benzene treated under N_2 with a soln. of Br_2 in the same solvent during several hrs. at such a rate that Br_2 was never in excess, and N_2 bubbling continued for 1–3 days until n.m.r. indicated completion of reaction → 4-bromo-3-oxopentanenitrile. Y 88%. F.e.s. B. Föhlisch et al., Chem. Ber. *115,* 355–80 (1982).

1-Halogenohydrazones from hydrazones CH=N–N< → C(Hal)=N–N<

s. *29,* 502; s.a. R.S. Tewari, P. Parihar, J. Chem. Eng. Data *26,* 418–20 (1981).

Irradiation s. under N-Bromosuccinimide

Electrolysis ϟ

Electrochemical aromatic halogenation H → Cl

s. 31, 482s*36;* *p*-chlorination of phenolethers s. Y. Matsuda, H. Hayashi, Chem. Letters *1981,* 661–2.

β,γ-Ethylenechlorides from ethylene derivs.
Regiospecific anodic chlorination

449.

A soln. of dehydrolinalyl acetate and 12 equivalents *NaCl* in 2:1 methylene chloride/water electrolysed using Pt foil electrodes at a constant current of 50 mA for 4 hrs. at room temp. → product. Y 91%. F.e. and limitations s. S. Torii et al., Tetrahedron Letters *22,* 2291–4 (1981); retention of 2-azetidinone ring s. ibid. *22,* 3193–6 (1981).

Sodium hydride s. under Hexachloroethane *NaH*

Sodium hydroxide/benzyltriethylammonium chloride/carbon tetrachloride $NaOH/[PhCH_2NEt_3]^+Cl^-/CCl_4$

Chlorination with carbon tetrachloride in alkaline medium

s. *25,* 385; phase transfer catalysis, also bromination with bromotrichloromethane, s. S.E. Lauritzen, C. Rømming, L. Skattebøl, Acta Chem. Scand. *35B,* 263–8 (1981).

n-*Butyllithium* n-*BuLi*

3,4-Oxidoalcohols from 3-ethylenealcohols via cyclic 2-bromocarbonic acid esters Stereospecific ring closure O

450.

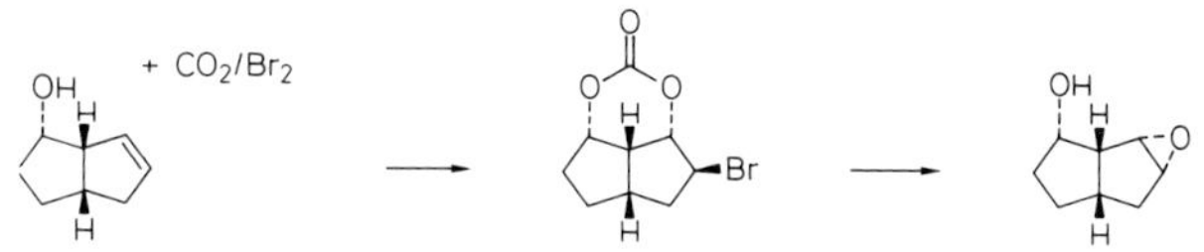

n-Butyllithium added to a soln. of startg. m. in tetrahydrofuran at –78° under argon, CO_2 bubbled through the mixture for 30 min., warmed to –20°, and a soln. of Br_2 in dichloromethane added during 30 min. while allowing the temp. to reach 0° → 3-bromo-2,8-dihydroxybicyclo[3.3.0]octane carbonate (Y 79%), stirred in methanol containing 2 *M* KOH at room temp. for 2 hrs. → 2,3-epoxybicyclo[3.3.0]octan-8-ol (Y 94%). F.e.s. M.F. Haslanger, S. Ahmed, J. Org. Chem. *46,* 4808–10 (1981); **cyclic 2-iodo carbonic acid esters** s. G. Cardillo et al., J. Chem. Soc. Chem. Commun. *1981,* 465–6.

Sodium hydrogen carbonate *$NaHCO_3$*

Iodolactonization

s. *35,* 351; γ-iodo-γ-lactones with added KI cf. L.-C. Yu, P. Helquist, Synth. Commun. *11,* 591–7 (1981); **asym. iodolactonization** s. W.C. Still, I. Galynker, J. Am. Chem. Soc. *104,* 1774–6 (1982); δ-iodo-γ-lactones from γ,δ-ethylenecarboxylic acids s. A.R. Chamberlin, M. Dezube, P. Dussault, Tetrahedron Letters *22,* 4611–4 (1981).

2-α-Iodo-O-heterocyclics from ethylenealcohols

s. *33,* 477; with $NaHCO_3$, also from ethyleneethers, cf. S.D. Rychnovsky, P.A. Bartlett, J. Am. Chem. Soc. *103,* 3963–4 (1981).

Potassium carbonate s. under t-*BuOCl* *K_2CO_3*

Potassium hydrogen carbonate s. under N-Bromosuccinimide *$KHCO_3$*

Lithium chloride *LiCl*

Replacement of hydrogen by bromine H → Br

with $NaBr/Cl_2$ cf. *20,* 417; with $LiCl/Br_2$, α-bromoketones, s. S.I. Zavyalov, G.I. Ezhova, Izv. Akad. Nauk SSSR Ser. Khim. *1980,* 2575–8.

Sodium bromide s. under Me_3SiCl *NaBr*

Tetrabutylammonium hydroxide s. under N-Bromosuccinimide *$Bu_4N^+OH^-$*

Triethylamine s. under Hexabromocyclopentadiene *Et_3N*

Pyridine (s.a. under PhSeHal) *C_5H_5N*

α-Halogenosulfoximines from sulfoximines via N-halogenosulfoximines H → Hal

451.

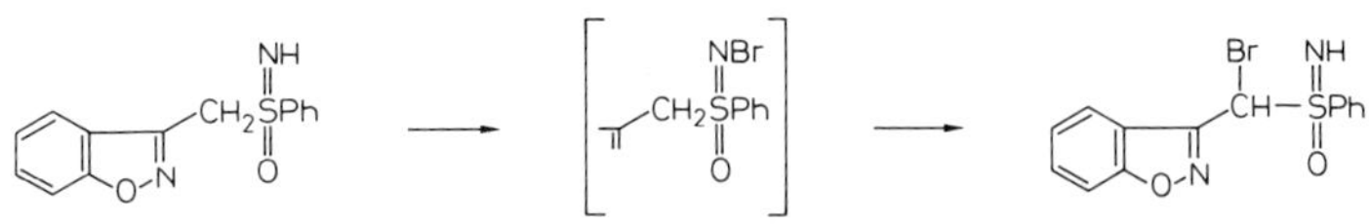

A soln. of phenyl[(1,2-benzisoxazol-3-yl)methyl]sulfoximine in pyridine treated with Br_2 for 2–3 hrs. → product, Y 80–90%, (startg. m. for 431). F.e., methods, also isolation of intermediate, s. T. Yoshida, S. Naruto, H. Nishimura, Tetrahedron Letters *21,* 2315–6 (1980); also chlorination s. Chem. Pharm. Bull. *30,* 1175–82 (1982); with *tert*-butyl hypochlorite/K_2CO_3 cf. C.R. Johnson, H.G. Corkins, J. Org. Chem. *43,* 4136–40 (1978); with hexachloroethane/NaH cf. C.R. Johnson, A. Tangerman, Synthesis *1982,* 286–8.

Cupric acetate *(MeCOO)₂Cu*

***o*-Iodination of phenols** H → I

with Na acetate cf. *16*, 584; with Cu(II) acetate s. C.A. Horiuchi, J.Y. Satoh, J. Chem. Soc. Chem. Commun. *1982*, 671–2.

Cupric bromide/tert-*butyl hydroperoxide* $CuBr_2$/t-*BuOOH*

Benzylic bromination H → Br

photochemically with N-bromosuccinimide cf. *20*, 648s*32*; with $CuBr_2$/*tert*-butyl hydroperoxide in polar medium (acetic anhydride) s. A. Chaintreau, G. Adrian, D. Couturier, Synth. Commun. *11*, 669–72 (1981).

Aluminium chloride $AlCl_3$

β-Bromo-γ-ketocarboxylic acid esters from γ-ketocarboxylic acid esters

452.

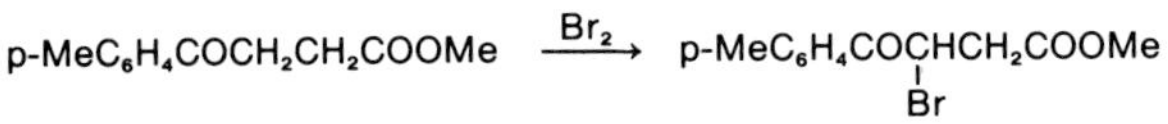

A stirred mixture of methyl 4-keto-4-(*p*-tolyl)butanoate and a little $AlCl_3$ in dry ether treated dropwise with a soln. of Br_2 in tetrachloromethane, and stirring continued for 5 min. → methyl 3-bromo-4-keto-4-(*p*-tolyl)butanoate. Y 100%. F.e.s. K.S. Bhat, A.S. Rao, Indian J. Chem. *20B*, 355–8 (1981).

Ammonium ceric nitrate/ *tetra*-n-*butylammonium iodide* $(NH_4)_2Ce(NO_3)_6$/n-$Bu_4N^+I^-$

Ar. iodination H → I

with $PhI(OCOCF_3)$ cf. *36*, 469; with $(NH_4)_2Ce(NO_3)_6$/*n*-$Bu_4N^+I^-$ s. T. Sugiyama, Bull. Chem. Soc. Japan *54*, 2847–8 (1981).

Methanol s. under SO_2Cl_2 CH_3OH

Amberlyst A 26 dichloroiodidate(I) $[Ⓟ\text{–}CH_2NMe_3]^+\ ICl_2^-$

Polymer-supported chlorination H → Cl

453.

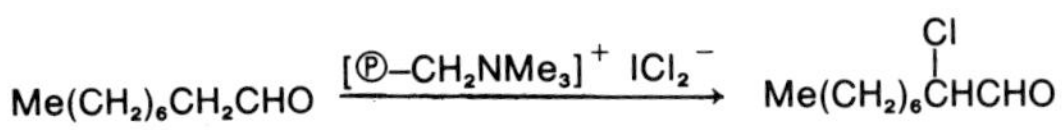

α-Chloraldehydes. Nonanal added to a slurry of Amberlyst A-26 (I^--form) containing absorbed Cl_2 in carbon tetrachloride, and refluxed with stirring for 20 min. → product. Y 88%. F.e., **also 1,2-dichlorides from ethylene derivs.,** s. A. Bongini et al., J. Chem. Soc. Chem. Commun. *1980*, 1278–9.

tert-*Butyl hydroperoxide s. under* $CuBr_2$ t-*BuOOH*

Lauroyl peroxide s. under CCl_4 *(RCOO)₂*

Benzoyl peroxide s. under N-Halogenosuccinimide *(PhCOO)₂*

Trichloroacetic acid CCl_3COOH

Aryl derivs. via aromatic thallation ←

with trifluoroacetic acid cf. *25*, 389; more safely and economically with trichloroacetic acid s. S. Uemura, O. Itoh et al., Bull. Chem. Soc. Japan *53*, 553–4 (1980).

Iodobenzene difluoride $PhIF_2$

Fluorination H → F

s. *24*, 574s*37*

Brominations with
hexabromocyclopentadiene/triethylamine C_5Br_6/Et_3N
***p*-Bromination** H → Br

454. 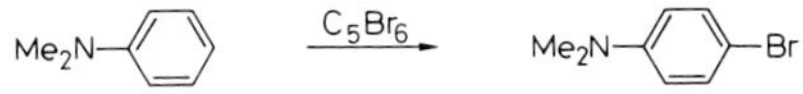

of anilines. Equimolar amounts of N,N-dimethylaniline and perbromocyclopentadiene in acetonitrile reacted at room temp. for 24 hrs. in the presence of triethylamine → *p*-bromodimethylaniline. Y 99%. F.e., also of ar. acylamines, phenols, phenolethers, and 5-membered arenes, s. B. Fuchs et al., J. Chem. Soc. Chem. Commun. *1982,* 778–9.

Hexachloroethane/sodium hydride C_2Cl_6/NaH
***α*-Chlorosulfoximines from sulfoximines** H → Cl
s. *37,* 451

Carbon tetrachloride (s.a. under NaOH) CCl_4
***α*-Chlorination of bicyclic amidines**

455. 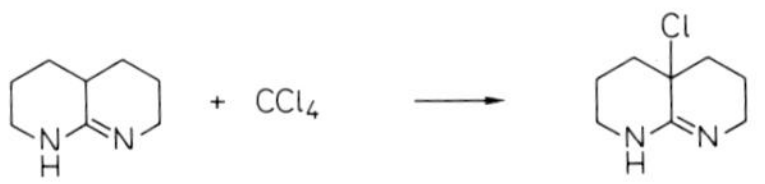

A soln. of startg. m. in carbon tetrachloride kept in the dark at room temp. in a degassed, sealed n.m.r. tube for a fortnight → product. Y 100%. F.e.s. S. Lofas, P. Ahlberg, J. Chem. Soc. Chem. Commun. *1981,* 998–9.

Carbon tetrachloride/dodecanoyl peroxide $CCl_4/(n\text{-}C_{11}H_{23}COO)_2$
***γ*-Chlorothiolic acid esters from 2-alkylthiotetrahydrofurans** C

456. 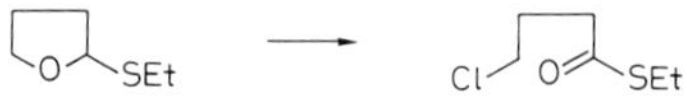

A soln. of 2-ethylthiotetrahydrofuran in carbon tetrachloride containing a little dodecanoyl peroxide heated at 80° for 2 hrs. → product. Y 85% (65% conversion). F.e.s. N.A. Batyrbaev et al., Khim. Geterotsikl. Soedin. *1981,* 992–3; **2-chlorothioloc acid esters from 1,3-oxathiolanes** cf. Zh. Org. Khim. *18,* 1334–5 (1982).

N-Bromosaccharin ←
Replacement of hydrogen by bromine H → Br
benzylic bromination s. *11,* 231s*32;* also *α*-bromoketones s. E.I. Sánchez, M.J. Fumarola, J. Org. Chem. *47,* 1588–90 (1982).

N-Halogenosuccinimide/benzoyl peroxide $\begin{matrix}CH_2\text{-}CO\\ |\\ CH_2\text{-}CO\end{matrix}\!\!>NHal/(PhCOO)_2$

Halogenation of aliphatics H → Hal
s. *10,* 427; **review** s. Y. Ogata, M. Kimua, Yakagaku *29,* 788–95 (1980) (Japan); Chem. Abstr. *94,* 83107 (1981).

N-Chlorosuccinimide $\begin{matrix}CH_2\text{-}CO\\ |\\ CH_2\text{-}CO\end{matrix}\!\!>NCl$

2-Chloro-3-silylenamines from 3-silylenamines H → Cl
s. *37,* 577

N-Bromosuccinimide $CH_2{-}CO \diagdown NBr$ / $CH_2{-}CO \diagup$

α-Bromosilanes from silanes H → Br
s. *37,* 197

N-Bromosuccinimide/irradiation $CH_2{-}CO \diagdown NBr$/⚡ / $CH_2{-}CO \diagup$

1-Halogenisothiocyanates from isothiocyanates
s. *36,* 471; 1,1-bromisocyanates from isocyanates s. Chem. Ber. *115,* 860–70 (1982).

N-Bromosuccinimide/tetrabutylammonium hydroxide/potassium hydrogen carbonate $CH_2{-}CO \diagdown NBr/Bu_4N^+OH^-/KHCO_3$ / $CH_2{-}CO \diagup$

2-Bromoenollactones from acetylenecarboxylic acids ○

457. 

N-Bromosuccinimide, $KHCO_3$, and 0.4 *M* tetrabutylammonium hydroxide added sequentially to a soln. of 2-phenyl-4-hexynoic acid in methylene chloride at 25°, and the mixture stirred vigorously for 30 min. → 3-phenyl-5(E)-(1-bromoethylidene)tetrahydro-2-furanone. Y 93%. F.e., **also enollactones** (cf. Synth. Meth. *35,* 85), s. G.A. Krafft, J.A. Katzenellenbogen, J. Am. Chem. Soc. *103,* 5459–66 (1981).

Chloramine-T/methanesulfonic acid $TsN(Cl)Na/MeSO_3H$

Chlorolactonization

458. 

δ-Chloro-γ-lactones. 4-Pentenoic acid and methanesulfonic acid added dropwise simultaneously to a suspension of chloramine T in benzene or chlorobenzene maintaining the temp. below 50°, and the mixture heated at 80° for 2 hrs. → product. Y 67%. F.e. incl. **ε-chloro-δ-lactones** s. B. Damin et al., J. Org. Chem. *46,* 3552–4 (1981).

Phosgene $COCl_2$

1-Chlorenamines from carboxylic acid amides via chlorimidium salts CHCON< → C=CCl(N<)
s. *25,* 398; s.a. Org. Synth. *59,* 26–34 (1979).

Benzeneselenyl halides/pyridine $PhSeHal/C_5H_5N$

α,β-Ethylene-α-halogeno- from α,β-ethylene-oxo compds. via α-arylseleno-α,β-ethylene-oxo compds. H → Hal

459. 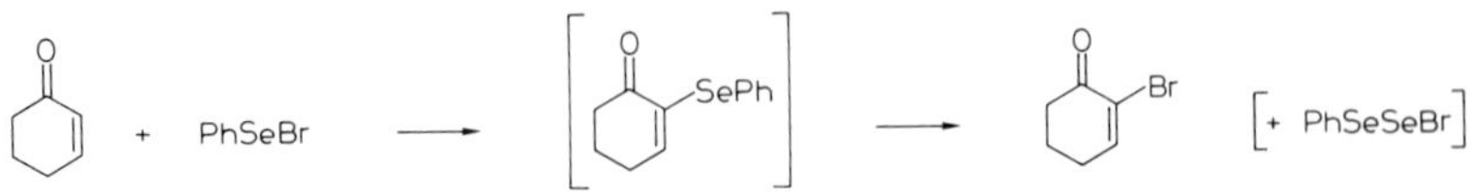

Startg. α,β-ethyleneketone treated with 2.5 equivalents each of phenylselenyl bromide and pyridine in dry methylene chloride at room temp. for 24 hrs. → product. Y 100%. F.e.s. S.V. Ley, A.J. Whittle, Tetrahedron Letters *22,* 3301–4 (1981).

Trimethylsilyl chloride *Me_3SiCl*

1,2-Chloroximes from oximes H → Cl

460. 

Trimethylsilyl chloride added to a soln. of startg. oxime in ether, and the crystals collected → 7*β*-chloro-5*α*-cholestan-6-one oxime. Y 75%. F.e.s. M. Husain, N.H. Khan, Syn. Commun. *11,* 185–207 (1981).

Trimethylsilyl chloride/sodium bromide *$Me_3SiCl/NaBr$*

***α*-Bromoketones from ketones** COCH → COCBr

s. *37,* 268

Trimethylsilyl bromide/chromium trioxide/sulfuric acid *$Me_3SiBr/CrO_3/H_2SO_4$*

***α*-Halogenokeones from oxido compds. via 2-siloxyhalides** $\triangledown_{O}$ → COCHal

461. 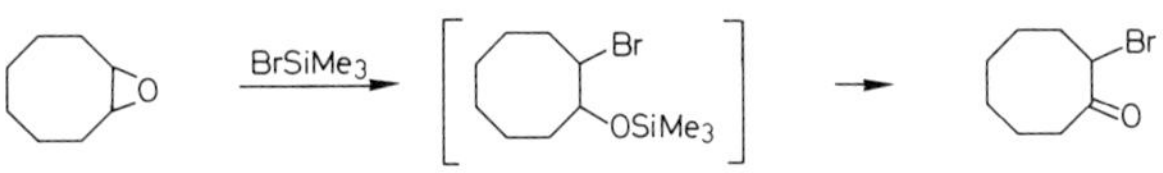

A soln. of oxidocyclooctene in dichloromethane added at 20° to a mixture of trimethylsilyl bromide and *triethylamine* in the same solvent, heated at 55° for 2.5 hrs., cooled to 0°, Jones reagent added, and stirring continued at this temp. for a further 0.75 hr. → 2-bromocyclooctanone. Y 91%. F.e., also with trimethyl silyl iodide, and isolation of intermediate (cf. Synth. Meth. *36,* 932), s. J.N. Denis, A. Krief, Tetrahedron Letters *22,* 1429–30 (1981); 1,2-siloxyhalides in the presence of triphenylphosphine cf. G.C. Andrews, T.C. Crawford, L.G. Contillo, Jr., ibid. *22,* 3803–6 (1981).

Lead tetraacetate *$(MeCOO)_4Pb$*

Oxidative fragmentation-type ring opening of bridgehead alcohols C

s. *37,* 948

Thionyl bromide *$SOBr_2$*

Carboxylic acid bromides from aldehydes CHO → COBr

s. *30,* 364s*37*

Sulfuryl chloride *SO_2Cl_2*

***α*-Chloro-*β*-ketocarboxylic acid derivs. from *β*-ketocarboxylic acid derivs.** H → Cl

s. *6,* 573; *α*-chloro-*β*-ketocarboxylic acid amides s. Z. Ejmocki, J. Korczak, Z. Eckstein, Przem. Chem. *59,* 597 (1980) (Pol.); Chem. Abstr. *94,* 191862 (1981).

Sulfuryl chloride/methanol *SO_2Cl_2/CH_3OH*

Chlorination

in the presence of silica gel cf. *33,* 484; in liq. SO_2 containing methanol, *α*-chloroketones, s. D. Masilamani, M.M. Rogić, J. Org. Chem. *46,* 4486–9 (1981).

Methanesulfonic acid s. under TsN(Cl)Na *$MeSO_3H$*

Sulfuric acid s. under Me_3SiBr *H_2SO_4*

Caesium fluoroxysulfate $CsSO_4F$

Replacement of hydrogen by fluorine H → F

462.

***o*-Fluorophenols.** A satd. soln. of $CsSO_4F$ in acetonitrile mixed with phenol, and allowed to react in the dark under N_2 until iodometric titration indicated the reaction to be completed → *o*-fluorophenol. Y 83%. – Also *o*-fluoroanisole s. D.P. Ip, E.H. Appelman, J. Am. Chem. Soc. *103,* 1964–8 (1981); ***α,β*-ethylenefluorides and cyclic *α*-fluoroketones** s. S. Stavber, M. Zupan, J. Chem. Soc. Chem. Commun. *1981,* 795–6.

Chromium trioxide s. under Me_3SiBr CrO_3

tert-*Butyl hypochlorite/potassium carbonate* t-$BuOCl/K_2CO_3$

***α*-Chlorosulfoximines from sulfoximines** H → Cl

s. *37,* 451

Calcium hypochlorite $Ca(OCl)_2$

***β,γ*-Ethylene-*α*-chloro-**
from *α,β*-ethylene-oxo compds. CH–C=C–CO → C=C–CCl–CO
2-Phase reaction

463.

A mixture of 70% $Ca(OCl)_2$ and mesityl oxide in 10:1 dichloromethane/water treated with dry ice during 3–4 hrs. → product. Y 71%. F.e. and aldehyde derivs. s. S.G. Hedge, J. Wolinsky, Tetrahedron Letters *22,* 5019–22 (1981).

Potassium bromate $KBrO_3$

Bromination H → Br

with $NaBrO_3$ cf. *3,* 447; with $KBrO_3$, ar. bromination, s. A. Banerjee et al., J. Indian Chem. Soc. *56,* 518–20 (1979); ibid. *58,* 985–8 (1981).

*Tetra-*n-*butylammonium iodide s. under $(NH_4)_2Ce(NO_3)_6$* n-$Bu_4N^+I^-$

Benzyltriethylammonium chloride s. under NaOH $[PhCH_2NEt_3]^+\ Cl^-$

Brominations with
3-bromoimidazo[1,2-b]pyridazinium tribromide ←

464.

Replacement of hydrogen by bromine. A suspension of 3-bromoimidazo[1,2-*b*]-pyridazinium tribromide in chloroform or acetic acid added to a stirred soln. of cyclopentanone in the same solvent, and stirring continued at room temp. for 30 min. → 2-bromocyclopentanone. Y 80%. F.e., **also 1,2-dibromides from ethylene derivs.,** and with the 6-chloro-analog s. B. Stanovnik, M. Tišler, I. Drnovšet, Synthesis *1981,* 987–9.

Phenyltrimethylammonium tribromide $[PhNMe_3]^+Br_3^-$

***α*-Halogenacetals from ketones** CHCO → $CHHalC(O–)_2$

with $CuHal_2$ cf. *31,* 165; *α*-bromacetals with phenyltrimethylammonium tribromide s. S. Visweswariah, S. Chandrasekaran et al., Synthesis *1982,* 309–10.

Via intermediates *v.i.*

2-Halogenoperoxides via peroxymercuration C=C → CHalC(OOR)
s. *30,* 350; 2-bromoperoxides s. J. Chem. Soc. Perkin Trans. I *1981,* 3258–64.

Fluorinations with *xenon difluoride* *XeF_2*
H → F
s. *24,* 574s*31;* α-fluoroketones from enoxysilanes, also with iodobenzene difluoride, s. T. Tsushima, K. Kawada, T. Tsuji, Tetrahedron Letters *23,* 1165–8 (1982); from enolesters, also α,α-difluoro-β-diketones from β-diketones in the presence of HF, s. B. Zajc, M. Zupan, J. Org. Chem. *47,* 573–5 (1982); in the presence of poly-4-vinylpyridine–boron fluoride complex s. J. Chem. Soc. Chem. Commun. *1980,* 759–60; N-fluorodicarboxylic from dicarboxylic acid imides s. Y.L. Yagupolsky, T.I. Savina, Zh. Org. Khim. *17,* 1330–31 (1981).

Oxygen ↑ HalC⇅O

Without additional reagents *w.a.r.*

Iodonium salts from iodosohydroxy tosylates ←
s. *27,* 574; β-amino-α,β-ethylene-α-iodonioketones s. Zh. Org. Khim. *17,* 2114–8 (1981).

Imidazol-1-ylsulfonyloxy as reactive leaving group ←
Nucleophilic substitution under mild conditions

465. 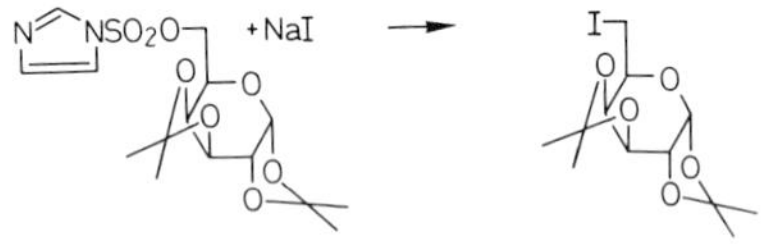

Iodides. Startg. imidazolylsulfonyl deriv. treated with a soln. of 2.5 equivalents NaI in dimethylformamide at room temp. for 6 hrs. → product. Y 81%. – The leaving group is resistant to hydrolysis (cf. triflates) and more suitable than conventional sulfonyloxy groups, which are normally displaced at much higher temperatures and are prone to elimination. F.e. and nucleophiles s. S. Hanessian, J.-M. Vatèle, Tetrahedron Letters *22,* 3579–82 (1981).

Lithium halides/hydrogen halides *LiHal/HHal*

Tert. halides from alcohols under mild conditions OH → Hal

466. t-BuOH → t-BuBr

HBr added dropwise to a mixture of LiBr and *tert*-butyl alcohol at 0°, and stirred for 2 hrs. → *tert*-butyl bromide. Y 85%. F.e., including iodides, and with Cd halides, s. H. Masada, Y. Murotani, Bull. Chem. Soc. Japan *53,* 1181–2 (1980).

Lithium bromide s. under CuBr *LiBr*

Sodium bromide s. under Me_3SiCl *NaBr*

Triethylamine (s.a. under Perfluoro-2-methyl-2-pentene) *Et_3N*

Triethylamine/triphenylphosphine dihalides *Et_3N/Ph_3PHal_2*

α,β-Ethylene-β-chloroketones from β-diketones COCHCO → CO–C=C–Cl
s. *31,* 494; f.e.s. E. Piers et al., Can. J. Chem. *60,* 210–23 (1982); with PPh_3/CCl_4 cf. M. Suama et al., Chem. Letters *1981,* 1459–62.

Cuprous bromide/lithium bromide *CuBr/LiBr*

Halogenallenes from sulfonyloxy-2-acetylenes $C(OSO_2R)C\equiv C \rightarrow C=C=CHal$

467. $HC\equiv C\text{–}CH(Ph)OSO_2Me \xrightarrow{LiCuBr_2} BrCH=C=CHPh$

A soln. of $LiCuBr_2$ in tetrahydrofuran added with stirring to the startg. mesylate in the same solvent at –60°, stirring continued for 1 hr., the mixture allowed to warm slowly to room temp., and stirred for a further 15 hrs. → product. Y 71%. F.e.s. M. Montury, J. Goré, Syn. Commun. *10,* 873–9 (1980).

Ethylmagnesium bromide *EtMgBr*

δ-Bromo-γ-lactones from γ,δ-oxidocarboxylic acid esters ○

s. *37,* 320

Cadmium halides *$CdHal_2$*

Tert. halides from alcohols OH → Hal

s. *37,* 466

Molecular sieve ←

1-Halogenothioethers from sulfoxides ←

s. *28,* 497; **review of 1-halogenothioethers** s. K. Arai, Yuki Gosei Kagaku Kyokaishi *39,* 374–84 (1981); Chem. Abstr. *95,* 79319 (1981).

Boron bromide *BBr_3*

Replacement of acetoxy groups by halogen OAc → Hal

s. *12,* 608s*31;* *β,γ*-ethylenebromides from formoxy-2-ethylenes with BBr_3 s. S. Wolfe, C. Shaw, Can. J. Chem. *60,* 144–53 (1982).

Thallous ethoxide/thionyl chloride *$TlOEt/SOCl_2$*

Glycosyl chlorides from aldoses OH → Cl

with $ZnCl_2$ cf. *25,* 396; with $SOCl_2$/thallous ethoxide s. A. Granata, A.S. Perlin, Carbohyd. Res. *86,* 305–8 (1980); with dimethylchloroformiminium chloride/collidine s. R.U. Lemieux et al., Can. J. Chem. *60,* 58–62; 63–7 (1982).

Diethyl azodicarboxylate s. under Ph_3P *EtOOC–N=N–COOEt*

Dimethylformamide (s.a. under PBr_3) *Me_2NCHO*

Dimethylformamide/thionyl chloride *$Me_2NCHO/SOCl_2$*

Carboxylic acid chlorides from the acids COOH → COCl

s. *15,* 441; *α*-alkylacrylic acid chlorides s. G.A. Marshalok, J.J. Yatchishin, D.K. Tolopko, Visn. L'vio. Politekh. Inst. *1981* (149), 146 (Russ.); Chem. Abstr. *96,* 162090 (1982).

Sulfonyl chlorides SO_2Cl

s. *15,* 441; *o*-naphthoquinonediazide sulfonyl chlorides s. B.Z. Yakovlev et al., Khim. Prom-st. *1981,* 505; Chem. Abstr. *96,* 34766 (1982).

Perfluoro-2-methyl-2-pentene/triethylamine *$(CF_3)_2C=CFC_2F_5/Et_3N$*

Replacement of hydroxyl groups by fluorine OH → F

468. $PhCOOH \xrightarrow{(CF_3)_2C=CFCF_2CF_3} PhCOF$

Carboxylic acid fluorides. A soln. of benzoic acid in acetone treated with triethylamine and perfluoro-2-methyl-2-pentene at room temp. for 3 hrs. → benzoyl fluoride. Y 86%. F.e., **also fluorides from alcohols,** and 2-fluoropyridine from 2-pyridone, s. S. Yanagida, Y. Noji, M. Okahara, Bull. Chem. Soc. Japan *54,* 1151–8 (1981).

1,1-Dihalides s. under $FeCl_3$ *R_2CHal_2*

Hexachloroacetone s. under Ph_3P *$(CCl_3)_2CO$*

Dimethylchloroformiminium chloride/collidine ←

Glycosyl chlorides from aldoses OH → Cl

s. *25,* 396s*37*

Trimethylsilyl halides *Me_3SiHal*

Glycosyl halides from acyl glycosides OAc → Hal

s. *19,* 523; glycosyl iodides with trimethylsilyl iodide, s. J. Thiem, B. Meyer, Chem. Ber. *113,* 3075–85 (1980); glycosyl bromides with trimethylsilyl bromide s. J.W. Gillard, M. Israel, Tetrahedron Letters *22,* 513–6 (1981).

Trimethylsilyl chloride/sodium bromide *$Me_3SiCl/NaBr$*

Bromides from alcohols OH → Br

with LiBr cf. *36,* 489; labelled bromides with $Na^{77}Br$, s. M.R. Kilbourn, K.D. McElvany, M.J. Welch, Intern. J. Appl. Radiation Isotopes *33,* 391–2 (1982).

Trimethylsilyl iodide *Me_3SiI*

Iodocarboxylic acid esters from lactones and alcohols ⊂

469.

Trimethylsilyl iodide added to a mixture of β-propiolactone and ethanol in dichloromethane at room temp., and the reaction continued for 30 min. with exclusion of moisture → product. Y >90%. – The method is general for β-, γ-, δ-, and ε-lactones. F.e.s. M. Kolb, J. Barth, Synth. Commun. *11,* 763–7 (1981).

Titanium tetrabromide *$TiBr_4$*

Replacement of acetoxy groups by halogen OAc → Hal

s. *12,* 608; glycosyl bromides s. H. Paulsen, A. Bünsch, Carbohyd. Res. *100,* 143–67 (1982).

Tributylfluorophosphonium bromide *$[Bu_3PF]^+Br^-$*

Halides from alkoxysilanes OSi⋛ → Hal

fluorides with $PhPF_4$ cf. *29,* 512s*36;* bromides via alkoxyphosphonium salts with $[Bu_3PF]^+Br^-$ s. R. Bartsch, O. Stelzer, R. Schmutzler, Synthesis *1982,* 326–9; bridgehead fluorides with SF_4 s. A.E. Sorochinskii et al., Zh. Org. Khim. *17,* 1642–8 (1981).

Triphenylphosphine/diethyl azodicarboxylate *$Ph_3P/EtOOC{-}N{=}N{-}COOEt$*

Iodides from alcohols OH → I

with PPh_3/CCl_4 cf. *8,* 616s*32,* with PPh_3/diethyl azodicarboxylate, isolation of intermediate alkoxyphosphonium iodides, inversion of configuration, carbohydrate derivs., s. H. Kunz, P. Schmidt, Tetrahedron Letters *1979,* 2123–4; Ann. Chem. *1982,* 1245–60.

Triphenylphosphine/hexachloroacetone *$Ph_3P/(CCl_3)_2CO$*

Replacement of hydroxyl groups by chlorine OH → Cl

β,γ-ethylenechlorides s. *33,* 492; also alkyl chlorides s. R.M. Magid, B.G. Talley, S.K. Souther, J. Org. Chem. *46,* 824–5 (1981).

Benzyltriphenoxyphosphonium chloride *$[(PhO)_3PCH_2Ph]^+Cl^-$*

Replacement of hydroxyl by halogen OH → Hal

iodides with methyltriphenoxyphosphonium iodide s. *8,* 616s*36;* chlorides with benzyltriphenoxyphosphonium chloride, *β,γ*-acetylenechlorides, s. A.N. Petel, Zh. Org. Khim. *16,* 2621 (1980).

Triphenylphosphine dihalides s. under Et_3N Ph_3PHal_2

Phosphorus tribromide/dimethylformamide PBr_3/Me_2NCHO

Halides from alcohols
chlorides s. *32,* 495; bromides with PBr_3 s. W.R. Dolbier, Jr., et al., J. Org. Chem. *47,* 2298–303 (1982).

Phosphorus pentachloride PCl_5

3,4-Dihydro-3-chromenes from 4-chromanones ←
s. *36,* 487; synthesis and reactions of **4-chromanones, review,** s. H.-J. Kabbe, A. Widdig, Angew. Chem. Intern. Ed. *21,* 247–56 (1982).

Thionyl chloride (s.a. under Me_2NCHO) $SOCl_2$

***o*-Chloracylamines from N-arylhydroxamic acids** ←

470.

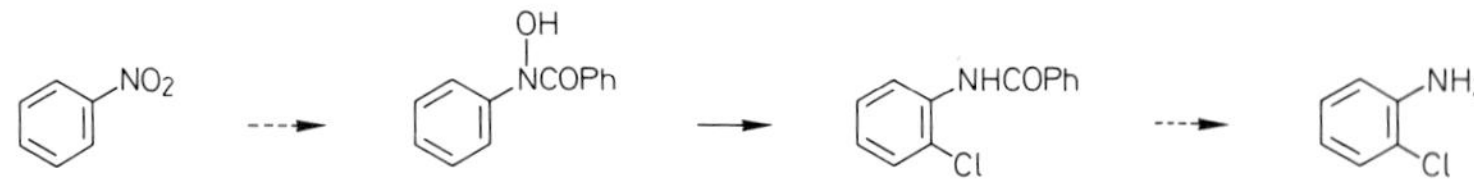

A soln. of $SOCl_2$ in dry benzene or ether added dropwise to a stirred soln. of N-phenylbenzohydroxamic acid in the same solvent at 0–5° for 30 min., followed by 1 hr. at 25° → 2-chlorobenzanilide. Y 84%. – This is seen as part of a multistep prepn. of ***o*-chloramines from ar. nitro compds.** F.e.s. N.R. Ayyangar, U.R. Kalkote, P.V. Nikrad, Tetrahedron Letters *23,* 1099–102 (1982).

Ring closures of azocarboxylic acids with intramolecular *o*-chlorination ○
4-(*o*-Chloroarylamino)-1,4-benzoxazin-3(2*H*)-ones

471.

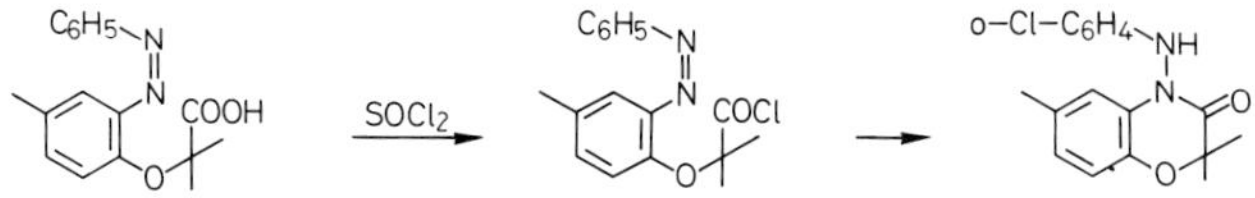

A soln. of 2-(2-phenylazo-4-methylphenoxy)-2-methylpropionic acid and $SOCl_2$ in carbon tetrachloride refluxed for 1 hr. → 4-(2-chloroanilino)-2,2,6-trimethyl-1,4-benzoxazin-3(2*H*)-one. Y 92%. F.e.s. M. Byers et al., J. Chem. Soc. Perkin Trans. I *1981,* 1092–5.

Thionyl bromide $SOBr_2$

Reactions with thionyl bromide CHO → $CHBr_2$
s. *30,* 364; 1,1-dibromides from aldehydes s. Org. Prep. Proc. Intern. *13,* 365–7 (1981); carboxylic acid bromides from aldehydes, also *α,β*-ethylene-*α*-bromoaldehydes, s. J. Prakt. Chem. *323,* 673–6 (1981).

Sulfuryl chlorofluoride $ClSO_2F$

Phosphinic halides from phosphinic acids P–OH → P–F
chlorides with PCl_5 s. *8,* 618; fluorides with $ClSO_2F$, also phosphorofluoridates, s. A. Łopusiński, J. Michalski, Angew. Chem. Intern. Ed. *21,* 294 (1982).

Sulfur tetrafluoride SF_4

Fluorides from alkoxysilanes OSi≡ → F
s. *29,* 512s*37*

Replacement of oxygen by fluorine CO → CF_2
1,1-Difluorides
s. *33,* 486; s.a. F.A. Bloshchitsa et al., Zh. Org. Khim. *17,* 1417–20 (1981).

Chlorine monofluoride *ClF*

1-Alkoxy-1,1-difluorides from carboxylic acid esters COOR → CF_2OR

472. $ClCH_2COOMe \xrightarrow{ClF} ClCH_2CF_2OMe$

Chlorine monofluoride passed into a soln. of methyl chloracetate in anhydrous HF at –60 to –40° during 18 min., and the mixture poured onto ice → 2,2-difluoro-1-chloro-3-oxabutane. Y 80%. F.e.s. L.S. Boguslavskaya, I.Y. Panteleeva, N.N. Chuvatkin, Zh. Org. Khim. *18,* 222–3 (1982).

*Tetra-*n*-butylammonium fluoride* n-$Bu_4N^+F^-$

Fluorides from mesylates OSO_2CH_3 → F

Carbohydrate derivs.

with KF cf. *14,* 606; with tetra-*n*-butylammonium fluoride s. L. Hough, A.A.E. Penglis, A.C. Richardson, Can. J. Chem. *59,* 396–405 (1981).

Hydrogen halides s. under LiHal *HHal*

Hydrogen bromide *HBr*

1-Bromothioethers from 1-hydroxythioethers OH → Br

1-Subst. cyclopropyl thioethers

473.

HBr slowly bubbled through a stirred soln. of 1-(methylthio)cyclopropanol in dichloromethane at 0° for 0.5 hr., and stirring continued for a further 0.5 hr. → product. Y 90–95%. F.e. and reactions s. R. Jorritsma, H. Steinberg, Th. J. de Boer, Rec. Trav. Chim. Pays-Bas *100,* 184–200 (1981).

Ferric chloride/1,1-dihalides $FeCl_3/R_2CHal_2$

Dihalogenomethyl compds. from aldehydes CHO → $CHHal_2$

s. *37,* 183

Via intermediates *v.i.*

Bromides from alcohols via selenides OH → SeAr → Br

Retention of configuration

474.

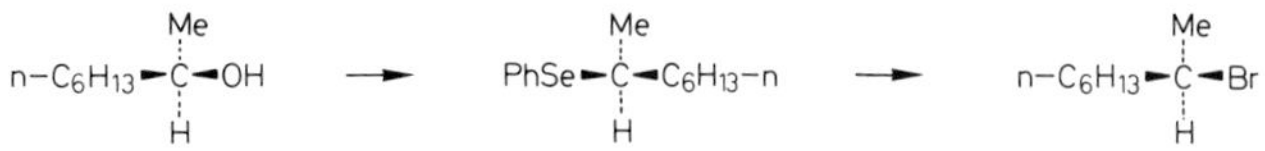

Sec. bromides. Startg. (R)-(–)-alcohol treated according to Synth. Meth. *32,* 599 → (S)-(+)-selenide (Y 74%) treated according to Synth. Meth. *32,* 513 → (R)-(–)-bromide (Y 90%). F.e. and method s. M. Sevrin, A. Krief, J. Chem. Soc. Chem. Commun. *1980,* 656–7.

Nitrogen ↑ HalC↓↑N

Without additional reagents *w.a.r.*

1,1-Difluorides from diazo compds. $C=N_2 \rightarrow CF_2$

475. $Ph_2C=N_2 \rightarrow Ph_2CF_2$

9:1 N_2/F_2 bubbled through a rapidly stirred soln. of diphenyldiazomethane in $CFCl_3$ at –70° for 15 min. → difluorodiphenylmethane. Y 71%. F.e.s. T.B. Patrick, J.J. Scheibel, G.L. Cantrell, J. Org. Chem. *46,* 3917–8 (1981).

Sodium nitrite/pyridine *$NaNO_2/C_5H_5N$*

Fluorides from prim. amines C(NH_2)COOH → CFCOOH
α-Fluorocarboxylic acids
s. *31,* 503; s.a. Helv. Chim. Acta *64,* 2528–30 (1981); J. Barber, R. Keck, J. Rétey, Tetrahedron Letters *23,* 1549–52 (1982).

Potassium iodide *KI*

Halides from diazonium fluoroborates N_2^+ → Hal
bromides with $CuBr_2$ cf. *26,* 549; iodides with KI/I_2 s. A. Citterio, A. Arnoldi, Synth. Commun. *11,* 639–42 (1981).

Pyridine s. under $NaNO_2$ *C_5H_5N*

Cyanogen bromide *CNBr*

Bromhydrins from *tert*-aminoalcohols NR_2 → Hal
s. *37,* 252

Trimethylsilyl halides *Me_3SiHal*

Aryl halides from triazenes N=N–NR_2 → Hal
fluorides s. *36,* 492; iodides and bromides in the presence of trimethylsilyl halides s. H. Ku, J.R. Barrio, J. Org. Chem. *46,* 5239–41 (1981).

Halogen ↑ HalC⇃↾Hal

Without additional reagents *w.a.r.*

Replacement of bridgehead bromine and iodine by fluorine Br → F
I → F

476.

A mixture of ca. 1.5% F_2 in N_2 slowly bubbled through a soln. of 1-bromoadamantane in $CFCl_3$ at –75° for 1–3 hrs. using an efficient vibromixer → 1-fluoroadamantane. Y 99%. – Similarly, 1-iodoadamantane → 1-fluoroadamantane. Y 74%. F.e.s. S. Rozen, M. Brand, J. Org. Chem. *46,* 733–6 (1981).

Sodium iodide *NaI*

Replacement of chlorine by iodine COCl → COI
Carboxylic acid iodides

477.

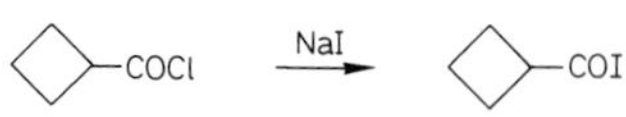

Startg. acid chloride added dropwise to a freshly prepd. soln. of NaI in acetonitrile under N_2, the mixture stirred at room temp. for 30 min., then extracted with pentane at low temp. *in vacuo* during 5 hrs. → product. Y 92%. F.e., selectivity, and description of apparatus, s. H.M.R. Hoffmann, K. Haase, Synthesis *1981,* 715–9; dicarboxylic acid iodides s. ibid. *1982,* 237–9.

Sodium iodide-^{125}I/dicyclohexyl-18-crown-6 polyether ←

Replacement of bromine by iodine Br → I*
with KI cf. *1,* 451s*32;* with $Na^{125}I$/dicyclohexyl-18-crown-6 polyether, I-labelled compds., s. B.L. Liu et al., Radioisotopes *30,* 391 (1981) (Eng.); Chem. Abstr. *95,* 203240 (1981).

Potassium fluoride *KF*

Replacement of chlorine by fluorine Cl → F

s. *4,* 513; with spray-dried KF, also replacement of bromine, s. N. Ishikawa et al., Chem. Letters *1981,* 761–4.

Potassium fluoride/trimethylamine KF/Me_3N

Trimethylamine as solid-liq. phase transfer catalyst

Replacement of chlorine by fluorine and azido groups

478. 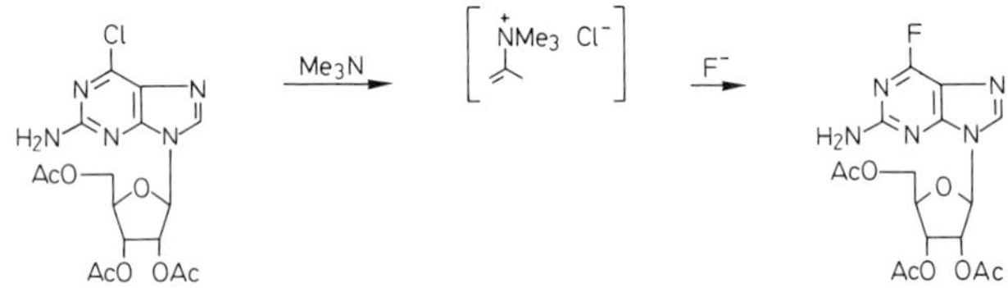

A suspension of KF in a solution of 9-(2,3,5-tri-O-acetyl-β-D-ribofuranosyl)-2-amino-6-chloropurine in dimethylformamide containing a little trimethylamine stirred vigorously at room temp. for 14 hrs. → 6-fluoro deriv. Y 93%. – Similarly, with NaN_3 in acetonitrile → 6-azido deriv. Y 100%. M.J. Robins, B. Uzmański, Can. J. Chem. *59,* 2601–7 (1981).

Sodium iodide *NaI*

Regiospecific cleavage of ethers under mild, neutral conditions C

479. 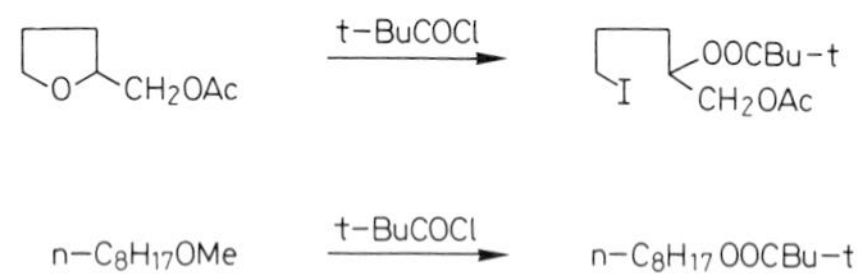

Acoxyiodides. A soln. of pivaloyl chloride in acetonitrile added to a mixture of 2-acetoxymethyltetrahydrofuran and NaI in the same solvent at 0°, stirred for 23 hrs. at room temp., and quenched with aq. $NaHSO_3$ → product. Y 71%. – Similarly, octyl methyl ether during 22 hrs. → octyl pivalate. Y 90%. – In contrast to Lewis acid-catalysed cleavage, reaction takes place at the less substituted carbon-oxygen bond. F.e.s. A. Oku, T. Harada, K. Kita, Tetrahedron Letters *23,* 681–4 (1982).

Rubidium fluoride-^{18}F $Rb^{18}F$

F-Labelled compds. F → F*

480. $Rb^{18}F + p\text{-}F\text{-}C_6H_4CN \rightarrow p\text{-}^{18}F\text{-}C_6H_4CN$

Ar. fluorides-^{18}F. 4-Cyanofluorobenzene added to a soln. of $Rb^{18}F$ (produced by heterogeneous exchange of RbCl with $^{18}F_2$) in dimethyl sulfoxide, and the mixture stirred at 165° for 4 hrs. → 4-cyanofluorobenzene-^{18}F. Exchange Y 94%. F.e.s. F. Cacace et al., J. Labelled Compounds Radiopharm. *18,* 1721–30 (1981).

Trimethylamine s. under KF Me_3N

Silver fluoride *AgF*

Replacement of bromine by fluorine Br → F*

s. *12,* 620; ^{18}F-labelled compds. with $Ag^{18}F$ s. M. Yagi, G. Izawa, M. Murano, Kakuriken Kenkyu Hokoku *14,* 62 (1981) (Japan); Chem. Abstr. *95,* 219716 (1981); Radiochem. Radioanal. Lett. *49,* 173 (1981); Chem. Abstr. *95,* 219719 (1981); with $Et_4N^{+}\,^{18}F^{-}/AgO$ s. S.J. Gatley, Intern. J. Appl. Radiation Isotopes *33,* 255–8 (1982).

Silver fluoroborate *$AgBF_4$*

α-Fluoroketones from 1,2-oxido-1-chlorides C

481. 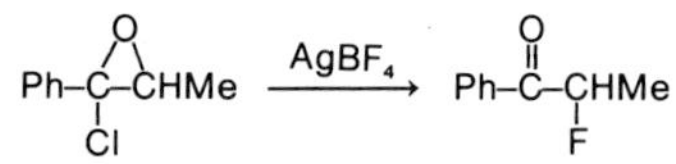

A soln. of $AgBF_4$ in ether treated at –30° with a soln. of startg. m. in the same solvent during 5 min., the mixture allowed to warm to room temp., and stirred for 30 min. → 2-fluoro-1-phenyl-1-propanone. Y 83%. F.e.s. K. Griesbaum et al., Chem. Ber. *114,* 1858–70 (1981).

Dicyclohexyl-18-crown-6 polyether s. under Sodium iodide ←

Methyl 2-pyridyl sulfoxide **as phase transfer catalyst** ←

Nucleophilic substitution of halogen Br → Cl

482. $PhCH_2Br + LiCl \rightarrow PhCH_2Cl$

Replacement of bromine by chlorine. Methyl 2-pyridyl sulfoxide added to a heterogeneous mixture of benzyl bromide and LiCl in benzene (or benzene/water), and stirred vigorously for 4 hrs. at 70° → benzyl chloride. Y 70%. – The catalyst is readily prepared, and can be recovered quantitatively by shaking the mixture with aq. HCl. Also replacement by cyano, thiocyanato, aryloxy, and arylthio groups s. N. Furukawa et al., Tetrahedron Letters *22,* 4409–12 (1981); f. 2-pyridyl sulfoxides and catalysts s. Heterocycles *16,* 1927–32 (1981).

^{123}Iodide *I^-*

I-Labelled compds. I → I*

s. *19,* 638; labelled iodocarboxylic acids s. Intern. J. Appl. Radiation Isotopes *33,* 499–504 (1982).

Trimethylsilyl iodide *Me_3SiI*

Iodides from fluorides F → I

483. 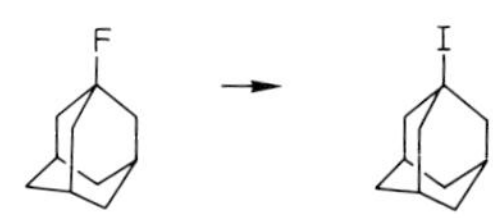

Bridgehead iodides. Trimethylsilyl iodide added to a stirred soln. of 1-adamantyl fluoride in dichloromethane at 25°, and stirred under N_2 for 16 hrs. → 1-adamantyl iodide. Y 87%. – The method is suitable for preparing prim., sec., and tert. iodides. F.e., also from tert. chlorides, and using trimethylsilyl iodide prepd. *in situ* s. G.A. Olah, S.C. Narang, L.D. Field, J. Org. Chem. *46,* 3727–8 (1981).

Antimony trifluoride *SbF_3*

Replacement of chlorine by fluorine Cl → F

s. *3,* 485; 1,1-difluorides from 1,1-dichlorides, s. H. Volz, W.D. Mayer, Ann. Chem. *1981,* 1407–14.

Antimony pentafluoride *SbF_5*

Trifluoromethyl from trichloromethyl groups $CCl_3 \rightarrow CF_3$

with KF cf. *28,* 516; with SbF_5, N-(trifluoromethyl)amines, s. L.M. Yagupolsky et al., Zh. Org. Khim. *16,* 2508–12 (1980).

Sulfur ↑ HalC ⇃↾ S

Sodium bromide *NaBr*

Halides from dithiourethans
via immoniodithiocarbonic acid esters
$RSC(SR')=\overset{+}{N}< \rightarrow RHal$

484. 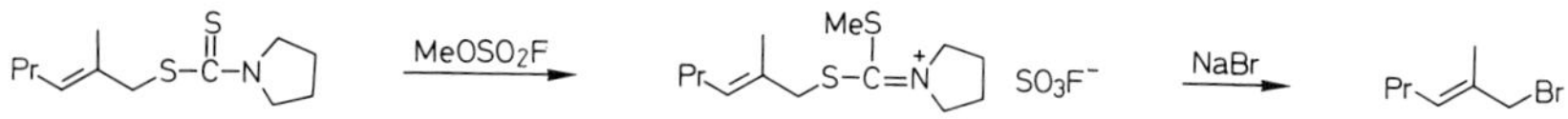

***β,γ*-Ethylenebromides.** A soln. of methyl fluorosulfonate in dichloromethane added with cooling to a soln. of startg. m. in the same solvent, the mixture stirred for 5 hrs., concd., a soln. of NaBr in dimethylformamide added, then stirred for 48 hrs. at room temp. → product. Y 62%. F.e.s. A. Sakurai, T. Hayashi, T. Oishi, Synthesis *1980,* 843–4; ***β,γ*-ethyleneiodides** s. ibid. *1978,* 370–2.

Triphenylphosphine (s.a. under SO_2Cl_2) Ph_3P

Iodides $SO_3H \rightarrow I$

485. $Me(CH_2)_4SO_3H \xrightarrow{I_2} Me(CH_2)_4I$

Alkyl iodides are simply prepared from a variety of organosulfur compds. by reaction with triphenylphosphine/I_2. E: **from sulfonic acids.** A soln. of startg. sulfonic acid and 5 equivalents triphenylphosphine in benzene treated with 1.5 equivalents I_2, and the mixture allowed to stand at room temp. for 26 hrs. under N_2 → product. Y 81%. F.e., **also from sulfonyl chlorides and esters, thiosulfonyl esters, sulfinic acids, mercaptans, and sym. disulfides,** s. S. Oae, H. Togo, Synthesis *1981,* 371–3.

Sulfuryl chloride SO_2Cl_2

Chloroformic acid esters
from monothiolcarbonic acid esters ROC(O)SR′ → ROC(O)Cl

486. $(O_2N)_3CCH_2OC(O)SEt \rightarrow (O_2N)_3CCH_2OCOCl$

A soln. of startg. S-ethyl thiocarbonate and excess SO_2Cl_2 in dichloromethane refluxed for 6 hrs. → product. Y 97%. F.e.s. W.H. Gilligan, S.L. Stafford, Synthesis *1979,* 600–2.

Sulfuryl chloride/triphenylphosphine SO_2Cl_2/Ph_3P

Replacement of sulfhydryl by chlorine SH → Cl
s. *21,* 619; with added triphenylphosphine or triphenylphosphite, also from disulfides, s. I.W.J. Still, G.W. Kutney, D. McLean, J. Org. Chem. *47,* 560–1 (1982).

Via intermediates *v.i.*

Halides from dithiourethans $RSC(S)N< \rightarrow RHal$
s. *37,* 484

Remaining Elements ↑ HalC ⇃↾ Rem

Without additional reagents *w.a.r.*

***α,β*-Ethylenehalides from enetellurium trichlorides** $C=CTeCl_3 \rightarrow C=CHal$
s. *37,* 446

Sodium chloride *NaCl*

Halides from tellurides via tellurium dihalides $R_2TeHal_2 \rightarrow RHal$

487. $n\text{-}C_{15}H_{31}Cl + LiCH_2TePh \dashrightarrow n\text{-}C_{15}H_{31}CH_2TePh \rightarrow n\text{-}C_{15}H_{31}CH_2Te(Ph)Cl_2 \rightarrow n\text{-}C_{15}H_{31}CH_2Cl$

A soln. of SO_2Cl_2 in carbon tetrachloride added slowly to a stirred soln. of 1-phenyltellurohexadecane in the same solvent at 0° under N_2, the mixture stirred at room temp. for 10 min., solvent and excess reagent removed, the residue mixed with NaCl and dimethylformamide, and heated with stirring at 100° for 1 hr. under N_2 → 1-chlorohexadecane. Y 92%. – This is part of a novel **homologization of halides.** F.e. incl. iodides, **also 1,1-dibromides and aldehydes from 1,1-ditellurides,** s. K. Chikamatsu et al., Chem. Letters *1982,* 1081–4.

Ammonium bromide-^{82}*Br/chloramine-T/hydrogen chloride* $NH_4Br/TsN(Cl)Na/HCl$

Bromides from stannanes Sn⋛ → Br*

with $CuBr_2$ cf. *28,* 511s*36;* labelled ar. bromides with $NH_4{}^{82}Br$ in the presence of chloramine-T and HCl s. J. Chem. Soc. Chem. Commun. *1982,* 625–6.

Phenyltrimethylammonium tribromide $[PhNMe_3]^+Br_3^-$

β-Bromoketones from 2-ethylene-3-siloxysilanes C(OSi⋛)=C–CSi⋛ → COCHCBr

s. *37,* 654

Carbon ↑ HalC⇂↑C

Without additional reagents *w.a.r.*

Chlorosulfinylcarboxylic acid chlorides from (acylthio)carboxylic acids ←

488. $AcSCH_2CH_2COOH \rightarrow ClS(O)CH_2CH_2COCl$

Cl_2 passed into a stirred soln. of startg. β-(acetylthio)carboxylic acid in methylene chloride at –30°, and warmed to ca. 20° → product. Y 83%. F.e. and with SO_2Cl_2 s. T.P. Vasileva et al., Izv. Akad. Nauk. SSSR Ser. Khim. *1981,* 159–65.

Elimination ⇑

Nitrogen ↑ HalC⇑N

Without additional reagents *w.a.r.*

Halides from N,N-dihalogenamines $NHal_2 \rightarrow Hal$

489.

Bridgehead halides. A 12% soln. of 1-(dichloroamino)adamantane in dichloromethane injected onto a g.l.c column of 15–30% SE-30 on Chromosorb W (injector temp. 250°; column temp. 155°) with He flow rate of 150 ml/min. → 1-chloroadamantane. Y 90% (by peak area). F.e., **also nitriles from prim. N,N-dihalogenamines,** s. J.T. Roberts et al., J. Org. Chem. *45,* 5239–43 (1980).

Sulfur ↑ **HalC⇑S**

Antimony pentafluoride SbF_5

Halides from sulfonyl halides $SO_2F \rightarrow F$

chlorides cf. *25,* 410; fluorides with SbF_5 s. A.F. Ermolov et al., Zh. Org. Khim. *17,* 2239–40 (1981).

Carbon ↑ **HalC⇑C**

Palladium–carbon *Pd–C*

Aryl chlorides from arylcarboxylic acid chlorides COCl → Cl

Decarbonylation

with $(Ph_3P)_3RhCl$ cf. *21,* 624; with Pd–C s. J.W. Verbicky, B.A. Dellacoletta, L. Williams, Tetrahedron Letters *23,* 371–2 (1982).

Formation of S–S Bond

Exchange ↓↑

Hydrogen ↑ **SS↓↑H**

Oxidations with

trimethylamine N-oxide/antimony pentachloride $Me_3NO/SbCl_5$

490. $$2\ PhSH \xrightarrow{Me_3NO.SbCl_5} PhS\text{–}SPh$$

A soln. of thiophenol and trimethylamine oxide-$SbCl_5$ (1:1 complex) in nitromethane allowed to stand at room temp. for 20 hrs. → diphenyl disulfide. Y 77%. – Oxidation is more facile and milder than with the corresponding uncomplexed N-oxides. F.e., **also oxo compds. from alcohols,** and with pyridine N-oxide-antimony pentachloride, s. J. Yamamoto et al., Bull. Chem. Soc. Japan *55,* 345–6 (1982).

Sym. disulfides from mercaptans 2 RSH → RSSR

N,N′-Thiobisphthalimide/aniline ←

Sym. bis(sulfonyl) sulfides from sulfinic acids $2\ RSO_2H \rightarrow RSO_2SSO_2R$

491.

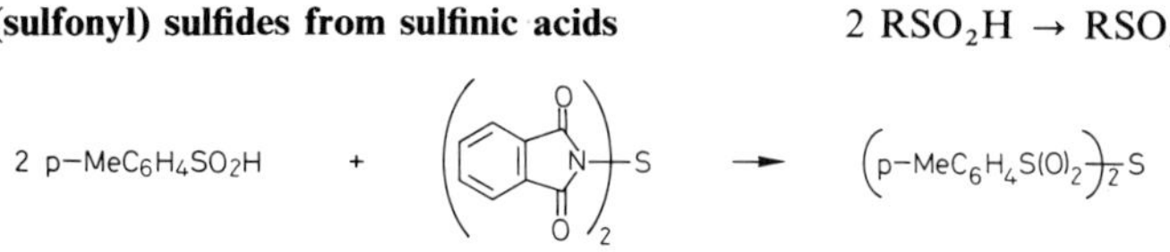

A mixture of N,N′-thiobisphthalimide, 2 equivalents *p*-toluenesulfinic acid, and a little aniline in dichloromethane refluxed for 1 hr. → bis(*p*-toluenesulfonyl) sulfide. Y 88%. F.e., and bases, **also N-sulfonylthiophthalimides** with 1 equivalent of sulfinic acid, s. Y. Abe et al., Bull. Chem. Soc. Japan *53,* 3678–82 (1980).

Oxidations with *bis(p-methoxyphenyl) selenoxide* Ar_2SeO
Sym. disulfides from mercaptans 2 RSH → RSSR

492. $$2\ PhSH \xrightarrow{(p\text{-}MeOC_6H_4)_2SeO} PhSSPh$$

A mixture of bis(*p*-methoxyphenyl) selenoxide and thiophenol in dichloromethane stirred at room temp. for 10 min. → diphenyl disulfide. Y 90%. F.e., **also sulfoxides from thioethers, phosphine oxides from phosphines, quinones from quinols,** and oxidative ring closures, s. F. Ogura et al., Bull. Chem. Soc. Japan *55,* 641–2 (1982).

Nitric oxide *NO*
with NO_2 s. *34,* 505; also with NO s. W.A. Pryor et al., J. Org. Chem. *47,* 156–9 (1982).

Antimony pentachloride s. under Me_3NO $SbCl_5$

Manganese dioxide MnO_2
1,1′-Disiloxydisulfides from 1-siloxymercaptans
s. *37,* 535

Oxygen ↑ SS↓↑O

Boron halides/potassium iodide/tetrabutylammonium iodide $BHal_3/KI/n\text{-}Bu_4N^+I^-$
Sym. disulfides from sulfonic acids $2\ RSO_3H \rightarrow RSSR$

493. $$2\ p\text{-}MeC_6H_4SO_3H \rightarrow (p\text{-}MeC_6H_4S)_2$$

KI suspended in a stirred soln. of BBr_3 in 1,2-dichloroethane containing a catalytic amount of tetra-*n*-butylammonium iodide, *p*-toluenesulfonic acid added, and the mixture refluxed for 16 hrs. under N_2 → *p*-tolyl disulfide. Y 96%. F.e., **also from sulfonic acid esters, chlorides, and fluorides,** s. G.A. Olah et al., J. Org. Chem. *46,* 2408–10 (1981); with polyphosphoric acid, ethyl polyphosphate, or P_2O_5 in place of boron halides cf. S. Oae, H. Togo, Synthesis *1982,* 152–5.

*Polyphosphoric acid derivs./potassium iodide/tetra-*n*-butylammonium iodide*
s. *37,* 493

Halogen ↑ SS↓↑Hal

Potassium iodide/pyridine KI/C_5H_5N
Thiosulfonyl esters from two sulfonyl chloride molecules $2\ RSO_2Cl \rightarrow RSO_2SR$
with $Fe(CO)_5/BF_3$ cf. *24,* 593; with KI/pyridine s. G. Palumbo, R. Caputo, Synthesis *1981,* 888–90.

Imidazoles/triethylamine ←
Sym. disulfides from sulfenyl halides 2 RSHal → RSSR
with KI cf. *16,* 369; with imidazole derivs. and triethylamine s. G. Sosnovsky, J.A. Krogh, Ann. Chem. *1982,* 121–36.

Sulfur ↑ SS⇃↾S

Without additional reagents *w.a.r.*

Disulfides from thiosulfonyl esters $RS–SO_2R' \rightarrow RS–SR''$
and mercaptans cf. *7,* 600; and sulfenyl chlorides s. J. Lázár, E. Vinkler, Acta Chim. Acad. Sci. Hung. *105,* 171–4 (1980).

Irradiation ⫻

Unsym. from sym. disulfides RSSR → RSSR′
s. *8,* 635; from two different disulfide molecules by irradiation cf. D. Gupta, Acta Chim. Acad. Sci. Hung. *108,* 31–5 (1981).

Carbon ↑ SS⇃↾C

Sulfuryl chloride SO_2Cl_2

Sym. phosphoryl disulfides from thiophosphoric esters $(RO)_2P(O)SSP(O)(OR)_2$

494. $(MeO)_3P{=}S + SO_2Cl_2 \rightarrow [(MeO)_2P(O){-}S]_2$

A mixture of SO_2Cl_2 and startg. phosphorothioate stirred in methylene chloride at room temp. for 2.5 hrs. → product. Y 95%. F.e.s. E. Krawczyk, A. Skowrońska, Phosphorus-Sulfur *9,* 189–92 (1980).

Bromine Br_2

Sym. disulfides from thiolic acid esters 2 RSAc → RSSR

495.

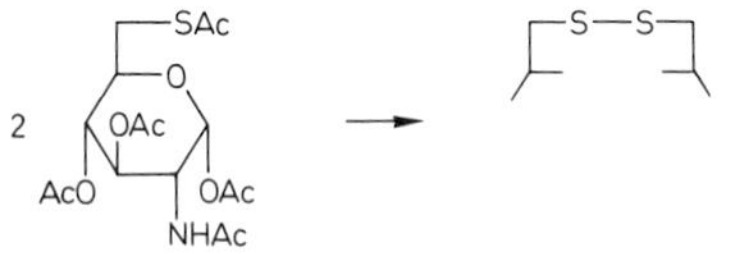

Br_2 in chloroform added at –10° with stirring to a soln. of 2-acetamido-1,3,4-tri-O-acetyl-6-S-acetyl-2-deoxy-6-thio-α-D-glucopyranose in the same solvent, and worked up after 2 min. → 6,6′-dithiobis-2-acetamido-1,3,4-tri-O-acetyl-2-deoxy-α-D-glucopyranose, Y 98%, (startg. m. for 501). **Also glycosyl disulfides,** and **sym. diselenides from isoselenouronium salts,** s. G.C. Chen et al., J. Chem. Soc. Perkin Trans. I *1980,* 2287–93.

Elimination ⥮

Sulfur ↑ SS⥮S

Tris(dimethylamino)phosphine $(Me_2N)_3P$

Disulfides from trisulfides RSSSR → RSSR

496. $PhCH_2S{-}S{-}SCH_2Ph \rightarrow PhCH_2S{-}SCH_2Ph$

A soln. of startg. trisulfide in ether treated with tris(dimethylamino)phosphine, and the mixture stirred for 2–3 hrs. at room temp. → product. Y 100%. F.e. and reagents s. D.N. Harpp, D.K. Ash, R.A. Smith, J. Org. Chem. *45,* 5155–60 (1980).

Formation of S–Rem Bond

Uptake ⇓

Addition to Sulfur and Sulfur SRem ⇓ SS

Irradiation ⇊

Insertion of silylenes into disulfide bonds RSSR → RS(Si⋜)SR
Bis(alkylthio)silanes

497. $EtS{-}SEt \xrightarrow{[Me_2Si:]} EtS{-}SiMe_2{-}SEt$

A soln. of dodecamethylcyclohexasilane and diethyl disulfide in cyclohexane irradiated in a quartz tube by a circular array of germicidal lamps for 2 hrs. → bis(ethylmercapto)dimethylsilane. Y 97%. F.e., also insertion into silicon-sulfur bonds in low yield, s. A. Chihi, W.P. Weber, J. Organometal. Chem. *210,* 163–8 (1981).

Exchange ⇃↾

Hydrogen ↑ SRem ⇃↾ H

Chlorotris(triphenylphosphine)rhodium(I) *$(Ph_3P)_3RhCl$*

Organothiostannanes from mercaptans and organotin hydrides RSH → RSSn⋛

498. $PhSH + HSnBu_3 \rightarrow PhSSnBu_3$

A mixture of thiophenol and tri-*n*-butyltin hydride in tetrahydrofuran treated under N_2 with a little chlorotris(triphenylphosphine)rhodium, and, when evolution of H_2 had ceased (ca. 10 min.), the soln. kept at room temp. for a further 20 min. → tri-*n*-butylstannyl phenyl sulfide. Y 95%. F.e.s. J.J. Talley, A. M. Colley, J. Organometal. Chem. *215,* C38–40 (1981).

Oxygen ↑ SRem ⇃↾ O

*2,4-Bis(*p*-methoxyphenyl)-1,3,2,4-dithiadiphosphetane-2,4-disulfide* ←

Replacement of P-oxygen by P-sulfur P(O) → P(S)

499. $Ph_2P(O)OMe \rightarrow Ph_2P(S)OMe$

A mixture of methyl diphenylphosphinate and Lawesson's reagent refluxed in benzene or toluene for 1 hr. → methyl diphenylphosphinothionate. Y 80%. F.e., also thiophosphonic acid derivs., thiophosphoric esters, and thiophosphonomonoamidates, s. L. Horner, H. Lindel, Phosphorus-Sulfur *12,* 259–61 (1982).

Halogen ↑ SRem ⇅ Hal

Sodium mercaptides *NaSR*

Replacement of halogen by alkylthio groups on group IV elements ≧SnHal → ≧SnSR

s. *28,* 520; organothiostannanes from halogenostannanes with Na-mercaptides cf. A.N. Nesmeyanov, D.N. Kravtsov, S.I. Pombrik, Izv. Akad. Nauk SSSR Ser. Khim. *1981,* 474; with *n*-BuLi s. R.M. Lequan, M. Lequan, J. Organometal. Chem. *201,* 371–5 (1980).

Caesium salt *Cs⁺*

Dithiocarboxylic acid silyl esters from caesium dithiocarboxylates C(S)SSi≦

500. $$PhCS_2^-Cs^+ + Me_3SiI \rightarrow PhC(S)SSiMe_3$$

Trimethylsilyl iodide added to a suspension of Cs dithiobenzoate in *n*-hexane at 0°, and the mixture stirred at 0–5° for 2 hrs. → trimethylsilyl dithiobenzoate. Y 94%. F.e., also conversion to **dithiocarboxylic acids-SD,** s. S. Kato, Synthesis *1982,* 457–8.

Remaining Elements ↑ SRem ⇅ Rem

Without additional reagents *w.a.r.*

Thioarsinous esters from disulfides RSSR → RSAs<

501. 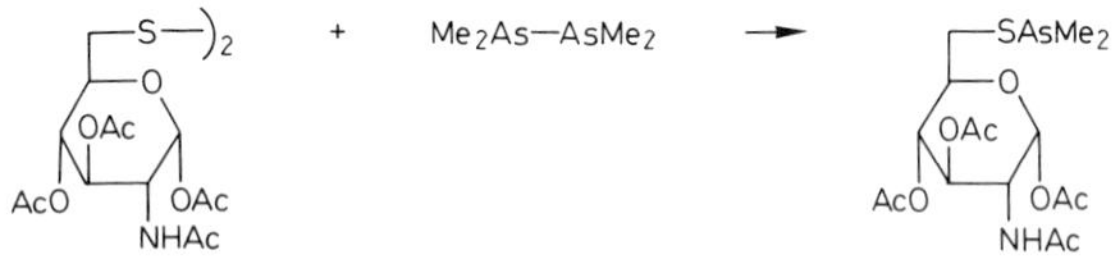

Tetramethyldiarsine added to 6,6′-dithiobis-2-acetamido-1,3,4-tri-O-acetyl-2-deoxy-*α*-D-glucopyranose (prepn. s. 495) in dichloromethane under N_2 at room temp. in a dry box, and the mixture left overnight → 2-acetamido-1,3,4-tri-O-acetyl-2-deoxy-6-S-dimethylarsino-6-thio-*α*-D-glucopyranose. Y 101%. F.e., **also selenoarsinous acid esters from diselenides,** s. G.C. Chen et al., J. Chem. Soc. Perkin Trans. I *1980,* 2287–93. **from halogenoarsines** cf. M.V. Rosenthal, R.A. Zingaro, Phosphorus-Sulfur, *9,* 107–16 (1980).

Carbon ↑ SRem ⇅ C

Triethylamine ←

Selective protection of sulfhydryl groups as diphenylthiophosphinic esters SH → $SP(O)Ph_2$

502. $$NH_2CH_2CH_2SH + NCP(O)Ph_2 \underset{F^-}{\overset{NEt_3}{\rightleftharpoons}} NH_2CH_2CH_2SP(O)Ph_2$$

A soln. of diphenylphosphinyl cyanide in chloroform added during 15 min. to a soln. of cysteamine hydrochloride and triethylamine in the same solvent, and stirred 2 hrs. at room temp. → S-diphenylphosphinyl deriv. Y 98% by n.m.r. – The protective group is readily removed with tetrabutylammonium or caesium fluoride. F.e. and protection with diphenylthiophosphinyl cyanide s. L. Horner, R. Gehring, H. Lindel, Phosphorus-Sulfur *11,* 349–56 (1981).

Formation of S–C Bond

Uptake ⇓

Addition to Hydrogen and Carbon SC⇓HC

Bis(trimethylsilyl) sulfate *$(Me_3SiO)_2SO_2$*

Reactions of bis(trimethylsilyl) sulfate H → SO_3H
Sulfonic acids
s. *36,* 512; f. reactions including trimethylsilylation s. M.G. Voronkov, V.K. Roman, E.A. Maletina, Synthesis *1982,* 277.

Sulfur monochloride *S_2Cl_2*

Replacement of hydrogen by sulfhydryl groups H → SH
s. *16,* 649; isolation of intermediate polysulfides cf. F. Bottino, R. Fradullo, S. Pappalardo, J. Org. Chem. *46,* 2793–5 (1981).

Addition to Oxygen and Carbon SC⇓OC

1,8-Diazabicyclo[5.4.0]undec-7-ene ←

1-Siloxythiolic acid esters RCHO → RCH(OSi⋛)SAc
from thionocarboxylic acid silyl esters and aldehydes

503. 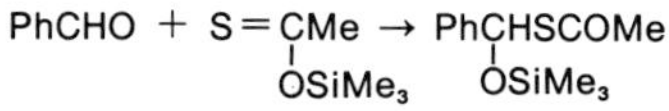

One drop of 1,8-diazabicyclo[5.4.0]undec-7-ene added to a stirred mixture of benzaldehyde and startg. silyl ester, and worked up after 15 min. at room temp. → product. Y 100%. F.e.s. T. Horii et al., Bull. Chem. Soc. Japan *54,* 790–2 (1981).

Zinc iodide *ZnI_2*

2-Hydroxythioethers from oxido compds.  → C(OH)C(SR)

and Na mercaptides cf. *28,* 524; and silyl sulfides, *trans*-derivs., in the presence of ZnI_2 s. Y. Guindon, R.N. Young, R. Frenette, Synth. Commun. *11,* 391–8 (1981).

Addition to Nitrogen and Sulfur SC⇓NS

Zinc *Zn*

Sulfinamides from thionyl amines N=SO → NHS(O)R
and Grignard compds. cf. *17,* 647; and organozinc bromides α-carbalkoxysulfinamides, s. A. De Blic et al., Synthesis *1982,* 281–3.

Addition to Nitrogen and Carbon SC⇓NC

Without additional reagents *w.a.r.*

1-Aminodithiocarboxylic acid esters from dithiocarboxylic acids and azomethines C=NR → C(NHR)SC(S)R′

504.

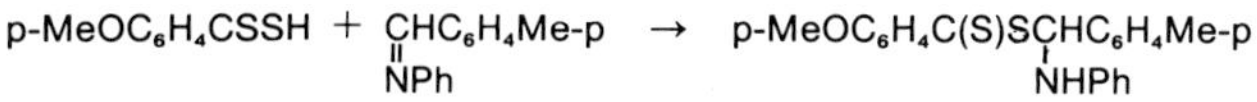

A soln. of freshly prepd. 4-methoxydithiobenzoic acid in ether added dropwise to a soln. of 4-methylbenzylideneaniline in the same solvent at 0° the solvent replaced by 1:4 ethyl acetate/*n*-hexane, cooled to –20°, and kept at the same temp. for 12 days → anilino(4-methylphenyl)methyl 4-methoxydithiobenzoate. Y 90%. F.e. and conversion to **2-ketodithiocarboxylic acid esters** s. M. Ishida, S. Kato, M. Mizuta, Z. Naturforsch. *36B,* 1047–9 (1981).

2-Aminothioethers from aziridines C
s. *8,* 643; f.e.s. M.G. Voronkov, V.I. Knutov, Khim. Geterotsikl. Soedin. *1981,* 921–3.

Sodium hydride *NaH*

2-Imino-1,3,4-thiadiazolidines from isothiocyanates and hydrazones O

505.

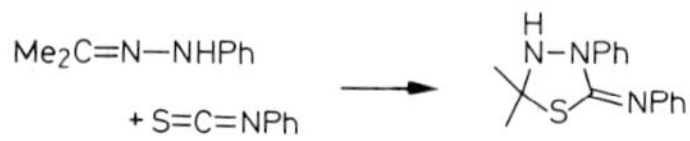

Acetone phenylhydrazone added to a suspension of NaH in dimethylformamide at room temp., phenyl isothiocyanate added dropwise after 10 min., and the mixture stirred for 1 hr. → 2,2-dimethyl-4-phenyl-5-phenylimino-1,3,4-thiadiazolidine. Y 90%. F.e., **also 1,3,4-thiadiazolidine-2-thiones** with carbon disulfide, s. Yamamoto I et al., J. Chem. Soc. Perkin Trans. I *1980,* 574–8.

Addition to Carbon and Carbon SC⇓CC

Without additional reagents *w.a.r.*

Immoniodithiocarbonic acid esters from dithiourethans RSC(=S)N< → RSC(SR′)=N⁺<
s. *37,* 484

Addition of sulfonyl thiocyanates to carbon-carbon multiple bonds ←
s. *31,* 86; *β,γ*-ethylene-*δ*-thiocyanatosulfones from 1,3-dienes s. M.M. Tanaskov et al., Zh. Org. Khim. *17,* 1800–7 (1981).

2-Chloro-1-silylsulfides from enesilanes and sulfenyl chlorides C=CSi≡ → C(Cl)C(Si≡)SAr
s. *37,* 548

***a*-Alkylthio-*a,β*-ethylenecarboxylic acid esters from 1,1-dialkoxyallenes** C=C=C(OR)$_2$ → C=C(SR′)COOR

506.

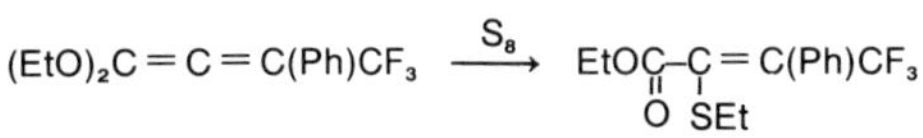

A stirred soln. of 1,1-diethoxy-4,4,4-trifluoro-3-phenyl-1,2-butadiene in dichloromethane treated under N_2 with sulfur at room temp., and the mixture left for 2 hrs. → ethyl 2-ethylthio-4,4,4-trifluoro-3-phenylcrotonate. Y 75%. F.e. and reactions s. R.W. Saalfrank et al., Chem. Ber. *115,* 57–64 (1982).

Without additional reagents *w.a.r.*

3,6-Dihydro-2*H*-1,3,4-thiadiazines from enazo compds. and thioketones ○

507. 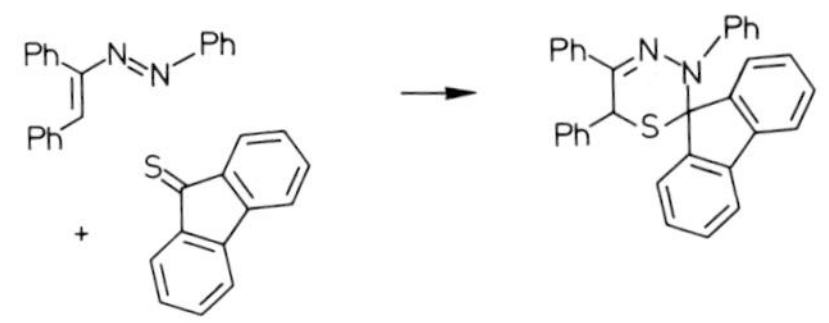

A soln. of fluorenethione in benzene added at room temp. to a stirred soln. of phenylazostilbene in the same solvent, and the mixture chromatographed after 5 hrs. → 3′,6′-dihydro-3′,5′,6′-triphenylspiro(fluorene-9,2′-[1,3,4]-thiadiazine). Y 91%. F.e., **also 5,6-dihydro-2*H*-1,2,3-thiadiazine 1-oxides from sulfines,** s. B.F. Bonini et al., J. Chem. Soc. Perkin Trans. I *1981,* 2322–7.

Irradiation ⚡

(E)-Ene-1,2-dithiocyanates from acetylene derivs. C≡C → C(SCN)=C(SCN)

508. $HC{\equiv}CPh + (SCN)_2 \rightarrow NCS{-}CH{=}C(Ph){-}SCN$

Equimolar amounts of thiocyanogen and phenylacetylene in benzene UV-irradiated under N_2 for 3 min. at 20–25° → (E)-1,2-dithiocyanato-1-phenylethylene. Y 98% (94% E-isomer). F.e.s. R.G. Guy et al., Tetrahedron *36,* 1839–42 (1980).

Potassium hydroxide/sodium sulfide *KOH/Na_2S*

Thiophenes from 1,3-diynes ○

with $NaOH/H_2S$ cf. *18,* 643; with KOH/Na_2S s. M.G. Voronkov et al., Khim. Geterotsikl. Soedin. *1981,* 1694.

Liq. ammonia *NH_3*

1-Alkylthio-1,3-enynes from 1,3-diynes C≡C → CH=C(SR)

s. *12,* 653s*36;* in liq. ammonia cf. Zh. Org. Khim. *18,* 269–74 (1982).

Quinine ←

Asymmetric synthesis thioethers from ethylene derivs. C=C → CHC(SR)

509. $PhCH{=}CHNO_2 + HSCH_2COOH \rightarrow Ph\overset{*}{C}HCH_2NO_2$ (SCH_2COOH on the starred C)

A soln. of (2-nitroethenyl)benzene in toluene added to a stirred mixture of thioglycolic acid and 1.09 equivalents quinine in the same solvent at room temp. under N_2, and stirring continued for 15 min. → (+)-product. Y 86% (enantiomeric excess 58%). F.e. and Cinchona alkaloids s. N. Kobayashi, K. Iwai, J. Org. Chem. *46,* 1823–8 (1981).

Chiral aminoalcohols ←

s. *34,* 513; with chiral aminoalcohols, 3-ketothioethers, s. T. Mukaiyama, A. Ikegawa, K. Suzuki, Chem. Letters *1981,* 165–8.

Calcium carbonate *$CaCO_3$*

2-Functionalized thioethers from ethylene derivs. and thiosulfonium salts via regio- and stereo-specific addition of nucleophiles ←

Oxysulfenylation

510. 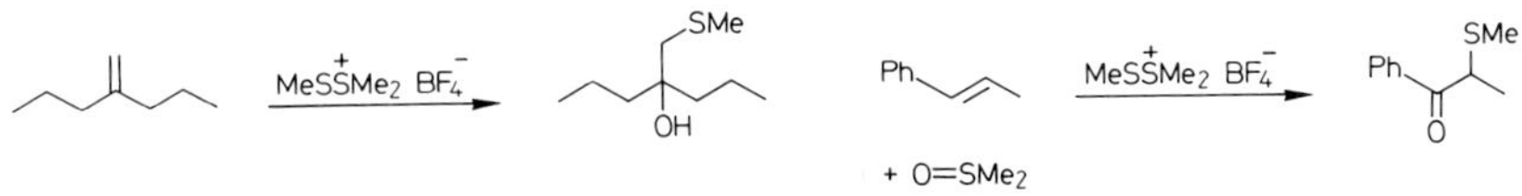

2-Hydroxythioethers. 2-Propyl-1-pentene

2-Ketothioethers. (E)-1-Phenylpropene

and dimethyl(methylthio)sulfonium fluoroborate mixed in methylene chloride

containing a little *dimethyl sulfide,* aq. $CaCO_3$ added at room temp., and worked up after 3–5 days → product. Y 97%. – Regiospecificity is reversed with acetate as nucleophile.

for 15 min., and dimethyl sulfoxide, HgO, and diisopropylethylamine added in rapid succession at room temp. → product. Y 81%.

F.e., also cyanosulfenylation with CN^- as nucleophile, and stereospecificity s. B.M. Trost, T. Shibata, J. Am. Chem. Soc. *104,* 3228–30 (1982); **azasulfenylation** s. ibid. *104,* 3225–8 (1982); *anti*-addition of tert. amines as evidence for **thiiranium salts as intermediates** cf. M.C. Caserio, J.K. Kim, ibid. *104,* 3231–3 (1982).

Mercuric oxide/dimethyl sulfoxide/ethyldiisopropylamine *$HgO/Me_2SO/i\text{-}Pr_2NEt$*

2-Ketothioethers from ethylene derivs. CH=C → C(O)C–SR

s. *37,* 510

Triethylaluminium *Et_3Al*

4-Arylthio-2-ethylenealcohols from 3-ethyleneoxido compds. C

Geospecific ring opening

511. 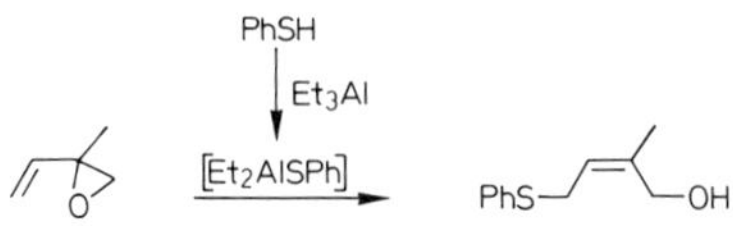

A mixture of 3,4-epoxy-3-methyl-1-butene and 2 equivalents diethylaluminium benzenethiolate (prepd. *in situ* from triethylaluminium and thiophenol) in benzene allowed to react at room temp. for 1 hr. → product. Y 92% (98% *cis*-isomer). F.e. and C-alkylation s. A. Yasuda, M. Takahashi, H. Takaya, Tetrahedron Letters *22,* 2413–6 (1981).

Azodiisobutyronitrile *$Me_2C(CN)N{=}NC(CN)Me_2$*

γ-Alkylthio-β-lactones from mercaptans C=C → CHC(SR)

s. *28,* 534; review of addition reactions to *exo*-methylene bond of diketene s. T. Kato, T. Chiba, Yuki Gosei Kagaku Kyokaishi *39,* 733–40 (1981) (Japan); Chem. Abstr. *95,* 168405 (1981).

Ethyldiisopropylamine s. under HgO *$(i\text{-}Pr)_2NEt$*

Alkaloids ←

Thiolic acid esters from ethylene derivs. C=C → CHCSCOR

with tri-*n*-butylamine cf. *21,* 641; with alkaloids, **asym. Michael addition** to α,β-ethyleneketones, s. J. Gawronski, K. Gawronska, H. Wynberg, J. Chem. Soc. Chem. Commun. *1981,* 307–8.

Dimethyl sulfoxide s. under HgO *Me_2SO*

Lithium perchlorate *$LiClO_4$*

2-Sulfoniothioethers from ethylene derivs. C=C → C(SR)C–$\overset{+}{S}$<
and thiosulfonium salts cf. *21,* 639; from thioethers and sulfenyl chlorides in the presence of $LiClO_4$ s. I.V. Bodrikov et al., Zh. Org. Khim. *17,* 1113–4 (1981).

Hydrogen chloride *HCl*

3-Oxoxanthates from *α,β*-ethyleneoxo compds. C=C → CHCSC(S)OR

512. EtOCSSK + CH_2=CHCHO → EtOC(S)SCH_2CH_2CHO

10% HCl added to a soln. of K O-ethylxanthate in water at 5°, acrolein added (pH 2), and the mixture stirred for 1.5 hrs. → product. Y 85%. F.e.s. S.V. Tsarenko et al., Izv. Vyssh. Uchebn. Zaved., Khim. Khim. Tekhnol. *23,* 836–8 (1980); Chem. Abstr. *94,* 30113 (1981). **3-Ketodithiocarboxylic acid esters** cf. J. Ramachandran, S.R. Ramadas, C.N. Pillai, Org. Prep. Proc. Intern, *13,* 71–9 (1981).

Rearrangement ∩

Hydrogen/Carbon Type SC∩HC

Sodium hydride *NaH*

3-(Alkylthio)thioenolethers from ketenemercaptals ←
1,3-Alkylthio group migration

513.

A soln. of startg. ketene dithioacetal in dry dimethylformamide added dropwise to a well stirred suspension of NaH in the same solvent, and the mixture stirred at 50–60° for 4 hrs. → product. Y 70%. F.e.s. S. Apparao et al., Tetrahedron Letters *23,* 971–4 (1982).

Lithium diisopropylamide *i-Pr_2NLi*

Epimerization of sulfoxides ←
s. *22,* 608; of sulfones with Li diisopropylamide cf. G. Sauer et al., Ann. Chem. *1982,* 448–58.

Oxygen/Sulfur Type SC∩OS

Without additional reagents *w.a.r.*

1,3-Dithiole-2-carboxaldehydes from 1,4-dithiin 1-oxides

514.

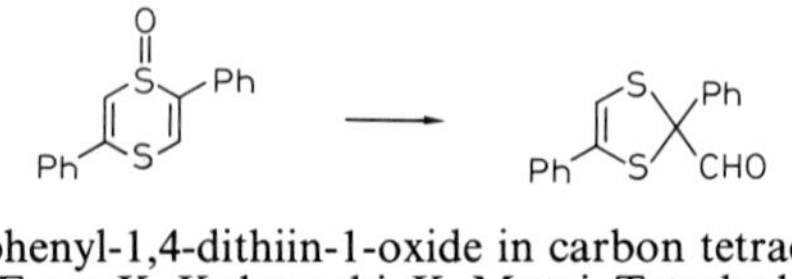

A dil. soln. of 2,5-diphenyl-1,4-dithiin-1-oxide in carbon tetrachloride heated at 75° → product. Y 100%. F.e.s. K. Kobayashi, K. Mutai, Tetrahedron Letters *22,* 5201–4 (1981).

Oxygen/Carbon Type **SC∩OC**

Sodium carbonate *Na_2CO_3*

2-Thiopyrones from 2-pyranthiones ○

515. 

A suspension of 4,6-diphenyl-2*H*-pyran-2-thione in methanol refluxed for 1 hr. with 10% aq. Na_2CO_3 → 4,6-diphenyl-2*H*-thiopyran-2-one. Y 85%. F.e.s. I.E. El-Kholy et al., J. Heterocyc. Chem. *18,* 105–10 (1981).

Trifluoroacetic acid *CF_3COOH*

Dithiolcarbonic acid esters from xanthates OCSSR → SCOSR
s. *18,* 649; in trifluoroacetic acid with inversion of configuration s. M.W. Fichtner, N.F. Haley, J. Org. Chem. *46,* 3141–3 (1981).

Nitrogen/Carbon Type **SC∩NC**

Sulfuric acid *H_2SO_4*

2-Thio-1,3-N,N-from 2-imino-1,3-S,N-heterocyclics ○
with MeCOOH cf. *20,* 443; with H_2SO_4, 2-amino-4*H*-3,1-benzothiazin-4-ones, s. M.R. Chaurasia, S.K. Sharma, Heterocycles *16,* 621–9 (1981).

Carbon/Carbon Type **SC∩CC**

Potassium tert-*butoxide* t-*BuOK*

1,2-Bis(alkylthio)-1,3-dienes from 1,4-bis(alkylthio)-2-acetylenes ←
1,3-Alkylthio group migration

516. $MeSCH_2{-}C{\equiv}C{-}CH_2SMe \rightarrow MeSCH{=}\underset{SMe}{C}{-}CH{=}CH_2$

A vigorously stirred soln. of K *tert*-butoxide in dimethyl sulfoxide treated with 1,4-bis(methylthio)-2-butyne maintaining the temp. below 35°, and the mixture stirred for 12 hrs. → 1,2-bis(methylthio)-1,3-butadiene. Y 85%. F.e. and conversion to **aromatic thioethers by diene synthesis with acetylene derivs.** s. M.E. Garst, P. Arrhenius, Syn. Commun. *11,* 481–7 (1981).

Exchange ↓↑

Hydrogen ↑ **SC↓↑H**

Without additional reagents *w.a.r.*

4-Imino-4*H*-1,3-oxathiin-2-ones
from *β*-ketocarboxylic acid thioamides ○

517. 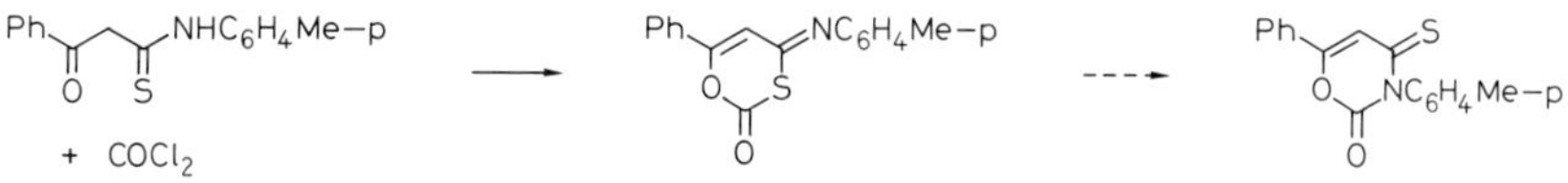

A 2% soln. of phosgene in toluene added gradually to a soln. of startg. benzoylthioacetanilide in the same solvent, and kept at 60–65° for 15–30 min. → 4-*p*-

tolylimino-6-phenyl-1,3-oxathiin-2-one. Y 90%. F.e., also rearrangement to **1,3-oxazin-2-one-4-thiones,** s. W. Zankowska-Jasinska, Z. Kamela, U. Zieba, Pol. J. Chem. *53,* 1251–7 (1979).

5-Subst. 1,3,4-thiadiazol-2(3*H*)-ones from phosgene
5-alkylthio-derivs. cf. *28,* 539; 5-amino derivs. from thiosemicarbazides s. H. Ogura, H. Takahashi, O. Sato, Chem. Pharm. Bull. *29,* 2188–92 (1981).

Thionyl chloride *$SOCl_2$*
Isothiazoles
s. *24,* 612; isothiazol-3-ones cf. R.J.S. Beer, D. Wright, Tetrahedron *37,* 3867–70 (1981).

Oxygen ↑ SC⇃↾O

Sodium methoxide *NaOMe*
Trithiocarbonic acid esters from acoxy compds. OAc → SC(S)SR

518.
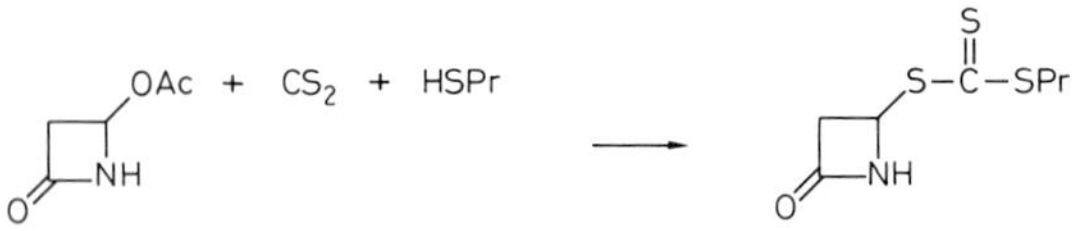

Propyl mercaptan and carbon disulfide added sequentially to a soln. of Na methoxide in methanol at 0°, stirred for 5 min., 4-acetoxy-2-azetidinone added, and stirring continued for 1 hr. at room temp. → 4-[(propylthio)thiocarbonyl]thio-2-azetidinone. Y 74%. F.e.s. S. Oida et al., Chem. Pharm. Bull. *28,* 3232–43 (1980).

Potassium tert-*butoxide* t-*BuOK*
Cyclic thioethers from methanesulfonic acid esters ○
with K thioacetate cf. *36,* 531; with K *tert*-butoxide s. Tetrahedron Letters *23,* 913–4 (1982).

Lithium diisopropylamide *i-Pr_2NLi*
Optically active sulfoxides S(O)OR → S(O)R′
from sulfinyl esters and organolithium compds.
1-sulfinylthioethers cf. *34,* 524; α-cyanosulfoxides with Li diisopropylamide, S→C chirality transfer, s. R. Annunziata et al., J. Chem. Soc. Perkin Trans. I *1981,* 614–7; β-amino-α,β-ethylenesulfoxides s. ibid. *1982,* 1183–6.

Potassium carbonate s. under $(EtO)_2P(O)CN$ *K_2CO_3*

Sodium acetate *MeCOONa*
Benzothiazolines from *o*-aminomercaptans and aldehydes ○
s. *8,* 657; with fused Na acetate cf. H.K. Gakhar, S. Chand, N. Kumar, Monatsh. Chem. *112,* 1083–90 (1981).

Lithium mercaptides *RSLi*
Thioenolethers from enol phosphates $C{=}COPO(OR)_2$ → C=CSR
β-Arylthio-α,β-ethyleneketones

519.

A soln. of Li benzenethiolate in tetrahydrofuran treated with a soln. of startg. enol phosphate in the same solvent at 0°, and stirred at room temp. for 12 hrs. → product. Y 88%. F.e., also synthesis of ethylene derivs. from alanes, s. N. Fukamiya, M. Oki, T. Aratani, Chem. Ind. *1981,* 606–7.

Sodium salt Na^+

2-Ethylenethioethers from allyl phosphates and mercaptans $OPO(OR)_2 \rightarrow SR'$

with Me_3Al cf. *32,* 537; with Na thiophenoxide and f. nucleophiles s. S. Araki, K. Minami, Y. Butsugan, Bull. Chem. Soc. Japan *54,* 629–30 (1981).

Triethylamine s. under Ph_3P, N-(Diphenoxyphosphoryl)-Δ^4-2-oxazolone, and 2,4,6-Trichlorobenzoyl chloride Et_3N

Pyridine s. under $Me_2NP(O)Cl_2$ C_5H_5N

4-Dimethylaminopyridine s. under 2,4,6-Trichlorobenzoyl chloride and $C_6H_{11}N{=}C{=}NC_6H_{11}$ ←

Ethylmagnesium bromide *EtMgBr*

Thioethers from oxo compds. and mercaptans SH → SR

s. *37,* 520

Trimethylaluminium Me_3Al

Benzothiazoles from *o*-aminomercaptans and carboxylic acid esters ○

s. *37,* 326

Pyridine-borane/trifluoroacetic acid $C_5H_5N.BH_3/CF_3COOH$

Thioethers from oxo compds. and mercaptans SH → SR

520. $$PhCHO + HS(CH_2)_6Me \rightarrow PhCH_2S(CH_2)_6Me$$

A soln. of 1-heptanethiol in methylene chloride added to a mixture of benzaldehyde and trifluoroacetic acid with cooling, stirred for 2–3 min., pyridine-borane added, and stirring continued for a further 5 min. → benzyl heptyl sulfide. Y 69%. F.e.s. Y. Kikugawa, Chem. Letters *1981,* 1157–8; with ethylmagnesium bromide as reagent cf. M.T. Cocco, A. Plumitallo, M. Secci, Gazz. Chim. Ital. *111,* 519–20 (1981).

Boron chloride s. under $R_3SnSSnR_3$ BCl_3

Dicyclohexylcarbodiimide/4-dimethylaminopyridine ←

Thiolic acid esters from carboxylic acids and mercaptans SH → SAc

with dimethylformamide cf. *17,* 663; in the presence of 4-dimethylaminopyridine, prevention of side reactions, s. B. Neises, W. Steglich, T. Van Ree, South African J. Chem. *34,* 58 (1981) (Eng.); Chem. Abstr. *95,* 186795 (1981).

Trifluoroacetic anhydride s. under CF_3SO_3H $(CF_3CO)_2O$

Formic acid *HCOOH*

1-Tosylcarbodiimides from sulfinic acids and thioureas via 1-tosylthioureas $S(O)OH \rightarrow SO_2CH_2NHCSN\lessdot$

521. $$p\text{-}MeC_6H_4S(O)OH + H_2NC(S)NHCPh_3 \xrightarrow{CH_2O} p\text{-}MeC_6H_4S(O)_2CH_2NHC(S)NHCPh_3 \rightarrow p\text{-}MeC_6H_4S(O)_2CH_2N{=}C{=}NCPh_3$$

A soln. of N-(triphenylmethyl)thiourea, 36% formalin, Na *p*-toluenesulfinate, and formic acid in 1,2-dimethoxyethane refluxed for 1.5 hrs., and kept at 0° for 30 min. → N-(tosylmethyl)-N'-(triphenylmethyl)thiourea (Y 70%), in acetone refluxed for 30 min. with HgO, a second portion of the latter added, and heating continued for a further 30 min. → N-(tosylmethyl)-N'-(triphenylmethyl)carbodiimide (Y 90%). F.e.s. A.M. van Leusen et al., J. Org. Chem. *46,* 2069–72 (1981).

Acetic acid s. under HCl *MeCOOH*

Trifluoroacetic acid s. under $C_5H_5N.BH_3$ *CF_3COOH*

Carbon tetrachloride s. under Ph_3P *CCl_4*

2,4,6-Trichlorobenzoyl chloride/triethylamine/4-dimethylaminopyridine ←

Thiolic acid esters from carboxylic acids and mercaptans SH → SAc

with ethyl chloroformate/N-ethylpiperidine in pyridine cf. *8,* 528; with 2,4,6-trichlorobenzoyl chloride/triethylamine/4-dimethylaminopyridine s. Y. Kawanami et al., Bull. Chem. Soc. Japan *54,* 943–4 (1981).

Hexamethyldisilthiane *$(Me_3Si)_2S$*

Replacement of carbonyl oxygen by sulfur CO → CS

s. *37,* 522

Hexacyclohexyldistannathiane/boron chloride *$(R_3Sn)_2S/BCl_3$*

522.

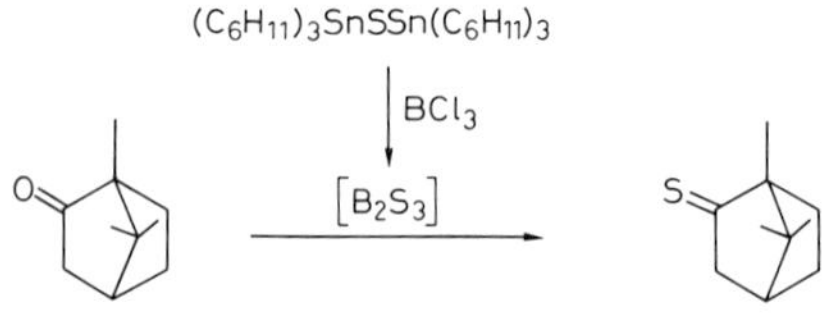

Hindered thioketones. 0.8 *M* BCl_3 in dry toluene injected into a stirred soln. of bis(tricyclohexyltin) sulfide and *DL*-camphor in the same solvent under N_2, and the mixture refluxed for 2.5 hrs. → *DL*-thiocamphor. Y 95%. F.e., incl. **thiolactams and thiolactones,** also with hexamethyldisilthiane, s. K. Steliou, M. Mrani, J. Am. Chem. Soc. *104,* 3104–6 (1982).

Triphenylphosphine (s.a. under $Pd(acac)_2$) *Ph_3P*

Triphenylphosphine/carbon tetrachloride/triethylamine *$Ph_3P/CCl_4/Et_3N$*

Δ^2-Thiazolines from 2-aminomercaptans and carboxylic acids ○

s. *37,* 332

Diethyl phosphorocyanidate/potassium carbonate *$(EtO)_2P(O)CN/K_2CO_3$*

Thiocyanates from sulfinic acids S(O)OH → SCN

523. $PhSO_2Na + (EtO)_2P(O)CN \rightarrow PhSCN$

A soln. of diethyl phosphorocyanidate in tetrahydrofuran added to Na phenylsulfinate and K_2CO_3, and the mixture refluxed for 1 hr. → phenyl thiocyanate. Y 74%. F.e.s. S. Harusawa, T. Shioiri, Tetrahedron Letters *23,* 447–8 (1982).

2,4-Bis(alkylthio)-1,3,2,4-dithiadiphosphetane-2,4-disulfides ←

Dithiocarboxylic acid esters from carboxylic acids COOH → CSSR

524.

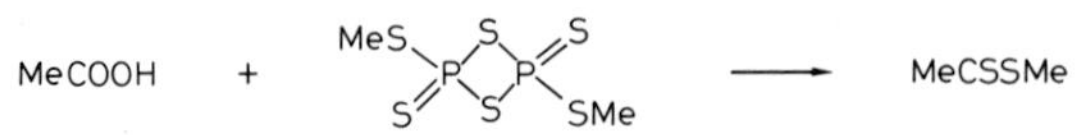

A mixture of 2 equivalents acetic acid and 1.1 equivalents 2,4-bis(methylthio)-1,3,2,4-dithiadiphosphetane-2,4-disulfide in 1,2,4-trichlorobenzene heated rapidly at 130° for 10 min. → methyl dithioacetate. Y 65%. F.e., also prepn. of reagent from P_4S_{10} and thiols, s. H. Davy, J. Chem. Soc. Chem. Commun. *1982,* 457–8.

Dimethylaminophosphoryl dichloride/pyridine *$Me_2NP(O)Cl_2/C_5H_5N$*
Thiolic acid esters from carboxylic acids and mercaptans SH → SAc
cf. *33,* 534s*34;* with pyridine instead of triethylamine s. H.-J. Liu, S.I. Sabesan, Can. J. Chem. *58,* 2645–8 (1980).

N-(Diphenoxyphosphoryl)-Δ^4-2-oxazolone/triethylamine ←
s. *37,* 333

Polyphosphoric acid *$(HPO_3)_n$*
with $POCl_3$/pyridine cf. *15,* 470; with polyphosphoric acid, hindered compds., s. T. Imamoto, M. Kodera, M. Yokoyama, Synthesis *1982,* 134–6.

Phosphorus pentasulfide *P_2S_5*
Replacement of oxygen by sulfur in N-heterocyclics CO → CS
s. *6,* 636s*36;* 9-acridanethiones s. R.R. Smolders et al., Synthesis *1982,* 493–4.
Carboxylic acid thioamides from amides CON⋜ → CSN⋜
Ultrasonication

525. $PhCONMe_2 \rightarrow PhCSNMe_2$

P_4S_{10} added to a stirred soln. of N,N-dimethylbenzamide in tetrahydrofuran, the mixture irradiated in an 80 W. Bransonic 12 bath sonicator for 20–30 min. at 30–40°, a further portion of P_4S_{10} added, and sonication continued with stirring for an additional 30–90 min. → N,N-dimethylthiobenzamide. Y 78%. – This method has the advantage of shorter reaction time and lower temperatures, and avoids a large excess of P_4S_{10} by comparison with the standard procedure (Synth. Meth. *8,* 670). F.e.s. S. Raucher, P. Klein, J. Org. Chem. *46,* 3558–9 (1981).
1,3,4-Thiadiazoles from N,N′-diacylhydrazines ○
s. *37,* 538

Trifluoromethanesulfonic acid/trifluoroacetic anhydride *$CF_3SO_3H/(CF_3CO)_2O$*
Triarylsulfonium salts from arenes and diaryl sulfoxides ←

526.

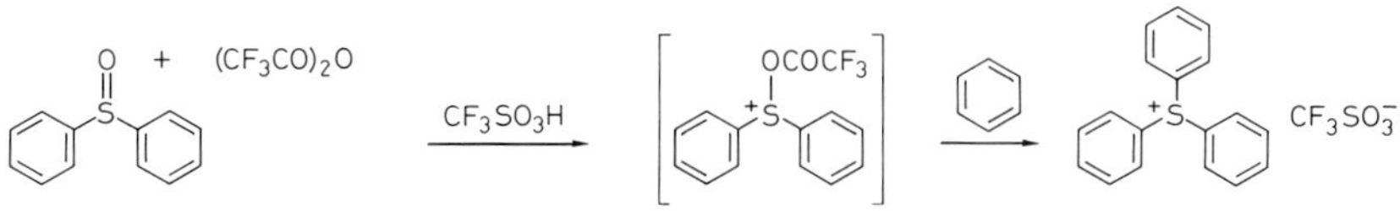

A soln. of diphenyl sulfoxide in benzene treated with 2 equivalents trifluoroacetic anhydride and 1 equivalent trifluoromethanesulfonic acid, the mixture stirred at 0° for 30 min., warmed to 20°, and the product separated off after 1 hr. → triphenylsulfonium triflate. Y 95%. F.e.s. Y Endo, K. Shudo, T. Okamoto, Chem. Pharm. Bull. *29,* 3753–5 (1982).

Sulfuric acid *H_2SO_4*
Mercaptals from acetals $C(OR)_2 \rightarrow C(SR)_2$
with *p*-toluenesulfonic acid cf. *36,* 540; with H_2SO_4 s. F.J.L. Aparicio, F.Z. Benitez, F.S. Gonzalez, Carbohyd. Res. *102,* 69–81 (1982).

Hydrogen chloride/acetic acid *HCl/MeCOOH*
Replacement of cyclic oxygen by sulfur ⊂O → ⊂S
s. *29,* 558; thiophenes from furans s. Zh. Org. Khim. *18,* 394–9 (1981).

Palladium(II) acetylacetonate/triphenylphosphine/triethylaluminium *Pd(acac)$_2$/Ph$_3$P/Et$_3$Al*

Sulfones from acoxy compds. and sulfinic acids OAc → SO_2R
β,γ-Ethylenesulfones

527.

A mixture of benzenesulfinic acid and 3-acetoxy-1,7-octadiene heated at 100° in toluene for 5 hrs. in the presence of a little Pd(II) acetylacetonate, triphenylphosphine, and triethylaluminium → 3-phenylsulfonyl-1,7-octadiene. Y 98%. F.e.s. U.M. Dzhemilev et al., Izv. Akad. Nauk SSSR Ser. Khim. *1981,* 696–7.

Via intermediates *v.i.*

Thioethers from alcohols via chlorides OH → SR
s. *28,* 547; with isolation of intermediate chlorides s. V.E. Eswarakrishnan, L. Field, J. Org. Chem. *46,* 4182 (1981).

Nitrogen ↑ SC⇃↾N

Without additional reagents *w.a.r.*

Carboxylic acid thioamides from amidines C(=N–)N< → C(S)N<
s. *13,* 604; thioformamides s. M. Tišler et al., Synthesis *1981,* 299–300.

Radical S-arylation ←
Thioiminoesters from carboxylic acid thioamides

528.
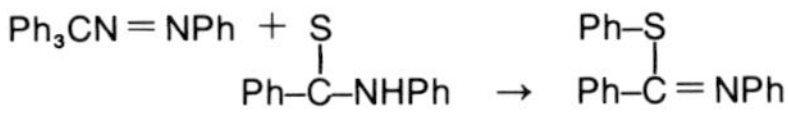

A soln. of startg. thioamide and phenylazotriphenylmethane in acetone refluxed with stirring in a stream of N_2 for 8 hrs., and left overnight → product. Y 85%. F.e.s. I.I. Kandror, I.O. Bragina, R.K. Freidlina, Dokl. Akad. Nauk SSSR *249,* 867–70 (1979); **1-aminothioenolethers** from N,N-disubst. amides cf. Izv. Akad. Nauk SSSR Ser. Khim. *1981,* 1167–9; **S-arylisothiosemicarbazides** s. ibid. *1981,* 1329–31.

Potassium thiocyanate *KSCN*

5-Imino-Δ²-1,3,4-thiadiazolines from 1,1-nitrohydrazones ○
s. *37,* 754

Cuprous chloride *CuCl*

Sulfonyl chlorides from amines NH_2 → SO_2Cl
with $CuCl_2$ cf. *12,* 671; *m*-trifluoromethylbenzenesulfonyl chloride with CuCl s. R.V. Hoffman, Org. Synth. *60,* 121–6 (1981).

Trifluoroacetic acid *CF_3COOH*

Diaryl sulfides from aromatic sulfenimides and arenes $ArSNAc_2$ → ArSAr′
Aromatic sulfenium salts as intermediates

529. $Ac_2NSC_6H_4\text{–}NO_2\text{-}p \xrightarrow{CF_3COOH} [CF_3CO_2^- \ \overset{+}{S}C_6H_4\text{–}NO_2\text{-}p] \xrightarrow{MeOPh} p\text{-}MeO\text{–}C_6H_4SC_6H_4\text{–}NO_2\text{-}p$

A soln. of N,N-diacetyl-*p*-nitrophenylsulfenimide in dichloromethane treated with anisole in the presence of 2.6 equivalents dry trifluoroacetic acid at room temp. →

4-methoxyphenyl 4′-nitrophenyl sulfide (Y 96%) and diacetamide (Y 99%). F.e., **also 2-acoxythioethers from ethylene derivs.,** s. R.A. Abramovitch. J. Pilski, J. Chem. Soc. Chem. Commun. *1981,* 704–5; ar. sulfenium salts from 1-arylthiopyridinium salts cf. ibid. *1981,* 703–4.

Hydrogen sulfide/sulfuric acid *H_2S/H_2SO_4*

Thionoformic acid esters from formiminoesters HC(=N–)OR → HC(=S)OR

s. *37,* 55

Rhodium(II) acetate *$(MeCOO)_4Rh_2$*

2-Ketothioethers from α-diazoketones and mercaptans $C=N_2$ → CHSR

530. $MeCOCH=N_2 + HSPh \rightarrow MeCOCH_2SPh$

A soln. of diazoacetone in dry benzene added slowly during 10 min. to a stirred mixture of thiophenol and a little rhodium(II) acetate in the same solvent at room temp., and stirring continued for a further 25 min. → product. Y 84%. – The method avoids regiospecific α-functionalization on which most other methods are based. F.e.s. M.A. McKervey, P. Ratananukul, Tetrahedron Letters *23,* 2509–12 (1982).

Via intermediates *v.i.*

Replacement of prim. amino groups by S-nucleophiles NH_2 → S–

thiocyanates s. *33,* 195s*36;* with 2-methylthio-4,6-diphenylpyridine as leaving group, **also thioethers and mercaptans,** cf. P. Molina et al., Synthesis *1982,* 472–5.

Halogen ↑ SC⇂↾Hal

Without additional reagents *w.a.r.*

Diaryl sulfides from ar. sulfenyl chlorides and arenes ArSHal → ArSAr′

with SiO_2 cf. *31,* 558; with $SnCl_4$ or without reagent s. F. Bottino, R. Fradullo, S. Pappalardo, J. Org. Chem. *46,* 2793–5 (1981).

2-Ethylenethioethers from ethylene derivs. and sulfenyl chlorides with double bond migration C=C–CH → RSC–C=C

531. $PhSCl + CH_2=CMe_2 \rightarrow [PhSCH_2C(Cl)Me_2] \rightarrow PhSCH_2C(Me)=CH_2$

Solns. of benzenesulfenyl chloride and 2-methyl-prop-1-ene in methylene chloride mixed at –20°, warmed to room temp. during 1–2 hrs., the solvent removed *in vacuo,* and the residue pyrolysed at 130° until evolution of HCl ceased → product. Y 76%. F.e. and regiospecificity s. B. Giese, P. Mazumdar, Chem. Ber. *114,* 2859–65 (1981).

Phenyllithium *PhLi*

1-Alkylthio-1,3-enynes from 3-iodothiophenes C

532.

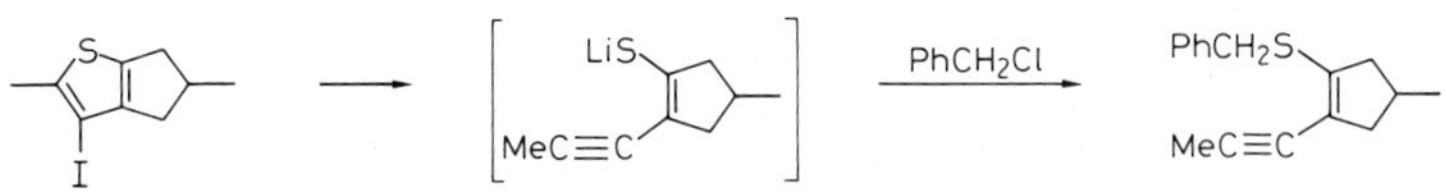

Phenyllithium in ether added via syringe to a stirred soln. of startg. 3-iodothiophene in the same solvent under N_2, after 15 min. benzyl chloride added, and left for 2 hrs. → 1-(benzylthio)-4-methyl-2-(1-propynyl)cyclopentene. Y 61%. F.e.s. T. Frejd, J.O. Karlsson, S. Gronowitz, J. Org. Chem. *46,* 3132–5 (1981); 1-alkylthio-4-silyl-1,3-enynes s. Chem. Scripta *18,* 192–4 (1981).

Lithium diisopropylamide *i-Pr₂NLi*

Thiodienolethers from 2,5-dihydrothiophene 1-oxides
with Mg cf. *24,* 636; with Li diisopropylamide, also 2,4-diene-sulfoxides and -sulfones, s. R.L. Crumbie, D.D. Ridley, Australian J. Chem. *34,* 1017–26 (1981).

Enetrithiocarbonic acid esters
from 1,3-dithiolane-2-thiones and halides

533.

A stirred soln. of Li diisopropylamide in tetrahydrofuran at –78° treated with 1,3-dithiolane-2-thione in the same solvent, the mixture kept at the same temp. for 40 min., methyl iodide added, and stirring continued at –78° for 2 hrs. → product. Y 87%. F.e.s. K. Hatanaka, S. Tanimoto, Tetrahedron Letters *23,* 425–6 (1982).

Sodium/liq. ammonia *Na/NH_3*

1,3-Dithioles from 1-(chloromethylthio)-1-acetylenes ○
s. *36,* 553; **review of 1,3-dithioles and 1,3-dithiolium salts** s. V.Y. Khodorvskii, O. Neilands, Latv. PSR Zinat. Akad. Vestis, Kim. Ser. *1982,* 131–58 (Russ.); Chem. Abstr. *97,* 23651 (1982).

Sodium isopropylmercaptide *$NaSCHMe_2$*

Ar. mercaptans from halides ArHal → ArSH
with Na methylmercaptide cf. *7,* 638s*36;* with Na isopropylmercaptide s. J. Org. Chem. *46,* 3070–3 (1981).

Sodium sulfide *Na_2S*

Cyclic thioethers from dibromides ○
High concentration method

534.

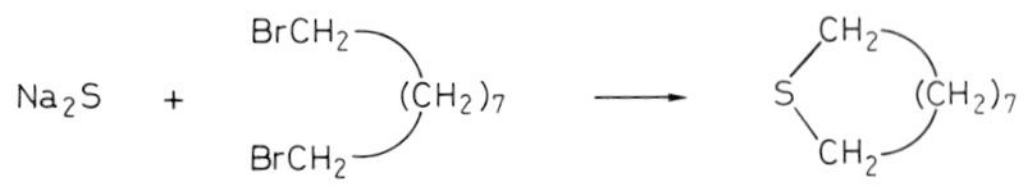

A soln. of 1,9-dibromononane in a small volume of ethanol treated with solid $Na_2S.9H_2O$, and the heterogeneous mixture stirred at room temp. for 12 hrs. → thiacyclodecane. Y 62%. – Previous syntheses of medium ring thiacycloalkanes have required high dilution techniques giving only moderate yields. F.e.s. A. Singh, A. Mahrotra, S.L. Regen, Syn. Commun. *11,* 409–11 (1981).

Sodium sulfide/triethylhexadecylammonium chloride *$Na_2S/[C_{16}H_{33}NEt_3]^+Cl^-$*

Thietanes from 1,3-dihalides
with H_2S cf. *31,* 557; with Na_2S and triethylhexadecylammonium chloride as phase transfer catalyst s. M. Lancaster, D.J.H. Smith, Synthesis *1982,* 582–3.

Sodium disulfide/
tributylhexadecylphosphonium bromide *$Na_2S_2/[C_{16}H_{33}P(C_4H_9)_3]^+Br^-$*

Sym. acyl disulfides from carboxylic acid chlorides 2 RCOCl → RCOS–SCOR
s. *5,* 454; with Na_2S_2 and tributylhexadecylphosphonium bromide as phase transfer catalyst cf. M. Kodomari, M. Fukuda, S. Yoshimoto, Synthesis *1981,* 637–8.

Sodium hydrogen sulfide *NaHS*

Partial replacement of chlorine by sulfhydryl Cl → SH
in liq. NH_3 cf. *24,* 634; in dimethyl sulfoxide s. L.S. Sologub, A.F. Pavlenko, T.I. Cherepenko, Fiziol. Akt. Veshchestva *12,* 83 (1980) (Russ.); Chem. Abstr. *95,* 6697 (1981).

Potassium salt/18-crown-6 polyether ←

Xanthates from halides Cl → SC(S)OR

with K salt cf. *20,* 459; with added 18-crown-6 polyether, also from alcohols (cf. *15,* 652), s. R. Chênevert, R. Paquin, A. Rodrigue, Synth. Commun. *11,* 817–21 (1981); also with Na salt s. S.V. Tsarenko et al., Izv. Vyssh. Uchebn. Zaved., Khim. Tekhnol. *24,* 284 (1981); Chem. Abstr. *95,* 96942 (1981).

Tri-n-butylamine s. under $K_2S_2O_5$ n-Bu_3N

Pyridine (s.a. under P_2S_5*)* C_5H_5N

1-Siloxymercaptans from aldehydes CHO → CH(SH)OSi≡

535. $$PrCHO + ClSiMe_3 \xrightarrow{H_2S} PrCH(SH)OSiMe_3$$

H_2S bubbled into a mixture of butanal, trimethylchlorosilane, and pyridine in dichloromethane, stirred vigorously below 15° until no further gas absorbed, and stirring continued at room temp. for 2 hrs. → product. Y 73%. F.e., also oxidation with MnO_2 to **1,1'disiloxydisulfides,** and conversion to **1-siloxythioethers** with ethylene derivs., s. T. Aida, T.-H. Chan, D.N. Harpp, Angew. Chem. Intern. Ed. *20,* 691–2 (1981).

Cuprous chloride/triethylamine hydrochloride *CuCl/*$Et_3N.HCl$

Sulfones from sulfonyl halides $SO_2Hal \rightarrow SO_2R$

with $SbHal_3$ cf. *30,* 411; with CuCl and triethylamine hydrochloride, 1,3-dienesulfones, s. M.M. Tanaskov, M.D. Stadnichuk, A.A. Petrov, Zh. Org. Khim. *17,* 1115 (1981).

Magnesium *Mg*

Sulfoxides from sulfinyl esters S(O)OR → S(O)R
Optically active compds.

s. *20,* 463; from carbohydrate derivs. s. D.D. Ridley, M.A. Smal, Australian J. Chem. *35,* 495–507 (1982); s.a. J. Chem. Soc. Chem. Commun. *1981,* 505–6; improved procedure s. J. Drabowicz, B. Bujnicki, M. Mikotajczyk, J. Org. Chem., *47,* 3325–7 (1982).

Sulfones from sulfonyl halides $SO_2Hal \rightarrow SO_2R$

from fluorides cf. *34,* 547; from chlorides s. E.V. Polunin, A.M. Moiseenkov, A.V. Semenovsky, Izv. Akad. Nauk SSSR Ser. Khim. *1981,* 1354–8.

Ethylaluminium dichloride $EtAlCl_2$

β,γ-Ethylenesulfoxides
from ethylene derivs. and sulfinyl chlorides CH–C=C → C=C–C–S(O)R
Ene reaction

536.

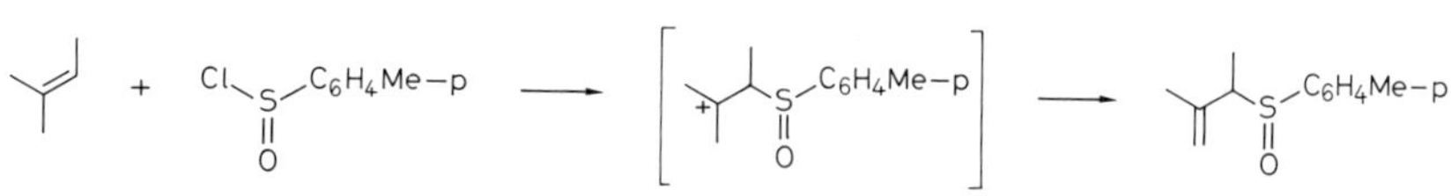

A soln. of ethylaluminium dichloride in heptane added to a soln. of 2-methyl-2-butene and *p*-toluenesulfinyl chloride in ether under N_2 at 0°, the soln. allowed to warm to 25°, stirred for 2 hrs., and worked up with aq. NaOH → product. Y 73%. F.e.s. B.B. Snider, J. Org. Chem. *46,* 3155–6 (1981).

Molecular sieve ←

Thiazoles from 1,2-oxido-1-halides and carboxylic acid thioamides ○

537.

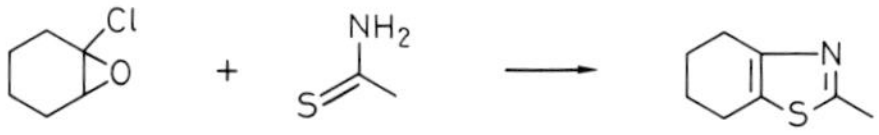

A soln. of thioacetamide in 4:1 dichloromethane-methanol added to a soln. of startg. epoxide in dichloromethane followed by molecular sieves 4Å, and the mixture heated at 40° for 24 hrs. → 2-methyl-4,5,6,7-tetrahydrobenzothiazole. Y 78%. F.e., **also 2-amino-thiazoles and -selenazoles from thioureas and selenoureas,** respectively, s. J. Gasteiger, C. Herzig, Tetrahedron *37,* 2607–11 (1981).

18-Crown-6 polyether s. under Potassium salt ←

Amberlyst A 26 mercaptide ←

Thioglycosides ←

with Na cf. *7,* 634; with Amberlyst A 26 mercaptide s. S.A. Abbas, J.J. Barlow, K.L. Matta, Carbohyd. Res. *100,* C33–4 (1982); from triflates cf. M. Blanc-Muesser, J. Defaye, H. Driguez, J. Chem. Soc. Perkin Trans. I *1982,* 15–8.

Tributylhexadecylphosphonium bromide s. under Na_2S_2 $[C_{16}H_{33}P(C_4H_9)_3]^+Br^-$

Phosphorus pentasulfide/pyridine P_2S_5/C_5H_5N

1,3,4-Thiadiazoles from acylhydrazines and carboxylic acid halides ○

538.

Startg. 4-subst. benzoyl chloride added slowly with stirring to a soln. of startg. 4-subst. benzoic acid hydrazide in pyridine at 0–5°, stirring continued at the same temp. for 45 min. and for further 45 min. at room temp., heated 3 hrs. at 80°, P_4S_{10} added at room temp., and heated again at 100° for 12 hrs. → product. Y 70–85%. F.e., **also from** the intermediate **N,N′-diacylhydrazines,** s. K. Dimitrowa et al., J. Prakt. Chem. *322,* 933–44 (1980).

*Potassium metabisulfite/tri-*n*-butylamine* $K_2S_2O_5$/n-Bu_3N

Sulfonic acids from halides Hal → SO_3H

with Na_2SO_3 cf. *1,* 488/9; *3,* 523; with $K_2S_2O_5$ in the presence of tert. amines, e.g. tri-*n*-butylamine, s. M. Gisler, H. Zollinger, Angew. Chem. Intern. Ed. *20,* 203–4 (1981).

Triethylamine hydrochloride s. under CuCl $Et_3N.HCl$

Triethylhexadecylammonium chloride s. under Na_2S $[C_{16}H_{33}NEt_3]^+Cl^-$

Nickel(II) bromide phosphine complex/sodium salt ←

Aryl thioethers from halides ArHal → ArSAr′

Diaryl sulfides

with $Pd(PPh_3)_4$/Na *tert*-butoxide cf. *36,* 561; with nickel(II) bromide phosphine complex and Na arylmercaptides s. H.J. Cristall et al., Synthesis *1981,* 892–4.

Via intermediates *v.i.*

Mercaptans from halides Hal → SH
via 3-alkylthio-1,2-benzisothiazole 1,1-dioxides

539. 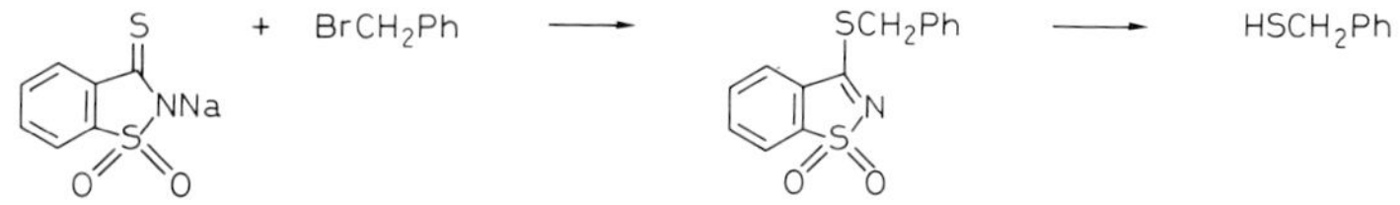

A soln. of benzyl bromide in dimethylformamide added to a stirred soln. of 1,2-benzisothiazol-3(2*H*)-thione 1,1-dioxide Na salt in the same solvent at room temp. under N_2, and the mixture stirred for 3 hrs. → 3-(benzylthio)-1,2-benzisothiazole 1,1-dioxide (Y 93%) in acetonitrile treated with piperidine at room temp. for 10 min. → benzyl mercaptan (Y 100%). F.e.s. K. Inomata, H. Yamada, H. Kotake, Chem. Letters *1981,* 1457–8.

Sulfur ↑ SC⇃↾S

Sodium hydroxide *NaOH*

1,3,4-Thiadiazoles from carboxylic acid thiohydrazides ○
s. *18,* 692; also isolation of intermediate Δ^2-1,3,4-thiadiazolines from oxo compds. cf. D.M. Evans, D.R. Taylor, J. Chem. Soc. Chem. Commun. *1982,* 188–9.

Δ^2-5-Thiazolones from (thioacylthio)acetic acids and α-aminocarboxylic acid esters

540. 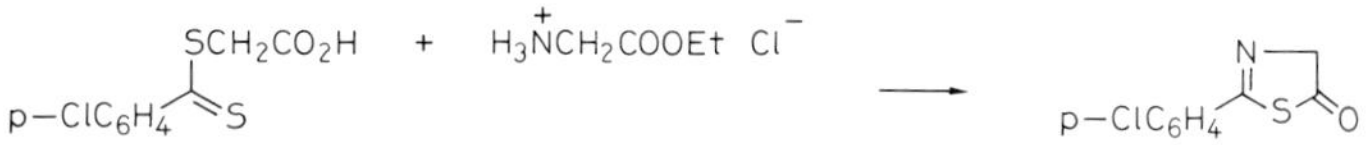

A soln. of startg. dithiobenzoate and ethyl glycinate hydrochloride in ethanol treated with 1 *M* aq. NaOH (pH ~9) at room temp., and the mixture stirred for 3 hrs. under slightly reduced pressure → 2-(*p*-chlorophenyl)-2-thiazolin-5-one. Y 82%. F.e.s. N. Suzuki, Y. Izawa, Org. Prep. Proc. Intern. *13,* 360–3 (1981).

Sodium ethoxide *NaOEt*

Thioethers from thiosulfonic acid esters H → SR
s. *25,* 456; α-(organothio)-carboxylic and -malonic acid esters s. K. Ogura et al., Bull. Chem. Soc. Japan *55,* 1216–20 (1982).

Potassium tert-*butoxide s. under* n-*BuLi* t-*BuOK*

Organolithium compd. *RLi*

Dithiocarbamic acid esters
from bisthiuram disulfides >N–C(S)S–]$_2$ → 2 >N–C(S)SR
and Grignard compds. cf. *26,* 620; and organolithium compds., also conversion to **arylmercaptans,** s. K.-Y. Jen, M.P. Cava, Tetrahedron Letters *23,* 2001–4 (1982).

n-*Butyllithium/potassium* tert-*butoxide* n-*BuLi*/t-*BuOK*

Electrophilic *o*-substitution of ethynylarenes H → SR
Dicarbanions as intermediates

541. 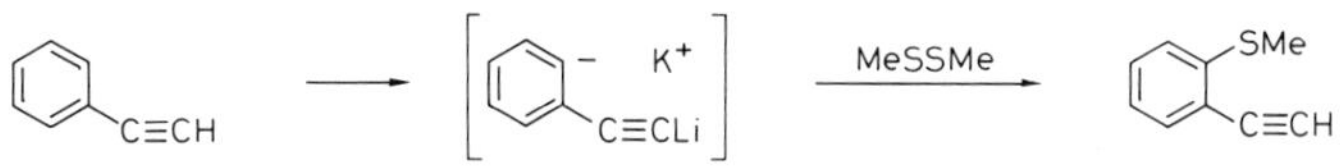

Phenylacetylene treated with 2 equivalents *n*-butyllithium in tetrahydrofuran and 1 equivalent K *tert*-butoxide in hexane, the mixture left at –20° for 1.5–2 hrs., cooled to −50° to −60°, and the dianion quenched with dimethyl disulfide → product. Y 78%. F. electrophiles, **also benzo[*b*]selenophene ring closure,** s. H. Hommes, H.D. Verkruijsse, L. Brandsma, J. Chem. Soc. Chem. Commun. *1981,* 366–7; *o*-halogenophenylacetylenes s. Tetrahedron Letters *22,* 2495–6 (1981).

Sodium amide *$NaNH_2$*

α-Alkylthiodithiocarboxylic acid esters from dithiocarboxylic acid esters

542.

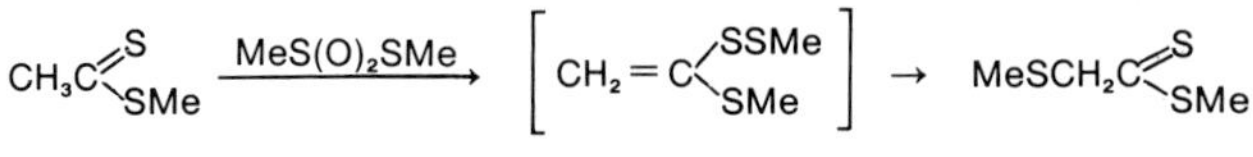

A well-stirred suspension of $NaNH_2$ in liq. NH_3 treated with startg. dithiocarboxylic acid ester, the soln. allowed to stand for 5 min., and poured cautiously into a stirred soln. of methyl methanethiosulfonate in the same solvent, the mixture heated at 50° until all NH_3 evaporated, and water added → product. Y 80%. F.e., and isolation of intermediate **1,2-ethylene-1,1-alkylthiodisulfides,** s. R.S. Sukhai et al., Rec. Trav. Chim. Pays-Bas *99,* 191–4 (1980).

Remaining Elements ↑ SC⇂↾Rem

Without additional reagents *w.a.r.*

Thiolic acid esters from trithiophosphites COCl → COSR

from carboxylic acids cf. *13,* 624; from carboxylic acid halides s. V.A. Alfonsov et al., Izv. Akad. Nauk SSSR Ser. Khim. *1982,* 1386–8.

Thioethers from stannanes and sulfenyl halides Sn⋚ → SR

543.

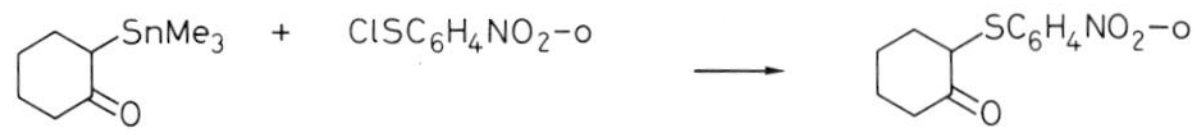

2-Ketothioethers. Startg. stannane and *o*-nitrophenylsulfenyl chloride in methylene chloride allowed to react in a sealed vessel for 10 min., and poured into water → 2-(*o*-nitrophenylthio)cyclohexanone. Y 97%. F.e. including **1-acetylene-1-thioethers** s. A.N. Kashin et al., Zh. Org. Khim. *16,* 2561–6 (1980).

Irradiation/trichloromethanesulfonyl chloride *$/Cl_3CSO_2Cl$*

Sulfonyl chlorides from alkyl cobaloximes ←

544.

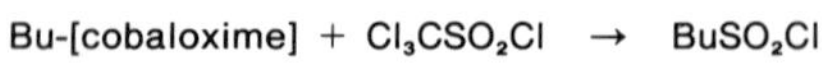

A mixture of startg. alkylcobaloxime and trichloromethanesulfonyl chloride in methylene chloride irradiated at ~10° in a Pyrex vessel using six 150 W tungsten spotlights for 210 min. → product. Y 95%. F.e.s. P. Bougeard, M.D. Johnson, G.M. Lampman, J. Chem. Soc. Perkin Trans. I *1982,* 849–54.

Sodium hydride s. under $NaBH_4$ *NaH*

Potassium fluoride/dicyclohexyl-18-crown-6 polyether ←

Thiolic acid esters from carboxylic acid halides and silyl sulfides COHal → COSR

s. *29,* 581; with KF and dicyclohexyl-18-crown-6 polyether cf. W. Ando et al., Chem. Letters *1982,* 885–6; without reagent cf. J.J. Talley, Synthesis *1981,* 549.

*Zinc iodide/tetra-*n*-butylammonium iodide* *ZnI_2/*n*-$Bu_4N^+I^-$*

Thioglycosides OR → SR

from alkylthiostannanes cf. *33,* 565; from silyl sulfides with ZnI_2/*n*-$Bu_4N^+I^-$ s. S. Hanessian, Y. Guindon, Carbohyd. Res. *86,* C3–6 (1980).

Sodium tetrahydridoborate/sodium hydride *$NaBH_4$/NaH*

Thietanes from α,β-ethyleneketones via β-(thiophosphorylthio)ketones C=C → CHCSP(S)(OR)$_2$

545. 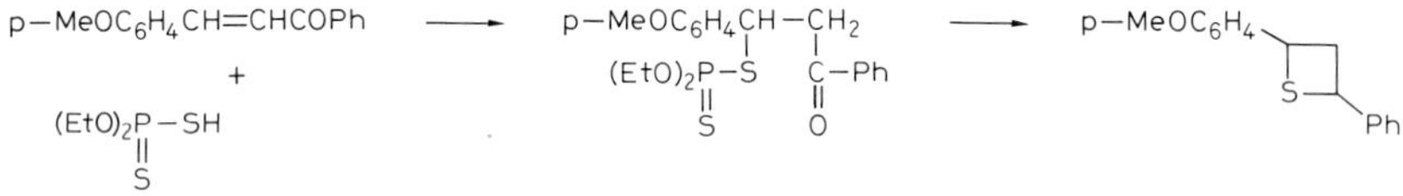

A soln. of startg. chalcone in benzene treated with a soln. of O,O-diethyl hydrogen phosphorodithioate in the same solvent at room temp., and the mixture stirred at 50–60° for 3 hrs. under N_2 → intermediate (Y 98%) in *tert*-butanol treated with $NaBH_4$, the mixture stirred at room temp. for 3 hrs. then at 60° for 1 hr. under N_2, NaH added at room temp., and stirring continued for 4 hrs. at the same temp. then for 1 hr. at 60° → product (Y 93%). F.e.s. Y. Ueno, L.D.S. Yadav, M. Okawara, Synthesis *1981,* 547–8.

Dicyclohexyl-18-crown-6 polyether s. under KF ←

Stannic chloride *$SnCl_4$*

β-Ketosulfoxides from enoxysilanes and sulfinyl chlorides C(OSi⋛)=C → COCS(O)R

546. 

A stirred soln. of 1-trimethylsilyloxycyclohexene in dichloromethane at –78° treated with a soln. of benzenesulfinyl chloride in the same solvent in one portion under argon, $SnCl_4$ added dropwise via syringe, and the mixture stirred at –78° for 30 min. → 2-phenylsulfinylcyclohexanone. Y 90%. – Introduction of a Lewis acid markedly promotes the reaction. F.e.s. N.A. Meanwell, C.R. Johnson, Synthesis *1982,* 283–4; **α-sulfinylcarbonyl compds.** s.a. V.N. Sergeev, G.S. Zaitseva, Y.I. Baukov, Zh. Obshch. Khim. *50,* 699–700 (1980).

Trimethylsilyl triflate *$Me_3SiOSO_2CF_3$*

Mercaptals from ketones and silyl sulfides CO → C(SR)$_2$

547. 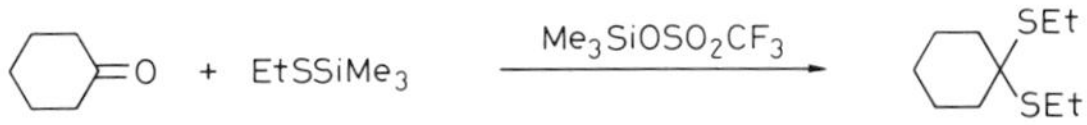

Treatment of cyclohexanone with 2 equivalents of ethylthiotrimethylsilane and a little trimethylsilyl triflate in methylene chloride at –78° for 10 hrs. → product. Y 84%. F. details of **reactions with trimethylsilyl triflate** s. R. Noyori, S. Murata, M. Suzuki, Tetrahedron *37,* 3899–910 (1981); **trialkylsilyl perfluoroalkanesulfonates as silylating agents and Lewis acids, review,** s. H. Emde et al., Synthesis *1982,* 1–26.

Trichloromethanesulfonyl chloride s. under ⫻ *Cl_3CSO_2Cl*

*Tetra-*n*-butylammonium iodide s. under ZnI_2* n-*$Bu_4N^+I^-$*

Via intermediates *v.i.*

Thioenolethers from enesilanes and sulfenyl chlorides via 2-chloro-1-silylsulfides C=CSi⋛ → C=CSAr

548. 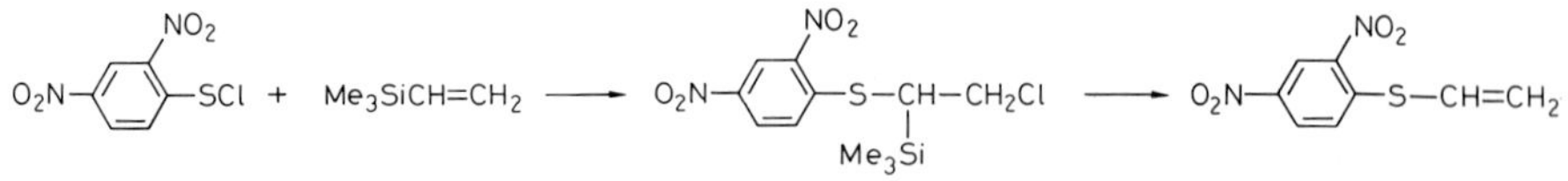

A suspension of 2,4-dinitrophenylsulfenyl chloride in dichloromethane treated with a soln. of trimethylvinylsilane in the same solvent at –76°, the mixture allowed to warm to 20°, and stirred for 15 hrs. → 2-chloro-1-(trimethylsilyl)ethyl 2,4-dinitrophenyl sulfide (95%), in 30% aq. ethanol stirred at 65° for 24 hrs. → 2,4-dinitrophenyl vinyl sulfide (Y 87%). F.e.s. F. Cooke et al., J. Org. Chem. *45,* 1046–53 (1980).

Carbon ↑ SC⇂↾C

Without additional reagents *w.a.r.*

1,4-Benzodithiins from 1,4,2-benzodithiazines and acetylene derivs. ○

549. 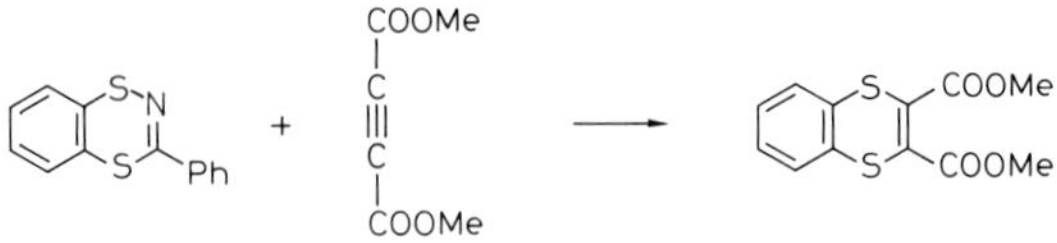

A mixture of 3-phenyl-1,4,2-benzodithiazine and 3 equivalents dimethyl acetylenedicarboxylate heated at 195° for 3 hrs. → 2,3-bis(methoxycarbonyl)-1,4-benzodithiin. Y 90%. F.e., **also from 2-azido-1,3-benzodithioles,** s. J. Nakayama et al., J. Chem. Soc. Chem. Commun. *1982,* 612–3.

Potassium hydroxide *KOH*

Cyclic 1-hydroxythioethers
s. *28,* 574; f. method from halogenoketones cf. G. Kavadias, S. Velkof, Can. J. Chem. *59,* 321–6 (1981).

Sodium methoxide *NaOMe*

Preferential S-dealkylation of aryl thioethers SR → SR′
s. *37,* 550

Sodium methylmercaptide *NaSMe*

550. 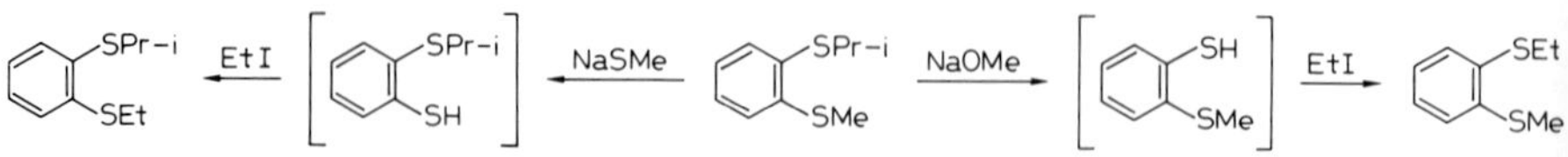

Transalkylation. A soln. of *o*-(isopropylthio)phenyl methyl sulfide in hexamethylphosphoramide stirred with

Na methylmercaptide at 100° for 6 hrs. | Na methoxide at 120° for 15 hrs.

under N_2, the mixture cooled, and quenched with excess ethyl iodide

→ *o*-(ethylthio)phenyl isopropyl sulfide. Y 95%. | → *o*-(methylthio)phenyl ethyl sulfide. Y 87%.

F.e.s. M. Tiecco et al., Synthesis *1982,* 478–80.

Potassium salt K^+

Thiolic acid esters COCl → COSR

from carboxylic acid chlorides and potassium O-alkyl xanthates

551.

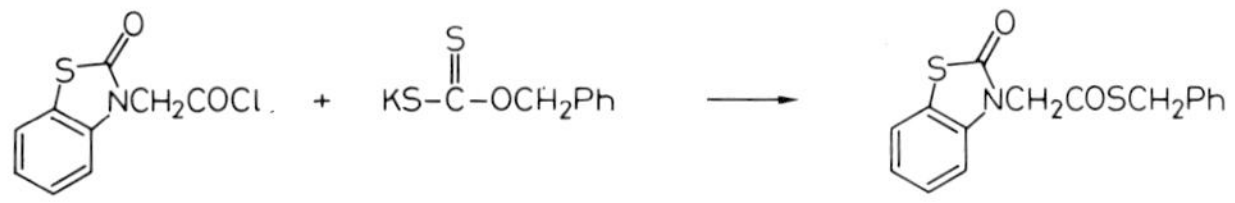

A stirred slurry of K benzyl dithiocarbonate dihydrate in acetone treated with 2-oxo-3-benzothiazolineacetyl chloride in one portion at 22°, the mixture stirred at 25–30° for 18 hrs., ice-water added, and stirring continued at 0–10° for 2 hrs. → product. Y 79%. F.e.s. J.J. D'Amico, T. Schafer, D. Jackson, Phosphorus Sulfur *7,* 301–4 (1979); **thiolcarbamic acid esters from carbamyl chlorides** s. ibid. *8,* 301–4 (1980).

Piperidine $C_5H_{11}N$ ○

2-Aminothiazoles from 2-ketoxanthates and cyanamides

552.

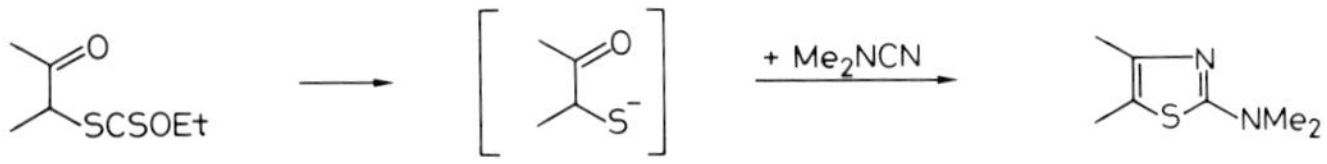

A soln. of startg. m. in ethanol treated with 1.1 equivalents piperidine at 20° for 30 min., N,N-dimethylcyanamide added, and the mixture refluxed for 3 hrs. → 2-(dimethylamino)-4,5-dimethylthiazole. Y 73%. F.e.s. M.D. Brown et al., J. Chem. Soc. Chem. Commun. *1982,* 444–5.

Mercuric chloride $HgCl_2$

1,3-Dithiolanes from 2-alkoxy-1,3-dithiolanes and oxo compds.

553.

A mixture of 2-ethoxy-1,3-dithiolane and *p*-nitrobenzaldehyde in dichloromethane treated with a little $HgCl_2$ at 0–5°, warmed to room temp., stirred for 24 hrs., and poured into ice water/ether → product. Y 91%. F.e.s. S. Jo et al., Bull. Chem. Soc. Japan *54,* 1434–6 (1981); **1,3-oxathiolanes from 2-alkoxy-1,3-oxathiolanes** s. ibid. *54,* 3237–8 (1981).

Sulfuryl chloride/thioacetic acid $SO_2Cl_2/MeCOSH$

1,2-Dithiolan-3-ones from β-(acylthio)carboxylic acids

554.

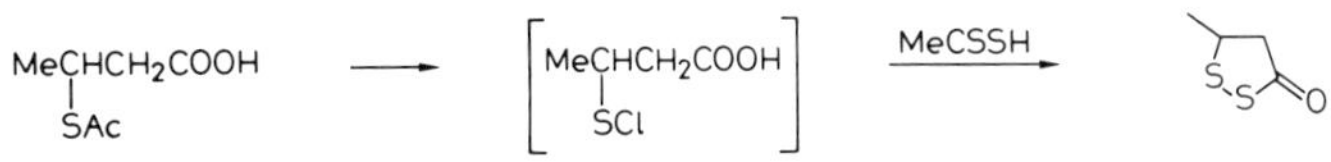

SO_2Cl_2 added at 20° to a soln. of startg. m. in carbon tetrachloride, left 1.5 hrs., solvent and acetyl chloride removed *in vacuo* (20–30°/10 mm), thioacetic acid added at 0°, left 1 hr., the excess acid removed *in vacuo* (40°/2 mm), and the residue heated at 140–200° at 2 mm → product. Y 78%. F.e. and methods s. T.P. Vasileva et al., Izv. Akad. Nauk SSSR Ser. Khim. *1980,* 2111–4.

Via intermediates *v.i.*

Transetherification of thioenolethers C=CSR → C=CSR′

s. *10,* 492; via *α,β*-ethylenesulfoxides cf. Y. Yamada et al., Synthesis *1981,* 64–5; s.a. K. Yamamoto et al., Tetrahedron Letters *23,* 897–900 (1982).

Elimination ⇑

Oxygen ↑ SC⇑O

Dicyclohexylcarbodiimide/1-hydroxybenzotriazole/triethylamine ←

Thiolactones from mercaptocarboxylic acids

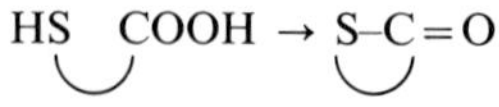

with dicyclohexylcarbodiimide s. *20,* 473; with added 1-hydroxybenzotriazole and triethylamine, macrocyclic thiolactones, s. S.A. Khan, B.W. Erickson, J. Am. Chem. Soc. *103,* 7374–6 (1981).

Trimethyl phosphite $(MeO)_3P$

Rearrangement of penicillin S-oxides ←

s. *26,* 629; Δ^2-oxazoline ring s. S. Yamamoto et al., Tetrahedron Letters *22,* 3089–92 (1981).

p-Toluenesulfonic acid *TsOH*

3,4-Dihydrothiocoumarins ○

37, 735

Nitrogen ↑ SC⇑N

Alumina Al_2O_3

Δ^3-Cephem-4-carboxylic acid esters via Δ^2-thiazoline ring opening ○

s. *32,* 569; f. method via acylaminodisulfides cf. S. Torii et al., Tetrahedron Letters *23,* 2495–8 (1982).

Halogen ↑ SC⇑Hal

Irradiation

1,3-S,N-Heterocyclics by dehydrohalogenative ring closure ○

benzothiazoles s. *35,* 414; 1,3-benzothiazin-4-ones s. T. Nishiwaki et al., Heterocycles *16,* 1203–4 (1981).

Sulfur ↑ SC⇑S

Ammonia NH_3

Ring closures with thiosulfonic acid esters ○

s. *31,* 576; Δ^3-cephems with NH_3 s. S. Torii et al., Tetrahedron Letters *23,* 2187–8 (1982).

Potassium cyanide *KCN*

Reviews of prepn. of thiocyanates RSCN

s. *21,* 676s*32;* s.a. S. Harusawa, T. Shioiri, Yuki Gosei Kagaku Kyokaishi *39,* 741 (1981) (Japan); Chem. Abstr. *95,* 167873 (1981).

Tributylphosphine *Bu_3P*

Ring closures via thioacylisocyanates ○

s. *32,* 575; s.a. Chem. Ber. *114,* 549–63, 808–9 (1981); 3-thioacyl-2-oxazolidones and 6,7-dihydro-1,5,3-oxathiazepin-2(2*H*)-ones s. ibid. *114,* 1972–5 (1981).

Remaining Elements ↑ SC⤊Rem

Sodium tetrahydridoborate/sodium hydride *$NaBH_4/NaH$*

Thietanes from β-(thiophosphorylthio)ketones ○

s. *37,* 545

Carbon ↑ SC⤊C

Sodium ethoxide *NaOEt*

Sulfido compds. from 1,3-oxathiolanes

s. *31,* 579; from 1,3-oxathiolan-2-ylideneammonium salts with ethanolic Na ethoxide, retention of configuration, cf. D. Hoppe, R. Follmann, Ann. Chem. *1980,* 1779–85.

Sodium hydride *NaH*

Sulfido compds. from 2-ketothioethers via 2-hydroxythioethers 

555.

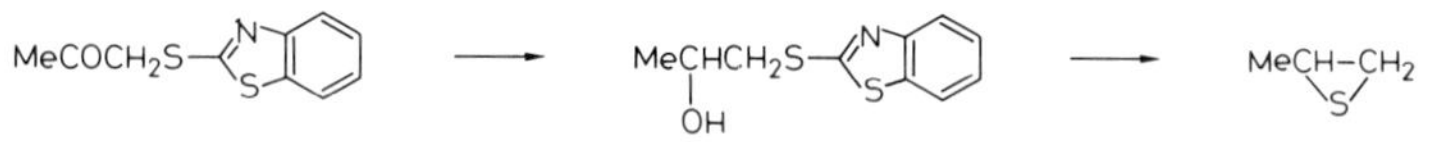

Startg. ketosulfide and $NaBH_4$ in methanol stirred at room temp. until t.l.c. indicated completion of reaction, the solvent replaced by tetrahydrofuran, NaH added, and the suspension stirred at room temp. for 2–5 hrs. → 2-methylthiirane. Y 95%. F.e.s. V. Calo, L. Lopez, G. Pesce, Gazz. Chim. Ital. *109,* 703–4 (1979).

Formation of Rem–Rem Bond

Uptake ⤋

Addition to Remaining Elements RemRem⤋Rem

Without additional reagents *w.a.r.*

Phosphine selenides ⋝PSe

s. *18,* 696; f. methods with Grignard compds. cf. N.A. Buina, F.G. Sibgatullina, I.A. Nuretdinov, Zh. Obshch. Khim. *49,* 2386–7 (1979).

Exchange ⇂↾

Nitrogen ↑ RemRem⇂↾N

N-Bromosuccinimide

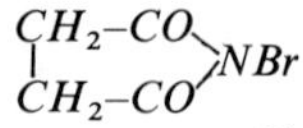

Sym. 2,2′-dibromodi(ene)diselenides from 1,2,3-selenadiazoles C

556. 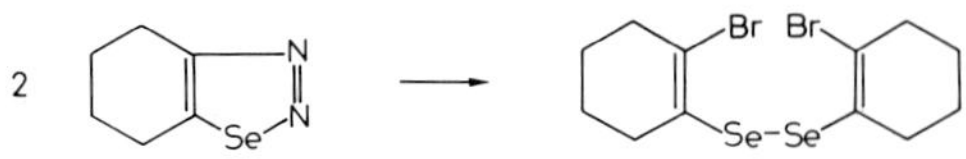

A soln. of startg. 1,2,3-selenadiazole in carbon tetrachloride treated with N-bromosuccinimide, and the mixture stirred at 20–40° in the presence of a little dibenzoyl peroxide until reaction complete by t.l.c. → product. Y 67%. F.e.s. H. Meier, J. Zountsas, H. Petersen, Nouveau J. Chem. *6,* 73–4 (1982).

Halogen ↑ RemRem⇂↾Hal

Electrolysis ϟ

Sym. disilanes from halogenosilanes ⋛Si–Si⋚
with Na,K cf. *8,* 695; by electrolysis s. E. Hengge, H. Firgo, J. Organometal. Chem. *212,* 155–61 (1981).

Lithium *Li*

Ultrasonication

557. $2\ Ph_3SiCl \rightarrow Ph_3Si{-}SiPh_3$

Triphenylchlorosilane added to 2.5:1 benzene/tetrahydrofuran and 2 equivalents of lithium wire in a flask partly submerged in a common ultrasound laboratory cleaner (150 W), and vigorously stirred for 10 hrs. → product. Y 73%. F.e., **also sym. distannanes,** s. P. Boudjouk, B.H. Han, Tetrahedron Letters *22,* 3813–4 (1981).

Formation of Rem–C Bond

Uptake ⇓

Addition to Oxygen and Carbon RemC⇓OC

Without additional reagents *w.a.r.*

1-Hydroxy-1,1-bis[phosphorus(V) compds.] from acylphosphorus(V) compds. CO → C(OH)P(O)⋜
s. *28,* 584; 1-hydroxy-1,1-bis(phosphine oxides) s. E. Lindner, M. Steinwand, Z. Naturforsch. *37B,* 407–11 (1982).

Sodium hydride *NaH*

Alternative generation of sodium phenylselenide C

558.

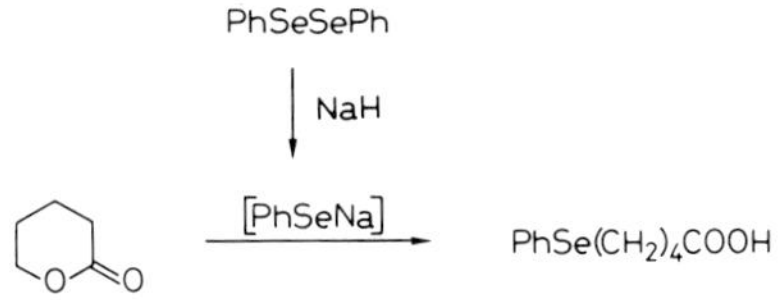

The conventional diselenide reductants, $NaBH_4$ and Na/liq. NH_3, may be replaced *by NaH* for generation of Na phenylselenide. E: A mixture of diphenyl diselenide and NaH in tetrahydrofuran refluxed for 45–60 min. under N_2, cooled to room temp., a little O_2-free hexamethylphosphoramide added by syringe, the orange soln. treated with δ-valerolactone, refluxed for 1 hr., and quenched with methanol → 5-(phenylseleno)valeric acid. Y 87%. F.e.s. P. Dowd, P. Kennedy, Synth. Commun. *11,* 935–41 (1981).

Sodium carbonate/benzyltriethylammonium chloride $Na_2CO_3/[PhCH_2NEt_3]^+Cl^-$

Abramov reaction

α-Hydroxyphosphonic acid esters from oxo compds. CO → C(OH)PO(OR)$_2$

s. *10,* 495; solid-liq. phase transfer catalysis with Na_2CO_3 and benzyltriethylammonium chloride in benzene/dimethyl sulfoxide cf. Y.A. Zhdanov, Z.I. Glebova, L.A. Uzlova, Zh. Obshch. Khim. *51,* 1675–6 (1981); with KF or CsF cf. F. Texier-Boullet, A. Foucaud, Syxnthesis *1982,* 165–6.

Potassium fluoride or caesium fluoride *KF or CsF*

s. *10,* 495s*37*

Triethylamine Et_3N

α-Hydroxyphosphinic acids and esters from ketones CO → C(OH)P(O)(OR)R

acids cf. *22,* 656; esters from phosphonous monoesters with triethylamine, carbohydrate derivs., s. M. Yamashita et al., Carbohyd. Res. *84,* 35–41 (1980).

α-(Selenoacylamino)ketones from oxazolium salts C

s. *37,* 573

Sodium tetrahydridoborate $NaBH_4$

2-Hydroxyselenides from oxido compds.  → C(OH)C(SeR)

s. *36,* 600s*37*

Selenoboric esters $(RSe)_3B$

by electrolysis cf. *36,* 600; with selenoboric esters s. A. Cravador, A. Krief, Tetrahedron Letters *22,* 2491–4 (1981); from diselenides with $NaBH_4$ cf. T. Arunachalam, E. Caspi, J. Org. Chem. *46,* 3415–20 (1981).

Tris(trimethylsilyl)aluminium $(Me_3Si)_3Al$

Nucleophilic C-silylation

1-Hydroxysilanes from oxo compds. CO → CH(OH)Si≡

559.

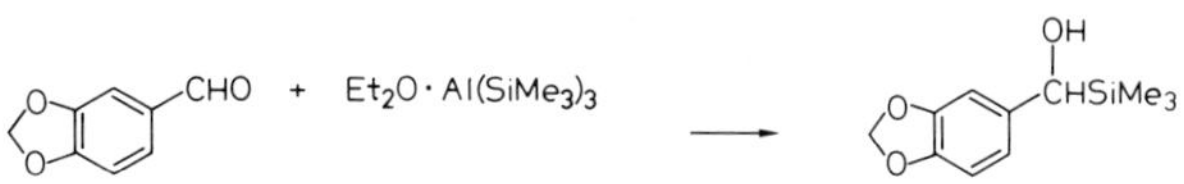

A soln. of piperonal in ether added to 0.33 equivalents tris(trimethylsilyl)aluminium etherate in the same solvent, the mixture refluxed for 4 hrs. under argon, and worked up with methanol/dil. HCl at room temp. → 3,4-methylenedioxy-α(trimethylsilyl)-

benzyl alcohol. Y 88%. F.e., **also 1-hydroxy-1,1-disilanes from carboxylic acid esters,** and silylation with Li tetrakis(trimethylsilyl)aluminate, s. L. Rösch, G. Altnau, W.H. Otto, Angew. Chem. Intern. Ed. *20,* 581–2 (1981); **(E)-enesilanes from acetylene derivs.** s. Tetrahedron Letters *21,* 4069–72 (1980).

Benzyltriethylammonium chloride s. under Na_2CO_3 $[PhCH_2NEt_3]^+Cl^-$

Addition to Nitrogen and Carbon RemC⇓NC

Without additional reagents *w.a.r.*

α-Aminophosphonic acids from azomethines N=C → NHC–$PO(OR)_2$
with phosphorous acid cf. *34,* 572; with tris(trimethylsilyl) phosphite s. J. Zón, Pol. J. Chem. *55,* 643–6 (1981).

α-Aminophosphonic esters N=C → NHC–$PO(OH)_2$
...from azomethines and dialkyl phosphites
with HCl cf. *25,* 465; without reagent and hydrolysis to the acids s. K. Issleib, K.-P. Döpfer, A. Balszuweit, Z. Naturforsch. *36B,* 1392–4 (1981).

...from aldehydes and amines NH_2 → NHC–$PO(OR)_2$
s. *23,* 477; from anilines and trialkyl phosphites s. B.E. Ivanov et al., Izv. Akad. Nauk SSSR Ser. Khim. *1982,* 178–80.

Addition to Carbon and Carbon RemC⇓CC

Without additional reagents *w.a.r.*

Peroxymercuration of ethylene derivs. C=C → C(OOR)C–Hg^+
s. *26,* 637; also demercuration to peroxides s. J. Chem. Soc. Chem. Commun. *1981,* 1117–8.

Dialkylchloroboranes from ethylene derivs. 2 C=C → $(CHC)_2BCl$
s. *37,* 664

1,1-Diboranes from acetylene derivs. C≡C → $CH_2C(BR_2)_2$
37, 732

1,4-Addition with dialkyl silyl phosphites
Cyclic β,γ-ethylene-γ-siloxyphosphonic esters from α,β-ethyleneketones COCH=CH → C(OSi≡)=C–C–$PO(OR)_2$

560.

A mixture of startg. enone and 1.8 equivalent of diethyl trimethylsilyl phosphite in acetonitrile heated 180° for 12 hrs. in a sealed tube → product. Y 100%. – The products are enone β-carbanion equivalents. F.e. and limitations s. D. Liotta, U. Sunay, S. Ginsberg, J. Org. Chem. *47,* 2227–9 (1982); **γ-alkoxy-β,γ-ethylene-γ-siloxyphosphonic acid esters from α,β-ethylenecarboxylic acid esters** s. Y. Okamoto, H. Sakurai, Synthesis *1982,* 497–9; Chem. Letters *1981,* 1265–8.

α,β-Acetylenesulfones C≡C → $C(SO_2R)$=CSeAr
from selenosulfonyl esters and terminal acetylene derivs.
via 1-ethylene-2-sulfonylselenides

561. $p\text{-}MeC_6H_4SO_2SePh$ + HC≡CPh → $p\text{-}MeC_6H_4SO_2CH$=C(Ph)SePh → $p\text{-}MeC_6H_4SO_2C$≡CPh

A soln. of startg. selenosulfonate and phenylacetylene in benzene refluxed for 16 hrs. → *trans*-β-(phenylseleno)vinyl sulfone (Y 86%), treated with 4 equivalents of 30%

H_2O_2 in tetrahydrofuran at room temp. → 1-phenyl-2-*p*-tolylsulfonylethyne (Y 93%). F.e.s. T. Miura, M. Kobayashi, J. Chem. Soc. Chem. Commun. *1982,* 438–9; with added azodiisobutyronitrile cf. T.G. Back, S. Collins, Tetrahedron Letters *22,* 5111–4 (1981).

Regiospecific oxytelluration of ethylene derivs. C=C → C(OR)–CTe⋚

562. 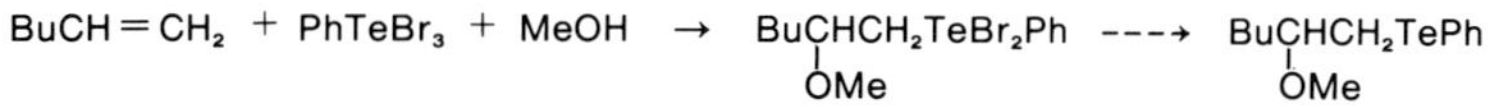

A soln. of phenyltellurium tribromide in methanol treated with 1-hexene, and the mixture refluxed with stirring for 3 hrs. → product. Y 63%. F.e. and stereospecificity, also reduction to **1,2-alkoxytellurides,** s. S. Uemura et al., Tetrahedron Letters *23,* 1177–80 (1982).

2-Chloroenetellurium trichlorides from acetylene derivs. C≡C → CCl=CTeCl_3
s. *37,* 446

Electrolysis ϟ

**α-Arylseleno-α,β-ethyleneoxo compds.
from 2-acetylenealcohols
via regiospecific hydroxyselenation** C(OH)C≡C → C=C(SeAr)CO

563. 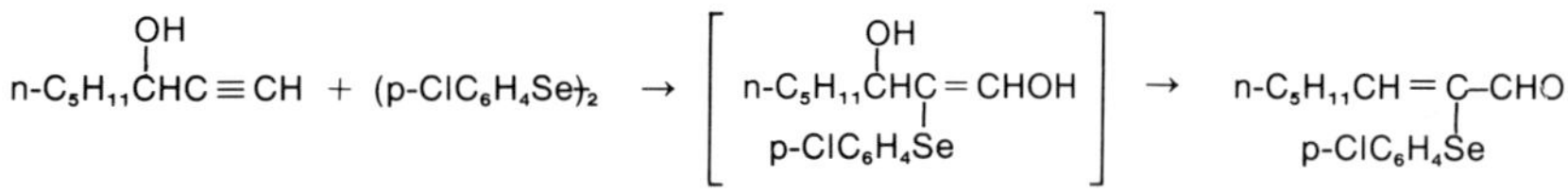

A mixture of startg. acetylenealcohol, *p*-chlorophenyl diselenide, and concd. H_2SO_4 in 6:1 acetonitrile/water electrolysed in the presence of tetraethylammonium perchlorate in an undivided cell using Pt foil electrodes at 65° with a constant current of 3.3 mA/cm² until 35 F/mol passed (14 hrs.) → product. Y 81%. F.e.s. K. Uneyama, K. Takano, S. Torii, Tetrahedron Letters *23,* 1161–4 (1982).

Lithium chloride *LiCl*

O-Heterocyclics by intramolecular oxytelluration ○

564. 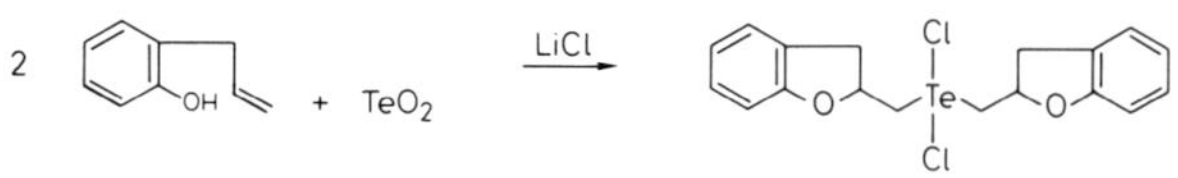

Coumarans. TeO_2, 2-allylphenol, and LiCl refluxed in acetic acid for 1.5 hrs. → bis{2-(2,3-dihydrobenzofuranyl)methyl}tellurium dichloride. Y 78%. F.e., incl. tetrahydrofurans and -pyrans, s. J. Bergman, L. Engman, J. Am. Chem. Soc. *103,* 5196–200 (1981).

Sodium acetate s. under Al_2Se_3 *MeCOONa*

Tris(trimethylsilyl)aluminium *$(Me_3Si)_3Al$*

(E)-Enesilanes from acetylene derivs. C≡C → CH=CSi⋚
s. *37,* 559

Aluminium selenide/sodium acetate *$Al_2Se_3/MeCOONa$*

Se-Heterocyclics from dienes ○

565. 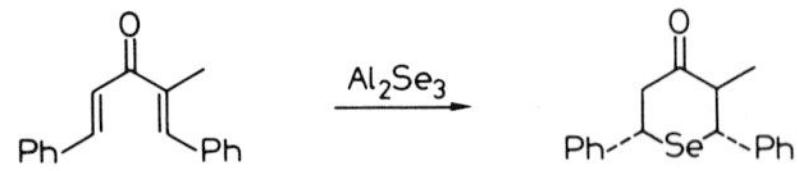

4-Selenanones. A boiling mixture of 2-methyl-1,5-diphenyl-1,4-pentadien-3-one and Na acetate in ethanol treated with Al_2Se_3, and heating continued for 6 hrs. → *r*-2,*cis*-6-diphenyl-*trans*-3-methyl-4-selenanone. Y 65%. F.e.s. P. Nanjappan et al., J. Org. Chem. *45,* 4622–9 (1980).

Aluminium chloride *$AlCl_3$*

Enesilanes from acetylene derivs. C≡C → CH=CSi⋲

with H_2PtCl_6 cf. *12,* 700s*26, 36;* with $AlCl_3$, vinylsilanes, s. M.G. Voronkov, S.N. Adamovich, V.B. Pukhnarevich, Zh. Obshch. Khim. *51,* 2385–6 (1981).

Azodiisobutyronitrile *$Me_2C(CN)N=NC(CN)Me_2$*

Stannanes from ethylene derivs. C=C → CHCSn⋲

without catalyst cf. *11,* 710; with added azodiisobutyronitrile without solvent, *β*-stannyl-carboxylic acid esters and -nitriles, s. J.C. Podesta, A.B. Chopa, A.D. Ayala, J. Organometal. Chem. *212,* 163–70 (1981); ibid. *229,* 223–8 (1982).

Nickel(II) phosphine oxide complexes ←

Hydrosilylation C=C → CHCSi⋲

with chlorotris(triphenylphosphine)rhodium(I) cf. *29,* 603; with nickel(II) phosphine oxide complexes s. N.A. Moldavskaya et al., Zh. Obshch. Khim. *51,* 1621–4 (1981).

Tetrakis(triphenylphosphine)palladium(0) *$Pd(PPh_3)_4$*

1,2-Ethylene-2,3-disilanes from allenes C=C=C → C=C(Si⋲)CSi⋲

566. 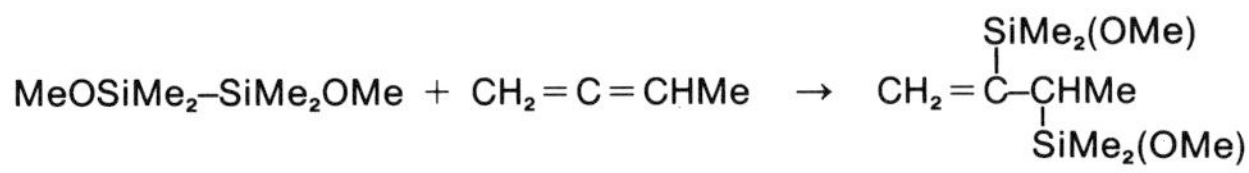

1,2-Butadiene distilled into a mixture of *sym*-dimethoxytetramethyldisilane and a little tetrakis(triphenylphosphine)palladium(0) contained in an evacuated stainless steel vessel, and heated with stirring at 120° for 15 hrs. → product. Y 88%. F.e.s. H. Watanabe et al., J. Chem. Soc. Chem. Commun. *1981,* 617–8; J. Organometal. Chem. *225,* 343–56 (1982).

Chloroplatinic acid *H_2PtCl_6*

Hydrosilylation C=C → CHC(Si⋲)

s. *12,* 700; 3-aminosilanes from 2-ethyleneamines s. E. Lukevits et al., Khim.-Farmatsevt. Zh. *15,* No. 4, 42–6 (1981).

Rearrangement ∩

Hydrogen/Carbon Type RemC∩HC

Lithium diisopropylamide/hexamethylphosphoramide *i-$Pr_2NLi/(Me_2N)_3PO$*

C-Arylseleno group migration ←

Isomerization of 2-ketoselenides

with NaH cf. *36,* 585; with *i*-Pr_2NLi/hexamethylphosphoramide s. D. Liotta, M. Saindane, D. Brothers, J. Org. Chem. *47,* 1598–600 (1982).

Oxygen/Remaining Elements Type RemC⌒ORem

Without additional reagents *w.a.r.*

α,β-Ethylene-γ-silylcarboxylic acid esters from O-silyl O-alkyl vinylketeneacetals ←
1,5 O → C Silyl group migration

567.

Startg. dieneacetal heated at 150–65° for 30 min. → product. Y 90%. F.e.s. G. Anderson et al., Tetrahedron Letters *22,* 4347–8 (1981).

Lithium diisopropylamide *i-Pr$_2$NLi*

***o*-Hydroxyphosphonic esters from phenols via monoaryl phosphates**

568.
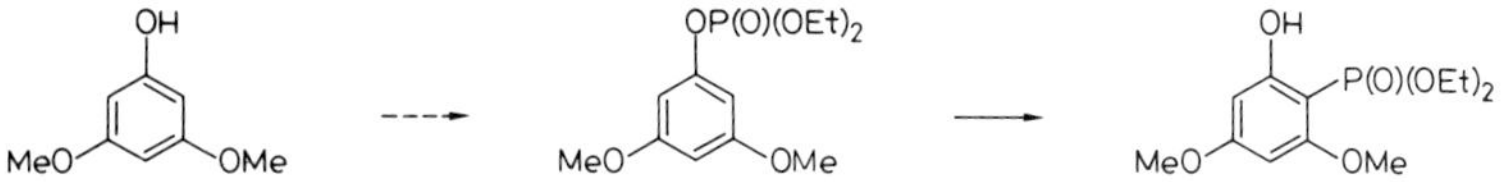

A soln. of diethyl 3,5-dimethoxyphenyl phosphate (prepd. from 3,5-dimethoxyphenol) in tetrahydrofuran added slowly to a freshly prepd. soln. of Li diisopropylamide in the same solvent at –78°, the mixture stirred at this temp. for 15 min., then at 0° for 30 min. → diethyl 2,4-dimethoxy-6-hydroxyphenylphosphonate. Y 93%. F.e. and bases s. L.S. Melvin, Tetrahedron Letters *22,* 3375–6 (1981); also from *o*-bromophosphoric acid esters with n-butyllithium cf. R.C. Cambie, B.D. Palmer, Australian J. Chem. *35,* 827–37 (1982).

Oxygen/Carbon Type RemC⌒OC

Ethyl iodide *C_2H_5I*

Phosphonic esters from phosphites ←
Arbuzov-Michaelis rearrangement s. *10,* 499; using a film reactor cf. M. Tandara, I. Pilipovic, Kem. Ind. *30,* 673 (1981) (Serbo-Croatian); Chem. Abstr. *96,* 162863 (1982).

Ethylene dibromide *$BrCH_2CH_2Br$*

Arbuzov-Michaelis rearrangement ←
s. *24,* 114; **review** s. A.K. Bhattacharya, G. Thyagarajan, Chem. Rev. *81,* 415–30 (1981).

Exchange ⇅

Hydrogen ↑ RemC⇅H

Without additional reagents *w.a.r.*

Deuteriation of phosphonium salts H → D
s. *20,* 482; f.e.s. M. Orfanopoulos, J. Chem. Res. (S) *1982,* 188–9.

Potassium carbonate K_2CO_3

Organogold compds. H → Au

569. $(Ph_3PAu)_3O^+BF_4^- + NCCH_2COOEt \rightarrow Ph_3PAuCH(CN)COOEt$

Ethyl cyanoacetate and water added to *tris(triphenylphosphinegold)oxonium fluoroborate* and K_2CO_3 in tetrahydrofuran, and stirred at 20° for 2 min. → triphenylphosphinegold(cyano)carbethoxymethane. Y 94%. F.e.s. E.I. Smyslova et al., J. Organometal. Chem. *215,* 269–79 (1981).

Sodium acetate CH_3COONa

Deuteriation H → D

with NaOEt/EtOD cf. *19,* 722; with Na acetate in D_2O, 1-deuteriation of nitro compds., also *a,a*-dideuteriocarbonyl compds., s. N. Ono et al., Chem. Letters *1982,* 1079–80.

Alumina Al_2O_3

570. $(EtO_2C)_2CH_2 \xrightarrow{D_2O/Al_2O_3} (EtO_2C)_2CD_2$

Deuteriation of acidic sites on deuteriated alumina or molecular sieves is quicker and easier than conventional techniques (e.g. K *tert*-butoxide/*tert*-butanol-OD). E: Diethyl malonate mixed with a slurry of D_2O–treated alumina in 1:1 ether/pentane for 5 min. → diethyl *a,a*-dideuteriomalonate. Y >96%. F.e. and with D_2O-treated molecular sieves s. G.W. Kabalka et al., J. Org. Chem. *46,* 1513–4 (1981).

Platinum *Pt*

s. *22,* 664; of nucleosides s. T. Kinoshita, K.H. Schram, J.A. McCloskey, J. Labelled Compds. Radiopharm. *9,* 525–34 (1982).

Oxygen ↑ RemC↓↑O

Without additional reagents *w.a.r.*

Acylphosphine oxides $HPOR_2 \rightarrow AcPOR_2$

from phosphinous acids and carboxylic acid anhydrides

571.

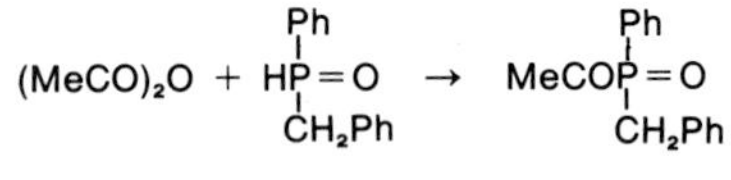

Acetic anhydride added to benzylphenylphosphine oxide at –10°, and the mixture heated for 30 min. at 40° → acetylbenzylphenylphosphine oxide. Y 100%. F.e., **also acylphosphines** from silylphosphines or sec. phosphines, s. E. Lindner, G. Frey, Chem. Ber. *113,* 3268–74 (1980).

Lithium disilylcuprates $LiCu(SiR_3)_2$

2-Ethylenesilanes from acoxy-2-ethylenes C(OAc)C=C → C=C–CSi≡

Regiospecific synthesis

572.

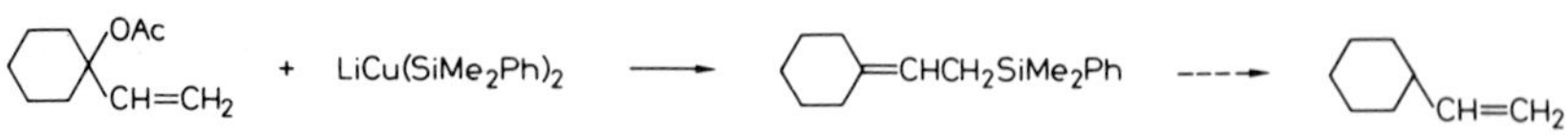

A soln. of startg. silylcuprate (prepd. from phenyldimethylsilyllithium and CuI) in tetrahydrofuran treated under N_2 with a soln. of startg. allyl acetate in the same solvent, and the mixture kept at 0° for 1 hr. then at room temp. overnight → product. Y 93%. F.e. and limitations, also conversion to **ethylene derivs.** restoring the position of the unsaturation, s. I. Fleming, D. Marchi, Synthesis *1981,* 560–1.

Via intermediates *v.i.*

Selenazolium from oxazolium salts via *α***-(selenoacylamino)ketones** ○

573. 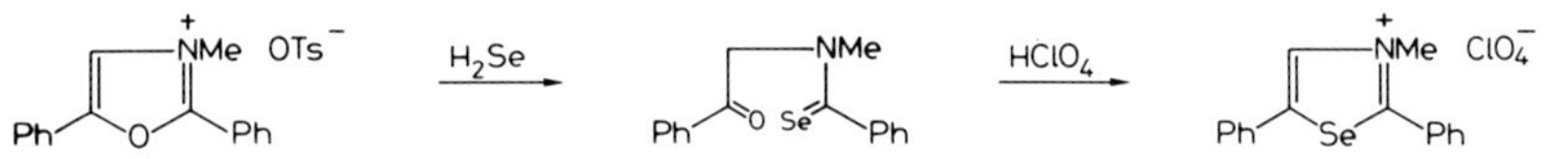

A mixture of startg. oxazolium salt and triethylamine in methanol satd. with H_2Se until a dark-red precipitate separated, and filtered after 10 min → selenoamide (Y 85%) suspended in acetic anhydride, dil. with ether, and treated dropwise while stirring with freshly prepd. 70% $HClO_4$ in acetic anhydride cooled to room temp. → selenazolium perchlorate (Y 95%). F.e. and conversion to **imidazoles** s. O.P. Shvaika, V.F. Lipnitskii, Zh. Obshch. Khim. *51,* 1842–7 (1981).

Nitrogen ↑ RemC↓↑N

Without additional reagents *w.a.r.*

Phosphine oxides from phosphinous acids and trisylhydrazones $R_2PHO \rightarrow R_2POR'$

574. 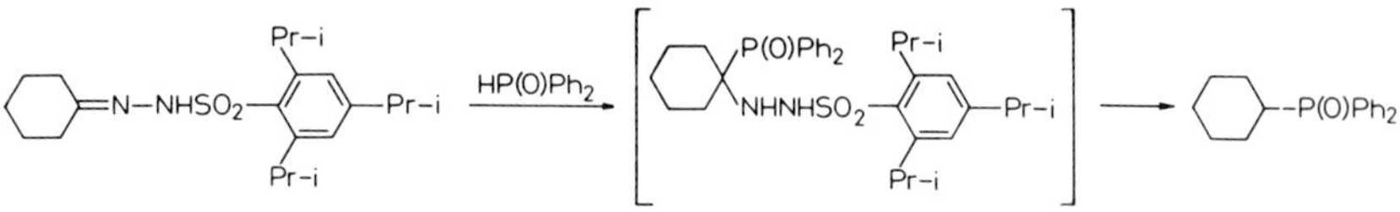

Cyclohexanone 2,4,6-tri-isopropylbenzenesulfonylhydrazone refluxed with diphenylphosphine oxide in tetrahydrofuran for 1 day → product. Y 83%. – The method avoids strong acids, bases, and oxidizing agents. – With tosylhydrazones, the intermediate *α***-tosylhydrazinophosphine oxides** are isolable. F.e.s. S.H. Bertz, G. Dabbagh, J. Am. Chem. Soc. *103,* 5932–4 (1981).

*α***-(Arylseleno)aldehydes from aldehydes under mild conditions** CHCHO → C(SeAr)CHO

575. 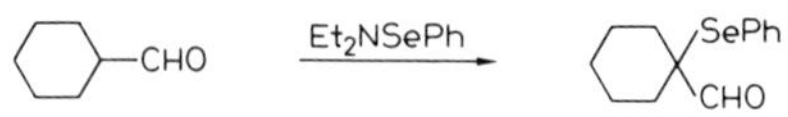

1.25 Equivalents N,N-diethylbenzeneselenamide added via syringe to a soln. of startg. aldehyde in methylene chloride, and the mixture stirred at room temp. for 6 hrs. → product. Y 87%. – The procedure, which requires neither strong acid or base, represents the first high-yielding method for the direct formation of *α*-phenylselenoaldehydes from unactivated aldehydes. F.e.s. M. Jefson, J. Meinwald, Tetrahedron Letters *22,* 3561–4 (1981); s.a. C. Paulmier, P. Lerouge, ibid. *23,* 1557–60 (1982).

Halogen ↑ RemC↓↑Hal

Sodium *Na*

Replacement of alkoxy by silyl groups OR → Si≡

2-Ethylenesilanes

with Li cf. *36,* 592; with Na, also benzyl silanes, s. D. Tzeng, W.P. Weber, J. Org. Chem. *46,* 265–7 (1981); via 2-ethylene(alkoxy)silanes cf. ibid. *46,* 693–6 (1981).

Sodium hydride/cuprous iodide *NaH/CuI*

Selenides from halides and selenols Hal → SeR

with Na in ethanol cf. *25,* 471; diaryl selenides with NaH/CuI s. H. Suzuki, H. Abe, A. Osuka, Chem. Letters *1981,* 151–2.

Potassium hydroxide *KOH*

Selenoynols as intermediates

1-Acetylene-1-selenides from 1,2,3-selenadiazoles and halides C

576. 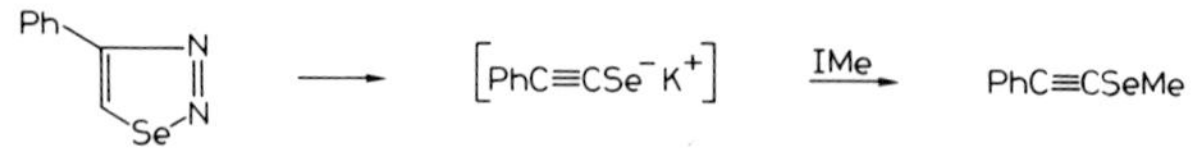

4-Phenyl-1,2,3-selenadiazole added to a soln. of KOH in 10:1 dioxane ethanol, left for 30 min. (after which evolution of N_2 ceased), treated with methyl iodide, and stirred for 2–3 hrs. → product. Y 71%. F.e., **also 2-imino-1,3-thiaselenoles and 1,3-thiaselenole-2-thiones** from isothiocyanates and CS_2, respectively, s. V.Z. Laishev, M.L. Petrov, A.A. Petrov, Zh. Org. Khim. *18,* 514–9 (1982).

Sodium methoxide *NaOMe*

α,β-Oxidophosphonic esters from dialkyl phosphites 

s. *32,* 593; with α-tosyloxyketones instead of α-halogenoketones, also α,β-oxidophosphinic esters, cf. S. Inokawa et al., Carbohyd. Res. *88,* 341–4 (1980).

Methyllithium *MeLi*

α-Sulfinylphosphines from sulfoxides S(O)CH → S(O)CPR_2

s. *37,* 87

n-*Butyllithium* n-*BuLi*

1-Nitroselenides from nitro compds. CHNO_2 → C(SeR)NO_2

s. *37,* 916

n-*Butyllithium/N,N,N′,N′-tetramethylethylenediamine* n-*BuLi/TMEDA*

Synthesis of α,β-ethyleneketones from enamines H → Si≡

via 3-silylenamines and 2-halogeno-3-silylenamines

One pot synthesis

577. 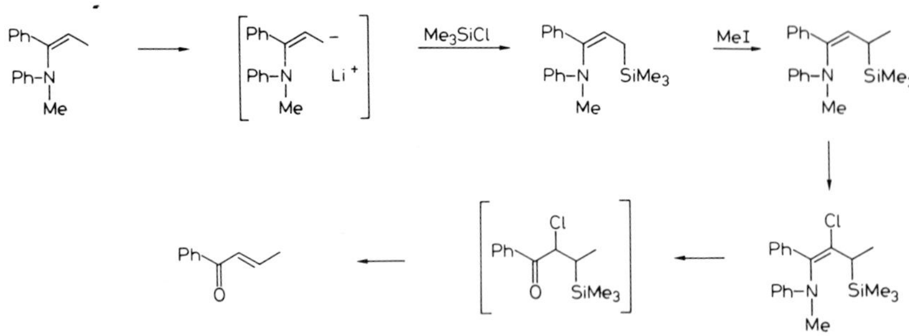

n-Butyllithium in hexane added dropwise to a soln. of startg. enamine in 10:1 tetrahydrofuran/tetramethylethylenediamine under argon at 0° the mixture stirred for 2 hrs., cooled to −78°, quenched with trimethylsilyl chloride, after stirring for 1 hr. a further equivalent of *n*-butyllithium in hexane added, stirred at 0° for 3 hrs., quenched with methyl iodide, stirring continued for 30 min., and allowed to warm to room temp. → crude alkylated silylenamine, in carbon tetrachloride stirred with N-chlorosuccinimide at room temp. for 5 hrs. → crude chloro deriv., cooled in an ice-bath, treated with trifluoroacetic acid, stirred at room temp. for 4 hrs., the mixture refluxed for 60 hrs. after addition of ether and 4 *M* HCl, the crude hydrolysate treated with NaF and 95% alcohol, and the mixture refluxed for 16 hrs. → product (Y 82%). F.e., **also via 3-sulfenylation of enamines,** s. C.S. Sudheendranath, H. Ahlbrecht, J. Indian Chem. Soc. *58,* 973–9 (1981).

sec-*Butyllithium/N,N,N',N'-tetramethylethylenediamine* s-*BuLi/TMEDA*

Enephosphines from trisylhydrazones CHCO → C=CPR_2

Regiospecific synthesis

578.

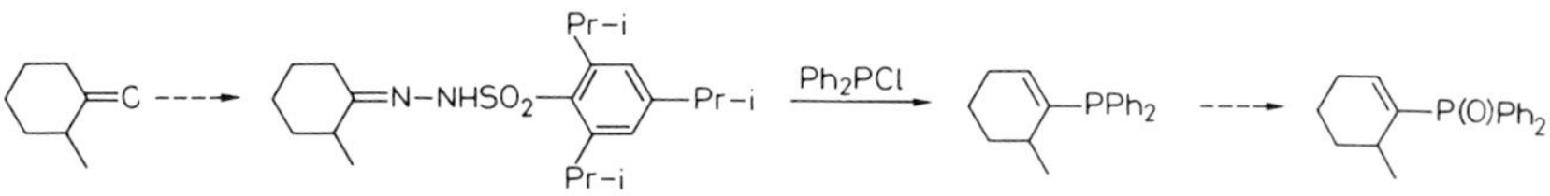

Startg. 2,4,6-triisopropylbenzenesulfonyl hydrazone added portionwise over 10–15 min. to 2 equivalents of *sec*-butyllithium in TMEDA at –45°, stirred for 1 hr., allowed to warm to 0°, kept at this temp. for 30 min., recooled to –45°, chlorodiphenylphosphine added, and stirring continued for 1 hr. at this temp., then overnight at room temp. → product. Y 92%. F.e. and conversion to **enephosphine oxides** s. D.G. Mislankar, B. Mugrage, S.D. Darling, Tetrahedron Letters *22,* 4619–22 (1981).

Lithium diisopropylamide *i-Pr_2NLi*

o-**Metalation of hetaromatic halides** ←

579.

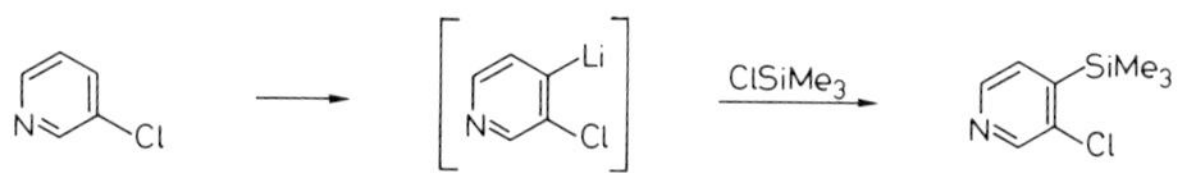

A soln. of 3-chloropyridine in tetrahydrofuran added over 15 min. with stirring to a soln. of Li diisopropylamide in the same solvent at –78° under N_2, the obtained 3-chloro-4-lithiopyridine stirred for 30 min., treated with chlorotrimethylsilane in tetrahydrofuran over 5–10 min. at the same temp., and allowed to warm to room temp. overnight → 3-chloro-4-trimethylsilylpyridine. Y 96%. F.e. and electrophiles s. G.W. Gribble, M.G. Saulnier, Tetrahedron Letters *21,* 4137–40 (1980).

C-*α*-Silylation of carboxylic acid esters

580.

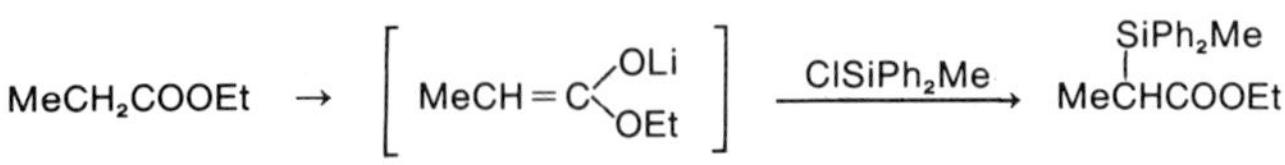

A soln. of ethyl propionate in tetrahydrofuran added via syringe to Li diisopropylamide in the same solvent at –78°, stirred at the same temp. for 30 min., diphenylmethylchlorosilane in tetrahydrofuran added dropwise via syringe, stirring continued for 1.5 hrs., warmed to 25° for 2 hrs., and hydrolysed with 1.5 *M* HCl → ethyl (diphenylmethylsilyl)propionate. Y 93%. – O-Silylation occurs exclusively with ketones. F.e., **also of lactones,** s. G.L. Larson, L.M. Fuentes, J. Am. Chem. Soc. *103,* 2418–9 (1981).

Sodium/liq. ammonia *Na/NH_3*

1,3-Thiaselenoles and 1,3-thiatelluroles ○

s. *36,* 553; 1,3-ditelluroles s. S.L. Bender, M.R. Detty, N.F. Haley, Tetrahedron Letters *23,* 1531–4 (1982).

Lithium cyanide *LiCN*

Cyano- from chloro-silanes R_3SiCl → R_3SiCN

s. *36,* 595s*37*

Potassium cyanide/potassium iodide *KCN/KI*
s. *36,* 595; with added KI cf. M.T. Reetz, I. Chatziiosifidis, Synthesis *1982,* 330; with LiCN in bis[2-(2-methoxyethoxy)ethyl] ether, prepn. of trimethylsilyl cyanide s. T. Livinghouse, Org. Synth. *60,* 126–32 (1981).

N,N,N′,N′-Tetramethylethylenediamine s. under n-*BuLi and* s-*BuLi* *TMEDA*

α-Picoline ←
Phospholes from phosphonous dihalides and 1,3-dienes ○

581.

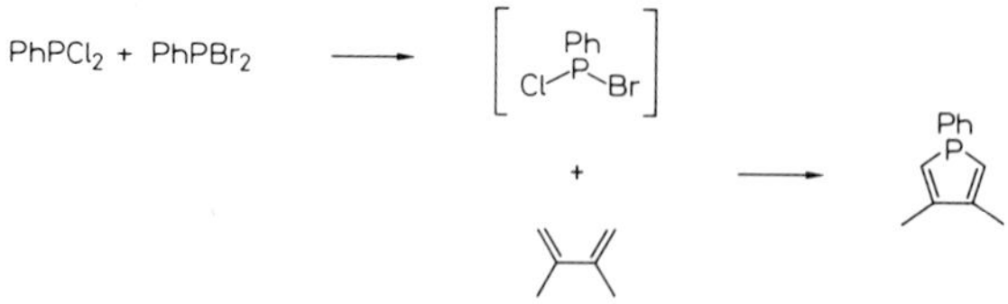

A mixture of dichlorophenylphosphine and dibromophenylphosphine allowed to stand at room temp. for 30 min. under N_2, 2,3-dimethylbutadiene added, cooled at 0° for 1 day, allowed to stand at room temp. for 8 days, the white crystalline mass crushed in 3:2 hexane/dichloromethane, a soln. of 2-methylpyridine in dichloromethane added dropwise at room temp. with vigorous stirring over 30 min., and worked up with 3 HCl after 5–20 hrs. → product. Y 88%. F.e. and bases s. A. Breque, F. Mathey, P. Savignac, Synthesis *1981,* 983–5.

Lithium disilylcuprates *$LiCu(SiR_3)_2$*
α,β-Ethylene-β-silyl from α,β-ethylene-β-chloro-ketones C=CCl → C=CSi≡

582.

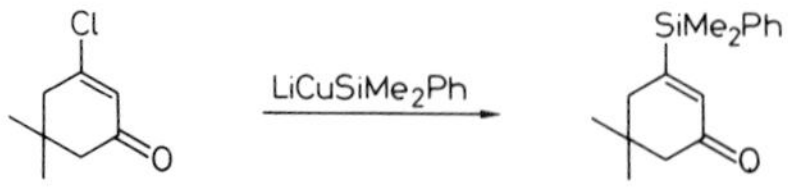

Phenyldimethylsilyllithium in tetrahydrofuran added slowly to a stirred suspension of CuCN in the same solvent at 0° under argon, after 20 min. cooled to –78°, startg. enechloride in tetrahydrofuran added dropwise, and the crude product in acetone treated with Jones reagent to remove undesired carbonyl adduct → enesilane. Y 99%. F.e. and method s. I. Fleming, D.A. Perry, Tetrahedron *37,* 4027–34 (1981).

Cuprous iodide s. under NaH *CuI*

Magnesium *Mg*
Silylacetylenes from terminal acetylene derivs. C≡CH → C≡C–Si≡
with ethylmagnesium bromide cf. *23,* 656; with magnesium s. M.G. Voronkov et al., Izv. Akad. Nauk SSSR Ser. Khim. *1981,* 897; with ethyl trimethylsilyl acetate cf. E. Nakamura, K. Hashimoto, I. Kuwajima, Bull. Chem. Soc. Japan *54,* 805–8 (1981).
Sym. dibromostannanes from Grignard compds. R_2SnHal_2

583. $2\ EtMgBr + (acac)_2SnBr_2 \rightarrow Et_2SnBr_2$

A soln. of bis(acetylacetonate)dibromotin(IV) in benzene treated with 2 molar proportions of ethylmagnesium bromide in ether/benzene at room temp., and refluxed for 2–3 hrs. → product. Y 93%. F.e.s. R.K. Chandrasekaran, S. Venkataraman, J.D. Koola, J. Organometal. Chem. *215,* C43–4 (1981).
Sym. diselenides from halides 2 RHal → RSeSeR
with K_2Se_2 cf. *11,* 719; with Mg/Se, prepn. of diphenyl diselenide, s. H.J. Reich, M.L. Cohen, P.S. Clark, Org. Synth. *59,* 141–7 (1979).

Magnesium/1,2-dibromoethane *Mg/BrCH₂CH₂Br*

Sym. distannanes from Grignard compds. ⋛SnSn⋚

584. $ArMgBr + SnCl_2 \rightarrow [Ar_3SnMgBr] \rightarrow Ar_3SnSnAr_3$

A soln. of 2,4,6-triethylphenyl magnesium bromide in tetrahydrofuran treated dropwise at –5° with a soln. of $SnCl_2$ in the same solvent, stirred for 2 hrs. at 20°, recooled to –5°, *1,2-dibromoethane as oxidant* added dropwise, the mixture stirred at 35° for 1 hr., and hydrolysed at 20° with 10% HCl → hexakis(2,4,6-triethylphenyl)distannane. Y 80%. F.e., **also from organotin hydrides** with azodiisobutyronitrile, s. H.-U. Buschhaus, W.P. Neumann, T. Apoussidis, Ann. Chem. *1981,* 1190–7.

Sulfur ↑ RemC⇅S

Lithium/naphthalene ←

1-Silylsulfides from mercaptals and chlorosilanes $C(SAr)_2 \rightarrow C(SAr)Si$⋚

585.

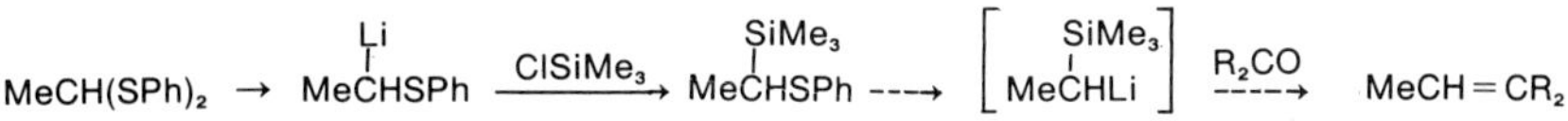

A soln. of Li naphthalenide in tetrahydrofuran treated at –78° with startg. mercaptal, the mixture stirred at the same temp. for 15 min., and quenched with trimethylsilyl chloride → product. Y 90%. F.e., also conversion to **ethylene derivs.** via f. reductive lithiation with Li naphthalenide (cf. Synth. Meth. *28,* 856), s. D.J. Ager, Tetrahedron Letters *22,* 2923–6 (1981).

Azodiisobutyronitrile *Me₂C(CN)N=NC(CN)Me₂*

2-Ethylenestannanes from β,γ-ethylenesulfones with allyl rearrangement $C=C\ CSO_2R \rightarrow$ ⋛$SnC-C=C$

586.

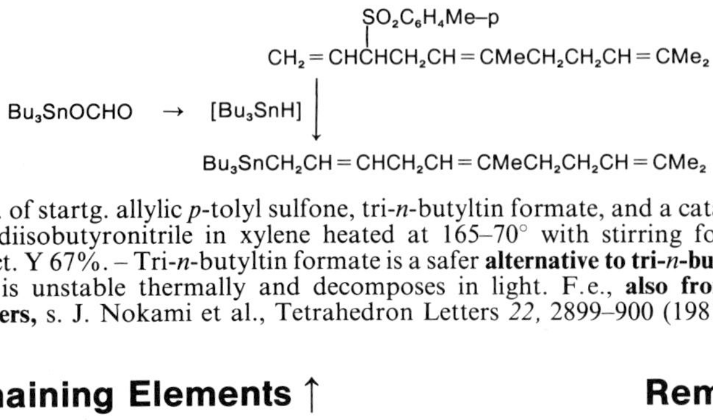

A soln. of startg. allylic *p*-tolyl sulfone, tri-*n*-butyltin formate, and a catalytic amount of azodiisobutyronitrile in xylene heated at 165–70° with stirring for 2.8 hrs. → product. Y 67%. – Tri-*n*-butyltin formate is a safer **alternative to tri-*n*-butyltin hydride,** which is unstable thermally and decomposes in light. F.e., **also from 2-ethylenethioethers,** s. J. Nokami et al., Tetrahedron Letters *22,* 2899–900 (1981).

Remaining Elements ↑ RemC⇅Rem

Without additional reagents *w.a.r.*

Heterocyclic phosphines from silanes Si⋚ → P⋚

587.

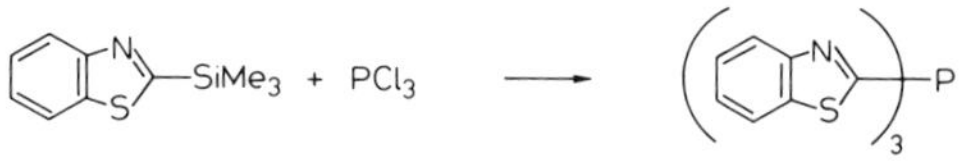

A mixture of PCl_3 and 2-(trimethylsilyl)benzothiazole stirred at room temp. for 2 hrs., heated at 60° for 30 min. with distillation of trimethylchlorosilane, and the solidified product heated at 100–10° for 10 hrs. → tris(benzothiazol-2-yl)phosphine. Y 83%. F.e.s. S.S. Moore, G.M. Whitesides, J. Org. Chem. *47,* 1489–93 (1982).

Lithium/1-dimethylaminonaphthalene ←

α-Silylketones from 2-ketoselenides via 1-ethylene-2-siloxyselenides SeAr → Si⋚

588. 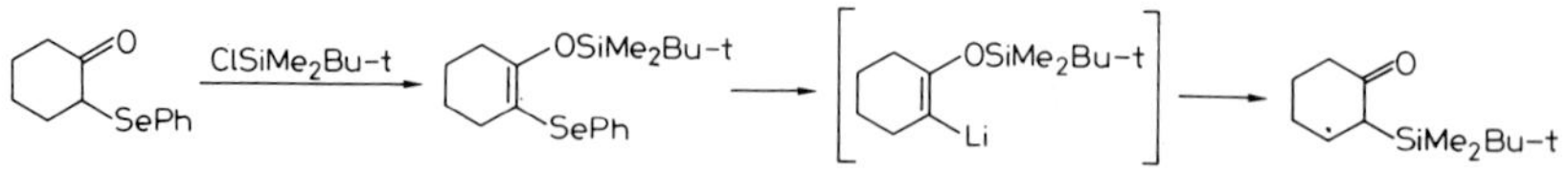

A soln. of 2-phenylselenocyclohexanone in tetrahydrofuran/hexamethylphosphoramide treated with Li diisopropylamide and *tert*-butyldimethylsilyl chloride at –78° → intermediate (Y 82%) in tetrahydrofuran treated with 6 equivalents of Li metal and 0.2 equivalents of 1-dimethylaminonaphthalene at 0°, and the mixture quenched with water after 1 hr. → 2-(*tert*-butyldimethylsilyl)cyclohexanone (Y 85%) and diphenyl diselenide (Y ca. 90%). F.e.s. I. Kuwajima, R. Takeda, Tetrahedron Letters *22,* 2381–4 (1981).

Sodium hydroxide/benzyltriethylammonium chloride $NaOH/[PhCH_2NEt_3]^+Cl^-$

Selenides from diselenides and halides RSeSeR → RSeR′

by electrolysis cf. *36,* 600; with NaOH and benzyltriethylammonium chloride as phase transfer catalyst s. J.V. Comasseto et al., J. Chem. Res. (S) *1982,* 212–3.

Sodium methoxide *NaOMe*

Reviews of crown polyethers ←

s. *31,* 592; *32,* 171; s.a. E. Weber, Kontakte (Darmstadt) *1982,* 24–35.

1-Dimethylaminonaphthalene s. under Li ←

Cuprous iodide s. under $NaBH_4$ *CuI*

Sodium tetrahydridoborate/cuprous iodide $NaBH_4/CuI$

Unsym. tellurides from ditellurides RTeTeR → RTeR′

with Mg cf. *18,* 719; with $NaBH_4$/CuI s. H. Suzuki et al., Chem. Letters *1981,* 1115–6.

tert-*Butyl hydroperoxide* t-*BuOOH*

2-Ketoselenides from ethylene derivs. and diselenides CH=C → CO–C–Se

s. *34,* 595; f.e.s. Bull. Chem. Soc. Japan *54,* 3100–5 (1981).

Stannous chloride $SnCl_2$

2-Ethyleneselenides from 2-ethylenesilanes and selenyl chlorides Si⋚ → SeAr

589. 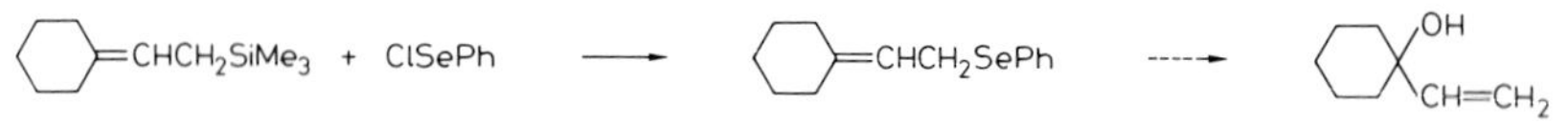

A soln. of phenylselenyl chloride in dichloromethane added to a soln. of startg. allylsilane in the same solvent at –78°, the mixture stirred for 10 min., warmed to 0°, a little anhydrous $SnCl_2$ added, and stirring continued for 1.5 hrs. → product. Y 80%. F.e., also oxidation to **2-ethylenealcohols** (cf. Synth. Meth. *34,* 143), and prepn. of **2-ethylenethioethers from sulfenyl chlorides with allyl rearrangement,** s. H. Nishiyama et al., Tetrahedron Letters *22,* 5285–8 (1981).

Phosphoryl chloride $POCl_3$

Phosphonic diamides from organoaluminium dichlorides $RAlCl_2 \rightarrow RPO(N\leqslant)_2$

590. $$EtAlCl_2 + POCl_3 \rightarrow [EtP(O)Cl_2.AlCl_3] \xrightarrow{HNEt_2} EtP(O)(NEt_2)_2$$

A soln. of ethylaluminium dichloride in benzene added slowly with stirring to a soln. of $POCl_3$ in the same solvent at room temp., refluxed for 15 hrs., diethylamine added at 5°, and refluxed again for 5 hrs. → product (isolated via the hydrochloride). Y 74%. F.e.s. G. Sonnek, H. Reinheckel, K.-G. Baumgarten, Z. Chem. *21,* 268–9 (1981).

Iodine I_2

Reactions with benzeneselenyl iodide prepd. *in situ* ○

Isocyclics from dienes

591.

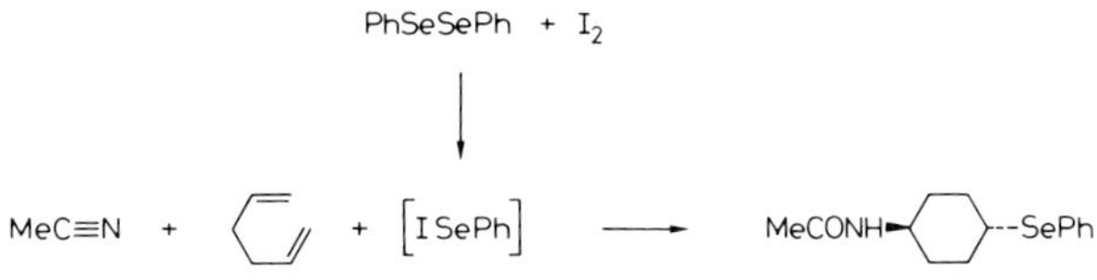

Cyclic 4-acylaminoselenides. Solid I_2 added to a soln. of diphenyl diselenide in acetonitrile, the mixture stirred for 10 min. at ambient temp., a soln. of hexa-1,5-diene in the same solvent added, and refluxed with stirring for 18 hrs. → 1-phenylseleno-4-acetamidocyclohexane. Y 75%. F.e.s. A. Toshimitsu, S. Uemura, M. Okano, J. Chem. Soc. Chem. Commun. *1982,* 87–9.

Benzyltriethylammonium chloride s. under NaOH $[PhCH_2NEt_3]^+Cl^-$

Tetrakis(triphenylphosphine)palladium(0) $Pd(PPh_3)_4$

Silanes from halides Hal → Si≡

with NaOMe cf. *31,* 592; from benzylic chlorides and methylchlorodisilanes with a little tetrakis(triphenylphosphine)palladium(0) s. H. Matsumoto et al., Chem. Letters *1982,* 399–400.

Iodo(phenyl)bis(triphenylphosphine)palladium $[Pd(Ph)I(PPh_3)_2]$

Stannanes from halides Hal → Sn≡

with bis(π-allylpalladium chloride) cf. *10,* 508s*36;* with iodo(phenyl)bis(triphenylphosphine)palladium s. Zh. Org. Khim. *17,* 905–10 (1981); with tetrakis(triphenylphosphine)palladium(0) cf. M. Kosugi et al., Chem. Letters *1981,* 829–30.

Carbon ↑ RemC↓↑C

Without additional reagents *w.a.r.*

Organobismuth compds. ←

s. *20,* 493; **review of preparation and reactions of organobismuth compds.** s. Chem. Rev. *82,* 59–76 (1982).

Sodium hydride *NaH*

Selenides from lactones C

Decarboxylation

592.

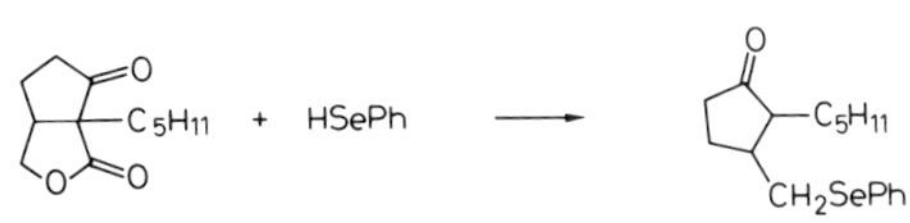

Startg. lactone treated with a mixture of phenylselenol and NaH in tetrahydrofuran/hexamethylphosphoramide at 25° for 2 hrs., and refluxed for 16 hrs. → product. Y 99%. Cf. Synth. Meth. *35,* 418. F.e. and conversion to ethylene derivs. s. D.J. Goldsmith, J.K. Thottathil, Tetrahedron Letters *22,* 2447–50 (1981).

Triethylamine s. under CuBr and CuHal₂ Et_3N

Cuprous bromide/triethylamine *CuBr/Et₃N*

Acetylene-1-selenides
from terminal acetylene derivs. and selenocyanates C≡CH → C≡CSeAr

593. PhSeCN + HC≡CBu → PhSeC→CBu

A mixture of 1-hexyne, phenyl selenocyanate, and 0.5 equivalents CuBr in dichloromethane treated with triethylamine at room temp. under argon, and the mixture stirred vigorously for 10 min. → product. Y 95%. F.e. and with CuCN s. S. Tomoda, Y. Takeuchi, Y. Nomura, Chem. Letters *1982,* 253–6.

Cupric halides/triethylamine *$CuHal_2/Et_3N$*

(E)-2-Halogeneneselenides C≡C → C(Hal)=CSeAr
from acetylene derivs. and selenocyanates

594.

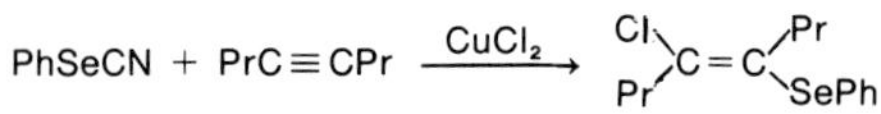

A little triethylamine added dropwise to a mixture of startg. alkyne, phenyl selenocyanate, and 2 equivalents $CuCl_2$ (or $CuBr_2$) in dichloromethane under argon at 26°, and stirred for 24 hrs. → product. Y 66–93%. F.e.s. S. Tomoda, Y. Takeuchi, Y. Nomura, Chem. Letters *1981,* 1715–8.

Zinc/trimethylsilyl chloride *Zn/Me_3SiCl*

α-Hydroxyphosphonic acid esters CO → $C(OH)PO(OR)_2$
from aldehydes and carbo-2,2,2-trichloroethoxyphosphonic esters

595. PhCHO + $CCl_3CH_2OCOP(O)(OEt)_2$ → $PhCH(OH)P(O)(OEt)_2$

Diethyl 2,2,2-trichloroethoxycarbonylphosphonate (prepared from 2,2,2-trichloroethyl chloroformate and triethyl phosphite at room temp. for 30 min.) treated with 1.1 equivalents of Zn powder and trimethylsilyl chloride in ether at room temp. for 2.5 hrs., benzaldehyde added, and worked up after 30 min. → product. Y 84%. F.e.s. M. Sekine, H. Yamagata, T. Hata, J. Chem. Soc. Chem. Commun. *1981,* 970–1.

Boron fluoride *BF_3*

α-Carbalkoxyaminophosphonic esters OR → $PO(OR)_2$
from 1,1-alkoxyurethans and phosphorous esters

596.

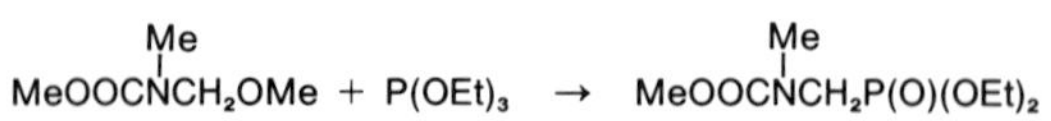

BF_3-etherate added dropwise to a stirred soln. of N-carbomethoxy-α-methoxydimethylamine and triethyl phosphite in methylene chloride at –20°, the mixture warmed to room temp., and stirred for 7 hrs. → product. Y 95%. F.e. and Lewis acids s. T. Shono, Y. Matsumura, K. Tsubata, Tetrahedron Letters *22,* 3249–52 (1981).

Acetic acid *MeCOOH*

α-Aminophosphonic acids
from aldehydes and urethans CHO → $C(NHR)–P(O)(OH)_2$
s. *36,* 608; **α-aminophosphinic acids** from dichlorophosphines cf. Monatsh. Chem. *113,* 59–71 (1982).

Tributylphosphine *Bu_3P*

Aromatic tellurides from alcohols and tellurocyanates OH → TeAr

597. n-$C_{12}H_{25}OH$ + PhTeCN → n-$C_{12}H_{25}$TePh

Tributylphosphine added in one portion to a stirred soln. of dodecyl alcohol and phenyl tellurocyanate in dichloromethane under N_2, and the mixture stirred for 30 min. at room temp. → dodecyl phenyl telluride. Y 62%. F.e., also prepn. of phenyl tellurocyanate, s. F. Ogura et al., Syn. Commun. *12,* 131–6 (1982).

*Tetra-*n*-butylammonium fluoride* n-$Bu_4N^+F^-$

Silylation with ethyl trimethylsilylacetate C≡CH → C≡CSi⋛

O-silylation s. *32,* 99; silylacetylenes from terminal acetylene derivs., also S-silylation, s. Bull. Chem. Soc. Japan *54,* 805–8 (1981).

Elimination ⇑

Oxygen ↑ RemC ⇑ O

Perchloric acid *$HClO_4$*

Selenazolium salts ○

s. *37,* 573

Halogen ↑ RemC ⇑ Hal

1,8-Diazabicyclo[5.4.0]undec-7-ene ←

Alkylidenephosphines PClCH → P=C

s. *34,* 599; f. eliminative methods cf. R. Appel, J. Peters, A. Westerhaus, Tetrahedron Letters *22,* 4957–60 (1981); **review of alkylidene- and alkylidyne-phosphines** s. Angew. Chem. Intern. Ed. *20,* 731–44 (1981).

Remaining Elements ↑ RemC ⇑ Rem

Without additional reagents *w.a.r.*

Alkylidynephosphines P≡C

598. Ph–C(SiMe₃)=P–Cl → Ph–C≡P

Chloro[phenyl(trimethylsilyl)methylene]phosphine pyrolysed on quartz wool at 700° in a flow system under high vacuum → phenylmethylidynephosphine which was condensed out at –196°. Y ca. 100%. F.e.s. R. Appel et al., Angew. Chem. Intern. Ed. *20,* 197 (1981).

Formation of C–C Bonds

Uptake ⇓

Addition to Hydrogen and Carbon CC⇓HC

tert-*Butyllithium* t-*BuLi*

Regiospecific *o*-metalation of methoxymethyl phenolethers ArH → ArCOOH
s. *32,* 601; f. electrophiles s. C.A. Townsend, L.M. Bloom, Tetrahedron Letters *22,* 3923–4 (1981); s.a. M.R. Winkle, R.C. Ronald, J. Org. Chem. *47,* 2101–8 (1982).

Addition to Oxygen and Carbon CC⇓OC

Electrolysis ϟ

Haloform condensation CO → C(OH)CCl_3
with KOH cf. *4,* 623; by electrolysis s. T. Shono et al., Tetrahedron Letters *22,* 871–4 (1981); f. electrophiles s. B. Strübing, P. Jeroschewski, A. Deutsch, Z. Chem. *20,* 442 (1980).

Potassium–silica gel *K–SiO_2*

Pinacols 2 CO → C(OH)C(OH)
s. *37,* 772

Sodium hydroxide/perchloric acid *NaOH/$HClO_4$*

1-β-Hydroxypyridinium from pyridinium salts and aldehydes via 2,3-dihydroxazolo[3,2-*a*]pyridines ←

599.

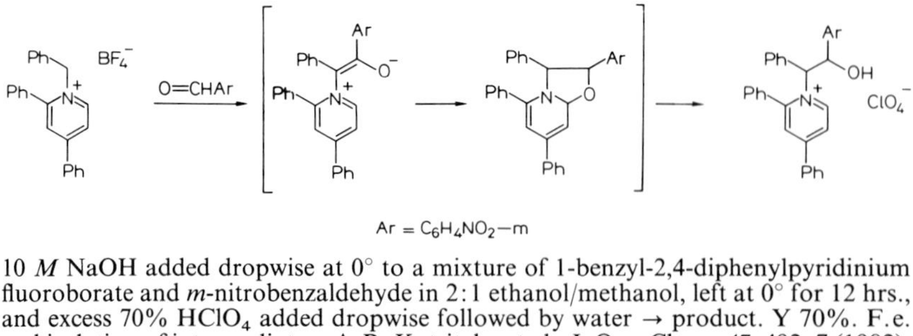

10 *M* NaOH added dropwise at 0° to a mixture of 1-benzyl-2,4-diphenylpyridinium fluoroborate and *m*-nitrobenzaldehyde in 2:1 ethanol/methanol, left at 0° for 12 hrs., and excess 70% $HClO_4$ added dropwise followed by water → product. Y 70%. F.e. and isolation of intermediate s. A.R. Katritzky et al., J. Org. Chem. *47,* 492–7 (1982).

Potassium hydroxide/1,2,5-trimethyl-4-piperidone ←

2-Acetylenealcohols from ketones CO → C(OH)C≡C
in liq. NH_3 cf. *19,* 734; in benzene containing 1,2,5-trimethyl-4-piperidone s. D.V. Sokolov et al., Izv. Akad. Nauk Kaz. SSR, Ser. Khim. *1981,* 62 (Russ.); Chem. Abstr. *95,* 42261 (1981).

Lithium 1,1-bis(trimethylsilyl)-3-methylbutoxide *$LiOC(SiMe_3)_2CCH_2CHMe_2$*

Directed aldol condensation CHO → CH(OH)C–CO

cf. *32,* 605; review s. T. Mukaiyama, Org. React. *28,* 203–331 (1982).

Sodium acetylide s. under Me_2AlCl *RNa*

Potassium tert-*butoxide* t-*BuOK*

Electrophilic substitution of aromatic silyl groups ←
Nucleophilic catalysis

600. 

Benzyloxysilanes. K *tert*-butoxide added to a stirred mixture of startg. silane and benzaldehyde in dimethylformamide at room temp., and the mixture allowed to react for 1 hr. → product. Y 80%. – With ar. silanes containing electron-donating groups, silyl substitution was achieved **via** the corresponding **tricarbonylchromium complexes** (p. 266). F.e. and electrophiles s. F. Effenberger, W. Spiegler, Angew. Chem. Intern. Ed. *20,* 265–6, 266–7 (1981).

Alkyllithium [s.a. under $(BuO)_3ZrCl$] *RLi*

Terminal ethylene derivs. from oxo compds. CO → $C(OH)CH_2CH_2Si\equiv$
via 3-hydroxysilanes
Synthesis with addition of two C-atoms

601. 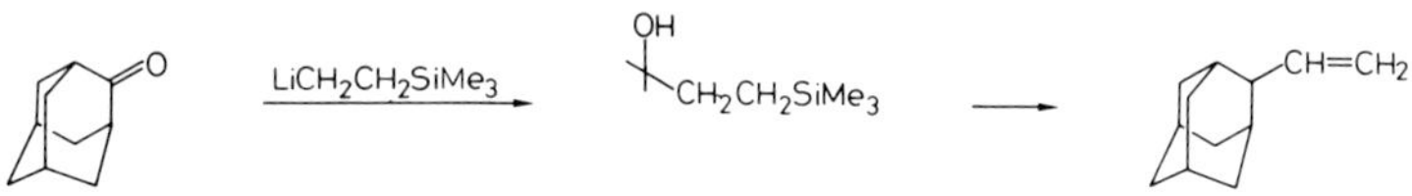

2-Adamantanone added to a soln. of [β-(trimethylsilyl)ethyl]lithium in ether at –78° → intermediate hydroxysilane (Y 72%), in dichloromethane treated with BF_3-acetic acid complex at 25° for 5 min. → product (Y 100%). F.e.s. S.R. Wilson, A. Shedrinsky, J. Org. Chem. *47,* 1983–4 (1982).

Synthesis of diols from lactones C
with Grignard compds. cf. *5,* 477; with organolithium compds. s. J. Lehmann, Arch. Pharm. *315,* 241–8 (1982).

Methyllithium/tert-*butyllithium* *MeLi*/t-*BuLi*

Aldol condensation with hindered oxo compds. CO → C(OH)C–CO
O,C-Dianions as intermediates

602. 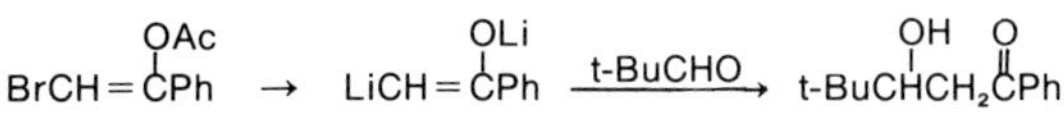

β-Hydroxyketones. Startg. bromoenol acetate in ether treated with methyllithium at 0° followed by *tert*-butyllithium at –78° to 0°, a slight excess of trimethylacetaldehyde added at –78°, and the soln. allowed to warm to 0° → product. Y 84%. – The dianions appear to be considerably more nucleophilic than simple enolates. F.e., **also styryl ketones from aromatic aldehydes,** s. C.J. Kowalski, K.W. Fields, J. Am. Chem. Soc. *104,* 1777–9 (1982).

n-*Butyllithium* n-*BuLi*

threo-**β-Hydroxycarboxylic acid thioamides** CO → CH(OH)C–C(S)N<
from carboxylic acid thioamides
s. *37,* 620

β-Hydroxysulfones from oxo compds. CO → C(OH)R
with ethylmagnesium bromide cf. *9,* 745; with *n*-butyllithium (cf. Synth. Meth. *29,* 796) s. J.-P. Fournier et al., Eur. J. Med. Chem. *17,* 53–8 (1982).

n-Butyllithium *n-BuLi*

Hydroxymethylation of sulfones $SO_2CH \rightarrow SO_2C\text{–}CH_2OH$

s. *37,* 961

β-Hydroxysulfoximines from ketones CO → C(OH)–C–S(=O)(=N–)

Asymmetric synthesis

s. *37,* 996

n-Butyllithium/chiral diamines ←

Sec. alcohols from aldehydes CHO → CH(OH)R

Asymmetric synthesis

s. *35,* 439; f. chiral diamines s. J.-P. Mazaleyrat, D.J. Cram, J. Am. Chem. Soc. *103,* 4585–6 (1981).

N,N′-Dimethyl-N,N′-propyleneurea (DMPU) as cosolvent ←

603. 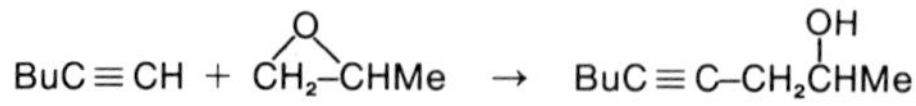

N,N′-Dimethyl-N,N′-propyleneurea (DMPU) can replace carcinogenic hexamethylphosphoramide as a safe cosolvent in many synthetically useful reactions occurring under strongly basic and nucleophilic conditions. E: A cold soln. of 1-hexyne in tetrahydrofuran treated dropwise with a soln. of butyllithium in hexane below 5°, the mixture stirred for 2 hrs. under ice-bath cooling, cooled to –76°, 50% by volume of N,N′-dimethyl-N,N′-propyleneurea added, the mixture stirred for 10 min., treated with methyloxirane, allowed to warm to room temp. during 6 hrs., stirred at this temp. for 12 hrs., and acidified with 10% acetic acid → 4-nonyn-2-ol. Y 59%. F. reactions s. T. Mukhopadhyay, D. Seebach, Helv. Chim. Acta *65,* 385–91 (1982).

3′-Metalation of 2-ethyleneamines ←

604. 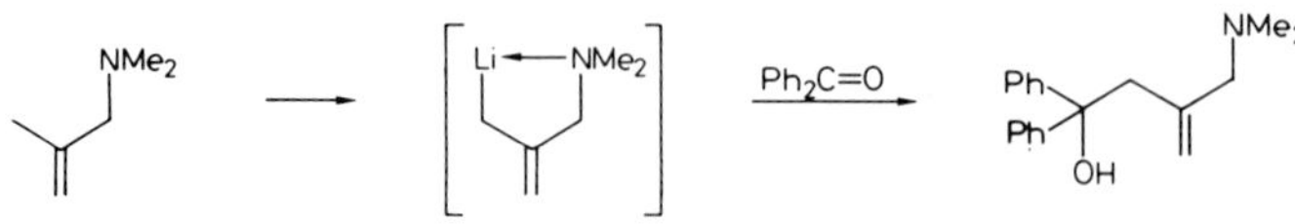

A 2.4 *M* soln. of *n*-butyllithium in hexane added dropwise to a stirred soln. of N,N-dimethylmethallylamine in dry tetrahydrofuran at 0° under N_2, the mixture stirred for 4 hrs. at 0°, benzophenone in tetrahydrofuran added in one portion, and stirring continued for a further 15 min. at 0° and 1.5 hrs. at room temp. → 4-(dimethylamino)-3-methylene-1,1-diphenylbutanol. Y 86%. F.e. and electrophiles, also conversion to 3-methylenetetrahydrofurans, s. J.J. Fitt, H.W. Gschwend, J. Org. Chem. *46,* 3349–52 (1981).

Oxazoles from lactones and isonitriles O

605. 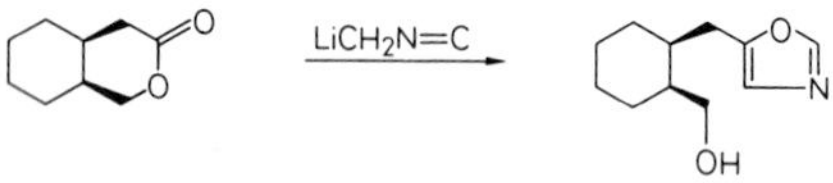

A soln. of startg. lactone in tetrahydrofuran added dropwise with efficient stirring to a cooled (–80°) suspension of lithiomethyl isocyanide (prepared by metalation of methyl isocyanide with *n*-butyllithium) in tetrahydrofuran-hexane, allowed to warm slowly to room temp. until an orange heterogeneous mixture resulted, stirred 1 hr., sufficient dry dimethylformamide added to make homogeneous, and stirring continued for a further 16 hrs. → 5-[[*cis*-2-(hydroxymethyl)cyclohexyl]methyl]oxazole. Y 64%. F.e.s. P.A. Jacobi, D.G. Walker, I.M.A. Odeh, J. Org. Chem. *46,* 2065–9 (1981).

Phthalides from carboxylic acid amides via *o*-lithiation

s. *19,* 738; heterocyclics via *o*-lithiated amides, review, s. V. Snieckus, Heterocycles *14,* 1649–76 (1980); *o*-lithiation of N,N-disubst. carboxylic acid amides with added TMEDA cf. P. Beak, R.A. Brown, J. Org. Chem. *47,* 34–46 (1982).

n-*Butyllithium/N,N,N',N'-tetramethylethylenediamine* n-*BuLi/TMEDA*

C-α-Metalation of benzyl ethers CO → C(OH)C(OR)Ar

606. 

Glycol monoethers. *n*-Butyllithium in hexane added dropwise via a rubber septum to a soln. of benzyl methyl ether and TMEDA in the same solvent at –10°, after 3 hrs. a soln. of benzophenone in *hexane* added dropwise at the same temp., and stirred for a further 3 hrs. → product. Y 76%. – In contrast to benzyl alcohols and amines, no *o*-lithiation was observed. F.e. and electrophiles s. M.K. Yeh, J. Chem. Soc. Perkin Trans. I *1981,* 1652–3. Review of allylic and benzylic carbanions substituted by heteroatoms s. J.F. Biellmann, J.-B. Ducep, Org. React. *27,* 1–344 (1982).

n-*Butyllithium/hexamethylphosphoramide* n-*BuLi/(Me₂N)₃PO*

3-Acetylenealcohols from oxido compds. (epoxide) → C(OH)C–C≡CH

s. *2,* 589s*37*

n-*Butyllithium/zinc chloride* n-*BuLi/ZnCl₂*

1,2-Addition to α,β-ethyleneketones C=C–CO → C=C–C(OH)R

of 1,3-dithianes cf. *33,* 639; of nitriles with added $ZnCl_2$ s. M.-C. Roux-Schmitt, L. Wartski, J. Seyden-Penne, J. Chem. Res. (S) *1980,* 346.

sec-*Butyllithium/N,N,N',N'-tetramethylethylenediamine* s-*BuLi/TMEDA*

2-sec-Aminoalcohols from oxo compds.
via metalation of N,N-disubst. 2,2-diethylbutyramides ←

607. 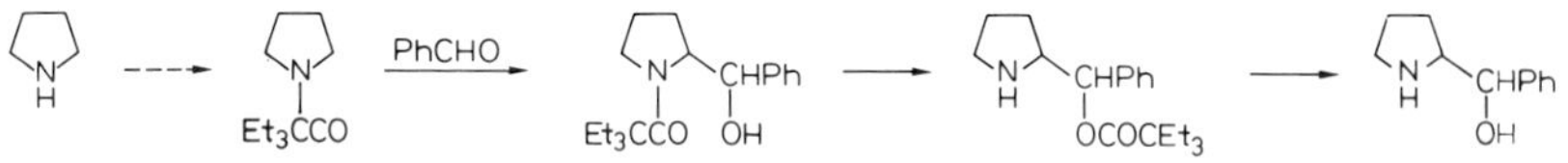

A soln. of the startg. amide in ether added to a soln. of TMEDA and *sec*-butyllithium in the same solvent at –78°, stirred for 45 min. at 0°, cooled to –78°, benzaldehyde added, allowed to warm to room temp., and the crude product refluxed in 2:1 methanol/concd. HCl for 17 hrs. → intermediate acoxyamine (Y 72%) heated for 35 hrs. with K *tert*-butoxide in *tert*-butanol → aminoalcohol (Y 91%). F.e.s. D.B. Reitz, P. Beak, A. Tse, J. Org. Chem. *46,* 4316–7 (1981).

tert-*Butyllithium* t-*BuLi*

α,β-Ethylenealdehydes from oxo compds. CO → C=CH–CHO
Synthesis with addition of two C-atoms

608. 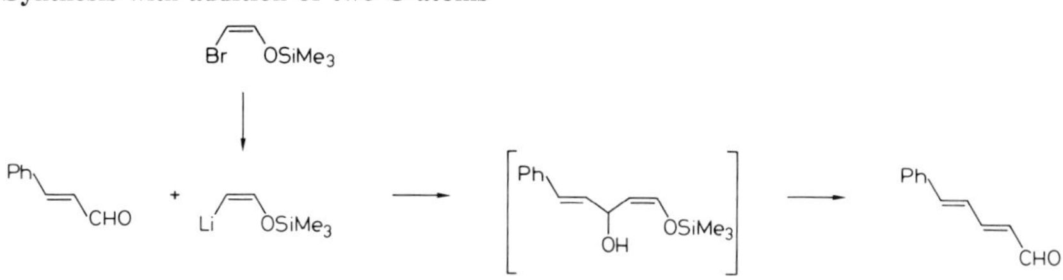

A soln. of *tert*-butyllithium in pentane added dropwise to a soln. of (Z)-2-bromo-1-(trimethylsiloxy)ethylene as **acetaldehyde carbanion equivalent** in ether at –70° under N_2, stirred for 90 min., cinnamaldehyde in the same solvent added during 5 min., after 30 min. at –75° the mixture allowed to warm to 0–5° during 20 min., treated with 1.5 *M* HCl in tetrahydrofuran, and worked up after 30 min. → 5-phenyl-2,4-pentadienal. Y 86%. F.e. with isolation of intermediates s. L. Duhamel, F. Tombret, J. Org. Chem. *46,* 3741–2 (1981).

tert-*Butyllithium* t-*BuLi*

Synthesis of sec. amines via metalation of formamidines CO → C(OH)C–N–CH(=N–)

609.

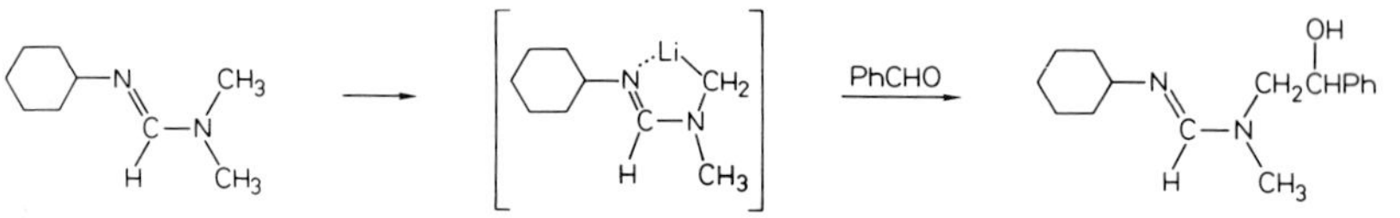

2-*sec*-Aminoalcohols. Startg. formamidine in tetrahydrofuran treated with *tert*-butyllithium at –78°, warmed to –25° for 1 hr., recooled to –78°, benzaldehyde added, the mixture warmed to room temp., and quenched with methanol/water at 20° over 15 hrs. → crude intermediate, treated with KOH in 2:1 methanol/water under reflux for 18 hrs. → product. Y 77%. F.e. and electrophiles s. A.I. Meyers, W. Ten Hoeve, J. Am. Chem. Soc. *102,* 7125–6 (1980).

tert-*Butyllithium/diethylaluminium chloride* t-*BuLi/*Et_2AlCl_3

Furans from alkoxy-2-acetylenes and aldehydes via 4-alkoxy-2,3-dienols CHO → CH(OH)C=C=COR

610.

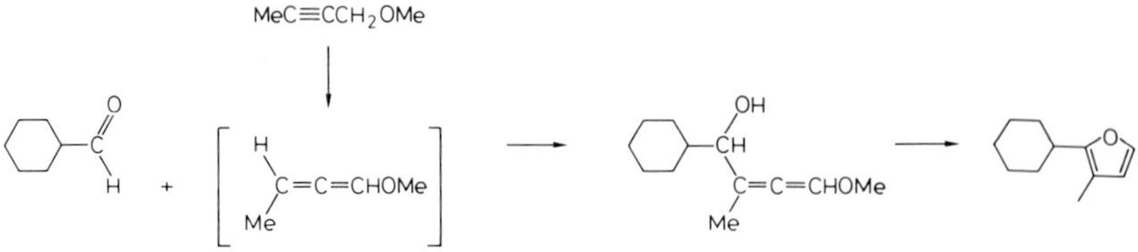

1.8 *M tert*-Butyllithium in pentane added dropwise to a stirred soln. of 1-methoxy-2-butyne in dry tetrahydrofuran under argon at –78°, stirring continued at this temp. for an additional 40 min., 1 *M* diethylaluminium chloride in hexane added dropwise, the mixture stirred for 10 min., freshly distilled cyclohexanecarbaldehyde added during 5 min., and stirring continued at –78° for 30 min. followed at 20° for 30 min. → crude intermediate, in dry methanol at 0° treated with 2 drops of HCl, and the mixture stirred at 20° for 1 hr. → 2-cyclohexyl-3-methylfuran. Y 82%. F.e., also with triisobutylaluminium, s. M. Ishiguro, N. Ikeda, H. Yamamoto, Chem. Letters *1982,* 1029–30.

Sodium acetylide *NaC≡CH*

Ethynylcarbinols from oxo compds. CO → C(OH)C≡CH

with Li acetylide/ethylenediamine cf. *20,* 499; with Na acetylide prepd. from acetylene and Na in liq. NH_3 s. A. Fisch, J.M. Coisne, H.P. Figeys, Synthesis *1982,* 211–4.

Lithium diisopropylamide [s.a. under $(C_5H_5)_2ZrCl_2$*]* *i-*Pr_2NLi

6-Metalation of uracils ←

611.

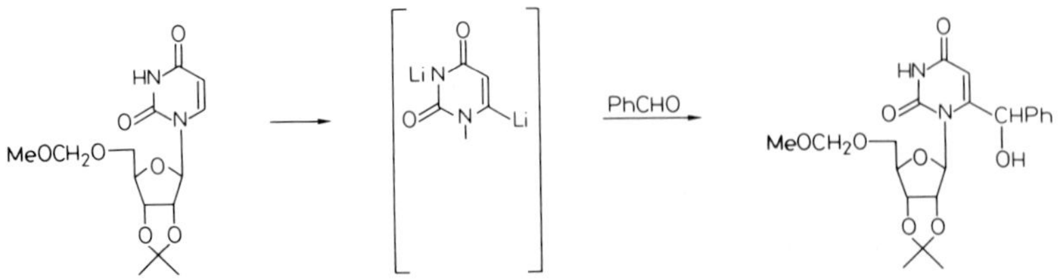

A soln. of 2′,3′-O-isopropylidene-5′-O-methoxymethyluridine in tetrahydrofuran added to a stirred soln. of Li diisopropylamide in the same solvent under nitrogen at –70°, the mixture stirred for 1 hr., freshly distilled benzaldehyde added dropwise maintaining the temp. below –70°, stirred at this temp. for 2 hrs., quenched with acetic acid, and allowed to warm to room temp. → 2′,3′-O-isopropylidene-5′-O-methoxymethyl-6-(α-hydroxybenzyl)uridine. Y 86%. F.e.s. H. Tanaka, H. Hayakawa, T. Miyasaka, Chem. Pharm. Bull. *29,* 3565–72 (1981).

α-Amino-β-hydroxynitriles
from α-aminonitriles and aldehydes CHO → CH(OH)C(N⊂)CN
s. *37,* 260

2-Hydroxythioethers from oxo compds. CO → C(OH)C(SR)
s. *29,* 262s*36;* α-arylthio-β-hydroxynitriles s. D. Morgans, G.B. Feigelson, J. Org. Chem. *47,* 1131–3 (1982); α-alkylthio-β-hydroxycarboxylic acids s. K. Tanaka, M. Terauchi, A. Kaji, Chem. Letters *1982,* 351–2.

Asymmetric synthesis of alcohols
from sulfoxides and oxo compds. CO → C(OH)CS(O)R
via 2-hydroxy-sulfoxides and -thioethers

612. 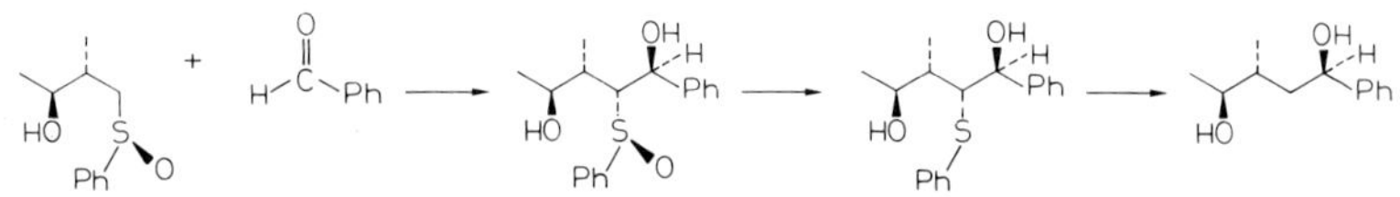

Optically active 1,4-diols. A soln. of startg. sulfoxide in tetrahydrofuran treated rapidly with 2 equivalents Li diisopropylamide in the same solvent at –78° for 1 min., benzaldehyde added, the mixture stirred for 30 sec., then aq. NH_4Cl added → intermediate hydroxysulfoxide (Y 77%), reduced with BH_3.THF in tetrahydrofuran at 25° for 24 hrs. → hydroxythioether (Y 95%), in ethanol treated with Raney nickel at 22° → product (Y 90–96%). This methodology affords both 1,2- and 1,3-asymmetric induction. F.e.s. D.R. Williams, J.G. Phillips, J.C. Huffman, J. Org. Chem. *46,* 4101–3 (1981).

1,2-Addition of enolates
of ester enolates cf. *33,* 641; of carboxylic acid enediolates to α,β-ethylenealdehydes, stereospecificity, s. J. Mulzer et al., Chem. Ber. *114,* 3701–24 (1981).

α-Hydroxyselenoacetals from selenoacetals and oxo compds. CO → C(OH)C(SeR)$_2$
s. *37,* 980

Lithium diisopropylamide/lithium dialkylcuprates *i-Pr$_2$NLi/LiCuR$_2$*

Syntheses via enolonium equivalents ←
α′-Subst. α,β-ethyleneketones

613. 

from α,β-oxidoketones. A soln. of startg. ketone in tetrahydrofuran added dropwise during 7 min. to a soln. of ca. 1.2 equivalents Li diisopropylamide in the same solvent at –78°, the mixture stirred at the same temp. for 0.5 hrs., 2.4 equivalents Li dimethylcuprate added, stirring continued at –23° for 2 hrs., and quenched with satd. NH_4Cl soln. at 0° → intermediate alcohol, refluxed in benzene with *p*-toluenesulfonic acid for 15 min. → product. Y 80%. F.e., also via enoxysilanes or enol phosphates, **and from α-halogenoketones,** s. P.A. Wender, J.M. Erhardt, L.J. Letendre, J. Am. Chem. Soc. *103,* 2114–6 (1981); **via enol phosphates** and reductive cleavage to cyclohex-2-en-1-ols s. J.P. Marino, H. Abe, J. Org. Chem. *46,* 5379–83 (1981); **via enoxysilanes** cf. J. Am. Chem. Soc. *104,* 3165–72 (1982); application to prostoglandin synthesis s. J. Org. Chem. *46,* 4389–93 (1981).

Lithium diisopropylamide/titanium tetrachloride/ titanium isopropoxide *i-$Pr_2NLi/TiCl_4/Ti(OPr-i)_4$*

Stereospecific Aldol condensation via titanium(IV)enolates CO → C(OH)C–CO

614.

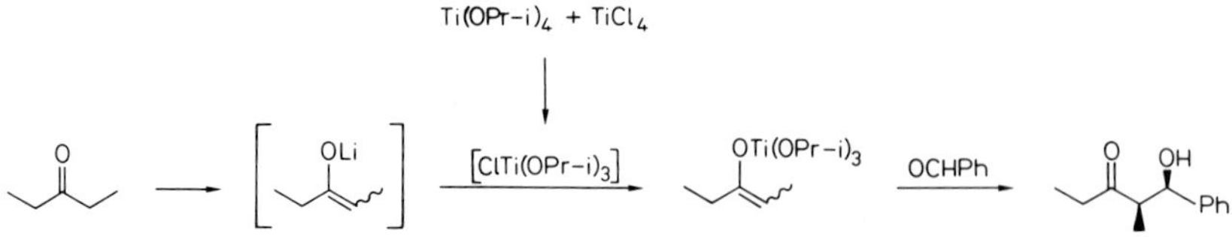

erythro-β-**Hydroxyketones.** A soln. of startg. Li enolate (prepared from the corresponding ketone and Li diisopropylamide) in tetrahydrofuran quenched with a mixture of $TiCl_4$ and titanium(IV) isopropoxide at –78°, the solvent removed at room temp., the Ti enolate obtained was dissolved in pentane, the soln. cooled to –120°, benzaldehyde added, stirring continued for 1 hr., and the mixture worked up with satd. aq. NH_4F → product. Y >70%. – The predominant *erythro*-selectivity is independent of geometry of the startg. enolate. F.e. and reagents s. M.T. Reetz, R. Peter, Tetrahedron Letters *22,* 4691–4 (1981).

Lithium dicyclohexylamide *R_2NLi*

α,β-Ethylenebromides from oxo compds. via 2,2-dibromalcohols
Synthesis with addition of one C-atom CO → C(OH)$CHBr_2$

615.

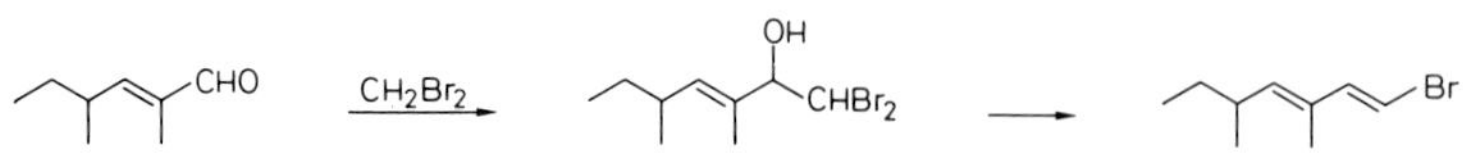

1-Bromo-1,3-dienes. A soln. of lithium dicyclohexylamide in tetrahydrofuran added to a mixture of dibromomethane and startg. carbonyl compd. in the same solvent at –78° → dibromoalcohol (Y 72%), refluxed with excess Zn dust and 2–6 equivalents of glacial acetic acid in methylene chloride for 4 hrs. → product (Y 93%, E/Z ratio 69:31). F.e.s. D.R. Williams et al., Tetrahedron Letters *22,* 3745–8 (1981).

Sodium iodide s. under $SnCl_2$ *NaI*

Sodium iodide/potassium cyanide/pyridine *$NaI/KCN/C_5H_5N$*

α-Siloxynitriles from oxo compds. CO → C(CN)OSi≡

s. *28,* 610; s.a. P.G. Gassman, J.J. Talley, Org. Synth. *60,* 14–20 (1981); with trimethylsilyl cyanide prepd. *in situ* from Me_3SiCl and KCN in the presence of NaI and pyridine s. F. Duboidin et al., Synthesis *1982,* 212–4.

N,N,N′,N′-Tetramethylethylenediamine s. under n-BuLi, s-BuLi *TMEDA*

N-Ethylpiperidine s. under Stannous triflate ←

Chiral diamines s. under Organolithium compd. ←

1,2,5-Trimethyl-4-piperidone s. under KOH ←

Pyridine s. under NaI *C_5H_5N*

Lithium dialkylcuprates s. under i-Pr_2NLi *$LiCuR_2$*

Cuprous iodide s. under Mg *CuI*

Calcium hydroxide *Ca(OH)₂*

C-Bis(hydroxymethylation) $CH_2 \rightarrow C(CH_2OH)_2$
with K_2CO_3 cf. *17,* 735; with $Ca(OH)_2$, simultaneous reduction of keto groups, s. D.L. Leland, M.P. Kotick, J. Med. Chem. *24,* 717–21 (1981).

Magnesium *Mg*

Grignard reaction in arenes CO → C(OH)R
s. *36,* 618; with 3-halogenamines in toluene s. A. Miodownik et al., Synth. Commun. *11,* 241–6 (1981).

Tert. alcohols from ketones
s. *1,* 680; carbohydrate derivs. s. M. Matsuzawa et al., Bull. Chem. Soc. Japan *54,* 3505–9 (1981).

Cyclic 1,2-divinylglycols from *o*-quinones CO → C(OH)CH=CH₂
s. *37,* 932

Stereospecific Grignard addition to carbohydrates C
Glycitols

616.

A soln. of 2,3,4-tri-O-benzyl-D-xylopyranose in tetrahydrofuran added dropwise to methylmagnesium iodide in ether at 0°, the mixture stirred for 1 hr. at room temp., refluxed for 1 hr., and worked up with satd. aq. NH_4Cl → 2,3,4-tri-O-benzyl-6-deoxy-D-iditol. Y 90%. F.e.s. Y. Tsuda, T. Nunozawa, K. Yoshimoto, Chem. Pharm. Bull. *28,* 3223–31 (1980).

Synthesis of 2-α-hydroxytetrahydrofurans from γ,δ-oxidoketones
Regiospecific intramolecular ring opening C O

617.

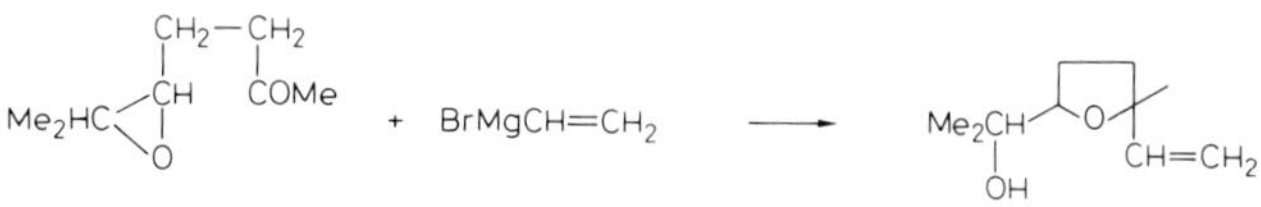

A soln. of the startg. oxidoketone in tetrahydrofuran treated dropwise with vinylmagnesium bromide in the same solvent at −15° under N_2, allowed to warm to room temp. slowly, and the mixture stirred for 12 hrs. → product. Y 75%. – The reaction produced no detectable trace of the tetrahydropyran isomer. F.e., also with organolithium compds., s. M. Chastrette, G.P. Axiotis, J. Organometal. Chem. *206,* 139–47 (1981).

Asym. synthesis with chiral diamines ←
s. *35,* 440; reactions with chiral diamines having a pyrrolidine ring s. Tetrahedron *37,* 4111–19 (1981).

Magnesium/cuprous iodide *Mg/CuI*

Synthesis of alcohols from oxido compds. epoxide → C(OH)CR

s. *35,* 441; 3-ethylene-5-silylalcohols s. H. Nishiyama, S. Narimatsu, K. Itoh, Tetrahedron Letters *22,* 5289–92 (1981).

2-Methylene-4-hydroxysilanes from oxido compds. ←
s. *37,* 198

Synthesis of carboxylic acids from β-lactones C
s. *36,* 620; ω-hydroxycarboxylic acids s. Chem. Letters *1982,* 569–70.

Magnesium/zirconocene dichloride *$Mg/(C_5H_5)_2ZrCl_2$*

Regio- and stereo-specific syntheses via 2-alkenylzirconium complexes
***threo*-3-Ethylenealcohols from aldehydes** CHO → CH(OH)C–C=C

618. 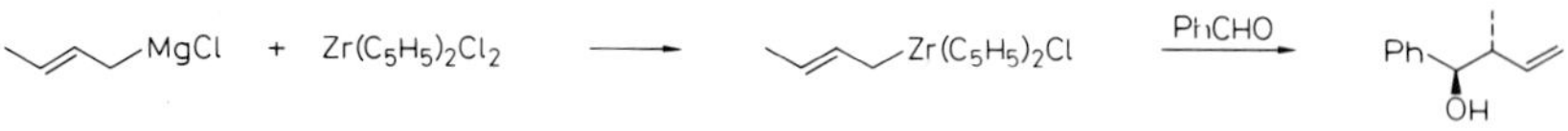

A soln. of startg. crotylzirconium complex (prepared from crotylmagnesium chloride and zirconocene dichloride) in tetrahydrofuran treated with benzaldehyde at –78°, and quenched with water at –30° to 0° → product. Y 95% (85% *threo*-isomer). F.e.s. Y. Yamamoto, K. Maruyama, Tetrahedron Letters *22,* 2895–8 (1981).

Zinc *Zn*

Synthesis with ultrasonically-dispersed zinc
2,2,2-Trifluoroalcohols from oxo compds. CO → $C(OH)CF_3$

619. $PhCHO + CF_3I \rightarrow PhCH(OH)CF_3$

A mixture of Zn powder, CF_3I, and benzaldehyde in dimethylformamide sonicated in the water bath of a 32 KHz ultrasound laboratory cleaner for 30 min. → phenyltrifluoromethylcarbinol. Y 72%. – No product was detected without sonication. F.e.s. T. Kitazume, N. Ishikawa, Chem. Letters *1981,* 1679–80.

Dimethylmagnesium *Me_2Mg*

Synthesis of alcohols from oxido compds.  → C(OH)CR

s. *16,* 729; improved prepn. of dialkylmagnesium compds. by hydromagnesation of ethylene derivs. s. B. Bogdanović, M. Schwickardi, P. Sikorsky, Angew. Chem. Intern. Ed. *21,* 199–200 (1982).

Isopropylmagnesium bromide *i-PrMgBr*

Stereospecific Aldol condensation with carboxylic acid thioamides CO → C(OH)–C–CSN$\lt$

620. 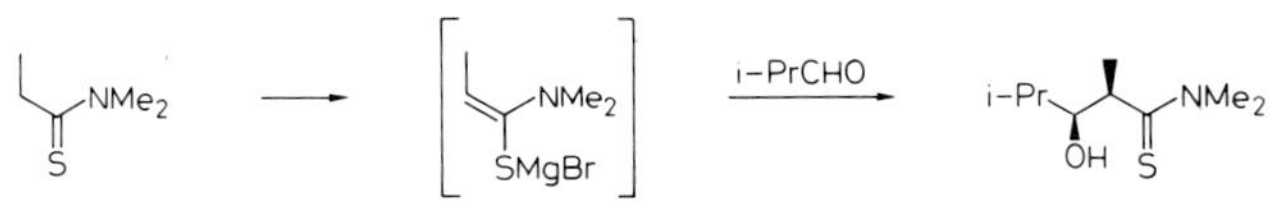

erythro-β-**Hydroxycarboxylic acid thioamides.** A soln. of startg. thioamide in tetrahydrofuran treated in one portion with isopropylmagnesium bromide in the same solvent at room temp., stirred for 4 hrs., a soln of isobutyraldehyde in tetrahydrofuran added at –78°, and the mixture stirred for 30 min. → *erythro*-product. Y 85–100%. F.e. and base, **also stereospecific thio-Claisen rearrangement,** s. Y. Tamaru et al., J. Am. Chem. Soc. *102,* 7806–8 (1980); *threo*-derivs. with *n*-butyllithium **via dianions** cf. ibid. *104,* 4018–9 (1982).

n-*Propylmagnesium bromide/titanocene dichloride* n-*$PrMgBr/(C_5H_5)_2TiCl_2$*

3-Ethylenealcohols from β,γ-ethylenehalides and aldehydes via 2-ethylenetitanium complexes
Regio- and stereo-specific synthesis CHO → CH(OH)C–C=C

621. 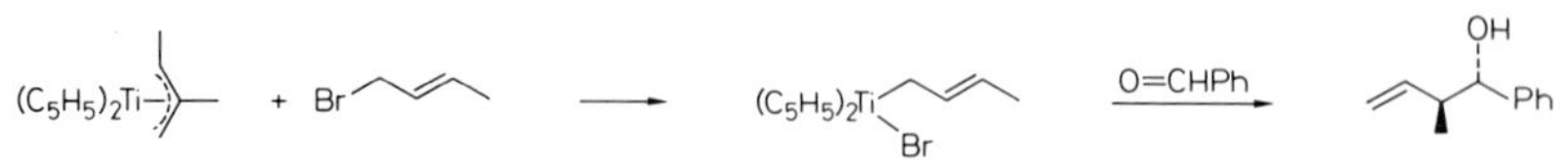

A mixture of startg. η^3-allyltitanium complex (prepared *in situ* from titanocene dichloride and propylmagnesium bromide in the presence of isoprene), 1-bromo-*trans*-

but-2-ene, and benzaldehyde in ether allowed to react at –30° to 35° under argon → *threo*-product. Y 92%. – Stereoselection is dependent on the halogen atom of the Ti complex rather than the bulk of the aldehyde. F.e.s. F. Sato et al., J. Chem. Soc. Chem. Commun. *1981,* 1140–1.

Dialkylzinc s. under $TiCl_4$ — ZnR_2

Zinc chloride s. under n-*BuLi* — $ZnCl_2$

Aluminium amalgam — *Al/Hg*

2-Silyl-2-allenealcohols from oxo compds. CO → C(OH)C(Si≡)=C=C

622. 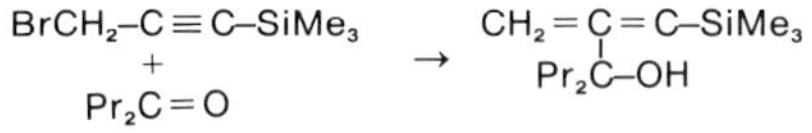

A soln. of startg. organoaluminium complex (prepd. by refluxing trimethylsilylpropargyl bromide with Al amalgam for 3–5 hrs. in tetrahydrofuran) treated with heptan-4-one at 0° for 1 hr., and quenched with satd. NH_4Cl → product. Y 81%. F.e., **also 3-acetylene-4-silylalcohols** with the corresponding Reformatskii reagent, s. R.G. Daniels, L.A. Paquette, Tetrahedron Letters *22,* No. 17, 1579–82 (1981).

Lithium tetrahydridoaluminate s. under $SnCl_2$, $CrCl_3$ — $LiAlH_4$

Dimethylaluminium chloride/sodium acetylide — Me_2AlCl/RNa

3-Acetylenealcohols from oxido compds.  → C(OH)C–C≡CH

with Na acetylide cf. *2,* 589; with ethynyldimethylaluminium etherate, prepd. from dimethylaluminium chloride and Na acetylide, s. T.F. Murray et al., J. Am. Chem. Soc. *103,* 7520–8 (1981); with *n*-butyllithium in the presence of hexamethylphosphoramide cf. J.-L. Gras, M. Bertrand, Compt. Rend. Ser. II *293,* 675–6 (1981).

Diethylaluminium chloride s. under t-*BuLi* — Et_2AlCl

Cerium amalgam — *Ce/Hg*

Synthesis of tert. alcohols from ketones and iodides CO → C(OH)R

623. 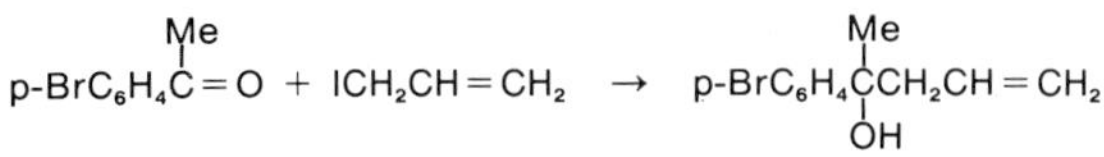

3-Ethylenealcohols. A mixture of allyl iodide and *p*-bromoacetophenone in tetrahydrofuran added at 0° under N_2 to Ce amalgam (prepared from Ce ingots and $HgCl_2$), and stirred at 0° for 5 hrs. then at room temp. for 1 hr. → 4-(*p*-bromophenyl)-4-hydroxy-1-pentene. Y 95%. – Esters, nitriles, and ar. bromides are unaffected. F.e., **also 2-arylalcohols** from benzyl iodide, s. T. Imamoto et al., Tetrahedron Letters *22,* 4987–8 (1981).

Cerium/iodine — Ce/I_2

Pinacols from two oxo compd. molecules 2 ArCHO → ArCH(OH)CH(OH)Ar

624. 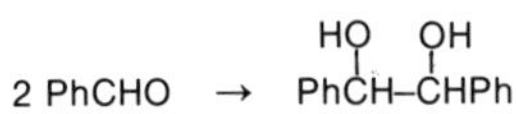

I_2 added to a mixture of freshly cut Ce and benzaldehyde in tetrahydrofuran at 0° under argon, and the mixture stirred at this temp. for 2 hrs. and at room temp. for 12 hrs. → product. Y 94%. – The reaction is applicable to aliphatic and aromatic carbonyl compounds, and leaves a range of functional groups, such as esters, nitriles, and vinylic halides, intact. F.e. and with other low-valent cerium halides s. T. Imamoto et al., Tetrahedron Letters *23,* 1353–6 (1982).

Ytterbium *Yb*

Syntheses with organoytterbium iodides CO → C(OH)R
Alcohols from oxo compds.

625.

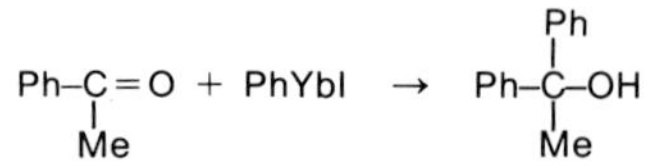

A soln. of phenylytterbium iodide (prepd. from powdered Yb and iodobenzene) in tetrahydrofuran treated with a soln. of acetophenone in the same solvent at –20°, the mixture stirred overnight at room temp., refluxed for 2 hrs., and hydrolysed with aq. 2 *M* HCl → product. Y 78%. F.e., reactions, and comparison with Grignard reactions, s. T. Fukagawa et al., Chem. Letters *1981,* 1771–4; **2-acetylenealcohols** from ytterbium acetylides s. G.B. Deacon, T.D. Tuong, J. Organometal. Chem. *205,* C4–6 (1981).

Cyclic dipeptide ←

Asym. induction with synthetic enzymes CHO → CH(OH)CN
Cyanohydrins from aldehydes

626.

A mixture of benzaldehyde, HCN, and a little cyclo(L-phenylalanyl-L-histidine) in benzene heated at 35° under N_2 with stirring for 1hr. → (R)-cyanohydrin. Y 76% (optical Y 76%). F.e.s. J. Oku, S. Inoue, J. Chem. Soc. Chem. Commun. *1981,* 229–30; Makromol. Chem. *183,* 579–86 (1982).

Thiazolium salts ←

Acyloins from two aldehyde molecules 2 CHO → CH(OH)CO
s. *29,* 630; s.a. H. Stetter, R.Y. Rämsch, Synthesis *1981,* 477–8.

Bis(2-imidazolidinylidenes) ←

Benzoins from two aldehyde molecules
in aprotic non-polar medium 2 ArCHO → ArCH(OH)COAr

627.

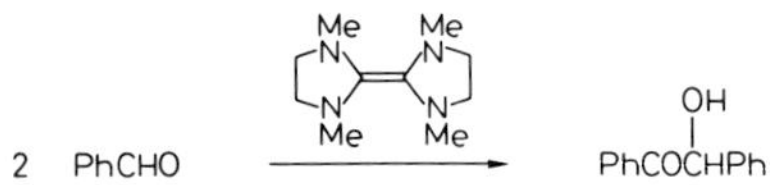

A little bis(1,3-dimethyl-2-imidazolidinylidene) and benzaldehyde heated in benzene at 80° for 30 min. under inert atmosphere → benzoin. Y 76%. F.e.s. M.F. Lappert, R.K. Maskell, J. Chem. Soc. Chem. Commun. *1982,* 580–1.

Cation exchanger ←

1,3-Dioxanes from ethylene derivs. and aldehydes O

628.

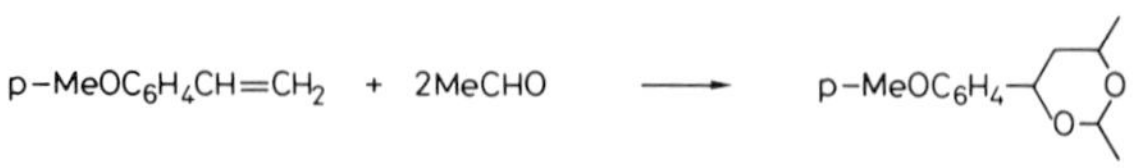

A mixture of sulfonic acid cation exchange resin, e.g. SP120 Lewatit, and toluene treated with startg. alkene and acetaldehyde, and stirred at 40° for 12 hrs. → product. Y 90%. – The method overcomes a major drawback of the Prins reaction: namely its limitation to formaldehyde (cf. Synth. Meth. *4,* 637). F.e.s. R. El Gharbi, M. Delmas, A. Gaset, Synthesis *1981,* 861–2; **3-phase acid catalysis** with cation exchangers cf. Tetrahedron Letters *22,* 723–4 (1981).

Tris(diethylamino)sulfonium trimethyldifluorosilicate *$[(Et_2N)_3S]^+[Me_3SiF_2]^-$*

β-Siloxyketones from enoxysilanes and aldehydes CHO → CH(OSi≡)C–CO
Stereospecific Aldol-type condensation
via tris(diethylamino)sulfonium enolates

629.

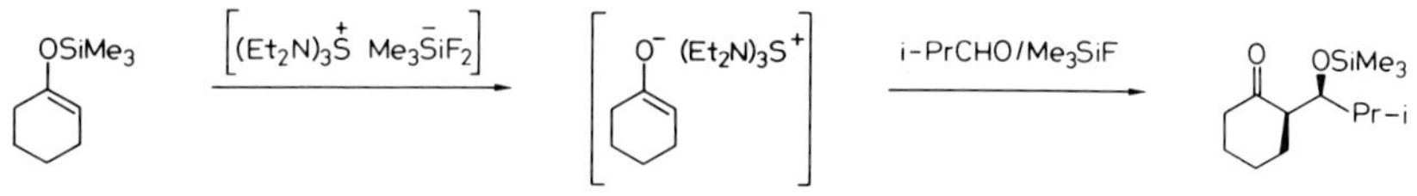

A mixture of startg. enoxysilane, 1.3 equivalents isobutyraldehyde, 4 equivalents trimethylfluorosilane, and a little tris(diethylamino)sulfonium trimethyldifluorosilicate in tetrahydrofuran allowed to stand at −78° for 8 hrs. → *erythro*-product. Y 67%. – The moderate to very high level of *erythro* selection is independent of the geometry of the startg. enolate. F.e., also new procedures for establishing *erythro/threo* **nomenclature,** s. R. Noyori, I. Nishida, J. Sakata, J. Am. Chem. Soc. *103,* 2106–8 (1981).

Silica gel s. under K *SiO_2*

Stannous triflate/N-ethylpiperidine ←
Stereospecific Aldol condensation via tin(II) enolates CO → C(OH)–C–CO

630.

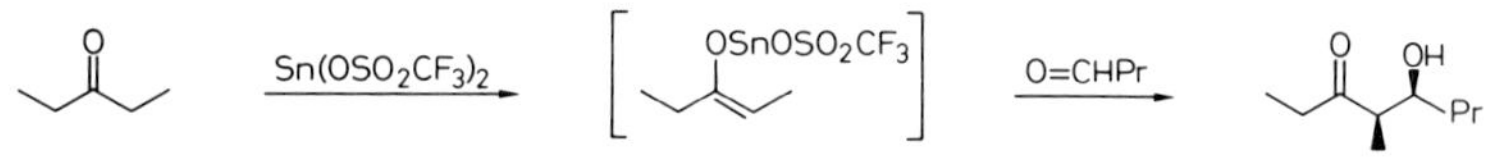

erythro-β-**Hydroxyketones.** A stirred suspension of tin(II) triflate and N-ethylpiperidine in dichloromethane treated dropwise with a soln. of 3-pentanone in the same solvent at −78° under argon, stirring continued for 30 min., a soln. of *n*-butyraldehyde in the same solvent added dropwise at −78°, allowed to stand for 2,5 hrs., and poured into a vigorously stirred mixture of dichloromethane and neutral phosphate buffer at 0° → 5-hydroxy-4-methyl-3-octanone. Y 86% (>91% *erythro*). The reaction proceeds **under** essentially **neutral conditions.** F.e., **also self-condensation of ketones,** s. T. Mukaiyama, R.W. Stevens, N. Iwasawa, Chem. Letters *1982,* 353–6; **via enoxystannanes** cf. Y. Yamamoto, H. Yatagai, K. Maruyama, J. Chem. Soc. Chem. Commun. *1981,* 162–5; **asym. aldol condensation** using chiral diamines s. Chem. Letters, *1982,* 1441–4.

Titanium(IV) isopropoxide s. under i-Pr_2NLi *$Ti(OPr\text{-}i)_4$*

Titanocene dichloride s. under n-*PrMgBr* *$(C_5H_5)_2TiCl_2$*

Tributoxyzirconium chloride/methyllithium *$(BuO)_3ZrCl/MeLi$*
Syntheses with organozirconium(IV) compds. CO → C(OH)R
Alcohols from oxo compds.

631.

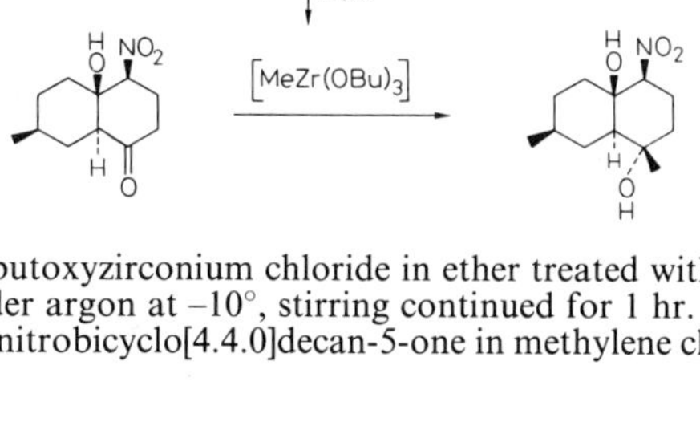

A stirred soln. of tributoxyzirconium chloride in ether treated with methyllithium in the same solvent under argon at −10°, stirring continued for 1 hr. at 0°, a soln. of 1-hydroxy-8-methyl-2-nitrobicyclo[4.4.0]decan-5-one in methylene chloride added, and the mixture warmed to room temp. with stirring overnight → 5,8-dimethyl-2-nitrobicyclo[4.4.0]decan-1,5-diol. Y 74%. – Compared with organo-lithium and -titanium compds., organozirconium reagents are of exceedingly low basicity and highly selective as aldehyde and ketone carbonylophiles. F.e. and selectivity s. B. Weidmann, C.D. Maycock, D. Seebach, Helv. Chim. Acta *64*, 1552–7 (1981).

Zirconocene dichloride (s.a. under Mg) $(C_5H_5)_2ZrCl_2$

Zirconocene dichloride/lithium diisopropylamide $(C_5H_5)_2ZrCl_2/i\text{-}Pr_2NLi$

Stereospecific Aldol condensation via zirconium(IV) enolates CO → C(OH)–C–CO

s. *36*, 625; **asym. synthesis** of β-hydroxycarboxylic acids s. J. Am. Chem. Soc. *103*, 2876–8 (1981).

Titanium trichloride $TiCl_3$

Glycols from oxo compds. C(OH)C(OH)

pinacols with $TiCl_4/Zn$ cf. *30*, 561; unsym. glycols with $TiCl_3$ s. A. Clerici, O. Porta, J. Org. Chem. *47*, 2852–6 (1982).

Titanium tetrachloride (s.a. under i-Pr_2NLi*)* $TiCl_4$

Titanium tetrachloride/dialkylzinc $TiCl_4/ZnR_2$

Stereospecific synthesis of alcohols from oxo compds. CO → C(OH)R

s. *36*, 624; with octahedral titanium(IV) reagents s. Synth. Commun. *11*, 647–54 (1981); **asymmetric synthesis** of sec. alcohols with chiral organotitanium compds. s. A.G. Olivero, B. Weidmann, D. Seebach, Helv. Chim. Acta *64*, 2485–8 (1981).

Stannous halides $SnHal_2$

3-Ethylenealcohols
from β,γ-ethyleneiodides and oxo compds. CO → C(OH)–C–C=C

632. $$CH_2{=}CHCH_2I + SnF_2 \rightarrow [CH_2{=}CHCH_2SnF_2I] \xrightarrow{O=CHPh} CH_2{=}CHCH_2CH(OH)Ph$$

A suspension of SnF_2, allyliodide, and benzaldehyde in 1,3-dimethyl-2-imidazolidone stirred for 1 hr. at room temp. → allylphenylcarbinol. Y 88%. F.e. and with $SnCl_2$ or $SnBr_2$ s. T. Mukaiyama, T. Harada, S. Shoda, Chem. Letters *1980*, 1507–10; **acoxy-3-ethylenes** with SnF_2 in the presence of acyl chlorides ibid. *1981*, 1109–10.

Stannous fluoride SnF_2

2,2,2-Tribromoalcohols from aldehydes CHO → $CH(OH)CBr_3$

633.

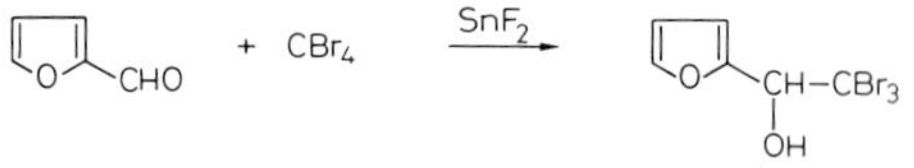

2 Equivalents of solid SnF_2 added in two portions at 5 min. intervals to a stirred soln. of furfural and 2.5 equivalents of carbon tetrabromide in dimethyl sulfoxide under argon at room temp., the mixture stirred for a further 5 min., dil. with dichloromethane and *M* HCl, and stirring continued for 30 min. → 1-(2-furyl)-2,2,2-tribromoethanol. Y 80%. – The reaction is selective and proceeds with high **asym. induction** in the case of aldehydes with an asymmetric centre at the α-carbon. F.e.s. T. Mukaiyama, M. Yamaguchi, J. Kato, Chem. Letters *1981*, 1505–8; **perfluoroalkyl carbinols** s. ibid. *1981*, 1337–8.

Stannous chloride/sodium iodide $SnCl_2/NaI$

2-Allenealcohols
from oxo compds. and 2-acetylenehalides CO → C(OH)–C=C=C

with $CrCl_3/LiAlH_4$ cf. *34*, 615; with $SnCl_2/NaI$ s. T. Mukaiyama, T. Harada, Chem. Letters *1981*, 621–4.

Stannous chloride/lithium tetrahydridoaluminate *$SnCl_2/LiAlH_4$*

β-Hydroxycarboxylic acid esters
from α-bromocarboxylic acid esters and oxo compds. CO → C(OH)C–COOR
Reformatskii-type synthesis under mild conditions

634.

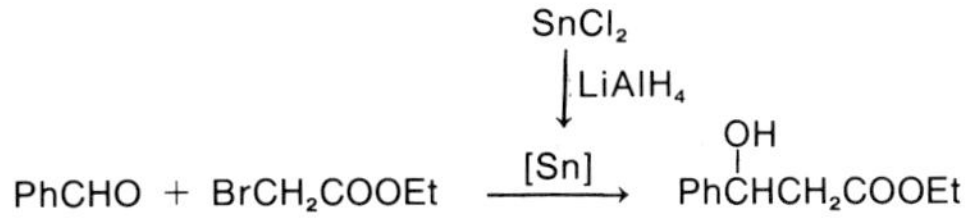

$LiAlH_4$ added portionwise to a suspension of anhydrous $SnCl_2$ in tetrahydrofuran under argon, the resulting suspension of *activated Sn* treated dropwise with a soln. of ethyl bromoacetate in tetrahydrofuran followed by benzaldehyde in the same solvent at room temp., and stirred for an additional 2 hrs. → ethyl 3-hydroxy-3-phenylpropionate. Y 83%. – The method overcomes one of the major problems associated with the classical Zn procedure, namely initiation and control of the subsequent exothermic reaction. F.e. solvents, and stereospecificity s. T. Harada, T. Mukaiyama, Chem. Letters *1982,* 161–4; *erythro-β*-**hydroxyketones** cf. ibid. *1982,* 467–70; **regiospecific synthesis of 3-ethylenealcohols from β,γ-ethyleneiodides** s. ibid. *1981,* 1527–8.

Hexamethylphosphoramide s. under n-*BuLi* *$(Me_2N)_3PO$*

Chromium(III) chloride/lithium tetrahydridoaluminate *$CrCl_3/LiAlH_4$*

3-Ethylenealcohols from oxo compds. CO → C(OH)–C–C=C

s. *34,* 614; stereochemistry s. M.D. Lewis, Y. Kishi, Tetrahedron Letters *23,* 2343–6 (1982).

Iodine s. under Ce *I_2*

Perchloric acid s. under NaOH *$HClO_4$*

Hydrogen chloride *HCl*

Cyanohydrins from aldehydes CHO → CH(OH)CN

s. *12,* 722; *α,β*-dihydroxynitriles s. W. Althoff, R. Karsdorf, P. Tinapp, Arch. Pharm. *314,* 518–24 (1981).

Cobalt carbonyl/carbonylhydridotris(triphenylphosphine)rhodium(I) *$Co_2(CO)_8/(Ph_3P)_3Rh(CO)H$*

α-Acylaminocarboxylic acids
from 2-ethylenealcohols C=C–CHOH → CHCHC(NHAc)COOH
One-pot synthesis

635.

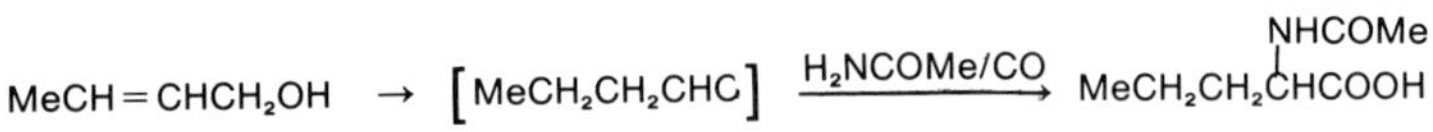

A mixture of crotyl alcohol and acetamide in dioxane treated with a little $Co_2(CO)_8$ and carbonylhydridotris(triphenylphosphine)rhodium in a stainless steel autoclave, the reactor charged with a 1:1 mixture of carbon monoxide and hydrogen to an initial pressure of 100 atm. at room temp., and heated at 110° for 17 hrs. with stirring → N-acetylnorvaline. Y 70%. – This procedure may be used to replace the two-step synthesis represented by Synth. Meth. *21,* 191 and *28,* 613. F.e. and co-catalysts s. K. Hirai, Y. Takahashi, I. Ojima, Tetrahedron Letters *23,* 2491–4 (1982).

Palladium(II) complex ←

Regio- and geo-specific synthesis of 2-ethylenealcohols from 3-ethyleneoxido compds. ←

636.

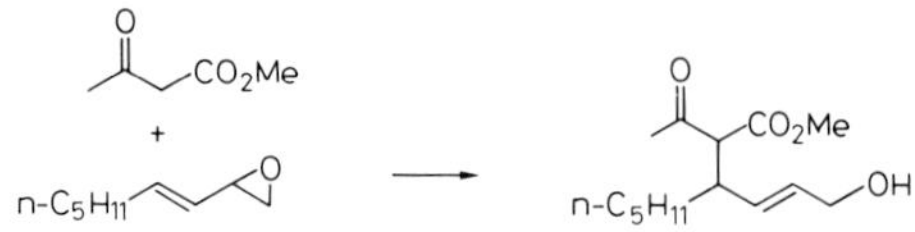

A mixture of startg. epoxide and methyl acetoacetate in tetrahydrofuran containing a little (tribenzylideneacetylacetonato)palladium(II) complex and 2-ethyl-2-(hydroxymethyl)-1,3-propanediol phosphite allowed to react at room temp. for 3 hrs. → trans-product. Y 65%. F.e., nucleophiles, and catalysts, s. J. Tsuji, H. Kataoka, Y. Kobayashi, Tetrahedron Letters *22,* 2575–8 (1981); with palladium(0) catalysis cf. B.M. Trost, G.A. Molander, J. Am. Chem. Soc. *103,* 5969–72 (1981).

Addition to Nitrogen and Carbon CC⇓NC

Without additional reagents *w.a.r.*

Reactions with stable sulfonium ylids ←

s. *24,* 674; **review** s. Y.V. Belkin, N.A. Polezhaeva, Russ. Chem. Rev. *50,* 481–97 (1981).

1-Carbamyllactones from aldehydocarboxylic acids and isonitriles ○
Intramolecular Passerini reaction

637.

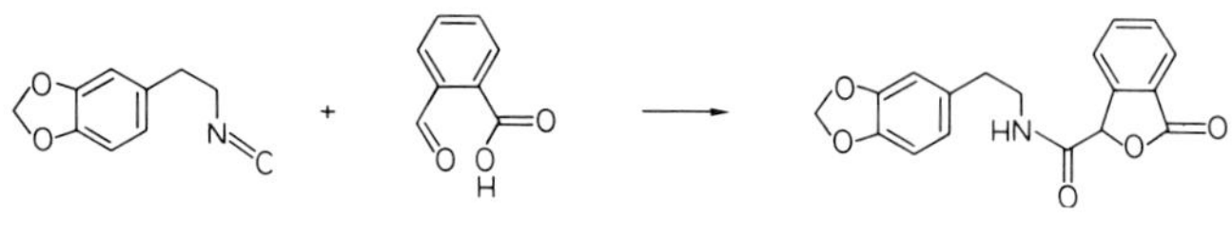

3-Carbamylphthalides. A soln. of 2-carboxybenzaldehyde in methylene chloride treated dropwise with a soln. of startg. isonitrile in the same solvent, and the mixture stirred for 3 hrs. → product. Y 78%. F.e.s. J.R. Flack, S. Manna, Tetrahedron Letters *22,* 619–20 (1981).

β-Diketones from nitriles and enoxyboranes via 1*H*-2,3,1-boroxazines

638.

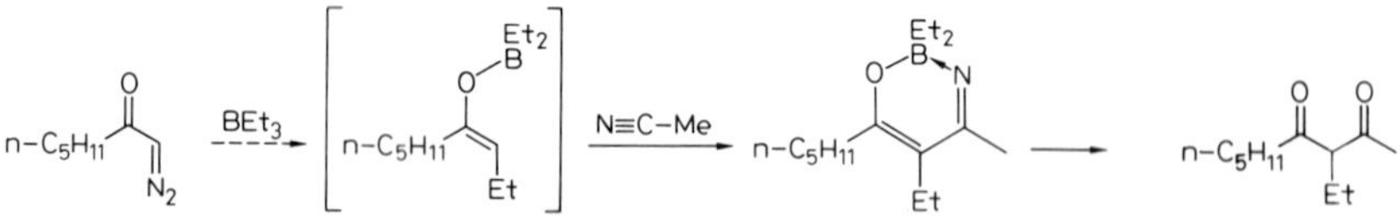

Acetonitrile added to a soln. of startg. enoxyborane (prepd. *in situ* from triethylborane and 1-diazo-2-heptanone) in tetrahydrofuran, and heated at 50° for 12 hrs. → 2,2,5-triethyl-6-methyl-4-pentyl-1*H*-2,1,3-boroxazine (Y 87%) in 4:1 ethanol/water treated with 3 drops of concd. HCl, and the soln. refluxed ca. 2 hrs. → 3-ethylnonane-2,4-dione (Y 96%). F.e.s. J. Hooz, J. Oudenes, Syn. Commun. *12,* 189–94 (1982).

Sodium alkoxide *NaOR*

2-Pyridones

s. *34,* 619; 4,6-diamino-derivs. s. E. Schmitz, S. Schramm, J. Prakt. Chem. *324,* 82–4 (1982).

Organolithium compd. *RLi*

Synthesis of tert. amines from azomethinium salts $C=\overset{+}{N}\lessdot \rightarrow CRN\lessdot$

s. *32,* 618; with unsatd. organometallics, producing e.g. 3-acetyleneamines, s. G. Courtois, M. Harama, P. Miginiac, J. Organometal. Chem. *218,* 1–15, 275–98 (1981).

Diisopropylamine s. under EtMgBr *i-Pr_2NH*

Ethyldiisopropylamine s. under $Bu_2BOSO_2CF_3$ *i-Pr_2NEt*

Pyridine *C_5H_5N*

α-Acoxycarboxylic acid amides from oxo compds. ←
Passerini reaction

s. *5,* 483; with pyridine as catalyst, and reduction to 2-aminoalcohols, s. W.C. Lumma, J. Org. Chem. *46,* 3668–71 (1981).

Zinc *Zn*

β-Aminocarboxylic acid esters N=C → NHC–C–COOR
from azomethines and α-bromocarboxylic acid esters

s. *29,* 638; β-carbalkoxyaminocarboxylic acid esters s. T.D. Truskanova, V.I. Panteleev, A.F. Kolomiets, Zh. Org. Khim. *18,* 671–2 (1982); α-aminocarboxylic acids cf. M. Bellassoued, R. Arous-chtara, M. Gaudemar, J. Organometal. Chem. *231,* 185–9 (1982).

Ethylmagnesium bromide/diisopropylamine *EtMgBr/i-Pr_2NH*

Synthesis of enamines from nitriles CN → $C(NH_2)=C$

with $ZnCl_2$/triethylamine cf. *36,* 633; with ethylmagnesium bromide/diisopropylamine s. T. Hiyama, K. Kobayashi, Tetrahedron Letters *23,* 1597–600 (1982).

Di-n-butylboryl triflate/ethyldiisopropylamine *$Bu_2BOSO_2CF_3$/i-Pr_2NEt*

β-Aminothiolic
from thiolic acid esters and azomethines C=NR → C(NHR)C–COSR′
via 1-alkylthioenoxyboranes

639.

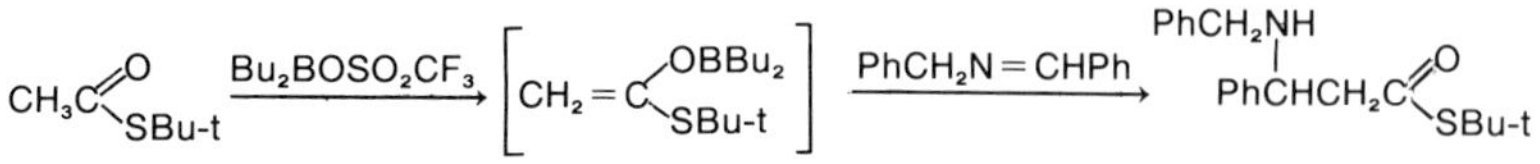

A soln. of startg. boryl enolate, prepd. *in situ* by reaction of di-*n*-butylboryl triflate and diisopropylethylamine with *tert*-butyl thioacetate in ether at 0°, treated with a soln. of the startg. azomethine in the same solvent, allowed to warm to room temp. over 30 min., and oxidized after a further hr. with 30% aq. H_2O_2 at 0° for 30 min. → product. Y 80%. F.e.s. M. Otsuka et al., Tetrahedron Letters *22,* 2109–12 (1981).

Polyphosphoric acid *$(HPO_3)_n$*

1,4-Dihydroisoquinol-3(2*H*)-one ring from α-arylnitriles and aldehydes ○

640.

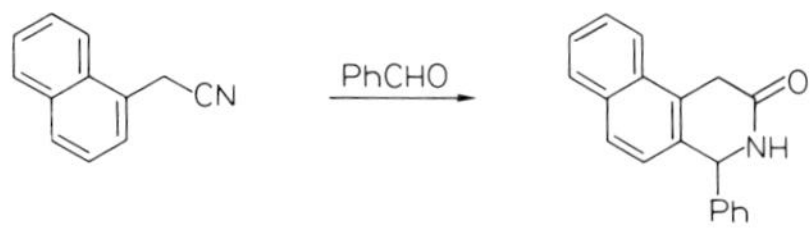

A soln. of 1-naphthylacetonitrile in polyphosphoric acid stirred at 90° for 30 min., benzaldehyde added dropwise, the mixture heated with continued stirring at 120° for 3 hrs., cooled to 80°, poured into water, and worked up with concd. NH_4OH → 1-phenyl-1,4-dihydro-3(2*H*)-benzo[*f*]isoquinolone. Y 69%. F.e.s. L. Hazai, G. Deák, M. Dóda, Acta Chim. Acad. Sci. Hung. *108,* 225–63 (1981).

Addition to Halogen and Carbon CC⇓HalC

Via intermediates *v.i.*

Homologization of halides RHal → RCH_2Hal

s. *37,* 487

Addition to Sulfur and Carbon CC⇓SC

Without additional reagents *w.a.r.*

Insertion of isonitriles into carbon-sulfur bonds ←

1-Carbalkoxyaminothioenolethers

from α-(organothio)carboxylic acid esters

641. 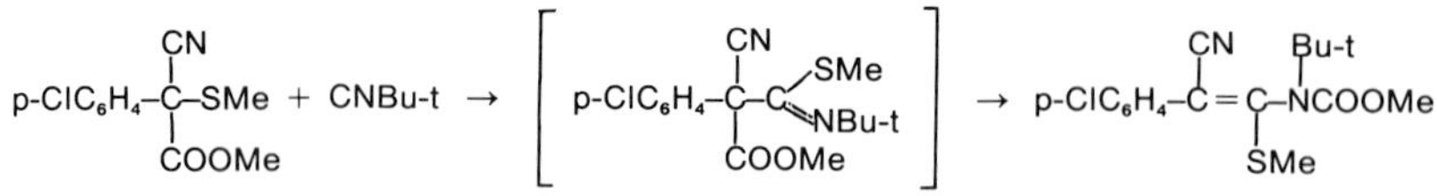

A mixture of *tert*-butyl isonitrile and startg. thioether in acetonitrile refluxed for 7 hrs. → product. Y 94%. F.e., also with isolation of intermediate thioiminoesters, s. G. Morel et al., Tetrahedron Letters *23,* 2023–6 (1982).

Sodium hydride *NaH*

Ketene mercaptals from active methylene groups $CH_2 \rightarrow C{=}C(SR)_2$

s. *19,* 758as*36* f. methods, ketoketene mercaptals, cf. R.K. Dieter, J. Org. Chem. *46,* 5031–3 (1981).

5-Aminothiazoles from isothiocyanates ○

642. 

A mixture of tosylmethyl isocyanide and methyl isothiocyanate in dimethoxyethane added to a stirred suspension of NaH in the same solvent during 30 min. at 0–5°, then stirred for a further 30 min. at room temp. → 5-(methylamino)-4-tosylthiazole, Y 72% (startg m. for 308). F.e., **also 5-mercaptoimidazoles** with e.g. *n*-butyllithium, s. S.P.J.M. van Nispen et al., Rec. Trav. Chim. Pays-Bas *101,* 28–34 (1982).

Addition to Remaining Elements and Carbon CC⇓RemC

Without additional reagents *w.a.r.*

Insertion of C-atoms into carbon-silicon bonds ←

with aldehydes cf. *27,* 883; with ketenes, prepn. of heterocyclic enoxysilanes, s. A. Medici, P. Pedrini, A. Dondoni, J. Chem. Soc. Chem. Commun. *1981,* 655–6.

Addition to Carbon–Carbon Bonds **CC⇓CC**

Without additional reagents *w.a.r.*

α,β-Ethylene-β-(organothio)carboxylic acid thioamides from ynamines C(S) → C=C–C(S)N<

643.

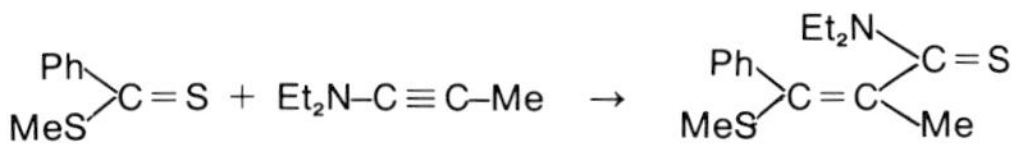

and dithiocarboxylic acid esters. A soln. of startg. dithioester in acetonitrile added under N_2 to a stirred soln. of 1-(diethylamino)propyne in the same solvent under N_2, and the mixture heated at 84° for 4 hrs. or at 20° for 24 hrs. → product. Y 70% (30% recovery of dithioester). F.e., **also α,β-ethylene-β,β-bis(organothio)carboxylic acid thioamides from trithiocarbonic acid esters,** s. V.H.M. Elferink, R.G. Visser, H.J.T. Bos, Rec. Trav. Chim. Pays-Bas *100,* 414–20 (1981).

Regio- and stereo-specific 1,2-cyanoselenation of carbon–carbon double bonds C=C → C(SeAr)CCN

α-Amino-β-arylselenonitriles from enamines

644.

A mixture of phenyl selenocyanate and startg. enamine in ethanol allowed to react at 22° for 5 min. under argon → product. Y 98%. F.e. S. Tomoda, Y. Takeuchi, Y. Nomura, Tetrahedron Letters *23,* 1361–4 (1982); **β-arylselenonitriles** with Lewis acids, especially $SnCl_4$, s. J. Chem. Soc. Chem. Commun. *1982,* 871–2; **review of** recent advances in **enamine chemistry** s. P.W. Hickmott, Tetrahedron, *38,* 1975–2050 (1982).

Cyclopropane ring from ethylene derivs. 

via Δ^1-pyrazoline ring

s. *33,* 629; azidocyclopropanes from enazides s. T. Hiyama, M. Kai, Tetrahedron Letters *23,* 2103–4 (1982).

[2+2]-Cycloaddition to bridgehead carbon–carbon double bonds 

645.

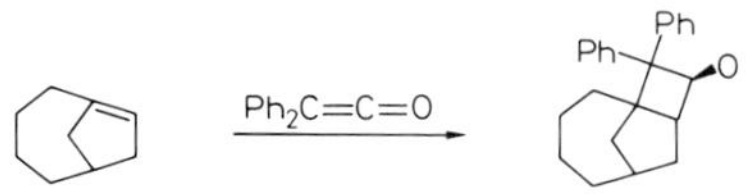

A soln. of bicyclo[4.2.1]non-1(8)-ene in benzene treated with diphenylketene at room temp. for 10 min. → 2,2-diphenyltricyclo[4.4.1.0^{1,4}]undecan-3-one. Y 97%. F.e.s. K.B. Becker, M.K. Hohermuth, G. Rihs, Helv. Chim. Acta *65,* 229–34, 235–42 (1982).

Azetidinones from α-diazoketones and azomethines 

s. *13,* 681; tri- and tetra-aryl derivs. s. K.N. Mehrotra, S.B. Singh, Bull. Chem. Soc. Japan *54,* 1838–40 (1981).

Diene synthesis 

with enolethers

review s. *17,* 750s*23;* also with enolesters (cf. *30,* 499) s. J. Tamariz, P. Vogel, Helv. Chim. Acta *64,* 188–97 (1981).

with thioenolethers

s. *33,* 626; with α-arylthio-α,β-ethyleneketones s. S. Knapp, R. Lis, P. Michna, J. Org. Chem. *46,* 624–6 (1981).

with enolesters
s. *30,* 499; with 3-acoxy-3-buten-2-ones s. R.J. Ardecky, D. Dominguez, M.P. Cava, J. Org. Chem. *47,* 409–12 (1982).

with 1,3-dienes prepd. *in situ*
s. *37,* 744

with siloxy-1,3-dienes
s. *31,* 622; **review** s. S. Danishefsky, Acc. Chem. Res. *14,* 400–6 (1981); **review of heterosubst. 1,3-dienes** s. M. Petrzilka, J.I. Grayson, Synthesis *1981,* 753–86; addition of 2-alkoxy-4-siloxy-1,3-dienes to quinones s. D.W. Cameron et al., Australian J. Chem. *34,* 2401–21 (1981).

with alkoxydienes
s. *30,* 500; with 1,4-dialkoxy-1,3-dienes s. R.K. Geivandov, E.I. Kovshev, Zh. Org. Khim. *17,* 543–52 (1981).

with dienamines
s. *26,* 696; with 1-acylamino-1,3-dienes cf. L.E. Overman et al., J. Am. Chem. Soc. *103,* 2816–22 (1981).

with cyclic dienones
with cyclones s. *33,* 776; also with cyclohexa-2,4-dienones, regiospecific addition, s. A. Oku et al., J. Org. Chem. *46,* 4152–7 (1981).

with allenes
s. *24,* 688; *16,* 733; with bis(trimethylsilyl)allenedicarboxylate s. T. Suzuki et al., J. Chem. Soc. Perkin Trans. I *1980,* 2801–4; with 1,2,4,5-tetraenes (cf. *15,* 531) s. G. Schön, H. Hopf, Ann. Chem. *1981,* 165–80; addition of 1,2,4,5-hexatetraene to acetylene derivs., and subsequent dimerization to **[2.2]paracyclophanes,** s. Org. Synth. *60,* 41–48 (1981).

with acetylene derivs.
s. *28,* 655s*36;* with acetylenedicarboxylic acid via the acid chloride s. G. Maier, W.A. Jung, Chem. Ber. *115,* 804–8 (1982).

with dihydropyridines
s. *34,* 627; *25,* 506s*29;* with N-carbalkoxy-1,2-dihydropyridines s. G.R. Krow et al., Tetrahedron Letters *23,* 2527–8 (1982).

with vinyl-N-heterocyclics

646.

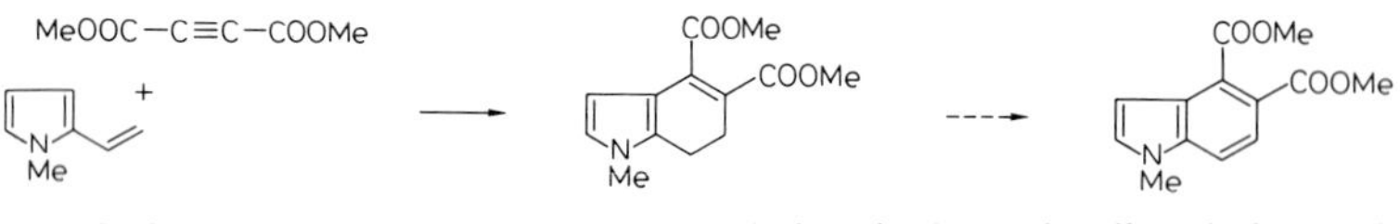

Dihydroindoles from vinylpyrroles. 1-Methyl-2-vinylpyrrole, dimethyl acetylenedicarboxylate, and a little hydroquinone in chloroform heated under reflux for 3 hrs. → dimethyl 1-methyl-6,7-dihydroindole-4,5-dicarboxylate. Y 67%. F.e., dienophiles, and aromatization to indoles (according to Synth. Meth. *10,* 639) s. A.R. Jones et al., J. Org. Chem. *45,* 4515–9 (1980); s.a. W.E. Noland et al., ibid. *45,* 4582–4 (1980); with 1-benzyl-3-vinylindole s. J.D. Lambert, Q.N. Porter, Australian J. Chem. *34,* 1483–90 (1981).

Asymmetric diene synthesis
with chiral 1,3-dienes s. *36,* 667; containing a carbohydrate group, allylic acoxy and methoxy as directing group, s. R.W. Franck et al., J. Am. Chem. Soc. *104,* 1106–7 (1982); with chiral acrylates s. W. Oppolzer et al., Tetrahedron Letters *22,* 2545–8 (1981); Helv. Chim. Acta *64,* 2802–7 (1981); **review** of asym. diene synthesis s. Y. Mori, Yuki Gosei Kagaku Kyokaishi *40,* 321–31 (1982) (Japan).

High pressure diene synthesis
s. *27,* 696; *34,* 626; with (1,3-dien)olethers, prepn. of 2-ethylenecarbohydrates, s. W. Abele, R.R. Schmidt, Tetrahedron Letters *22,* 4807–10 (1981); with 2,4-dienecarboxylic acid esters s. W.G. Dauben, W.R. Baker, Tetrahedron Letters *23,*

2611–4 (1982); bicyclo[4.2.0]oct-3-enes s. A.V. Kamernitsky et al., Izv. Akad. Nauk SSSR Ser. Khim. *1980,* 1902; with furans s. J. Jurczak et al., Helv. Chim. Acta *65,* 1021–4 (1982); H. Kotsuki et al., Bull. Chem. Soc. Japan *55,* 496–9 (1982).

2-Alkoxy-3,4-dihydropyrans from α,β-ethyleneketones and enolethers
s. *20,* 294; *6,* 696; with triformylmethane s. L.-F. Tietze et al., Tetrahedron Letters *23,* 1147–50 (1982); also from 1,2-dimethoxyethene s. R.R. Schmidt, M. Maier, ibid. *23,* 1789–92 (1982); s.a. H.K. Hall, Jr. et al., ibid. *23,* 603–6 (1982); 2,2-dialkoxy-derivs. from keteneacetals s. C.G. Baker, J.W. Scheeren, R.J.F. Nivard, Rec. Trav. Chim. Pays-Bas *100,* 13–20 (1981).

Dihydro-Δ²-thiopyran ring by heterodiene synthesis
s. *33,* 654; with 2-thioacyl-furans and -thiophenes s. H. Ohmura, S. Motoki, Chem. Letters *1981,* 235–8; with α,β-ethylenethioketones prepd. *in situ* s. H. Yamaguchi et al., Tetrahedron Letters *23,* 1263–6 (1982).

Thiopyran ring from acetylene derivs. and α,β-ethylenethiocarbonyl compds.
s. *30,* 470; 4-amino-derivs. s. J.B. Rasmussen, R. Shabana, S.-O. Lawesson, Tetrahedron *37,* 3693–8 (1981); ibid. *38,* 1705–9 (1982).

3-Sulfonyl-Δ²-isoxazolines from ethylene derivs.
s. *35,* 477; improved generation of sulfonylnitrile oxides s. P.A. Wade, M.K. Pillay, J. Org. Chem. *46,* 5425–7 (1981); from bis(benzenesulfonyl)furoxan cf. R.A. Whitney, E.S. Nicholas, Tetrahedron Letters *22,* 3371–4 (1981).

High pressure [3+2]-cycloaddition with nitrones
2-alkoxyisoxazolidine ring s. *32,* 628; 5-alkoxyisoxazolidine ring from enolethers s. C.M. Dicken, P. DeShong, J. Org. Chem. *47,* 2047–51 (1982).

Anthracenes from isobenzofurans
via 1,4,9,10-diepoxy-1,2,3,4,4a,9,9a,10-octahydroanthracenes

647.

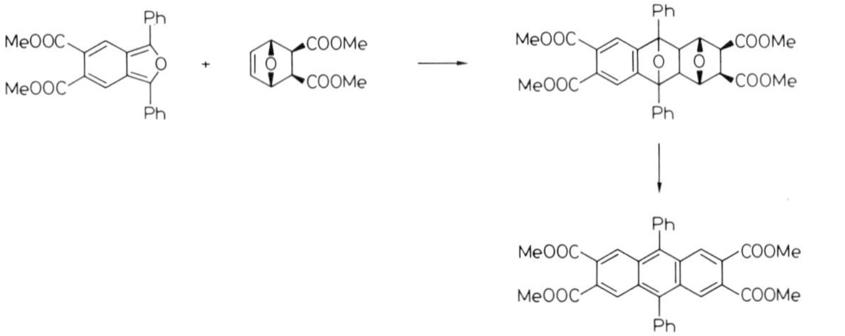

A soln. of dimethyl 1,3-diphenylisobenzofuran-5,6-dicarboxylate and *cis-exo*-7-oxabicyclo[2.2.1]hept-5-ene-2,3-dicarboxylate in chloroform refluxed under N_2 for 2 days → intermediate (Y 90%), in dichloromethane containing H_2SO_4 stirred under ultrasonic agitation below 10° for 6 hrs. → tetramethyl 9,10-diphenylanthracene 2,3,6,7-tetracarboxylate (Y 65%). F.e.s. C. Schmitz, J.M. Aubry, J. Rigaudy, Tetrahedron *38,* 1425–30 (1982).

3,4,7-Triazabicyclo[3.3.0]octa-2,7-dienes
from 2,5-dihydro-1,2,4-triazines and acetylene derivs.

648.

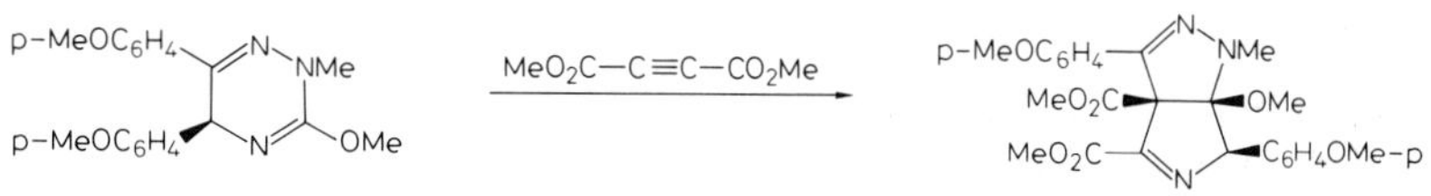

A mixture of startg. dihydrotriazine and dimethyl acetylenedicarboxylate in toluene heated at 110° under argon for 6 hrs. → product. Y 80%. F.e.s. T. Sasaki, K. Minamoto, K. Harada, J. Org. Chem. *45,* 4587–93 (1980).

Without additional reagents *w.a.r.*

2-Amino-2*H*-benzimidazo[2,1-*b*][1,3]thiazine-4(3*H*)-thiones
from *o*-diisothiocyanates and enamines
Double ring closure

649.

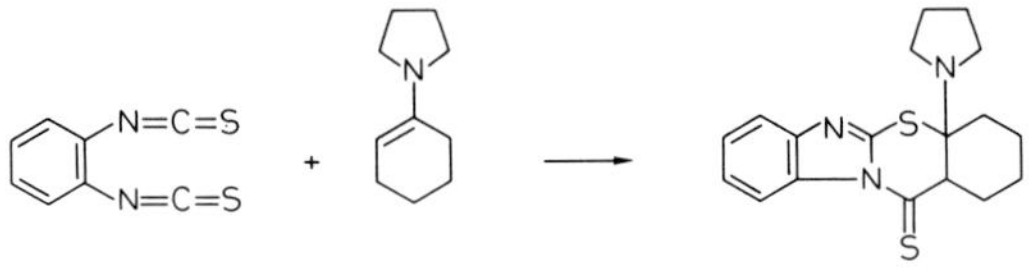

A soln. of 1-pyrrolidin-1-ylcyclohexene in ether added to a soln. of *o*-phenylene diisothiocyanate in the same solvent, and the mixture stirred for 4 hrs. at ambient temp. → 6a-pyrrolidin-1-yl-6a,7,8,9,10,10a-hexahydrobenzimidazo[2,1-*b*][1,3]benzothiazine-11-thione. Y 84%. F.e. and deamination s. A.W. Faull et al., J. Chem. Soc. Perkin Trans. I *1980,* 1, 2587–90.

Ring expansion by two C-atoms with acetylene derivs.
s. *18,* 759s*36;* of 1-*β*-styryl-1,2-dihydropyridine s. L. Yerino, M.E. Osborn, P.S. Mariano, Tetrahedron *38,* 1579–91 (1982); of thebaine cf. K. Hayakawa et al., J. Am. Chem. Soc. *103,* 4605–6 (1981); A. Singh et al., J. Org. Chem. *47,* 752–4 (1982); of imidazole ring s. N. Abe, T. Nishiwaki, Heterocycles *16,* 537–8 (1981).

Irradiation (s.a. under 1-Cyanonaphthalene)
Photochemical cycloaddition of *α,β*-ethyleneketones to ethylene derivs.

s. *22,* 705s*36;* effect of micellar solns. as media cf. N. Berenjian et al., Can. J. Chem. *60,* 425–36 (1982).

Regiospecific [2+2]-cycloaddition with enolesters
s. *37,* 244

1,4-Di(siloxy)bicyclo[2.2.0]hexane ring from 1,2-di(siloxy)cyclobut-1-enes
s. *37,* 256

Electrolysis
Reductive dimerization of ethylene derivs. 2 C=C → CH–C–C–CH
s. *21,* 731; prepn. of tetraethyl 1,2,3,4-butanetetracarboxylate s. D.A. White, Org. Synth. *60,* 58–62 (1981).

Sodium hydroxide/benzyltriethylammonium chloride $NaOH/[Et_3NCH_2Ph]^+Cl^-$
Anionic [3+2]-cycloaddition
Pyrrolidines
with Li diisopropylamide cf. *26,* 712; with NaOH and benzyltriethylammonium chloride as phase transfer catalyst s. V. Dryanska, K. Popandova, C. Ivanov, Synth. Commun. *12,* 343–50 (1982); with NaOMe on alumina in dimethyl sulfoxide s. I.P. Beletskaya et al., Zh. Org. Khim. *18,* 1129–31 (1982); with excess of KOH cf. A.T. Malkhasian et al., Arm. Khim. Zh. *33,* 956 (1980).

Sodium methoxide s. under $H_2BCl.SMe_2$ *NaOMe*

Sodium methoxide–alumina $NaOMe–Al_2O_3$
Anionic [3+2]-cycloaddition
s. *26,* 712s*37*

Potassium tert-*butoxide/chiral crown polyethers* ←
Asym. Michael addition C=C → CHCR
with alkaloids cf. *12,* 737s*31;* with K *tert*-butoxide in the presence of chiral crown ethers s. D.J. Cram, G.D.Y. Sogah, J. Chem. Soc. Chem. Commun. *1981,* 625–28;

review of chiral crown heterocyclics s. S.T. Jolley, J.S. Bradshaw, R.M. Izart, J. Heterocyc. Chem. *19,* 3–19 (1982); **high-pressure asym. Michael addition** with alkaloids cf. K. Matsumoto, T. Uchida, Chem. Letters *1981,* 1673–6; addition of chiral perhydro-1,4-oxazepine-5,7-diones to 1-nitroethylene derivs. cf. T. Takeda, T. Hoshiko, T. Mukaiyama, ibid. *1981,* 796–800.

Organolithium compd. (s.a. under CuI) *RLi*

Syntheses with styrenetricarbonylchromium(0) complexes
Benzylic carbanions as intermediates

650. 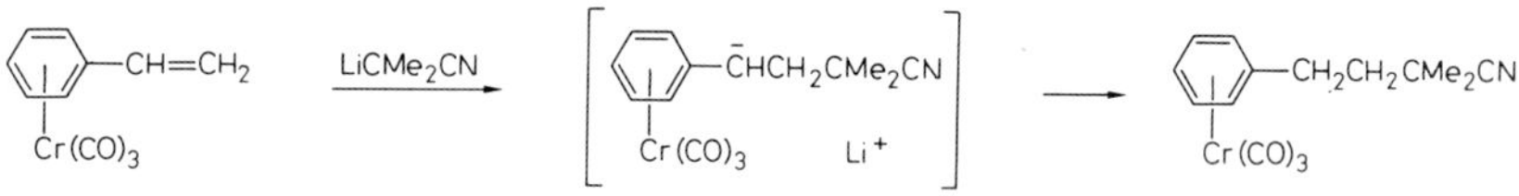

2-Lithio-2-methylpropionitrile added slowly to η^6-styrenechromium tricarbonyl at –30° in tetrahydrofuran, and quenched with aq. NH_4Cl after 3 hrs. → product. Y 92%. F.e., electrophiles, stereospecificity, and limitations, s. M.F. Semmelhack, W. Seufert, L. Keller, J. Am. Chem. Soc. *102,* 6584–6 (1980).

n-*Butyllithium* n-*BuLi*

1,4-Addition to α,β**-ethylenecarboxylic acid thioamides**
s. *35,* 556s*36;* also 1,4- and 1,6-addition to 2,4-dienecarboxylic acid thioamides s. Y. Tamaru, Tetrahedron Letters *23,* 2383–6 (1982).

Michael-type addition of lithium trialkyl-1-alkynylborates ←
s. *36,* 671; synthesis of ketones from electron-deficient ethylene derivs. without $TiCl_4$ cf. A. Pelter, L. Hughes, J.M. Rao, J. Chem. Soc. Perkin Trans. I *1982,* 719–22.

Lithium diisopropylamide *i-*Pr_2NLi

Michael addition with dithiocarboxylic acid esters C=C → CHCR
δ**-Ketodithiocarboxylic acid esters**
from α,β**-ethyleneketones**

651. 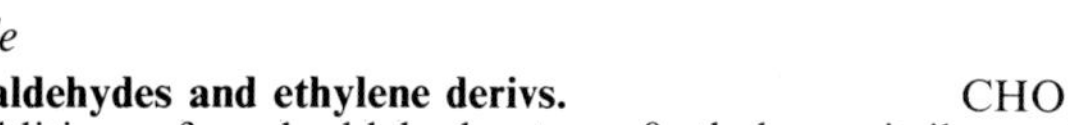

A soln. of methyl dithioacetate in tetrahydrofuran treated with Li diisopropylamide at –50°, followed by addition of cyclopent-2-enone, and quenched with aq. NH_4Cl after 15 min. → product. Y 78%. F.e.s. P. Metzner, J. Chem. Soc. Chem. Commun. *1982,* 335–6.

Potassium carbonate s. under $(MeCOO)_2Pd$ K_2CO_3

Sodium cyanide *NaCN*

Ketones from aldehydes and ethylene derivs. CHO → CO–C–CH
s. *28,* 648; addition of aryl aldehydes to α,β-ethylene nitriles s. H. Stetter, H. Kuhlmann, G. Lorenz, Org. Synth. *59,* 53–8 (1979).

Sodium nitrite/acetic acid $NaNO_2/MeCOOH$

trans-**2-Ene-1,6-diones from two** α,β**-ethyleneketone molecules** ←

652. 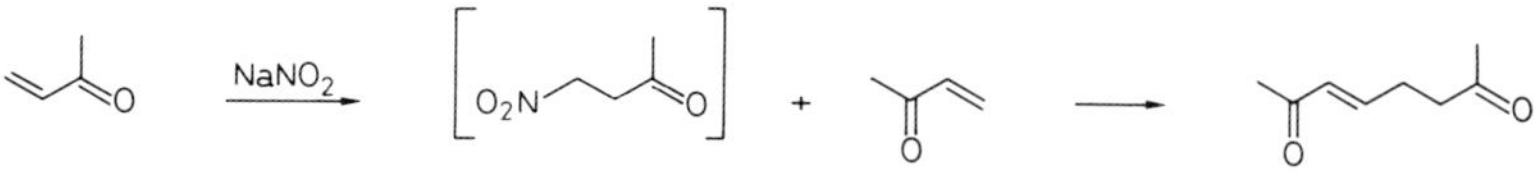

Acetic acid added to a stirred mixture of $NaNO_2$ and 5 equivalents of 3-buten-2-one in dimethyl sulfoxide at 20–25°, and stirring continued for an additional 16 hrs. → *trans*-3-octene-2,7-dione. Y 86%. F.e.s. T. Miyakoshi, S. Saito, J. Kumanotani, Chem. Letters *1982,* 83–4.

Potassium bromide s. under $(CF_3COO)_2Hg$ *KBr*

Diethylamine s. under Pd phosphine complexes *Et_2NH*

Triethylamine *Et_3N*

4-Ethylene-Δ^2-5-oxazolones from Δ^2-5-oxazolones ←

s. *36,* 661; **review of Δ^2-5-oxazolone chemistry** s. A.K. Mukerjee, P. Kumar, Heterocycles *16,* 1995–2034 (1981).

Pyrrolidine ring from azomethinium ylids ○

with dimethylformamide cf. *35,* 473; with triethylamine s. G.A. Kraus, J.O. Nagy, Tetrahedron Letters *22,* 2727–30 (1981).

Triethylenediamine ←

Michael addition with dithiolmalonic acid esters C=C → CHCR

653. 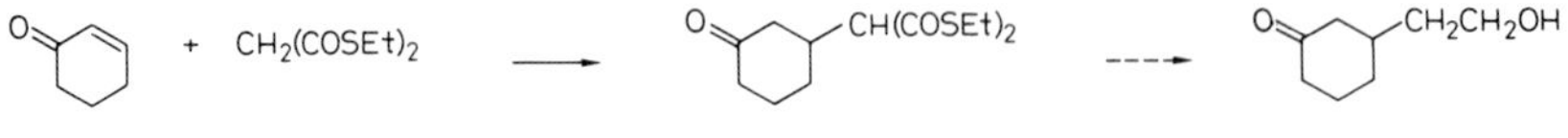

A soln. of S,S′-diethyl dithiomalonate, 1,4-diazabicyclo[2.2.2]octane, and 2-cyclohexenone in 1,2-dimethoxyethane stirred at room temp. for 48 hrs. under argon, and the mixture acidified with 1 *M* aq. HCl → product. Y 93%. F.e. and Raney nickel reduction to **δ-hydroxycarbonyl compds.** (cf. Synth. Meth. *36,* 789) s. H.-J. Liu, I.V. Oppong, Can. J. Chem. *60,* 94–6 (1982).

1,8-Diazabicyclo[5.4.0]undec-7-ene ←

Michael addition to α,β-ethylenesulfoxides SO–C=C → SO–CHCR

with Li diisopropylamide cf. *31,* 634; with 1,8-diazabicyclo[5.4.0] undec-7-ene, γ-nitrosulfoxides, s. N. Ono et al., Tetrahedron Letters *23,* 2957–60 (1982).

Pyridine N-oxide s. under $HgCl_2$ ←

Lithium diorganocuprates *$LiCuR_2$*

1,4-Addition C=C → CHCR

to 1-nitroethylene derivs.

s. *4,* 658s*31;* carbohydrate derivs. s. H.H. Baer, Z.S. Hanna, Carbohyd. Res. *85,* 136–42 (1980).

to α,β-ethylenesulfones

s. *23,* 693s*28;* with Li divinylcuprates s. G. De Chirico et al., J. Chem. Soc. Chem. Commun. *1981,* 523–4.

Lithium dialkylcuprates *$LiCuR_2$*

Synthesis of α,β-ethyleneketones from α,β-ethylene-β-silylketones COC=C → C(OSi⋮)=C–CR

654. 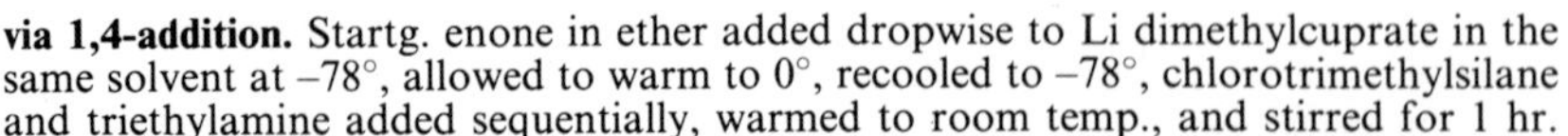

via 1,4-addition. Startg. enone in ether added dropwise to Li dimethylcuprate in the same solvent at –78°, allowed to warm to 0°, recooled to –78°, chlorotrimethylsilane and triethylamine added sequentially, warmed to room temp., and stirred for 1 hr. → intermediate enoxysilane (Y 82%), in tetrahydrofuran treated dropwise while stirring with trimethylphenylammonium tribromide in the same solvent at room temp., filtered, the solvent replaced by carbon tetrachloride, a slow stream of HBr introduced at reflux for 1 hr., and excess HBr removed with N_2 → crude β-bromoketone, in carbon tetrachloride treated with a soln. of 1,5-diazabicyclo[5.4.0]undec-5-ene in the same solvent → product. F.e., **also via 1,4-addition-alkylation, and 1,4-addition to β-silyl-α,β-acetyleneketones,** s. I. Fleming, D.A. Perry, Tetrahedron *37,* 4027–34 (1981).

Dilithium pentaalkyltricuprates *$Li_2Cu_3R_5$*
1,4-Addition to α,β-ethylenealdehydes C=C → CHCR

655. 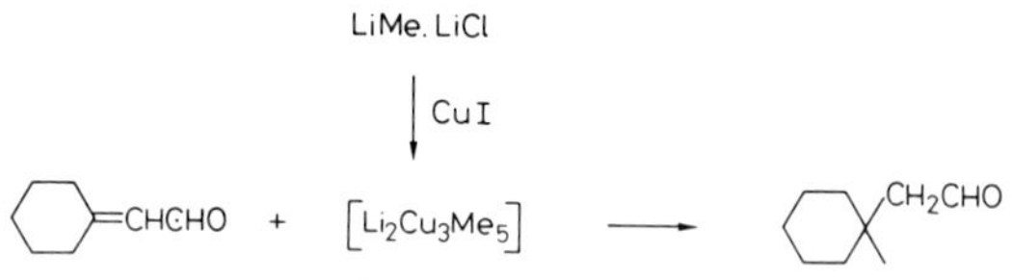

5 Equivalents of ethereal MeLi.LiCl added to a slurry of 3 equivalents CuI in ether at 0°, a soln. of 1 equivalent startg. aldehyde in the same solvent added dropwise at –75°, after 1–2 hrs. the mixture warmed to –40°, recooled to –75°, then quenched with acetic acid → product. Y 90%. – The reaction proceeds with <1.5% 1,2-addition, in contrast with Li dimethylcuprate (13% 1,2-addition). F.e.s. D.L.J. Clive, V. Farina, P. Beaulieu, J. Chem. Soc. Chem. Commun. *1981,* 643–4; J. Org. Chem. *42,* 2572–82 (1982); 1,4-addition with magnesium dialkylcuprate chloride cf. M. Bourgain-commerçon, J.-P. Foulon, J. François, J. Organometal. Chem. *228,* 321–6 (1982).

Methylcopper(I)/aluminium chloride *$MeCu/AlCl_3$*
Stereospecific 1,4-addition to α,β-ethyleneketones

656. 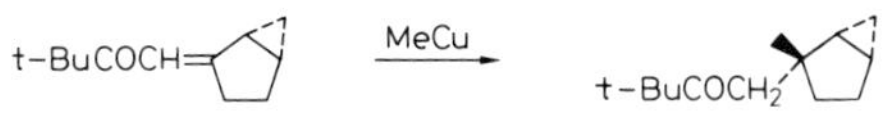

β-Cyclopropylketones. A stirred suspension of methylcopper in ether at –70° treated dropwise under argon with a soln. of *$AlCl_3$* in the same solvent, the mixture stirred for 1 hr., a soln. of startg. enone in ether added at –70°, and the temp. allowed to rise to –50° → product. Y 89%. – In the absence of $AlCl_3$ no reaction takes place, whereas using Gilman's reagent (R_2CuLi) a competing cyclopropane ring opening occurs. F.e.s. T. Ibuka, E. Tabushi, J. Chem. Soc. Chem. Commun. *1982,* 703–4; **γ-acoxyketones** s. ibid. *1980,* 1193–4.

Cuprous halides (s.a. under Mg) *CuHal*

*Cuprous iodide/*n*-butyllithium* *CuI/*n*-BuLi*
Synthesis of *trans*-ethylene from acetylene derivs. C≡C → CH=CR
s. *31,* 639; stereospecific synthesis of ethylene derivs. by carbometallation of acetylene derivs., **review,** s. J.F. Normant, A. Alexakis, Synthesis *1981,* 841–70.

Cuprous iodide/phenyllithium/boron fluoride *$CuI/PhLi/BF_3$*
Synthesis of carboxylic acids via asym. 1,4-addition C=C → CHCR

657. 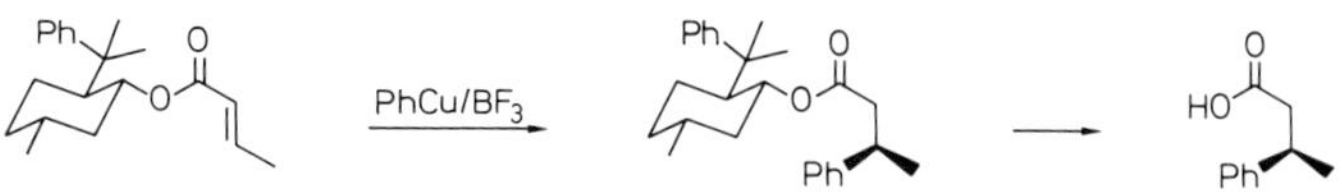

to α,β-ethylenecarboxylic acid esters. Phenyllithium added to purified CuI in ether at –40° to –20°, the resulting dark soln. cooled to –70°, BF_3-etherate and startg. (–)-8-menthyl α,β-ethylenecarboxylate slowly added sequentially, the mixture stirred for 10 min. at this temp., allowed to warm to 25° slowly, and quenched with NH_4Cl → satd.

ester (Y 76%), heated in a mixture of 11.6 *M* Na methoxide in methanol and water at 75° for 24 hrs. → (R)-(–)-3-phenylbutyric acid (99% enantiomeric purity). F.e. with ca. 100% recovery of (–)-8-menthol s. W. Oppolzer, H.J. Löher, Helv. Chim. Acta *64,* 2808–11 (1981); **with chiral lithium dialkylcuprates** cf. H. Malmberg, M. Nilsson, C. Ullenius, Acta Chem. Scand. *35B,* 625–9 (1981).

Magnesium *Mg*

658. 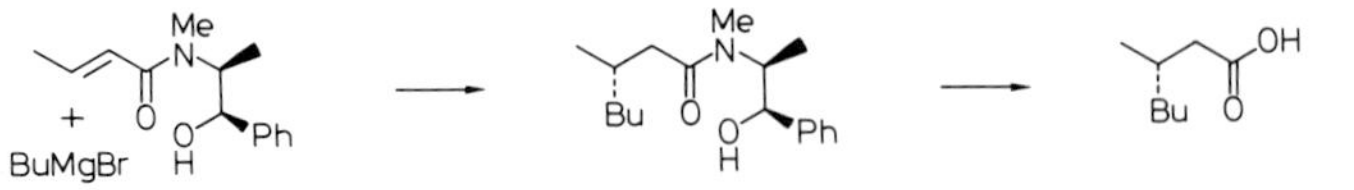

...to α,β-ethylenecarboxylic acid amides. A soln. of *n*-butylmagnesium bromide in ether added to methacrylic acid *l*-ephedrine amide in the same solvent at –78° under argon, stirred for 48 hrs. at –40°, and treated with phosphate buffer (pH 7) → 1,4-adduct (Y 63%), refluxed in 2:1 3*M* H_2SO_4/acetic acid for 3 hrs. → (S)-3-methylheptanoic acid (Y 84%, enantiomeric excess 85%). F.e.s. T. Mukaiyama, N. Iwasawa, Chem. Letters *1981,* 913–6.

Magnesium/cuprous halides *Mg/CuHal*

Synthesis of ethylene from acetylene derivs. C≡C → CH=CR

with CuI cf. *31,* 641; with $CuBr(LiBr)_3$ complex s. H. Westmijze et al., Rec. Trav. Chim. Pays Bas *100,* 98–102 (1981); with $AgBr(LiBr)_3$ complex, 2-subst. 1,3-enynes, cf. ibid. *100,* 337–41 (1981).

Stereospecific 1,4-addition to α,β-acetylenecarboxylic acid esters

s. *23,* 693s*36;* synthesis of maleic acid esters s. H. Nishiyama, M. Sasaki, K. Itoh, Chem. Letters *1981,* 905–8.

Magnesium/cuprous chloride *Mg/CuCl*

1,4-Addition to 5-alkylidene-1,3-dioxane-4,6-diones C=C → CHCR

s. *9,* 770s*36;* with added CuCl cf. X. Huang, C.-C. Chan, Q.-L. Wu, Tetrahedron Letters *23,* 75–6 (1982).

Magnesium/cuprous iodide/boron fluoride *Mg/CuI/BF₃*

4-Alkylation of pyridines via 1-carbalkoxypyridinium salts and 1-carbalkoxy-1,4-dihydropyridines ←

659. 

A soln. of CuI in tetrahydrofuran treated with butylmagnesium bromide in the same solvent at –20° under N_2, the mixture stirred at this temp. for 30 min., cooled to –78°, BF_3-etherate added slowly, stirring continued for 10 min., the resulting soln. added to a suspension of N-ethoxycarbonylpyridinium chloride (prepd. from pyridine and ethyl chloroformate) in tetrahydrofuran, the mixture allowed to warm slowly to room temp., and stirring continued for 3 hrs. → intermediate 1,4-dihydropyridine (Y 89%) stirred for 6.5 hrs. under a slow stream of O_2, and worked up with 5 *M* aq. HCl → 4-butylpyridine (Y 68%). F.e.s. K. Akiba, Y. Iseki, M. Wada, Tetrahedron Letters *23,* 429–32 (1982).

Zinc (s.a. under $HgCl_2$) *Zn*

γ-Arylthio-α,α-dichloro-γ-lactones from α,β-ethylenesulfoxides ○
Stereospecific cycloaddition

660.

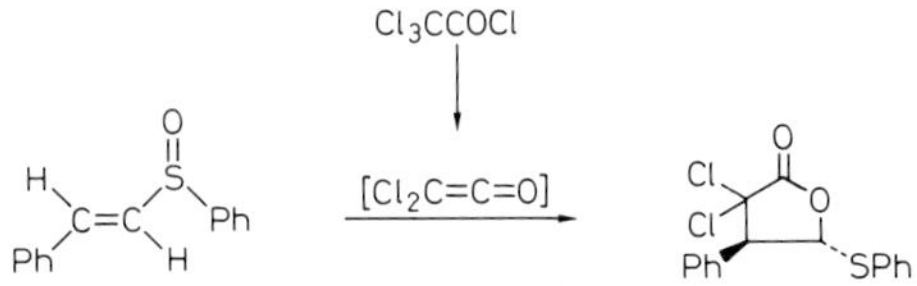

A soln. of trichloroacetyl chloride in ether added dropwise over 15 min. to a refluxing suspension of zinc metal and (E)-β-styryl phenyl sulfoxide in the same solvent under N_2, and heating continued for a further 15 min. → *trans*-2,2-dichloro-3-phenyl-4-(phenylthio)-γ-butyrolactone. Y 65%. – Reaction is thought to proceed via intramolecular addition to a vinyloxysulfonium cation. F.e.s. J.P. Marino, M. Neisser, J. Am. Chem. Soc. *103,* 7687–9 (1981).

Mercuric acetate s. under B_2H_6 *$(MeCOO)_2Hg$*

Mercuric trifluoroacetate/potassium bromide *$(CF_3COO)_2Hg/KBr$*

Bicyclic lactones from dienecarboxylic acids
Double ring closure by intramolecular oxymercuration
s. *37,* 680

Mercuric chloride/pyridine N-oxide/zinc ←

α-Chloro-β-hydroxycarboxylic acid esters
from aldehydes CHO → CH(OH)CH(Cl)COOR
Synthesis with addition of two C-atoms

661. $$\text{PhCHO} + \text{HC}\equiv\text{COEt} \xrightarrow{\text{HgCl}_2} \text{PhCH(OH)–CH(Cl)COOEt}$$

A soln. of pyridine N-oxide in tetrahydrofuran added to a soln of $HgCl_2$ in the same solvent at room temp. under argon, treated with ethoxyacetylene in tetrahydrofuran at 0°, the resulting mixture stirred at this temp. for 1 hr., allowed to warm to room temp., a soln. of benzaldehyde in the same solvent and Zn dust added successively, and stirred for a further 3 hrs. → ethyl 2-chloro-3-hydroxy-3-phenylpropionate. Y 81%. F.e.s. T. Mukaiyama, M. Murakami, Chem. Letters *1981,* 1129–32.

Zinc chloride *$ZnCl_2$*

1,4-Addition of halides to 1,3-dienes ←
s. *21,* 737; in the absence of HCl, 1-halogen-2,6-enynes from 2-acetylenehalides s. H. Mayr, H. Klein, J. Org. Chem. *46,* 4097–100 (1981).

Tetrahydrofuran-borane/mercuric acetate/
sodium tetrahydridoborate *$H_3B.THF/(MeCOO)_2Hg/NaBH_4$*

Reductive codimerization of electron-rich and electron-deficient ethylene derivs. under mild conditions

662. $$\text{EtMeC}{=}\text{CH}_2 + \text{THF.BH}_3 \rightarrow [(\text{EtMeCHCH}_2)_3\text{B}] \xrightarrow{\text{Hg(OAc)}_2} [\text{EtMeCHCH}_2\text{HgOAc}] \xrightarrow{\text{CH}_2=\text{CHCN}} \text{EtMeCHCH}_2\text{CH}_2\text{CH}_2\text{CN}$$

A soln. of $THF.BH_3$ in tetrahydrofuran added to 2-methylbut-1-ene in the same solvent under N_2 at 0°, after 45 min. Hg acetate added, the mixture allowed to react at 20° for 30 min., dil. with methylene chloride, mixed with acrylonitrile, a suspension of $NaBH_4$ in water added rapidly, and worked up after 5 min. → product. Y 65%. F.e.s. B. Giese, G. Kretzschmar, Angew. Chem. Intern. Ed. *20,* 965–6 (1981).

Sodium tetrahydridoborate *$NaBH_4$*
γ-Nitrocarboxylic $NO_2C–C–C–COOR$
from α,β-ethylenecarboxylic acid esters and 1-nitroethylene derivs.

663. 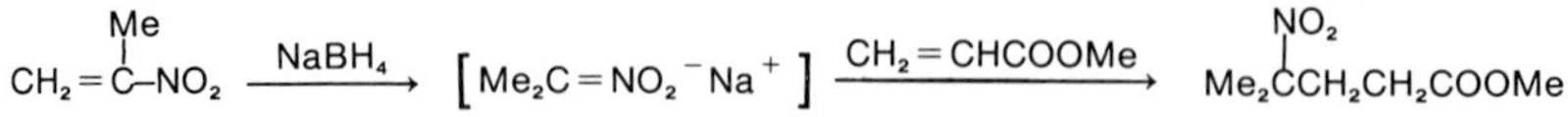

A mixture of methyl acrylate and startg. nitroalkene in methanol treated with powdered $NaBH_4$, and stirred for 24 hrs. at room temp. → product. Y 62%. F.e., **also with simultaneous transesterification,** s. T.-L. Ho, Syn. Commun. *12,* 339–41 (1982).

Lithium tetrahydridoaluminate s. under $H_2BCl.SMe_2,TiCl_3$ *$LiAlH_4$*

Alumina s. under NaOMe *Al_2O_3*

Monochloroborane-dimethylsulfide/ lithium tetrahydridoaluminate/ iodine/sodium methoxide *$H_2BCl.SMe_2$/ $LiAlH_4$/ I_2/NaOMe*
Synthesis of *cis*-ethylene from acetylene derivs. via dialkylchloroboranes C≡C → CH=CR

664. 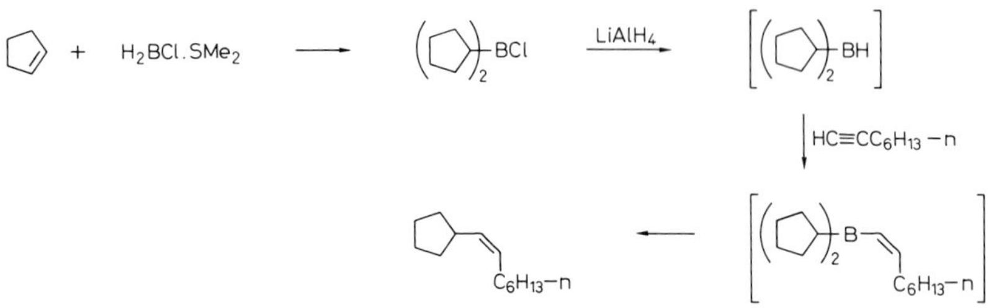

Cyclopentene added to a soln. of monochloroborane-dimethyl sulfide in methylene chloride at 0° under N_2, stirred for 2 hrs. at room temp., solvent replaced by tetrahydrofuran at 0°, 1-octyne added, the vigorously stirred mixture treated dropwise with a soln. of $LiAlH_4$ in the same solvent, stirring continued for 2 hrs. at 0° then 0.5 hrs. at room temp., cooled to –78°, solns. of Na methoxide in methanol and I_2 in tetrahydrofuran added sequentially, stirring continued at –78° for 3 hrs., and excess I_2 removed with aq. $Na_2S_2O_3$ → *cis*-1-cyclopentyl-1-octene. Y 62%. F.e.s. S.U. Kulkarni, D. Basavaiah, H.C. Brown, J. Organometal. Chem. *225,* Cl-5; isolation of dialkylchloroboranes s. ibid. *225,* 63–9 (1982); synthesis of **trisubst. (Z)-ethylene derivs.** s. J. Org. Chem. *47,* 171–3 (1982); modified Zweifel ethylene derivs. and **ketones from acetylenehalides** (cf. Synth. Meth. *23,* 864) s. ibid. *47,* 754–6 (1982); novel prepn. of monochloroborane-dimethyl sulfide s. W.E. Paget, K. Smith, J. Chem. Soc. Chem. Commun. *1980,* 1169–70.

Lewis acids ←
Regiospecific diene synthesis O
s. *28,* 655s*36;* with 5-silylcyclopentadienes s. W.R. Roush, T.E. D'Ambra, J. Org. Chem. *46,* 5045–7 (1981); with 2-trimethylsilylmethyl-1,3-butadiene s. A. Hosomi et al., Tetrahedron Letters *23,* 551–4 (1982).

Boron fluoride (s.a. under CuI) *BF_3*

3-Ethylenealcohols from ethylene derivs. and aldehydes ←

Ene synthesis

with Me_2AlCl cf. *36,* 665; with BF_3, stereospecific ene synthesis of steroid side chains, s. A.D. Batcho et al., Helv. Chim. Acta *64,* 1682–7 (1981).

Dimethylaluminium chloride *Me_2AlCl*

Cyclohexanol ring from enisocyclics and α,β-ethyleneoxo compds. ○

Sequential ene syntheses

665.

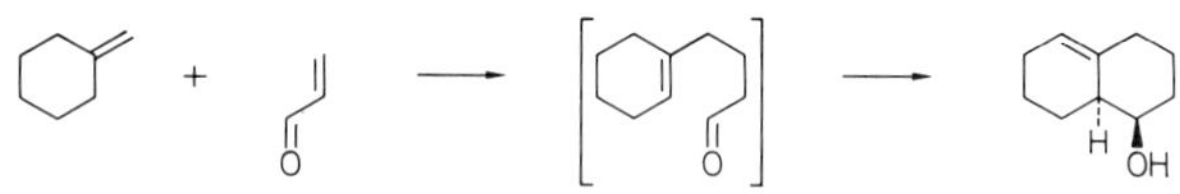

Methylenecyclohexane added to a soln. of acrolein and dimethylaluminium chloride in methylene chloride at 0°, and stirred for 20 min. at 0° → product. Y 63%. F.e.s. B.B. Snider, E.A. Deutsch, J. Org. Chem. *47,* 745–7 (1982).

Diethylaluminium chloride *Et_2AlCl*

γ-Lactones from α,β-ethyleneoxo compds.

via β,γ-ethylene-γ-iminolactones and γ-iminolactones

666.

A soln. of diethylaluminium chloride in tetrahydrofuran added with stirring to a soln. of pulegone and methyl isocyanide in the same solvent at 5–10°, and stirred at room temp. for 12 hrs. → unsatd. iminolactone (Y 85%) in acetic acid hydrogenated with 10% Pd on carbon under 10 atm. H_2 at 50° for 15 hrs. → satd. iminolactone (Y 83%), hydrolysed in aq. oxalic acid at reflux for 24 hrs. → *cis*-γ-lactone (Y 80%). F.e.s. Y. Ito, H. Kato, T. Saegusa, J. Org. Chem. *47,* 741–3 (1982).

Ethylaluminium dichloride *$EtAlCl_2$*

Ene synthesis

with $AlCl_3$ cf. *29,* 676; with $EtAlCl_2$, stereo- and regio-specificity, s. J. Am. Chem. Soc. *104,* 1930–6 (1982).

7-Hydroxybicyclo[2.2.1]heptan-2-ones from α,β-ethylenealdehydes ○

via diene synthesis with 1-arylthio-2-siloxy-1,3-dienes

667.

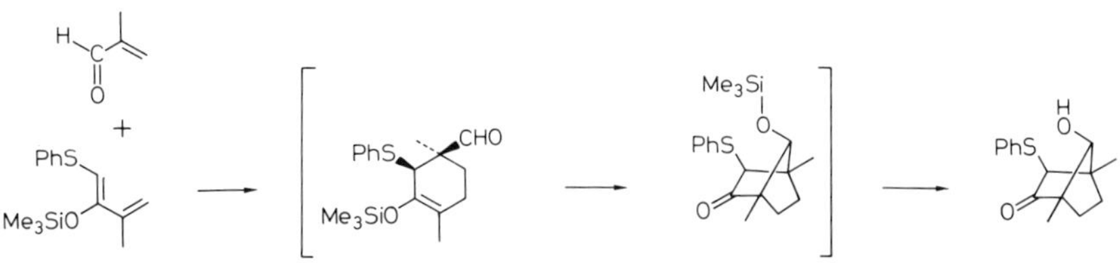

A mixture of (Z)-3-methyl-1-(phenylthio)-2-[(trimethylsilyl)oxy]-1,3-butadiene and methacrolein in methylene chloride treated with ethylaluminium dichloride at –78° for 15 min., worked up with bicarbonate, and hydrolysed with 5% HCl in acetone → product. Y 90%. F.e. and isolation of [4+2]-adducts s. A.P. Kozikowski, E.M. Huie, J. Am. Chem. Soc. *104,* 2923–5 (1982).

Aluminium chloride (s.a. under MeCu) $AlCl_3$

α-Fluoro-β-ketocarboxylic acid esters from carboxylic acid halides
Synthesis with addition of two C-atoms

COCl → COCHFCOOR

668. $$MeCOCl + FCH{=}CF_2 \rightarrow MeCOCHFCF_2Cl \xrightarrow{EtOH} MeCOCHFCOOEt$$

A mixture of $AlCl_3$, acetyl chloride, and dichloromethane in a pressure vessel cooled in a dry ice/acetone bath, trifluoroethene introduced, the mixture stirred at room temp. for 48 hrs., and poured onto ice/concd. HCl → intermediate (Y 64%), mixed with Na ethoxide in ethanol, and stirred for 2 hrs. at room temp. → product (Y 88%). F.e. and method s. N. Ishikawa et al., Chem. Letters *1980,* 1107–10.

Diene synthesis with push–pull ethylene derivs. as ketene equivalents ○

669.

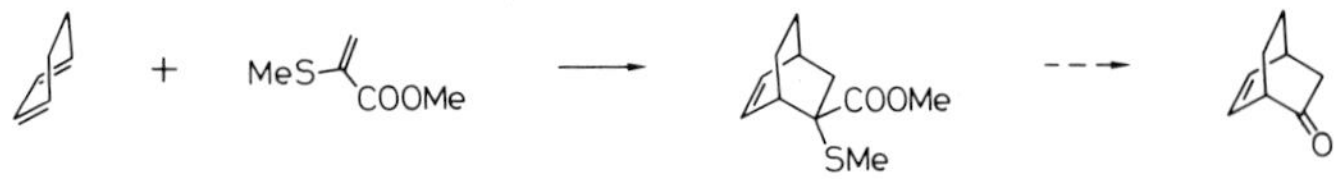

A mixture of 5 equivalents 1,3-cyclohexadiene and 1 equivalent startg. olefin allowed to react at 20° for 4 hrs. in the presence of 0.5 equivalent $AlCl_3$ → product. Y 81%. – The **3-cyclohexenone ring** may be obtained by subsequent treatment according to Synth. Meth. *33,* 236. F.e. and Lewis acid catalysts s. L. Stella, J.-L. Boucher, Tetrahedron Letters *23,* 953–6 (1982).

Water s. under Rhodium carbonyl H_2O

Chiral crown polyethers s. under KOBu-t ←

Benzyltriethylammonium chloride s. under NaOH $[Et_3NCH_2Ph]^+Cl^-$

Polymer-based thiazolium salts ←

Ketones from aldehydes and ethylene derivs. CHO → CO–C–CH

with thiazolium salts cf. *28,* 648s*34;* with polymer-supported thiazolium salts s. C.S. Sell, L.A. Dorman, J. Chem. Soc. Chem. Commun. *1982,* 629–30; unsym. γ-polyketones, 4,7,10-trioxo-esters, and 4,7,10-trioxo-nitriles, s. H. Stetter, A. Mertens, Ann. Chem. *1981,* 1550–60.

1-Cyanonaphthalene/irradiation ←

Photochemical C-alkylation of 5-membered heterocyclics with ethylene derivs. H → R

670.

2-Alkylfurans. A soln. of 1,1-diphenylethylene and 10 equivalents of 2-methylfuran in acetonitrile containing a little 1-cyanonaphthalene irradiated with a high-pressure mercury-arc using a Pyrex filter (> 280 nm) at room temp. → 5-(2,2-diphenylethyl)-2-methylfuran. Y 87%. F.e., incl. pyrrole and thiophene analogs, s. K. Mizuno, M. Ishii, Y. Otsuji, J. Am. Chem. Soc. *103,* 5570–2 (1981).

Tetramethylguanidine ←

Michael addition to α,β-ethyleneketones C=C → CHCR

with KOH cf. *20,* 521; with tetramethylguanidine, also to α,β-ethylenecarboxylic acid esters, s. N. Ono et al., Tetrahedron Letters *23,* 2957–60 (1982).

Acetic acid s. under $NaNO_2$ *MeCOOH*

Titanium tetrachloride *$TiCl_4$*

1-Silylcyclopentenes from allenesilanes and ethylene derivs. with silyl group migration
Regiospecific[3+2]-cycloaddition ○

671.

1.5 Equivalents of $TiCl_4$ added rapidly to a soln. of methyl vinyl ketone and the startg. allenylsilane in methylene chloride at −78°, and the mixture stirred at this temp. for 1 hr. → product. Y 68–75%. – The reaction fails with *β,β*-disubst. *α,β*-ethyleneketones. F.e.s. R.L. Danheiser, D.J. Carini, A. Basak, J. Am. Chem. Soc. *103,* 1604–6 (1981).

Titanium trichloride/lithium tetrahydridoaluminate *$TiCl_3/LiAlH_4$*

Reductive dimerization of ethylene derivs. 2 C=C → CH−C−C−CH
with $TiCl_4/LiAlH_4/(MeCOO)_2Cu$ cf. *35,* 471; with $TiCl_3/LiAlH_4$, 1,6-diketones, s. J. Janssen, W. Lüttke, Chem. Ber. *115,* 1234–43 (1982).

Stannous chloride s. under $PdCl_2$ *$SnCl_2$*

Stannic chloride *$SnCl_4$*

***β*-Arylselenonitriles from ethylene derivs.** C=C → C(SeAr)C−CN
s. *37,* 644

Triphenylphosphine s. under Bis(1,5-cyclooctadiene)nickel(0), Ph_3P $(MeCOO)_2Pd$, and $PdCl_2$

Iodine s. under $H_2BCl.SMe_2$ *I_2*

*Tetra-*n*-butylammonium fluoride* n-*$Bu_4N^+F^-$*

Michael addition C=C → CHCR
with KF cf. *23,* 700; with tetra-*n*-butylammonium fluoride s. K. Matsumoto, Angew. Chem. Intern. Ed. *20,* 770–1 (1981).

High pressure Michael addition
with NEt_3 cf. *36,* 660; with *n*-$Bu_4N^+F^-$ s. Angew. Chem. Intern. Ed. *20,* 770–1 (1981).

Insertion of four C-atoms with *α,β*-ethyleneketones ←
***ε*-Nitro-*β*-diketones from *α*-nitroketones**

672.

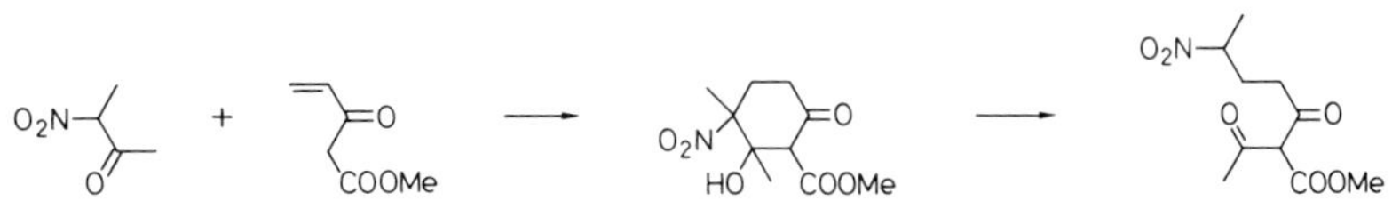

A soln. of 3-nitro-2-butanone in tetrahydrofuran treated simultaneously at 0° under argon with solns. of startg. *α,β*-ethyleneketone and tetrabutylammonium fluoride in the same solvent during 30 min., further $Bu_4N^+F^-$ in tetrahydrofuran added during 1 hr., the mixture stirred for 30 min., treated again with $Bu_4N^+F^-$ in tetrahydrofuran during 30 min., the mixture warmed to 20°, and stirring continued at this temp. for 3 hrs. → 2-acetyl-6-nitro-3-oxoheptanoic acid methyl ester. Y 82%. F.e. and isolation of the intermediate **4-nitrocyclohexan-3-ol-1-ones, also 1,3,7-triketones from *β*-diketones,** s. A. Lorenzi-Riatsch, Y. Nakashita, M. Hesse, Helv. Chim. Acta *64,* 1854–65 (1981).

Hydrogen chloride *HCl*

γ-Nitroketones
from 1-nitroethylene derivs. and enamines $NO_2C=C \rightarrow NO_2CHCR$
Stereospecific Michael addition

673.

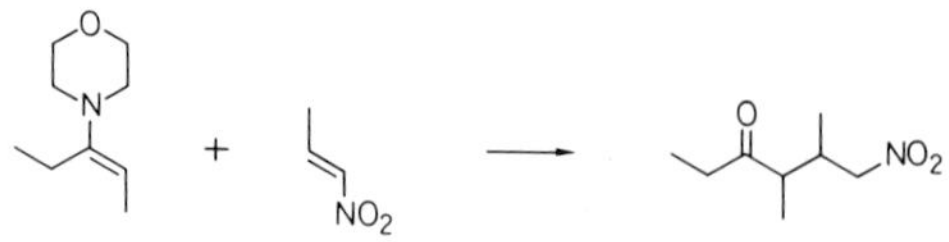

A soln. of N-(2-penten-3-yl)-morpholine and 1-nitro-1-propene in ether kept at room temp. for several days, solvent replaced by 1:1 10% aq. HCl/ethanol, and stirred at room temp. for 1 hr. → 4,5-dimethyl-6-nitro-3-hexanone. Y 81% (91% diasteriomeric purity). F.e., and elaboration of a **topological rule** for carbon-carbon bond formation between prochiral centres, s. D. Seebach, J. Golinski, Helv. Chim. Acta *64,* 1413–23 (1981).

Dicarbonylcyclopentadienylcobalt $(C_5H_5)_2Co(CO)_2$

Pyridine ring from acetylenenitriles and acetylene derivs. ○
Double ring closure

674.

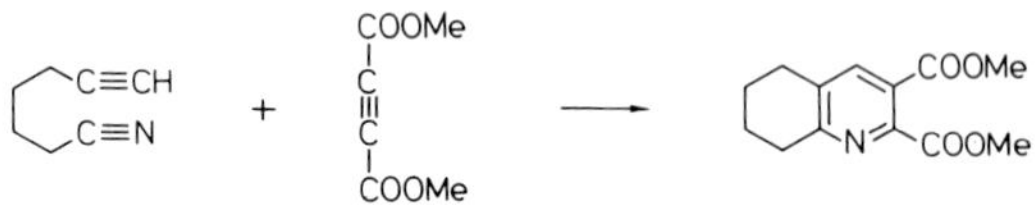

5,6,7,8-Tetrahydroquinolines. A mixture of startg. acetylenenitrile, dimethyl acetylenedicarboxylate, and a little dicarbonyl(cyclopentadienyl)cobalt in *m*-xylene added during 4–8 hrs. via syringe pump to refluxing xylene irradiated by a visible light GE–ENH 250 W. slide projector lamp → product. Y 83%. Cf. Synth. Meth. *31,* 658s*33.* F.e., incl. tri- and penta-methylene analogs, and regiospecificity, s. D.J. Brien, A. Naiman, K.P.C. Vollhardt, J. Chem. Soc. Chem. Commun. *1982,* 133–4.

Bis(1,5-cyclooctadiene)nickel(0)/triphenylphosphine ←

[2+2+2]-Cycloaddition with two acetylene deriv. molecules and heterocumulenes with carbodiimides cf. *36,* 675; with isocyanates, prepn. of 2-pyridones, s. Synthesis *1982,* 324–5.

Polymer-supported nickel(II) acetylacetonate ←

Michael addition under neutral conditions $C=C \rightarrow CHCR$
with nickel(II) acetylacetonate cf. *35,* 476; with polymer-supported nickel(II) acetylacetonate and Fe catalysts s. C.P. Fei, T.H. Chan, Synthesis *1982,* 467–8.

Dihydridotetrakis(trialkylphosphine)ruthenium(II) $(R_3P)_4RuH_2$

trans-**1,3-Enynes from terminal acetylene derivs. and 1,3-dienes** $C \equiv C-C=C$

675.

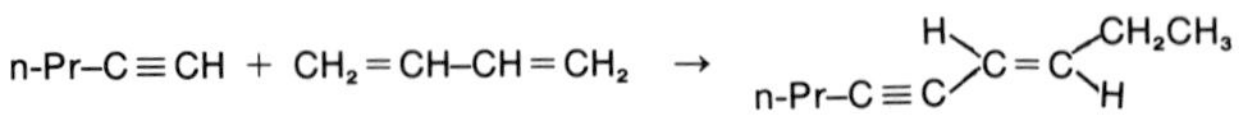

A mixture of dihydridotetrakis(tri-*n*-butylphosphine)ruthenium(II), benzene, 1,3-butadiene, and 1-pentyne heated at 60° in an ampoule under argon for 2 hrs. → product. Y 96%. F.e. and catalyst s. T. Mitsudo et al., J. Chem. Soc. Chem. Commun. *1981,* 496–7.

Rhodium oxide/iron pentacarbonyl $Rh_2O_3/Fe(CO)_5$

Amines from ethylene derivs. $C{=}C \rightarrow CHC{-}CH_2N{<}$

Synthesis with addition of one C-atom

s. *27,* 723; f. Rh catalysts, scope and pathway s. F. Jachimowicz, J.W. Raksis, J. Org. Chem. *47,* 445–7 (1982).

Rhodium carbonyl/ tetramethyl-1,3-diaminopropane/water $Rh_6(CO)_{16}/H_2N(CH_2)_3NH_2/H_2O$

Prim. alcohols from terminal ethylene derivs. $CH{=}CH_2 \rightarrow CH_2CH_2CH_2OH$

Reductive carbonylation

676. $$n\text{-}C_6H_{11}CH{=}CH_2 + CO + H_2O \rightarrow n\text{-}C_6H_{11}CH_2CH_2CH_2OH$$

A mixture of 1-octene, water, a little $Rh_6(CO)_{16}$, and 3 equivalents N,N,N′,N′-tetramethyl-1,3-propanediamine in ethoxyethanol heated under an initial pressure of 5 atm. CO with stirring at 80° for 5 hrs. → 1-nonanol. Y 84%. The hydroformylation occurs without the addition of molecular hydrogen. F.e., **also oxo from** α,β**-ethyleneoxo compds.,** s. K. Kaneda et al., Chem. Letters *1981,* 2, 1763–6.

Chlorotris(triphenylphosphine)rhodium(I) $(Ph_3P)_3RhCl$

Double ring closure of diynes with acetylene derivs. ○

with dicarbonylcyclopentadienylcobalt cf. *33,* 658s*36;* with chlorotris(triphenylphosphine)rhodium(I), condensed benzene ring, s. R. Grigg, R. Scott, P. Stevenson, Tetrahedron Letters *23,* 2691–2 (1982).

Silica-supported rhodium-phosphine complex ←

Hydroformylation $C{=}C \rightarrow CHC(CHO)$

s. *4,* 667; *20,* 55; with a silica-supported rhodium-phosphine catalyst cf. F. Jiao et al., Ranliao, Huaxue Huebao *9* (1), 47 (1981) (Ch); Chem. Abstr. *96,* 52483 (1982); **asymmetric hydroformylation** with $PtCl_2$-$SnHal_2$ containing a chiral diphosphine s. T. Hayashi et al., Bull. Chem. Soc. Japan *54,* 3438–46 (1981); **of vinyl esters** with chiral rhodium-phosphine complex cf. W.S. Knowles et al., J. Org. Chem. *46,* 4422–7 (1981); **review of hydroformylation** s. C. Botteghi et al., Chim. Ind. (Milan) *63,* 29–37 (1981).

Water-soluble palladium phosphine complex/diethylamine ←

1,3-Dien-5-ynes from 3 acetylene deriv. molecules ←

Homogeneous oligomerization in aq. medium

677. $$3\ Me_2C(OH)C{\equiv}CH \xrightarrow{PdCl_2(Ph_2PC_6H_4SO_3H\text{-}p)_2} Me_2C(OH)CH{=}CH{-}CH{=}\underset{}{C}(CMe_2OH){-}C{\equiv}CC(OH)Me_2$$

A mixture of 3-methyl-1-butyn-3-ol, water, and a little water-soluble dichlorobis(triarylphosphine)palladium(II) heated at 90° for 2 hrs. in the presence of diethylamine → product. Y 84%. F.e. and Rh complex s. G.A. Chukhadzhyan, T.S. Elbakyan, L.I. Sagradyan, Arm. Khim. Zh. *34,* 163–5 (1981).

Palladium(II) acetylacetonate/triphenylphosphine $Pd(acac)_2/Ph_3P$

Palladium-catalysed telomerization of 1,3-dienes ←

with 2-aminoalcohols cf. *27,* 724s*34;* with 2-ethyleneamines, f. catalysts, s. C. Moberg, Tetrahedron Letters *22,* 4827–30 (1981).

Palladium(II) acetate/triphenylphosphine/ potassium carbonate $(MeCOO)_2Pd/PPh_3/K_2CO_3$

Addition of polyhalides to carbon–carbon double bonds ←

s. *18,* 776s*36; 19,* 776s*29;* with Pd(II) acetate, Ph_3P, and K_2CO_3 cf. J. Tsuji, K. Sato, H. Nagashima, Chem. Letters *1981,* 1169–70.

Palladium(II) chloride/stannous chloride/triphenylphosphine *$PdCl_2/SnCl_2/PPh_3$*

α-Methylene-γ-lactones from 3-acetylenealcohols ○

s. *30,* 502; with catalytic amounts of $PdCl_2$, $SnCl_2$, and triphenylphosphine in acetonitrile cf. T.F. Murray et al., J. Am. Chem. Soc. *103,* 7520–8 (1981).

Rearrangement ∩

Hydrogen/Oxygen Type CC∩HO

Potassium hydride *KH*

Ketones from ethylenealcohols ←

Oxy-Cope rearrangement

s. *31,* 664; of 1,2-dialkenylcyclobutanols to cyclooctenones s. R.C. Gadwood, R.M. Lett, J. Org. Chem. *47,* 2268–75 (1982); bicyclo[5.3.1]undec-7-ene ring with hexamethylphosphoramide as co-solvent or with added 18-crown-6 polyether s. S.F. Martin, J.B. White, R. Wagner, J. Org. Chem. *47,* 3190–2 (1982).

Macroisocyclic ketones via sigmatropic rearrangement

Ring expansion by eight C-atoms

678.

3,7,9-Trienones. Startg. m. treated with a soln. of KH in tetrahydrofuran at room temp. for 1 hr. → product. Y 90%. – The reaction proceeds either by a [5.5]-sigmatropic rearrangement or two consecutive Cope rearrangements. P.A. Wender, S. McN. Sieburth, Tetrahedron Letters *22,* 2471–4 (1981).

Bridgehead carbon-carbon double bonds by oxy-Cope rearrangement

679.

Startg. vinylcarbinol refluxed with KH in tetrahydrofuran for 4 hrs. (cf. Synth. Meth. *31,* 664) → product. Y 79%. F.e.s. S.G. Levine, R.L. McDaniel, J. Org. Chem. *46,* 2199–200 (1981).

Via intermediates *v.i.*

Stereospecific double ring closure by oxymercuration-demercuration ○

Bicyclic lactones from dienecarboxylic acids

680.

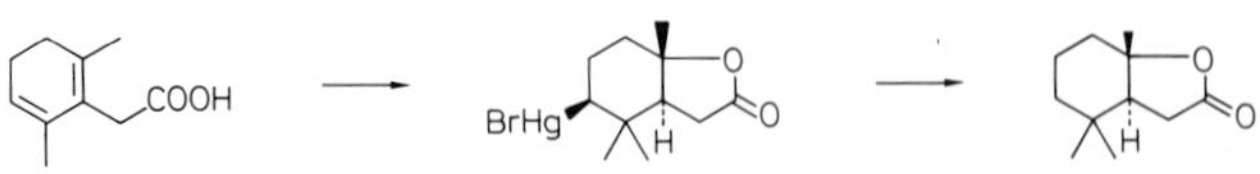

A soln. of Hg trifluoroacetate in nitromethane added to a stirred soln. of startg. diene in the same solvent at room temp. under N_2, after 20 min. excess satd. aq. KBr added, and the mixture stirred for 14 hrs. → intermediate organomercury bromide, in 1:1 dichloromethane/95% ethanol treated dropwise under argon with a soln. of $NaBH_4$ in 1.4 *M* NaOH at room temp. for ca. 1 hr. → product. Y 88%. F.e., **also bicyclic enolethers and ethers from dienones and dienols, respectively,** s. T.R. Hoye, A.J. Caruso, M.J. Kurth, J. Org. Chem. *46,* 3550–2 (1981).

Hydrogen/Carbon Type CC∩HC

Without additional reagents *w.a.r.*

Intramolecular ene reaction O

dihydronaphthalenes cf. *28,* 667; [3.3.3]propellanes s. H. Schostanez, L.A. Paquette, Tetrahedron *37,* 4431–5 (1981); J. Am. Chem. Soc. *103,* 722–24 (1981).

Stereospecific intramolecular ene reaction

s. *36,* 682; α-cyclopropyloxo compds. from 2,5-dienols s. F.-G. Klärner, W. Rüngeler, W. Maifeld, Angew. Chem. Intern. Ed. *20,* 595–6 (1981); review of regio- and stereospecific intramolecular cycloaddition and ene reactions s. W. Oppolzer, Pure Appl. Chem. *53,* 1181–201 (1981).

(E)-Thioenolethers from 1-vinylcyclopropyl thioethers C

681. 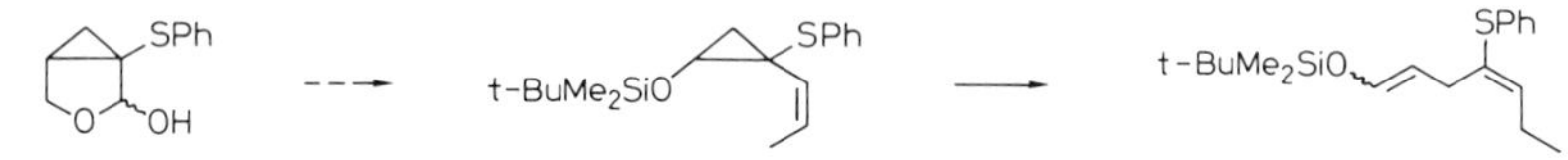

Startg. m. flash vacuum pyrolysed at 545–650° and 0.05–0.30 torr → product (E,E/Z,E = 9:1). Y 96% (from startg. lactol). – The obtained **4-arylthio-1-siloxy-1,4-dienes** are synthetically useful **as** chemodifferentiated **1,4-dioxo compd. equivalents.** F.e.s. B.M. Trost, P.L. Ornstein, J. Org. Chem. *47,* 748–51 (1982).

Irradiation (s.a. under $AlCl_3$, Me_2CO and Methyl 2-naphthyl ketone ⫻

α-Oxo-N-heterocyclics from enacylamines O

s. *25,* 522; 5,6-dihydro-5-spiro-2-pyridone ring s. J.-C. Gramain, Y. Troin, D. Vallee, J. Chem. Soc. Chem. Commun. *1981,* 832–3.

Carbostyrils from α,β-ethylenecarboxylic acid anilides

s. *19,* 787; photochemically with dihydrofuran ring opening, 3-(β-hydroxy)carbostyrils, s. S.P. Rajendran et al., Synthesis *1982,* 160–2.

Sodium hydride *NaH*

Isomerization of cyclic β-ketothiolic acid esters by skeletal rearrangement

682. 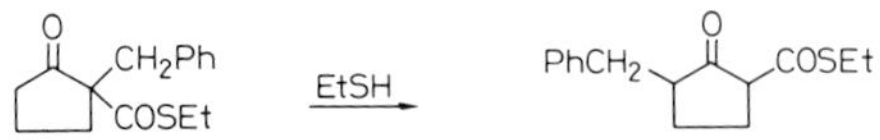

A soln. of startg. m., NaH, and ethanethiol in 1,2-dimethoxyethane refluxed for 4 hrs. → product. Y 93%. F.e.s. H.-J. Liu, H.K. Lai, Syn. Commun. *11,* 65–9 (1981).

Lithium diisopropylamide *$i\text{-}Pr_2NLi$*

Epimerization of cyano groups ←

with KOBu-*t* cf. *24,* 718; with Li diisopropylamide s. G. Sauer et al., Ann. Chem. *1982,* 448–58.

Potassium carbonate *K_2CO_3*

Intramolecular Michael addition O

with KH cf. *33,* 662s*36;* with K_2CO_3 and f. methods s. I.A. Blair et al., Australian J. Chem. *34,* 1887–98 (1981); s.a. T. Imanishi et al., Tetrahedron Letters *22,* 4001–4 (1981).

Calcium oxide *CaO*

O-Propenyl from O-allyl derivs. O–CH–C=C → O–C=C–CH

trans-derivs. with iridium-phosphine complex cf. *34,* 668; *cis*-derivs. with CaO s. H. Matsuhashi, H. Hattori, K. Tanabe, Chem. Letters *1981,* 341–2.

Alumina Al_2O_3

2E,4Z-Diene- from β-allene-carboxylic acid esters CH–C=C=CH → C=CH–CH=C

s. *23,* 801s*37*

Aluminium chloride/irradiation $AlCl_3/h\nu$

trans-cis-**Rearrangement** ←

with acetophenone as sensitizer cf. *28,* 669; in the presence of $AlCl_3$, *cis-α,β*-ethylenecarboxylic acid esters, s. F.D. Lewis, J.D. Oxman, J. Am. Chem. Soc. *103,* 7345–7 (1981); E → Z rearrangement of ethylene derivs. with Ni cf. M. Prashad, M. Seth, A.P. Bhaduri, Indian J. Chem. *19B,* 393 (1980).

Hydroquinone ←

Intramolecular diene synthesis with vinylheterocyclics ○

683.

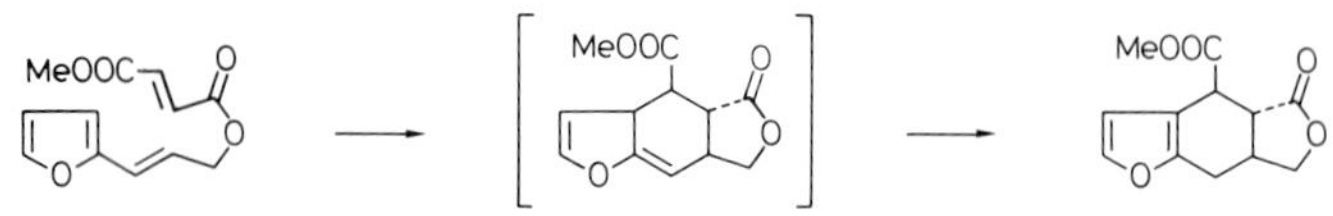

A soln. of startg. m. in toluene containing a small amount of hydroquinone heated at 200° for 20 hrs. in a sealed tube under N_2 → product. Y 73%. F.e. and aromatization to **benzofuran ring** s. H. Kotsuki et al., Chem. Letters *1981,* 917–20.

Acetone/irradiation $Me_2CO/h\nu$

Skeletal rearrangement of bicyclo[2.2.2]oct-2-enes ←

684.

Tricyclo[3.3.0.0^{2,8}]octan-3-ones. A soln. of startg. m. in acetone irradiated under argon in a water-cooled quartz vessel with a Rayonet RPR-208 photoreactor for 72 hrs. → tricyclo[3.3.0.0^{2,8}]octan-3-one. Y 85%. – The products are synthons for polycyclopentanoid terpenes and prostacyclins. F.e.s. M. Demuth et al., Helv. Chim. Acta *63,* 2434–9 (1980); **bicyclo[3.2.1]oct-2-ene ring** cf. D. Bravetti, R.M. Bettolo, A. Lupi, ibid. *65,* 371–6 (1982).

Methyl 2-naphthyl ketone/irradiation ←

1,5-C-Acyl group migration ←

acyl-*p*-quinols cf. *36,* 686; 7-acyloxepin-2(3*H*)-ones, photochemical method with methyl 2-naphthyl ketone as sensitizer, s. N. Hoshi, H. Uda, J. Chem. Soc. Chem. Commun. *1981,* 1163–4.

Azodiisobutyronitrile/thiophenol $Me_2C(CN)N{=}NC(CN)Me_2/PhSH$

1,3-Dienes from allenes CH–C=C=CH → C=CH–CH=C

2,4-dienones s. *23,* 801; 2,4-dienecarboxylic acids with azodiisobutyronitrile/thiophenol s. T. Sato et al., Tennen Yuki Kagobutsu Toronkai Koen Yoshishu, 24th, *1981,* 425 (Japan); Chem. Abstr. *96,* 199032 (1982); 2E,4Z-dienecarboxylic acid esters with Al_2O_3 s. S. Tsuboi et al., Tetrahedron Letters *23,* 209–12 (1982).

Zirconium(IV) n-*propoxide* *Zr(OPr*-n*)*$_4$

Stereospecific intramolecular Michael addition with subsequent Aldol condensation ○

685. 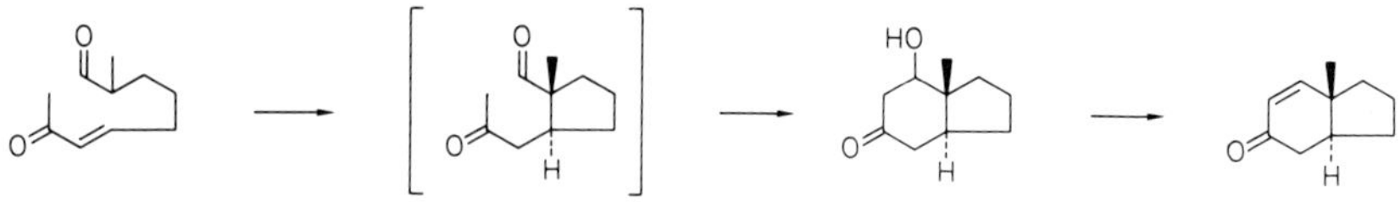

trans-**Hydrindenones.** A soln. of the startg. m. in dry benzene treated with 1 equivalent Zr(IV) *n*-propoxide, dil. with methanol after 1 hr. at room temp., 0.5 *M* aq. LiOH added, and stirred overnight → intermediate bicyclic ketol, in dry methylene chloride containing a crystal of 4-(N,N-dimethylamino)pyridine treated with 3 equivalents of trifluoroacetic anhydride and 5 equivalents of diazabicycloundecene at –40° under argon, allowed to warm to 0° over 1 hr., further diazabicycloundecene added, and the mixture allowed to warm to room temp. → product. Overall Y 90% (*trans*:*cis* 40:1). – The *trans*-fused product can normally be obtained only via the *cis*-fused derivs. G. Stork, C.S. Shiner, J.D. Winkler, J. Am. Chem. Soc. *104,* 310–2 (1982).

Stannic chloride *SnCl*$_4$

Cyclic ketones from ethyleneketones
Intramolecular C-*a*-*tert*-alkylation

686. 

Cyclic *β*-ketocarboxylic acid esters. $SnCl_4$ added slowly to a stirred soln. of methyl 7-methyl-3-oxo-6-octenoate in dichloromethane at 0°, warmed to room temp., then stirred for 14 hrs. → methyl 2,2-dimethyl-6-oxocyclohexanecarboxylate. Y 96%. F.e. and *β*-diketone analogs s. M.T. Reetz, I. Chatziiosifidis, K. Schwellnus, Angew. Chem. Intern. Ed. *20,* 687–9 (1981).

Sulfuric acid *H*$_2$*SO*$_4$

***a,β*-Ethyleneketones**
from *γ,δ*-ethyleneketones C=C–CH–C–CO → CH–C–C=C–CO

687. 

Startg. m. treated with 5% H_2SO_4 in acetic acid → product. Y 96%. F.e.s. S.R. Wilson, R.N. Misra, J. Org. Chem. *45,* 5079–81 (1980).

Iodine *I*$_2$

Migration of carbon-carbon double bonds ←

s. *19,* 789; into the [3.3.3]propellane ring s. H. Schostarez, L.A. Paquette, J. Am. Chem. Soc. *103,* 722–4 (1981).

Hydrogen chloride *HCl*

...into rings
Cyclopent-2-enones

s. *31,* 680; s.a. A. Barco et al., Synthesis *1981,* 199–200; 2-alkylindenes on irradiation cf. H. Morrison, D. Giacherio, J. Org. Chem. *47,* 1058–63 (1982).

Nickel *Ni*

E → Z Rearrangement ←

s. *28,* 669s*37*

Chiral rhodium phosphine complex ←

(E)-Enamines from 2-ethyleneamines C=C–CH–N⋜ → CH–C=C–N⋜

Asym. induction

688. 

A mixture of N,N-diethylnerylamine and a little cationic chiral Rh-phosphine complex in tetrahydrofuran heated at 40° for 23 hrs. → (3R)-N,N-diethylcitronellal-(E)-enamine. Y ca. 100% (94% enantiomeric excess). – The method is of potential for preparing **optically active aldehydes.** F.e. and catalysts s. K. Tani et al., J. Chem. Soc. Chem. Commun. *1982,* 600–1.

Bis(benzonitrile)dichloropalladium(II) *$PdCl_2(PhCN)_2$*

Isomerization of cyclopentenes ←

s. *37,* 106

Bis(benzonitrile)dichloroplatinum(II) *$PtCl_2(PhCN)_2$*

Enolethers from alkoxycyclopropanes C

689. 

γ-Alkoxy-β,γ-ethylenecarboxylic acid esters. A mixture of startg. cyclopropyl ether and a little bis(benzonitrile)dichloroplatinum(II) heated at 100° for 10 min. → product. Y 98%. F.e. and catalysts, incl. copper bronze, s. M.P. Doyle, D. Van Leusen, J. Am. Chem. Soc. *103,* 5917–9 (1981).

Via intermediates *v.i.*

Geoisomerization of ethylene derivs. via 1,3-dithiolane-2-thiones ←

690. 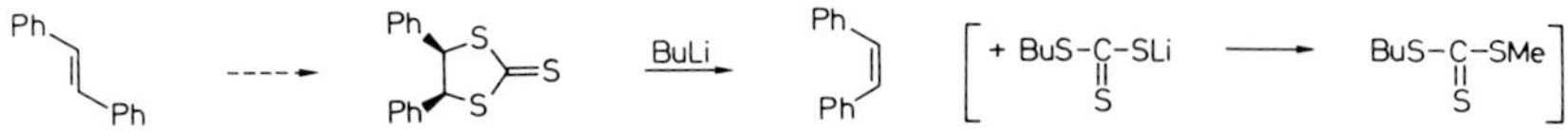

A soln. of startg. *cis*-4,5-diphenyl-1,3-dithiolane-2-thione [prepd. from *trans*-stilbene by the method of G.G. Overberger and A. Drucker, J. Org. Chem. *29,* 360 (1964)] in tetrahydrofuran at –78° treated with a soln. of *n*-butyllithium in hexane under N_2, and the mixture stirred at the same temp. for 45 min. with exclusion of daylight → *cis*-stilbene. Y 99%. F.e. and prepn. of **trithiocarbonic acid esters** by reaction of the Li trithiocarbonates with alkyl halide, s. K. Hatanaka et al., Tetrahedron Letters *22,* 5195–6 (1981).

Oxygen/Carbon Type CC∩OĊ

Without additional reagents *w.a.r.*

Cyclic α-aminoketones from 2-aminoenolethers RO–C=CH–N⋜ → CO–CH–N⋜

2-Aminocyclohexanones

691. 

Startg. m. heated in the vapor phase under an argon flow at 450–470° in a quartz reactor → 2-morpholinocyclohexanone. Y 70%. F.e.s. R.A. Karakhanov, M.M. Vartanyan, R.B. Apandiev, Izv. Akad. Nauk SSSR Ser. Khim. *1981,* 694–5; **2-aminocyclopentanones** s. ibid. *1980,* 1951.

Irradiation

Cyclic alcohols from ketones
3-Azetidinols ○
s. *26,* 744; condensed derivs. s. J.-C. Gramain, L. Ouazzani-Chahdi, Y. Troin, Tetrahedron Letters *22,* 3185–8 (1981).

Electrolysis

Oxo from oxido compds.
Isomerization
with BF_3 cf. *9,* 794; electrochemically s. J. Delaunay et al., J. Chem. Soc. Chem. Commun. *1982,* 387–8.

Sodium alkoxide *NaOR*

Cyclic β-diketones from enollactones ←
with Na methoxide in benzene cf. *21,* 764; with ethanolic Na ethoxide, cyclopentene-1,3-diones from 5-alkylidene-2(5*H*)-furanones, s. R.A. Massy-Westropp, M.F. Price, Australian J. Chem. *34,* 2369–83 (1981).

n-*Butyllithium* n-*BuLi*

1,5-Dien-3-ols from diallyl ethers ←
Regio- and geo-specific Wittig rearrangement

692. 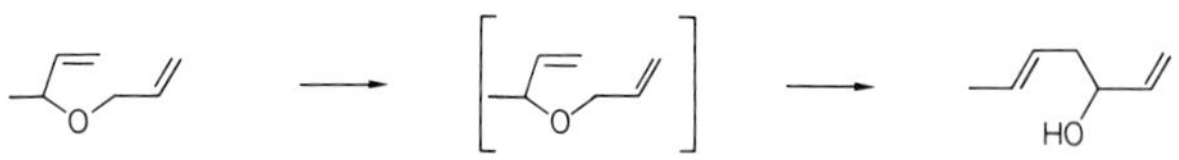

A soln. of *n*-butyllithium in hexane added dropwise to a soln. of startg. dienol in tetrahydrofuran at –85° under N_2, the mixture stirred at the same temp. for 5–8 hrs., allowed to warm to 0°, and quenched with HCl → product. Y 79% (>95% E isomer). F.e.s. T. Nakai et al., J. Am. Chem. Soc. *103,* 6492–4 (1981).

Lithium diisopropylamide *i-Pr₂NLi*

α-Aminoketones from iminoesters

693. $p\text{-}MeC_6H_4N{=}C(Ph)OCH_2Ph \rightarrow p\text{-}MeC_6H_4NHCH(Ph)COPh$

Startg. iminoester treated with a soln. of Li diisopropylamide in tetrahydrofuran at –5 to 0° → product. Y 46%. – This appears to be the first example of an **imino-Wittig rearrangement.** F.e.s. A.R. Katritzky, N.K. Ponkshe, Tetrahedron Letters *22,* 1215–6 (1981).

'Homo-Claisen' rearrangement ←
δ,ε-Ethylenecarboxylic acids from 2-acoxy-3-ethylenesilanes
via γ,δ-ethylene-ε-silylcarboxylic acids

694. 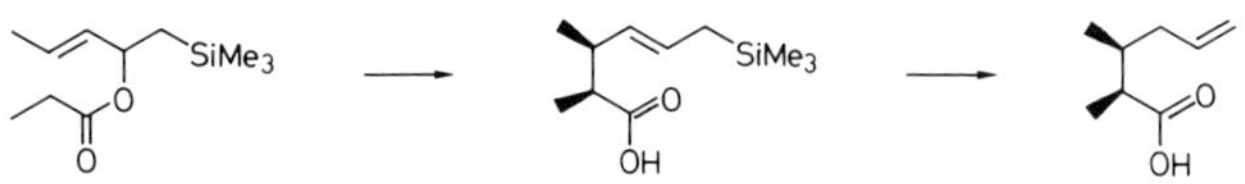

Startg. m. treated with Li diisopropylamide between –78° and room temp. (cf. Synth. Meth. *28,* 683) → intermediate (Y 61%, 90% *trans-erythro*), treated with BF_3-acetic acid in methylene chloride at 25° for 5 min. → product (Y 100%). F.e.s. S.R. Wilson, M.F. Price, J. Am. Chem. Soc. *104,* 1124–6 (1982).

Lithium bromide *LiBr*

Cyclobutanones from alkylidenecyclopropanes via oxaspiropentanes

695. 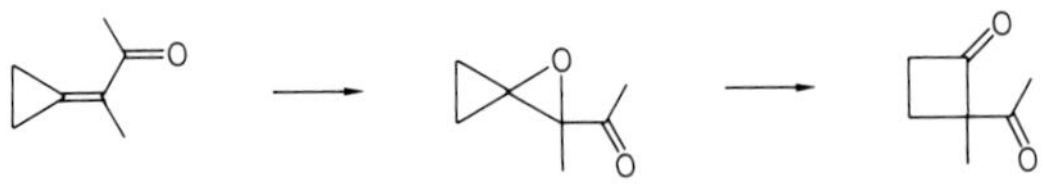

2-Acylcyclobutanones. A mixture of startg. m., *m*-chloroperbenzoic acid, and Na_2-HPO_4 in methylene chloride allowed to react at 30° → intermediate (Y 63%) and LiBr in tetrachloromethane heated at 60° for 27 hrs. → product. (Y 100%). F.e., also with *a*-cyclopropylideneacetals, s. F. Huet, A. Lechevallier, J.-M. Conia, Chem. Letters *1981,* 1515–8.

Dialkylaluminium mercaptides *R_2AlSPh*

Cyclic *β*-diketones from enollactones ←

696. 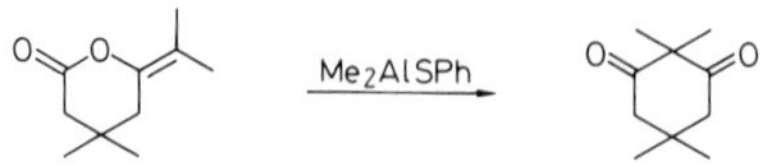

A soln. of startg. *δ*-lactone in tetrahydrofuran added to dimethylaluminium phenylthiolate (prepd. from thiophenol and trimethylaluminium), and the mixture refluxed for 1 hr. → product. Y 91%. Cf. Synth. Meth. *21,* 764. F.e. and reagents s. H. Tomioka, K. Oshima, H. Nozaki, Tetrahedron Letters *23,* 99–100 (1982).

Boron fluoride *BF_3*

***β*-Alkoxyaldehydes from 1-vinyloxyethers** ←

s. *37,* 936

Cation exchanger ←

***a*-Vinyloxyketones from 4-methylene-1,3-dioxolanes**

697. 

A mixture of 2-ethyl-4-methylene-1,3-dioxolane and a little Lewatit CNP 80, a *weakly acidic* carboxy-type cation exchanger, heated with stirring at 170° for 8 hrs. → 1-(1-propenyloxy)-2-propanone. Y 68%. F.e. and catalysts s. J. Mattay, W. Thünker, H.-D. Scharf, Ann. Chem. *1981,* 1105–17.

Nitrogen/Sulfur Type CC∩NS

Triethylamine *Et_3N*

***o*-Aminomercaptals from mercaptal mono-S-arylimines** ←
Stereospecific rearrangement

698. 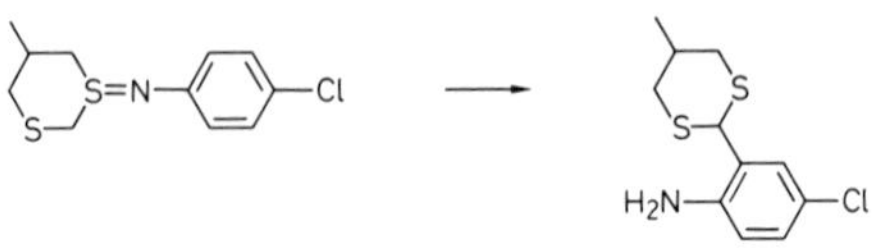

A soln. of startg. sulfimide in 1:1 triethylamine/benzene heated at 75–80° for 10–15 hrs. → product. Y 75%. F.e.s. J. Bailer, P.K. Claus, F.W. Vierhapper, Tetrahedron *36,* 901–11 (1980).

Nitrogen/Carbon Type CC∩NC

Without additional reagents *w.a.r.*

***o*-Amino-Claisen rearrangement of quaternary 2-ethyleneanilinium salts** ←
s. *34,* 680; s.a. Chem. Pharm. Bull. *29,* 2465–77 (1981).

***m*-Amino-Claisen rearrangement of quarternary 2-ethyleneanilinium salts** ←

699.

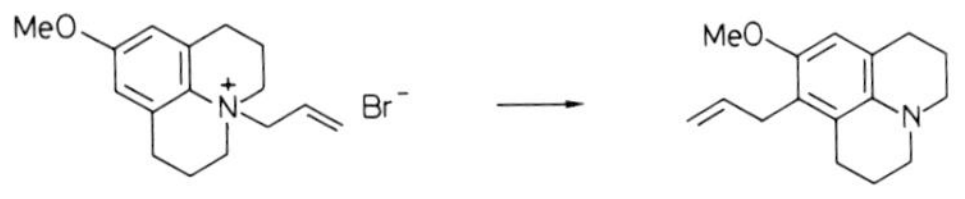

A soln. of 4-allyl-9-methoxyjulolidinium bromide heated in 2:1 glycerol/water at 140° for 4 hrs. under N_2 → 8-allyl-9-methoxyjulolidine. Y 91%. F.e., also in the presence of $NaHCO_3$, and ***p*-amino-Claisen rearrangement,** s. H. Katayama, J. Chem. Soc. Chem. Commun. *1980,* 1009–10.

Irradiation

Cyclopropane from Δ^1-pyrrol-5-one ring ▽
s. *26,* 753; s.a. Org. Synth. *59,* 132 (1979).

Potassium hydroxide *KOH*

2-Alleneamines from quaternary 2-acetyleneammonium salts ←
Stevens rearrangement

700.

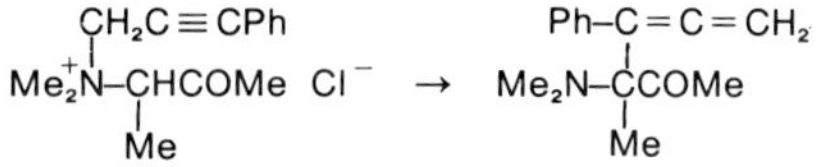

Powdered KOH and dimethyl sulfoxide added to a soln. of startg. salt in ether, and the mixture stirred occasionally until the exothermic reaction ceased → product. Y 75%. F.e.s. S.T. Kocharyan et al., Arm. Khim. Zh. *34,* 429–31 (1981); **3-ethyleneamines from quaternary 2–ethyleneammonium salts** s. ibid. *34,* 409–16 (1981); Zh. Org. Khim. *18,* 1861–4 (1982); **intramolecular pericyclic reactions of acetylene derivs. review,** s. A. Viola, J.J. Collins, N. Filipp, Tetrahedron *37,* 3765–811 (1981).

Sodium alkoxide *NaOR*

Stereospecific 1,5-electrocyclic ring closure of azomethines ○
3-Spiroindolines

701.

A soln. of the startg. m. in toluene added to a soln. of Na ethoxide in ethanol at room temp., and quenched with water after 4 min. → *trans*-product. Y 90%. – With Na *tert*-butoxide in *tert*-butanol → *cis*-product. Y 71%. F.e.s. W.N. Speckamp et al., J. Am. Chem. Soc. *103,* 4643–5, 4645–6 (1981); **asym. induction** in the presence of (–)-menthol s. J. Chem. Soc. Chem. Commun. *1982,* 369–70.

Lithium 2,2,6,6-tetramethylpiperidide ←

Indoles from ar. isonitriles via *o*-α-metalation
s. *33,* 691; **review** s. Kagaku no Ryoiki *36,* 91 (1982) (Japan); Chem. Abstr. *96,* 199436 (1982).

Potassium fluoride *KF*

Stevens rearrangement ←

with NaOH cf. *11,* 795; with KF in dimethyl sulfoxide s. S.T. Kocharyan, T.L. Razina, A.T. Babayan, Arm. Khim. Zh. *33,* 684–5 (1980).

Sodium iodide *NaI*

Enamines from aziridines C

with HCl cf. *7,* 756; with NaI in acetone, dehydroalanine-containing peptides s. K. Okawa et al., Bull. Chem. Soc. Japan *55,* 174–6 (1982).

Methylammonium sulfite *$[MeNH_3]_2^+SO_3^{2-}$*

Anilines from pyridinium salts O

702.

Startg. pyridinium salt heated with an aq. soln. of methylammonium sulfite at 150° for 30 hrs. → product. Y 96%. F.e.s. A.N. Kost et al., Dopov. Akad. Nauk Ukr. RSR Ser. B *1980,* 59–62; Chem. Abstr. *93,* 71157 (1980); 1-aminonaphthalenes from isoquinolinium salts s. Khim. Geterotsikl. Soedin. *1979,* 1386–9; s.a. ibid. *1981,* 508–10; 2-aminodiaryls s. Org. Prep. Proc. Intern. *13,* 203–7 (1981); **review of pyridine ring nucleophilic recyclizations** s. Tetrahedron, *37,* 3423–54 (1981).

Sulfur/Carbon Type CC∩SC

Without additional reagents *w.a.r.*

γ,δ-Ethylene- and N-allyl-carboxylic acid thioamides from S-allylthioiminoesters ←

703.

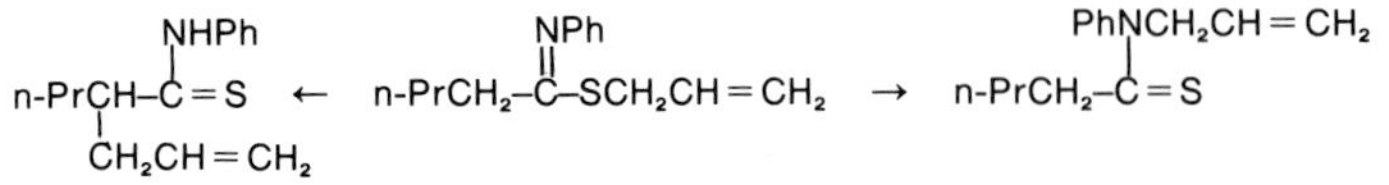

Startg. thioiminoester

heated at 200° for 1 hr. under argon under slightly reduced pressure in a Kugelrohr apparatus → N-phenyl-2-allylthiopentanoylamide. Y 76%.

in tetrahydrofuran containing a little bis(benzonitrile)dichloropalladium(II) refluxed for 6 hrs. under argon → N-phenyl-N-allylthiopentanoylamide. Y 80%.

F.e.s. Y. Tamaru, M. Kagotani, Z. Yoshida, Tetrahedron Letters *22,* 4245–8 (1981).

Sodium methoxide *NaOMe*

C-α-Allylation of thiolactams via thio-Claisen rearrangement ←

704.

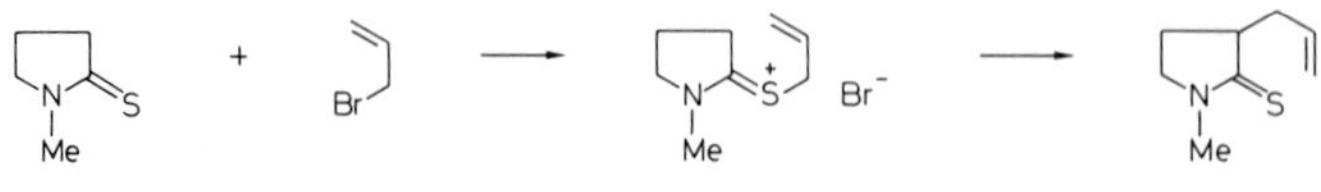

A soln. of startg. thiolactam in acetonitrile stirred with allyl bromide at room temp. under N_2 for 2 days → intermediate crude sulfonium salt, in tetrahydrofuran stirred with Na methoxide at 0° for 15 min. followed by 15 hrs. at room temp. → product. Y 90.3%. F.e. and geospecificity, also *α,β*-diallyl derivs. by repeating the reaction, s. S. Takano, M. Hirama, K. Ogasawara, Chem. Letters *1982,* 529–32.

Potassium tert-*butoxide* t-*BuOK*

Ring expansion of cyclic 2-ethylenesulfonium ylids by three C-atoms
s. *34,* 690s*36;* also with *i*-Pr_2NLi, stereochemical control, s. J. Org. Chem. *46,* 3315–21 (1981).

Lithium diisopropylamide *i-Pr_2NLi*

[2.3]-Sigmatropic rearrangement of unsatd. ethylenethioethers ←
s. *29,* 709s*30;* 3-ethylenemercaptans from 2-ethylenethioethers s.a. K. Tanaka, M. Terauchi, A. Kaji, Chem. Letters *1981,* 315–8 (1981).

Lithium N-isopropylcyclohexylamide ←

Sulfinylcyclopropanes from thietane 1-oxides
Retention of configuration

705.

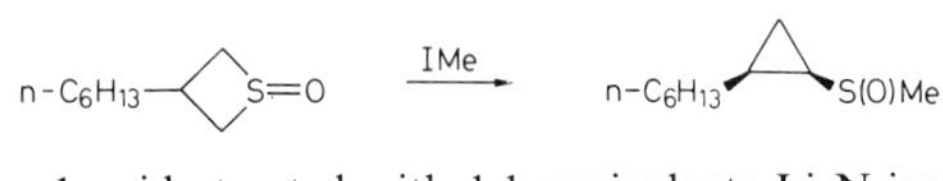

cis-3-Hexylthietane 1-oxide treated with 1.1 equivalents Li N-isopropylcyclohexylamide in tetrahydrofuran at –20° for 15 min., followed by addition of methyl iodide → *cis*-2-hexyl-1-(methylsulfinyl)cyclopropane. Y 89%. F.e.s. D.N. Jones et al., J. Chem. Soc. Chem. Commun. *1982,* 589–91.

Remaining Elements/Carbon Type CC∩RemC

Without additional reagents *w.a.r.*

Siloxycarbenes as intermediates CHCOSi≡ → C=CH–OSi≡
s. *31,* 704; **review of carbenes and carbenoids with neighbouring heteroatoms** s. K.G. Taylor, Tetrahedron *38,* 2751–72 (1982).

Carbon/Carbon Type CC∩CC

Without additional reagents *w.a.r.*

Cope rearrangement
cycloheptadiene ring s. *31,* 709; 4,5-dihydrooxepins s. J.-P. Beny, J.-C. Pommelet, J. Chuche, Bull. Chem. Soc. France II *1981,* 377–86.

Intramolecular diene synthesis

with furans
s. *36,* 724; substituent and solvent effects s. D.D. Sternback, D.M. Rossana, Tetrahedron Letters *23,* 303–6 (1982); as part of a gibberellin synthesis s. W.M. Grootaert, P.J. De Clercq, ibid. *23,* 3291–4 (1982); as part of a corticosteroid synthesis s. ibid. *23,* 3283–6 (1982).

Hydrindene ring
s. *26,* 758s*34, 36; trans*-hydrindene ring s. J. Am. Chem. Soc. *104,* 1033–41 (1982); *cis*-hydrindene ring cf. ibid. *104,* 3216–7 (1982).

Decalin ring
s. *34,* 693; from 1,2,4-trienes cf. E.A. Deutsch, B.B. Snider, J. Org. Chem. *47,* 2682–4 (1982); *cis*-decalin ring from methyl undeca-2,8,10-trienoates s. W.R. Roush, S.E. Hall, J. Am. Chem. Soc. *103,* 5200–11 (1981); *a*-eudesmol ring s. D.F. Taber, S.A. Saleh, Tetrahedron Letters *23,* 2361–4 (1982).

Highly condensed polyisocyclics
s. *26,* 758; iceane skeleton s. D.P.G. Hamon, P.R. Spurr, J. Chem. Soc. Chem. Commun. *1982,* 372–3.

Without additional reagents *w.a.r.*

Asym. total steroid synthesis via *o*-quinodimethanes

s. *32,* 711s*36;* (+)-chenodeoxycholic acid s. J. Org. Chem. *47,* 2331–42 (1982); 19-norsteroids s. G. Quinkert et al., Ann. Chem. *1981,* 2335–71.

7-Azabicyclo[4.2.0]octa-2,4-dien-8-ones ←
from 2-azabicyclo[4.2.0]octa-4,7-dien-3-ones

706. 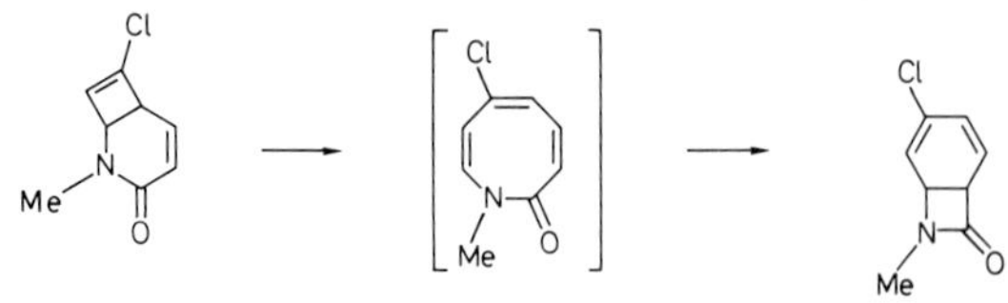

A soln. of startg. m. in 1,2,4-trichlorobenzene heated at 120° for 5 hrs. → product. Y 73%. F.e.s. K. Somekawa et al., Bull. Chem. Soc. Japan *54,* 1112–6 (1981).

Cyclohexenes from vinylcyclobutanes

s. *35,* 478; 1-siloxycyclohexene ring s. T. Sano et al., Heterocycles *16,* 1463–8 (1981).

Irradiation

2-Azabicyclo[2.2.0]hex-5-en-3-ones from 2-pyridones

707. 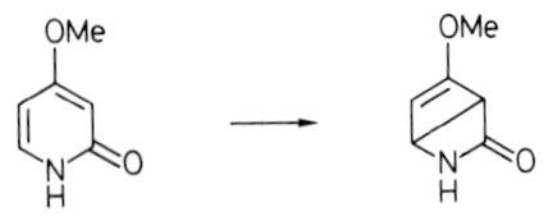

A soln. of 4-methoxy-2-pyridone in ether irradiated through Pyrex with a Toshiba 400P high-pressure Hg-lamp → 5-methoxy-2-azabicyclo[2.2.0]hex-5-en-3-one. Y 83%. F.e., also thermal rearrangement of product to **6-alkoxy-2-pyridones,** s. C. Kaneko et al., J. Chem. Soc. Chem. Commun. *1980,* 1177–8.

Photochemical intramolecular [2+2]-cycloaddition

s. *22,* 761; 2-heterobicyclo[3.2.0]heptane and -[4.2.0]octane ring s. M. Ikeda et al., Chem. Pharm. Bull. *29,* 2062–8 (1981); s.a. Bull. Chem. Soc. Japan *54,* 3492–4 (1981); dibenzo[4.4.2.2]buttaflane ring s. J. Chem. Soc. Chem. Commun. *1982,* 82–3; [4.4.3]propellanes and spirocyclics from terminal allenes s. D. Becker et al., J. Org. Chem. *47,* 3297–306 (1982); tricyclo[2.2.0.0^{2,6}]hexanes s. A. Padwa et al., J. Am. Chem. Soc. *103,* 7202–10 (1981); 1,2-bridged quadricyclanes s. H. Hart, C. Peng, Isr. J. Chem. *22,* 53–5 (1982); quadricyclanes as energy storage compds., s.a. K. Maruyama, K. Terada, Y. Yamamoto, J. Org. Chem. *46,* 5294–306 (1981); limitations s. A. Behr et al., J. Chem. Tech. Biotech. *32,* 627–30 (1982); f.e.s. Synth. Meth. *36,* 311.

n-*Butyllithium* n-*BuLi*

2,4-Dienesulfones from 3-*a*-sulfonylcyclobutenes
Carbanion-promoted ring opening

708. 

A soln. of *n*-butyllithium in hexane added at –78° under N_2 to a soln. of 3,3-(ethylenedioxy)-7-(*p*-toluenesulfonyl)methylbicyclo[4.2.0]oct-1(6)-ene in tetrahydrofuran, the temp. raised to –30°, and the mixture stirred for 10 min. → product. Y 90%. F.e., **also 2,4-dienesulfoxides with and without sulfinyl group migration,** and via *a*-alkylation at –78°, s. T. Kametani et al., J. Am. Chem. Soc. *103,* 1256–8 (1981).

n-*Butyllithium/N,N,N′,N′-tetramethylethylenediamine* n-*BuLi/TMEDA*

Cyclopentene from vinylcyclopropane ring

s. *31,* 708s*37*

Ethyldiisopropylamine *i-Pr₂NEt*

Stereospecific intramolecular diene synthesis via indole-2,3-quinodimethanes ○

709. 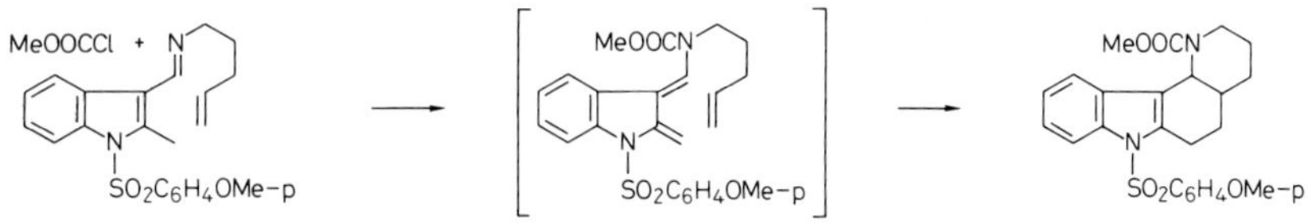

with N-carbalkoxylation. Methyl chloroformate added to a soln. of N-([(4-methoxybenzenesulfonyl)-2-methyl-1*H*-indol-3-yl]methylene)-4-pentenylamine and ethyldiisopropylamine in chlorobenzene at 0°, and heated to 130° for 3.5 hrs. → 1-carbomethoxy-7-(4-methoxybenzenesulfonyl)-1,2,3,4,4a*a*,5,6,11c*a*-octahydro-7*H*-pyrido[3,2-*c*]carbazole. Y 88%. F.e.s. T. Gallagher, P. Magnus, Tetrahedron *37,* 3889–97 (1981); **1,4-dihydrocarbazoles** from acetylene derivs. s. J. Chem. Soc. Chem. Commun. *1982,* 613–4; s.a. J. Am. Chem. Soc. *104,* 1140–1 (1982).

N,N.N′,N′-Tetramethylethylenediamine s. under n-*BuLi* *TMEDA*

*Lithium hydridotri-*sec-*butylborate* *Li[*s-*Bu₃BH]*

Cyclohex-1-en-4-ols from 2-vinylcyclobutanones via 2-vinylcyclobutanols

710. 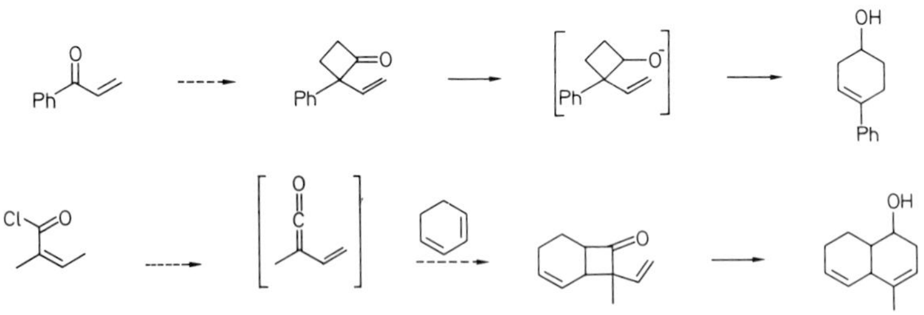

A soln. of startg. cyclobutanone in tetrahydrofuran treated with 2 equivalents Li-hydridotri-*sec*-butylborate in the same solvent at −78° and allowed to warm to 25° during 2 hrs. → 4-phenyl-3-cyclohexenol. Y 65–71%. – This is part of a cyclohexenol [3+3]-annelation starting **from *a,β*-ethyleneketones;** a [4+2]-annelation can be achieved by combining with a 2-vinylcyclobutanone synthesis from vinylketenes and activated ethylene derivs. F.e.s. R.L. Danheiser, C. Martinez-Davila, H. Sard, Tetrahedron *37,* 3943–50 (1981).

Menthyloxyaluminium dichloride ←

Stereospecific intramolecular diene synthesis ○

711. 

Menthyloxyaluminium dichloride added to a soln. of startg. triene in toluene/dichloromethane, and the mixture allowed to react at 23° for 48 hrs. → product. Y 79–83% (100% *trans*-fused). F.e., also with $AlCl_3$ and organoaluminium chlorides, s. W.R. Roush, H.R. Gillis, J. Org. Chem. *45,* 4267–8 (1980); s.a. ibid. *46,* 1510–1 (1981); **also asymmetric synthesis** from chiral trienes s. J. Am. Chem. Soc. *104,* 2269–83 (1982).

Propylene oxide ←

Intramolecular diene synthesis with arynes

712.

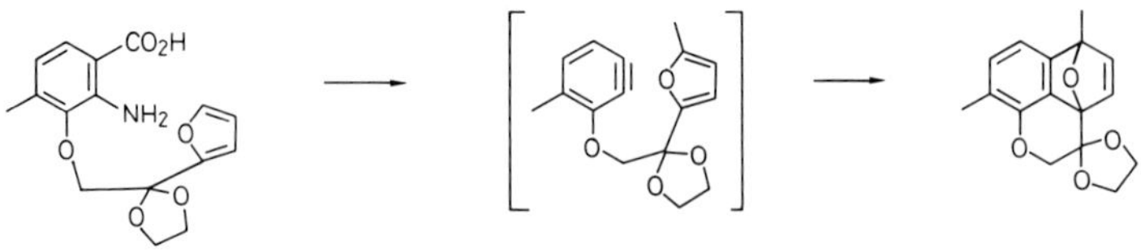

Startg. m. in ethanol treated with isoamyl nitrite and HCl at 0° for 30 min., and the diazonium salt heated in 1,2-dichloroethane containing propylene oxide → product. Y 86%. F. method s. W.M. Best, D. Wege, Tetrahedron Letters *22,* 4877–80 (1981).

Trifluoroacetic acid CF_3COOH

Acid-catalysed Cope rearrangement ←

713.

of 1,5-dien-3′-ones. A soln. of startg. m. in methylene chloride containing 1 equivalent trifluoroacetic acid allowed to react at room temp. for 15 min. → product. Y 74%. F.e., also with BF_3-etherate, s. W.G. Dauben, A. Chollet, Tetrahedron Letters *22,* 1583–6 (1981).

Polymer-based stannous chloride/phosphine complex ←

Dienes by double cyclopropane ring opening ⊂

s. *30,* 538; with polymer-based stannous chloride/phosphine complex, norbornadienes, s. M.E. Landis, D. Gremaud, T.B. Patrick, Tetrahedron Letters *23,* 375–8 (1982).

Tetrakis(triphenylphosphine)palladium(0) $Pd(PPh_3)_4$

Cyclopentene from vinylcyclopropane ring ◯

s. *31,* 708; with $Pd(PPh_3)_4$, 3-vinylcyclopentenes, s. Y. Morizawa, K. Oshima, H. Nozaki, Tetrahedron Letters *23,* 2871–4 (1982); with *n*-butyllithium/TMEDA cf. T.V. Akhachinskaya et al., Zh. Org. Khim. *18,* 458–9 (1982); 1-silylcyclopentene ring s. L.A. Paquette et al., Tetrahedron Letters *23,* 263–6 (1982).

Exchange ⇂↾

Hydrogen ↑ CC⇂↾H

Without additional reagents *w.a.r.*

Aminomethylation with methyleneammonium salts COCH → COC–CH_2NR_2

Mannich reaction

with chlorides s. *32,* 715; with trifluoroacetates, regiospecificity, s. M. Gaudry, Y. Jasor, T.B. Khac, Org. Synth. *59,* 153–8 (1979).

Electrolysis/sodium iodide ↯*/NaI*

Anodic dimerization 2 RH → R–R

with added Li diisopropylamide cf. *28,* 703s*30;* with NaI, tetramethyl 1,1,2,2-ethanetetracarboxylate, s. D.A. White, Org. Synth. *60,* 78–80 (1981); **review of anodic and cathodic carbon–carbon bond formation** s. H.J. Schäfer, Angew. Chem. Intern. Ed. *20,* 911–34 (1981).

Glow discharge ϟ

Replacement of hydrogen by cyano groups H → CN
with acetonitrile cf. *36*, 726; with cyanogen s. J. Am. Chem. Soc. *103*, 4204–9 (1981).

Sodium tert-*pentoxide s. under HCSSK* $NaOCH_2CMe_3$

n-*Butyllithium* n-*BuLi*

Cyclic 2-benzylamino-1,1-sulfinylthioethers from N-benzylcyclimmonium salts ←
s. *37*, 964

Lithium diisopropylamide/p-*toluenesulfonyl cyanide* $i\text{-}Pr_2NLi/TsCN$

α-Cyanoketones from ketones H → CN
Kinetic cyanation

714. 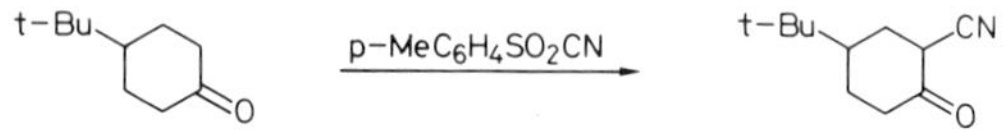

A soln. of 4-*tert*-butylcyclohexanone in tetrahydrofuran added to a stirred soln. of Li diisopropylamide in the same solvent at –78° under N_2, the mixture stirred for 20 min., added by syringe during 30 sec. to a soln. of *p*-toluenesulfonyl cyanide in tetrahydrofuran at –78° under N_2, the mixture quenched with concd. NH_4OH, and allowed to warm to room temp. → product. Y 80%. F.e., also from enolesters, s. D. Kahne, D.B. Collum, Tetrahedron Letters *22*, 5011–4 (1981).

Potassium dithioformate/sodium tert-*pentoxide* $HCSSK/NaOCH_2CMe_3$

C-Thioformylation H → CH=S

715. 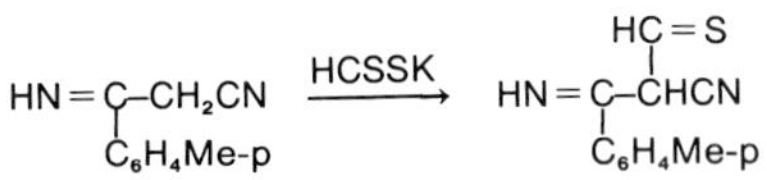

α-Cyanothioaldehydes. A mixture of freshly prepared β-aryl-β-iminopropionitrile, K dithioformate, and Na *tert*-pentoxide in tetrahydrofuran stirred at room temp. for 24 hrs., water added, and acidified with 2 *M* HCl to pH ca. 2 → product. Y 96%. – The thioaldehydes were clearly monomeric (by osmometry), very stable, and had distinct melting points. F.e.s. M. Muraoka, T. Yamamoto, T. Takeshima, Chem. Letters *1982*, 101–4.

Sodium iodide s. under ϟ *NaI*

Tributylamine s. under $Bu_2BOSO_2CF_3$ Bu_3N

Ethyldiisopropylamine $i\text{-}Pr_2NEt$

C-α-Dialkoxymethylation H → $CH(OR)_2$
β-Ketoacetals from ketones

716. 

A stirred slurry of *diethoxycarbenium fluoroborate* (prepd. *in situ* from ethyl orthoformate and BF_3-etherate) in methylene chloride at –78° treated under N_2 with cyclohexanone followed by dropwise addition of ethyldiisopropylamine during 30 min., stirring continued at the same temp. for 1 hr., the mixture poured into satd. aq. $NaHCO_3$ at 25°, more solvent added, and stirred vigorously for 10 min. at room temp. → 2-(diethoxymethyl)cyclohexanone. Y 87%. – The products can be easily hydrolysed to β-ketoaldehydes, the overall method being simpler and milder than conventional C-α-formylations in acidic media. F.e. and limitations s. W.L. Mock, H.-R. Tsou, J. Org. Chem. *46*, 2557–61 (1981).

Pyridine s. under $(MeCOO)_4Pb$ C_5H_5N

Silver acetate s. under $(MeCOO)_4Pb$ *MeCOOAg*

Mercuric bis(trimethylsilyl)amide $Hg\{N(SiMe_3)_2\}_2$

Oxidative dimerization

Sym. 1,2-dinitriles from nitriles 2 RH → R–R

717.

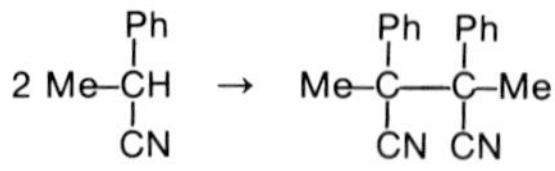

A mixture of 2-phenylpropionitrile and Hg bis(trimethylsilyl)amide in tetrahydrofuran heated 8 hrs. at ca. 30° in a sealed tube → *meso*-2,3-dimethyl-2,3-diphenylsuccinonitrile. Y 97%. F.e., **also sym. succinic acid esters,** s. O.A. Kruglaya et al., Zh. Obshch. Khim. *51,* 716 (1981).

Dibutylborinyl triflate/tributylamine $Bu_2BOSO_2CF_3/Bu_3N$

1-β-Keto-1,2-dihydroisoquinolines from isoquinolinium salts and ketones ←

Enoxyboranes as intermediates

718.

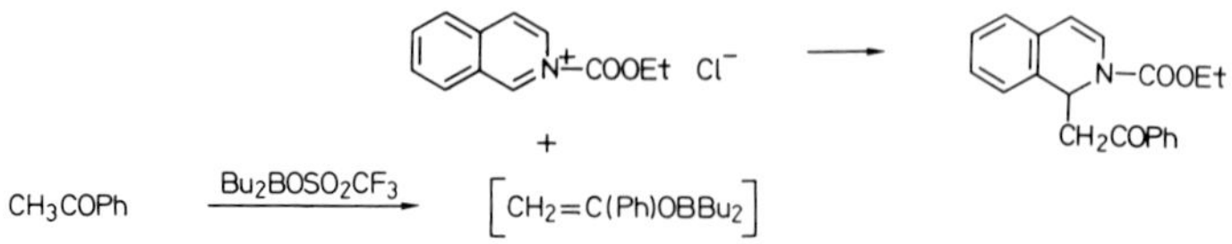

A suspension of startg. isoquinolinium salt in ether treated via syringe under N_2 with a soln. of startg. boron enolate (prepared *in situ* at –78° from acetophenone, dibutylboron triflate, and tributylamine) in the same solvent, the mixture stirred at room temp. for 3 hrs., and worked up with 3 *M* NaOH → product. Y 98%. F.e.s. K. Akiba et al., Tetrahedron Letters *22,* 4961–4 (1981).

Aluminium chloride s. under Acetone cyanohydrin $AlCl_3$

Cerium(IV) salt *Ce(IV)*

Aromatic α-nitroalkylation ArH → Ar–C(NO_2)

with manganese(III) acetate cf. *33,* 714; nitromethylation with Ce(IV) salts, e.g. Ce(IV) acetate, s. J. Org. Chem. *46,* 4672–6 (1981).

Acetone cyanohydrin/aluminium chloride $Me_2C(OH)CN/AlCl_3$

Aromatic formylation ArH → ArCHO

719.

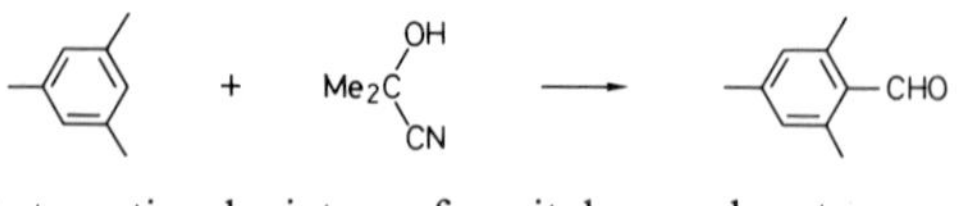

$AlCl_3$ added slowly to a stirred mixture of mesitylene and acetone cyanohydrin in 1,2-dichloroethane at 0°, allowed to warm to room temp., refluxed for 24 hrs., and hydrolysed → 2,4,6-trimethylbenzaldehyde. Y 48%. – The method has the same limitations as Gatterman's procedure, but avoids handling HCN. F.e.s. A. Rahm, R. Guilhemat, M. Pereyre, Syn. Commun. *12,* 485–7 (1982).

Trimethylsilyl iodide Me_3SiI

Modified Mannich reaction CO → COC–CH_2NR_2

β-Aminoketones from enoxysilanes

s. *32,* 715; with aminomethyl ethers in the presence of trimethylsilyl iodide or triflate cf. A. Hosomi, S. Iijima, H. Sakurai, Tetrahedron Letters *23,* 547–50 (1982).

Lead tetraacetate/silver acetate/pyridine *$(MeCOO)_4Pb/MeCOOAg/C_5H_5N$*
Diketones from cyclic alcohols C
Oxidative dimerization

720.

A mixture of startg. alcohol, Pb tetraacetate, Ag(I) acetate, and pyridine in benzene stirred 10 hrs. at 80° → dodecane-2,11-dione. Y 57% (conversion 77%). F.e.s. G.I. Nikishin, S.S. Spektor, N.I. Kapustina, Izv. Akad. Nauk SSSR Ser. Khim. *1981,* 1172–5.

Chlorosulfonyl isocyanate *$ClSO_2NCO$*
Replacement of hydrogen by cyano groups H → CN
s. *23,* 521; 2-cyanopyrroles in acetonitrile s. C.E. Loader, H.J. Anderson, Can. J. Chem. *59,* 2673–6 (1981).

p-*Toluenesulfonyl cyanide s. under i-Pr_2NLi* *TsCN*

Palladium(II) acetate *$(MeCOO)_2Pd$*
C-Arylation of *p*-quinones H → Ar

721.

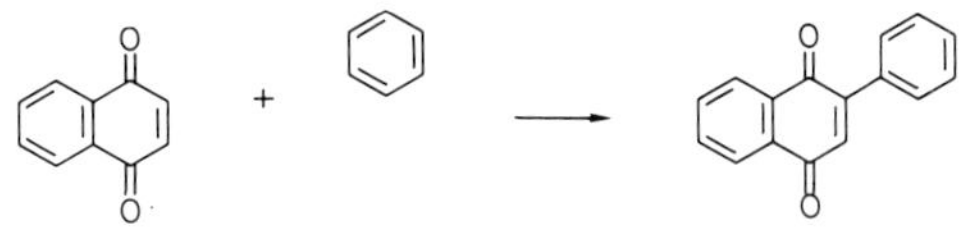

A soln. of 1,4-naphthoquinone and palladium(II) acetate in 1:1 acetic acid/benzene stirred at 80° under N_2 for 14 hrs. → 2-phenyl-1,4-naphthoquinone. Y 85%. F.e. and regiospecificity s. T. Itahara et al., J. Chem. Soc. Chem. Commun. *1981,* 859–60.

Oxygen ↑ CC⇃↾O

Without additional reagents *w.a.r.*
Aminomethylene compds. from amines CH_2 → C=C–N<
s. *31,* 728; f.e.s. Chem. Ber. *114,* 3471–84 (1981).

4(1*H*)-Pyrimidinethiones from acylisothiocyanates ○
and enamines s. *19,* 820; 6-alkylthio derivs. from 1-aminothioenolethers s. V. Aggarwal, H. Ila, H. Junjappa, Synthesis *1982,* 65–8.

Isocoumarin ring

722.

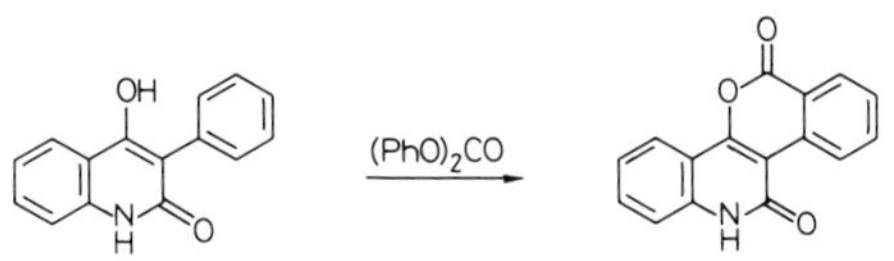

Startg. m. and diphenyl carbonate heated under reflux for 0.5–3 hrs. → 6*H*-[2]benzopyrano[4,3-*c*]quinoline-6,11-dione. Y 82%. F.e.s. W. Stadlbauer, N. Ghobrial, T. Kappe, Z. Naturforsch. *35B,* 892–5 (1980).

Without additional reagents *w.a.r.*

Pyrrolidine ring from 3-ethylene-2-hydroxyamines and aldehydes with simultaneous isocyclic ring expansion
Stereospecific 2-azonia-[3.3]-sigmatropic rearrangement ○ ◑

723.

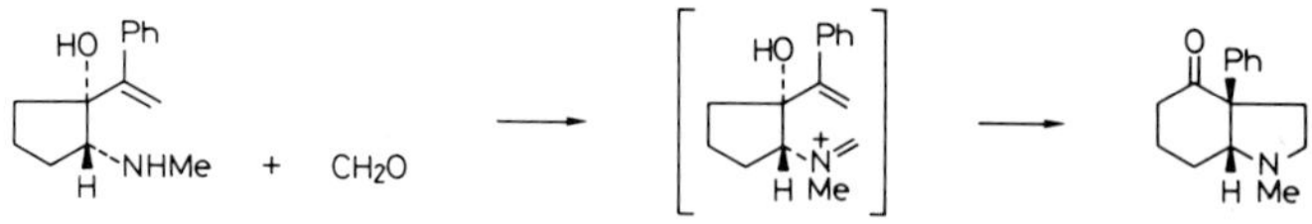

7-Azabicyclo[4.3.0]nonan-2-ones. Startg. 3-ethylene-2-hydroxyamine and paraformaldehyde refluxed in ethanol for 20 hrs. → *cis*-product. Y 78%. F.e.s. L.E. Overman, L.T. Mendelson, J. Am. Chem. Soc. *103,* 5579–81 (1981); from N-cyanomethyl derivs. cf. Tetrahedron Letters *23,* 2741–4 (1982).

Lithium (s.a. under CuI and $TiCl_3$) *Li*

Ketones from carboxylic acids COOH → COR

s. *5,* 532; phenyl ketones, reagents for removal of excess PhLi, s. D.E. Nicodem, M.L.P.F.C. Marchiori, J. Org. Chem. *46,* 3928–9 (1981).

Sodium hydroxide/polyethylene glycol dimethyl ether 400 ←

***α,β*-Ethylenenitriles from aldehydes** CHO → CH=C–CN

with NaOR cf. *11,* 821; phase transfer catalysis with NaOH/polyethylene glycol dimethyl ether 400 s. B. Zupančič, M. Kokalj, Synthesis *1981,* 913–4.

Sodium hydroxide/benzyltriethylammonium chloride *$NaOH/[PhCH_2NEt_3]^+Cl^-$*

2-Acoxy-1,1,1-trichlorides from acylals $C(OAc)_2$ → $C(OAc)–CCl_3$

Synthesis with addition of one C-atom

724.

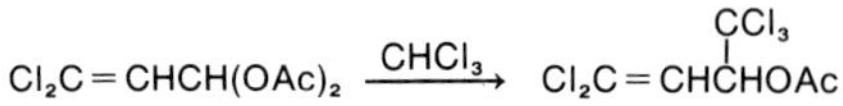

50% aq. NaOH and methylene chloride added dropwise with vigorous stirring and ice-cooling to a mixture of startg. acylal, chloroform, and a little benzyltriethylammonium chloride at –5 to 10°, and stirred for a further 1 hr. at room temp. → product. Y 65%. F.e.s. G.V. Kryshtal et al., Izv. Akad. Nauk SSSR Ser. Khim. *1981,* 2820–3.

2-Pyridones from *β*-dicarbonyl compds. ○

725.

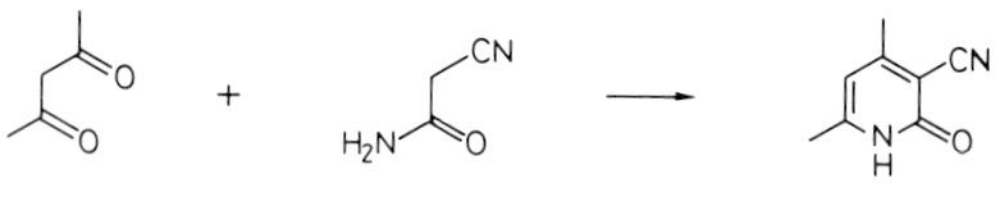

A mixture of startg. *β*-dicarbonyl compd., cyanoacetamide, and a little triethylbenzylammonium chloride in 50% NaOH stirred for 4 hrs. → product. Y 95%. – The method is advantageous with respect to yield, ease of operation, and absence of organic solvent. F.e., **also 6-amino- and 6-hydroxy-2-pyridones** from ethoxymethylenemalonic acid derivs., s. V.S. Hawalder, S.V. Sunthankar Indian J. Chem. *19B,* 151–2 (1980).

Sodium enolate [s.a. under $Pd(PPh_3)_4$] ←

3-(Arylthio)coumarins from *o*-hydroxyaldehydes and thiolacetic acid aryl esters
Perkin-Oglialore reaction

726.

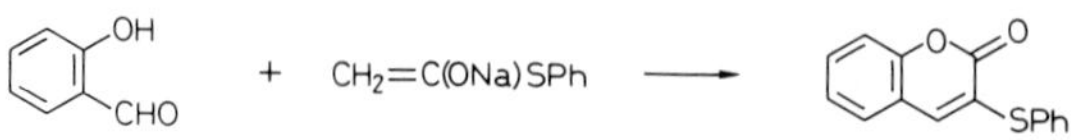

A mixture of startg. *o*-hydroxyaldehyde and S-phenyl sodiothioacetate in acetic anhydride refluxed at 170–80° for 6–7 hrs. → product. Y 60–80%. F.e.s. J.R. Merchant, P.J. Shah, J. Heterocyc. Chem. *18,* 441–2 (1981).

Lithium enolate ←

2-Nitroallylation $H \rightarrow CH_2C(NO_2){=}CH_2$

727. CH_3COOEt

↓

$CH_2{=}C(NO_2){-}CH_2OOCBu\text{-}t$ + $[CH_2{=}C(OLi)OEt]$ → $CH_2{=}C(NO_2){-}CH_2CH_2COOEt$

γ,δ-Ethylene-γ-nitrocarboxylic acid esters. A 0.2 *M* soln. of 3-pivaloyloxy-2-nitro-1-propene in tetrahydrofuran added to a soln. of startg. Li enolate in the same solvent at −78°, stirred for 1–3 hrs., warmed to −30° during 10 min., and worked up with 2% acetic acid → product. Y 87%. F.e.s. P. Knochel, D. Seebach, Nouveau J. Chim. *5,* 75–7 (1981); Tetrahedron Letters *22,* 3223–6 (1981).

Methyllithium s.a. under 9-Borabicyclo[3.3.1]nonane *MeLi*

n-*Butyllithium s. under* $TiCl_3$ n-*BuLi*

n-*Butyllithium/N,N,N',N'-tetramethylethylenediamine* n-*BuLi/TMEDA*

3-Hydroxy-2-thioenolethers $CO \rightarrow C(OH)C(SR){=}C$
from 2-alkoxythioethers and oxo compds.
1-Metalated thioenolethers as intermediates

728. $MeOCH_2{-}CH_2SMe$

↓

$[CH_2{=}C(SMe)Li]$ + $O{=}CHPh$ → $CH_2{=}C(SMe)CH(OH)Ph$

A hexane soln. of *n*-butyllithium added to a soln. of 2-methoxyethyl methyl sulfide and TMEDA in tetrahydrofuran at −30°, the mixture stirred for 1 hr., a soln. of benzaldehyde in the same solvent added, and warmed to room temp. → 3-hydroxy-2-methylthio-3-phenylpropene. Y 80%. F. e. and methods, s. T. Takeda, H. Furukawa, T. Fujiwara, Chem. Letters *1982,* 593–6.

Polymer-based trityllithium ←

Reviews of polymer supports in organic synthesis ←
s. *33,* 724; of functionalized polymers in organic synthesis s. A. Akelah, D.C. Sherrington, Chem. Rev. *81,* 557–87.

Lithium diethylamide Et_2NLi

β-Ketosulfoxides $S(O){-}CH \rightarrow S(O){-}C{-}CO$
from sulfoxides and carboxylic acid esters
with K *tert*-butoxide cf. *19,* 827; with Li diethylamide, **asym. synthesis,** s. N. Kunieda, A. Suzuki, M. Kinoshita, Bull. Chem. Soc. Japan *54,* 1143–50 (1981).

Lithium diisopropylamide $i\text{-}Pr_2NLi$

Review of syntheses via carboxylic acid dianions ←
s. *28,* 733s*32;* review of reactions of dianions of carboxylic acids and ester dienolates s. N. Petragnani, M. Yonashiro, Synthesis *1982,* 521–78.

Synthesis of β-ketocarboxylic acids $CO{-}C{-}COOH$
s. *28,* 733; **review of synthesis and uses of β-ketocarboxylic acids** s. L. Oshry, S.M. Rosenfeld, Org. Prep. Proc. Intern. *14,* 251–64 (1982).

2-Azetidinones from carboxylic acid esters and azomethines 

s. *36,* 740; 3-acylamino-3-alkoxy-2-azetidinones s. A.K. Bose, M.S. Khajavi, M.S. Manhas, Synthesis *1982,* 407–9.

Lithium diisopropylamide *i-Pr₂NLi*

2(5*H*)-Furanones from α,β-ethylenecarboxylic acids and oxo compds. O,C-Dianions as intermediates ○

729. 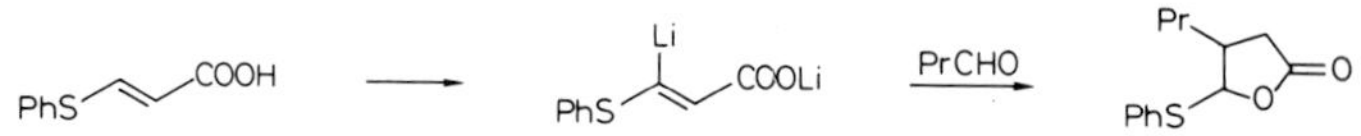

4-Arylthio-2(5*H*)-furanones. A soln. of startg. acid in tetrahydrofuran at −80° treated with 2 equivalents Li diisopropylamide in the same solvent, butanal added, allowed to react for 3 hrs. between −80 and −40°, and the product treated with *p*-toluenesulfonic acid in benzene → lactone. Y 98%. F.e.s. Y. Takahashi et al., Heterocycles *15*, 225–8 (1981).

Sodium carbonate s. under $RuH_2(PPh_3)_4$ Na_2CO_3

Potassium carbonate s. under $(PhO)_2P(O)N_3$ K_2CO_3

Sodium acetate s. under $POCl_3$ *MeCOONa*

Sodium cyanide (s.a. under Me_3SiCl*)* *NaCN*

Michael addition of cyanide ion with subsequent Dieckmann condensation Cyclic α-cyanoketones

730. 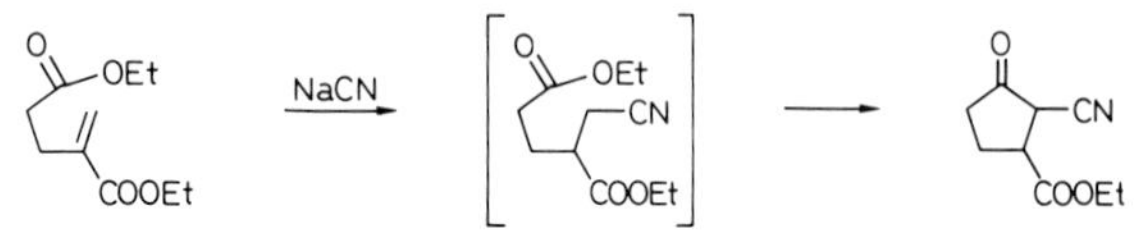

Startg. ester added slowly to a soln. of NaCN in dimethyl sulfoxide at 80°, and the mixture stirred for 90 min. → 2-cyano-3-oxo-1-cyclopentanecarboxylic acid ethyl ester. Y 84%. F.e.s. H. Stetter, K. Marten, Ann. Chem. *1982,* 240–9; bicyclic α-cyanoketones s. ibid. *1982,* 250–3.

Sodium iodide s. under Me_3SiCl *NaI*

Triethylamine s. under Cyanuric chloride, $(EtO)_2P(O)CN$*, and Bis(2-oxazolidon-3-yl)phosphoryl chloride* Et_3N

N,N,N′,N′-Tetramethylethylenediamine s. under n-*BuLi* *TMEDA*

Piperidine $C_5H_{11}N$

Thioalkylation by Mannich-type reaction H → CH_2S–

with CH_2O cf. *33,* 727; with aldehydes, 2-pyrone derivs., s. P. de March et al., J. Heterocyc. Chem. *19,* 335–6 (1982); regiospecific thiomethylation of nucleoside bases s. S.S. Jones, C.B. Reese, A. Ubasawa, Synthesis *1982,* 259–60.

Morpholine ←

2*H*-[1]Benzopyrano[2,3-*d*]pyrimidine-2,4(3*H*)-diones from *o*-hydroxyaldehydes Double ring closure ○

731. 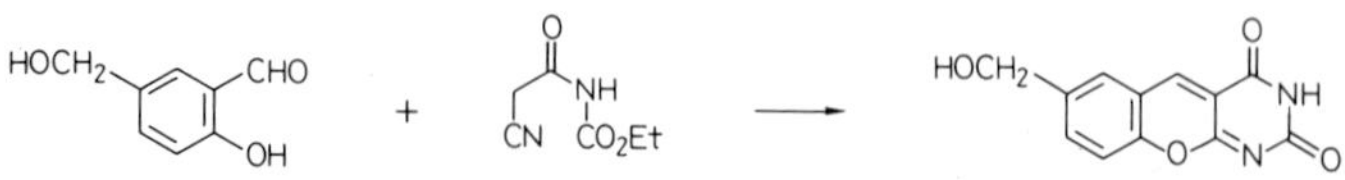

A soln. of 5-hydroxymethylsalicylaldehyde and N-cyanoacetylurethan in ethanol treated with morpholine, and refluxed for 3–4 hrs. → 7-hydroxymethyl-2*H*-[1]benzopyrano[2,3-*d*]pyrimidine-2,4(3*H*)-dione. Y 75%. F.e.s. D.J. Blythin et al., Heterocycles *16,* 203–7 (1981).

Cuprous iodide/lithium *CuI/Li*

Synthesis of allenes C=C=C−R

from tosyloxy-2-acetylenes s. *24,* 784s*31;* and mixed lithium vinylcuprates s. R. Baudouy, J. Goré, J. Chem. Res. (S) *1981,* 278–9.

Zinc/copper (s.a. under $TiCl_3$) *Zn/Cu*

6-Hydroxychromans from *p*-quinones and 2-ethylenealcohols ○

with Zn/formic acid cf. *22,* 793; with Cu,Zn s. M. Kajiwara, O. Sakamoto, S. Ohta, Heterocycles *14,* 1995–8 (1980).

Ethylmagnesium bromide *EtMgBr*

1,1-Diaryl derivs. from acetals $C(OR)_2 \rightarrow CAr_2$

s. *16,* 825; 1,1-bis-(*o*-hydroxyaryl)alkanes from phenols with EtMgBr cf. G. Casiraghi et al., J. Chem. Soc. Perkin Trans. I *1982,* 805–8.

Bromomagnesium dialkylamides *$BrMgNR_2$*

Claisen condensation ←

s. *37,* 924

Sodium tetrahydridoaluminate s. under $NbCl_5$ *$NaAlH_4$*

9-Borabicyclo[3.3.1]nonane/methyllithium ←

Ketones from acetylene derivs. and carboxylic acid esters via enoxyboranes $C{\equiv}C \rightarrow CH_2CHCOR$

732.

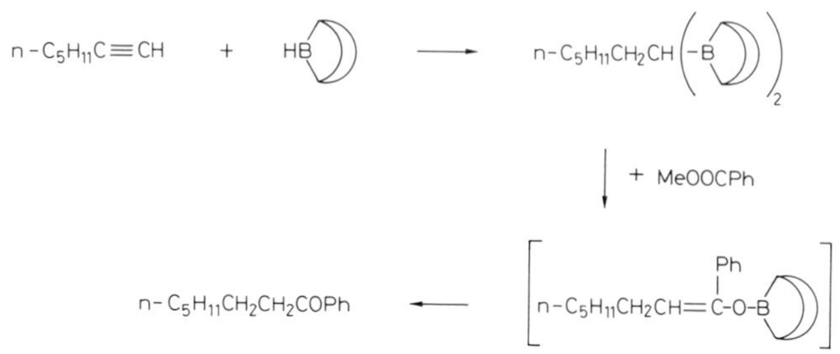

A soln. of 2 equivalents 9-borabicyclo[3.3.1]nonane in tetrahydrofuran added under argon to a soln. of 1-heptyne in the same solvent at 0°, the mixture stirred for 3 hrs. at room temp., solvent removed, residue dissolved in tetrahydrofuran, a soln. of 2.5 equivalents methyllithium in ether added at −78°, the mixture allowed to warm to room temp. after 10 min., treated with a soln. of methyl benzoate in tetrahydrofuran, stirred for 3 hrs. at room temp., and worked up with water → 1-phenyl-1-octanone. Y 93%. F.e., **also β-hydroxyketones** by stereospecific cross-aldol condensation with the intermediate enoxyboranes (cf. *36,* 622), s. T. Mukaiyama et al., Chem. Letters *1981,* 1193–6.

Boron fluoride *BF_3*

C-Allylation of phenolethers with allyl phosphates $H \rightarrow CH_2CH{=}CH_2$

s. *29,* 747; f.e.s. Chem. Letters *1982,* 797–800.

Syntheses via cobalt carbonyl-stabilized carbonium ions ←

s. *34,* 724; (E)-1,3-enynes from 1,4-enyn-3-ols with BF_3 s. Tetrahedron Letters *23,* 2555–8 (1982).

Aluminium chloride *$AlCl_3$*

1-Indanones from cyclopropanecarboxylic acid esters and arenes ○

733. 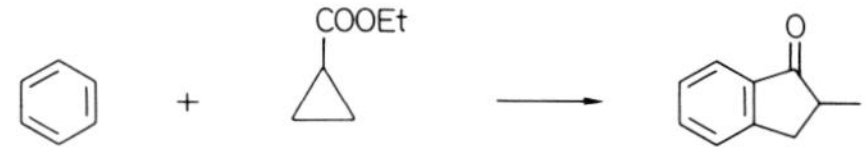

A soln. of ethyl cyclopropanecarboxylate in dry benzene treated portionwise with 3.6 equivalents of dry $AlCl_3$ over 20 min., and the mixture refluxed for 16 hrs. → 2-methyl-1-indanone. Y 93%. F.e. and regiospecificity s. H.W. Pinnick et al., J. Org. Chem. *46,* 3758–60 (1981).

Polyethylene glycol dimethyl ether 400 s. under NaOH ←

Hydroquinone s. under TsOH ←

Ammonium acetate *$MeCOONH_4$*

Coumarins from β-ketocarboxylic acid esters
with H_2SO_4 cf. *1,* 591; with NH_4 acetate s. T. Kappe, C. Mayer, Synthesis *1981,* 524–6.

4-Chromanones from *o*-hydroxyketones and oxo compds.
with pyrrolidine cf. *35,* 523; with ammonium acetate, 3-nitroflavanones, s. C. Paparao, K.V. Rao, V. Sundaramurthy, Synthesis *1981,* 236–7; f. method with isolation of intermediate s. Indian J. Chem. *20B,* 144–6 (1981).

Benzyltriethylammonium chloride s. under NaOH *$[PhCH_2NEt_3]^+Cl^-$*

N-Methylanilinium trifluoroacetate *$[PhNH_2Me]^+CF_3CO_2^-$*

α-Methylenation of ketones $COCH_2 \rightarrow COC{=}CH_2$
s. *32,* 726; s.a. Org. Synth. *60,* 88–91 (1981).

Acetone cyanohydrin *$Me_2C(OH)CN$*

α-Aminonitriles from oxo compds. CO → C(N⊂)CN
Synthesis with addition of one C-atom
with NaCN cf. *27,* 781; with acetone cyanohydrin, α-amino-β,γ-ethylenenitriles, s. G.M. Zhdankina et al., Izv. Akad. Nauk SSSR Ser. Khim. *1982,* 346–50.

Trifluoroacetic anhydride (s.a. under H_3PO_4) *$(CF_3CO)_2O$*

3-Ethylenethioethers from sulfoxides and ethylene derivs.
Geospecific ene reaction with thionium salts ←

734. 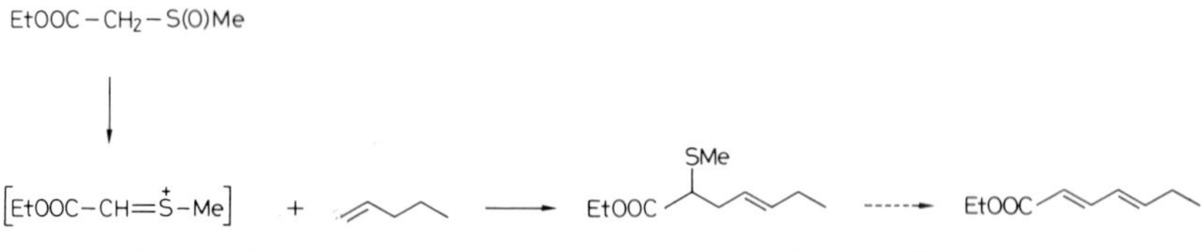

α-Alkylthio-γ,δ-ethylenecarboxylic acid esters. A stirred soln. of ethyl α-methylsulfinylacetate in trifluoroacetic acid at 0° treated sequentially with trifluoroacetic anhydride and 1-pentene, and stirring continued at the same temp. for 1 hr. → ethyl 2-methylthio-E-4-heptenoate. Y 75%. F.e. and conversion to **(E),(E)-2,4-dienecarboxylic acid esters** s. Y. Tamura et al., Tetrahedron Letters *22,* 1343–4 (1981).

Ion-exchanger ←

4-Alkylidene-5-oxazolones from aldehydes
Azlactone synthesis ○
with Na acetate cf. *2,* 654; with ion-exchange resins s. A. Gaset, J.P. Gorrichon, Synth. Commun. *12,* 71–9 (1982).

Pivalic acid *Me_3CCOOH*

Orthoester Claisen rearrangement ←

s. *27,* 795s*34;* with pivalic acid, *γ,δ*-ethylene-*α*-chlorocarboxylic acid esters, s. W.H. Moos, R.D. Gless, H. Rapoport, J. Org. Chem. *46,* 5064–74 (1981).

3,4-Dihydrothiocoumarins ←
from arylmercaptans and *α,β*-ethyleneorthocarboxylic acid esters
via monothioorthoester Claisen rearrangement

735. 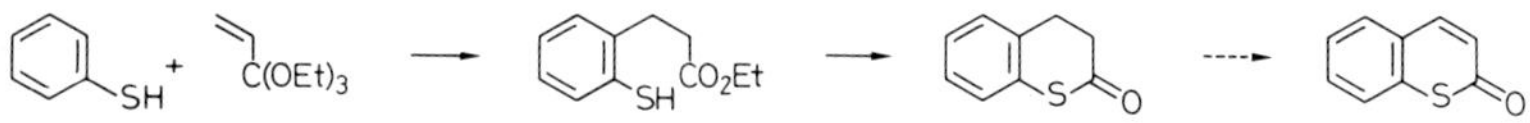

A soln. of thiophenol, triethyl orthoacrylate, and pivalic acid in *p*-cymene refluxed for 20 hrs. → ethyl 2-mercaptodihydrocinnamate (Y 94%), in diphenyl ether refluxed for 24 hrs., a little *p*-toluenesulfonic acid added, and refluxing continued for 18 hrs. → dihydrothiocoumarin (Y 73%). F.e., also dehydrogenation with Pd–C to **thiocoumarins,** and prepn. of the latter **from *α,β*-acetyleneorthocarboxylic acid esters** via *o*-mercaptocinnamic acid esters, s. J.A. Panetta, H. Rapoport, J. Org. Chem. *47,* 2626–8 (1982).

*Bis(*p*-methoxyphenyl) telluroxide* *R_2TeO*

Aldol condensation under mild, neutral conditions CO–CH_2 → CO–C=C
***α,β*-Ethyleneoxo compds.**

736. 

A mixture of *α*-tetralone, benzaldehyde, and a little bis(*p*-methoxyphenyl) telluroxide in toluene refluxed for 2 hrs. → 2-benzylidene-*α*-tetralone. Y 96%. – Relatively poor yields were obtained with very acidic methylene compds. F.e.s. L. Engman, M.P. Cava, Tetrahedron Letters *22,* 5251–2 (1981).

Cyanuric chloride/triethylamine *$(CNCl)_3/Et_3N$*

2-Azetidinones from carboxylic acids and azomethines 

s. *22,* 445s*37*

Silicon tetraisothiocyanate *$Si(NCS)_4$*

Reviews of silicon compds. in organic synthesis

s. *33,* 735s*36;* s.a. L.A. Paquette, Science *217,* 793–800 (1982); P. Magnus, Aldrichimica Acta *13,* 43–51 (1980).

Trimethylsilyl cyanide/stannic chloride *$Me_3SiCN/SnCl_4$*

Replacement of alkoxy by cyano groups OR → CN

737. 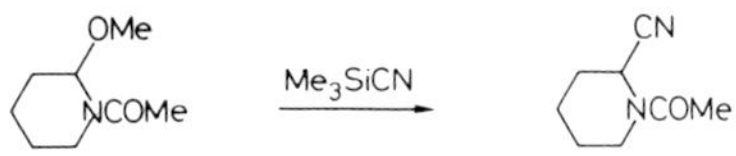

Cyclic *α*-acylaminonitriles. N-Acetyl-2-methoxypiperidine treated with trimethylsilyl cyanide and a little $SnCl_4$ → N-acetyl-2-cyanopiperidine. Y 95%. F.e. and catalysts s. V. Asher et al., Tetrahedron Letters *22,* 141–4 (1981); **1-cyanoacetals and *α*-alkoxynitriles** with BF_3 cf. K. Utimoto et al., ibid. *22,* 4279–80 (1981).

Trimethylsilyl chloride/sodium iodide/sodium cyanide *$Me_3SiCl/NaI/NaCN$*

Nitriles from alcohols under mild conditions OH → CN

738. $PhCH_2OH \xrightarrow{NaCN} PhCH_2CN$

A deaerated soln. containing benzyl alcohol, 2 equivalents NaCN, and a little NaI in 1:1 *acetonitrile*/dimethylformamide treated with 2 equivalents chlorotrimethylsilane under argon at room temp., and the mixture heated with stirring at 60–65° for 2 hrs. → benzyl cyanide. Y 98%. – Reaction is facile with prim., sec., and tert. alcohols, and proceeds **with inversion of configuration.** F.e.s. R. Davis, K.G. Untch, J. Org. Chem. *46,* 2985–7 (1981).

Titanium trichloride/zinc, copper *$TiCl_3/Zn,Cu$*

Sym. ethylene derivs. from two molecules of oxo compd. 2 >CO → C=C

with $TiCl_4$/Zn cf. *30,* 561; with $TiCl_3$/Zn,Cu s. L. Castedo et al., J. Org. Chem. *46,* 4292–4 (1981).

Titanium trichloride/lithium *$TiCl_3/Li$*

Ethylene derivs. from two ketone molecules

s. *30,* 561s*32,* sym. ethylene derivs. s. Org. Synth. *60,* 113–7 (1981); halogoaromatic subst. olefins s. W.H. Richardson, Synth. Commun. *11,* 895–9 (1981).

*Titanium trichloride/*n*-butyllithium* *$TiCl_3$/*n*-BuLi*

Reductive coupling of alcohols 2 R–OH → R–R

with MeLi cf. *23,* 793; with *n*-BuLi s. K. Schlögl, W. Weissensteiner, Synthesis *1982,* 50–3.

Titanium tetrachloride *$TiCl_4$*

α-Aminocarboxylic acids from 1-alkoxyurethans and isonitriles OR → CONHR
via α-(carbalkoxyamino)carboxylic acid amides

739.

A soln. of the startg. α-methoxyurethan in methylene chloride added to a stirred soln. of $TiCl_4$ in the same solvent at –70°, stirred for 5 min., phenyl isocyanide added, and the mixture kept at the same temp. for 3 hrs. → intermediate (Y 82%) in 47% HBr refluxed for 4 hrs. → DL-valine (Y 63%). F.e.s. T. Shono, Y. Matsumara, K. Tsubata, Tetrahedron Letters *22,* 2411–2 (1981).

Stannic chloride s. under Me_3SiCN *$SnCl_4$*

Triphenylphosphine s. under $Pd(PPh_3)_4$ *Ph_3P*

Diethyl phosphorocyanidate/triethylamine *$(EtO)_2P(O)CN/Et_3N$*

C-Acylation with carboxylic acids COOH → COR

s. *34,* 730; α-nitroketones s. Chem. Pharm. Bull. *29,* 259–61 (1981).

o-Cyano-N-heterocyclics from cyclic N-oxides ←
Modified Reissert-Henze reaction

with PhCOCl/KCN cf. *26,* 813; with diethyl phosphorocyanidate/triethylamine s. S. Harusawa, Y. Hamada, T. Shiori, Heterocycles *15,* 981–4 (1981).

Diphenyl phosphorazidate/potassium carbonate *$(PhO)_2P(O)N_3/K_2CO_3$*

Oxazoles from isonitriles and carboxylic acids ○
Oxazole-4-carboxylic acid esters

740.

A soln. of methyl isocyanoacetate in dimethylformamide added to a stirred suspension of the startg. carboxylic acid and $2K_2CO_3.3H_2O$ in the same solvent at room temp., cooled to 0° after 5 min., diphenyl phosphorazidate added, and stirring continued at 0–5° for 2 hrs. then at room temp. for 14 hrs. → product. Y 53%. F.e.s. Y. Hamada, T. Shioiri, Tetrahedron Letters *23,* 235–6 (1982).

Bis(2-oxazolidon-3-yl)phosphoryl chloride/triethylamine ←

2-Azetidinones from carboxylic acids and azomethines 

with $POCl_3$ cf. *21,* 445; with bis(2-oxazolidon-3-yl)phosphoryl chloride s. D.R. Shridhar, B. Ram, V.L. Narayana, Synthesis *1982,* 63–5; with cyanuric chloride/triethylamine s. M.S. Manhas, A.K. Bose, M.S. Khajavi, Synthesis *1981,* 209–11.

Phosphoric acid H_3PO_4

Aromatic chloromethylation $ArH \rightarrow ArCH_2Cl$

s. *3,* 645; s.a. M.S. Ogii, T.D. Novobranets, N.J. Danina, Vopr. Khim. Khim. Tekhnol. *1980,* 60, 20; Chem. Abstr. *95,* 186736 (1981).

Chromans from phenols and 2-ethylenealcohols 

with BF_3 cf. *27,* 790*a*; with phosphoric acid, chroman-6-carboxaldehydes, s. V.K. Ahluwalia, K.K. Arora, R.S. Jolly, Synthesis *1981,* 527–9.

Phosphoric acid/trifluoroacetic anhydride $H_3PO_4/(CF_3CO)_2O$

Heteroaromatic C-acylation $COOH \rightarrow COAr$

s. *33,* 741s*34;* 3-acylation of indoles s. Y. Murakami et al., Heterocycles *14,* 1939–41 (1980).

Phosphoryl chloride/sodium acetate $POCl_3/MeCOONa$

C-γ-Methylenation of α,β-ethyleneketones $CH_2 \rightarrow C{=}CH_2$

6-Methylene-Δ⁴-3-ketosteroids

741. 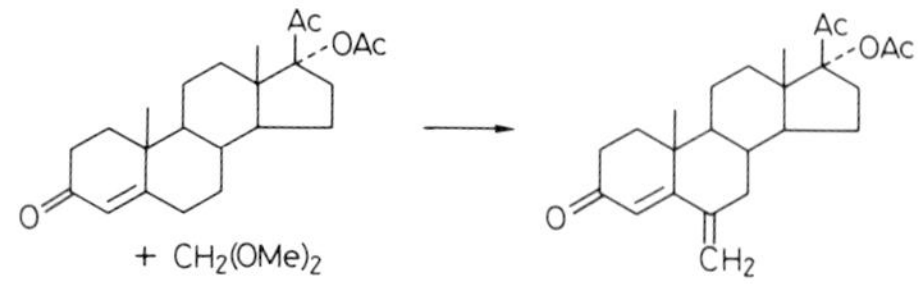

A suspension of Na acetate in chloroform containing methylal and $POCl_3$ stirred at reflux for 1 hr., startg. ketosteroid added, the mixture treated dropwise with further $POCl_3$ during 2.5 hrs., stirring continued at reflux for 1.5 hrs., the suspension cooled, and satd. aq. $NaHCO_3$ added dropwise with vigorous stirring during ca. 1 hr. until the aq. layer became alkaline → product. Y 76%. F.e. and conditions s. K. Annen et al., Synthesis *1982,* 34–40.

Niobium pentachloride/sodium tetrahydridoaluminate $NbCl_5/NaAlH_4$

Reductions with low-valent niobium $2 \gtrdot CO \rightarrow C{=}C$

Ethylene derivs.

742. 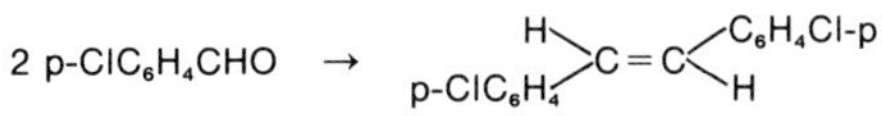

from 2 oxo compd. molecules. A soln. of $NaAlH_4$ in tetrahydrofuran added to 2 equivalents of $NbCl_5$ in 40:1 benzene/tetrahydrofuran at 0° under argon, after 10 min. the black suspension treated dropwise with a soln. of *p*-chlorobenzaldehyde in benzene, and the mixture refluxed for 3 hrs. → *p,p′*-dichlorostilbene. Y 95% (E/Z = >20:1). F.e., **also from oxido compds., (Z)-ethylene from acetylene derivs.,** and coupling of allylic or benzylic alcohols and halides, s. M. Sato, K. Oshima, Chem. Letters *1982,* 157–60.

Trimethylsilyl triflate/hydrogen peroxide $CF_3SO_2OSiMe_3/H_2O_2$

Electrophilic α-substitution of α,β-ethyleneketones ←
via 3-siloxy-2-ethyleneselenides

743. 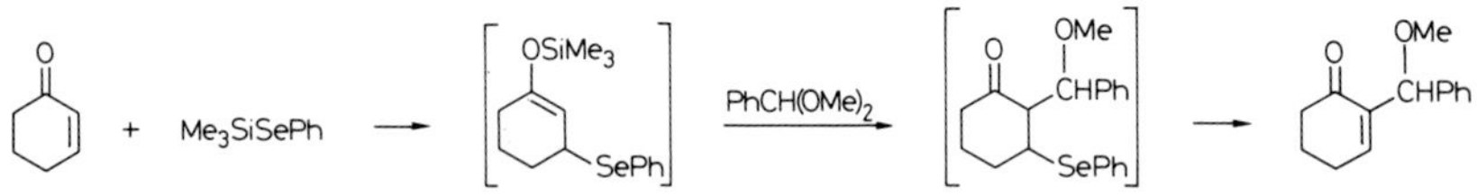

A soln. of trimethylsilyl triflate in dichloromethane treated sequentially with phenyl trimethylsilyl selenide and 2-cyclohexenone at –78° under argon, the mixture stirred at the same temp. for 30 min., benzaldehyde dimethyl acetal added, warmed to –25°, kept at the same temp. for 40 min., pyridine and 30% H_2O_2 added, and warmed slowly to 15° → 2-methoxyphenylmethyl-2-cyclohexenone. Y 75%. – Subsequent 1,4-addition affords the corresponding *α,β*-**disubst. ketones** without contamination by *α′,β*-disubst. isomers which may accompany the conventional 1,4-addition-alkylation. F.e. and with orthoesters as electrophile s. M. Suzuki, T. Kawagishi, R. Noyori, Tetrahedron Letters *22,* 1809–12 (1981).

p-*Toluenesulfonic acid* p-$CH_3C_6H_4SO_3H$

Friedländer quinoline ring synthesis ○

s. *28,* 757; modified method from *o*-aminoazomethines cf. A.S. Kende, D.P. Lorah, R.J. Boatman, J. Am. Chem. Soc. *103,* 1271–3 (1981), **review** s. C.C. Cheng, S.-J. Yan, Org. Reactions *28,* 37–201 (1982).

2-Pyridones from two β-ketocarboxylic acid amide molecules

with $ZnCl_2$ cf. *26,* 800; with *p*-toluenesulfonic acid, 2-pyridone-5-carboxylic acid amides, s. J.B. Pierce, Z.S. Ariyan, G.S. Ovenden, J. Med. Chem. *25,* 131–6 (1982).

p-*Toluenesulfonic acid/hydroquinone* ←

Diene synthesis with 1,3-dienes prepared *in situ*
Cyclohexene ring from 2-ethylenealcohols

744. 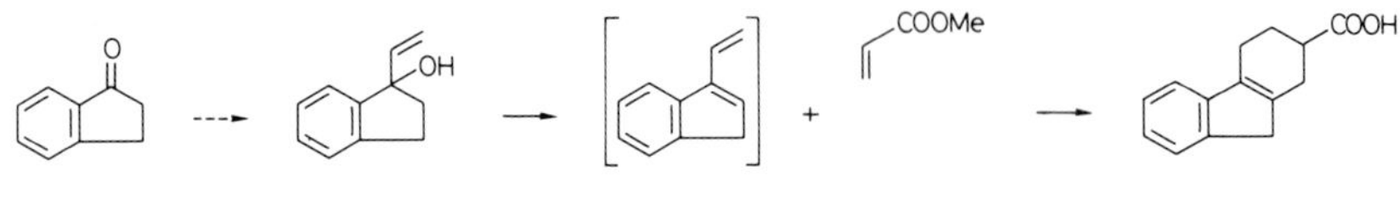

A soln. of crude allyl alcohol (prepd. from 1-indanone and vinylmagnesium bromide) in benzene treated under N_2 with a mixture of methyl acrylate and catalytic amounts of *p*-toluenesulfonic acid and hydroquinone, the mixture refluxed azeotropically using a Dean-Stark trap for 10 hrs., cooled in an ice-water bath, HCl passed through for 30 min., the mixture allowed to stand for a further 20 min., and the crude ester adduct hydrolysed with a 10% soln. of KOH in 10:1 methanol/water under reflux for 2 hrs. → 1,2,3,4-tetrahydrofluorene-2-carboxylic acid. Y 51% based on startg. ketone. F.e.s. B.C. Ranu, U.R. Ghatak et al., J. Chem. Soc. Perkin Trans. I *1982,* 865–73.

Bromine s. under $Na[C_5H_5Fe(CO)_2]$ Br_2

Hydrochlorides ←

Aminomethylation of ketones H → $CH_2N\lt$

β-Aminoketones

s. *14,* 479; with formaldehyde-d_2, *β,β*-dideuterio-*β*-aminoketones, s. J.-P. Denhez, G. Girault, P. Dizabo, J. Labelled Compounds Radiopharm. *18,* 759–63 (1981).

Hydrogen chloride *HCl*

Döbner-Miller quinoline synthesis ○

s. *9,* 851; from dimethyl 2-oxoglutaconate s. E.J.. Corey, A. Tramontano, J. Am. Chem. Soc. *103,* 5599–600 (1981).

Sodium dicarbonylcyclopentadienylferrate(1-)/bromine *$Na[C_5H_5Fe(CO)_2]/Br_2$*

Carboxylic acid esters from tosylates by carbonylation OTs → COOR

Homologous glycuronic acid esters

745.

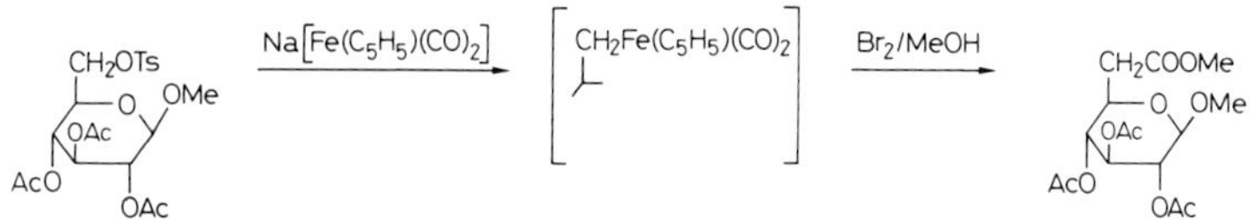

A mixture of methyl 2,3,4-tri-O-acetyl-6-O-*p*-tolylsulfonyl-*β*-D-glucopyranoside and Na dicarbonylcyclopentadienylferrate(1-) in tetrahydrofuran stirred for 10 min., cooled to 0° to −5°, a stream of CO passed through the soln., methanol and Br_2 added sequentially, and the mixture left for a few min. → methyl (methyl 2,3,4-tri-O-acetyl-6-deoxy-*β*-D-glucoheptopyranoside)uronate. Y 81%. F.e. and isolation of the intermediate, **also from halides,** s. H.H. Baer, H.R. Hanna, Carbohyd. Res. *102,* 169–83 (1982).

Cobalt(II) chloride *$CoCl_2$*

C-*a*-Benzylation of *β*-dicarbonyl compds. H → R

s. *22,* 782s*37*

Nickel acetylacetonate *$Ni(acac)_2$*

Pyrimidines from *β*-dicarbonyl compds. ○

746.

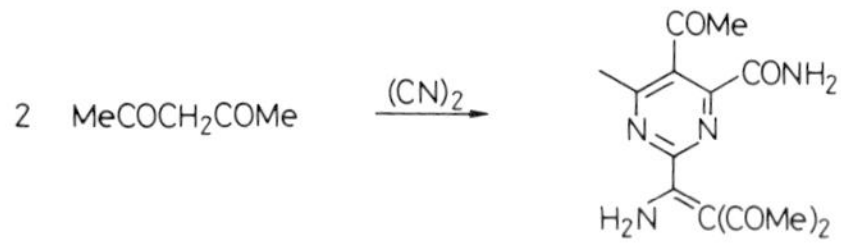

Ni acetylacetonate added with stirring to a 0.6 *M* soln. of *cyanogen* in 1,2-dichloroethane, treated with startg. *β*-diketone, and the mixture stirred slowly at 20° for 140 hrs. → product. Y 100%. F.e. and with Cu(II) acetylacetonate s. B. Corain, M. Basato, H.-F. Klein, Angew. Chem. Intern. Ed. *20,* 972–3 (1981).

Dihydrotetrakis(triphenylphosphine)ruthenium(II)/ sodium carbonate *$RuH_2(PPh_3)_4/Na_2CO_3$*

C-*a*-Alkylation of nitriles with alcohols H → R

with NaOR cf. *22,* 782; with Na_2CO_3 in the presence of $RuH_2(PPh_3)_4$ s. R. Grigg, T.R.B. Mitchell, S. Sutthivaiyakit, Tetrahedron Letters *22,* 4107–10 (1981); of *β*-dicarbonyl compds. with benzyl alcohols using $CoCl_2$ cf. J. Marquet, M. Moreno-Mañas, Chem. Letters *1981,* 173–6.

Hexadecacarbonylhexarhodium/pyridine *$Rh_6(CO)_{16}/C_5H_5N$*

N-Subst. maleimides ○

s. *28,* 766; **review** s. G.M. Gherasim, I. Zugravescu, Rev. Chim. (Bucharest) *32,* 1047–54 (1981) (Rom); Chem. Abstr. *96,* 142574 (1982).

Tetrakis(triphenylphosphine)palladium(0) *$Pd(PPh_3)_4$*

Synthesis of ethylene derivs. from acoxy-2-ethylenes C=C–C(OAc) → C=C–CR

s. *26,* 827s*36; a,β*-ethylene-carboxylic acids and -nitriles with tetrakis(triphenylphosphine)palladium(0) s. J. Tsuji et al., Tetrahedron Letters *22,* 2573–4 (1981); 2-ethylenealcohols s. J.P. Genet, M. Balabane, Y. Legras, Tetrahedron Letters *23,* 331–4 (1982).

Tetrakis(triphenylphosphine)palladium(0)/triphenylphosphine — *$Pd(PPh_3)_4/Ph_3P$*

Nitro-3-ethylene derivs. from nitronates and acoxy-2-ethylenes — H → C–C=C

747. $PhCH{=}NO_2^-\ Li^+ + AcOCH_2CH{=}CHPh \rightarrow PhCH(NO_2)CH_2CH{=}CHPh$

Tetrahydrofuran, a little tetrakis(triphenylphosphine)palladium(0), triphenylphosphine, and cinnamyl acetate added to startg. Li nitronate in water, and the mixture refluxed for 2 hrs. → product. Y 84% (79% regio-purity). F.e.s. P.A. Wade, S.D. Morrow, S.A. Hardinger, J. Org. Chem. *47,* 365–7 (1982), also with 2-ethylenealcohols and aryloxy-2-ethylenes s. P. Aleksandrowicz, H. Piotrowska, W. Sas, Tetrahedron, *38,* 1321–7 (1982).

Tetrakis(triphenylphosphine)palladium(0)/ triphenylphosphine/organosodium compd. — *$Pd(PPh_3)_4/PPh_3/RNa$*

Regiospecific synthesis of enesilanes from 1-acoxy-2-ethylenesilanes — H → C–C=CSi≡

748. $(EtOOC)_2CHNa + CH_2{=}CHCH(OAc)SiMe_3 \rightarrow (EtOOC)_2CHCH_2CH{=}CHSiMe_3$

A mixture of 1-trimethylsilylallyl acetate triphenylphosphine, and a little tetrakis(triphenylphosphine)palladium(0) treated with Na diethyl malonate for 9 hrs. at room temp. → product. Y 72%. (E:Z ratio = ca. 1:4). F.e., **also from 3-acoxyenesilanes,** s. T. Hirao et al., Tetrahedron Letters *22,* 3079–80 (1981).

Nitrogen ↑ — CC↓↑N

Without additional reagents — *w.a.r.*

C-Alkylation with diazo compds. — CH → CR

indoles s. *23,* 816; 4-alkylation of 5,6-dihydro-2-pyrones s. G. Falsone, B. Spur, Arch. Pharm. *315,* 597–602 (1982).

Reactions via nitrilium salts — H → COR

749.

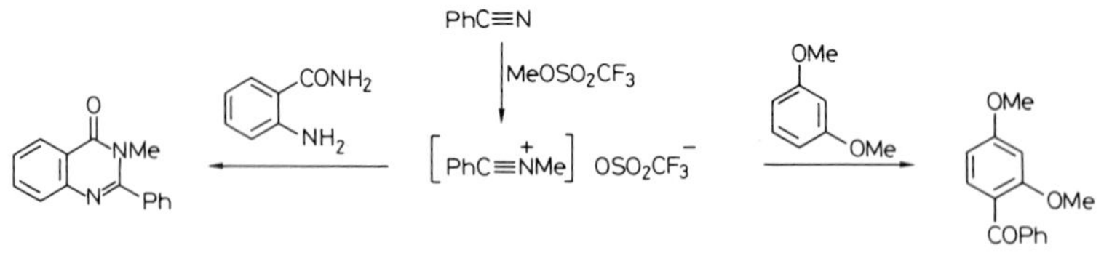

4(3*H*)-Quinazolones from *o*-aminocarboxylic acid amides | **C-Acylation.**

Benzonitrile added dropwise to a soln. of methyl triflate in benzene → N-methylbenzonitrilium triflate (Y 96%),

in nitromethane treated with startg. *o*-aminocarboxylic acid amide at 20° for 24 hrs. → product. Y 100%.

in benzene treated with *m*-dimethoxybenzene at 20° for 24 hrs., and the intermediate ketimine hydrolysed with methanolic HCl → product. Y 86%.

F.e. and reactions s. B.L. Booth, K.O. Jibodu, M.F. Proença, J. Chem. Soc. Chem. Commun. *1980,* 1151–3; C-acylation of phenolethers, s.a. J. Chem. Soc. Perkin Trans. I *1980,* 2894.

Without additional reagents *w.a.r.*

Enaminolactams from enamines
2-Piperidone ring

750. 

A stirred soln. of startg. enamine in dioxane treated with acrylamide, and the mixture refluxed for 3 hrs. → a mixture of 3,4,5,6,7,8-hexahydro-2(1*H*)-quinolone and 3,4,4a,5,6,7-hexahydro-2(1*H*)-quinolone. Y 100%. F.e.s. A.A. El-Barbary, S. Carlsson, S.-O. Lawesson, Tetrahedron *38,* 405–12 (1982).

4(1*H*)-Pyridazones from *α*-azoketones

751. 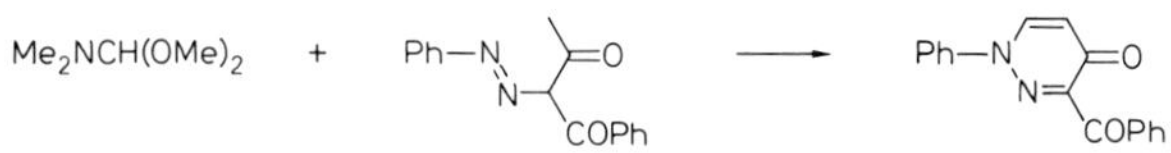

3-Acyl-4(1*H*)-pyridazones. A suspension of startg. azo compd. in dimethylformamide dimethyl acetal refluxed for 30 min., and the mixture kept at –10° overnight → product. Y 70%. F.e.s. S. Plescia et al., J. Heterocyc. Chem. *18,* 333–4 (1981).

Pyridazines from 1,2,4,5-tetrazines
and oxo compds. s. *35,* 541; and acetylene derivs., 4-silylpyridazines, s. L. Birkofer, E. Hänsel, A. Steigel, Chem. Ber. *115,* 2574–85 (1982); and amidines, 4-aminopyridazines, s. H.P. Figeys, A. Mathy, A. Dralants, Synth. Commun. *11,* 655–62 (1981); **1,2,4-triazines from amidines** cf. Tetrahedron Letters *22,* 1393–6 (1981); 4-*ω*-aminopyridazines from cyclic azomethines s. U. Berger, G. Dannhardt, R. Obergrusberger, Arch. Pharm. *315,* 428–37 (1982); **review of cycloaddition to 1,2,4,5-tetrazines** (Carboni-Lindsey reaction) s. E.G. Kovalev et al., Khim. Geterotsikl. Soedin. 1462–78 (1981).

Regiospecific diene synthesis with iron tricarbonyl complexes
3,9*a*-Dihydropyridazino[2,3,-*d*]azepines from 1,2,4,5-tetrazines

752. 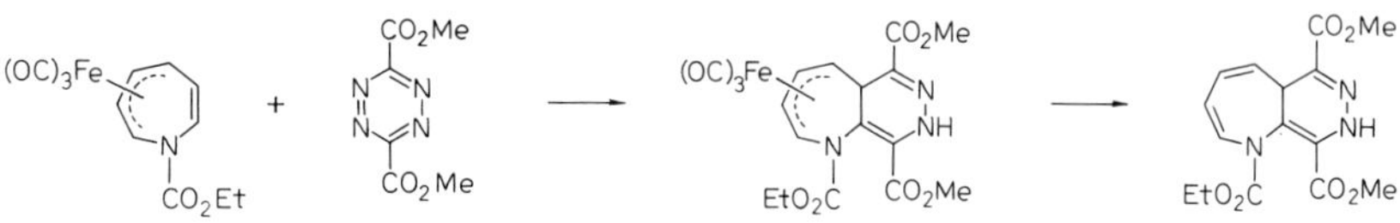

A soln. of 3,6-bis(methoxycarbonyl)-1,2,4,5-tetrazine and tricarbonyl[N-(ethoxycarbonyl)azepine]iron in methylene chloride stirred at room temp. for 7 hrs. → intermediate (Y 92%) in chloroform treated with *o*-chloranil at room temp. for 10 min. with stirring → product. (Y 71%). – The tricarbonyliron group preferentially activates the C(2)–C(3) double bond of seven-membered unsatd. rings. F.e. and aromatization. T. Ban et al., J. Org. Chem. *47,* 110–6 (1982).

Irradiation

Solid state insertion with diazo compds. $CH_2 \rightarrow CHCH_2R$
s. *33,* 756; insertion of carbenes into hydrogen-carbon bonds, temperature dependence, s. Bull. Chem. Soc. Japan *55,* 492–5 (1982); cyclopropanes from ethylene derivs., stereospecificity, s. Tetrahedron Letters *18,* 1917–20 (1982).

Organosodium compd. s. under Pd(PPh$_3$)$_4$ *RNa*

Sodium methoxide *NaOMe*

Pyrroles from 1-nitroethylene derivs. and ketones via anomalous Michael addition ○

753. 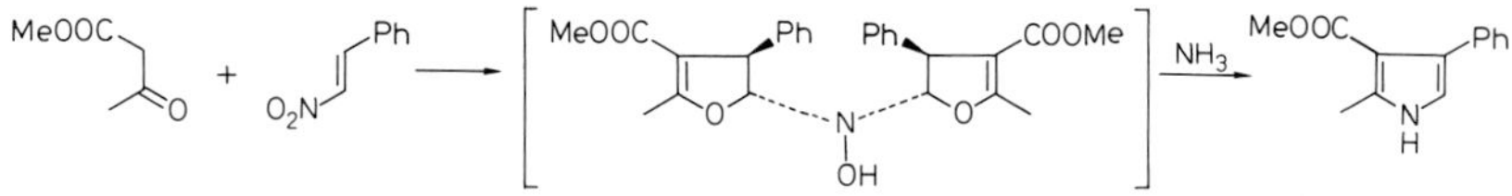

A soln. of startg. *β*-dicarbonyl compd. and *β*-nitrostyrene in 0.72 *M* methanolic Na methoxide stirred for 5 min., the suspension obtained dil. with ice-cold methanol, and ammonia bubbled through the mixture until satd. → methyl 2-methyl-4-phenylpyrrole-3-carboxylate. Y 83%. F.e. and isolation of intermediate **N,N-bis(2,3-dihydrofuran-2-yl)hydroxylamines** s. A. Gómez-Sánchez et al., J. Chem. Soc. Perkin Trans. I *1982,* 441–7; from enamines cf. ibid. *1980,* 1199–205.

Ring closures with vinylogous amidinium salts
s. *25,* 581s*31;* 2-*a*-aldehydo-4-chromanones from *o*-hydroxyketones s. R. Kirchlechner, Tetrahedron Letters *22,* 1497–500 (1981).

Sodium ethoxide *NaOEt*

Ring closures with 1-nitrohydrazones

754. 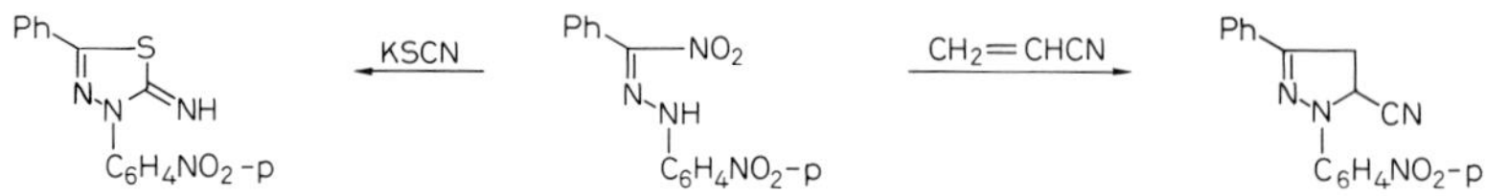

5-Imino-Δ^2-1,3,4-thiadiazolines
A soln. of KSCN in ethanol added to a soln. of startg. nitrohydrazone in the same solvent, and the mixture refluxed for 15 min. → product. Y 70–85%.

Δ^2-Pyrazolines from ethylene derivs.
A mixture of Na ethoxide, acrylonitrile, and startg. nitrohydrazone in ethanol refluxed for 90 min. → product. Y 75%.

F.e., prepn. of startg. m., and conversion to **1-subst. hydrazones** with nucleophiles, s. A.S. Shawali, H.M. Hassaneen, S.M. Sherif, J. Heterocyc. Chem. *17,* 1745–9 (1980).

Sodium isopropoxide *NaOPr-i*

Aminomethylenation under mild conditions $CH_2 \rightarrow C{=}CH{-}N<$
Vinylogous carboxylic acid amides from ketones
s. *36,* 369; with Na isopropoxide instead of Na methoxide s. Synth. Commun. *12,* 35–41 (1982).

n-*Butyllithium* n-*BuLi*

2-Ethylenealcohols from oxo compds. $CH{-}C({=}N{-}NHSO_2R) \rightarrow C{=}C{-}C(OH)R$
Shapiro reaction
s. *33,* 761; *a*-alkylidene-*β*-hydroxycarboxylic acid amides and f. electrophiles s. R.M. Adlington, A.G.M. Barrett, J. Chem. Soc. Chem. Commun. *1981,* 65–7.

Lithium diisopropylamide *i-Pr₂NLi*

***a*-Ketocarboxylic acid esters from cyanoorthoformic acid esters** RLi → RCOCOOR′

755. 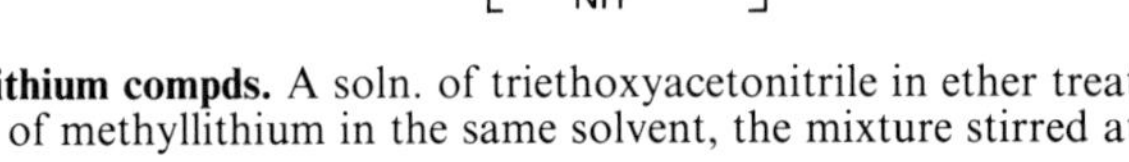

and organolithium compds. A soln. of triethoxyacetonitrile in ether treated under N_2 with a soln. of methyllithium in the same solvent, the mixture stirred at room temp.

for 1–2 hrs., and hydrolysed with 1 *M* HCl → ethyl pyruvate. Y 80–89%. F.e., **also carboxylic acid esters from Grignard compds.,** s. G.P. Axiotis, Tetrahedron Letters *22,* 1509–10 (1981).

Lithium diisopropylamide/hydrogen chloride *i-Pr_2NLi/HCl*

Quinolines from *o*-aminoketones and oximes
O,C-Dianions as intermediates O

756. 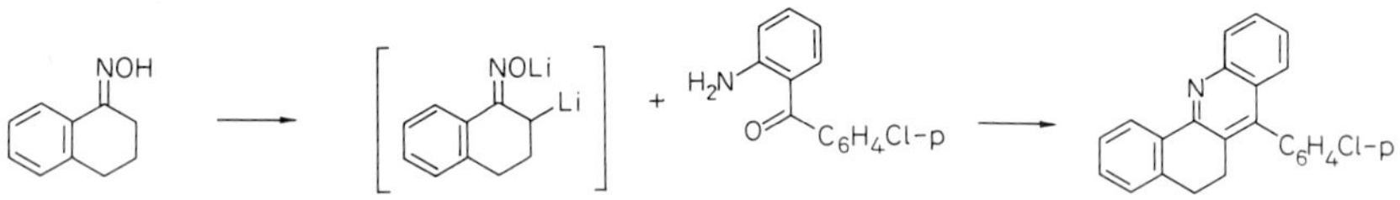

A soln. of the startg. oxime in tetrahydrofuran added to 3 equivalents of Li diisopropylamide in the same solvent at 0°, stirred for 45 min., treated with a soln. of startg. 2-aminobenzophenone in the same solvent, stirred for a further 60 min., 3 *M* HCl added, and the 2-phase mixture heated with stirring under reflux for ca. 1 hr. → 7-(4-chlorophenyl)-5,6-dihydrobenz[*c*]acridine. Y 76%. F.e.s. D.J. Park, T.D. Fulmer, C.F. Beam, J. Heterocyc. Chem. *18,* 649–51 (1981).

Sodium acetate *MeCOONa*

Pyrroles from 1,2-diaminocyclopropanes and aldehydes

757. 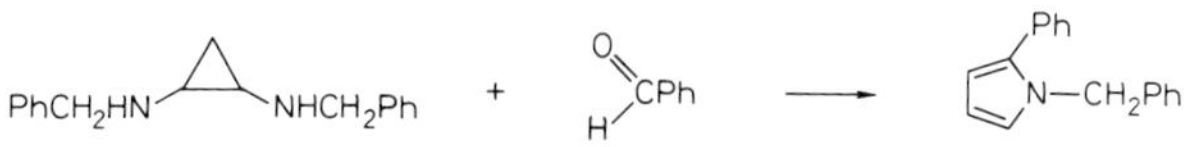

2-Arylpyrroles. *trans*-1,2-Dibenzylaminocyclopropane hydrobromide and 2 equivalents benzaldehyde in ethanol allowed to react in the presence of 13 equivalents Na acetate for 15 hrs. → 1-benzyl-2-phenylpyrrole. Y 78%. – Reaction is independent of the configuration of the startg. diamine. F.e.s. H. Quast, W. von der Saal, J. Stawitz, Angew. Chem. Intern. Ed. *20,* 588–9 (1981).

Lithium bromide *LiBr*

Ketones from aldehydes and diazo compds. CHO → COR
Deoxybenzoins
with ZnI_2 cf. *7,* 816s*36,* with LiBr s. C.A. Loeschorn, M. Nakajima, J.-P. Anselme, Bull. Soc. Chim. Belges *90,* 985–6 (1981).

Potassium salt s. under $CuCO_3$ *K^+*

Copper *Cu*

Insertion of diazomethane into carbon-selenium bonds
2-Ketoselenides from selenolic acid esters COSeR → $COCH_2SeR$

758. $MeCOSePh + CH_2N_2 \rightarrow MeCOCH_2SePh$

A mixture of startg. m. and a little Cu powder in excess ethereal alcohol-free diazomethane stirred until reaction complete by t.l.c. (ca. 4–12 hrs.) with periodic addition of diazomethane soln. to maintain excess of the reagent at all times → product. Y 65%. F.e., **also α-arylselenocarboxylic acid esters from selenolcarbonic acid esters,** s. T.G. Back, R.G. Kerr, Tetrahedron Letters *23,* 3241–4 (1982).

Cupric carbonate/potassium salt $CuCO_3/K^+$

2-Ethylenedithiophosphoric acid O,O-dialkyl esters from 1,3-dienes and diazonium salts ←

1,4-Addition

759. $PhN_2{}^+BF_4{}^- + CH_2{=}CH{-}CH{=}CH_2 + KSP(S)(OMe)_2 \rightarrow PhCH_2CH{=}CHCH_2SP(S)(OMe)_2$

Benzenediazonium fluoroborate added dropwise to a mixture of 1,3-butadiene, K O,O-dimethyl dithiophosphate, and basic $CuCO_3$ in acetone at –15 to –5°, and worked up when evolution of N_2 ceased → product. Y 80%. F.e. and conversion to **2-ethylenemercaptans** s. N.I. Ganushchak et al., Zh. Obshch. Khim. *50*, 2459–62 (1980).

Cuprous chloride *CuCl*

Intramolecular Meerwein reaction

2,3,7,8-Dihydroanthra[9,1-*b*,*c*:10,5-*b*′,*c*′]dipyrans ○

760.

CuCl added in small portions during 30 min. to a stirred suspension of 1,5-anthraquinone bis(diazonium hydrogen sulfate) and acrylonitrile in aq. dimethyl methanephosphonate at 25°, and the mixture stirred for an additional 30 min. when the exothermic (60°) evolution of N_2 ceased → product. Y 65%. F.e.s. H. Fritz et al., J. Heterocyc. Chem. *18*, 1571–80 (1981).

Boron fluoride s.a. under $MeSO_2Cl$ BF_3

Boron fluoride/acetic anhydride BF_3/Ac_2O

Chromones from *o*-hydroxyketones

Isoflavones

s. *31*, 789s*37*

Fluoroborates $[BF_4]^-$

Cinnolines from diazonium fluoroborates and enamines

761.

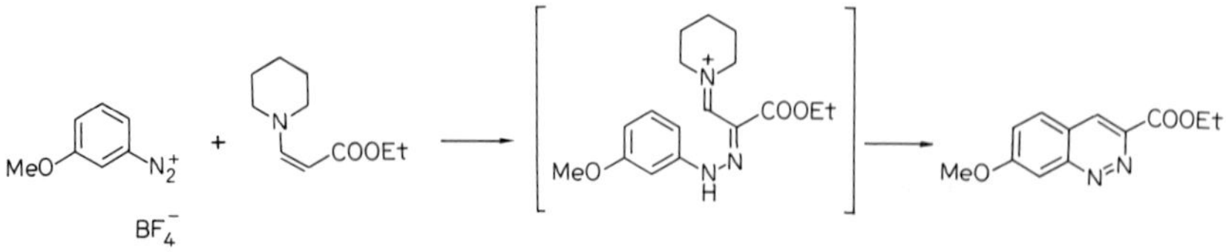

Cinnoline-3-carboxylic acid esters. Powdered *m*-methoxybenzenediazonium fluoroborate added to a soln. of startg. ester in acetonitrile, and the mixture stirred for 2 hrs. at room temp. → ethyl 7-methoxycinnoline-3-carboxylate. Y 72%. F.e., also cinnoline-3-carboxamides, and isolation of the intermediate *a*-**hydrazonoazomethinium salts,** s. C.B. Kanner, U.K. Pandit, Tetrahedron *37*, 3513–8 (1981); imidazole ring closure with base s. ibid. 3519–23 (1981).

Ammonium acetate *$MeCOONH_4$*

Pyridines from 2-ketopyridinium salts and α,β-ethyleneketones

s. *18,* 859; s.a. R.S. Tewari et al., J. Indian Chem. Soc. *57,* 1035–7 (1980).

Acetic anhydride s. under BF_3 *$(CH_3CO_2)O$*

Triphenylphosphine s. under $Pd(PPh_3)_4$ *Ph_3P*

Phosphorus trichloride *PCl_3*

2,3-Disubst. indoles from ketones and hydrazines
Fischer indole ring closure under mild conditions

762.

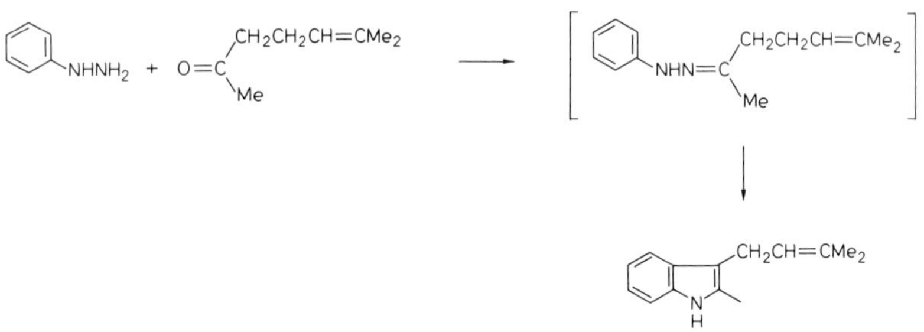

A soln. of startg. ketone and phenylhydrazine in benzene treated with 1 equivalent PCl_3, and the mixture allowed to react at room temp. for a few min. → product. Y 90%. F.e., also from the intermediate hydrazones, s. G. Baccolini, P.E. Todesco, J. Chem. Soc. Chem. Commun. *1981,* 563.

Phosphoryl chloride *$POCl_3$*

Vilsmeier aldehyde synthesis H → CHO

s. *31,* 105; with $Me_2N^{13}CHO$ s. G. Buldain, A. Valasinas, J. Labelled Compounds Radiopharm. *19,* 1–5 (1982).

Methanesulfonyl chloride/boron fluoride *$MeSO_2Cl/BF_3$*

Chromones from *o*-hydroxyketones ○

s. *31,* 789; s.a. J.M. Clough, B.K. Snell, K. Brian, Phytochemistry *20,* 1752 (1981), Chem. Abstr. *96,* 19915 (1982); with acetic anhydride instead of $MeSO_2Cl$, isoflavones, cf. H.C. Jha, F. Zilliken, E. Breitmaier, Angew. Chem. Intern. Ed. *20,* 102–3 (1981).

Hydrogen chloride (s.a. under i-Pr_2NLi) *HCl*

γ-Nitroketones from 1-nitroethylene derivs. and enamines $NO_2C=C \rightarrow NO_2CHCR$
Stereospecific Michael addition

s. *37,* 673

Rhodium(II) acetate *$(MeCOO)_4Rh_2$*

Sym. ethylene derivs. from diazo compds. $2\ C=N_2 \rightarrow C=C$
***cis*-Stilbenes**

with LiBr cf. *18,* 862s*36;* with Rh(II) acetate s. B.K.R. Shankar, H. Shechter, Tetrahedron Letters *23,* 2277–80 (1982).

Hexadecacarbonylhexarhodium *$Rh_6(CO)_{16}$*

Cyclopropane ring from diazo compds. and ethylene derivs. ▽

s. *23,* 819s*37*

Hexadecacarbonylhexarhodium *$Rh_6(CO)_{16}$*

Ylids from diazo compds. ←

3- from 2-Ethylenethioethers via [2.3]-sigmatropic rearrangement

763.

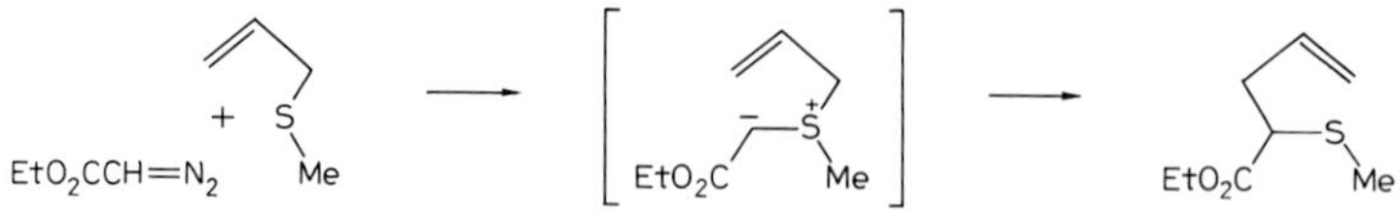

A soln. of ethyl diazoacetate in benzene added with stirring over 24–30 hrs. to a little $Rh_6(CO)_{16}$ and 5–10 equivalents of the startg. allyl sulfide at 60° under N_2 → ethyl 2-(thiomethyl)-4-pentenoate. Y 96%. – With rhodium(II) acetate, Y 91%. F.e., **also 3-from 2-ethylene-*tert*-amines, and γ,δ-from β,γ-ethyleneiodides,** s. M.P. Doyle, W.H. Tamblyn, V. Bagheri, J. Org. Chem. *46,* 5094–102 (1981).

Palladium(II) acetate *$(MeCOO)_2Pd$*

Cyclopropane ring from diazo compds. and ethylene derivs.

s. *23,* 819s*27;* from terminal ethylene derivs. s. M. Suda, Synthesis *1981,* 714; vinylcyclopropanes from 1,3-dienes, regiospecificity with Rh and Pd catalysts, s. M.P. Doyle et al., Tetrahedron Letters *23,* 2261–4 (1982); f. Rh catalysts s. Synthesis *1981,* 787–9; with $Rh_6(CO)_{16}$ s. Tetrahedron Letters *22,* 1783–6 (1981); **asym. synthesis** of cyclopropanecarboxaldehydes s. H. Abdallah, R. Grée, R. Carrié, Tetrahedron Letters *23,* 503–6 (1982).

Tetrakis(triphenylphosphine)palladium(0)/ triphenylphosphine/organosodium compd. *$Pd(PPh_3)_4/Ph_3P/RNa$*

Palladium-catalysed allylation with 3-nitroethylenes H → C–C=C

Synthesis of ethylene derivs.

764.

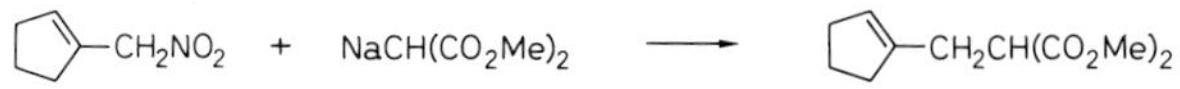

A soln. of dimethyl sodiomalonate in dimethyl formamide added to a mixture of startg. nitro compd. and catalytic amounts of tetrakis(triphenylphosphine)palladium(0) and triphenylphosphine in the same solvent, and heated at 70° for 3 hrs. → product. Y 70%. F.e., **also 2-ethyleneamines** with amines in place of carbanions, s. R. Tamura, L.S. Hegedus, J. Am. Chem. Soc. *104,* 3727–9 (1982); **with double bond migration** cf. N.Ono, I. Hamamoto, A. Kaji, J. Chem. Soc. Chem. Commun. *1982,* 821–2.

Halogen ↑ CC↓↑Hal

Without additional reagents *w.a.r.*

β-Keto-α-phosphoranylidenenitriles ⋛P=CH → ⋛P=C–COR

s. *37,* 966

Review of malononitrile in organic chemistry ←

s. *33,* 774; **reactions of malononitrile derivs., review,** s. F. Freeman, Synthesis *1981,* 925–54.

Irradiation [s.a. under KNH_2, $KOCMe_3$, and $Co_2(CO)_8$]

Photoarylation of heterocyclics H → Ar
Aryl- from chloro-pyridines

765. 

A soln. of 2-chloropyridine in *benzene* irradiated with a 60 W low-pressure Hg lamp for 2 hrs. → product. Y 93%. F.e., also 3- and 4-aryl analogs, s. M. Terashima, K. Seki, C. Yoshida, Heterocycles *15,* 1075–8 (1981).

Electrolysis

α,β-Ethyleneketones from ketones and 1,1,1-trichlorides CHCO → C=C−COR

766. 

A soln. of benzotrichloride and 2 equivalents of cyclopentanone in tetrahydrofuran containing 0.5 *M* $LiClO_4$ electrolysed in the presence of *NaH* at 0° in a divided cell with Hg cathode and Pt anode at a potential of −0.8 V until 100% conversion → 1-cyclopentenyl phenyl ketone. Y 64%. F.e.s. M. Steiniger, H.J. Schäfer, Angew. Chem. Intern. Ed. *21,* 79–80 (1982).

β,γ- from α,β-Acetylenecarboxylic acid esters
Deconjugative C-dialkylation $CH_2{-}C{\equiv}C{-}COOR \rightarrow C{\equiv}C{-}CR'_2COOR$

767. $MeCH_2C{\equiv}CCO_2Et + BuI \rightarrow MeC{\equiv}CC(Bu)_2CO_2Et$

A mixture of ethyl pent-2-ynoate and *n*-butyl iodide in dimethylformamide electrolysed using a Pt cathode in the presence of tetra-*n*-butylammonium iodide at a constant current of 0.1 A/cm² until passage of 5 F/mol → product. Y 82%. F.e. and alkylation of propiolic acid s. M. Tokuda, O. Nishio, J. Chem. Soc. Chem. Commun. *1980,* 188–9.

Ene-1,2-diolesters from
four carboxylic acid halide molecules 4 RCOHal → RC(OCOR)=C(OCOR)R

768. 4 PhCOCl → PhOCOC(Ph)=C(Ph)OCOPh

A stirred soln. of benzoyl chloride in acetone containing $LiClO_4$ electrolysed at 18° using a Hg pool cathode and a Pt plate anode separated by a porous glass diaphragm at a potential of −1.45 V until 1 F/mol passed → 1,2-diphenyl-1,2-ethenediol dibenzoate. Y 95%. F.e. with aroyl chlorides s. A. Guirado et al., J. Org. Chem. *47,* 142–4 (1982).

Electrochemical S_{RN}1-reactions
Indoles from *o*-halogenamines and oxo compds.

769. 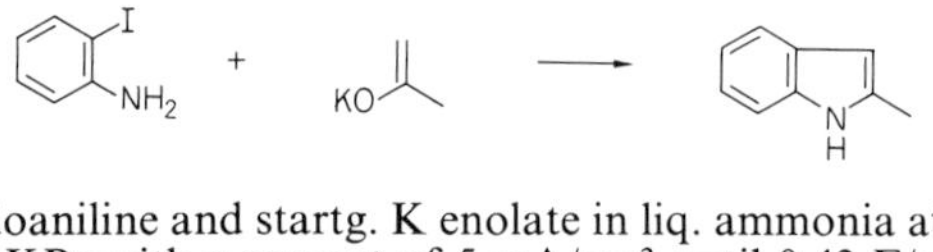

A mixture of *o*-iodoaniline and startg. K enolate in liq. ammonia at −33° electrolysed in the presence of KBr with a current of 5 mA/cm² until 0.43 F/mol electricity had passed (ca. 1 hr.) → product. Y 87%. Yields are higher than with the photochemical technique (cf. *36,* 795). F.e.s. K. Boujlel et al., Tetrahedron Letters *23,* 173–6 (1982).

Electrolysis/dibromobis(triphenylphosphine)nickel(II) ϟ/$NiBr_2(Ph_3P)_2$

Arylcarboxylic acids from halides Hal → COOH

with Ca cf. *18,* 891; electrochemically in the presence of $NiBr_2(PPh_3)_2$ s. M. Troupel et al., Nouveau J. Chem. *5,* 621–5 (1981).

Lithium (s.a. under CuI and Mg) *Li*

Coupling of halides 2 R–Cl → R–R
Ultrasonication

770. $$2\ C_6H_5\text{–}Cl \xrightarrow{Li} C_6H_5\text{–}C_6H_5$$

Sym. diaryls. A flask containing a mixture of chlorobenzene and Li wire in tetrahydrofuran immersed to the solvent level in an ultrasound laboratory cleaner for 12 hrs. → biphenyl. Y 70%. F.e., also hydrocarbons, **and α-diketones from 2 carboxylic acid chloride molecules,** s. B.H. Han, P. Boudjouk, Tetrahedron Letters *22,* 2757–8 (1981).

Aldehydes from halides RHal → RCHO
Synthesis with addition of one C-atom
Improved Bouveault reaction by ultrasonication

771. $$n\text{-}BuBr + Me_2NCHO \rightarrow n\text{-}BuCHO$$

A mixture of *n*-butyl bromide, dimethylformamide, and Li sand (mineral oil suspension) in tetrahydrofuran at 10° sonicated with an ultrasound cleaner at 40 kHz for 5 min. → product. Y 88% by g.l.c. The reaction proceeds rapidly at ambient temp. without side-reactions, a major limitation of the standard Bouveault reaction (cf. Synth. Meth. *11,* 869). F.e. incl. ar. aldehydes and **dialdehydes** s. C. Pétrier, A.L. Gemal, J.-L. Luche, Tetrahedron Letters *23,* 3361–4 (1982).

Synthesis of 1,2,4-trienes from 5-chlor-3-en-1-ynes ←
with Mg cf. *28,* 812; with Li, 1-silyl-1,2,4-trienes, s. J.-P. Dulcere, J. Grimaldi, M. Santelli, Tetrahedron Letters *22,* 3179–80 (1981).

Lithium/cuprous iodide/ *Li/CuI/*
N-methyl-N-phenylaminotributylphosphonium iodide $[Bu_3PNMePh]^+I^-$

Synthesis of allenes from 2-acetylenealcohols ←
with Mg/CuI/2-fluoropyridinium salt cf. *34,* 808; with mixed Li organocuprates in the presence of N-methyl-N-phenylaminotributylphosphonium iodide s. Y. Tanigawa, S.-I. Murahashi, J. Org. Chem. *45,* 4536–8 (1980).

Reactions with *potassium–silica gel* $K\text{–}SiO_2$

Coupling of aryl halides 2 Ar–Hal → Ar–Ar

772. $$2\ C_6H_5\text{–}Br \rightarrow C_6H_5\text{–}C_6H_5$$

K dispersed on silica gel allowed to react with a soln. of bromobenzene in ether/petroleum ether → biphenyl. Y 82%. F.e., **also pinacols from 2 ketone molecules,** s. J. Levy et al., Angew. Chem. Intern. Ed. *20,* 1033 (1981).

Sodium hydride (s.a. under n-*BuLi)* *NaH*

C-α-Alkylation of α-acylaminoketones H → R
s. *37,* 238

2-Hydroxysulfoxides from γ-chloroketones with cyclopropane ring closure ▽

773. 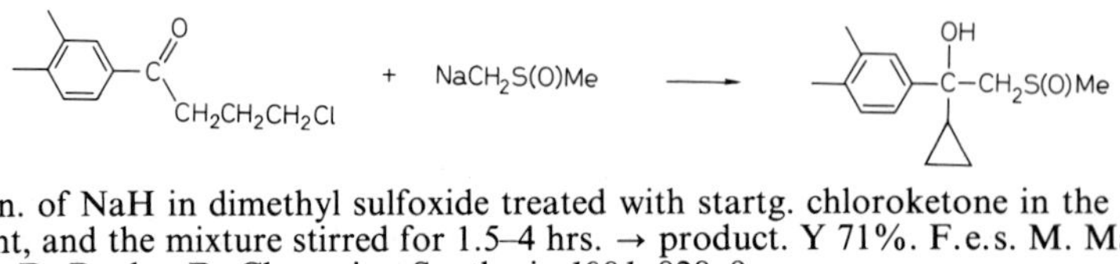

A soln. of NaH in dimethyl sulfoxide treated with startg. chloroketone in the same solvent, and the mixture stirred for 1.5–4 hrs. → product. Y 71%. F.e.s. M. Madesclaire, D. Roche, D. Chatonier, Synthesis *1981,* 828–9.

Sodium hydride/diethyl phosphite *NaH/(EtO)₂P(O)H*

(E)-α,β-Ethylene-
from α-halogeno-carboxylic acids and oxo compds. CO → C=C–COOH

774. 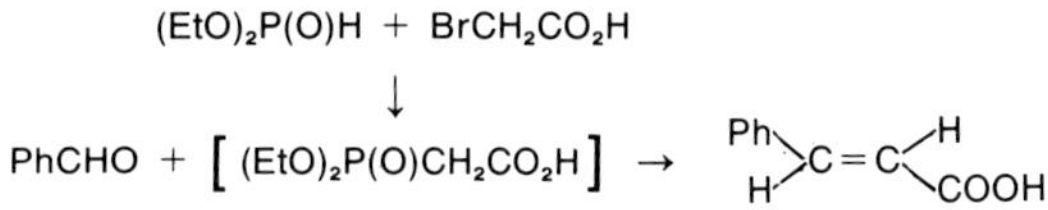

A soln. of bromoacetic acid in 1,2-dimethoxyethane added to a mixture of diethyl phosphite and NaH in the same solvent under N_2, stirred until evolution of H_2 ceased, benzaldehyde added, stirred for 1 hr., and quenched with ethanol → (E)-cinnamic acid. Y 85%. – This *in situ*-**variant of the Horner synthesis** (cf. Synth. Meth. *29,* 874) is operationally simple, and utilizes readily available and inexpensive startg. ms. F.e., also with Na alkoxide, and limitations s. D.R. Brittelli, J. Org. Chem. *46,* 2514–20 (1981).

Lithium hydroxide *LiOH*

α-Alkylation of β-ketocarboxylic acid esters H → R
s. *25,* 591; *21,* 433; with LiOH or K_2CO_3, α-alkylation of dialkyl acetonedicarboxylates, s. T. Kitahara et al., Agr. Biol. Chem. *46,* 1369–75 (1982).

Sodium hydroxide [s.a. under $Co_2(CO)_8$] *NaOH*

C-α-Alkylation of sulfones SO_2CH → SO_2CR
with *n*-BuLi cf. *29,* 796; with NaOH s. R.R. Regis, A.M. Doweyko, Tetrahedron Letters *23,* 2539–42 (1982).

Reimer-Tiemann reaction H → CHO
s. *2,* 714; **review** s. H. Wynberg, E.W. Meijer, Org. Reactions *28,* 1–36 (1982).

1,1-Dichlorocyclopropanes from ethylene derivs.
Ultrasonication

775. 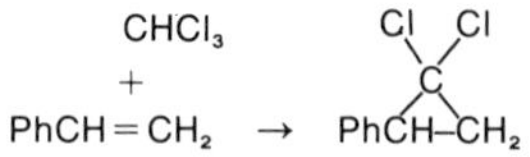

A stirred mixture of powdered NaOH and styrene in chloroform irradiated in a 45 kHz (35 W) ultrasonic bath-type sonicator at 40° for 1.5 hrs. → 1,1-dichloro-2-phenylcyclopropane. Y 95%. – The method avoids use of phase transfer catalysts. F.e.s. S.L. Regen, A. Singh, J. Org. Chem. *47,* 1587–8 (1982).

Acyloins as acyl carbanion equivalents ←
s. *36,* 793; compilation of references of formyl and acyl carbanion equivalents s. T.A. Hase, J.K. Koskimies, Aldrichimica Acta *14,* 73–7 (1981).

Sodium hydroxide/benzyltriethylammonium chloride *NaOH/[$PhCH_2NEt_3]^+Cl^-$*

Ion pair extraction ←
s. *34,* 762; recirculation and recovery of quaternary ammonium salts used to activate anions s. A. Braendstroem, J. Mol. Catal. *10,* 241 (1981); Chem. Abstr. *95,* 23608 (1981).

Sodium hydroxide/Aliquat 336 ←

Synthesis of 1-sulfonylthioethers from halides H → R
s. *37,* 191

Potassium hydroxide/polyethylene glycol ←

α-Alkylation of nitriles
with NaOH/benzyltriethylammonium chloride cf. *23,* 832s*31;* with KOH and polyethyleneglycols or their ethers s. K. Sukata, Yuki Gosei Kagaku Kyokaishi *39,* 1131–2 (1981) (Eng); Chem. Abstr. *97,* 5926 (1982).

Sodium methoxide *NaOMe*

Cyclopropanes from α,β-ethylenehalides

s. *24,* 830; s.a. J.H. Babler, B.J. Invergo, Tetrahedron Letters *22,* 2743–6 (1981).

2-Aminothiazoles from aldehydes

776.

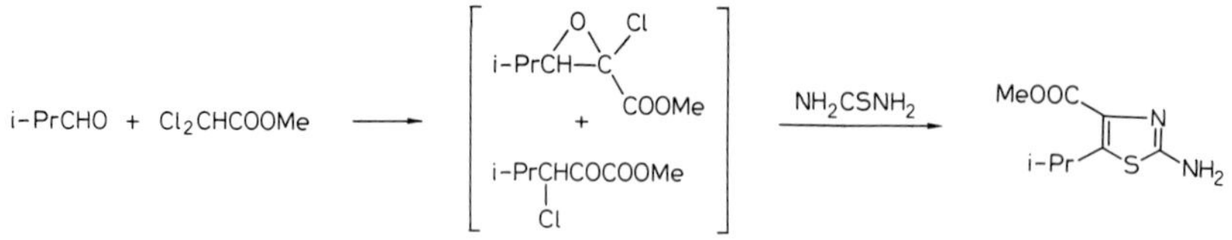

A soln. of Na methoxide added with vigorous stirring during 45 min. to a soln. of 20 g methyl dichloroacetate and 14 ml isobutyraldehyde in ether at 0°, after 1 hr. further solvent and brine added, the organic layer worked up, the crude mixed esters dissolved in methanol containing 8.5 g thiourea, and refluxed for 4 hrs. → 17.3 g methyl 2-amino-5-isopropylthiazole-4-carboxylate. F.e.s. A. Barton et al., J. Chem. Soc. Perkin Trans. I *1982,* 159–64.

Sodium ethoxide *NaOEt*

Synthesis of ketones via alkylation of β-ketosulfoxides CO–C–SOR → CO–CR′

with NaH in dimethyl sulfoxide cf. *22,* 835; with Na ethoxide s. K. Sato, S. Inoue, T. Sakamoto, Synthesis *1981,* 796–8.

Potassium tert-*butoxide s. under K/liq.* NH_3 *KOBu-t*

Potassium tert-*butoxide/liq. ammonia/irradiation* *KOBu-t*/NH_3/hν

Ring closures via $S_{RN}1$-reaction

indoles cf. *36,* 795; 1,2-dihydroisoquinolines s. Tetrahedron Letters *23,* 2313–4 (1982); 1*H*-pyrrolo[3,2-*b*]pyridines s. R. Fontan, C. Galvez, P. Viladoms, Heterocycles *16,* 1473–4 (1981).

Alkyllithium compd. *RLi*

α-Subst. β,γ-ethyleneketones ←
from carboxylic acid esters and 1,3-dichloroethylenes

777.

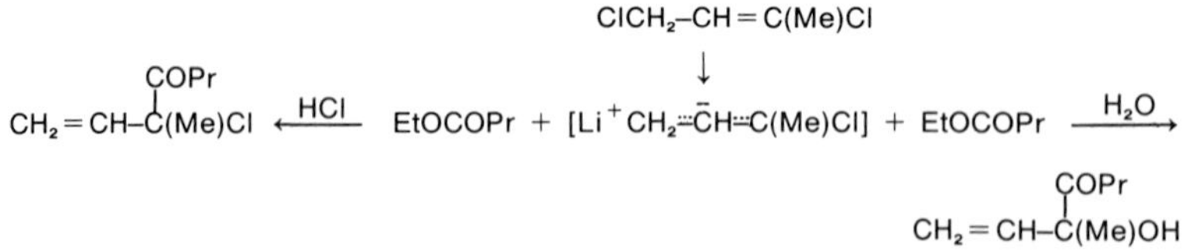

β,γ-Ethylene-α-chloroketones | **β,γ-Ethylene-α-hydroxyketones**

Startg. ester added dropwise to a soln. of *gem*-chloro(methyl)allyllithium in tetrahydrofuran at –90°, stirred at the same temp. for 30 min., treated with

3 *M* aq. HCl, | water,

and allowed to warm to room temp. → product.

Y 68%. | Y 71%.

F.e.s. B. Mauze, A. Doucoure, L. Miginiac, J. Organometal. Chem. *215,* 1–8 (1981).

Butyllithium *BuLi*

Synthesis of cyclopropenes from 1,1-difluorocyclopropanes ←

778.

A soln. of 1,1-difluoro-2-phenylcyclopropane in tetrahydrofuran treated with 2 equivalents butyllithium in the same solvent at −70° → 1-butyl-2-phenylcyclopropene. Y 91%. F.e.s. M. Suda, Tetrahedron Letters *21,* 4355–8 (1980).

4-Alkylation of pyridines
via 1-carbalkoxy-1,4-dihydropyridine-4-phosphonic esters H → R

779.

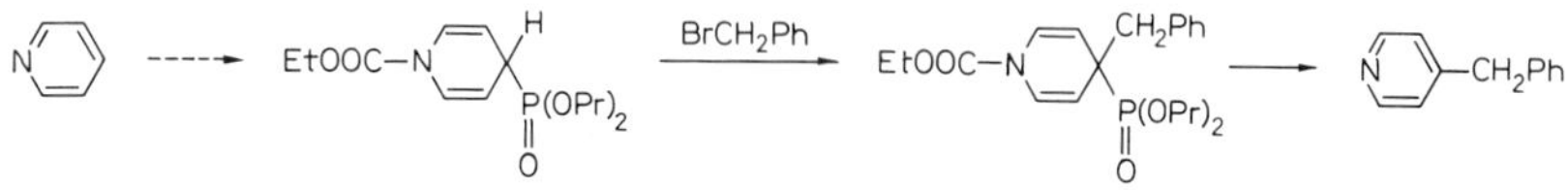

Startg. dihydropyridinephosphonate (obtained from pyridine, ethyl chloroformate and tri-isopropyl phosphite) treated under N_2 with one equivalent of butyllithium in tetrahydrofuran at −78°, and the resulting anion quenched with benzyl bromide → intermediate (Y 87%) in hexane treated with ca. 2 equivalents butyllithium in the same solvent between 0° and room temp. for 1 hr. → 4-benzylpyridine (Y 86%). F.e.s. K. Akiba, H. Matsuoka, M. Wada, Tetrahedron Letters *22,* 4093–6 (1981); A-alkylation **of quinolines** s. ibid. *23,* 1709–12 (1982).

n-*Butyllithium (s.a. under CuBr)* n-*BuLi*

Metalation of allenes C=C=CH → C=C=CLi
s. *31,* 806; f. electrophiles, incl. Me_3SiCl, s. Synthesis *1981,* 875–8.

Synthesis of acetylene derivs. H → C≡CH
via 1,1-dichloro-2-fluoroethylenes cf. *32,* 806; via 1,1,2-trichloroethylenes, alkynylcyclopropanes, s. T. Liese, A. de Meijere, Angew. Chem. Intern. Ed. *21,* 65–6 (1982).

Synthesis of 2-ethylenesilanes from 2-ethylene-3-iodosilanes C=CI → C=CR
s. *37,* 889

C-α-Alkylation of carboxylic acid thioamides via dianions H → R

780.

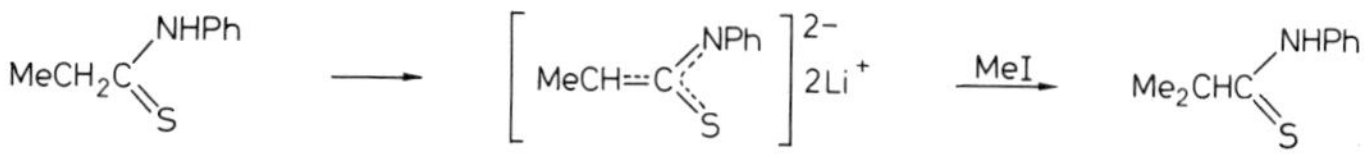

A soln. of the startg. thioamide in tetrahydrofuran treated with 2.2 equivalents of *n*-butyllithium at 0° for 1 hr., the dianion quenched with methyl iodide, warmed to room temp., and worked up with NH_4Cl → product. Y 83%. F.e., also 1,4-addition to *α,β*-ethylenecarboxylic acid thioamides, and prepn. of the dianion **from isothiocyanates** and Grignard compds., s. Y. Tamura et al., Tetrahedron Letters *22,* 3413–6 (1981).

Synthesis of β-subst. α,β-ethyleneketones
from β,γ-ethylene-β-halogenoketals ←

781.

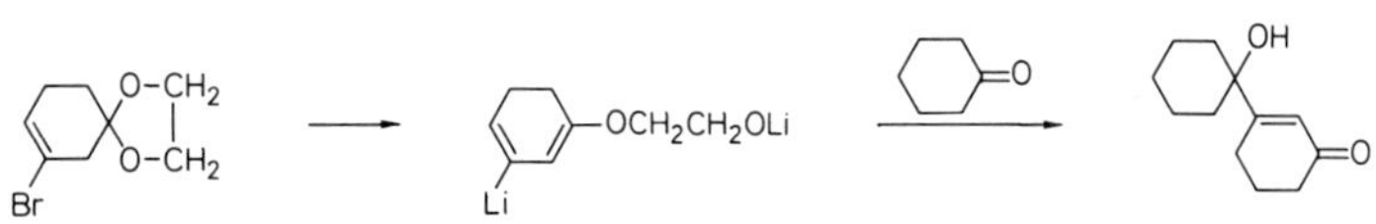

2.2 Equivalents of 1.59 *M* *n*-butyllithium added dropwise to a soln. of startg. m. in dry tetrahydrofuran at −78°, the mixture stirred for 0.5 hrs., cyclohexanone added to

the cloudy yellow suspension, and stirring continued at –78° for 1 hr. followed by 1 hr. at room temp. → product. Y 80%. F.e. and electrophiles s. C. Shih, J.S. Swenton, Tetrahedron Letters *22,* 4217–20 (1981).

2-Chloro-1-siloxycyclopropanes from enoxysilanes ▽

s. *37,* 970

7-Subst. benzoxazoles from *m*-fluoroacylamines ○
Arynes as intermediates

782. 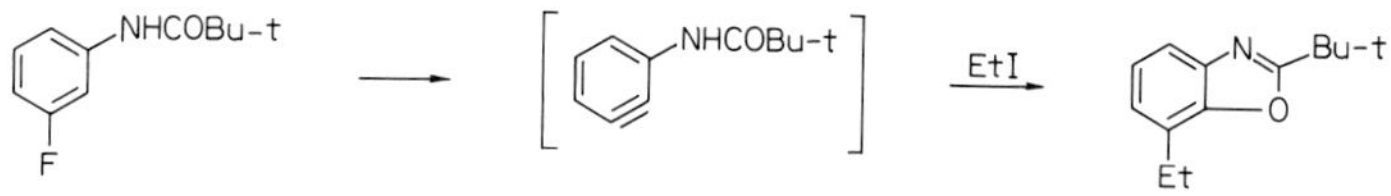

A soln. of the startg. m. in tetrahydrofuran treated dropwise with 2.5 equivalents of 1.6 *M* *n*-butyllithium in hexane at –20°, allowed to warm to 0° over 1 hr., cooled to –50°, ethyl iodide added, and allowed to warm to room temp. → product. Y 89%. F.e. and electrophiles s. R.D. Clark, J.M. Caroon, J. Org. Chem. *47,* 2804–6 (1982).

Bisarynes as intermediates
with Mg cf. *14,* 866; with *n*-butyllithium, phenanthrenes **via bisarynes** s. H. Hart, S. Shamouilian, J. Org. Chem. *46,* 4874–6 (1981).

Δ^2-5-Imidazolones from α-isocyanocarboxylic acid amides
with NaH cf. *33,* 779; with *n*-BuLi, also ring opening to optically active α-aminocarboxylic acids, s. U. Schöllkopf et al., Ann. Chem. *1981,* 439–58.

n-*Butyllithium/sodium hydride* n-*BuLi/NaH*

Reactions with stepwise-prepd. dicarbanions H → R
γ-Alkylation of β-ketocarboxylic acid esters
s. *26,* 861; γ-hydroxyalkylation s. L. Novák et al., Ann. Chem. *1982,* 1173–82.

n-*Butyllithium/N,N,N',N'-tetramethylethylenediamine* n-*BuLi/TMEDA*

3-Alkylation of oxindoles
with NaH cf. *12,* 841; via N,C-dianions with *n*-BuLi/TMEDA s. A.S. Kende, J.C. Hodges, Synth. Commun. *12,* 1–10 (1982).

Syntheses with 2-ethyleneurethans as homoenolate equivalents ←
s. *36,* 816; oxo compds. as electrophiles with *n*-butyllithium/TMEDA as base s. Angew. Chem. Intern. Ed. *20,* 1024–6 (1981).

n-*Butyllithium/morpholine* n-*BuLi/HN(CH$_2$CH$_2$)$_2$O*

Syntheses via protection of aldehydes *in situ* as lithium 1-aminoalkoxides Hal → R

783. 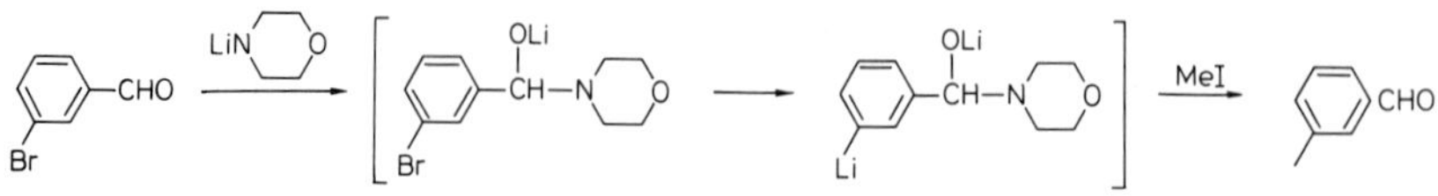

m-Bromobenzaldehyde treated with Li morpholide in tetrahydrofuran at –78°, after 15 min. *n*-butyllithium added, the intermediate dianion treated with methyl iodide, and the mixture poured into cold 10% aq. HCl → *m*-tolualdehyde. Y 90%. F.e., also with Li N-methyl-2-pyridylamide/trimethylchlorosilane for reactions of aldehydocarboxylic acid esters, s. D.L. Comins, J.D. Brown, Tetrahedron Letters *22,* 4213–6 (1981).

sec-Butyllithium s-*BuLi*

Electrophilic 2-substitution of N-heterocyclics via metalation of N-formimidoyl-N-heterocyclics ←

784. 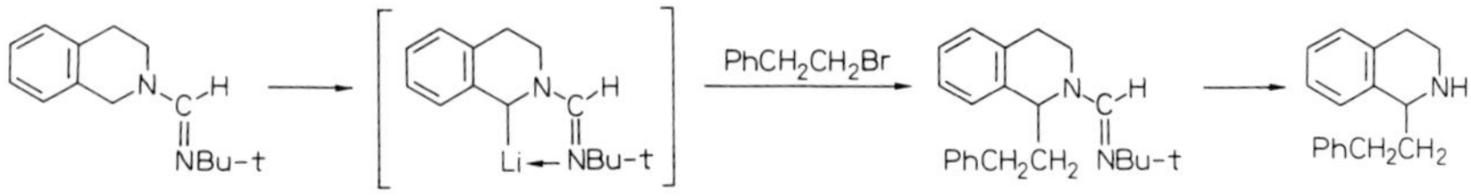

1-Alkylation of 1,2,3,4-tetrahydroisoquinolines. A soln. of startg. N-formimidoyl deriv. in tetrahydrofuran treated with *sec*-butyllithium at –78°, a soln. of phenethyl bromide in the same solvent added, the mixture allowed to warm to ca. –20° over 2–3 hrs., and quenched with water → crude intermediate refluxed in 1:1 10% aq. KOH/methanol for 24 hrs. → product (Y 61%). F.e., methods, and electrophiles, s. A.I. Meyers, S. Hellring, W. Ten Hoeve, Tetrahedron Letters *22,* 5115–8 (1981); **2-substitution of 1,2,3,4-tetrahydroquinolines and indolines,** also synthesis of sec. ar. amines, s. ibid. *22,* 5119–22 (1981).

Asymmetric synthesis with chiral leaving groups Ar–Hal + Ar′–OR → Ar–Ar′
Diaryls from phenolethers and aryl halides

785. 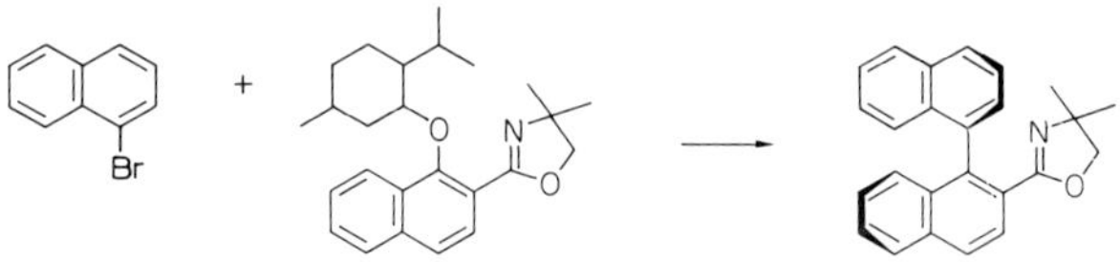

A soln. of α-bromonaphthalene in tetrahydrofuran at –78° treated with *sec*-butyllithium, the mixture allowed to warm to –42°, a soln. of startg. (–)-α-menthoxy-β-oxazolinonaphthalene in the same solvent added, and stirred at the same temp. for 1 hr. → (S)-product. Y 80% (67% enantiomeric excess). F.e. and chiral leaving groups, also via arylmagnesium bromides, s. J.M. Wilson, D.J. Cram, J. Am. Chem. Soc. *104,* 881–4 (1982); via **activation with chiral Δ^2-oxazolines** cf. A.I. Meyers, K.A. Lutomski, ibid. *104,* 879–81 (1982).

tert-Butyllithium (s.a. under i-Pr$_2$NLi) t-*BuLi*

Metalation of eneformamidines H → R
s. *37,* 71

1-Metalation of 2-aminothioenolethers

786. 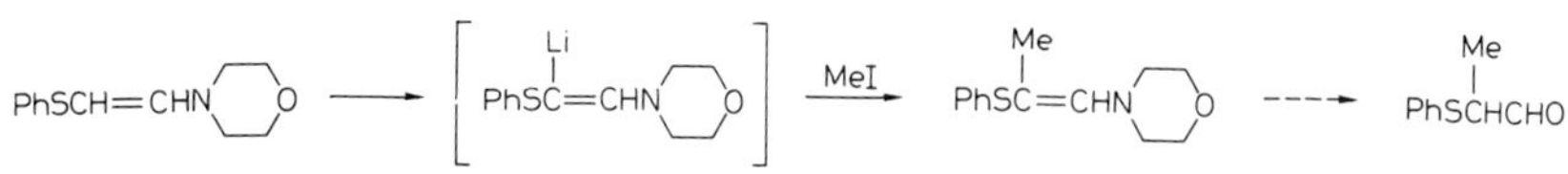

A soln. of the startg. enamine in 9:1 tetrahydrofuran/hexamethylphosphoramide treated with a soln. of *tert*-butyllithium in pentane under N_2 at –65°, stirred for 1.5 hrs. at this temp., a soln. of methyl iodide in tetrahydrofuran added at –65°, stirring continued for 1 hr., and the mixture allowed to warm to room temp. → product. Y 100%. F.e. and hydrolysis to α-**(arylthio)-aldehydes and -ketones** s. L. Duhamel, J. Chauvin, A. Messier, J. Chem. Res. (M) *1982,* 621–40.

tert-*Butyllithium (s.a. under i-Pr$_2$NLi)* t-*BuLi*

Ketones from carboxylic acids and halides RCOOH → RCOR′

N^4-Sulfonylamidrazones as acyl carbanion equivalents

787.

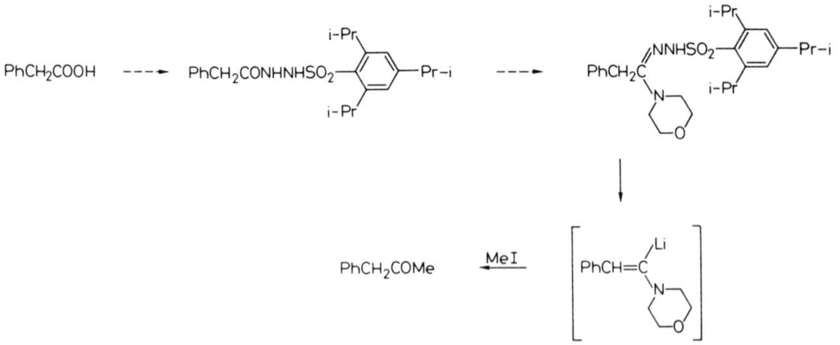

A soln. of startg. amidrazone (readily available in 2 steps from phenylacetic acid) in tetrahydrofuran at –78° treated with 2.3 equivalents *tert*-butyllithium for 30 min., warmed to 10° during 5 min., recooled to –78°, methyl iodide added, and the mixture gradually warmed to room temp. → phenylacetone. Y 80%. F.e.s. J.E. Baldwin, J.C. Bottaro, J. Chem. Soc. Chem. Commun. *1981,* 1121–2.

tert-*Butyllithium/N,N,N′,N′-tetramethylethylenediamine* t-*BuLi/TMEDA*

Electrophilic 1-substitution of 1,2,3,4-tetrahydroisoquinolines H → R

N-Pivaloyl as activating group

788.

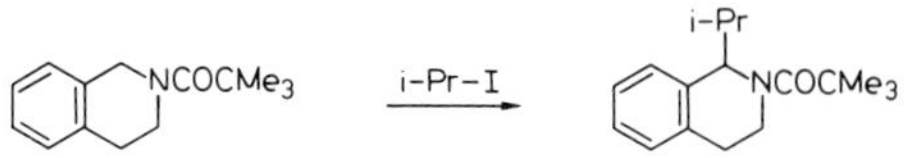

1-Alkylation. Startg. N-pivaloyl deriv. treated with *tert*-butyllithium and TMEDA for 5 min., and the orange red soln. quenched with 2-iodopropane → product. Y 90%. – **Removal of the activating group** was achieved with Na dihydridobis(2-methoxyethoxo)aluminate in benzene. F.e. and electrophile s. J.-J. Lohmann et al., Angew. Chem. Intern. Ed. *20,* 128–9 (1981).

tert-*Butyllithium/magnesium bromide/*
[1,4-bis(diphenylphosphino)butane]dichloropalladium(II) ←

2-Arylation of pyrroles and indoles H → Ar

789.

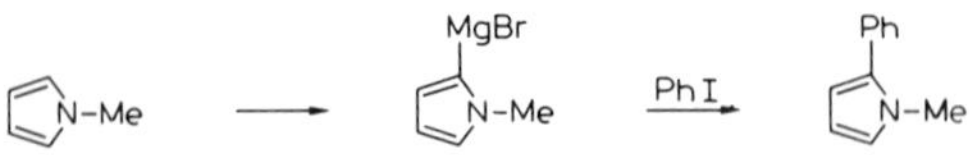

A soln. of 1-methyl-2-pyrrolyllithium (prepd. from 1-methylpyrrole and *tert*-butyllithium) in tetrahydrofuran added to $MgBr_2$ (or dried $ZnCl_2$) in the same solvent at room temp., the mixture stirred for 1 hr., added to a stirred mixture of iodobenzene and a little [1,4-bis(diphenylphosphino)butane]dichloropalladium(II) at room temp. under argon, and stirred for a further 20 hrs. → 1-methyl-2-phenylpyrrole. Y 78%. – Similarly, 1-methylindole → 1-methyl-2-phenylindole. Y 79%. F.e., **also hetarylation,** s. A. Minato et al., Tetrahedron Letters *22,* 5319–22 (1981).

Lithium diisopropylamide [s.a. under (i-C$_4$H$_9$)$_2$AlH] *i-Pr$_2$NLi*

α-Lateral metalation of oxazoles H → R

s. *37,* 251

C-α-Alkylation of β,γ-ethylenechlorides

790. $CH_2=C(Me)CH_2Cl \rightarrow [CH_2=C(Me)CH(Cl)Li] \xrightarrow{Br(CH_2)_4Br} CH_2=C(Me)CHCl(CH_2)_4Br$

A soln. of Li diisopropylamide in tetrahydrofuran added slowly via syringe to a soln. of methallyl chloride and 1,4-dibromobutane in the same solvent at –78° under inert atmosphere, and stirred at the same temp. for 15 min. → 7-bromo-3-chloro-2-methyl-1-heptene. Y 71%. – Sec. bromides were ineffective as alkylating agents. F.e.s. T.L. Macdonald, B.A. Narayanan, D.E. O'Dell, J. Org. Chem. *46,* 1504–6 (1981).

...of α-hydroxycarboxylic acids via 1,3-dioxolan-4-ones
Retention of configuration

791.
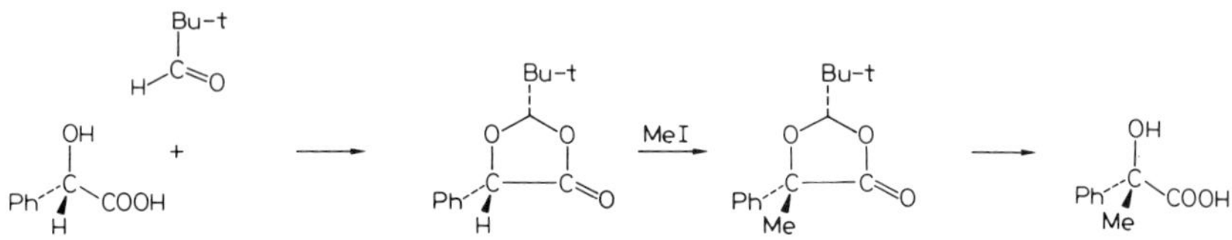

A soln. of S(+)-mandelic acid in benzene containing *p*-toluenesulfonic acid treated with pivalaldehyde at reflux → *cis*-1,3-dioxolan-4-one deriv. (Y 74%), in tetrahydrofuran treated with a mixture of Li diisopropylamide and hexamethylphosphoramide at –70°, methyl iodide added, and warmed to 0° during 10 min. → methyl deriv. (Y 94%; *cis/trans* ratio 93/7) hydrolysed with KOH in aq. methanol at 60° for 10 min. → S(+)-atrolactic acid (enantiomeric excess 85%). F.e.s. G. Frater, U. Muller, W. Gunther, Tetrahedron Letters *22,* 4221–4 (1981).

Asym. synthesis of ketones via C-alkylation of chiral azomethines ←
s. *19,* 896s*34; 30,* 461; α-alkyl- and α,α'-dialkyl cyclic ketones s. J. Org. Chem. *103,* 3081–7 (1981).

Synthesis of prim. amines
via N-α-alkylation of iminoesters $CH{-}N{=}C(OR) \rightarrow CR'{-}NH_2$

792. $PhCH_2N=C(Ph)OMe \xrightarrow{EtI} PhCH(Et)N=C(Ph)OMe \rightarrow PhCH(Et)NH_2.HCl$

Methyl N-benzylbenzenecarboximidate added with stirring to a soln. of Li diisopropylamide in tetrahydrofuran at –20° under N_2, after 30 min. treated with a soln. of ethyl iodide in the same solvent at 0°, and stirring continued 2–4 hrs. at 20° → methyl N-(α-ethylbenzyl)benzenecarboximidate (Y 86%) suspended in 1 *M* HCl, and after ca. 30 min. the homogeneous soln. heated with stirring for 1.5 hrs. at 60° → α-ethylbenzylamine hydrochloride (Y 90%). F.e.s. A.H. Schulthess, H.-J. Hansen, Helv. Chim. Acta *64,* 1322–36 (1981).

Reactions with α-siloxyphosphonic esters H → C–Hal
Enol phosphates from 1,1-dihalides

793.
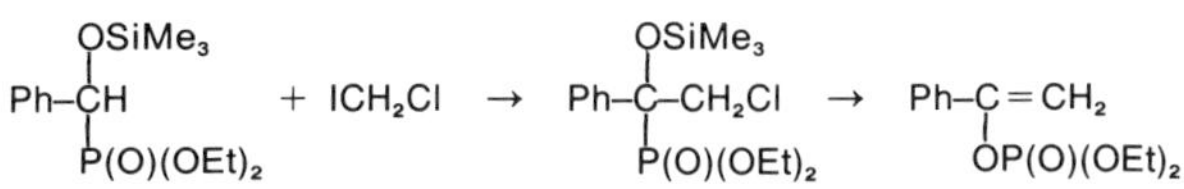

Diethyl α-[(trimethylsilyl)oxy]benzylphosphonate added dropwise to a soln. of Li diisopropylamide in tetrahydrofuran at –78°, the mixture stirred for 30 min., chloroiodomethane added dropwise, stirring continued for 30 min., poured into a mixture of 1 *M* NH_4Cl and dichloromethane, and treated with a little $Na_2S_2O_3$ → diethyl α-(chloromethyl)-α-[(trimethylsilyl)oxy]benzylphosphonate (Y 85%), in toluene refluxed for 13 hrs. → product (Y 73%). F.e., **also cyclic ene-1,2-diol phosphates from carboxylic acid halides,** s. M. Sekine, M. Nakajima, T. Hata, J. Org. Chem. *46,* 4030–4 (1981).

Lithium diisopropylamide (s.a. under $(i\text{-}C_4H_9)_2AlH$) *$i\text{-}Pr_2NLi$*

Carboxylic acid esters from halides via α,α-dialkoxynitriles Hal → COOR

794. $HC(OMe)_2CN \rightarrow [LiC(OMe)_2CN] \xrightarrow{n\text{-}C_8H_{17}Br} n\text{-}C_8H_{17}C(OMe)_2CN \rightarrow n\text{-}C_8H_{17}COOMe$

A soln. of dimethoxyacetonitrile **as carbomethoxy carbanion equivalent** in 4:1 tetrahydrofuran/hexamethylphosphoramide added to a stirred soln. of Li diisopropylamide in tetrahydrofuran at –78°, the mixture stirred for a further 90 min., 1-bromooctane added, and stirring continued at –78° for 90 min. followed by 30 min. at room temp. → 2,2-dimethoxydecanenitrile (Y 55%) in 5:1 acetone/water refluxed with *p*-toluenesulfonic acid for 3 hrs. → methyl nonanate (Y 96%). K. Utimoto et al., Tetrahedron Letters *22,* 4279–80 (1981).

β-Diketones
from ketones and carboxylic acid chlorides CO–CH + ClCOR → CO–C–COR
Inverse addition of lithium enolates

795. 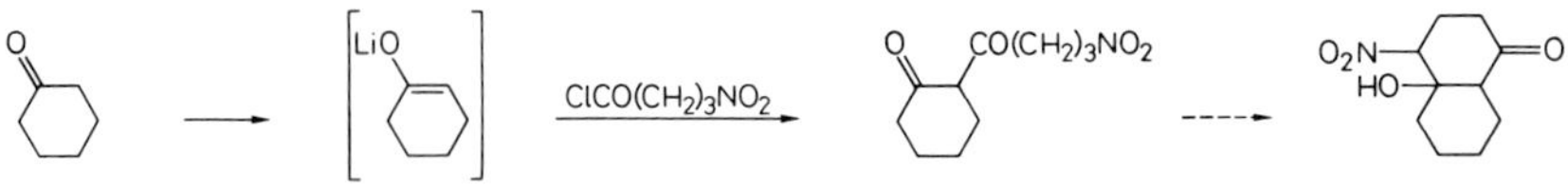

ε-Nitro-β-diketones. Cyclohexanone added quickly by syringe to a soln. of Li diisopropylamide in tetrahydrofuran under argon at –78°, after 30 min. the enolate soln. transferred via a connecting Teflon tube into a soln. of startg. acyl chloride in the same solvent at –78° during 15–30 min., a little tetrahydrofuran added, and stirred for a further 60 min. at –100 to –60° → 2-(4′-nitrobutyryl)cyclohexanone. Y 75%. F.e. and limitations, **also from enoxysilanes,** s. D. Seebach et al., Helv. Chim. Acta *64,* 716–35 (1981); ring closure to **4-nitrocyclohexan-3-ol-1-ones** with satd. $NaHCO_3$ in tetrahydrofuran s. ibid. *64,* 736–59 (1981).

Lithium diisopropylamide/hexamethylphosphoramide *$i\text{-}Pr_2NLi/(Me_2N)_3PO$*

Deconjugative C-α-alkylation
of α,β-ethylenecarboxylic acid esters CHC=C–COOR → C=C–CR′COOR
s. *29,* 810; stereochemistry, also deconjugative protonation and aldol condensation, s. A.S. Kende, B.H. Toder, J. Org. Chem. *47,* 163–7 (1982).

*Lithium dicyclohexylamide/*tert*-butyllithium* *R_2NLi/t-BuLi*

Preparation of ynolates ***in situ***
via Hofmann-type rearrangement COOR → CH_2COOR
One-step homologization of carboxylic acid esters

796. 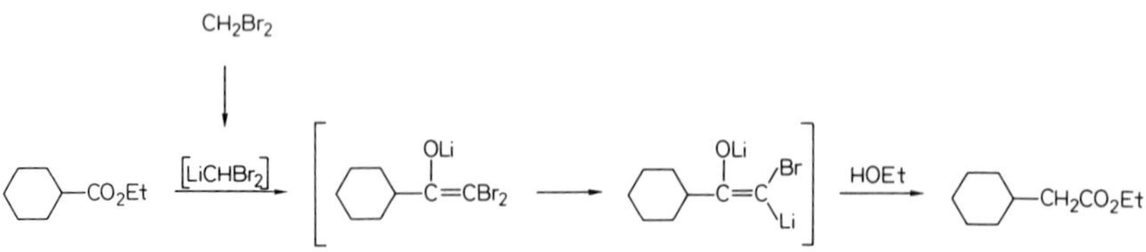

A mixture of ethyl cyclohexanecarboxylate and 2.4 equivalents dibromomethane in tetrahydrofuran treated sequentially at –78° with 2.4 equivalents Li dicyclohexylamide and 7.5 equivalents *tert*-butyllithium, allowed to warm to room temp., and the formed ynolate quenched with 10% H_2SO_4 in ethanol → ethyl cyclohexaneacetate. Y 63%. – The method holds decided advantages over the classical multistep Arndt-Eistert procedure based on the hazardous diazomethane. F.e. and method, **also α,β-ethylenecarboxylic acids from aldehydes,** s. C.J. Kowalski, K.W. Fields, J. Am. Chem. Soc. *104,* 321–3 (1982).

Sodium amide $NaNH_2$

Synthesis of acetylene derivs. from halides C≡CH → C≡CR

s. *9,* 888; methods s. D. Mesnard, F. Bernadou, L. Miginiac, J. Chem. Res. (S) *1981,* 270–1.

Potassium/liq. ammonia/potassium tert-*butoxide* $K/NH_3/KOBu$-t

Reductive C-alkylation of arenes ←

with Li/liq. NH_3 cf. *27,* 839; of *o*-methoxycarboxylic acids with K in liq. NH_3 containing K *tert*-butoxide s. J.M. Hook, L.N. Mander, M. Woolias, Tetrahedron Letters *23,* 1095–8 (1982).

Potassium amide/irradiation KNH_2/hν

α-Arylation of ketones H → Ar

S_{RN}1-reaction

s. *32,* 800s*34;* α-hetarylation with 2-chloroquinoxaline and 4-chloroquinazolines s. D.R. Carver, J.S. Hubbard, J.F. Wolfe, J. Org. Chem. *47,* 1036–40 (1982).

Sodium carbonate Na_2CO_3

3-Acyl-6-alkoxy-5,6-dihydro-4*H*-1,2-oxazines

from enolethers and β-halogeno-α-isonitrosoketones ○

s. *37,* 337

Potassium carbonate/potassium iodide K_2CO_3/KI

2-Methylene-2,3-dihydrofurans from 3-haloacetylenes

with CuBr/Na ethoxide cf. *34,* 798; with K_2CO_3/KI s. R. Chênevert et al., Synthesis *1982,* 75–7.

Potassium carbonate/benzyltriethylammonium chloride $K_2CO_3/[PhCH_2NEt_3]^+Cl^-$

C,C-Dialkylation $CH_2 \rightarrow CR_2$

with NaH cf. *18,* 872; with K_2CO_3 and benzyltriethylammonium chloride as phase transfer catalyst, 1,3-dioxane-4,6-diones, s. C.-C. Chan, X. Huang, Synthesis *1982,* 452–4.

Sodium hydrogen carbonate s. under $ZnCO_3$ $NaHCO_3$

Sodium cyanide $NaCN$

Review of small-ring propellanes ○

s. *23,* 840s*27;* review of synthesis of propellanes s. Y. Odaira, Y. Tobe, Yuki Gosei Kagaku Kyokaishi *39,* 1163–71 (1981).

Potassium cyanide/18-crown-6 polyether ←

2-Subst. Δ^2-thiazolines from 1,2-iodisothiocyanates

797.

2-Cyano-Δ^2-thiazolines. A mixture of *trans*-1-iodo-2-isothiocyanatocyclohexane, KCN, and 18-crown-6 polyether in 4:1 chloroform/water refluxed for 92 hrs. → 2-cyano-*cis*-3a,4,5,6,7,7a-hexahydrobenzothiazole. Y 100%. F. carbanions s. R.C. Cambie et al., J. Chem. Soc. Perkin Trans. I *1981,* 40–51; with S-nucleophiles s. ibid. *1981,* 52–7.

Potassium cyanide/benzyltriethylammonium chloride *$KCN/[PhCH_2NEt_3]^+Cl^-$*
α,β-Oxidonitriles from α-halogenoketones COCHal → 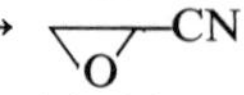

with $Et_4N^+CN^-$ cf. *35,* 581; with KCN and benzyltriethylammonium chloride as phase transfer catalyst s. K. Takahashi, T. Nishizuka, H. Iida, Synth. Commun. *11,* 757–62 (1981).

Caesium fluoride s. under SnF_2 *CsF*

Potassium iodide s. under K_2CO_3 *KI*

Trialkylamine s. under Pd *R_3N*

Triethylamine {s.a. under CuCl, CuI, $AgBF_4$, $SnCl_4$, $Ni(CO)_2(Ph_3P)_2$, Pd, $(MeCOO)_2Pd$, and Dichloro[1,1′-bis(diphenylphosphino)ferrocene]palladium(II)} *Et_3N*
Diazomethyltrimethylsilane as diazomethane substitute COCl → CH_2CO–
Modified Arndt-Eistert synthesis
of ketones s. *36,* 824; also prepn. of carboxylic acid amides s. Chem. Pharm. Bull. *29,* 3249–55 (1981); ketones s.a. ibid. *30,* 119–24 (1982).

α-Diazo-β-ketosulfones
from diazomethyl sulfones and carboxylic acid chlorides COCl → $COC(N_2)SO_2R$

798. 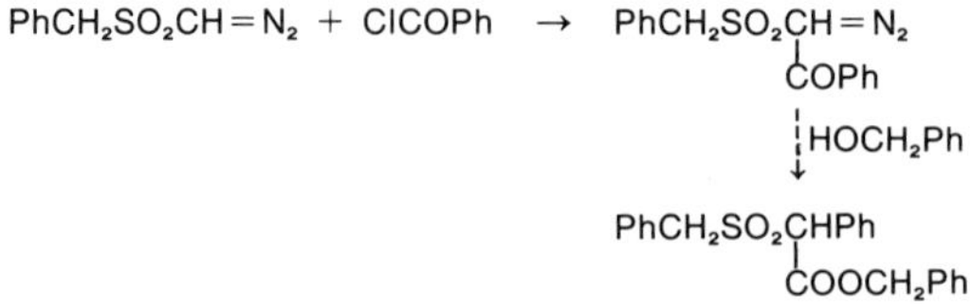

Benzoyl chloride added dropwise at 0° or room temp. to a mixture of benzyl sulfonyldiazomethane and triethylamine in acetonitrile, and stirred at room temp. for 24 hrs. → product. Y 87%. F.e., and thermal Wolff rearrangement to **α-sulfonylcarboxylic acid esters** with alcohols in refluxing toluene, s. Y.-C. Kuo, T. Aoyama, T. Shioiri, Chem. Pharm. Bull. *30,* 526–33 (1982).

β,γ-Ethylene-β-siloxycarboxylic
acid silyl esters CH–COCl → C=C(OSi≡)C–COOSi≡
from ketene disilyl acetals

799. 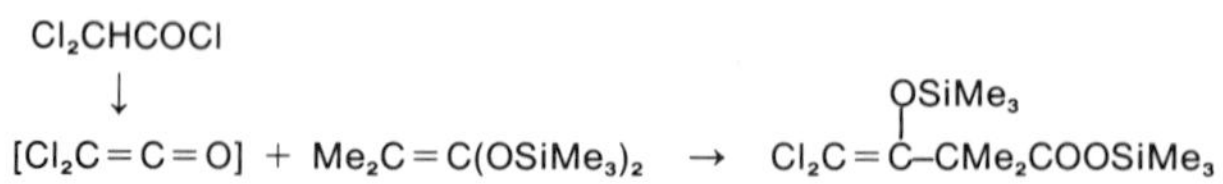

A soln. of dichloroacetyl chloride in hexane added during 2 hrs. to a stirred mixture of dimethylketene bis(trimethylsilyl)acetal and triethylamine in the same solvent at 22–24° under N_2, and the mixture stirred overnight → trimethylsilyl 4,4-dichloro-2,2-dimethyl-3-(trimethylsiloxy)-3-butenoate. Y 80%. F.e. and reactions of chloroketenes with ketene acetals s. W.T. Brady, R.D. Watts, J. Org. Chem. *46,* 4047–50 (1981).

1,2-Oxathiene dioxides
Sulfene as intermediate ○
s. *21,* 867; 1,2-oxathiino[5,6-*d*]-1-benzoxepin derivs. s. G. Menozzi et al., J. Heterocyc. Chem. *19,* 937–9 (1982).

Pyrazole ring from 1-halogenohydrazones and enamines
s. *34,* 796; isolation of intermediate 5-amino-Δ^2-pyrazolines cf. H. Biere et al., Eur. J. Med. Chem. *17,* 27–34 (1982).

Heterocyclics from 1-halogenohydrazones
review s. *34,* 796s*35;* from 2-keto-1-halogenohydrazones s. N.F. Eweiss, A. Osman, J. Heterocyc. Chem. *17,* 1713–7 (1980).

3,3-Dichloro-1,7-dithiacycloalk-5-en-2-ones
from cyclic *a,β*-ethylenemercaptals
Ring expansion by four C-atoms with dichloroketene

800. 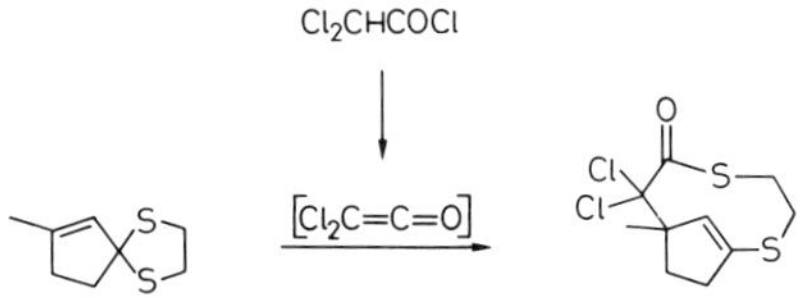

A mixture of 7-methyl-1,4-dithiaspiro[4.4]non-6-ene and dichloroacetyl chloride in hexane stirred at room temp. under N_2 while a soln. of triethylamine in the same solvent was added slowly during 1 hr., and stirred for 3–4 hrs. → 2,2-dichloro-1-methyl-4,7-dithiabicyclo[6.2.1]undec-8(11)-en-3-one. Y 85%. F.e.s. G. Rosini et al., J. Org. Chem. *46,* 2228–30 (1981).

*Tri-*n-*butylamine s. under $(MeCOO)_2Pd$* n-Bu_3N

N,N,N′,N′-Tetramethylethylenediamine s. under n-*BuLi and* t-*BuLi* *TMEDA*

Morpholine s. under n-*BuLi* ←

Pyridine C_5H_5N
Perfluoroalkylation with perfluoroalkyliodonium triflates ←

801. 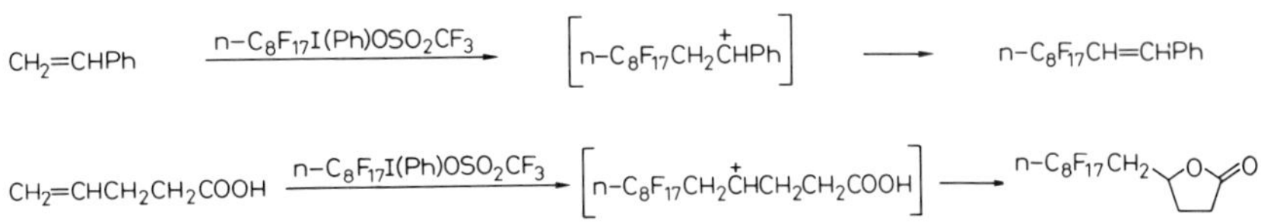

of ethylene derivs. A soln. of styrene and perfluorooctylphenyliodonium triflate in methylene chloride refluxed for 0.5 hrs. in the presence of one equivalent pyridine → *trans-β*-(perfluorooctyl)styrene. Y 73%. – The intermediate carbonium ion can also be trapped by nucleophiles in both inter- and intra-molecular processes. E: Pent-4-enoic acid → *γ*-lactone deriv. Y 69%. F.e.s. T. Umemoto, Y. Kuriu, S. Nakayama, Tetrahedron Letters *23,* 1169–72 (1982); **ar. perfluoroalkylation,** and with perfluoroalkyliodonium hydrogen sulfates, s. Chem. Letters *1981,* 1663–6; coupling with Grignard and organolithium compds. cf. Tetrahedron Letters *22,* 5197–200 (1981).

Copper *Cu*
(Polyfluoroalkyl)arenes from aryl iodides ArHal → $ArCF_3$
s. *26,* 867; trifluoromethylarenes with bis(trifluoromethyl)mercury s. N.V. Kondratenko, E.P. Vechirko, L.M. Yagupolsky, Synthesis *1980,* 932–3.

Silver(I) oxide *Ag_2O*

Thiolic acid esters from carboxylic acid chlorides COCl → CH_2COSR
Arndt-Eistert synthesis

802. 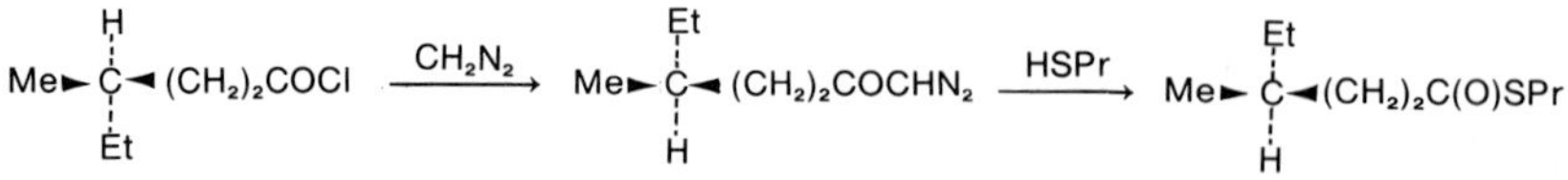

Optically active compds. S-(+)-4-Methylcaproyl chloride added to a soln. of diazomethane in ether, and allowed to stand at 20–25° for 3–4 hrs., solvent removed, treated with *n*-propyl mercaptan and Ag_2O, and refluxed for 2 hrs. on a water bath → S-(+)-*n*-propyl 5-methylthiolenanthate. Y 48%. F.e.s. I.G. Vasi, N.T. Nanavati, J. Indian Chem. Soc. *57,* 744–5 (1980); Indian J. Chem. *20B,* 509–10 (1981).

Lithium diorganocuprates *$LiCuR_2$*

Regiospecific synthesis of enoxysilanes CHalC(OSi⋮)=C → C=C(OSi⋮)CR
from β,γ-ethylene-β-siloxyhalides

803. 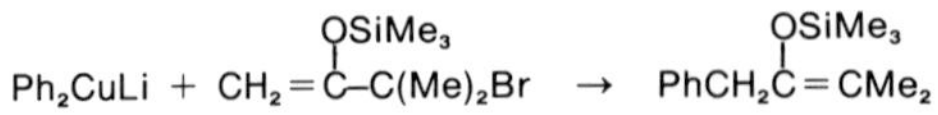

A soln. of startg. 2-siloxyallyl halide in ether added slowly to a soln. of Li diphenylcuprate in ether at 0°, and stirred at this temp. for 5 hrs., and at room temp. for 2 hrs. → product. Y 71%. F.e.s. H. Sakurai et al., Tetrahedron Letters *21,* 2325–8 (1980).

Lithium dialkylcuprates

Replacement of halogen by alkyl groups Hal → R
s. *23,* 831; introduction of labelled methyl groups s. B. Långström, S. Sjöberg, J. Labelled Compounds Radiopharm. *18,* 671–82 (1981).

Cuprous cyanide *CuCN*

Nitriles from halides Hal → CN
from chlorides s. *2,* 728; from bromides, 1-cyanoisochroman, s. A.G. Samodurova, E.A. Markaryan, Sint. Geterotsikl. Soedin. *1981* (12) 63; Chem. Abstr. *95,* 168923 (1981).

Acylcyanides from carboxylic acid halides COHal → COCN
with Me_3SiCN cf. *35,* 583; from acyl iodides with CuCN s. K. Hasse, H.M.R. Hoffmann, Angew. Chem. Intern. Ed. *21,* 83–4 (1982).

Cuprous chloride s.a. under Zn *CuCl*

Cuprous chloride/triethylamine *$CuCl/Et_3N$*

Synthesis of 1,3-diynes from 1-bromoacetylenes C≡CBr → C≡C–C≡C
s. *15,* 634; carbohydrate derivs. s. J.M.J. Tronchet, A.-P. Bonenfant, Helv. Chim. Acta *64,* 1893–901 (1981).

*Cuprous bromide/*n-*butyllithium* *CuBr/*n-*BuLi*

1,2-Dien-4-ynes from allenes and α,β-acetyleneiodides $C{\equiv}CI \rightarrow C{\equiv}C{-}C{=}C{=}C$

804. $$CH_2{=}C{=}CH_2 \rightarrow [CH_2{=}C{=}CHCu] \xrightarrow{IC\equiv CPh} CH_2{=}C{=}CH{-}C{\equiv}CPh$$

A soln. of *n*-butyllithium in hexane added to a stirred soln. of allene in tetrahydrofuran at –60°, stirred for 1 hr., CuBr added, allowed to warm to –20°, startg. 1-iodo-1-alkyne added at –40°, warmed to 0° during 30 min., and poured into satd. aq. NH_4Cl → product. Y 89%. F.e.s. K. Ruitenberg et al., J. Organometal. Chem. *217,* 267–71 (1981).

Cuprous bromide/dimethyl sulfoxide *CuBr/Me₂SO*

Cyclopropanes from ethylene derivs. and 1,1-dihalides 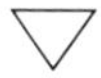

with Zn/CuCl cf. *26,* 876; with CuBr in dimethyl sulfoxide, 1,1-dicarbalkoxycyclopropanes, s. N. Kawabata et al., Bull. Chem. Soc. Japan *54,* 2539–40 (1981).

Cuprous iodide s.a. under Li and $PdCl_2$ *CuI*

Cuprous iodide/lithium *CuI/Li*

2-Allenealcohols from acetyleneoxido compds. ←

s. *30,* 594; 5-amino-2-allenealcohols s. A. Doutheau, A. Saba, J. Gore, Tetrahedron Letters *23,* 2461–4 (1982); with organozinc chlorides and a little $Pd(Ph_3P)_4$ s. H. Kleijn et al., Rec. Trav. Chim. Pays-Bas *101,* 97–101 (1982).

Cuprous iodide/triethylamine *CuI/Et_3N*

α,β-Acetyleneketones from carboxylic acid chlorides $C{\equiv}CH \rightarrow C{\equiv}C{-}COR$

with CuCl/triethylamine cf. *33,* 830; with CuI/triethylamine s. A.S. Zanina et al., Izv. Akad. Nauk SSSR Ser. Khim. *1981,* 1158–9.

Magnesium *Mg*

Coupling of halides 2 RHal → R–R

s. *4,* 780; disaccharide derivs. s. M. Ariatti, J. Žemlička, J. Org. Chem. *46,* 5204–12 (1981).

Aldehydes from Grignard compds. RHal → RCHO

Synthesis with addition of one C-atom

805. 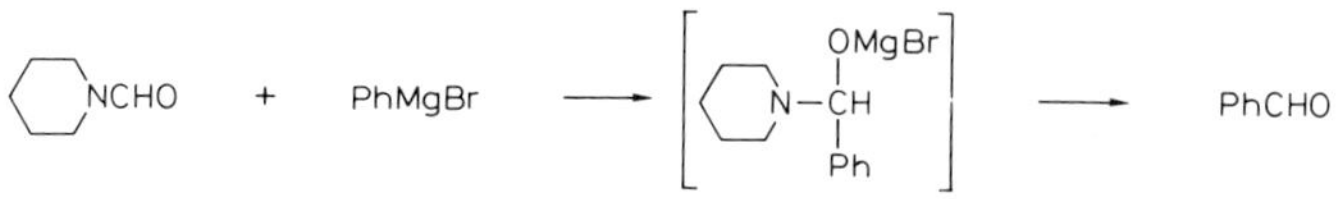

A soln. of N-formylpiperidine in ether slowly added during 2 min. to a stirred soln. of freshly prepared Grignard compd. in the same solvent, the mixture cooled to 0°, stirred at room temp. for 15 min., and worked up with 3 *M* HCl → benzaldehyde. Y 96%. The reaction is mild, inexpensive, and generally applicable. F.e., **also from organolithium compds.,** s. G.A. Olah, M. Arvanaghi, Angew. Chem. Intern. Ed. *20,* 878–9 (1981).

Oxo compds. from Δ^2-imidazolinium salts and halides Hal → CO

aldehydes s. *35,* 571; ketones, also from organolithium compds., s. M.W. Anderson, R.C.F. Jones, J. Saunders, J. Chem. Soc. Chem. Commun. *1982,* 282–3.

Ketones from carboxylic acid chlorides COCl → COR

s. *2,* 735; highly-hindered ketones s. J.-E. Dubois, B.L. Zhang, C. Lion, Tetrahedron *37,* 4189–94 (1981).

Synthesis of ethylene derivs. from ketones CH–CO → C=C–R

s. *12,* 856; 4-vinylphenols s. R.D. Bindal et al., Synthesis *1982,* 405–7.

Magnesium *Mg*

Ketones from hydroxamic acid esters CON(OR) → COR′

806.

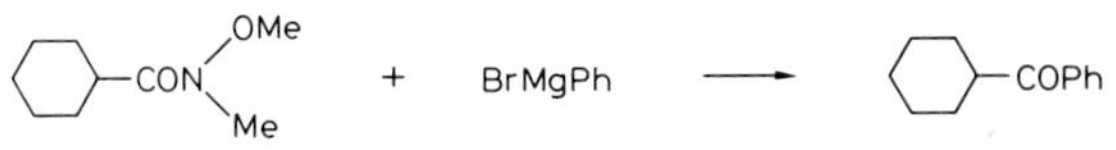

and Grignard compds. 6 Equivalents of phenylmagnesium bromide added to a soln. of N-methoxy-N-methylcyclohexanecarboxamide in dry tetrahydrofuran at low temp., and stirred at 65° for 1 hr. → cyclohexyl phenyl ketone. Y 100%. F.e., also from organolithium compds., and prepn. of **aldehydes** with excess of $LiAlH_4$ or diisobutylaluminium hydride, s. S. Nahm, S.M. Weinret, Tetrahedron Letters *22,* 3815–8 (1981).

Synthesis of acetals from mixed orthoformic acid esters Hal → $CH(OR)_2$
s. *25,* 605; also from organoaluminium halides s. F. Barbot et al., J. Chem. Res. (S) *1981,* 343.

***α,β*-Ethylenealdehydes from vinylogous amidinium salts** RHal → RC=C–CHO

807.

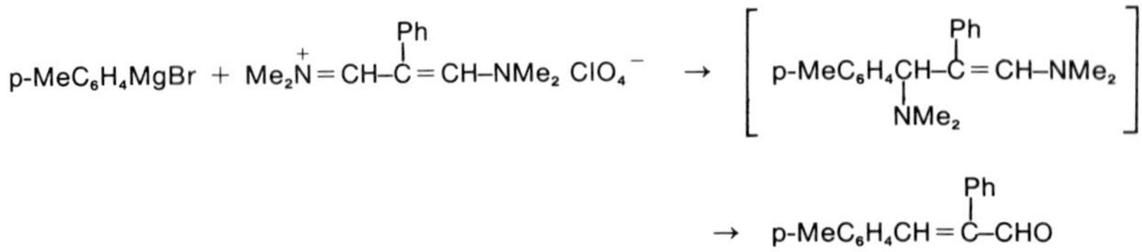

and Grignard compds. [3-(Dimethylamino)-2-phenyl-prop-2-en-1-ylidene]dimethylammonium perchlorate added in one portion to an ice-cold soln. of *p*-toluylmagnesium bromide in tetrahydrofuran, heated for 3 hrs. at reflux, stirred overnight at room temp., the mixture cooled in an ice-water bath, treated with 10% HCl, and stirring continued for 30 min. → 2-phenyl-3-*p*-toluylacrolein. Y 100%. F.e., also with organolithium compds., s. J.T. Gupton, C.M. Polaski, Syn. Commun. *11,* 561–70 (1981).

***α,β*-Acetylenecarboxylic acid esters**
from acetylene derivs. C≡CH → C≡C–COOR
s. *27,* 847; s.a. R.A. Earl, L.B. Townsend, Org. Synth. *60,* 81–7 (1981).

1,5-Dien-3-ols from *α*-halogenoketones ←
via *β,γ*-ethyleneketones

808.

A soln. of 2-chlorocyclohexanone in tetrahydrofuran treated with 1 equivalent of vinylmagnesium bromide at 0° for 2 hrs., propen-2-ylmagnesium bromide added, and the mixture refluxed for 3 hrs. → product. Y 63% (90:10 *cis:trans*). F.e. incl. **1,2-divinylalcohols** s. D.A. Holt, Tetrahedron Letters *22,* 2243–6 (1981).

Prepn. of highly reactive metal powders ←
magnesium s. *27,* 856; also zinc cf. R.D. Rieke et al., J. Org. Chem. *46,* 4323–4 (1981).

Prepn. of Grignard reagents from activated magnesium ←
review s. *14,* 863s*36;* by reduction of $MgCl_2$ with K/KI s. R.D. Rieke et al., Org. Synth. *59,* 85–94 (1979).

Magnesium/lithium *Mg/Li*

Regiospecific synthesis of ethylene derivs. from α-chlorocarboxylic acid chlorides ClC–COCl → C=CR$_2$

809. $ClCH_2COCl + 2\ BrMgCH_2CH=CH_2 \rightarrow CH_2=C(CH_2CH=CH_2)_2$

A soln. of chloroacetyl chloride in tetrahydrofuran added under argon at –60° during 15 min. to an ethereal soln. of allylmagnesium bromide and tetrahydrofuran, stirred for 1.5 hrs., Li powder added, stirring continued at the same temp. for a further 8 hrs., and allowed to warm to room temp. overnight → product. Y 92%. F.e., also less efficiently from α-chlorocarboxylic acid esters, and prepn. of **ethylene and deuterioethylene derivs. from α-chlorooxo compds.** F.e.s. J. Barluenga et al., J. Org. Chem. *46,* 2721–6 (1981).

Magnesium/lithium/liq. ammonia *Mg/Li/NH$_3$*

Reductive arylation of 2,4-dienones C=C–CO → CH–CH–CHAr

810.

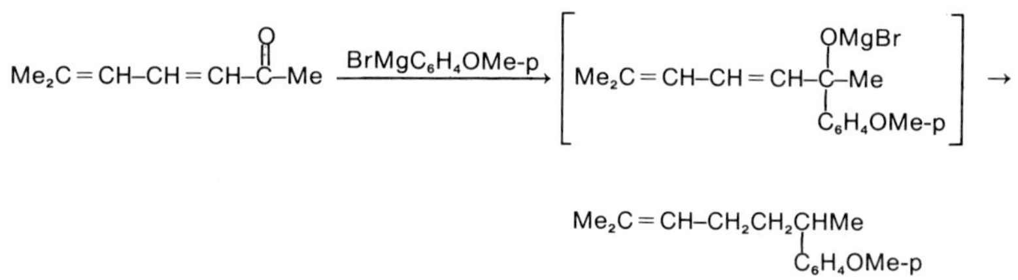

A soln. of *p*-bromoanisole in tetrahydrofuran added slowly at room temp. to active Mg (prepared by refluxing a stirred mixture of $MgCl_2$ and 2 equivalents of K in the same solvent for 2 hrs.) cooled to –60° after 45 min., a soln. of 6-methyl-3,5-heptadien-2-one in the same solvent added dropwise during ca. 3 min., stirring continued for 20 min. at this temp., allowed to warm to room temp. over ca. 45 min., NH_3 distilled into the mixture followed quickly by Li wire, after 15 min. ethanol slowly added during ca. 5 min., vigorous stirring continued for 2–3 hrs., and reaction quenched with excess NH_4Cl → 2-methyl-6-(*p*-methoxyphenyl)-2-heptene. Y 92%. F.e., **also reductive 1,4-cyclohexadien-1-ylation** with organolithium compds., s. J.S. Ryan Zilenovski, S.S. Hall, J. Org. Chem. *46,* 4139–42 (1981).

*Magnesium/*cis-*1,4-dichloro-2-butene* *Mg/ClCH$_2$CH=CHCH$_2$Cl*

Sym. diaryls from Grignard compds. 2 ArHal → Ar–Ar

811. $2\ C_6H_5MgBr \xrightarrow{ClCH_2CH=CHCH_2Cl} C_6H_5\text{–}C_6H_5\ [+CH_2=CH\text{–}CH=CH_2]$

A soln. of *cis*-1,4-dichloro-2-butene in ether added dropwise over 10 min. to a refluxing soln. of phenylmagnesium bromide in the same solvent, and heating continued for a further 3 hrs. with stirring → biphenyl. Y 92%. F.e. and promoters s. S.K. Taylor et al., J. Org. Chem. *46,* 2194–6 (1981).

Magnesium/iron carbonyl *Mg/Fe(CO)$_5$*

Aldehydes from halides Hal → CHO

Synthesis with addition of one C-atom

with $[FeH(CO)_4]^-$ cf. *34,* 822; from Grignard compds. with $Fe(CO)_5$ s. M. Yamashita et al., Bull. Chem. Soc. Japan *55,* 1663–4 (1982).

Magnesium/nickel acetylacetonate *Mg/Ni(acac)$_2$*

Cross-coupling of phenols via monoaryl phosphates Ar–OH → Ar–R

812.

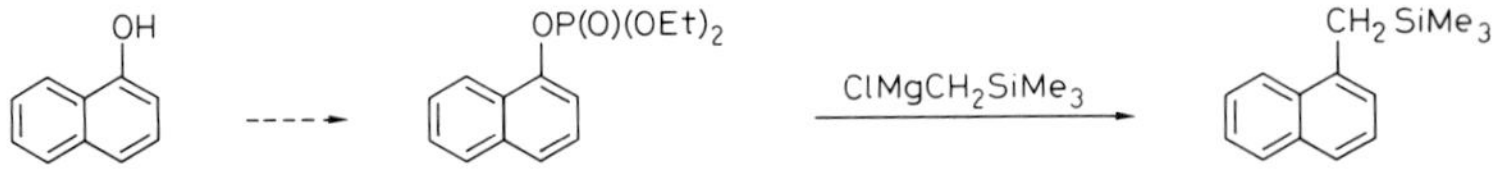

with Grignard compds. 1-Naphthyl diethyl phosphate added to a little Ni acetylacetonate and trimethylsilylmethylmagnesium chloride in ether, and the mixture stirred at room temp. for 10 hrs. → 1-trimethylsilylmethylnaphthalene. Y 79%. F.e., **also with alanes,** s. T. Hayashi et al., Tetrahedron Letters *22,* 4449–52 (1981).

Magnesium/{bis(diphenylphosphino)alkane}dichloronickel(II) ←

Transition metal-catalysed cross-coupling
with Grignard compds. RHal + HalR′ → R–R′
s. *26,* 875s*36;* 2-(ω-hydroxyalkyl)pyridines with {bis(diphenylphosphino)alkane}dichloronickel(II) s. O. Piccolo, T. Martinengo, Synth. Commun. *11,* 497–504 (1981).

Magnesium/{1,2-bis(diphenylphosphino)ethane}dichloronickel(II) ←

Aromatic cross-coupling ArHal → ArAr′
s. *32,* 828s*36;* with {1,2-bis(diphenylphosphino)ethane}dichloronickel(II), arylfurans, s. J. Org. Chem. *47,* 1590–2 (1982); 2-aryl-1,3-dienes with tetrakis(triphenylphosphine)palladium(0) s. S. Nunomoto, Y. Kawakama, Y. Yamashita, Bull. Chem. Soc. Japan *54,* 2831–2 (1981).

Magnesium/{1,1′-bis(diphenylphosphino)ferrocene}dichloronickel(II) ←

Magnesium/{1,1′-bis(diphenylphosphino)ferrocene}dichloropalladium(II) ←

Regiospecific synthesis of ethylene derivs. ←
by cross-coupling of Grignard compds.

813.

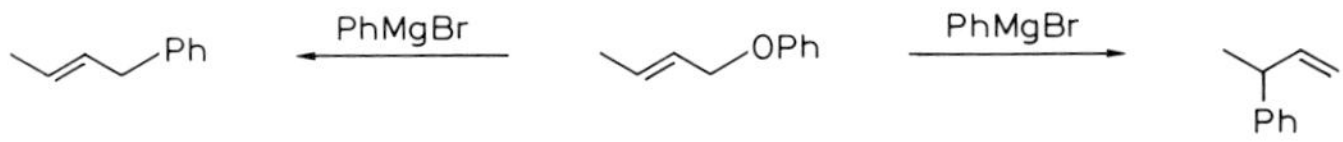

with 2-ethyleneethers. A soln. of phenylmagnesium bromide in ether added to a mixture of startg. allyl ether and a little

{1,1′-bis(diphenylphosphino)ferrocene} dichloropalladium(II),	{1,1′-bis(diphenylphosphino)ferrocene} dichloronickel(II),

and stirred at room temp. for 20 hrs. → product. Y >80%. F.e., **also with β,γ-ethylenehalides, siloxy-2-ethylenes, and 2-ethylenealcohols,** s. T. Hayashi et al., J. Chem. Soc. Chem. Commun. *1981,* 313–4.

Zinc (s.a. under Pd(PPh$_3$)$_4$) *Zn*

Cyclopropanes from ethylene derivs. ▽

Ultrasonication

814.

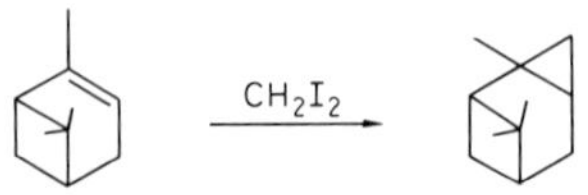

Mossy Zn in 1,2-dimethoxyethane ultrasonically irradiated for 2 hrs. under N_2, (–)-α-pinene added quickly, the mixture heated to reflux, diiodomethane added with stirring, and refluxed for ca. 4 hrs. with continued ultrasonic irradiation → product.

Y 90%. – The method is advantageous in that no chemical activation of Zn is needed, there is no delayed exotherm, crude Zn can be used, and the reaction proceeds smoothly, rapidly, and with high yields. F.e.s. O. Repič, S. Vogt, Tetrahedron Letters *23,* 2729–32 (1982).

Diene synthesis via *o*-quinodimethanes ○
Ultrasonication

815.

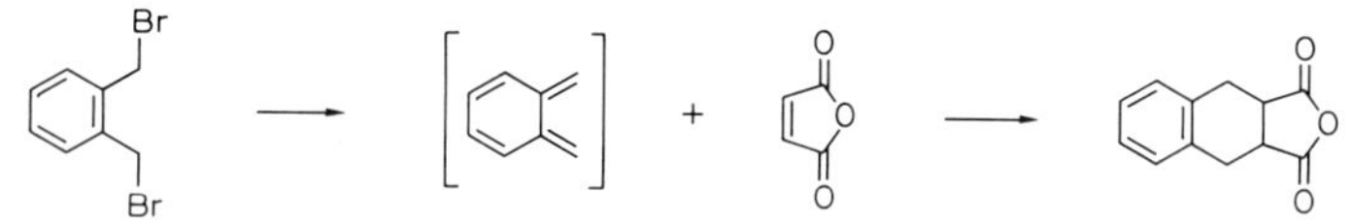

A mixture of *α,α'*-dibromo-*o*-xylene, maleic anhydride, and activated Zn added to peroxide-free dioxane under N_2, and the reaction vessel partially immersed in a common ultrasound laboratory cleaner (150 W) at 20–25° for 15 hrs. → product. Y 89%. – No reaction occurs without sonication. F.e.s. B.H. Han, P. Boudjouk, J. Org. Chem. *47,* 752–4 (1982).

Zinc/cuprous chloride *Zn/CuCl*

Aziridines from azomethines

s. *28,* 818; review of synthesis of 3-membered heterocyclics by carbene insertion s. P.J. Stang, Isr. J. Chem. *21,* 119–27 (1981).

Zinc/iodine *Zn/I_2*

***β*-Acoxy-*γ*-lactones**
from *α*-acoxyketones and *α*-bromocarboxylic acid esters ○
Reformatskii synthesis with 1,3-O-acyl group migration

816.

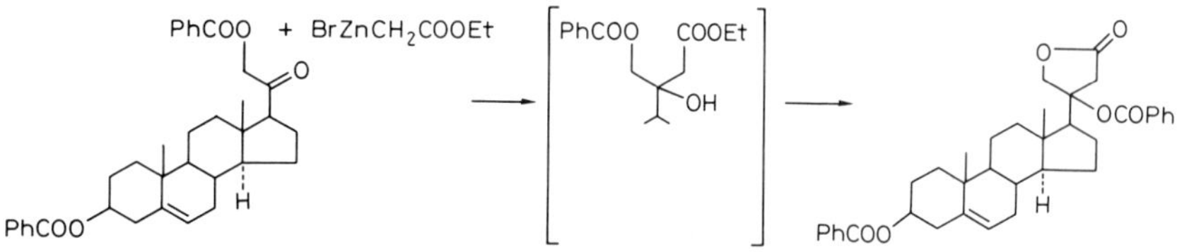

A soln. of 3*β*,21-dibenzoyloxy-5-pregnen-20-one in 1:1 benzene/ether treated with Zn powder and a few crystals of I_2, the mixture heated with stirring at 50°, ethyl bromoacetate added dropwise maintaining a gentle reflux, and refluxing continued for 1 hr. → 3*β*,20*ξ*-dibenzoyloxy-14*α*-card-5-enolide. Y 84%. F.e.s. A.M. Seldes, E.G. Gros, Steroids *39,* 181–90 (1982).

Zinc/nickel chloride *Zn/$NiCl_2$*

Synthesis of ketones from halides and 2-acoxypyridines RHal → RCOR'

817.

A soln. of 2-[6-(2-methoxyethyl)pyridyl] 4-phenylbutanoate and butyl iodide in dimethylformamide added to Zn dust and a little $NiCl_2$ under argon, and the mixture stirred at 50° for 5 hrs. → 1-phenyloctan-4-one. Y 87%. F.e.s. M. Onaka, Y. Matsuoka, T. Mukaiyama, Chem. Letters *1981,* 531–4.

Zinc/tetrakis(triphenylphosphine)palladium(0) *Zn/Pd(PPh$_3$)$_4$*

Regiospecific synthesis of allenes from 3-haloacetylenes C(Hal)C≡C → C=C=CR

818. $Me_2C(Br){-}C{\equiv}CH + PhZnCl \rightarrow Me_2C{=}C{=}CHPh$

Arylallenes. A mixture of startg. m., phenylzinc chloride, and a little tetrakis(triphenylphosphine)palladium(0) in tetrahydrofuran allowed to react at room temp. → product. Y 80–95%. F.e., **also from halogenallenes,** and synthesis of 1,2,4-trienes, 1,2,4,5-tetraenes, and 1,2,4-dienynes, s. K. Ruitenberg et al., Tetrahedron Letters *22,* 1451–2 (1981).

Diethylzinc *Et$_2$Zn*

3-Ethylene- from 2-ethylene-thioethers and 1,1-diiodides via [2.3]-sigmatropic rearrangement of 2-ethylenesulfonium ylids ←

819.

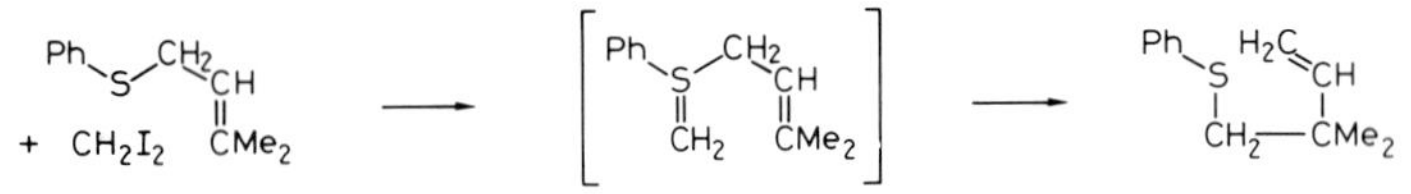

Startg. thioether injected into a stirred soln. of diethylzinc in oxygenated, anhydrous benzene at 55°, diiodomethane added, stirring continued at this temp. for 24 hrs., the mixture cooled to room temp., quenched with acetaldehyde, and worked up after 5 min. → product. Y 80%. – No cyclopropanation was observed. F.e. and Se-analog, **also with 1,3-chirality transfer,** s. Z. Kosarych, T. Cohen, Tetrahedron Letters *23,* 3019–22 (1982).

Organozinc chlorides/tetrakis(triphenylphosphine)palladium(0) *RZnCl/Pd(PPh$_3$)$_4$*

2-Allenealcohols from acetyleneoxido compds. ←

s. *30,* 594s*37*

Zinc carbonate/sodium hydrogen carbonate *ZnCO$_3$/NaHCO$_3$*

Coumarin ring from β-bromo-α,β-ethylenecarboxylic acid esters ○

s. *26,* 878; **synthesis of 3,4-fused coumarins, review,** s. M. Darbarwar, V. Sundaramurthy, Synthesis *1982,* 337–88.

Magnesium bromide s. under t-*BuLi* *MgBr$_2$*

Zinc bromide/tetrakis(triphenylphosphine)palladium(0) *ZnBr$_2$/Pd(PPh$_3$)$_4$*

Cross-coupling via lithium divinylcuprates ←

1,3-dienes cf. *33,* 806s*36;* styrenes s. Tetrahedron Letters *22,* 3851–2 (1981).

Diisobutylaluminium hydride/lithium diisopropylamide *(i-C$_4$H$_9$)$_2$AlH/i-Pr$_2$NLi*

Synthesis of aldehydes from nitriles CH–CN → CR–CHO

One pot conversion

820. $BuCH_2CN + HAl(Bu\text{-}i)_2 \rightarrow [BuCH_2CH{=}NAl(Bu\text{-}i)_2] \xrightarrow{BrC_5H_{11}\text{-}n} BuCH(C_5H_{11}\text{-}n)CHO$

A soln. of diisobutylaluminum hydride in hexane added to a soln. of hexanenitrile in ether at –10° under N$_2$, stirred for 30 min., Li diisopropylamide in ether/hexane, and hexamethylphosphoramide, added sequentially, stirring continued for 1 hr. at room temp., the mixture cooled to –10°, treated with 1-bromopentane, stirring continued at the same temp. for 5 hrs. then at room temp. for 15 hrs., refluxed for 1 hr., treated with 20% H$_2$SO$_4$, and steam distilled → 2-*n*-butylheptanal. Y 85%. F.e., incl. reaction with sec. nitriles, s. H.L. Goering, C.C. Tseng, J. Org. Chem. *46,* 5250–2 (1981).

Silver fluoroborate/triethylamine $AgBF_4/Et_3N$

4-Cycloheptenones from 1,3-dienes ○

and α,α'-dibromoketones cf. *26,* 890; and α-bromoketones with $AgBF_4$/triethylamine s. J. Mann, A.A. Usmani, J. Chem. Soc. Chem. Commun. *1980,* 1119.

Samarium diiodide SmI_2

Sym. α-diketones from carboxylic acid chlorides 2 RCOCl → RCOCOR

821. $$2\ PhCOCl \xrightarrow{SmI_2} PhCOCOPh$$

A soln. of benzoyl chloride in tetrahydrofuran added slowly under N_2 to 2 equivalents of a 0.1 *M* soln. of SmI_2 in the same solvent at room temp., and the yellow mixture worked up after 2 min. → benzil. Y 78%. F.e.s. P. Girard, R. Couffignal, H.B. Kagan, Tetrahedron Letters *22,* 3959–60 (1981).

Polyethyleneglycol ←

18-Crown-6 polyether s. under KCN ←

Polymer-based crown ethers ←

3-Phase catalysis ←

s. *31,* 799s*36;* with immobilized benzo-18-crown-6 cf. A. van Zon, F. de Jong, Y. Onwezen, Rec. Trav. Chim. Pays-Bas *100,* 429–32 (1981); with polymer-based phosphonium salts s.a. S.L. Regen, D. Bolikal, C. Barcelon, J. Org. Chem. *46,* 2511–4 (1981); kinetic study s. M. Tomoi, W.T. Ford, J. Am. Chem. Soc. *103,* 3821–8 (1981); f. catalysts s. V. Janout, P. Čefelín, Collection Czech. Chem. Commun. *47,* 1818–25 (1982).

Quaternary ammonium halides s. under $Co_2(CO)_8$ $R_4N^+X^-$

Benzyltriethylammonium chloride s. under NaOH, K_2CO_3 *and KCN* $[PhCH_2NEt_3]^+Cl^-$

Aliquat 336 s. under NaOH ←

Dodecyltrimethylammonium chloride s. under Tricarbonylnitrosylcobalt $[C_{12}H_{25}NMe_3]^+Cl^-$

Benzoyl peroxide $(PhCOO)_2$

γ-Lactones from ethylene derivs. and α-halogenocarboxylic acids ○

with $RuCl_2(Ph_3P)_3$ cf. *34,* 824; with benzoyl peroxide s. T. Nakano et al., Chem. Letters *1981,* 415–8.

cis-*1,4-Dichloro-2-butene s. under Mg* $ClCH_2CH{=}CHCH_2Cl$

Dimethyl sulfoxide s. under CuBr Me_2SO

Silica gel s. under K SiO_2

Trimethylsilyl cyanide/stannic chloride $Me_3SiCN/SnCl_4$

Tert. nitriles from halides Hal → CN

822.

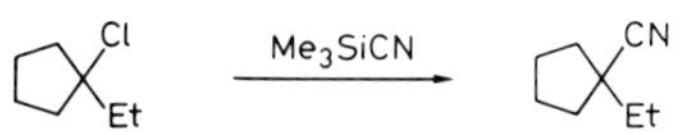

$SnCl_4$ added slowly to a stirred soln. of 1-chloro-1-ethylcyclopentane and trimethylsilyl cyanide in methylene chloride at room temp. under N_2, and allowed to stand for 24–38 hrs. → 1-ethylcyclopentacarbonitrile. Y 82%. F.e.s. M.T. Reetz, I. Chatziiosifidis, Angew. Chem. Intern. Ed. *20,* 1017–8 (1981).

Chlorobis(cyclopentadienyl)hydridozirconium s. under $PdCl_2(PhCN)_2$ — $(C_5H_5)_2Zr(Cl)H$

Stannous fluoride/caesium fluoride — SnF_2/CsF

α,β-Oxido from α,α-dibromo-ketones and aldehydes
Stereospecific ring closure

CO CHal$_2$ → 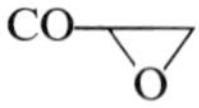

823.

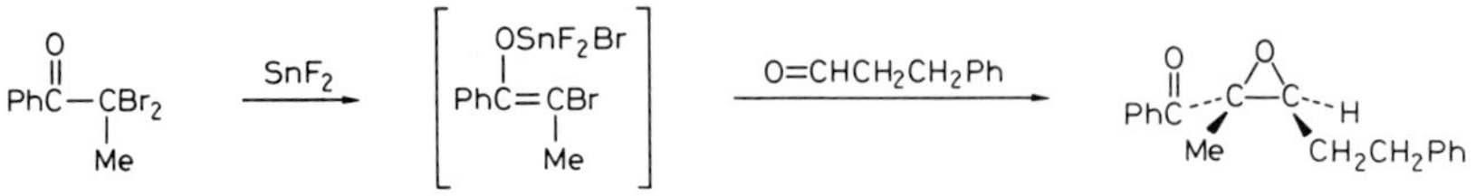

A suspension of SnF_2, α,α-dibromopropiophenone, and 3-phenylpropanal in dimethylformamide stirred at –45 to –55° for 15 hrs., CsF added, and stirring continued at 80° for 2 hrs. → 1,5-diphenyl-2-methyl-*trans*-2,3-epoxypentan-1-one. Y 81%. F.e.s. S. Shoda, T. Mukaiyama, Chem. Letters *1981,* 723–4.

Stannic chloride (s.a. under Me_3SiCN) — $SnCl_4$

Friedel-Crafts 2-acylation of eneurethans — H → COR

824.

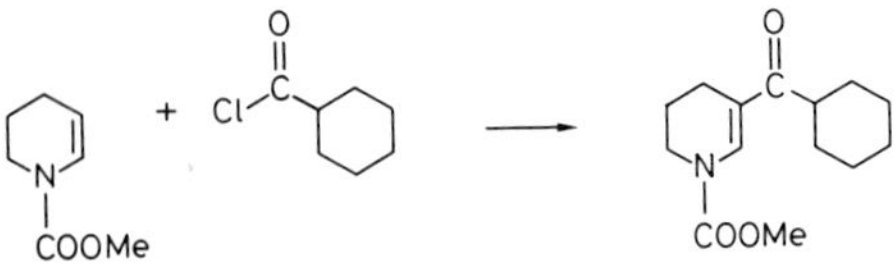

A stirred soln. of startg. acylhalide in methylene chloride treated dropwise with $SnCl_4$ during 5 min. at –60° under N_2, the mixture stirred for 15 min., a soln. of startg. eneurethan in the same solvent added dropwise during 30 min., and allowed to warm to room temp. during 14.5 hrs. → product. Y 85%. F.e., incl. **Vilsmeier formylation,** s. T. Shono et al., Tetrahedron Letters *23,* 1201–4 (1982).

Stannic chloride/triethylamine — $SnCl_4/Et_3N$

Lactams from α-alkylthio-α-chlorocarboxylic acid chlorides and amines via α-(alkylthio)lactams — ○

825.

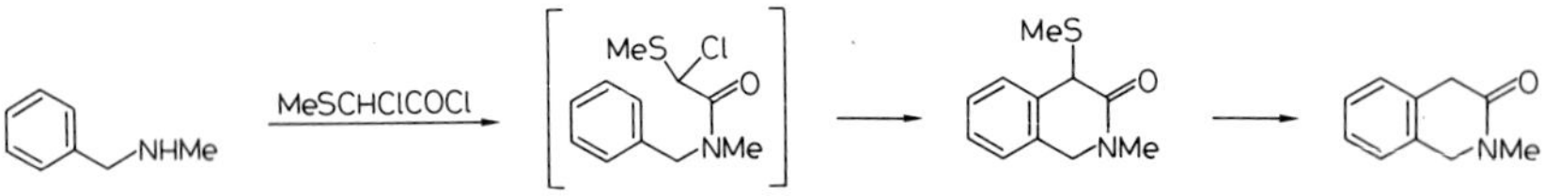

1,4-Dihydroisoquinol-3(2*H*)-ones. A soln. of α-chloro-α-(methylthio)acetyl chloride in dichloromethane added dropwise to a stirred soln. of startg. amine and triethylamine in the same solvent at 0°, stirring continued at room temp. for 15 min., treated with $SnCl_4$, and the mixture stirred for a further 30–50 min. → intermediate (Y 91%) in ethanol heated under reflux with Raney-Ni for 2 hrs. → product (Y 83%). F.e. and **oxindoles, also from α-(alkylthio)carboxylic acid chlorides via Pummerer rearrangement,** s. Y. Tamura et al., Synthesis *1981,* 534–6.

Triphenylphosphine s. under $(MeCOO)_2Pd$ and $PdCl_2(PPh_3)_2$ — Ph_3P

Tri-o-tolylphosphine s. under $(MeCOO)_2Pd$ — Ar_3P

N-Methyl-N-phenylaminotributylphosphonium iodide s. under Li — $[Bu_3PNMePh]^+I^-$

Hexamethylphosphoramide s. under i-Pr_2NLi — $(Me_2N)_3PO$

Diethyl phosphite s. under NaH *$(EtO)_2P(O)H$*

Chromium(II) chloride *$CrCl_2$*

Coupling of halides 2 RHal → R–R

s. *35,* 549; with $CrCl_2$ in stoichiometric amount, non-electrochemical method, cf. R. Sustmann, R. Altevogt, Tetrahedron Letters *22,* 5167–70 (1981).

Iodine s. under Zn *I_2*

Iron pentacarbonyl s. under Mg *$Fe(CO)_5$*

Sodium tetracarbonylferrate(-II) *$Na_2Fe(CO)_4$*

Oxo compds. from halides RHal + R′Hal′ → RCOR′

ketones s. *27,* 869; also aldehydes s. R.G. Finke, T.N. Sorrell, Org. Synth. *59,* 102–12 (1979).

Cobalt carbonyl/sodium hydroxide/ quaternary ammonium halides/irradiation *$Co_2(CO)_8/NaOH/R_4N^+X^-$////*

Carboxylic acids from halides Hal → COOH

Carbonylation

s. *12,* 867s*37*

Tricarbonylnitrosylcobalt/ dodecyltrimethyl ammonium chloride *$Co(CO)_3(NO)/[C_{12}H_{25}NMe_3]^+Cl^-$*

s. *12,* 867s*37*

Dicarbonylbis(triphenylphosphine)nickel/triethylamine *$Ni(CO)_2(Ph_3P)_2/Et_3N$*

Lactones from bromhydrins ○

α-Methylenelactones

with $Ni(CO)_4$ cf. *34,* 823; with dicarbonylbis(triphenylphosphine)nickel/triethylamine s. M.F. Semmelhack, S.J. Brickner, J. Org. Chem. *46,* 1723–6 (1981); with tetrakis(triphenylphosphine)palladium(0) s. L.D. Martin, J.K. Stille, J. Org. Chem. *47,* 3630–3 (1982).

Nickel acetylacetonate s. under Mg *$Ni(acac)_2$*

{Bis(diphenylphosphino)alkane}dichloronickel s. under Mg ←

{1,1′-Bis(diphenylphosphino)ferrocene}dichloronickel(II) s. under Mg ←

Dibromobis(triphenylphosphine)nickel(II) s. under ϟ *$NiBr_2(Ph_3P)_2$*

Nickel chloride s. under Zn *$NiCl_2$*

Palladium/trialkylamine *Pd/R_3N*

Arylation of ethylene derivs. C=CH → C=CAr

s. *27,* 871s*37*

*Palladium(II) acetate/triphenylphosphine/ tri-*n*-butylamine* *$(MeCOO)_2Pd/Ph_3P/$n-Bu_3N*

Lactams from halogenamines ○

Carbonylation

s. *36,* 844; *34,* 826; f.e. incl. carbostyrils s. M. Mori et al., Heterocycles *13* (Special Issue), 329–32 (1979); 1,4-benzodiazepine derivs. s. J. Org. Chem. *47,* 2456–61 (1982).

Palladium(II) acetate/tri-o-tolylphosphine/trialkylamine *$(MeCOO)_2Pd/Ar_3P/R_3N$*

Arylation of ethylene derivs. C=CH → C=CAr

s. *27,* 871s*36;* of N-alkenylphthalimides s. P.Y. Johnson, J.Q. Wen, J. Org. Chem. *46,* 2767–71 (1981); with Pd–graphite/tri-*n*-butylamine, (E)-styrenes and (E),(E)-1,3-dienes, cf. D. Savoia et al., J. Chem. Soc. Chem. Commun. *1981,* 541–2; with palladium black/triethylamine, hetarylation with 4-chloropyrimidines, s. T. Sakamoto et al., Heterocycles *16,* 965–8 (1981); arylation of polyfluoroalkenes with $PdCl_2$/KOAc s. T. Fuchikami, M. Yatabe, I. Ojima, Synthesis *1981,* 365–6.

review of palladium-catalysed vinylation of halides s. R.F. Heck, Org. Reactions *27,* 345–90 (1982).

Tetrakis(triphenylphosphine)palladium(0) s.a. under Zn, RZnCl, and $ZnBr_2$ *$Pd(PPh_3)_4$*

Tetrakis(triphenylphosphine)palladium(0)/zinc *$Pd(PPh_3)_4/Zn$*

β,γ-Ethylenecarboxylic acid esters from α,β-ethylenehalides

Retention of configuration C=C–Hal → C=C–C–COOR

826.

A suspension of tetrakis(triphenylphosphine)palladium(0) in ether treated sequentially with startg. vinylbromide, a soln. of Reformatskii reagent in methylal, and hexamethylphosphoramide, refluxed for 3 hrs., allowed to cool to ambient temp., and hydrolysed with satd. NH_4Cl → product. Y 69% (88% *trans* isomer). F.e., also with a little tetrakis(triphenylphosphine)nickel(0), s. J.F. Fauvarque, A. Jutand, J. Organometal. Chem. *209,* 109–114 (1981).

Palladium(II) chloride/cuprous iodide *$PdCl_2/CuI$*

Acetylene derivs. from halides C≡CH → C≡CR

with $Pd(OAc)_2/PPh_3$ cf. *27,* 851s*36;* with $PdCl_2$ and CuI, *cis*-stilbenes, s. E.B. Merkushev, T.S. Skorozhodova, Zh. Org. Khim. *18,* 355–8 (1982); with a little iodo(phenyl)bis(triphenylphosphine)palladium in the presence of tetra-*n*-butylammonium iodide cf. N.A. Bumagin, I.O. Kalinovsky, I.P. Beletskaya, Izv. Akad. Nauk SSSR Ser. Khim. *1981,* 2836.

Bis(benzonitrile)dichloropalladium(II)/chlorobis(cyclopentadienyl)hydridozirconium *$PdCl_2(PhCN)_2/(C_5H_5)_2Zr(Cl)H$*

Synthesis of ethylene from acetylene derivs. ←

trans-styrenes *cf.* 33, 828; (E)-5-alkenyl-2′-deoxyuridines with bis(benzonitrile)dichloropalladium(II) as catalyst s. P. Vincent, J.-P. Beaucourt, L. Pichat, Tetrahedron Letters *23,* 63–4 (1982).

Dichlorobis(triphenylphosphine)palladium(II)/triphenylphosphine *$PdCl_2(PPh_3)_2/Ph_3P$*

Carboxylic acids from halides Hal → COOH

Carbonylation

with $PdCl_2(PPh_3)_2$ s. *12,* 867s*32;* with added triphenylphosphine in 3:1 water/tributylamine, anthranilic acids, s. D. Valentine, J.W. Tilley, R.A. LeMahieu, J. Org. Chem. *46,* 4614–7 (1981); with tricarbonylnitrosylcobalt and dodecyltrimethylammonium chloride as catalyst (cf. *12,* 867s*36*); arylacetic acids, s. S. Gambarotta, H. Alper, J. Organometal. Chem. *212,* C23–6 (1981); aryl- and α,β-ethylene-carboxylic acids with a little $Co_2(CO)_8$, phase transfer catalysis under irradiation, cf. J.-J. Brunet, C. Sidot, P. Caubere, Tetrahedron Letters *22,* 1013–6 (1981); s.a. F. Francalanci, M. Foà, J. Organometal. Chem. *232,* 59–70 (1982); f. Pd catalysts s. Organometallics *1,* 775–8 (1982).

Dichlorobis(diphenylmethylphosphine)palladium(II) $PdCl_2(MePPh_2)_2$

N,N-Disubst. α-ketocarboxylic acid amides from halides and sec. amines RHal + HN< → RCOCON<

Biscarbonylation

827. $PhBr + 2\,CO + HNEt_2 \rightarrow PhCOCONEt_2$

A mixture of bromobenzene, diethylamine, and a little dichlorobis(diphenylmethylphosphine)palladium(II) contained in a stainless-steel autoclave stirred at 100° for 40 hrs. under CO at an initial pressure of 10 atm. (at room temp.) → product. Y 82%. F.e. and limitations s. F. Ozawa et al., Tetrahedron Letters *23,* 3383–6 (1982); with a stoichiometric amount of alkylpalladium(II) phosphine complex cf. Chem. Letters *1982,* 865–8; s.a. T. Kobayashi, M. Tanaka, J. Organometal. Chem. *233,* C64–6 (1982).

{1,4-Bis(diphenylphosphino)butane}dichloropalladium(II) s. under t-*BuLi* ←

{1,1'-bis(diphenylphosphino)ferrocene}dichloropalladium(II) s.a. under Mg ←

{1,1'-bis(diphenylphosphino)ferrocene}dichloropalladium(II)/triethylamine ←

α,β-Acetyleneketones from terminal acetylene derivs. and halides RHal + HC≡CR' → RCOC≡CR'

Carbonylation

828. $PhI + HC{\equiv}CPh \xrightarrow{CO} PhCOC{\equiv}CPh$

A mixture of iodobenzene, phenylacetylene, triethylamine, and a little [1,1'-bis(diphenylphosphino)ferrocene]dichloropalladium(II) heated with stirring in a stainless steel autoclave under 20 atm. CO at 120° for 4 hrs. → 1,3-diphenylprop-2-ynone. Y 85.6%. F.e.s. T. Kobayashi, M. Tanaka, J. Chem. Soc. Chem. Commun. *1981,* 333–4.

Via intermediates *v.i.*

4-Alkylation of pyridines H → R

s. *37,* 659; *37,* 779

β-Alkylation of α,β-ethylenecarbonyl compds. via α-sulfinylcarbonyl compds.

829.

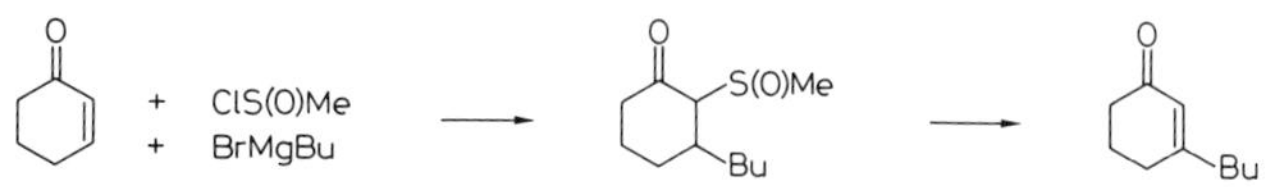

β-Subst. α,β-ethyleneketones. 2-Cyclohexenone treated with butylmagnesium bromide and a little CuCl in ether at –30° for 40 min., the soln. obtained added to methanesulfinyl chloride in ether at –78°, and allowed to warm slowly to 0° → 3-butyl-2-methanesulfinylcyclohexanone, refluxed in toluene with $CaCO_3$ for 7 hrs. → 3-butyl-2-cyclohexenone. Overall Y 76%. F.e., also β-alkylation of α,β-ethylenecarboxylic acid esters, s. T. Fujisawa et al., Chem. Letters *1981,* 1159–60.

Replacement of chlorine by cyano via ammonium groups Cl → CN

830.

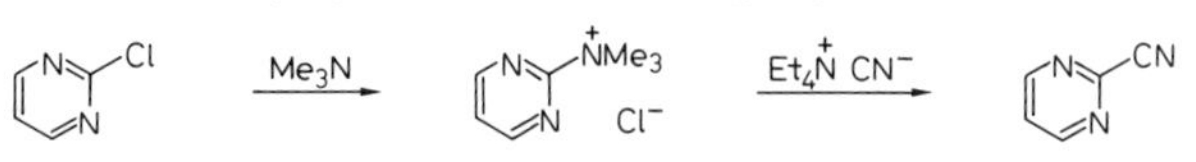

Cyanopyrimidines. An ice-cooled soln. of 2-chloropyrimidine in benzene treated with 20% trimethylamine in the same solvent during 1 hr., and warmed to room temp. → intermediate (Y 96%), suspended in methylene chloride, a soln. of tetraethylammonium cyanide in the same solvent added, and stirred at room temp. for 15–30 min.

→ 2-cyanopyrimidine (Y 84%). F.e. and **cyanoquinazolines** s. K. Hermann, G. Simchen, Ann. Chem. *1981,* 333–41; *α,β*-ethylene nitriles s. H.E. Zimmerman, R.J. Pasteris, J. Org. Chem. *45,* 4864–76 (1980).

Δ^2-Isoxazolines from nitrile oxides and ethylene derivs.

s. *16,* 888; 5-arylthio-Δ^2-isoxazolines s. G. Bianchetti et al., Gazz. Chim. Ital. *111,* 315–23 (1981).

Sulfur ↑ CC⇂↾S

Without additional reagents *w.a.r.*

Cyclopropane ring from ethylene derivs. and S-ylids

1,2-Diacylcyclopropanes

s. *23,* 869s*36;* s.a. Indian J. Chem. *20B,* 168–9 (1981).

trans-**Chrysanthemic acid esters**

s. *23,* 869; **review of synthesis of pyrethroid acids** s. D. Arlt, M. Jautelat, R. Lantzsch, Angew. Chem. Intern. Ed. *20,* 703–22 (1981).

1-Vinyl-Δ^2-1,2,3-triazolines from enazides

831.

Startg. vinyl azide added to 2 equivalents of dimethylsulfoxonium methylid in dimethyl sulfoxide, and the mixture stirred at room temp. for 12 hrs. → product. Y 95%. F.e., also flash vacuum pyrolysis to **1-vinylaziridines,** s. A. Hassner et al., Tetrahedron Letters *22,* 1863–6 (1981).

Azetidines from aziridines

and sulfonium ylids cf. *32,* 843; and dimethylsulfoxonium methylid, N-sulfonylazetidines, s. U.K. Nadir, V.K. Koul, J. Chem. Soc. Chem. Commun. *1981,* 417–8.

*Sodium hydroxide/tetra-*n*-butylammonium bromide* *NaOH/*n*-Bu$_4$N$^+$ Br$^-$*

2-Aminooxazoles from 1-sulfonylcarbodiimides and aldehydes

832.

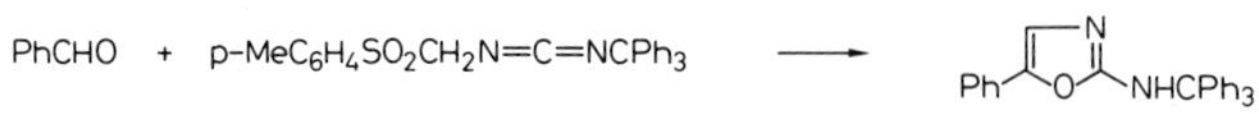

30% aq. NaOH added to a soln. of N-(tosylmethyl)-N′-(triphenylmethyl)carbodiimide, benzaldehyde, and tetra-*n*-butylammonium bromide in dichloromethane, and the mixture stirred for 1 hr. at room temp. → 5-phenyl-2-[(triphenylmethyl)amino]-oxazole. Y 78%. F.e. and methods s. A.M. van Leusen et al., J. Org. Chem. *46,* 2069–72 (1981).

Potassium tert-*butoxide* *KOBu*-t

Sulfonylcyclopropanes from *α,β*-ethylenesulfones and sulfonium salts

s. *26,* 898; f.e.s. D.B. Reddy, B. Sankaraiah, T. Balaji, Indian J. Chem. *19B,* 563–6 (1980).

4-Alkylthiopyridines from *α*-ketoketene mercaptals and ketones

833.

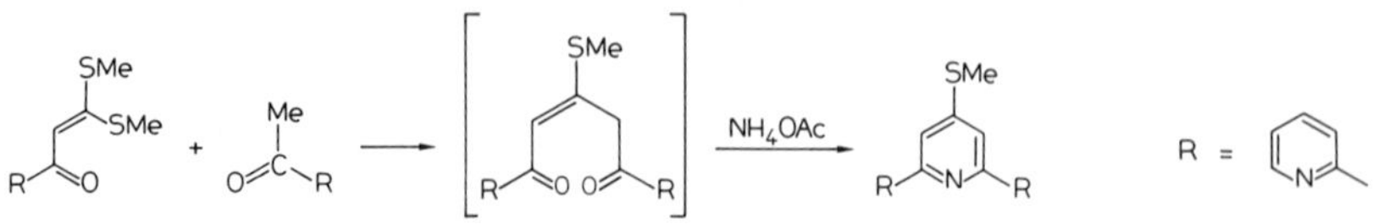

3,3-Bis(methylthio)-1-(2-pyridyl)-2-propen-1-one added to a soln. of 2-acetylpyridine and K *tert*-butoxide in tetrahydrofuran, stirred for 3 hrs. at room temp., ammonium acetate and glacial acetic acid added, and heated for 2 hrs. with continuous removal of solvent → 2,6-di(2-pyridyl)-4-(methylthio)pyridine. Y 79%. F.e. and **pyrylium salts** s. K.T. Potts et al., J. Am. Chem. Soc. *103,* 3585–6 (1981); isolation of the intermediate **3-alkylthio-2-ene-1,5-diones** s. ibid. *103,* 3584–5 (1981), s.a. J. Org. Chem. *47,* 3027–38 (1982).

Lithium diisopropylamide *i-Pr₂NLi*

Synthesis of 1,3-dienes from α-cyclopropylsulfones
Regio- and geo-specific ring opening

834.

(E,E)-2,4-Dienecarboxylic acid esters. A soln. of startg. cyclopropylsulfone in 10:1 tetrahydrofuran/hexamethylphosphoramide added at –78° tp 0.1 equivalent to Li diisopropylamide in tetrahydrofuran, the mixture stirred at the same temp. for 30 min., warmed to –40°, a 4-fold excess of methyl iodide added after 10 min., and stirring continued for 30–60 min. → product. Y 76%. F.e., also without alkylation, **and from 1-cyclopropylthioethers,** s. A. Tomažič, E. Ghera, Tetrahedron Letters *22,* 4349–52 (1981).

Cyanocyclopropanes from α,β-ethylenesulfones and nitriles
Stereospecific ring closure

835.

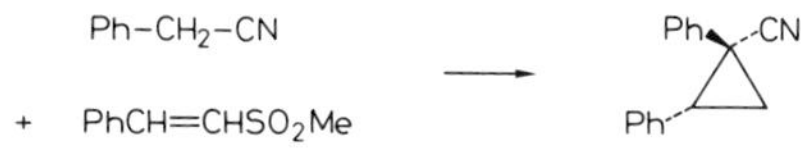

Startg. nitrile added to a soln. of Li diisopropylamide in tetrahydrofuran at 0–5°, the mixture stirred at –60 to –70° for 2–3 hrs., the vinyl sulfone added, the reaction allowed to continue for 4 hrs., kept at room temp. overnight, and refluxed for 5 hrs. → *r*-1-cyano-1,*c*-2-diphenylcyclopropane. Y 83%. F.e.s. T. Agawa et al., J. Chem. Soc. Perkin Trans. I *1981,* 751–5.

1,4-Dimethoxynaphthalenes from 3-sulfonylphthalides
s. *35,* 596; application to daunomycinone synthesis s. J. Am. Chem. Soc. *103,* 6378–86 (1981); anthraquinone ring s. R.A. Russell, R.N. Warrener, J. Chem. Soc. Chem. Commun. *1981,* 108–10.

Lithium diocyclohexylamide/mercuric perchlorate *R₂NLi/Hg(ClO₄)₂*

2-Alkylthio-1,3-benzodithioles CO → C(OH)COOR
as carbalkoxyl carbanion equivalents
α-Hydroxycarboxylic acid esters from ketones

836.

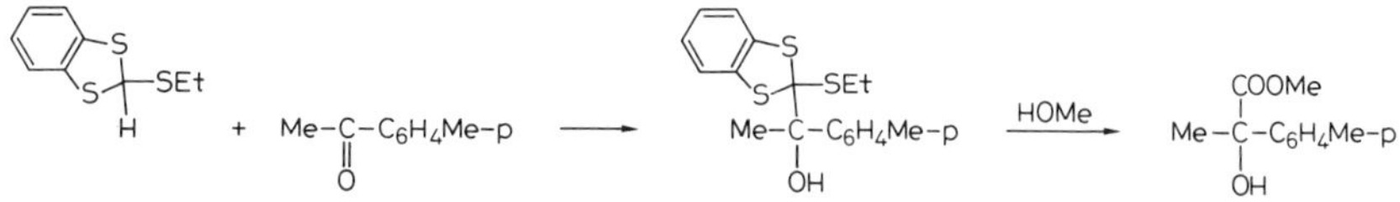

A soln. of 4-methylbenzophenone in dry tetrahydrofuran added to a stirred soln. of Li dicyclohexylamide and 2-ethylthio-1,3-benzodithiole in the same solvent, and the mixture stirred for 2 hrs. at –78° under N_2 → intermediate adduct, treated with $Hg(ClO_4)_2$ in methanol for 25 min. at room temp. with stirring → methyl 4-methylbenzilate. Y 91%. F.e. and limitations s. K. Fuji et al., Synth. Commun. *11,* 209–15 (1981).

Potassium cyanide/dicyclohexyl-18-crown-6 polyether ←

Polarity reversal of electron-deficient ethylene derivatives $CH-CSO_2R \rightarrow C(CN)=CH$

***α,β*-Ethylene-nitriles from -sulfones with position interchange**

837.

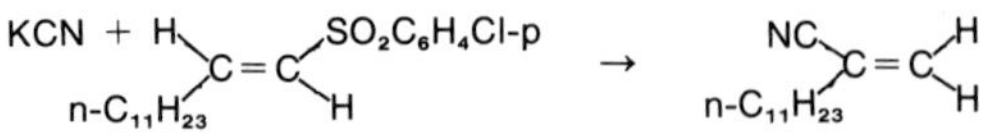

***α*-Methylenenitriles.** A mixture of startg. sulfone, a little dicyclohexyl-18-crown-6 polyether, powdered KCN, and a little *methylene blue* as radical inhibitor in *tert*-butyl alcohol refluxed with stirring under argon for 2.5 hrs. → product. Y 68%. – This is part of a 2-step synthesis of *α*-methylenenitriles **from aldehydes.** F.e. and trisubst. ethylene derivs. s. D.F. Taber, S.A. Saleh, J. Org. Chem. *46,* 4817–9 (1981).

Piperidine $C_5H_{11}N$

***α,β*-Ethylene-*γ*-hydroxynitriles from oxo compds. via vinylogous Pummerer rearrangement** $CHCO \rightarrow C(OH)-C=CH-CN$

Synthesis with addition of two C-atoms

838.

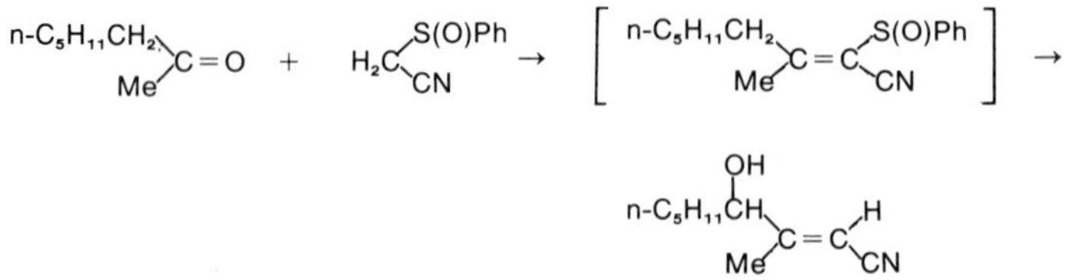

Phenylsulfinylacetonitrile treated with 1.2 equivalents 2-octanone and 1 equivalent piperidine in methanol at room temp. for 10 hrs. → 4-hydroxy-3-methyl-2-nonenenitrile. Y 84%. F.e. and reductive ring closure to furans s. J. Nokami et al., Tetrahedron Letters *22,* 4489–90 (1981).

Lithium dialkylcuprates $LiCuR_2$

Synthesis of ketones from thiolic acid 2-pyridyl esters $COSR \rightarrow COR'$

s. *37,* 188

Magnesium/nickel acetylacetonate $Mg/Ni(acac)_2$

Synthesis of ethylene derivs. from ethylenesulfones and Grignard compds. $C=C-SO_2R \rightarrow C=CR'$

from *β,γ*-ethylenesulfones with Cu acetylacetonate as catalyst cf. *36,* 857; from *α,β*-ethylenesulfones, geospecific synthesis of trisubst. ethylene derivs. with Ni acetylacetonate, s. Tetrahedron Letters *23,* 2469–72 (1982).

Magnesium/{bis(diphenylphosphino)alkane}dichloronickel(II) ←

Synthesis of ethylene derivs. from thioenolethers $C=C-SR \rightarrow C=CR'$

with $NiCl_2(PPh_3)_2$ cf. *29,* 854s*36;* with {bis(diphenylphosphino)alkane}dichloronickel(II) s. V. Fiandanese et al., J. Chem. Soc. Chem. Commun. *1982,* 647–9.

Magnesium/{1,1′-bis(diphenylphosphino)ferrocene}dichloropalladium(II) ←

Cross-coupling of 2-alkylthio-Δ^2-oxazolines with Grignard compds. $SR \rightarrow Ar$

2-Aryl-Δ^2-oxazolines

839.

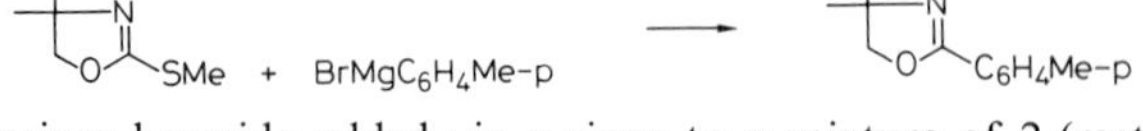

p-Tolylmagnesium bromide added via syringe to a mixture of 2-(methylthio)-4,4-dimethyl-2-oxazoline and a little [1,1′-bis(diphenylphosphino)ferrocene]dichloropalladium(II) in dry ether under N_2, and stirred at room temp. for 3 hrs. → 2-(4-methylphenyl)-4,4-dimethyl-2-oxazoline. Y 94%. F.e. and catalysts s. L.N. Pridgen, L.B. Killmer, J. Org. Chem. *46,* 5402–4 (1981).

Zinc iodide s. under Me_3SiCN — ZnI_2

Dicyclohexyl-18-crown-6 polyether s. under KCN — ←

*Tetra-*n-*butylammonium bromide s. under NaOH* — n-$Bu_4N^+Br^-$

Trimethylsilyl cyanide/zinc iodide — Me_3SiCN/ZnI_2

Replacement of sulfinyl by cyano groups — S(O)R → CN

Cyanomethyl ethers

840. n-$C_6H_{13}CH(Me)OCH_2S(O)Me$ + Me_3SiCN → n-$C_6H_{13}CH(Me)OCH_2CN$

A soln. of 2-octanol methylsulfoxymethyl ether in dichloromethane treated with trimethylsilyl cyanide and ZnI_2, and the mixture allowed to react at 25° for 3 hrs. → 2-octanol cyanomethyl ether. Y 95%. F.e.s. J.A. Schwindeman, P.D. Magnus, Tetrahedron Letters *22,* 4925–8 (1981).

Mercuric perchlorate s. under Lithium dicyclohexylamide — $Hg(ClO_4)_2$

Nickel acetylacetonate s. under Mg — $Ni(acac)_2$

{Bis(diphenylphosphino)alkane}dichloronickel(II) s. under Mg — ←

{1,1′-Bis(diphenylphosphino)ferrocene}dichloropalladium(II) s. under Mg — ←

Remaining Elements ↑ — CC⇂↿Rem

Without additional reagents — *w.a.r.*

Ketones from carboxylic acid chlorides and stannanes — COCl → COR

s. *34,* 862; *α,β*-acetyleneketones without catalyst s. G. Himbert, L. Henn, Tetrahedron Letters *22,* 2637–40 (1981); 2-acyl-quinolines and -pyridines s. Y. Yamamoto, A. Yanagi, Heterocycles *19,* 41–4 (1982); with bis(π-allylpalladium chloride) cf. Zh. Org. Khim. *18,* 1131–6 (1982); Izv. Akad. Nauk. SSSR Ser. Khim. *1981,* 1433.

Wittig synthesis of enolethers from formic acid esters — OCHO → OCH=C

841. CH_2=$CHCH_2OCHO$ + Ph_3P=C(Me)COOEt → CH_2=$CHCH_2OCH$=C(Me)COOEt

Allyl enolethers. A mixture of (carbethoxyethylidene)triphenylphosphorane and allylformate in toluene heated at 90° for 14 hrs. → allyl 2-carbethoxy-1-propenyl ether. Y 60%. F.e. and sigmatropic rearrangement of product s. M. Suda, Chem. Letters *1981,* 967–70.

Thio-Wittig synthesis — CS → C=C

enelactams cf. *33,* 849; enacylamines s. M. Slopianka, A. Gossauer, Ann. Chem. *1981,* 2258–65.

Ketenimines from alkylidenephosphoranes and isocyanates — N=C=O → N=C=C

s. *30,* 611; N-acylketenimines s. L. Capuano, K. Djokar, Chem. Ber. *114,* 1976–9

(1981); vinylketenimines s. Ann. Chem. *1982,* 80–6; **from phosphonic esters** with NaH, organothioketenimines, cf. J. Motoyoshiya, I. Yamamoto, H. Gotoh, J. Chem. Soc. Perkin Trans. I *1981,* 2727–30.

Claisen rearrangement of allyl enolethers under mild conditions ←
Synthesis of 4-ethylenealcohols from trialkylalanes

842.

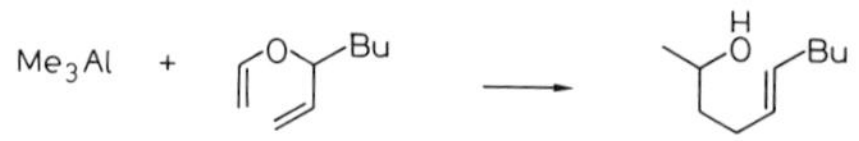

A soln. of trimethylaluminium in hexane added under argon to a soln. of 1-butyl-2-propenyl vinyl ether in 1,2-dichloroethane at 25°, and the mixture stirred for 15 min. → product. Y 91% (E:Z 47:53). F.e., **also reductive Claisen rearrangement** with triisobutylaluminium, and prepn. of **γ,δ-ethylenealdehydes** with diethylaluminium phenylmercaptide or diethylaluminium chloride-triphenylphosphine, s. K. Takai et al. Tetrahedron Letters *22,* 3985–8 (1981).

C-Glycosides by Wittig synthesis ←
s. *32,* 856; f.e. and comparison with the Knoevenagel-Doebner reaction s. P.M. Collins, W.G. Overend, T.S. Shing, J. Chem. Soc. Commun. *1982,* 297–8.

Cyclopropanes from ethylene derivs. and metal-carbene complexes ▽
with a tungsten-carbene complex cf. *33,* 873; with an iron-carbene complex, stereospecificity, s. M. Brookhart et al., J. Am. Chem. Soc. *102,* 7802–3 (1980).

1,1-Dichlorocyclopropenes from acetylene derivs.

843.

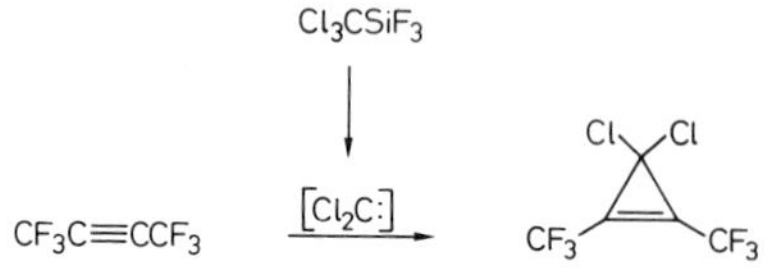

A mixture of hexafluorobut-2-yne and trifluoro(trichloromethyl)silane as *source of dichlorocarbene* heated at 6.5 atm. in a sealed Pyrex tube at 140° for 12 hrs. → 3,3-dichloro-1,2-bis(trifluoromethyl)cyclopropene. Y 93%. F.e.s. J.M. Birchall et al., J. Chem. Soc. Perkin Trans. I *1981,* 2080–6.

Benzene ring from 2-pyrones and acetylene derivs. ○
s. *22,* 877; also via acetylene derivs. prepd. *in situ* from 1,2,3-selenadiazoles s. H. Meier et al., Ann. Chem. *1982,* 914–23.

Pyrrolizin-3-ones from 2-acylpyrroles and cumulated ylids
with phosphoranylideneketenes cf. *31,* 865; with arsoranylideneketenes s. H.J. Bestmann, R.K. Bansal, Tetrahedron Letters *22,* 3839–42 (1981).

Irradiation s. under Li^+ ⫻

Sodium hydride *NaH*

Ketenimines from isocyanates and phosphonic esters N=C=O → N=C=C
Organothioketenimines
s. *30,* 611s*37*

Carboxylic acids from aldehydes CHO → CH_2COOH
via α-amino-α,β-ethylenephosphonic esters
Synthesis with addition of one C-atom
s. *24,* 875; f.e., incl. benzothiophene-2-acetic acid, s. C.R. Degenhardt, Synth. Commun. *12,* 415–21 (1982).

Sodium hydride *NaH*

α-Amino-α,β-ethylenephosphonic esters CHO → CH=C(NR_2)–P(O)($OR)_2$
from 1-amino-1,1-di-phosphonic esters and aldehydes
Horner synthesis

844. 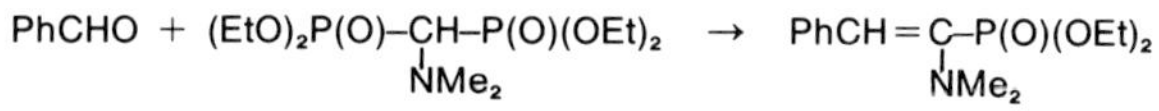

The startg. bis(phosphonic ester) added dropwise with stirring to a suspension of NaH in absolute dioxane at room temp. until reaction complete, benzaldehyde added in one portion, and the mixture heated for 1–2 hrs. at 100° → product. Y 67%. F.e.s. B. Costisella, I. Keitel, H. Gross, Tetrahedron *37,* 1227–32 (1981).

(E)-2-Ethyleneamines
from enephosphonium salts and aldehydes CHO → CH=C–C–NR_2
via (E)-2-ethylenephthalimides

845. 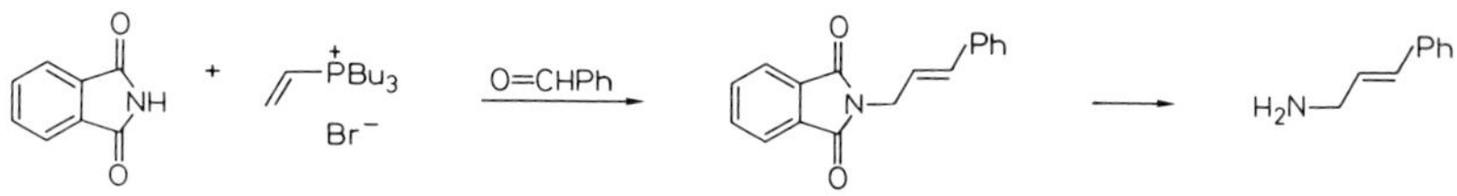

Benzaldehyde, vinyltributylphosphonium bromide, and phthalimide added sequentially to a suspension of NaH in tetrahydrofuran, and the mixture stirred at 25–65° for 3–20 hrs. → intermediate (Y 83%) treated with hydrazine in the usual manner → (E)-product (Y 95%). F.e.s. A.I. Meyers, J.P. Lawson, D.R. Carver, J. Org. Chem. *46,* 3119–23 (1981).

Oxido from oxo compds. and selenonium salts CO → (epoxide)

846. 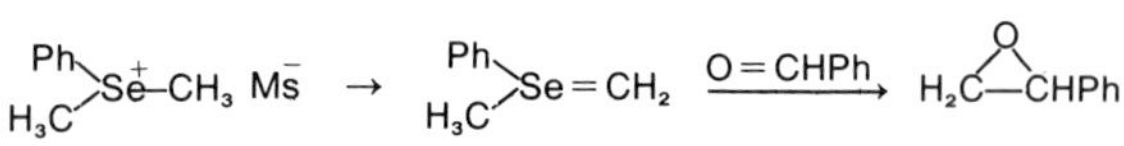

A mixture of dimethyl(phenyl)selenonium methanesulfonate and benzaldehyde in dry dimethylformamide added dropwise during 30 min. to a stirred suspension of NaH in dry tetrahydrofuran under N_2 at 0°, warmed to room temp. for 4 hrs., and heated at 50° for 30 min. → product. Y 94%. F.e.s. K. Takaki, M. Yasumura, K. Negoro, Angew. Chem. Intern. Ed. *20,* 671–2 (1981).

Sodium hydroxide s.a. under PhLi *NaOH*

Sodium hydroxide/tetrabutylammonium hydrogen sulfate/triphenylphosphine *NaOH/$Bu_4N^+HSO_4^-$/PPh_3*

Synthesis of ethylene derivs. from π-allyl palladium chloride complexes ←
with NaH cf. *33,* 715; with NaOH and tetrabutylammonium hydrogen sulfate as phase transfer catalyst in the presence of triphenylphosphine s. B. Åkermark, A. Ljungqvist, M. Panunzio, Tetrahedron Letters *22,* 1055–8 (1981).

Sodium hydroxide/benzyltriethylammonium chloride *NaOH/[$PhCH_2NEt_3]^+Cl^-$*

Eneselenides
from α-(arylseleno)phosphonic esters and aldehydes CHO → CH=CSeR
Horner synthesis

847. PhCHO + $(EtO)_2P(O)CH_2SePh$ → PhCH=CHSePh

A mixture of the startg. phosphonate, benzaldehyde, 50% aq. NaOH, and a little benzyltriethylammonium chloride in methylene chloride stirred at room temp. for 2.3 hrs. → product. Y 63% (*trans:cis* 9:1). F.e., **also from α-(arylseleno)phosphonium salts** by Wittig synthesis, s. J.V. Comasseto, C.A. Brandt, J. Chem. Res. (S) *1982,* 56–7.

Potassium hydroxide/benzyltriethylammonium chloride *KOH/[PhCH₂NEt₃]⁺Cl⁻*

Carboxylic acids from oxo compds.
via α-amino-α,β-ethylenenitriles CO → C=C(NR₂)CN
Synthesis with addition of one C-atom

848. $PhCHO + (EtO)_2P(O)CH(NMe_2)CN \rightarrow PhCH{=}C(NMe_2){-}CN \rightarrow PhCH_2COOH$

A soln. of diethyl 1-dimethylamino-1-cyanomethanephosphonate and benzaldehyde in methylene chloride added to a vigorously stirred mixture of 50% KOH, methylene chloride, and a little benzyltriethylammonium chloride, and heated under reflux for 1–3 hrs. → cyanoenamine (Y 64%), heated briefly with concd. HCl → phenylacetic acid (Y 80%). F.e. and methods s. B. Costisella, H. Gross, Tetrahedron *38,* 139–45 (1982).

Sodium ethoxide s. under Pd(Ph₃P)₄ *NaOEt*

Potassium tert-*butoxide* *KOBu*-t

Horner synthesis
with α,β-ethylene-β-phosphonylmercaptals CHO → CH=C–C=C(SR)₂
Vinylketene mercaptals from aldehydes

849. $PhCHO + (EtO)_2P(O)CH{=}CH{-}CH(SEt)_2 \rightarrow PhCH{=}CH{-}CH{=}C(SEt)_2$

$\dashrightarrow PhCH_2CH{=}CHCHO$

A soln. of 3-diethylphosphonoacrolein diethyl dithioacetal in tetrahydrofuran added dropwise to a cooled (dry ice/CCl_4), stirred suspension of K *tert*-butoxide in the same solvent under inert atmosphere, after 15 min. a soln. of benzaldehyde in tetrahydrofuran added, and stirring continued for 15 min. → product. Y 75%. – This is part of a multistep **synthesis of α,β-ethylenealdehydes with addition of three C-atoms.** F.e.s. G. Just, P. Potvin, G.H. Hakimelahi, Can. J. Chem. *58,* 2780–3 (1980).

Alkyllithium/tetramethylethylenediamine *RLi/TMEDA*

Synthesis of sulfines from enesilanes C=S=O
s. *34,* 843; review of sulfines s. Rec. Trav. Chim. Pays-Bas *101,* 1–27 (1982).

Methyllithium *MeLi*

Sequential Michael additions ○
with subsequent intramolecular Wittig synthesis
One-pot total steroid synthesis
from 1-tetralones, cyclopent-2-enones, and enephosphonium salts

850.

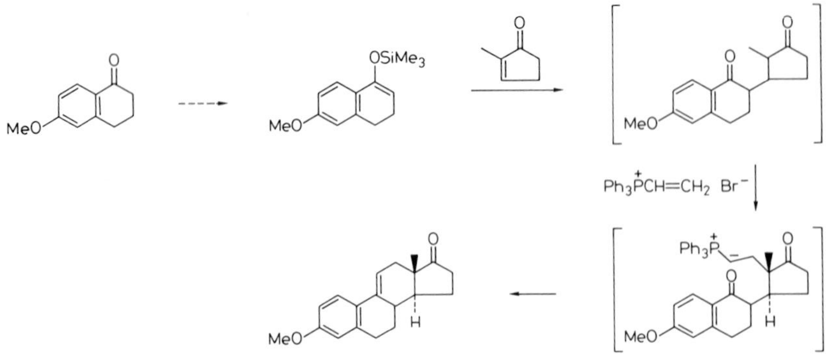

Methyllithium in ether added via syringe to a soln. of startg. tetralone silyl ether in tetrahydrofuran at 0° under N_2, stirred at room temp. for 60 min., cooled to –30°, treated dropwise with 2-methyl-2-cyclopentenone in tetrahydrofuran, after 30 min. dil. with further tetrahydrofuran and dimethylformamide, warmed to room temp., vinyltriphenylphosphonium bromide in dimethylformamide added dropwise over 4–6 hrs., and stirring continued overnight → (±)-9,11-dehydro-8-epiestrone methyl ether. Y 21%, based on reacted tetralone. G.H. Posner, J.P. Mallamo, A.Y. Black, Tetrahedron *37,* 3921–6 (1981).

Butyllithium *BuLi*

Horner synthesis with 1,2-dihydroisoquinoline-1-phosphonic esters CO → C=C

851.

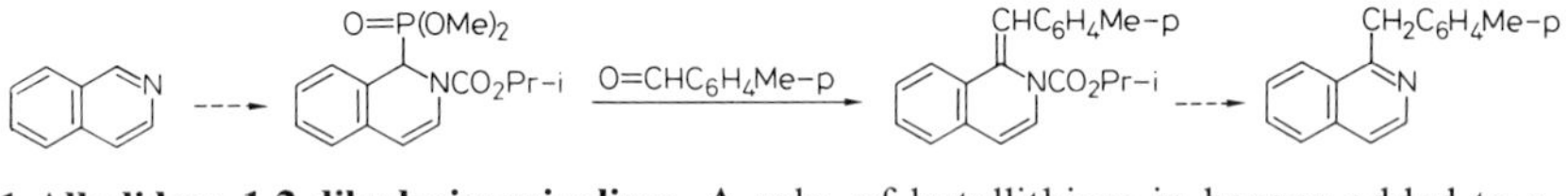

1-Alkylidene-1,2-dihydroisoquinolines. A soln. of butyllithium in hexane added to a soln. of dimethyl 2-isopropoxycarbonyl-1,2-dihydroisoquinoline-1-phosphonate in tetrahydrofuran below –70°, *p*-tolualdehyde added with stirring, and the mixture allowed to warm to room temp. → product. Y 63%. – This is the key step in a multistage **1-alkylation of isoquinolines.** F.e.s. K. Akiba et al., Tetrahedron Letters *22,* 4977–80 (1981).

n-*Butyllithium* n-*BuLi*

Synthesis of hindered aldehydes from ketones via enazomethines CO → CRCHO

852.

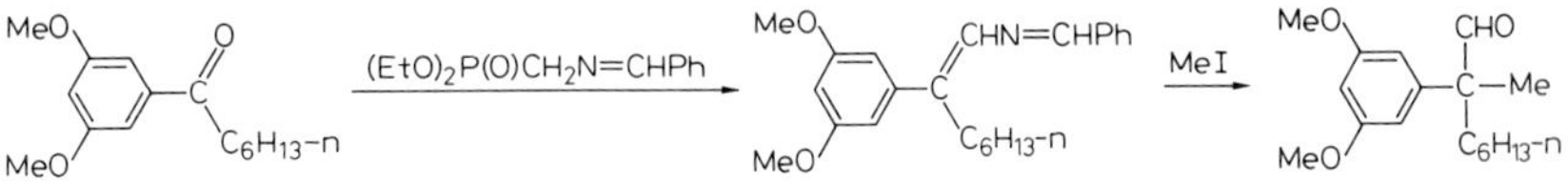

A soln. of diethyl N-benzylideneaminomethanephosphonate in tetrahydrofuran added at –78° to a stirred soln. of *n*-butyllithium in tetrahydrofuran/hexane, after 1 hr. a soln. of startg. ketone in tetrahydrofuran added, warmed to room temp., refluxed for 3 hrs., recooled to –78°, a soln. of *n*-butyllithium in hexane added slowly, stirred for 2 hrs., treated with methyl iodide, after 10 min. stirred for 2 hrs. at room temp., then mixed vigorously for 2 hrs. with aq. 1 *M* HCl → 2-(3,5-dimethoxyphenyl)-2-methyloctanal. Y 80%. – This one-pot method is useful for constructing **quaternary carbon atoms.** F.e.s. V. Singh, V.V. Kane, A.R. Martin, Synth. Commun. *11,* 429–37 (1981).

Horner synthesis with α**-sulfonylphosphonic esters** CO → C=C

α,β**-Ethylenesulfones from aldehydes**

s. *23,* 879s*27;* 1,3-diene-sulfones and -sulfoxides s. B.E. de Jong, H. de Koning, H.O. Huisman, Rec. Trav. Chim. Pays-Bas *100,* 410–4 (1981); **review of Horner synthesis of heterosubst. ethylene derivs.** s. H. Gross, I. Keitel, Z. Chem. *22,* 117–26 (1982).

Ketones from trialkylboranes and fluorosulfonyl esters R_3B → RCOR′

Synthesis with addition of two C-atoms

853.

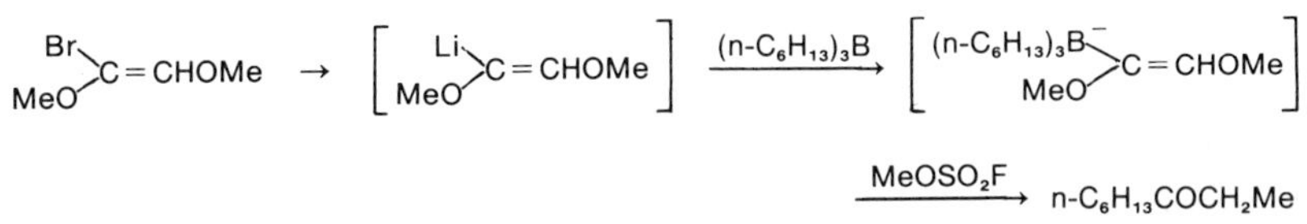

Trihexylborane in tetrahydrofuran added to a soln. of 1,2-dimethoxyethenyllithium (prepd. from 1-bromo-1,2-dimethoxyethene and *n*-butyllithium) in ether at –78°, after stirring for 10 min. methyl fluorosulfonate added, stirring continued for a further 30 min. at –78°, allowed to warm to room temp., stirred overnight, 3 *M* HCl added at 0°, then finally stirred at room temp. for 6 hrs. → 3-nonanone. Y 91%. F.e.s. T. Yogo, J. Koshino, A. Suzuki, Chem. Letters *1981,* 1059–60.

n-*Butyllithium* n-*BuLi*

Ethylene derivs.

from phosphine oxides and oxo compds. CHP(O)R_2 + OC → C=C

s. *32,* 866; 2,4-dienols with *n*-butyllithium s. E. Vedejs et al., J. Org. Chem. *47,* 1534–46 (1982).

Terminal ethylene derivs. from 2-pyridyl thioethers CHSR → C=CH_2

854.

1,3-Dienes. A soln. of *n*-butyllithium in hexane added to a soln. of startg. thioether in tetrahydrofuran at –78°, stirred for 30 min., a soln. of tri-*n*-butylstannylmethyl iodide in tetrahydrofuran added very slowly to the resultant red soln., and stirring continued at the same temp. for 20 min. followed by 1.5 hrs. at 0° → 1-phenylbutadiene. Y 73%. F.e., **also styrenes,** and synthesis from *β,γ*-ethylenesulfones, s. M. Ochiai et al., Tetrahedron Letters *23,* 2205–8 (1982); review of new possible applications of heavy main-group elements in organic synthesis s. T. Kauffmann, Angew. Chem. Intern. Ed. *21,* 410–29 (1982).

n-*Butyllithium/hydrogen peroxide/sodium hydroxide* n-*BuLi/H_2O_2/NaOH*

Lithium borates as intermediates ←

s. *36,* 872; review of organoborates in new synthetic reactions s. A. Suzuki, Acc. Chem. Res. *15,* 178–84 (1982).

sec-*Butyllithium* s-*BuLi*

γ-Diketones

from azomethines and *β,γ*-ethylene-*β*-siloxychlorides C=NR → CO–C–C–CO

Electrophilic acetonylation

855.

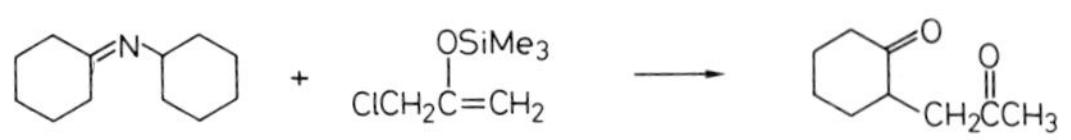

A soln. of 1.5 *M* *sec*-butyllithium in pentane added to a soln. of N-(cyclohexylidene)cyclohexylamine in tetrahydrofuran at –78°, 3-chloro-2-(trimethylsiloxy)-1-propene as **acetonyl carbonium equivalent** added at room temp., the mixture stirred for 30 min., and carefully hydrolysed with aq. oxalic acid → product. Y 82%. F.e. and regiospecificity, **also from hydrazones,** and acetonylation of CH-acidic compds., s. A. Hosomi et al., J. Org. Chem. *46,* 4631–3 (1981).

Phenyllithium/hydrogen peroxide/sodium hydroxide *PhLi/H_2O_2/NaOH*

Sec. alcohols from trialkylboranes and sulfones CH_2SO_2 → CH(OH)R

856.

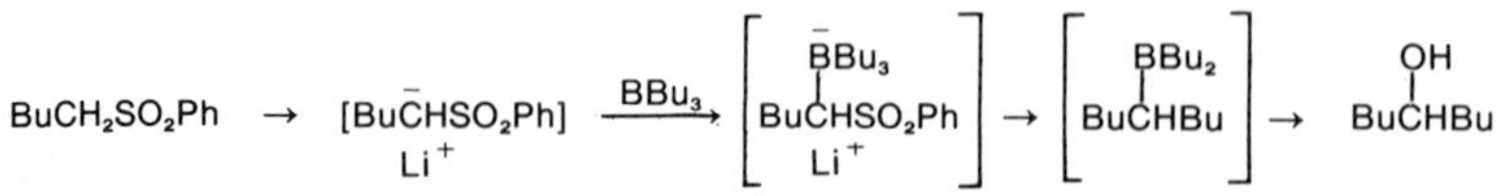

A soln. of startg. lithiosulfone (prepd. by metalation of amyl phenyl sulfone with phenyllithium) in tetrahydrofuran treated rapidly with tri(*n*-butyl)borane in the same solvent at –78°, the mixture stirred at the same temp. for 15 min., allowed to warm to room temp., stirring continued at 25–30° for 48 hrs., the mixture cooled to –20°, 3 *M* NaOH added quickly followed slowly by 30% H_2O_2, and stirred at 30–40° overnight → product. Y 84%. F.e.s. D. Uguen, Bull. Soc. Chim. France II *1981,* 99–102.

Lithium diisopropylamide *i-Pr_2NLi*

***(E)-α,β*-Ethyleneacylsilanes**
from halides and aldehydes CHO → CH=C(R)COSi≡
Synthesis with addition of two C-atoms

857.

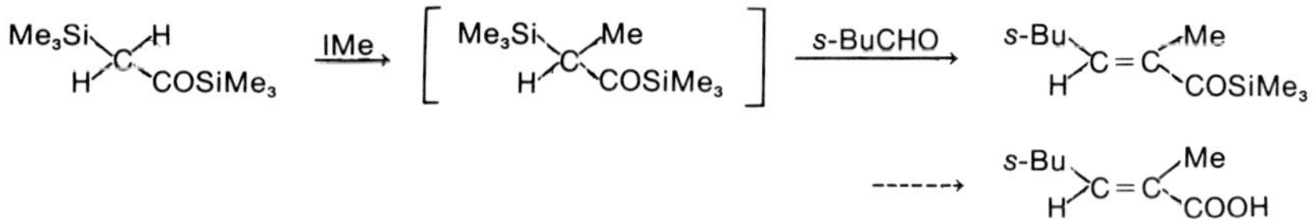

A soln. of Li diisopropylamide in tetrahydrofuran treated at –78° with a soln. of [(trimethylsilyl)acetyl]trimethylsilane in tetrahydrofuran, warmed to 0–5°, stirred for 30 min., recooled to –25°, a soln. of methyl iodide in tetrahydrofuran added, stirred for 4 hrs., warmed to 25°, the mixture added to a soln. of Li diisopropylamide in tetrahydrofuran at 0–5°, stirred at 25° for 1 hr., cooled to –78°, 2-methylbutyraldehyde in the same solvent added during 30 min., stirred at –78° for 15 min., warmed to 25°, then poured into 1 *M* HCl → product. Y 85%. F.e. and oxidation to ***α,β*-disubst. (E)-acrylic acids** s. J.A. Miller, G. Zweifel, J. Am. Chem. Soc. *103,* 6217–9 (1981).

Cyclopropanes from enephosphonium salts and halides 
Cyclopropyl ketones

858.

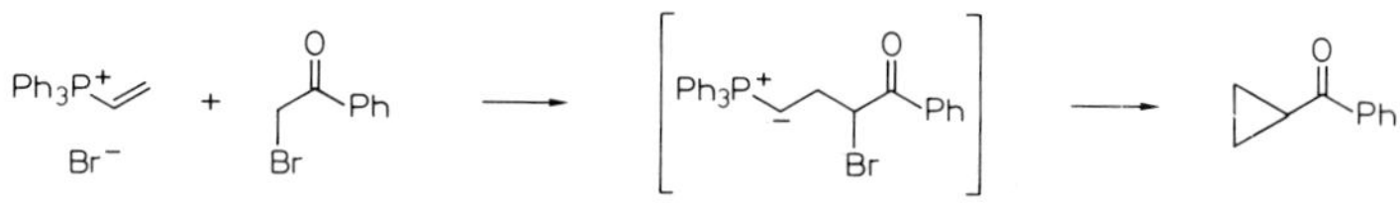

α-Bromoacetophenone in tetrahydrofuran added dropwise via syringe to a soln. of Li diisopropylamide in the same solvent at –78° under N_2, stirred for 1 hr., vinyltriphenylphosphonium bromide in dimethylformamide added dropwise with vigorous stirring, warmed slowly to 25° during 1–2 hrs., and stirring continued for 24 hrs. → cyclopropyl phenyl ketone. Y 70%. F.e.s. G.H. Posner, J.P. Mallamo, A.Y. Black, Tetrahedron *37,* 3921–6 (1981).

Lithium diisopropylamide/magnesium iodide *i-Pr_2NLi/MgI_2*

Synthesis of ethylene derivs. from silanes and oxo compds. CO → C=C
Peterson olefination

s. *28,* 856s*31; α,β*-ethylenenitriles from aldehydes, synthesis with addition of two C-atoms in the presence of MgI_2, s. Y. Yamakado et al., J. Am. Chem. Soc. *103,* 5568–70 (1981).

Lithium bis(trimethylsilyl)amide *$(Me_3Si)_2NLi$*

Wittig synthesis

with Na bis(trimethylsilyl)amide cf. *22,* 871s*31;* with Li bis(trimethylsilyl)amide, (E)-ethylenecarboxylic acids, s. B.E. Maryanoff, B.A. Duhl-Emswiler, Tetrahedron Letters *22,* 4185–8 (1981).

C-Acylation with acylphosphonic esters CO–CH → CO–C–CO
***β*-Diketones from ketones**

859. $PhCOP(O)(OEt)_2 + CH_3COPh \rightarrow PhCOCH_2COPh$

Acetophenone added to a cooled soln. of lithium bis(trimethylsilyl)amide in tetrahydrofuran at –78° under argon, the mixture stirred for 30 min., 2 equivalents diethyl benzoylphosphonate added, and stirring continued for 30 min. at –78° → product. Y 86%. F.e., also with NaH as base, s. M. Sekine et al., Chem. Letters *1981,* 1087–90.

Potassium bis(trimethylsilyl)amide — $(Me_3Si)_2NK$

trans-**Oxido from oxo compds. and arsonium ylids** — CO → 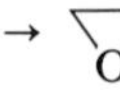

860. 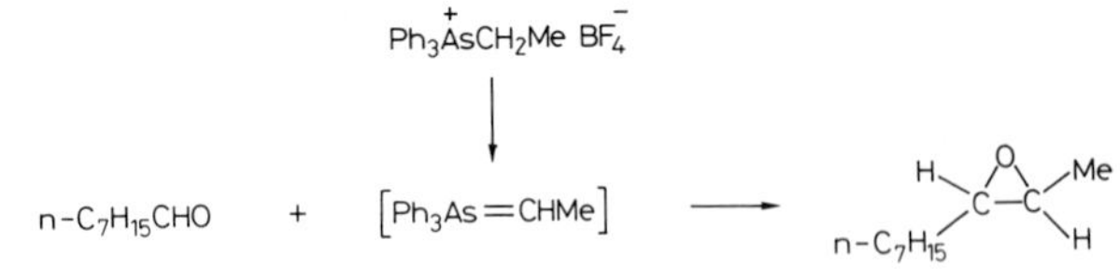

Ethyltriphenylarsonium fluoroborate treated with K bis(trimethylsilyl)amide in 10% hexamethylphosphoramide/tetrahydrofuran at −10° for 10 min., the mixture cooled to −78°, octanal added, and the mixture allowed to warm slowly to room temp. → *trans*-epoxide. Y 80%. F.e., also prepn. of startg. arsonium salts, s. W.C. Still, V.J. Novack, J. Am. Chem. Soc. *103,* 1283–5 (1981).

Sodium carbonate s. under $Pd(PPh_3)_4$ — Na_2CO_3

Potassium carbonate — K_2CO_3

Heterogeneous Wittig synthesis — CO → C=C

with NaOH in aq. benzene cf. *29,* 868; with K_2CO_3 in dioxane s. Y. Le Bigot, M. Delmas, A. Gaset, Synth. Commun. *12,* 107–12 (1982).

Potassium fluoride/18-crown-6 polyether — ←

2-Arylalcohols from aldehydes and benzylsilanes — CO → C(OH)R

s. *30,* 624s*37*

Ketones from acylsilanes and halides — COSi≡ + RHal → COR

861. $PhCOSiMe_3 + BrCH_2Ph \rightarrow PhCOCH_2Ph$

A mixture of benzoyl(trimethyl)silane and benzyl bromide in mesitylene heated at 160° for 3 hrs. in the presence of 3 equivalents KF and a little 18-crown-6 → product. Y 90%. F.e., **also azomethines from imidoylsilanes,** s. A. Degl'Innocenti et al., J. Chem. Soc. Chem. Commun. *1980,* 1201–2.

Lithium salt — Li^+

Tert. nitro compds. — $RTl(OAc)_2 \rightarrow RC(NO_2)<$

from sec. nitronates and monoorganothallium(III) compds.

862. $PhCH{=}CH{-}Tl(OAc)_2 + Me_2C{=}NO_2^-\ Li^+ \rightarrow PhCH{=}CHCMe_2NO_2$

A mixture of startg. nitronate and organothallium(III) acetate in dimethyl sulfoxide allowed to react at room temp. under nitrogen for 3 hrs. → product. Y 99%. F.e., also in methanol and with irradiation, s. H. Kurosawa, M. Sato, H. Okada, Tetrahedron Letters *23,* 2965–8 (1982).

Lithium salt/irradiation — Li^+/≋

...from sec. nitronates and organomercury halides — $RHgCl \rightarrow RC(NO_2)<$

863. $PhCH_2HgCl + Me_2C{=}NO_2^-\ Li^+ \rightarrow PhCH_2CMe_2NO_2$

Benzylmercury chloride added to a stirred soln. of lithiated 2-nitropropane in dimethyl sulfoxide, and the mixture irradiated in a Pyrex vessel with a 275 W sunlamp at 35–40° for 2 hrs. → 2-methyl-2-nitro-1-phenylpropane. Y 100%. F.e.s. G.A. Russell, J. Hershberger, K. Owens, J. Organometal. Chem. *225,* 43–56 (1982); **organomercury compds. in organic synthesis, review,** s. R.C. Larock, Tetrahedron *38,* 1713–54 (1982).

Triethylamine *Et_3N*

Syntheses with benzylamine-type palladium complexes ←
s. *32,* 876; *o-α*-aminoketones s. R.A. Holton, K.J. Natalie, Tetrahedron Letters *22,* 267–70 (1981).

Ethanolamine *$HOCH_2CH_2NH_2$*

Allylboration of carbonyl compds. CO → C(OH)C–C=C
2-Ethylenealcohols from oxo compds. with allyl rearrangement
s. *33,* 865s*36;* 2-alkoxy-3-ethylenealcohols s. R.W. Hoffmann, B. Kemper, Tetrahedron Letters *22,* 5263–6 (1981); *23,* 845–8 (1982).

Pyridine *C_5H_5N*

4-Hydroxyenesilanes
from 2-ethylene-1-boranyl-1-silanes and aldehydes C(OH)–C–C=C–Si⋛
Stereo- and regio-specific synthesis

864.

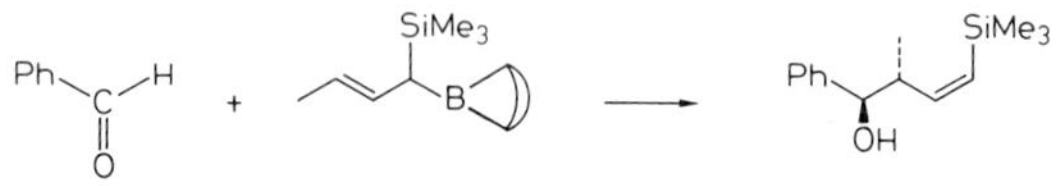

2 Equivalents pyridine added to a soln. of 9-(*α*-trimethylsilylcrotyl)-9-borabicyclo[3.3.1]nonane in ether at –78° under N_2, stirred for a while, benzaldehyde added, after 30 min. quenched with methanol/ethanolamine at –70°, and warmed to room temp. → product. Y 90% (92% (Z)-*threo*-isomer). – Stereochemistry is controlled over 4 contiguous C-atoms in one operation. F.e., **also 4-hydroxyenestannanes from 2-ethylene-1-boranyl-1-stannanes,** s. Y. Yamamoto, H. Yatagai, K. Maruyama, J. Am. Chem. Soc. *103,* 3229–31 (1981).

Copper(II) acetylacetonate/triphenylphosphine *$Cu(acac)_2/Ph_3P$*

Ketones from carboxylic acid chlorides and trialkylaluminium COCl → COR
s. *37,* 879

Dilithium trialkylcuprates *Li_2CuR_3*

Cross-coupling with arylmercury chlorides RHgCl → R–R′

865. C_6H_5–HgCl + Li_2CuMe_3 → C_6H_5–Me

Alkylarenes. Phenylmercuric chloride treated with 5 equivalents of dilithium trimethylcuprate in diethyl ether at –78° for 1 hr., warmed to 0°, and quenched with excess methyl iodide and oxygen → toluene. Y 92%. F.e., **also hydrocarbons** from alkylmercury chlorides, **and ethylene derivs. from enemercury chlorides,** s. R.C. Larock, D.R. Leach, Tetrahedron Letters *22,* 3435–8 (1981); with diiodo(methyl)-bis(triphenylphosphine)rhodium(III) in place of cuprate cf. ibid. *22,* 2443–6 (1981); J. Organometal. Chem. *225,* 31–41 (1982); s.a. Organometallics, *1* 74–81 (1982).

Cupric sulfate *$CuSO_4$*

Reactions of *α*-diazocarboxylic acid esters with enoxysilanes ←
γ-ketocarboxylic acid esters s. *28,* 858; *γ,γ*-dialkoxycarboxylic acid esters in the presence of $CuSO_4$ cf. K. Saigo, S. Okagawa, H. Nohira, Bull. Chem. Soc. Japan *54,* 3603–4 (1981).

Cuprous bromide *CuBr*

Allenes from acoxy-2-acetylenes
and trialkylaluminium C(OAc)C≡C → C=C=CR

866.

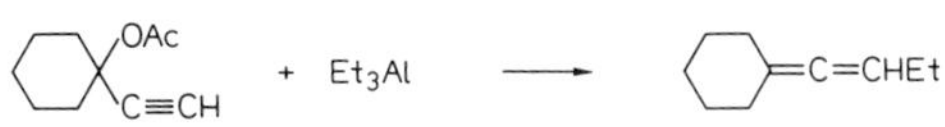

Solns. of CuBr and triethylaluminium in ether added sequentially with stirring to a soln. of startg. acetylene deriv. in the same solvent at 20°, the mixture heated 2 hrs., and hydrolysed with water → product. Y 68%. F.e.s. G.A. Tolstikov, T.Y. Romanova, A.V. Kuchin, Izv. Akad. Nauk SSSR Ser. Khim. *1981,* 1438–9.

Magnesium *Mg*

Ketones from acylcobalt(III) complexes ←

867.

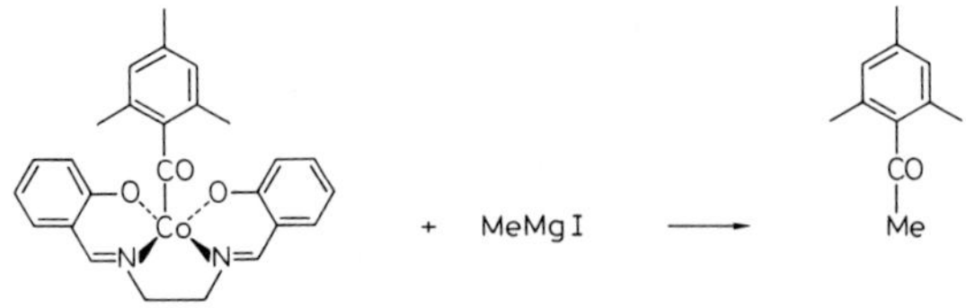

A soln. of methylmagnesium iodide in ether added dropwise to a stirred soln. of startg. acyl{N,N′-ethylenebis(salicylideneiminato}cobalt(III) complex in tetrahydrofuran, the mixture stirred for a further 3 hrs., and quenched with 1 *M* HCl → methyl mesityl ketone. Y 91%. F.e.s. I. Tabushi, K. Seto, Y. Kobuke, Tetrahedron *37,* 863–5 (1981).

Magnesium/dichlorobis(triphenylphosphine)nickel(III) *$Mg/NiCl_2(PPh_3)_2$.*

Cross-coupling of Grignard compds. with tellurides C=CTeR → C=CR′

Ethylene derivs. from enetellurides

with retention of configuration

868.

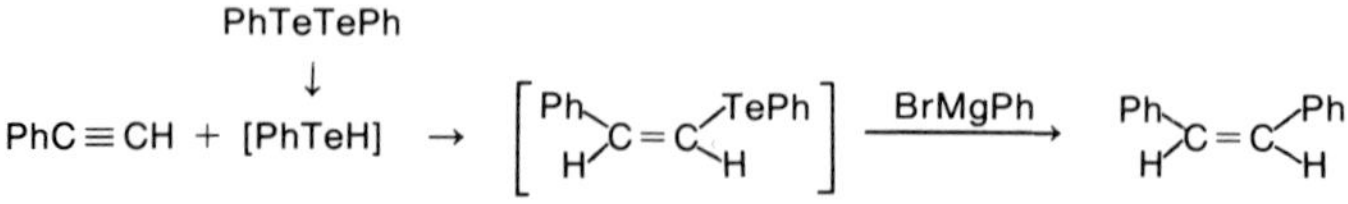

A soln. of phenylmagnesium bromide in tetrahydrofuran added to startg. enetelluride (prepared from diphenyl ditelluride and the appropriate acetylene deriv.) in the same solvent under N_2, and the mixture refluxed for 3 hrs. in the presence of a little dichlorobis(triphenylphosphine)nickel(II) → (Z)-stilbene. Y 100%. – Low yields were obtained with alkylmagnesium bromides. F.e. and catalysts, **also ar. cross-coupling,** s. S. Uemura, S. Fukuzawa, Tetrahedron Letters *23,* 1181–4 (1982).

Zinc/copper/isopropyl iodide *Zn/Cu/i-PrI*

Cyclic β-diketones from 1,2-di(siloxy)cycloalk-1-enes

via α-alkoxymethyl-α-hydroxyketones

Ring expansion by one C-atom

869.

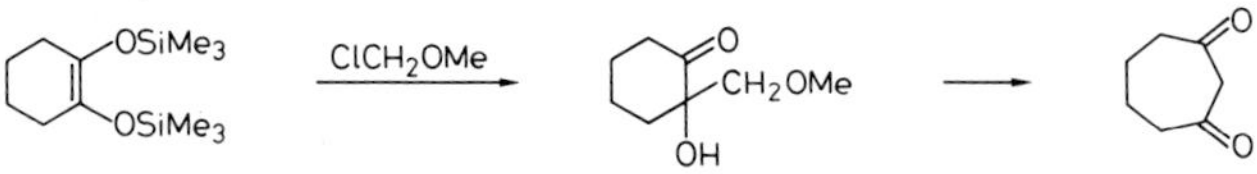

Isopropyl iodide added to a soln. of Zn-Cu couple (prepd. from Zn dust and CuCl) in methylene chloride, refluxed for 30 min. with stirring under N_2, 1,2-bis(trimethylsiloxy)cyclohexene added at 0–5° followed dropwise by a soln. of chloromethyl methyl ether in methylene chloride at 0–2°, and stirring continued at 0° for 1 hr. then at room temp. for 2 hrs. → intermediate (Y 67%) added dropwise to a little $KHSO_4$ at 170–180° and 20–25 mmHg → 1,3-cycloheptadione (Y 75%). F.e.s. I. Nishiguchi et al., Chem. Letters *1981,* 551–4.

Mercuric acetate/sodium tetrahydridoborate *$(MeCOO)_2Hg/NaBH_4$*

Synthesis of ketones

from siloxycyclopropanes and activated ethylene derivs.

870.

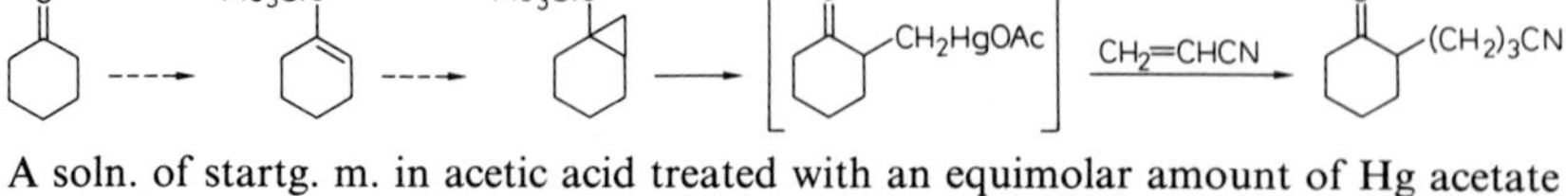

A soln. of startg. m. in acetic acid treated with an equimolar amount of Hg acetate at 20° for 1 hr., the solvent replaced by dichloromethane acrylonitrile added, and reduced with aq. $NaBH_4$ → product. Y 68%. – This is part of a multistep **synthesis of ε-keto-nitriles and -carbonyl compds. from ketones.** F.e.s. B. Giese, H. Horler, W. Zwick, Tetrahedron Letters *23,* 931–4 (1982).

Magnesium halide s. under $Pd(PPh_3)_4$ *$MgHal_2$*

Magnesium iodide s. under $i\text{-}Pr_2NLi$ *MgI_2*

Zinc bromide *$ZnBr_2$*

5-Subst. 2(5*H*)-furanones from 2-acoxyfurans ←
s. *37,* 911

Zinc iodide *ZnI_2*

Synthesis of ketones
from enoxysilanes and acetoxy compds. C=C–OSi⋶ → RC–CO

871.

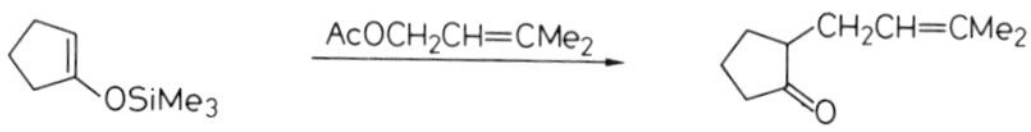

***γ,δ*-Ethyleneketones.** 1-Trimethylsiloxycyclopentene and prenyl acetate added to flame-dried ZnI_2 in dichloromethane under N_2, and the suspension stirred for 4 hrs. at room temp. → 2-(3-methyl-2-butenyl)cyclopentanone. Y 81%. F.e.s. M.T. Reetz, S. Hüttenhain, F. Hübner, Syn. Commun. *11,* 217–22 (1981); **4-*β*-keto-2-azetidinones** with $ZnCl_2$ cf. P.J. Reider, R. Rayford, E.J.J. Grabowski, Tetrahedron Letters *23,* 379–82 (1982); with trimethylsilyltriflate, **also carboxylic and thiolic acid esters,** cf. A.G.M. Barrett, P. Quayle, J. Chem. Soc. Chem. Commun. *1981,* 1076–7; carbohydrate derivs. from acyl glycosides s. Y.S. Yokoyama et al., Tetrahedron Letters *23,* 2673–6 (1982).

Sodium tetrahydridoborate s. under $(MeCOO)_2Hg$ *$NaBH_4$*

Diisobutylaluminium hydride *$(i\text{-}C_4H_9)_2AlH$*

Synthesis of sec. amines from O-sulfonyloximes and organoaluminium compds. via Beckmann rearrangement ◌

872.

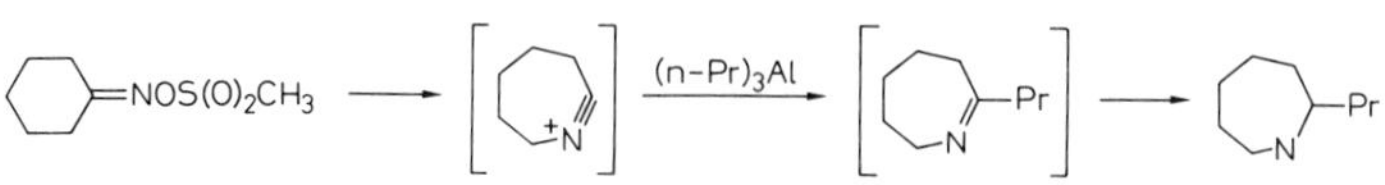

Ring expansion. A soln. of tri-*n*-propylaluminium in hexane added to cyclohexanone oxime methanesulfonate in methylene chloride at –78°, after 5 min. the soln. warmed to 0° and stirred for 1 hr., diisobutylaluminium hydride in hexane added at the same temp., stirring continued for 1 hr., quenched by diluting with methylene chloride followed by successive treatment with NaF and water, and the suspension stirred vigorously at 0° for 30 min. → 2-propylazacycloheptane. Y 64%. F.e., incl. acyclic *α*-alkylated sec. amines, regiospecificity, and methods, s. K. Hattori et al., J. Am. Chem. Soc. *103,* 7368–70 (1981).

Thallium(III) trifluoroacetate *$(CF_3COO)_3Tl$*

Electrophilic allylation of arenes H → C–C=C

873.

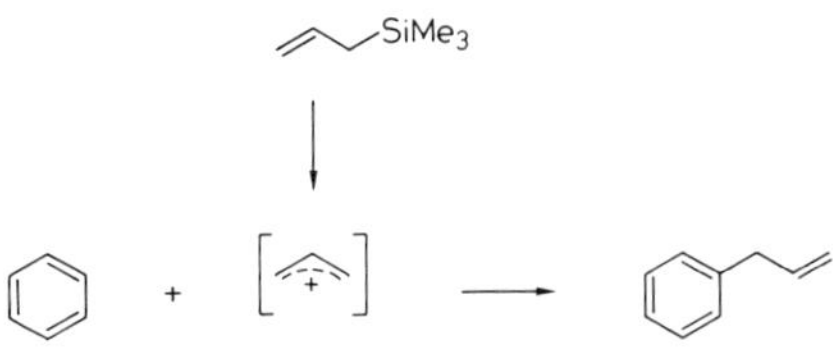

Thallium(III) trifluoroacetate reverses the conventional polarity of 2-ethylene-silanes, -stannanes, and -germanes affording reactive allyl cations. E: Benzene added at 0° to a suspension of thallium(III) trifluoroacetate in dichloromethane, allyltrimethylsilane added dropwise with stirring at the same temp., and stirring continued for 5 hrs. at 0°, then at room temp. for 30 min. → allylbenzene. Y 56%. F.e. including furan derivs. s. M. Ochiai, M. Arimoti, E. Fujita, Tetrahedron Letters *22,* 4491–4 (1981).

Boron fluoride *BF_3*

C-Allylation of *p*-quinones with 2-ethylenestannanes H → C–C=C

s. *34,* 853; 1,4-addition to functionalized *p*-quinones cf. Tetrahedron Letters *22,* 5221–4 (1981); with 2-ethylenesilanes cf. M. Ochiai, M. Arimoto, E. Fujita, J. Chem. Soc. Chem. Commun. *1981,* 450–1; s.a. Tennen Yuki Kagobutsu Toronkai Koen Yoshishu, 24th *1981,* 505 (Japan); Chem. Abstr. *97,* 55539 (1982).

Reductive C-allylation with 2-ethylenesilanes OH → C–C=C

874. $Ph_2CHOH + Me_3SiCH_2CH{=}CH_2 \rightarrow Ph_2CHCH_2CH{=}CH_2$

of benzyl alcohols. A soln. of benzhydrol and allyltrimethylsilane in methylene chloride at 0–5° treated with BF_3 gas, and the mixture stirred under N_2 until complete consumption of startg. m. by t.l.c. or g.c. → 4,4-diphenyl-1-butene. Y 100%. F.e., **also reductive C-arylation with ar. silanes,** and prepn. of **sym. 1,6-dienes from ketals,** s. J.A. Cella, J. Org. Chem. *47,* 2125–30 (1982).

3-Ethylenealcohols from 2-ethylenestannanes CO → C(OH)C–C=C

Stereospecific synthesis

s. *36,* 879; review of stereospecific synthesis of 3-ethylenealcohols from allylic organometallic compds. s. Y. Yamamoto, K. Maruyama, Yuki Gosei Kagaku Kyokaishi *40,* 332–42 (1982).

(Z)-*a*,*β*-Ethylenenitriles from aldehydes CHO → CH=C(CN)Si≡

via (E)-*a*,*β*-ethylene-*a*-silylnitriles

875.

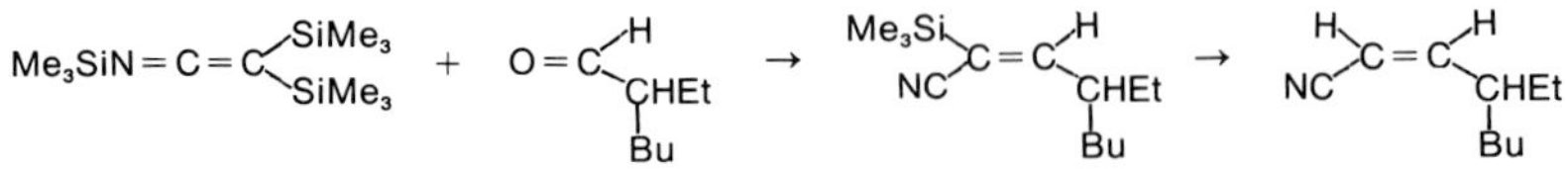

A soln. of tris(trimethylsilyl)ketenimine in benzene added dropwise to an ice-cooled soln. of startg. aldehyde and BF_3–etherate in the same solvent, the mixture stirred at 25° for 2 hrs. then refluxed for 1 hr. → crude intermediate (Y 74%), in ether at 0° treated with 1 *M* NaOH in methanol, and stirred for 1 hr. → product (Y 91%). F.e.s. Y. Sato, Y. Niinomi, J. Chem. Soc. Chem. Commun. *1982,* 56–7.

Synthesis with addition of two C-atoms

Nucleophilic 1-substitution of urethans via 1-alkoxyurethans OR → C–CO

***a*-Carbalkoxyaminoketones**

876.

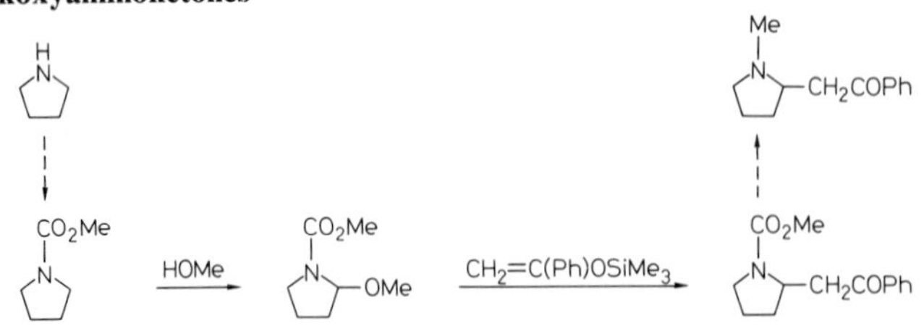

from enoxysilanes. A soln. of 1-(carbomethoxy)pyrrolidine in methanol containing tetraethylammonium *p*-toluenesulfonate electrolysed with a constant current of 3 A at carbon electrodes until 2.1 F/mol passed (cf. Synth. Meth. *23,* 186s*32*) → 1-(carbomethoxy)-2-methoxypyrrolidine (Y 80%), in dichloromethane added dropwise during 5 min. to a stirred soln. of BF_3-etherate in the same solvent under N_2 at –75°, stirred 5 min., startg. enoxysilane added dropwise over 5–10 min., stirring continued for 2 hrs., and allowed to warm to room temp. → 1-(carbomethoxy)-2-

phenacylpyrrolidine (Y 92%). – This is part of a multistep **synthesis of tert. methylamines from sec. amines.** F.e. and nucleophiles s. T. Shono, Y. Matsumura, K. Tsubata, J. Am. Chem. Soc. *103,* 1172–6 (1981).

Functionalized ethylene derivs. from 1,2-oxidosilanes
Enolesters

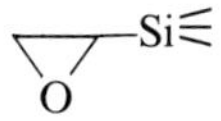 → C=C−OAc

s. *32,* 881; also enolethers with isolation of intermediates, ω-acetylenic vinylic esters and ethers, s. M.C. Croudace, N.E. Schore, J. Org. Chem. *46,* 5357–63 (1981).

1-Allyllactams from 1-acoxylactams ←
s. *37,* 893

Fluoroborates BF_4^-

β-Ketomercaptals from enoxysilanes C(OSi≡)=C → CO−C−C(SR)$_2$

877.

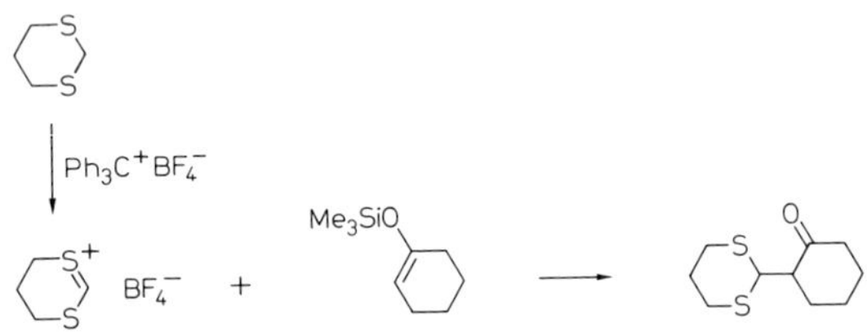

A soln. of 1,3-dithiane and trityl fluoroborate in methylene chloride refluxed for 2 hrs. → **1,3-dithienium fluoroborate** (Y 81%) **as formyl carbonium equivalent** in nitromethane added dropwise over 1 min. to a stirred soln. of startg. enoxysilane in methylene chloride at −78° under argon, and worked up after 10 min. → ketomercaptal (Y 92%). F.e., also ester and lactone derivs., s. I. Paterson, L.G. Price, Tetrahedron Letters *22,* 2829–32 (1981); **β,γ-ethylene-δ-ketomercaptals from (1,3-dien)oxysilanes** s. ibid. *22,* 2833–6 (1981).

Silver fluoroborate $AgBF_4$

Quinones from acetylene derivs. and cobaltacyclopent-3-ene-2,5-diones ○

878

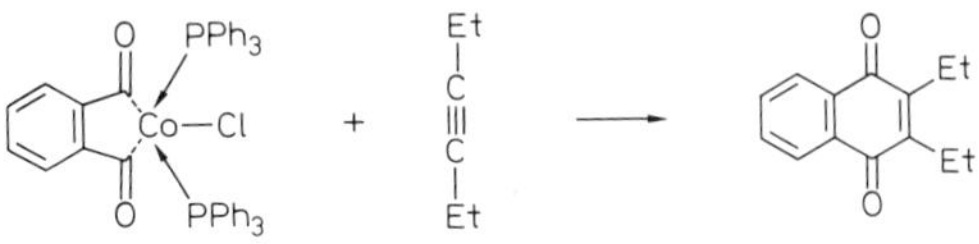

Naphthoquinones. A stirred mixture of cobalt complex, 3-hexyne, and silver fluoroborate in acetonitrile heated under N_2 in a sealed tube at 110° for 40 hrs. → 2,3-diethyl-1,4-naphthoquinone. Y 90%. F.e. and from the corresponding iron complexes s. L.S. Liebeskind, S.L. Baysdon, M.S. South, J. Am. Chem. Soc. *102,* 7397–8 (1980); prepn. of complexes s. J. Organometal. Chem. *202,* C73–6 (1980), s.a. Organometallics, *1,* 771–5 (1982).

Aluminium chloride $AlCl_3$

Ketones from carboxylic acid chlorides and aluminium trialkyls COCl → COR

879. MeCOCl + (i-Bu)$_3$Al → MeCOBu-i

A soln. of triisobutylaluminium in dichloromethane added over 1 hr. to a vigorously stirred mixture of $AlCl_3$ and acetyl chloride in the same solvent at −30 to −20°, and stirring continued at room temp. for 1 hr. → methyl isobutyl ketone. Y 70%. F.e. and catalysts s. G.A. Tolstikov, F.K. Valitov, A.V. Kuchin, Zh. Obshch. Khim. *50,* 2715–20 (1980); f. method cf. ibid. *52,* 1328–34 (1982); with copper(II) acetylacetonate/triphenylphosphine cf. K. Takai, K. Oshima, H. Nozaki, Bull. Chem. Soc. Japan *54,* 1281–2 (1981).

Aluminium chloride *$AlCl_3$*

4-Ketothioenolethers COCl → CO–C–C=CSR

from 2-ethylene-1-silylsulfides and carboxylic acid chlorides

880. 

A soln. of startg. silyl sulfide in dichloromethane treated with 1.2 equivalents propanoyl chloride in the presence of $AlCl_3$ at –78° for 7 hrs. → product. Y 63%. – This is part of a 2-step synthesis **from 2-ethylenethioethers with double bond migration.** F.e.s. K. Hiroi, L.-M. Chen, J. Chem. Soc. Chem. Commun. *1981,* 377–8.

Syntheses with 2-ethylene-3-chlorosilanes COCl → CO–CCl–C=C

β,γ-Ethylene-α-chloroketones from carboxylic acid chlorides

881. 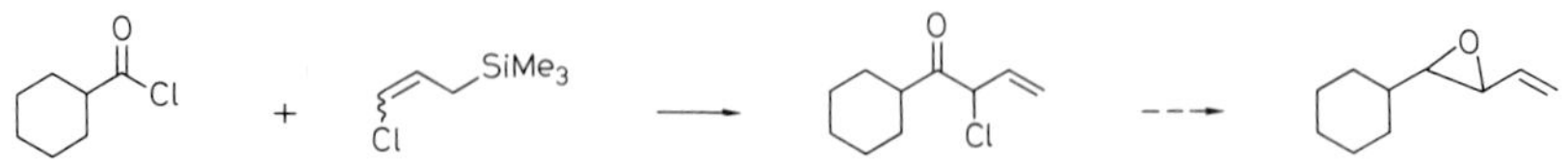

Cyclohexanecarboxylic acid chloride added dropwise to a stirred suspension of $AlCl_3$ in dichloromethane at –20° under N_2, the mixture stirred for 1 hr., a soln. of 3-chloroallyltrimethylsilane in the same solvent added dropwise, and stirring continued at –20° for 2 hrs. → product. Y 74%. F.e. and conversion to **3-ethyleneoxido compds.** s. M. Ochiai, E. Fujita, Tetrahedron Letters *21,* 4369–72 (1980); **1-alkoxy-3-ethylene-2-chlorides** from aldehydes s. J. Chem. Soc. Chem. Commun. *1980,* 1118.

α,β-Ethylenecarboxylic acid esters from aldehydes CHO → CH=CHCOOR

via β-hydroxy-α-silylcarboxylic acid esters

Geospecific synthesis with addition of two C-atoms

882. 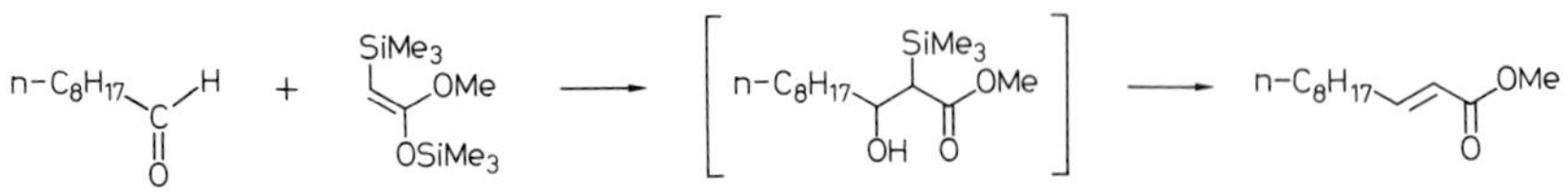

A mixture of nonanal, 1 equivalent $AlCl_3$, and 2 equivalents O-methyl-C,O-bis(trimethylsilyl)ketene acetal in carbon tetrachloride refluxed for 8 hrs. → product. Y 89% (92% (E)-isomer). F.e., also (Z)-isomer with $TiCl_4$, and isolation of intermediate s. I. Matsuda, Y. Izumi, Tetrahedron Letters *22,* 1805–8 (1981).

Aluminium bromide *$AlBr_3$*

Synthesis of hydrocarbons from halides and tetraalkylsilanes Cl → R

883. 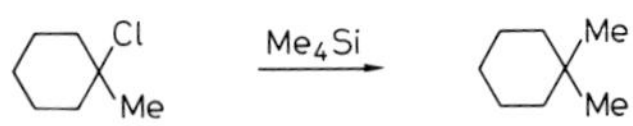

Tetramethylsilane, methylene chloride, and a little $AlBr_3$ added to 1-chloro-1-methylcyclohexane cooled in an acetone-dry ice bath, and stirred at room temp. for 15 min. → 1,1-dimethylcyclohexane. Y 80%. F.e., incl. replacement of bridgehead halogen, also from prim. halides with rearrangement, s. G.I. Bolestova et al., Zh. Org. Khim. *17,* 1357–63 (1981); **from ethylene derivs.** cf. J. Chem. Soc. Chem. Commun. *1980,* 748.

Gallium chloride *$GaCl_3$*

Synthesis of alcohols from oxo compds. and silanes CO → C(OH)R

s. *30,* 624; 2,2,2-trichloralcohols from chloral s. G. Déléris et al., Eur. J. Med. Chem. *16,* 533–7 (1981); 2-arylalcohols with KF/18-crown-6-polyether s. A. Ricci et al., Tetrahedron Letters *23,* 577–8 (1982).

18-Crown-6 polyether s. under KF ←

Tetrabutylammonium chloride s. under $PdCl_2$ $Bu_4N^+Cl^-$

Tetrabutylammonium hydrogen sulfate s. under NaOH $Bu_4N^+HSO_4^-$

Benzyltriethylammonium chloride s. under NaOH and KOH $[PhCH_2NEt_3]^+Cl^-$

Cation exchanger ←

Polymer-supported Horner synthesis CO → C=C

884. 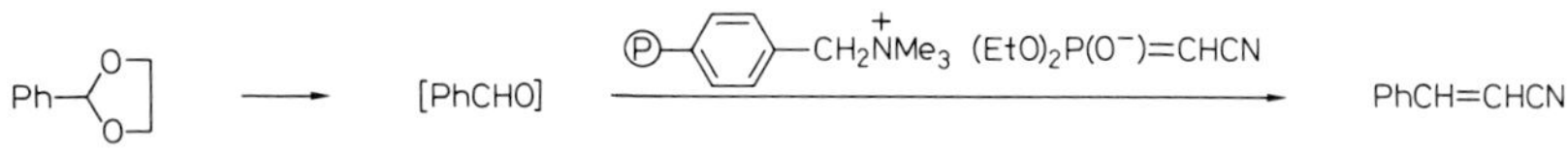

with 1,3-dioxolanes. A soln. of 2-phenyl-1,3-dioxolane in 9:1 tetrahydrofuran/water containing Amberlyst 15 H (H^+ form) stirred for 1 hr. at room temp. with startg. phosphonate bound on Amberlyst A-26 for 1 hr. → cinnamonitrile. Y 98%. F.e.s. G. Cainelli et al., J. Chem. Soc. Perkin Trans. I *1980,* 2516–9.

Maleic anhydride ←

Regiospecific synthesis of 1,4-dienes ←

from 1-alkenylzirconium complexes and π-allyl palladium chloride complexes

885. 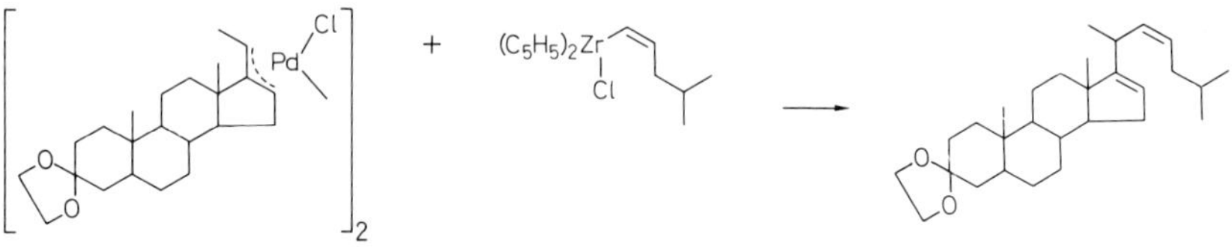

A soln. of chlorobis(cyclopentadienyl)(4′-methylpent-1′-en-1′-yl)zirconium in tetrahydrofuran added during 2 hrs. at −78° to an equilibrated mixture of maleic anhydride and startg. Pd π-allyl complex in the same solvent under N_2, and allowed to warm to room temp. overnight → product. Y 78%. F.e.s. J.S. Temple, J. Schwartz, J. Am. Chem. Soc. *102,* 7381–2 (1980); ibid. *104,* 1310–5 (1982).

Trifluoroacetic acid/acetic acid $CF_3COOH/MeCOOH$

Syntheses with azomethine-type palladium complexes ←

o-subst. azomethines from organolithium compds. s. *31,* 868; *o*-vinylaldehydes from ethylene derivs. with $CF_3COOH/MeCOOH$ s. J.R. Girdling, D.A. Widdowson, Tetrahedron Letters *23,* 1957–60 (1982).

Isopropyl iodide s. under Zn/Cu *i-PrI*

2-Chloroethoxytrimethylsilane $Me_3SiOCH_2CH_2Cl$

Wittig synthesis with retention of siloxy groups CO → C=C

886. 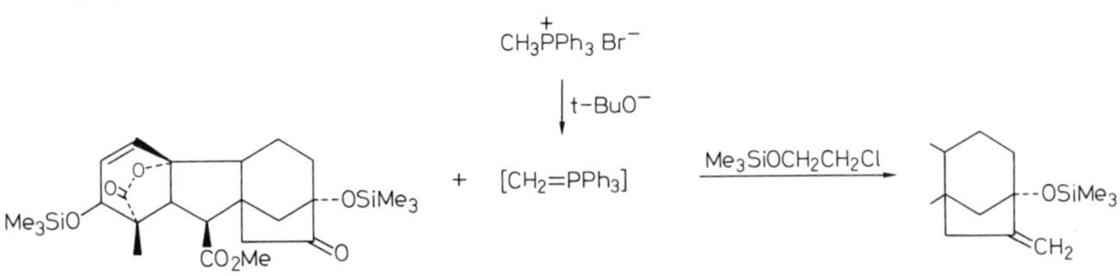

A stirred soln. of startg. ketone in tetrahydrofuran containing 20 equivalents 2-chloroethoxy(trimethyl)silane treated dropwise under N_2 at 20° with triphenylphosphine methylid until pale yellow, then further ylid added during 5 min. followed portionwise at hourly intervals until reaction completed → product. Y ca. 100%. 2-Chloroethoxy(trimethyl)silane, acting as a strong electrophile, intercepts any traces of *tert*-butoxide ion (in equilibrium with the ylid) or adventitious base which normally cause epimerization of the trimethylsiloxy groups. F.e., **also with replacement of benzoyloxy by trimethylsiloxy groups,** s. L.N. Mander, J.V. Turner, Tetrahedron Letters *22,* 4149–52 (1981).

Chloroaluminium-titanium complex ←

Methylene from carbonyl compds. $CO \rightarrow C{=}CH_2$

with chloroaluminium-titanium complex s. *35,* 622; f. methylene transfer complexes, terminal ethylene derivs. from aldehydes, s. T. Kauffmann et al., Tetrahedron Letters *22,* 5031–4 (1981).

Titanium tetraisopropoxide s. under $TiCl_4$ *$Ti(OPr\text{-}i)_4$*

Zirconium tetrabutoxide *$Zr(OBu)_4$*

2-Ethylation of terminal ethylene derivs. $CH{=}CH_2 \rightarrow CEt{:}CH_2$

887.

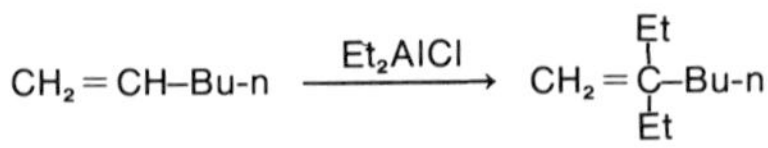

Startg. α-olefin added dropwise during 1 hr. under argon with stirring to a soln. of diethylaluminium chloride and a little Zr(IV) butoxide in heptane, and stirred for 1 hr. at 20° → 3-methyleneheptane. Y 89%. F.e.s. Y.M. Zzhemilev et al., Izv. Akad. Nauk SSSR Ser. Khim. *1979,* 2626–7.

Titanocene dichloride *$(C_5H_5)_2TiCl_2$*

Syntheses via 1,1-dimetalloalkenes $CO \rightarrow C{=}C{=}C$

Allenes from oxo compds. and alkynyl aluminium derivs.

888.

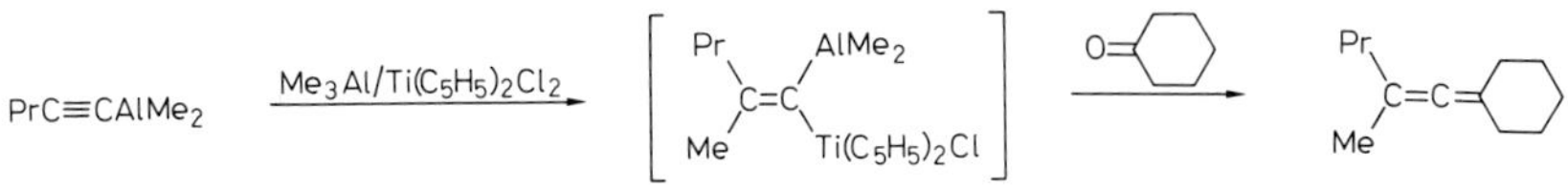

1-Pentynyldimethylaluminium treated with an equimolar amount of trimethylaluminium and titanocene dichloride in methylene chloride at room temp., and allowed to react with cyclohexanone for 30 min. at 0° → product. Y 83%. F.e., **also 1,2-ethylene-1,1-diiodides and α,β-ethyleneketones** using zirconocene dichloride, s. T. Yoshida, E. Negishi J. Am. Chem. Soc. *103,* 1276–7 (1981).

Zirconocene dichloride *$(C_5H_5)_2ZrCl_2$*

2-Ethylenesilanes from silyl-2-acetylenes $C{\equiv}C \rightarrow C(I){=}CR$

via 2-ethylene-3-iodosilanes

Regio- and stereo-specific synthesis

889.

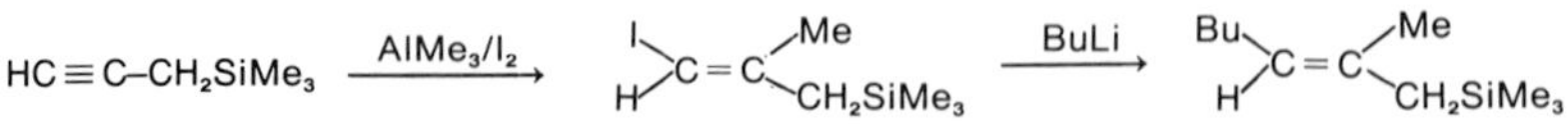

3-Trimethylsilyl-1-propyne added to a soln. of trimethylaluminium and a little zirconocene dichloride in 1,2-dichloroethane at 0°, the mixture stirred at room temp. for 2 hrs., and treated sequentially with a soln. of I_2 in tetrahydrofuran at –30°, ice-water, and 3 *M* HCl → intermediate 1-iodo-2-methyl-3-trimethylsilylprop-1-ene (Y 63%) in tetrahydrofuran treated with *n*-butyllithium at –50° for 3 hrs. → product (Y 78%; >98% stereochemical purity). F.e. and methods s. E. Negishi, F. Luo, C.L. Rand, Tetrahedron Letters *23,* 27–30 (1982); **α,β-ethyleneiodides** from dialkylaluminium chlorides cf. T. Yoshida, Chem. Letters *1982,* 293–4.

Titanium tetrachloride *$TiCl_4$*

2-Azetidinones from azomethines and O-silyl O-alkyl keteneacetals 

s. *33,* 879; s.a. I. Ojima, S. Inaba, M. Nagai, Synthesis *1981,* 545–7.

Bridgehead functionalization
with 2-ethylenesilanes and heteroanalogs ←

s. *35,* 623; from 1-adamantyl acetate, chloride, and silyl ether s. J. Org. Chem. *47,* 3219–24 (1982).

Titanium tetrachloride/titanium tetraisopropoxide *$TiCl_4/Ti(OPr\text{-}i)_4$*

γ-Ketocarboxylic acid esters
from 1-nitroethylene derivs.
and O-silyl O-alkyl keteneacetals $C(NO_2)=C \rightarrow CO-C-C-COOR$

890. 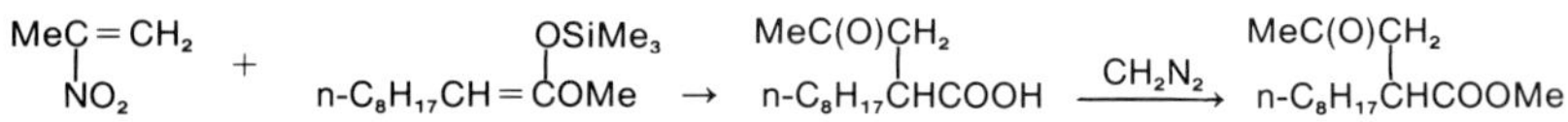

2-Nitropropene and startg. silyl acetal in dichloromethane containing $TiCl_4$ and Ti tetraisopropoxide stirred at −78° for 1 hr., the crude product hydrolysed in refluxing 70% aq. 1,2-dimethoxyethane, and esterified with diazomethane → product. Y 84%. F.e.s. M. Miyashita, T. Kumazawa, A. Yoshikoshi, Chem. Letters *1980,* 1043–4.

Stannic chloride *$SnCl_4$*

γ-Diketones
from enoxysilanes and 1-nitroethylene derivs. $C(OSi\equiv)=C \rightarrow CO-C-C-CO$

s. *32,* 888; s.a. Org. Synth. *60,* 117–20 (1981).

3-Ketothioethers from enoxysilanes and mercaptals $C(OSi\equiv)=C \rightarrow CO-C-C-SR$

891. 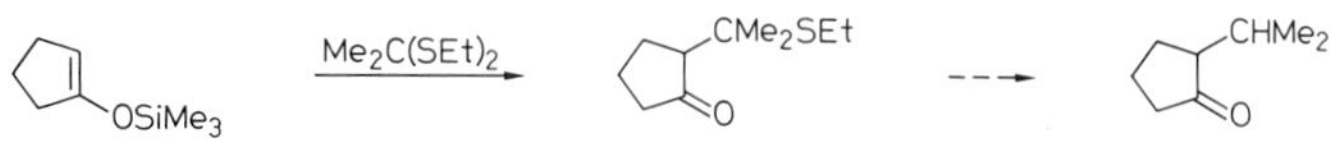

A suspension of $SnCl_4$ in methylene chloride at −60° treated with a cold soln. (−60°) of 2,2-bis(ethylthio)propane in the same solvent, immediately 1-trimethylsiloxycyclopentene in methylene chloride at the same temp. added during 1 min., and the mixture left to stand for 45 min. → 2-(2-ethylthio-2-propyl)cyclopentanone. Y 80%. – This is the key step of a **regiospecific** *α-sec-***alkylation of ketones.** F.e.s. M.T. Reetz, A. Giannis, Syn. Commun. *11,* 315–22 (1981).

Alkoxy-3-ethylenes
from 2-ethylenesilanes and monothioacetals $RO-C-SR' \rightarrow RO-C-C-C=C$

892. $PhCH_2CH_2OCH_2SMe + Me_3SiCH_2CH{=}CH_2 \longrightarrow PhCH_2CH_2OCH_2CH_2CH{=}CH_2$

A mixture of startg. silane, thioacetal, and $SnCl_4$ in dichloromethane reacted together at room temp. for 30 min. → product. Y 70%. F.e. and Lewis acids, also **3-ethylenethioethers** by C–O bond cleavage (preferably with $TiCl_4$), and **3-vinyl-O-heterocyclics** by intramolecular reaction, s. H. Nishiyama et al., J. Chem. Soc. Chem. Commun. *1982,* 459–60.

Stannic chloride *$SnCl_4$*

Replacement of alkoxy and acoxy by allyl groups

893. 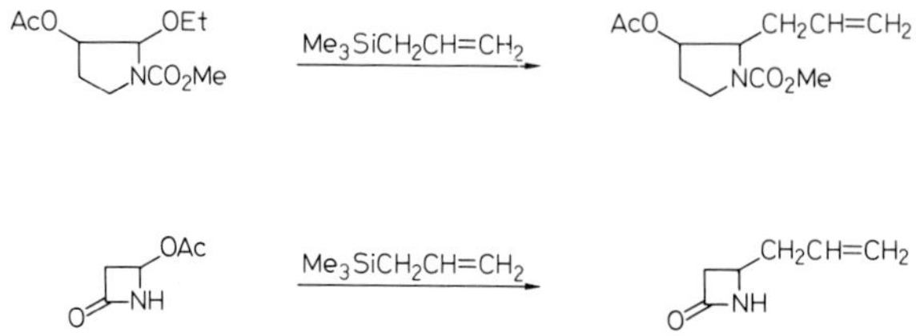

$SnCl_4$ added to a soln. of startg. carbamate and 2 equivalents allyltrimethylsilane in methylene chloride at room temp., and poured into brine when t.l.c. indicated completion of reaction → product. Y 99%. – Similarly 4-acetoxy-2-azetidinone with BF_3-etherate → 4-allyl-2-azetidinone. Y 63%. F.e.s. G.A. Kraus, K. Neuenschwander, J. Chem. Soc. Chem. Commun. *1982,* 134–5.

Triphenylphosphine s. under NaOH and Cu(acac)$_2$ *Ph_3P*

Hydrogen peroxide s. under PhLi *H_2O_2*

Trimethylsilyl triflate *$Me_3SiOSO_2CF_3$*

Synthesis of ketones
from enoxysilanes and acetoxy compds. C(OSi⋲)=C → CO–CR
s. *37,* 871

Molybdenum pentachloride *$MoCl_5$*

Sym. 1,4-Dienes 2 C=C–Si⋲ → C=C–C–C=C
from two enesilane molecules and acetals

894. 

A suspension of $MoCl_5$ in dichloromethane treated with a soln. of (E)-β-styryltrimethylsilane in the same solvent, benzaldehyde diethyl acetal added dropwise at –78°, and the mixture stirred at the same temp. for 2.5 hrs. then at –20° for 1.5 hrs. → 1,3,5-triphenyl-(E,E)-1,4-pentadiene. Y 61%. F.e. and Lewis acid catalysts s. T. Hirao et al., Tetrahedron Letters *22,* 3633–6 (1981).

*Tetra-*n-*butylammonium fluoride* n-$Bu_4N^+F^-$

Prepn. of *p*-quinodimethanes *in situ* from quaternary silylammonium salts ○
1,6-Elimination

895. 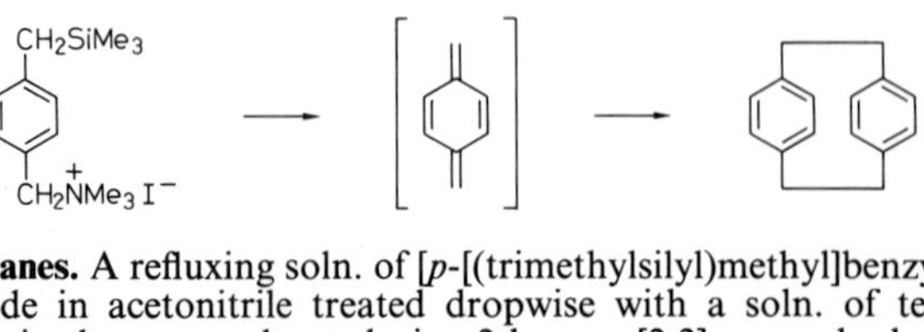

[2.2]Paracyclophanes. A refluxing soln. of [*p*-[(trimethylsilyl)methyl]benzyl]trimethylammonium iodide in acetonitrile treated dropwise with a soln. of tetrabutylammonium fluoride in the same solvent during 2 hrs. → [2.2]paracyclophane. Y 56%. Also furan- and thiophene-analog s. Y. Ito et al., J. Org. Chem. *46,* 1043–4 (1981).

Hydrogen chloride (s.a. under $PdCl_2$) *HCl*

Pyrogallol ring from 1,1,2,3-tetrasiloxy-1,3-dienes ○

896.

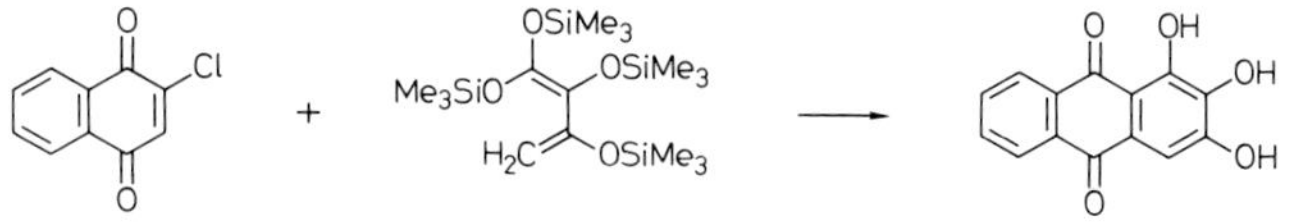

Anthragallols. A mixture of 2-chloronaphthoquinone and startg. siloxydiene heated at 100° for 36 hrs., and hydrolysed with 1:1 tetrahydrofuran/2% aq. HCl at 50° for a few min. → 1,2,3-trihydroxyanthraquinone. Y 82%. F.e.s. G. Roberge, P. Brassard, Synthesis *1981,* 381–3.

Catechol monoesters from acetylene derivs.

897.

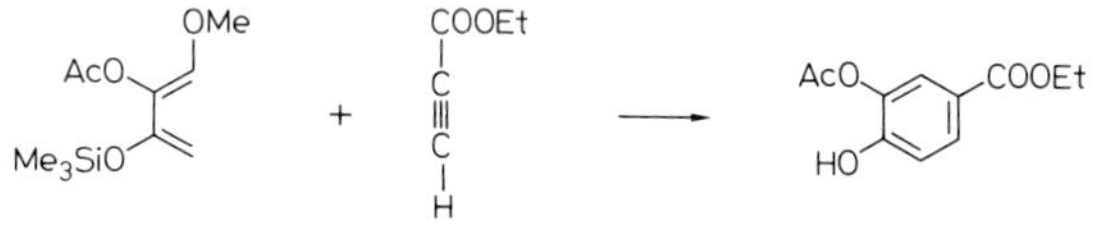

A soln. of ethyl propiolate and 1-methoxy-2-acetoxy-3-trimethylsilyloxybutadiene in benzene refluxed for 20 hrs. under argon, cooled to room temp., volatiles removed *in vacuo,* and the residue treated with 0.5 *M* HCl in tetrahydrofuran at room temp. for 30 min. → ethyl 3-acetoxy-4-hydroxybenzoate. Y 87%. F.e.s. S. Danishefsky, T.A. Craig, Tetrahedron *37,* 4081–6 (1981).

Diene synthesis with 1,3-disiloxy-2-aza-1,3-dienes
2-Pyridones from acetylene derivs.

898.

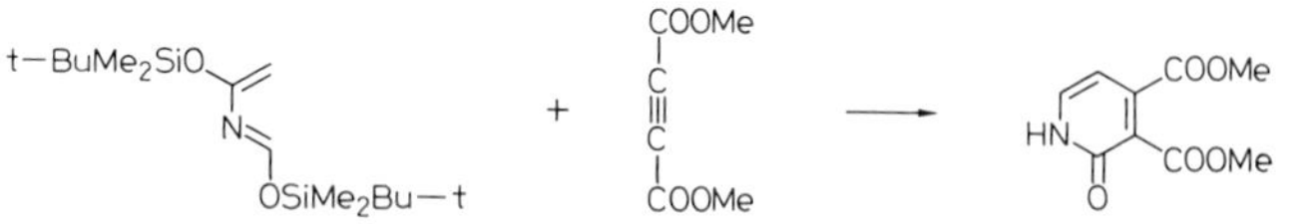

A mixture of the startg. disiloxyazadiene and 2 equivalents of dimethyl acetylenedicarboxylate refluxed in chloroform, and hydrolysed with a little concd. HCl → product. Y 64%. F.e., **also 3,4-dihydro-2-pyridones** from ethylene derivs., **and polycyclic heteroarenes from *p*-quinones,** s. F. Sainte et al., J. Am. Chem. Soc. *104,* 1428–30 (1982).

(2,2'-bipyridyl)diethylnickel *Ni(bipy)(Et)₂*

Syntheses of ketones COCl → COR

899. EtCOCl + (bipy)Ni(Et)$_2$ → EtCOEt

from carboxylic acid chlorides. Propionyl chloride added to a soln. of 2,2'-bipyridyl)diethylnickel complex in tetrahydrofuran, and the dark purple soln. worked up with hexane → diethyl ketone. Y 80%. F.e., **also from carboxylic acid esters or anhydrides,** and reactions s. T. Yamamoto, T. Kohara, A. Yamamoto, Bull. Chem. Soc. Japan *54,* 2010–6 (1981).

Dichlorobis(triphenylphosphine)nickel(II) s. under Mg *$NiCl_2(PPh_3)_2$*

Alkylrhodium(III) complex ←

Cross-coupling with monoorganomercury chlorides RHgCl → R–R'

s. *37,* 865

Palladium(II) acetate $(MeCOO)_2Pd$

Ketones from stannanes $ArN_2^+ + R_4Sn \rightarrow ArCOR$
3-Component synthesis
with halides s. *36*, 896; with diazonium salts in the presence of palladium(II) acetate cf. K. Kikukawa et al., Chem. Letters *1982*, 35–6.

Tetrakis(triphenylphosphine)palladium(0) $Pd(PPh_3)_4$

Regio- and geo-specific cross-coupling of alkenylaluminium alkyls with halides $C{=}C{-}Al{<} \rightarrow C{=}C{-}R$

900. 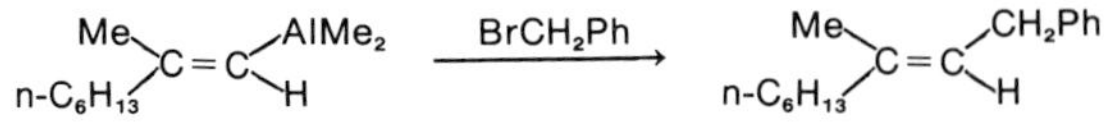

Allylarenes. (E)-2-Methyl-1-octenyldimethylaluminium (prepared from 1-octyne, trimethylaluminium, and zirconocene dichloride in 1,2-dichloroethane) added to a mixture of benzyl bromide and a little tetrakis(triphenylphosphine)palladium(0) in tetrahydrofuran, and stirred at room temp. for 3 hrs. → product. Y 91%. F.e., also by cross-coupling of α,β-ethylenehalides and organozinc halides, and **synthesis of diarylmethanes,** s. E. Negishi, H. Matsushita, N. Okukado, Tetrahedron Letters *22*, 2715–8 (1981); **1,4-dienes** s. J. Am. Chem. Soc. *103*, 2882–4 (1981).

Synthesis of ketones from 2-(organothio)enol phosphates via thioenolethers $C{=}COPO(OR)_2 \rightarrow C{=}CR'$

901. 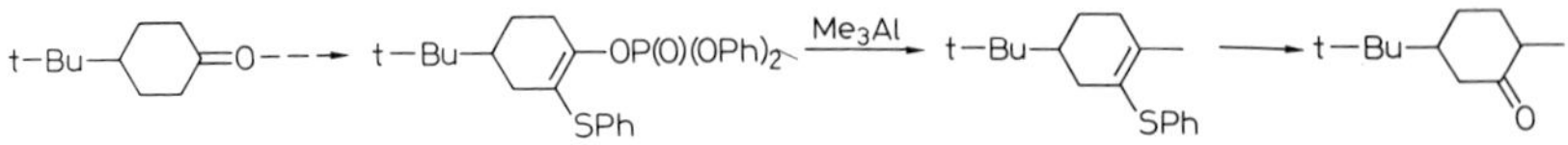

A soln. of the startg. enol phosphate (prepd. in 2 steps from 4-*tert*-butylcyclohexanone) in 1,2-dichloroethane treated with trimethylaluminium and a little tetrakis(triphenylphosphine)palladium(0), and the mixture heated at 80° for 4 hrs. → thioenolether (Y 80%), hydrolysed with $TiCl_4$ in water/methylene chloride → product (Y 78%). – This is part of a new **alkylative 1,2-transposition of keto groups.** F.e., also alkylative 1,3-transposition and synthesis of ketones **from thiolic acid esters via 1-(organothio)enol phosphates,** s. M. Sato et al., Tetrahedron Letters *22*, 1609–12 (1981).

Tetrakis(triphenylphosphine)palladium(0)/sodium ethoxide $Pd(PPh_3)_4/NaOEt$

Cross-coupling with eneboranes $ArHal \rightarrow ArC{=}C$
s. *36*, 893; **synthesis and applications of eneboranes, review,** s. H.C. Brown, J.B. Campbell, Aldrichimica Acta *14*, 3–11 (1981).

Tetrakis(triphenylphosphine)palladium(0)/sodium carbonate $Pd(PPh_3)_4/Na_2CO_3$

Aromatic cross-coupling with boronic acids $ArHal + Ar'B(OH)_2 \rightarrow Ar{-}Ar'$

902. $p\text{-}ClC_6H_4Br + (HO)_2B{-}C_6H_5 \rightarrow p\text{-}ClC_6H_4{-}C_6H_5$

Diaryls. A 10% excess of phenylboronic acid in ethanol added to a stirred mixture of *p*-chlorobromobenzene, a little tetrakis(triphenylphosphine)palladium(0), benzene, and 2 *M* Na_2CO_3 under N_2, refluxed for 6 hrs. with vigorous stirring, and residual boronic acid oxidized with 30% H_2O_2 at room temp. for 1 hr. → *p*-chlorobiphenyl. Y 89%. F.e.s. N. Miyaura, T. Yanagi, A. Suzuki, Syn. Commun. *11*, 513–9 (1981).

Tetrakis(triphenylphosphine)palladium(0)/magnesium halide $Pd(PPh_3)_4/MgHal_2$

1,3-Dienes by cross-coupling ←
with lithium divinylcuprates cf. *33*, 806s*36;* with vinylcopper(I) magnesium halide complexes s. Tetrahedron Letters *23*, 1589–92 (1982).

Palladium(II) chloride/tetrabutylammonium chloride/ hydrogen chloride *$PdCl_2/Bu_4N^+Cl^-/HCl$*

2-Chromanols from α,β-ethyleneketones and o-chloromercuriphenols ○

903. 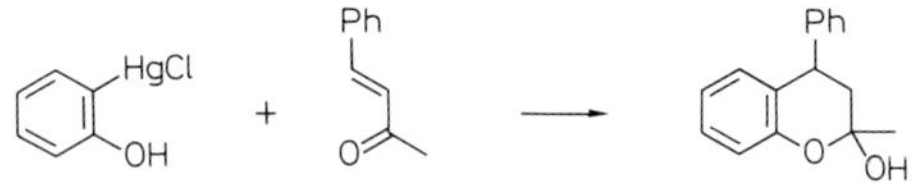

A stirred soln. of benzalacetone in dichloromethane treated with 3 *M* HCl, catalytic amounts of tetrabutylammonium chloride and $PdCl_2$, and (2-hydroxyphenyl)mercury chloride, and stirring continued for 4 hrs. at room temp. → 4-phenyl-2-methylchroman-2-ol. Y 86%. F.e.s. S. Cacchi, D. Misiti, G. Palmieri, J. Org. Chem. *47,* 2995–9 (1982).

Benzylchlorobis(triphenylphosphine)palladium(0) *$PhCH_2PdCl(PPh_3)_2$*

Ketones from carboxylic acid chlorides and stannanes COCl → COR

s. *34,* 862; with benzylchlorobis(triphenylphosphine)palladium(0) and further Pd catalysts, α,β-acetyleneketones, s. M.W. Logue, K. Teng, J. Org. Chem. *47,* 2549–53 (1982).

Via intermediates *v.i.*

Enamines from oxo compds. and α-aminophosphine oxides CO → C=C–N<

Synthesis with addition of one C-atom

from aldehydes s. *35,* 631; also from ketones s. Tetrahedron Letters *21,* 2671–4 (1980).

Carbon ↑ CC⇃↾C

Without additional reagents *w.a.r.*

C-Formylation with bis-2-imidazolidinylidenes via imidazolidines H → CHO

s. *22,* 874; N-(2,4-diformyl-5-hydroxyphenyl)acetamide s. J. Hocker, H. Giesecke, Org. Synth. *60,* 49–52 (1981).

Ethylene derivs. from α-carbalkoxyphosphonium salts and oxo compds. Wittig synthesis CO → C=C

904. 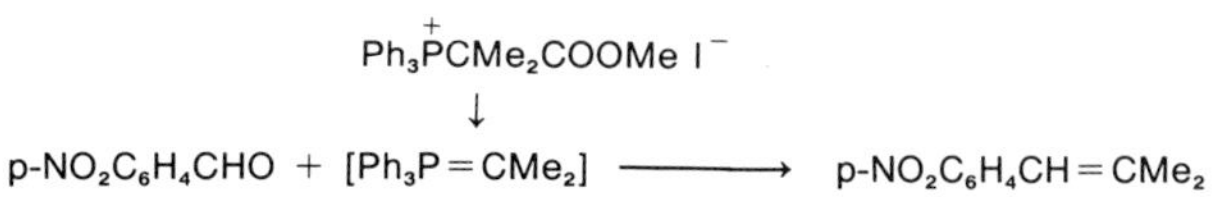

A mixture of startg. phosphonium salt and *p*-nitrobenzaldehyde under N_2 immersed in a preheated oil bath at 105°, after 3 min. the orange semi-solid heated to 119° during 8 min., and maintained at 119–122° for 4 min. → product. Y 90%. – The methods avoid handling sensitive alkylidenephosphoranes. F.e. and limitations s. J.L. Belletire, D.R. Walley, M.J. Bast, Synth. Commun. *12,* 469–75 (1982).

Reviews of interchange of ethylene derivs. ←

s. *30,* 467s*31;* s.a. R.L. Banks, Catalysis (London) *1981,* 100–29.

Improved prepn. of fulminic acid ○

Δ^2-Isoxazolines from ethylene derivs.

905. 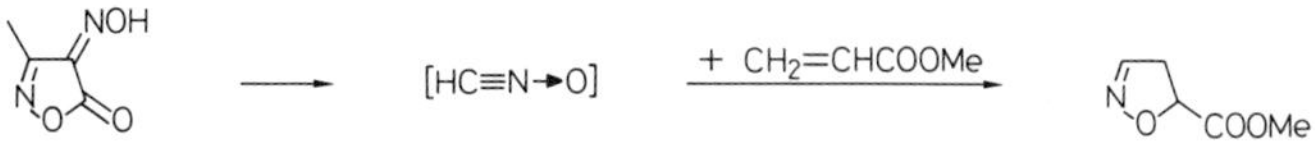

Startg. 4-oximino-isoxazol-5(4*H*)-one at 100–120° sublimed into a horizontal quartz tube at 450° and 10^{-3} to 10^{-4} torr during 10 hrs., the mixture allowed to attain atm.

pressure, and treated with methyl acrylate at –196° → product. Y 99%. – The method provides a safe and facile synthesis of chemically pure fulminic acid in essentially quantitative yield. F.e.s. C. Wentrup, B. Gerecht, H. Briehl, Angew. Chem. Intern. Ed. *18,* 467–8 (1979).

Isocoumarins from homophthalic acids
and carboxylic acid anhydrides cf. *21,* 894; and carboxylic acid halides, 3-arylisocoumarins, s. K. Nozawa et al., Chem. Pharm. Bull. *29,* 2491–5 (1981).

Pyrazoles from sydnones
s. *23,* 885; 1-amino- and 5-arylthio-pyrazoles s. H. Gotthardt, F. Reiter, Chem. Ber. *114,* 2450–64 (1981).

Electrolysis ϟ

Kolbe synthesis of dicarboxylic acid esters 2 RCOOH → R–R
s. *1,* 716; dimethyl decanedioate s. D.A. White, Org. Synth. *60,* 1–5 (1981).

Sodium hydride *NaH*

Pyrroles from α-acylaminonitriles and enephosphonium salts via intramolecular Wittig synthesis ○

906.

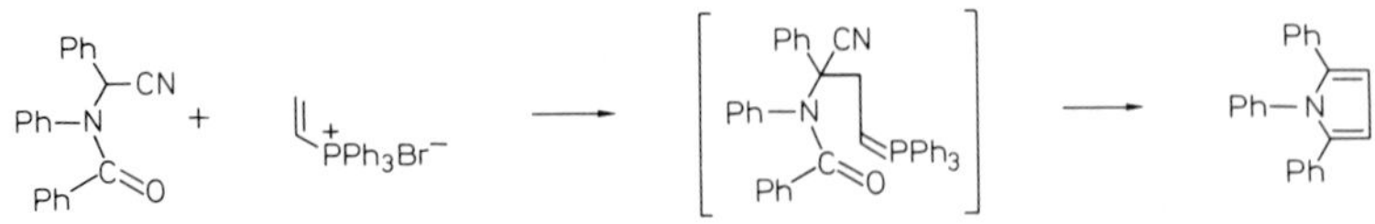

A soln. of startg. open-chain Reissert compd. in dimethylformamide added with stirring to NaH during 10 min. under N_2, after evolution of H_2 ceased the mixture treated with a soln. of vinyltriphenylphosphonium bromide in the same solvent, and heated at 100° for 24 hrs. → product. Y 100%. F.e.s. J.V. Cooney, W.E. McEwen, J. Org. Chem. *46,* 2570–3 (1981).

Potassium hydroxide *KOH*

***o*-Cyanoureas from aromatic nitro compds. and sec. amines** ←

907.

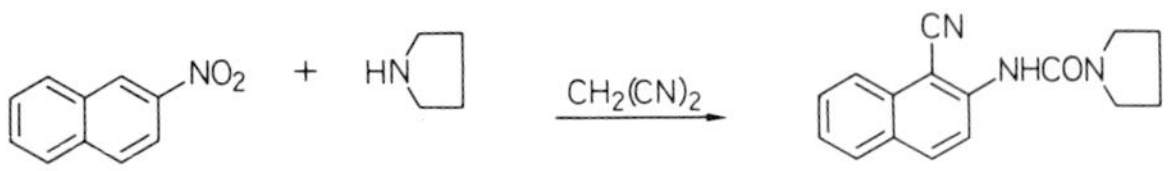

A stirred soln. of malononitrile, KOH, and pyrrolidine in dimethylformamide treated portionwise with 2-nitronaphthalene at room temp., and stirring continued for 24 hrs. → product. Y 58%. F.e., also with 6-nitroquinoline, s. T. Ohshima et al., Chem. Pharm. Bull. *29,* 1328–34 (1981); ***o*-acylaminonitriles** s. ibid. *29,* 1286–91, 1292–8 (1981).

Sodium ethoxide *NaOEt*

Uracil ring interchange via retro-Michael reaction ○

908.

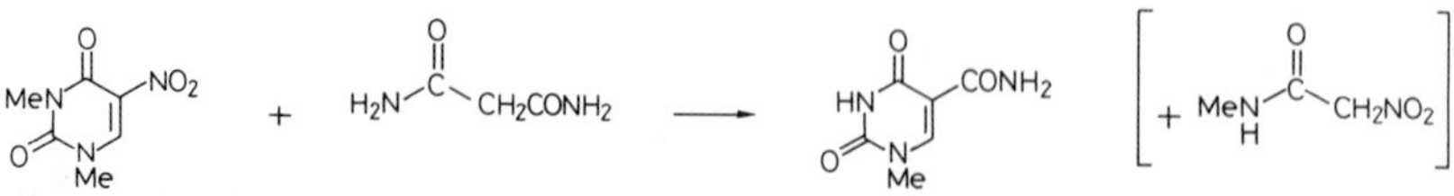

A soln. of 1,3-dimethyl-5-nitrouracil in ethanol refluxed with malonamide and Na ethoxide for 1 hr. → 5-carbamoyl-1-methyluracil. Y 84%. F.e.s. K. Hirota, Y. Kitade, S. Senda, Tetrahedron Letters *22,* 2409–10 (1981).

Sodium tert-*butoxide* *NaOBu*-t

Geospecific Wittig synthesis with cyanohydrins C(OH)CN → C=C

909. 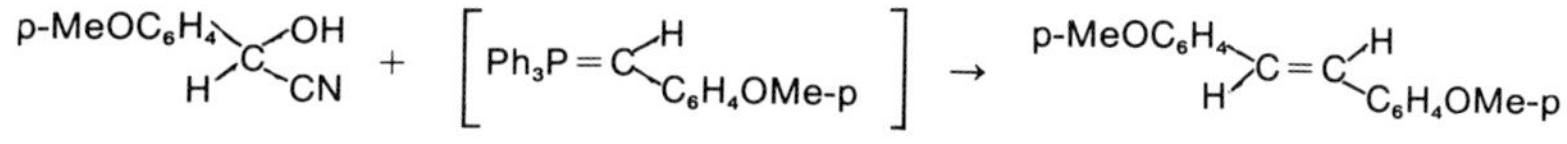

Stilbenes. A mixture of startg. cyanohydrin and *p*-methoxybenzyltriphenylphosphonium bromide in dioxane stirred under N_2 for 30 min., a soln. of Na *tert*-butoxide in the same solvent added, stirring continued for 2 hrs. under gentle reflux, the mixture cooled during 15 min., filtered, and the filtrate worked up → 4,4′-dimethoxystilbene. Y 82% (95% E-isomer). F.e.s. G. Kossmehl, R. Nuck, Chem. Ber. *114,* 2914–20 (1981).

Methyllithium/tert-*butyllithium* *MeLi*/t-*BuLi*

Indoles from *o*-bromacylamines and *α*-chloroketones ○

N,C-Dianions as intermediates

910. 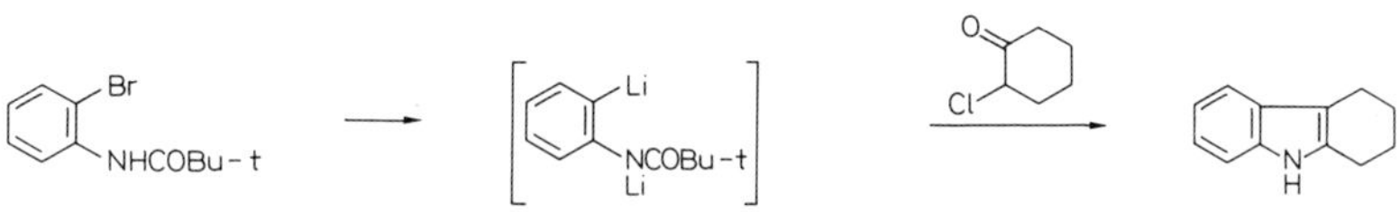

A soln. of 2′-bromo-2,2-dimethylpropionanilide in tetrahydrofuran treated with methyllithium at −78°, stirred for 10 min., 2 equivalents *tert*-butyllithium added during 5 min., the soln. kept at −78° for 1 hr., warmed slowly to −10° during 20 min., 2-chlorocyclohexanone added during 10 min., the mixture treated with K *tert*-butoxide and water at −10°, and quenched with 5% HCl at −78° → tetrahydrocarbazole. Y 77%. F.e., **also from *α,β*-ethylenealdehydes,** and regiospecificity, s. P.A. Wender, A.W. White, Tetrahedron Letters *22,* 1475–8 (1981).

Lithium chloride/*hexamethylphosphoramide* *LiCl*/$(Me_2N)_3PO$

Decarbalkoxylative C-alkylation ←

s. *32,* 903; spirocyclics by intramolecular process s. R.G. Eilerman, B.J. Willis, J. Chem. Soc. Chem. Commun. *1981,* 30–2.

Magnesium *Mg*

Synthesis of amines from *α*-aminonitriles C–CN → C–R

s. *7,* 858; aminosugar derivs. s. H. Böshagen, W. Geiger, B. Junge, Angew. Chem. Intern. Ed. *20,* 806 (1981).

Titanium tetrachloride $TiCl_4$

5-Subst. 2(5H)-furanones from 2-acoxyfurans ←

Hexamethylphosphoramide s. under LiCl $(Me_2N)_3PO$

911. 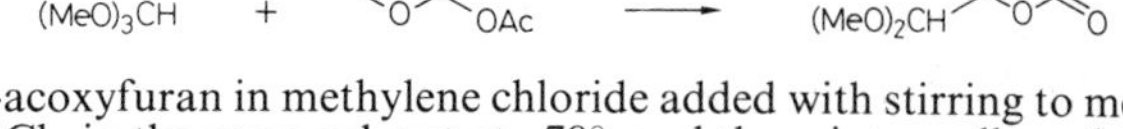

A soln. of 2-acoxyfuran in methylene chloride added with stirring to methyl orthoformate and $TiCl_4$ in the same solvent at −78°, and the mixture allowed to stand for an additional 2 hrs. → product. Y 89%. F.e. and with aldehydes as electrophile s. T. Shono, Y. Matsumura, S. Yamane, Tetrahedron Letters *22,* 3269–72 (1981); from 2-siloxyfurans with $ZnBr_2$ cf. A. Pelter, R. Al-Bayati, W. Lewis, ibid. *23,* 353–6 (1982); with $SnCl_4$ s. M. Asoaka et al., ibid. *22,* 4269–70 (1981).

p-*Toluenesulfonic acid* *TsOH*
Pyridines from two enamine molecules ○

912. 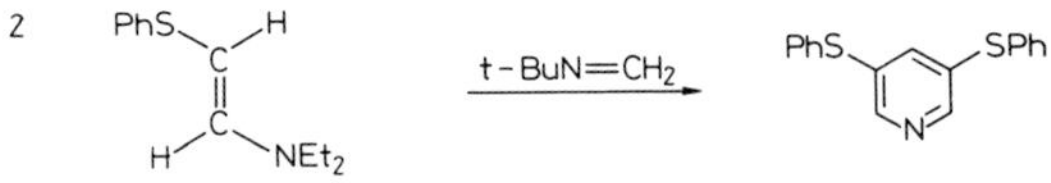

A 1:2 mixture of N-methylene-*tert*-butylamine and startg. enamine in benzene heated at 200° in the presence of a little *p*-toluenesulfonic acid for 9 hrs. → product. Y 87% by n.m.r. F.e.s. M. Komatsu et al., Angew. Chem. Intern. Ed. *21,* 213–4 (1982).

Elimination ⇑

Hydrogen ↑ CC⇑H

Irradiation s. under O_2 ⇈

Sodium ethoxide s. under t-*BuOCl* *NaOEt*

Mercuric acetate $(MeCOO)_2Hg$
Phenols from 2-cyclohexenones ←
with Pd–charcoal cf. *28,* 721; with mercuric acetate, *m*-aminophenols, s. H. Iida, Y. Yuasa, C. Kibayashi, Synthesis *1982,* 471–2.

Trifluoroacetic anhydride s. under Ph_2SeO $(CF_3CO)_2O$

Dimethyl sulfide s. under t-*BuOCl* Me_2S

Diphenyl diselenide/m-*iodoxybenzoic acid* $PhSeSePh/ArIO_2$
α,β-Ethyleneketones from ketones COCHCH → COC=C
Steroidal cross-conjugated dienones

913. 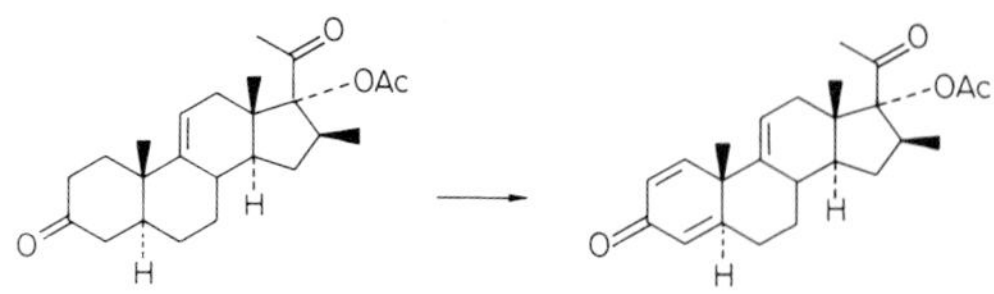

A mixture of *m*-iodoxybenzoic acid and a little diphenyl diselenide in toluene refluxed for 15 min., startg. m. added, and heating continued for a further 2.5 hrs. → product. Y 86%. – The method, involving **formation of benzeneseleninic anhydride** *in situ,* is inexpensive and avoids side reactions often associated with the use of benzeneseleninic anhydride in equivalent amount. F.e. and catalyst s. D.H.R. Barton et al., J. Chem. Soc. Chem. Commun. *1981,* 1044–5.

Diphenyl selenoxide/trifluoroacetic anhydride $Ph_2SeO/(CF_3CO)_2O$
Isoquinolines from 3,4-dihydroisoquinolines ←
s. *37,* 416

Chloranil ←
Dehydrogenation of steroids CH–CH → C=C
$\Delta^{4,6}$-Ketosteroids
s. *12,* 878; s.a. Z. Chen, Y. Wen, Y. Chen, Yaoxue Xuebao *17,* 299 (1982) (Ch.); Chem. Abstr. *97,* 92624 (1982).

2,3-Dichloro-5,6-dicyanoquinone ←

o-Quinones from cyclohex-3-ene-1,2-diol ring ←

914.

A soln. of startg. diol in dioxane treated with 12 equivalents 2,3-dichloro-5,6-dicyanoquinone, and the mixture refluxed for 24 hrs. → product. Y 63%. F.e.s. K.L. Platt, F. Oesch, Tetrahedron Letters *23,* 163–6 (1982).

Oxygen/irradiation O_2/*ℋ*

Cyclodehydrogenation ○

s. *28,* 890; 12*H*-[1]benzopyrano-[2,3-*h*]isoquinolin-12-ones and -[2,3-*f*]quinolin-12-ones s. I. Yokoe et al., J. Chem. Soc. Chem. Commun. *1981,* 442–3; naphthacenes s. W. Verboom, J.J.M. Brants, H.J.T. Bos, Rec. Trav. Chim. Pays-Bas *101,* 68–77 (1982).

tert-*Butyl hypochlorite/dimethyl sulfide/sodium ethoxide* t-*BuOCl/*Me_2S*/NaOEt*

Indoles from indolines under mild conditions ←

Indolin-1-ylsulfonium salts as intermediates

915.

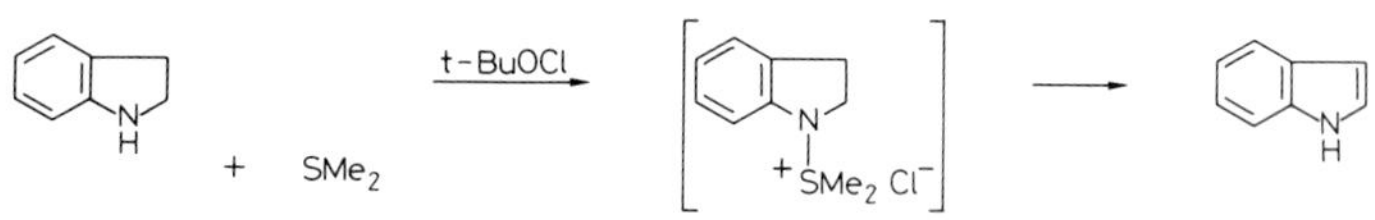

A soln. of *tert*-butyl hypochlorite in methylene chloride added dropwise to a vigorously stirred mixture of indoline, dimethyl sulfide, and *triethylamine* in the same solvent at –65° for 1 hr., stirred for 2 hrs. at this temp., ethanolic Na ethoxide added over 10 min., and allowed to warm to room temp. for 2 hrs. → indole. Y 82%. – The method is applicable to substrates containing groups sensitive to acids, strong bases, and elevated temp. F.e.s. Y. Kikugawa, M. Kawase, Chem. Letters *1981,* 445–6.

Via intermediates *v.i.*

1-Nitroethylene derivs. from nitro compds.
via 1,1-nitroselenides $CHCHNO_2 \rightarrow C{=}CNO_2$
One-pot synthesis

916.

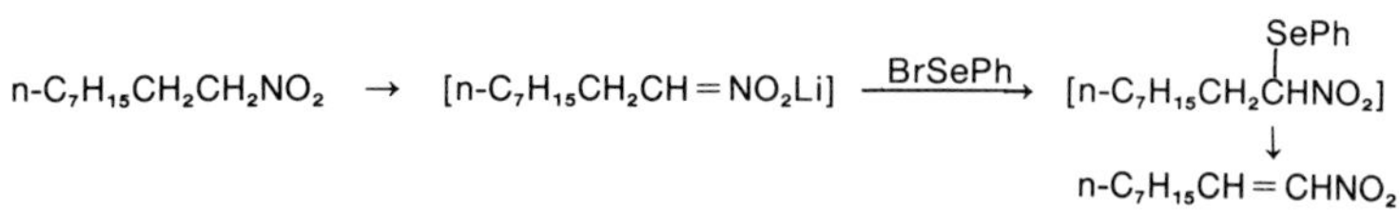

A soln. of phenylselenyl bromide in tetrahydrofuran added to a mixture of *n*-butyllithium and startg. nitroalkane in the same solvent at 0° under N_2, and treated with 35% aq. H_2O_2 → product. Y 83%. F.e., regiospecificity, and isolation of the intermediate, s. T. Sakakibara et al., J. Chem. Soc. Chem. Commun. *1981,* 261–2; with K nitronates and reactions of 1-nitroselenides cf. V.I. Erashko et al., Izv. Akad. Nauk SSSR Ser. Khim. *1982,* 161–5; prepn. and hydroxymethylation of the intermediates s. T. Sakahibara, S. Ikuta, R. Sudoh, Synthesis *1982,* 261–3.

Oxygen ↑ CC⇑O

Without additional reagents *w.a.r.*

Ethylene derivs. from urethans CH–C–OCON< → C=C

s. *19,* 930; f.e.s. R.F. Atkinson et al., J. Org. Chem. *46,* 2804–6 (1981).

N-Condensed 1,2,3,4-tetrahydropyridine ring
by intramolecular heterodiene synthesis ○

917. 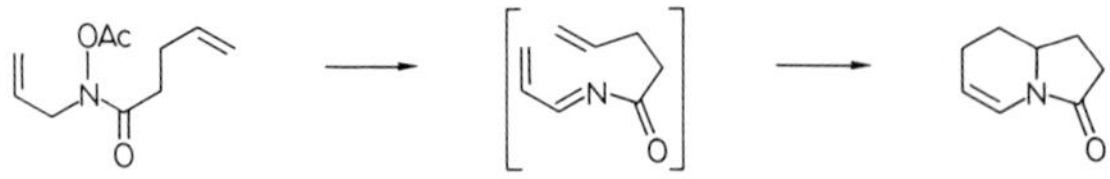

Startg. m. evaporated through a hot tube, the pyrolysate dissolved in ether, and passed through a column of K_2CO_3 to remove the formed acetic acid → product. Y 73%. F.e.s. Y.-S. Cheng, F.W. Fowler, A.T. Lupo, J. Am. Chem. Soc. *103,* 2090–1 (1981); **indolizidines** via N-acylformimines cf. N.A. Khatri et al., J. Am. Chem. Soc. *103,* 6387–93 (1981); N-condensed tetrahydropyridine ring **from N-acylimines,** stereospecificity, s. B. Nader et al., ibid. *103,* 7573–80 (1981).

Pyridine from 1,3-oxazin-4-one ring

918. 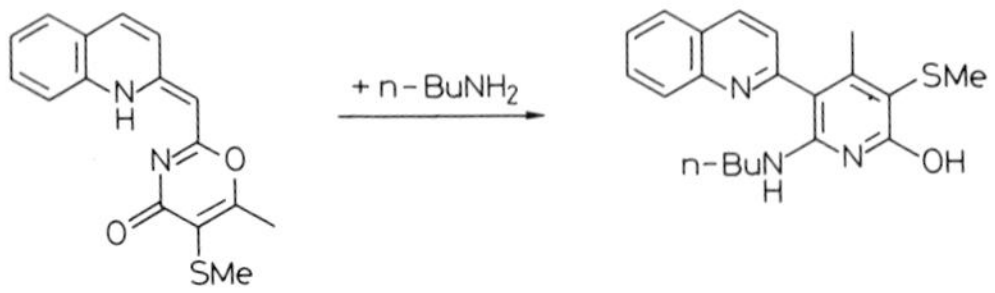

2-Amino-6-hydroxypyridines. 6-Methyl-5-methylthio-2-[2(1*H*)-quinolylidene]methyl-1,3-oxazin-4-one treated with *n*-butylamine, and the mixture kept at room temp. for 10 min. → product. Y 85%. F.e., **also 2,6-pyridiones,** s. T. Kinoshita et al., Chem. Pharm. Bull. *28,* 2892–9 (1980).

Electrolysis ϟ

Intramolecular reductive 3-alkylation of 2-cyclohexenones
Cathodic ring closure

919. 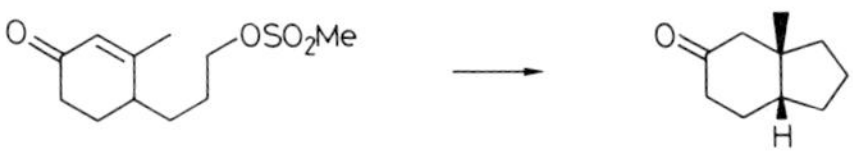

Bicyclo[4.3.0]nonan-3-ones. Startg. m. electrolysed at –2.20 V in dry dimethylformamide containing 0.1 *M* tetra-*n*-butylammonium perchlorate or tetrafluoroborate as the supporting electrolyte using a Pt-gauze anode and a Hg-pool cathode → 1-methyl-*cis*-bicyclo[4.3.0]nonan-3-one. Y 82%. Also prepn. of **bicyclo[4.1.0]heptan-3-ones** s. P.G. Gassman et al., J. Org. Chem. *46,* 5455–7 (1981).

Sodium hydride (s.a. under Bis{1,2-bis(diphenylphosphino)ethane}palladium(0)) *NaH*

1-Nitroethylene derivs. from 2-nitroalcohols $C(OH)CHNO_2 \rightarrow C{=}CNO_2$

920. 

Startg. *β*-nitroalcohol in tetrahydrofuran added dropwise with stirring at room temp. to a suspension of NaH in the same solvent under N_2, and stirring continued for 45 min. → product. Y 85%. – This is part of a multistep regiospecific prepn. of 1-nitroethylene derivs. **from ketones.** F.e.s. P. Dampawan, W.W. Zajac, Tetrahedron Letters *23,* 135–8 (1982).

Vinylcyclopropanes from acoxy-2-ethylenes

921.

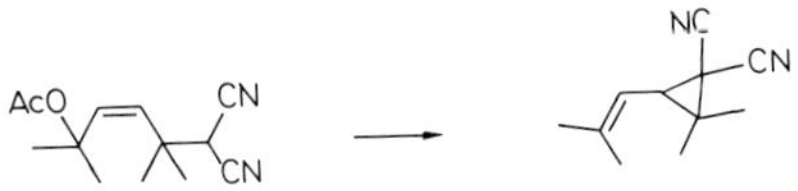

A soln. of startg. m. in tetrahydrofuran added at room temp. to a suspension of NaH in the same solvent, and the mixture stirred at 65° until t.l.c. indicated completion → 1,1-dicyano-2,2-dimethyl-3-(2-methyl-1-propenyl)cyclopropane. Y 75%. F.e. and stereospecificity in the presence of tetrakis(triphenylphosphine)palladium(0) s. J.P. Genet, F. Piau, J. Org. Chem. *46,* 2414–7 (1981).

Potassium hydroxide *KOH*

Phenol ring from α,β-ethylene-δ-ketoaldehydes ○

s. *37,* 319

Sodium ethoxide *NaOEt*

Carbostyrils from o-acylaminoketones

with K *tert*-butoxide s. *21,* 912; with NaOEt, 3-sulfonylamino derivs., s. F. Bahr, H. Usbeck, Pharmazie *36,* 668–70 (1981).

Potassium tert-*butoxide* *KOBu*-t

Intramolecular aldol condensation

with ZnO cf. *31,* 915; with K *tert*-butoxide in *tert*-butanol, cyclic α,β-ethyleneketones, s. A.B. Smith, B.A. Wexler, J. Slade, Tetrahedron Letters *23,* 1631–4 (1982).

Potassium 2,6-di-tert-*butylphenoxide* *KOAr*

Acetylene derivs. from enol triflates $CH=C(OSO_2CF_3) \rightarrow C\equiv C$

with pyridine cf. *29,* 920; with K 2,6-di-*tert*-butylphenoxide, simultaneous cleavage of silyl groups, s. M. Ladika, P.J. Stang, J. Chem. Soc. Chem. Commun. *1981,* 459–60.

Ethyllithium *EtLi*

Cumulenyl ethers from 4-siloxy-2-acetyleneethers $C=C=C=COR$

922.

$$Me_2C(OSiMe_3)-C\equiv C-CH_2OMe \rightarrow Me_2C=C=C=CHOMe$$

A well-stirred soln. of startg. m. in ether treated dropwise with ethyllithium in the same solvent during ca. 30 min. at –50°, the temp. allowed to rise to 0°, and a satd. soln. of NH_4Cl added with vigorous stirring and cooling below 0° → product. Y 87%. F.e.s. R.G. Visser, H.J.T. Bos, L. Brandsma, Rec. Trav. Chim. Pays-Bas *100,* 34–6 (1981); **from α,β-acetylene-γ-halogenacetals** cf. A. Gorgues, A. Le Coq, Tetrahedron Letters *21,* 5007–10 (1980).

n-Butyllithium (s.a. under WCl_6) n-*BuLi*

Indoles from carboxylic acid anilides ○

Improved Madelung synthesis

923. 

A stirred soln. of N-(2-methylphenyl)benzamide in tetrahydrofuran treated dropwise under N_2 at 20° with 3 equivalents of *n*-butyllithium in hexane, stirring continued at room temp. for 16 hrs., and the ice-cooled mixture treated dropwise with 2 *M* HCl → 2-phenylindole. Y 90%. F.e. and with Li diisopropylamide s. W.J. Houlihan, V.A. Parrino, Y. Uike, J. Org. Chem. *46,* 4511–5 (1981).

Potassium carbonate/dicyclohexyl-18-crown-6 polyether ←

1-Nitroethylene derivs. from 2-acoxynitro compds. $C(OAc)CHNO_2 \rightarrow C{=}CNO_2$

with KOH cf. *14,* 912; with K_2CO_3/dicyclohexyl-18-crown-6 polyether s. G. Just, H. Oh, Can. J. Chem. *59,* 2729–36 (1981).

Sodium acetate *MeCOONa*

Decarboxylative ring closure ○

s. *32,* 988; thiochroman-3-ones s. P.D. Clark, D.M. McKinnon, Can. J. Chem. *60,* 243–8 (1982).

Potassium hydrogen sulfate *$KHSO_4$*

Cyclic β-diketones from α-alkoxymethyl-α-hydroxyketones ○

s. *37,* 869

Sodium iodide s. under Me_3SiCl *NaI*

Trimethylamine s. under Me_3SiI *Me_3N*

Morpholine/camphoric acid ←

Cyclic α,β-ethylenealdehydes from dialdehydes ○

with piperidinium acetate cf. *15,* 168; with morpholine/camphoric acid, regiospecificity, s. T. Harayama et al., Chem. Pharm. Bull. *29,* 766–72 (1981).

Zinc/copper s. under $TiCl_3$ *Zn/Cu*

Bromomagnesium dialkylamides *$BrMgNR_2$*

Dieckmann cyclization

924. 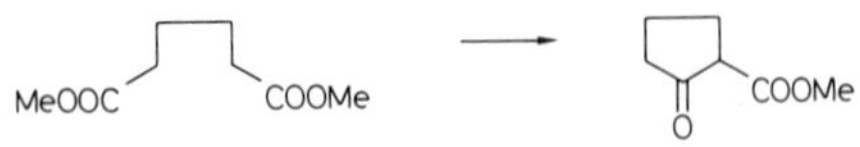

A soln. of startg. diester in dry ether added to bromomagnesium diethylamide in the same solvent at ice-bath temp. with continuous stirring under N_2, allowed to stand for 2–3 hrs., and refluxed for 30 min. → product. Y 75–80%. F.e. and with Li dialkylamides, **also Claisen condensation,** s. P.K. Singh, K. Rajeshwari, K. Ranganayakulu, Indian J. Chem. *19B,* 823–4 (1980).

Zinc chloride *$ZnCl_2$*

Total steroid synthesis by cationic double ring closure

with $SnCl_4$ cf. *29,* 929; with $ZnCl_2$ s. M.B. Groen, B. Hindriksen, F.J. Zeelen, Rec. Trav. Chim. Pays-Bas *101,* 148–55 (1982).

Zinc chloride/acetic acid *$ZnCl_2/MeCOOH$*

Anthracene ring from 3-arylphthalides

925.

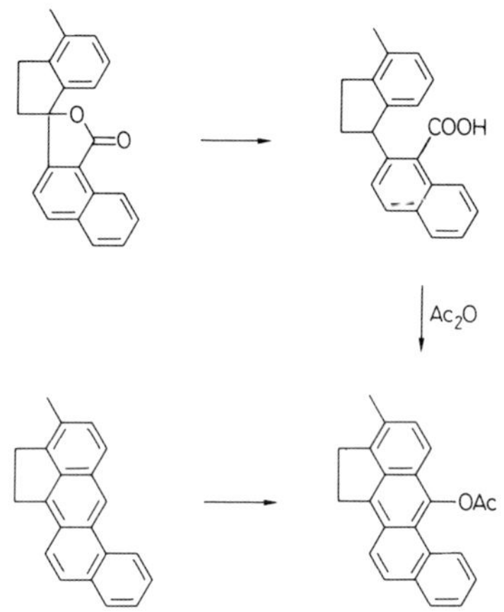

Polyarenes. A soln. of startg. lactone in pyridine added to a suspension of activated Zn dust in aq. methanolic KOH, and the mixture refluxed with stirring for 2.5 hrs. → acid (Y 91%), in glacial acetic acid treated with $ZnCl_2$, and refluxed with stirring for 20 min. → 6-acetoxy-3-methylcholanthrene (Y 90%), refluxed with 56% HI and 50% H_3PO_2 in propionic acid for 10 min. → 3-methylcholanthrene (Y 93%). F.e., also direct one-pot conversion with limitations, s. R.G. Harvey, C. Cortez, S.A. Jacobs, J. Org. Chem. *47,* 2120–5 (1982).

Sodium tetrahydridoborate/trifluoroacetic acid *$NaBH_4/CF_3COOH$*

Aromatization by elimination of O-bridges ←
with Zn/acetic acid cf. *14,* 866; with $NaBH_4/CF_3COOH$ s. G.W. Gribble, W.J. Kelly, M.P. Sibi, Synthesis *1982,* 143–5; with $Fe_2(CO)_9$ s. W.M. Best et al., Australian J. Chem. *35,* 843–8; with $TiCl_4/LiAlH_4$ s. Y.D. Xing, N.Z. Huang, J. Org. Chem. *47,* 140–2 (1982).

Lithium tetrahydridoaluminate s. under $TiCl_4$ *$LiAlH_4$*

Aluminium chloride *$AlCl_3$*

Indoles from α-arylaminoketones ○
with $ZnCl_2$ cf. *3,* 729; f. catalysts, especially $AlCl_3$, s. H. Galons et al., J. Heterocyc. Chem. *18,* 561–3 (1981).

Dicyclohexyl-18-crown-6 polyether s. under K_2CO_3 ←

Diethyl azodicarboxylate s. under Ph_3P *EtOOC−N=N−COOEt*

Phthalic anhydride ←

1-Nitroethylene derivs. from 2-nitroalcohols $C(OH)CHNO_2 \rightarrow C{=}CNO_2$
with $MeSO_2Cl/Et_3N$ cf. *31,* 930; with phthalic anhydride, 2-nitropropene, s. M. Miyashita, T. Yanami, A. Yoshikoshi, Org. Synth. *60,* 101–3 (1981).

Trifluoroacetic anhydride *$(CF_3CO)_2O$*

Ring closures via Pummerer rearrangement ○

Cyclic α-(alkylthio)ketones from β-ketosulfoxides
with CF_3COOH cf. *29,* 926; using $(CF_3CO)_2O$ with migration of carbon-carbon double bonds s. L.N. Mander, P.H.C. Mundill, Synthesis *1981,* 620–2; 3-alkylthio-

4-vinyl-2-pyrrolidones s. Y. Tamura et al., Tetrahedron Letters *22,* 4301–4 (1981); 1-alkylthio-3,4-dihydrophenanthrene-2(1*H*)-ones s. Australian J. Chem. *34,* 1235–42 (1981); 1,2,3,4-tetrahydrocarbazoles s. Y. Oikawa et al., Chem. Pharm. Bull. *29,* 1606–14 (1981).

Anion-cation exchanger ←

Simultaneous acid-base catalysis
Cyclopent-2-enones from γ-ketoacetals via γ-ketoaldehydes

926. 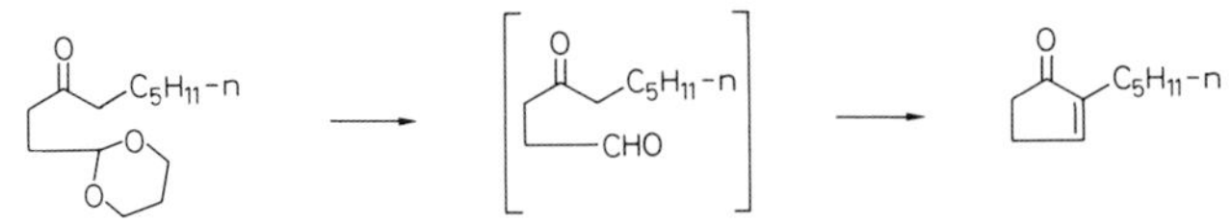

A mixture of 2-(3-oxononyl)-1,3-dioxane and R-276 Rexyn 300 ion-exchange resin in methanol refluxed with stirring for 1–2 hrs. → 2-*n*-pentyl-2-cyclopenten-1-one. Y 56%. – The ion-exchange resin consists of both sulfonic acid and quaternary ammonium hydroxide beads, and catalyses both steps of the reaction concurrently. F.e.s. J.C. Stowell, H.F. Hauck, J. Org. Chem. *46,* 2428–9 (1981).

Acetic acid (s.a. under $ZnCl_2$) *MeCOOH*

Thiasteroids ○

s. *36,* 929; review of partial and total synthesis of thiasteroids s. S.R. Ramadas et al., Heterocycles *19,* 861–933 (1982).

Stearic acid/palmitic acid *RCOOH*

Ethylenealcohols from diols
Partial dehydration C(OH)CH → C=C

927. $HOCH_2CH_2(CH_2)_{12}OH \rightarrow CH_2=CH(CH_2)_{12}OH$

A mixture of stearic acid, palmitic acid, and 1,14-tetradecanediol heated with distillation for 19 hrs. at 330–40° and 700 mmHg in a constant volume backmix flow reactor equipped with a fractionating column for recycling of unreacted startg. m. → 13-tetradecen-1-ol. Y 83% (based on 91% conversion). F.e.s. T. Yamanaka, T. Imai, Bull. Chem. Soc. Japan *54,* 1585–6 (1981).

Camphoric acid s. under Morpholine ←

Trifluoroacetic acid (s.a. under $NaBH_4$) *CF_3 COOH*

Transannular bicyclo[4.1.0]heptan-2-ol ring opening C ○
with simultaneous isocyclic ring closure
Condensed 6,7-dihydrotropolone ethers

928. 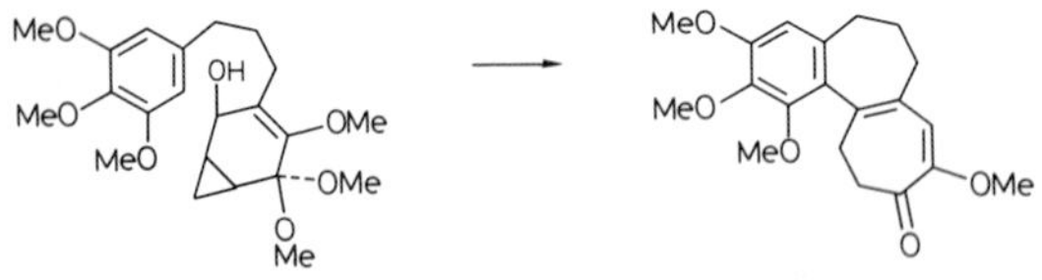

A soln. of 4,5,5-trimethoxy-2-[3-(3,4,5-trimethoxyphenyl)prop-1-yl]bicyclo[4.1.0]-hept-3-en-2-ol in trifluoroacetic acid stirred at room temp. for 18 hrs. → 11,12-dihydrodesacetamidoisocolchicine. Y 68%. F.e.s. D.A. Evans, S.P. Tanis, D.J. Hart, J. Am. Chem. Soc. *103,* 5813–21 (1981).

Hexamethyldisilazane s. under Me_3SiI *$(Me_3Si)_2NH$*

Trimethylsilyl chloride/sodium iodide *Me_3SiCl/NaI*

Ethylene derivs. from glycols under mild, neutral conditions C(OH)C(OH) → C=C

929. 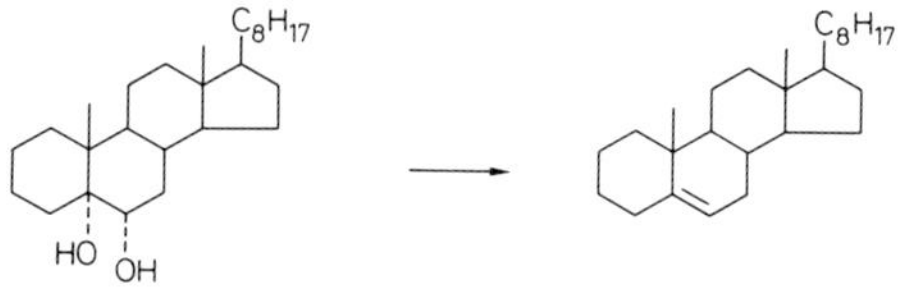

A soln. of cholestan-5α,6α-diol in acetonitrile treated with a soln. of NaI in the same solvent at room temp., the mixture stirred for 5 min., trimethylsilyl chloride added, and stirring continued for 30 min. → cholest-5-ene. Y 96%. F.e., incl. reaction of *trans*-glycols, s. N.C. Barua, R.P. Sharma, Tetrahedron Letters *23,* 1365–6 (1982).

Bicyclo[3.3.0]oct-1(5)-en-2-ones from bicyclo[4.2.0]octane-2,5-diones ○ ○

Deoxygenative skeletal rearrangement

930. 

A mixture of startg. m., trimethylsilyl chloride, and dried NaI in acetonitrile stirred at room temp. for 5 hrs. → product. Y 91%. F.e. and methods s. K. Sasaki, M. Oda, Tetrahedron Letters *23,* 2117–20 (1982).

Trimethylsilyl iodide/trimethylamine *Me_3SiI/Me_3N*

Divinyl sulfides from 1,1′-disiloxy-1-thioethers via 1,1′-diiodo-1-thioethers CH–C(OSi≡)SR → C=CSR

931. 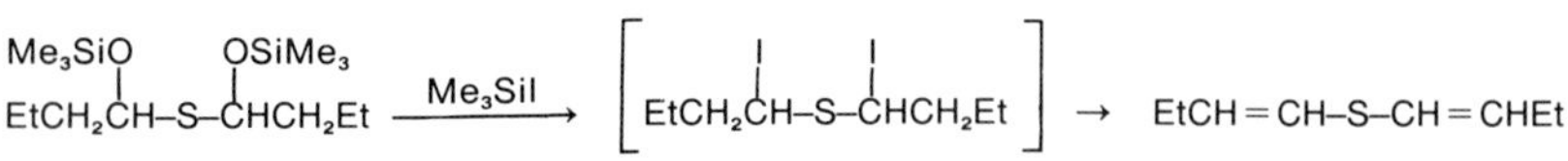

Trimethylsilyl chloride added dropwise to a stirred soln. of the startg. bis(trimethylsiloxy)sulfide in dry methylene chloride at –78° under N_2, warmed to 0° after 10 min., trimethylamine added dropwise after a further 10 min., and stirring continued at room temp. overnight → product. Y 88%. F.e.s. T. Aida, T.H. Chan, D.N. Harpp, Tetrahedron Letters *22,* 1089–92 (1981).

Trimethylsilyl iodide/hexamethyldisilazane *$Me_3SiI/(Me_3Si)_2NH$*

Enolethers from acetals $CHC(OR)_2$ → C=C(OR)

with $AlCl_3/Et_3N$ cf. *36,* 927; with trimethylsilyl iodide and hexamethyldisilazane s. R.D. Miller, D.R. McKean, Tetrahedron Letters *23,* 323–6 (1982).

Zirconium dioxide *ZrO_2*

Dehydration of alcohols CHC(OH) → C=C

with Al_2O_3 cf. *14,* 922; with ZrO_2, terminal ethylene derivs., comparison with Al_2O_3 and H_2SO_4, s. B.H. Davis, J. Org. Chem. *47,* 900–2 (1982).

Titanium trichloride/zinc/copper *$TiCl_3/Zn/Cu$*

Cyclic ethylene derivs. from dioxo compds. ○

s. *33,* 930; macrocyclic polyenes s. Tetrahedron Letters *23,* 1777–80, 2723–4 (1982).

Titanium tetrachloride/lithium tetrahydridoaluminate *$TiCl_4/LiAlH_4$*

Aromatization by elimination of O-bridges ←

s. *14,* 866s*37*

Phosphorus s. under HI *P*

Triphenylphosphine/diethyl azodicarboxylate $PPh_3/EtOOC-N=N-COOEt$

Dehydration of sec. alcohols CHC(OH) → C=C

with $PPh_3/CCl_4/Et_3N$ cf. *28,* 916s*32;* with PPh_3/diethyl azodicarboxylate, dehydroaminoacid derivs., s. R. Andruszkiewicz, J. Grzybowska, H. Wojciechowska, Pol. J. Chem. *55,* 657–71 (1981).

1,3-Dimethyl-2-phenyl-1,3,2-diazaphospholidine ←

Ethylene derivs. from 1,3-dioxolane-2-thiones C

s. *37,* 937

Phosphoric acid H_3PO_4

Dehydration of cyclic 1-acylaminoalcohols O

with Ac_2O cf. *31,* 920; with phosphoric acid, 1,4-dihydroisoquinol-3(2*H*)-ones by **cyclodehydration,** s. Y. Watanabe, Y. Kamochi, T. Miyazaki, Heterocycles *16,* 609–11 (1981).

Phosphoryl chloride $POCl_3$

Benzene ring from *o*-**quinones via 1,2-divinylglycols**
Polyarenes

932. 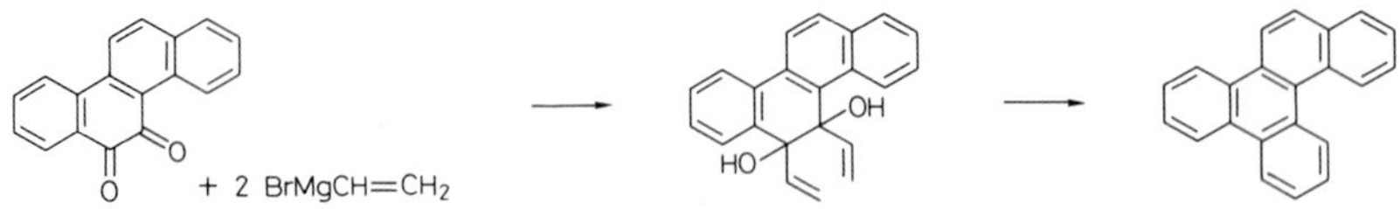

A soln. of vinylmagnesium bromide in tetrahydrofuran added dropwise to a stirred soln., of chrysene-5,6-dione in the same solvent at room temp. under N_2, and worked up after 2 hrs. → 5,6-dihydroxy-5,6-divinyl-5,6-dihydrochrysene (Y 56%, 60% conversion) in pyridine heated with $POCl_3$ at reflux for 10 min. → benzo[*g*]chrysene (Y 74%). F.e. and method s. K.B. Sukumaran, R.G. Harvey. J. Org. Chem. *46,* 2740–5 (1981).

Niobium pentachloride/sodium tetrahydridoaluminate $NbCl_5/NaAlH_4$

Ethylene derivs. from oxido compds. ⩗O → C=C

s. *37,* 742

Methanesulfonic acid $MeSO_3H$

Cyclic ketones from carboxylic acids O

with *β*-naphthalenesulfonic acid cf. *22,* 924; with $MeSO_3H$, 1-indanones and 1-tetralones, s. V. Premasagar, V.A. Palaniswamy, E.J. Eisenbraun, J. Org. Chem. *46,* 2974–6 (1981).

1,2-Dihydronaphthalene from 3-phenylcyclobutanol ring

933. 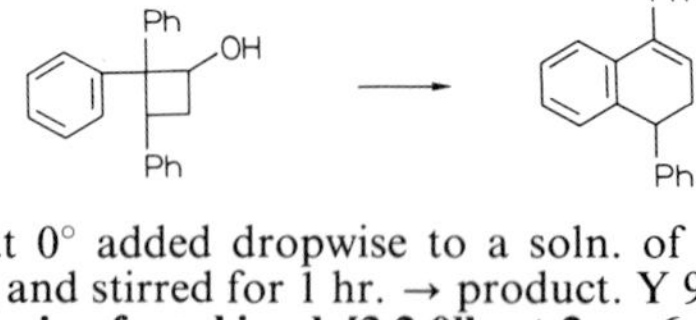

Methanesulfonic acid at 0° added dropwise to a soln. of startg. alcohol in ether, warmed to room temp., and stirred for 1 hr. → product. Y 95%. F.e. and with P_2O_5 in benzene, **also benzene ring from bicyclo[3.2.0]hept-2-en-6-ols,** s. E. Lee-Ruff et al., Can. J. Chem. *60,* 154–8 (1982); **2-tetralones** by cycloisomerization of 2-arylcyclobutanones s. ibid. *59,* 1675–9 (1981).

Sulfuric acid H_2SO_4

Cyclic ketones from carboxylic acids O

fluorenones cf. *16,* 181; thioxanthones (cf. *4,* 837) s. K. Sindelář, Collection Czech. Chem. Commun. *45,* 3166–81 (1980).

Sulfuric acid *H_2SO_4*

Cyclopent-2-enone ring from 2-acetylene-1,4-diols

with $MeSO_3H/P_2O_5$ cf. *31,* 932; with H_2SO_4 in methanol, regiospecificity, s. T. Hiyama et al., Bull. Chem. Soc. Japan *54,* 2747–58 (1981).

Aromatization by elimination of two O-bridges ←

Anthracenes

s. *37,* 647

*Tungsten hexachloride/*n*-butyllithium* *WCl_6/*n*-BuLi*

Ethylene derivs. from oxido compds.  → C=C

s. *27,* 932; s.a. M. Umbreit, K.B. Sharpless, Org. Synth. *60,* 29–34 (1981).

Hydrogen chloride *HCl*

Dehydration of alcohols with cleavage of ketals CHC(OH) → C=C

Cyclopent-2-enones

with H_2SO_4 cf. *21,* 933; with HCl in *n*-butanol s. T. Iwahashi, F. Matubara, Y. Yoshihiro, Nippon Kagaku Kaishi *1981,* 1121 (Japan); Chem. Abstr. *95,* 186677 (1981).

Ring closure with enolethers ○

Carbostyrils

with H_2SO_4 cf. *26,* 954; with HCl s. M. Tominaga et al., Chem. Pharm. Bull. *29,* 2161–5 (1981).

2-Cyclohexenones from 2-alkoxy-3,4-dihydro-2*H*-pyrans via regiospecific aldol condensation C ○

934. 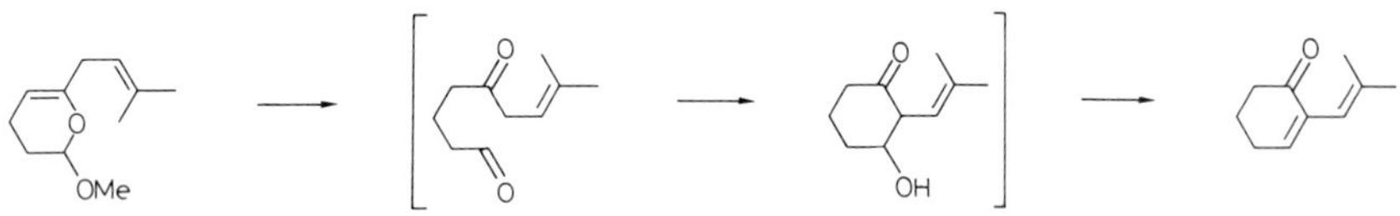

A soln. of startg. 2-methoxy-3,4-dihydro-2*H*-pyran in tetrahydrofuran treated with ca. 2 *M* HCl, and heated to 45° with stirring for 4 hrs. → 2-(2-methyl-1-propen-1-yl)-2-cyclohexen-1-one. Y 82%. F.e. and method, also isolation of intermediate **1,5-dioxo compds.** on mild hydrolysis, s. R.K. Boeckman, K.J. Bruza, Tetrahedron *37,* 3997–4006 (1981).

Hydrogen iodide/phosphorus *HI/P*

Reductive ring closure of *o*-ketocarboxylic acids ○

Anthracenes

s. *34,* 922; benz[*a*]anthracenes and dibenzanthracenes s. K.L. Platt, F. Oesch, J. Org. Chem. *46,* 2601–3 (1981).

Pentacarbonyltrimethylsilylmanganese *$Me_3SiMn(CO)_5$*

Enolethers from ketals under mild, neutral conditions $CHC(OR)_2$ → C=C(OR)

935. 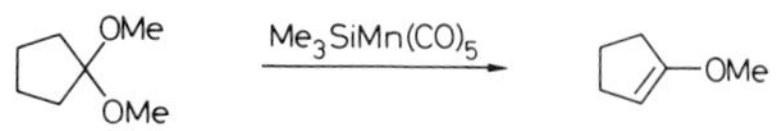

A mixture of the startg. ketal and 1.0–1.3 equivalents pentacarbonyltrimethylsilylmanganese in acetonitrile heated at 50° under N_2 for 2–4 hrs. → product. Y 93% by g.l.c. F.e. and limitations s. M. Marsi, J.A. Gladysz, Tetrahedron Letters *23,* 631–4 (1982).

Diiron nonacarbonyl *$Fe_2(CO)_9$*

Aromatization by elimination of O-bridges ←

s. *14,* 866s*37*

Palladium(II) acetate/chiral phosphinoferrocenyl complex ←

Regiospecific ring closure of aryloxy-2-ethylenes ○

s. *36*, 938; **asym. induction** with a chiral phosphinoferrocenyl complex in the presence of Pd(II) acetate, also from 2-ethylenecarbonic acid esters, cf. Tetrahedron Letters *23*, 3089–92 (1982).

Bis[1,2-bis(diphenylphosphino)ethane]palladium(0)/sodium hydride ←

Spirocyclics via palladium π-allyl complexes

s. *37*, 246

Bis(acetonitrile)dichloropalladium(II) $PdCl_2(MeCN)_2$

α,β-Ethyleneoxo compds. from 1-vinyloxyethers with 1,3-alkyl migration ←

936.

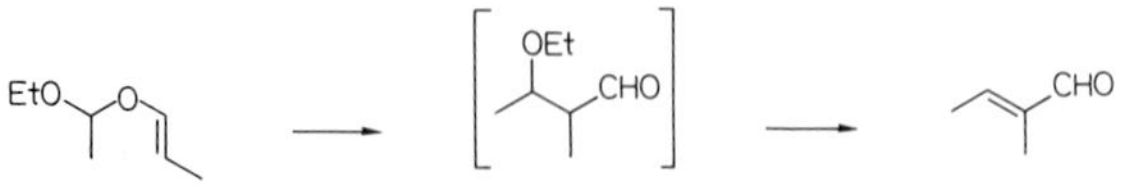

A catalytic amount of bis(acetonitrile)dichloropalladium(II) added to a soln. of acetaldehyde ethyl 1-propenyl acetal in dichloromethane, and the mixture stirred at room temp. for ca. 24 hrs. → 2-methyl-2-butenal. Y 71%. F.e.s. M. Takahashi et al., Chem. Letters *1981*, 1361–2; isolation of intermediate **β-alkoxyaldehydes** with BF_3-etherate s. ibid. *1981*, 1435–8.

Via intermediates *v.i.*

Ethylene derivs. from glycols via 1,3-dioxolane-2-thiones C(OH)C(OH) → C=C

Geospecific conversion under mild conditions

937.

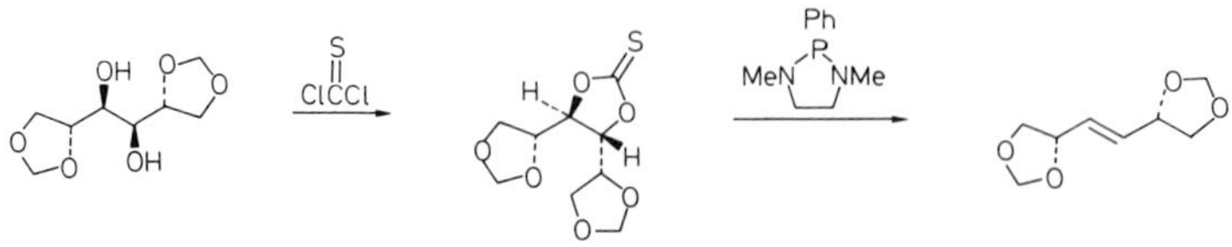

A soln. of thiophosgene in carbon tetrachloride added to a stirred soln. of 1,2:5,6-di-O-isopropylidene-D-mannitol and 4-(dimethylamino)pyridine in dry methylene chloride at 0° under argon, the mixture stirred for 1 hr., silica gel added, and allowed to warm to 25° → 1,2:5,6-di-O-isopropylidene-D-mannitol-3,4-thionocarbonate (Y 93%), suspended in 1,3-dimethyl-2-phenyl-1,3,2-diazaphospholidine, and stirred at 40° for 20 hrs. under argon → *trans*-3,4-didehydro-3,4-dideoxy-1,2:5,6-di-O-isopropylidene-D-*threo*-hexitol. Y 88%. F.e.s. E.J. Corey, P.B. Hopkins, Tetrahedron Letters *23*, 1979–82 (1982).

Nitrogen ↑ CC⇑N

Without additional reagents *w.a.r.*

Ethylene derivs. from disulfimides $CH{-}CN(SO_2R)_2 \rightarrow C{=}C$

Geospecific conversion

938.

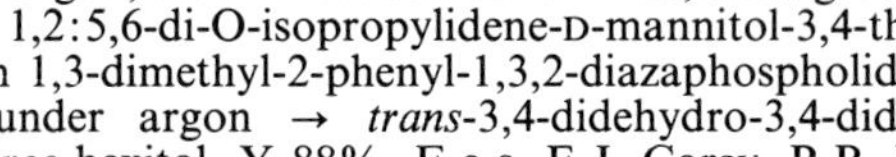

N-Cyclooctyl-N,N-di-(*p*-nitrobenzene)sulfonamide heated at 175° for 30 min. → *cis*-cyclooctene. 91%. – The method is milder than conventional pyrolytic ones, and expensive reagents are avoided. F.e.s. V.A. Curtis et al., Tetrahedron Letters *22*, 199–202 (1981).

Without additional reagents *w.a.r.*

3-Cyano-2-azetidinones from 4-azido-Δ^3-pyrrol-2-ones

939. 

A soln. of 4-azido-5-(benzyloxy)-3-chloro-1-isopropyl-2-pyrrolinone in chlorobenzene heated at 115° for 4 hrs. under N_2 → 4-(benzyloxy)-3-chloro-3-cyano-1-isopropyl-2-azetidinone. Y 90%. F.e. and limitations, also by irradiation, s. H.W. Moore et al., J. Am. Chem. Soc. *103,* 1769–77 (1981).

Irradiation

Pyrroles from 5-amino-1-vinyl-Δ^2-1,2,3-triazolines

940. 

2-Alkylpyrroles. A soln. of 4-methyl-1-(1-phenylvinyl)-5-(1-pyrrolidinyl)-4,5-dihydro-1*H*-1,2,3-triazole in methanol irradiated with a 100 W high-pressure Hg-lamp through a Pyrex vessel at 0° for 45 min. → 2-methyl-5-phenylpyrrole. Y 83%. – The expected 3-alkylpyrroles were not detected. F.e.s. M.M. Ito et al., Chem. Letters *1981,* 1519–22; **pyrrolenines** cf. Bull. Chem. Soc. Japan *54,* 2779–85 (1981).

Photoreactions of tetrazoles

s. *34,* 928; **review of the chemistry of tetrazoles** s. G.I. Koldobskii, V.A. Ostrovskii, V.S. Poplavskii, Khim. Geterotsikl. Soedin. *1981,* 1299–326.

Potassium tert-*butoxide* *KOBu*-t

Cyclopropane ring from tosylhydrazones

with LiH cf. *28,* 934; with K *tert*-butoxide, cage compds., s. K. Hirao, Y. Ohuchi, O. Yonemitsu, J. Chem. Soc. Chem. Commun. *1982,* 99–100.

*Tri-*n*-butyltin hydride* n-Bu_3SnH

Review of regio- and stereo-specificity in radical reactions ←

s. *36,* 943; rules determining regiospecificity of radical substitution and addition reactions s. J.M. Tedder, Angew. Chem. Intern. Ed. *21,* 401–10 (1982).

*Tri-*n*-butyltin hydride* n-$Bu_3SnH/$
azodiisobutyronitrile $Me_2C(CN)N{=}NC(CN)Me_2$

Ethylene derivs. from 2-isocyanoxanthates $C(NC)COCS_2R \rightarrow C{=}C$

941. 

A soln. of methyl 4,6-O-benzylidene-2-deoxy-2-isocyano-*α*-D-altropyranoside and a little azodiisobutyronitrile in toluene added dropwise to a refluxing soln. of tri-*n*-butyltin hydride in the same solvent, and heating continued for 1 hr. under N_2 → product. Y 90%. D.H.R. Barton et al., J. Chem. Soc. Perkin Trans. I *1980,* 2657–64.

Phosphorus pentoxide s. under $MeSO_3H$ P_2O_5

Phosphorus trichloride *PCl_3*

Fischer indole ring closure ○

s. *37,* 762

Methanesulfonic acid *$MeSO_3H$*

Intramolecular carbalkoxyaminoalkylation with 1,1-diurethans

942. 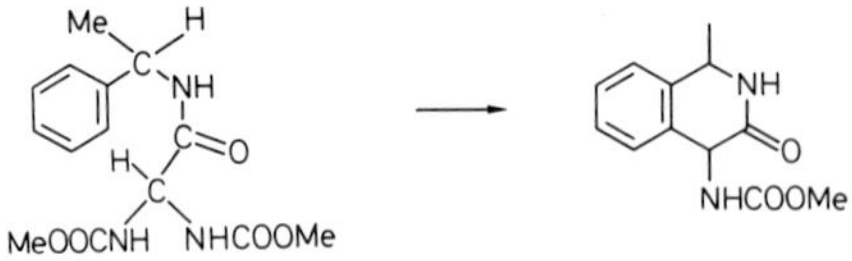

4-Carbalkoxyamino-1,4-dihydro-3(2*H*)-isoquinolones. Startg. m. in toluene treated with methanesulfonic acid at room temp. → product. Y 83%. F.e., including **3-carbalkoxyaminooxindoles and 1-carbalkoxyamino-4,5-dihydro-1*H*-3-benzazepin-3(2*H*)-ones,** s. D. Ben-Ishai, N. Peled, I. Sataty, Tetrahedron Letters *21,* 569–72 (1980); 2-pyrrolidone and γ-lactone derivs. s. J. Chem. Soc. Chem. Commun. *1980,* 687–8.

Methanesulfonic acid/phosphorus pentoxide *$MeSO_3H/P_2O_5$*

Cyclopent-2-enones from γ-nitrocarboxylic acids

943. 

Startg. nitrocarboxylic acid treated with a 67-fold excess of P_2O_5/methanesulfonic acid at room temp. for up to 40 hrs. → 3-methylcyclopent-2-enone. Y 52%. F.e., also with polyphosphoric acid at 100°, s. T.-L. Ho, J. Chem. Soc. Chem. Commun. *1980,* 1149–50.

p-*Toluenesulfonic acid* *TsOH*

Cyclopent-2-enone ring from β,γ-ethylene-α′-diazoketones with skeletal rearrangement ○ ◌

944. 

Condensed bicyclo[3.3.0]oct-3-en-2-ones. A soln. of startg. diazoketone in benzene containing *p*-toluenesulfonic acid boiled for 5 min. → product. Y 95%. F.e.s. G.O.S.V. Satyanarayana, P.R. Kanjilal, U.R. Ghatak, J. Chem. Soc. Chem. Commun. *1981,* 746–7.

Ethylenebis(salicylideneiminato)cobalt(II) ←

Acetylene derivs. from 1,2-di(hydrazones) under mild conditions C(=N−NH₂)C(=N−NH₂) → C≡C

945. 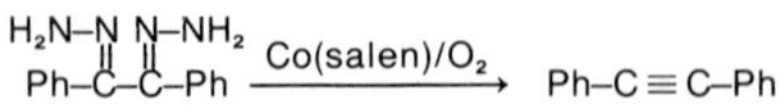

A soln. of startg. m. in 1:2 benzene/methanol added dropwise at room temp. during 15 min. with vigorous stirring to a little ethylenebis(salicylideneiminato)cobalt(II)suspended in methanol under O_2, and O_2 bubbled through the mixture for an additional 2 hrs. → diphenylacetylene. Y 98%. – Reaction appears limited to tolans. F.e. and with ethylenebis(3-methoxysalicylideneiminato)cobalt(II), **also azo from hydrazo compds.,** s. A. Inada, Y. Nakamura, Y. Morita, Chem. Pharm. Bull. *30,* 1041–2 (1982).

Rhodium(II) acetate *$(MeCOO)_4Rh_2$*

Ethylene derivs. from diazo compds. CH–CN_2 → C=CH

***cis-α,β*-Ethylenecarboxylic acid esters**

946. 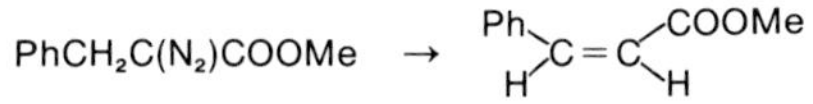

A soln. of startg. *α*-diazocarboxylic acid ester in benzene treated with a catalytic quantity of rhodium(II) acetate at room temp., and worked up after 30 min. when evolution of N_2 ceased → *cis*-product. Y 100%. F.e.s. N. Ikota et al., Tetrahedron Letters *22,* 4163–6 (1981).

Halogen ↑ CC⇑Hal

Without additional reagents *w.a.r.*

Porphyrins via dideoxybiladienes-*ac* ○

s. *21,* 948; **review of** advances in the **synthesis of porphyrins** s. R.P. Evstigneeva, Pure Appl. Chem. *53,* 1129–40 (1981).

Phenol ring by flash vacuum pyrolysis

s. *36,* 946; **review of flash vacuum pyrolysis.** Part I s. U.E. Wiersum, Rec. Trav. Chim. Pays-Bas *101,* 317 (1982); Part II s. ibid. *101,* 365–81 (1982).

Benzene from alkylidenecyclobutene ring by dehydrohalogenative ring expansion

947. 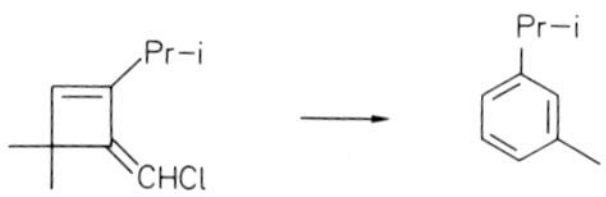

Startg. m. heated at 375° for 6 sec. → product. Y 100%. F.e., also from dihalogeno-1,2-bis(alkylidene)cyclobutanes, s. R. Maurin, C. Piscot, Z. Charrouf-Chafchaouni, Tetrahedron Letters *21,* 2425–8 (1980).

Irradiation (s.a. under i-Pr_2NLi, n-Bu_3SnH)

Photochemical cyclodehydrohalogenation ○

s. *28,* 892; *22,* 944; intramolecular arylation of N-(haloaryl)ethyl-*β*-enaminones s. T. Tiner-Harding, P.S. Mariano, J. Org. Chem. *47,* 482–5 (1982); **photochemistry and photocyclization of aryl halides, review,** s. J. Grimshaw, A.P. de Silva, Chem. Soc. Rev. *10,* 181–203 (1981).

Sodium hydroxide/ *NaOH/*
*tetra-*n-*butylammonium hydrogen sulfate* n-$Bu_4N^+HSO_4^-$

Acetylene derivs. from 1,2-dihalides CHalCHal → C≡C

s. *35,* 670s*36;* s.a. A. Le Coq, A. Gorgues, Org. Synth. *59,* 10–5 (1979).

Potassium hydroxide *KOH*

Polycyclic ketones from bridgehead alcohols
Skeletal rearrangement via cyclic iodoketones

948. 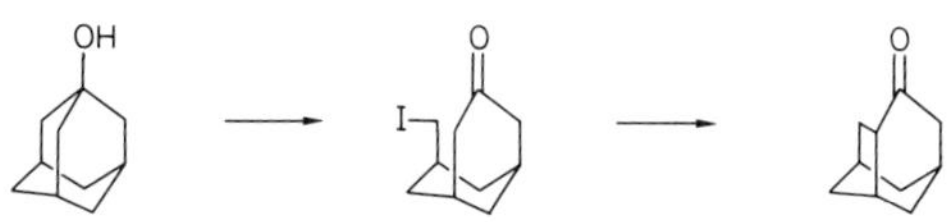

A stirred mixture of Pb tetraacetate, iodine, and 1-adamantanol in benzene heated to 80° during 20 min., allowed to cool to 70–75°, and stirring continued for 2 hrs. at this temp. then for an additional 1 hr. with cooling to room temp. → *endo*-7-

iodomethylbicyclo[3.3.1]nonan-3-one, treated with a soln. of KOH in methanol, and the mixture refluxed for 3 hrs. with stirring → 4-protoadamantanone. Y 71–82% (based on 1-adamantanol). F.e.s. Z. Majerski, Z. Hameršak, Org. Synth. *59,* 147–53 (1979).

2-Fluoro-1,3-dienes from 1,1-difluorocyclopropanes C
Geospecific ring opening

949.

Startg. 1,1-difluorocyclopropane treated with KOH in aq. tetrahydrofuran at room temp. for 5 hrs. → product (95% (E), (E)-isomer). Y quantitative. F.e. and conditions s. Y. Kobayashi et al., Tetrahedron Letters *22,* 5297–300 (1981).

Potassium hydroxide/tetraoctylammonium bromide $KOH/(n\text{-}C_8H_{17})_4N^+Br^-$
Acetylene derivs. from dihalides CHHalCHHal → C≡C
with K *tert*-butoxide/18-crown-6 polyether cf. *23,* 948s*36;* with KOH/tetraoctylammonium bromide and catalysts, s. Tetrahedron *37,* 1653–8 (1981).

Potassium tert-*butoxide* t-*BuOK*
Cyclopropane ring from halides ▽

with KOH cf. *8,* 930; with K *tert*-butoxide, iminocyclopropanes, s. H. Quast et al., Ann. Chem. *1980,* 1814–35.

n-*Butyllithium* n-*BuLi*
Benzocyclobutenes from dihalides □

s. *34,* 942; s.a. C.K. Bradsher, K.J. Edgar, J. Org. Chem. *46,* 4600–2 (1981); benzo[1,2:4,5]dicyclobutenes s. ibid. 4608–10 (1981).

Ring closure of dihalides ○
s. *32,* 966s*32; 34,* 942; benzo-condensed cyclic ethers s. C.K. Bradsher, D.C. Reames, J. Org. Chem. *46,* 1384–88 (1981).

Lithium diisopropylamide/irradiation *i-Pr₂NLi/*⚡

Oxindoles by cyclodehydrochlorination

950.

of o-halogenoacylamines. A soln. of N-propionyl-*o*-chloroaniline in tetrahydrofuran added to Li diisopropylamide in the same solvent under argon at –78°, allowed to warm to 25°, irradiated for 3 hrs. in a Rayonet Model RPR-240 photoreactor with 4 12.5 W lamps, and quenched with water → product. Y 73%. F.e. and **azaoxindoles** s. J.F. Wolfe, M.C. Sleevi, R.R. Goehring, J. Am. Chem. Soc. *102,* 3646–7 (1980); **of α-halogenocarboxylic acid anilides** cf. T. Hamada et al., Chem. Pharm. Bull. *29,* 128–36 (1981).

Sodium bis(trimethylsilyl)amide *$(Me_3Si)_2NNa$*

Intramolecular C-alkylation of protected cyanohydrins

951.

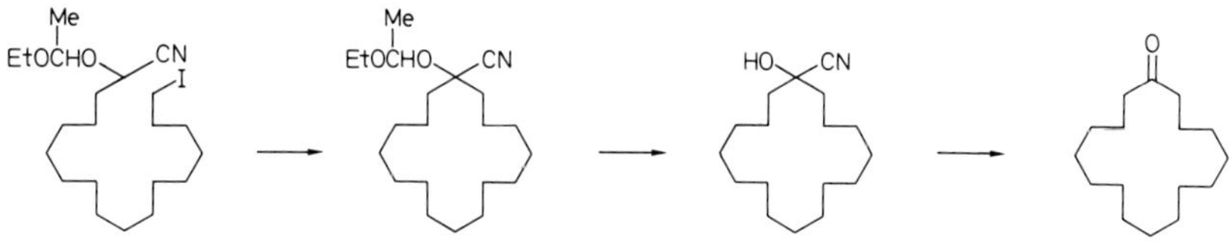

Macrocyclic ketones. A soln. of startg. m. in tetrahydrofuran added slowly during 2 hrs. to a soln. of Na bis(trimethylsilyl)amide in the same solvent at 40° under N_2 → intermediate (Y 71%) in tetrahydrofuran treated at 0° with 3 *M* HCl → cyclohexadecanone cyanohydrin in ether washed vigorously with 5% NaOH → cyclohexadecanone (Y 85% based on first intermediate). F.e. of medium- and large-ring ketones s. T. Takahashi, T. Nagashima, J. Tsuji, Tetrahedron Letters *22,* 1359–62 and 1363–6 (1981).

Potassium amide *KNH_2*

Ring closure via pyridynes
s. *32,* 964; **review of hetarynes** s. M.G. Reinecke, Tetrahedron *38,* 427–98 (1982).

Potassium bis(trimethylsilyl)amide *$(Me_3Si)_2NK$*

Intramolecular C-alkylation
s. *35,* 673; cyclic aminonitriles s. R. Achini, Helv. Chim. Acta *64,* 2203–16 (1981).

Potassium carbonate *K_2CO_3*

Acetylene derivs. from α,β-ethylenebromides C=CBr → C≡C
s. *23,* 955; aryloxy-2-acetylenes s. P.F. Shuda, M.R. Heimann, J. Org. Chem. *47,* 2484–7 (1982).

Sodium sulfide *Na_2S*

Ethylene derivs. from 1,2-dihalides CHalCHal → C=C
with Na_2Se cf. *23,* 956; *25,* 618; with Na_2S s. K. Fukunaga, H. Yamaguchi, Synthesis *1981,* 879–80.

Triethylamine *Et_3N*

Halogenoketenimines from 2,2-dihalogenoaziridines C

952.

Triethylamine added to a suspension of 1,3-diphenyl-2,2-dibromoaziridine in acetonitrile, and the mixture stirred for 7 hrs. at room temp. under N_2 → N-phenyl(phenylbromo)ketenimine. Y 86%. F.e.s. T. Hirao et al., Chem. Letters *1981,* 1197–8.

Pyridine *C_5H_5N*

5-Alkylidene-2(5*H*)-furanones from α,α-dichloro-γ-lactones ←

953.

A mixture of 2,2-dichloro-4-butyl-4-butanolide and pyridine refluxed for 1 hr. → 4-butylidenebut-2-enolide. Y 64% (82% (Z)-isomer). F.e.s. T. Nakano, Y. Nagai, J. Chem. Soc. Chem. Commun. *1981,* 815–6.

Zinc/copper *Zn/Cu*

Condensed α-methylene-γ-lactones by double ring closure ○

954.

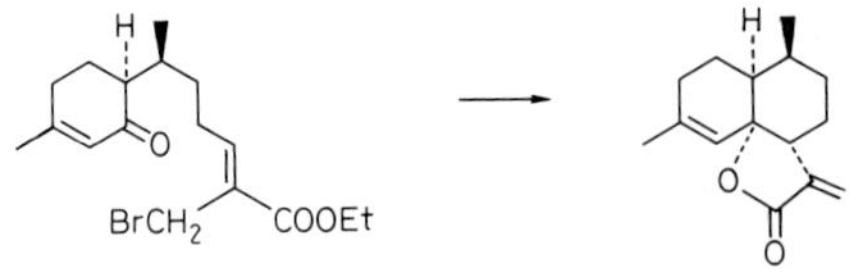

Zn/Cu couple added to a soln. of startg. bromoketoester in tetrahydrofuran, and stirred under N_2 for 14 hrs. → (1′R, 4′S, 4a′S, 8a′S)-2-(4′,7′-dimethyl-1′,2′,3′,4′,4a′,5′,6′,8a′-octahydro-1′-naphthyl)propen-8a′-olide. Y 63%. F.e.s. O. Goldberg et al., Helv. Chim. Acta *63,* 2455–68 (1981).

Zinc/acetic acid *Zn/MeCOOH*

α,β-Ethylenebromide from 2,2-dibromoalcohols $C(OH)CBr_2 \rightarrow C{=}CBr$

s. *37,* 615

Zinc/thiourea *Zn/$(NH_2)_2CS$*

Ethylene derivs. from 1,2-dibromides CBrCBr → C=C

with Zn/acetic acid cf. *22,* 178; with Zn/thiourea s. R.N. Majumdar, H.J. Harwood, Synth. Commun. *11,* 901–7 (1981).

Tetraoctylammonium bromide s. under KOH $(n\text{-}C_8H_{17})_4N^+Br^-$

*Tri-*n*-butyltin hydride/irradiation* n-Bu_3SnH/≀≀≀

Regiospecific ring closure with vinyl radicals ○

955.

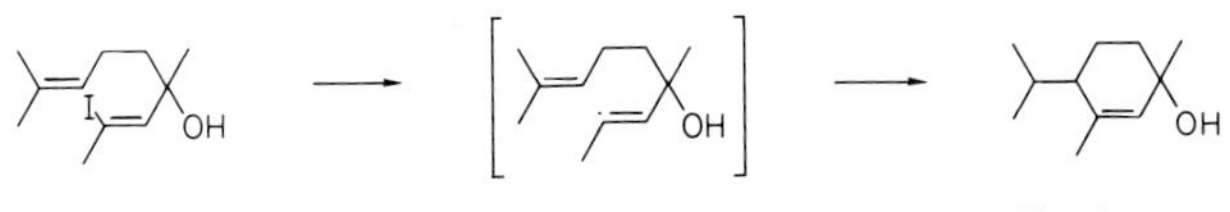

A soln. of the startg. m. in dry benzene containing a little azodiisobutyronitrile and 1.1 equivalents of tri-*n*-butyltin hydride irradiated with a GE 275 W sunlamp under reflux for 0.5–1 hr., the solvent removed, and the residue stirred rapidly for 1 hr. with ether and satd. aq. KF → product. Y 72%. F.e.s. G. Stork, N.H. Baine, J. Am. Chem. Soc. *104,* 2321–3 (1982).

Reductions with *iron–graphite* *Fe–C*

Ethylene derivs. from 1,2-dibromides CBrCBr → C=C

956. $n\text{-}C_9H_{19}CHBrCH_2Br \rightarrow n\text{-}C_9H_{19}CH{=}CH_2$

Deaerated water dropped into a slurry of iron–graphite (prepared from $FeCl_3$ and potassium graphite) in tetrahydrofuran under argon, the mixture stirred 5 min. at 30°, 1,2-dibromoundecane and decane (internal standard) added, and vigorously stirred at 70° for 3 hrs. → 1-undecene. Y 95%. – Similarly, *erythro*-1,2-dibromo-1,2-diphenylethane → *trans*-stilbene. Y 98%. F.e., **also ketones from α-bromoketones,** and cyclopent-2-enone and cyclohept-4-enone ring from *α,α′*-dibromoketones, s. D. Savoia et al., J. Org. Chem. *47,* 876–9 (1982).

Sulfur ↑ CC⇑S

Without additional reagents *w.a.r.*

Ethylene derivs. from 1-halogenothioethers CH–S–CHal → C=C
Dehydrohalogenative elimination of sulfur
Gas-phase pyrolysis

957. $$CH_3S\text{–}CCl_2\text{–}COOEt \rightarrow CH_2{=}CCl\text{–}COOEt$$

***α,β*-Ethylene-*α*-halogenocarboxylic acid esters.** Startg. ester vaporized at the top of an electrically heated quartz column filled with quartz beads, and pyrolysed at 490° at a pressure of 0.02–16 torr → product. Y 62%. F.e., **also *α,β*-ethylenecarboxylic acid esters and *α,β*-ethylene-*α*-halogenonitriles,** s. J.-C. Pommelet et al., Angew. Chem. Intern. Ed. *20,* 585–6 (1981).

Condensed N-heterocyclics from cyclic sulfones
via stereospecific intramolecular diene synthesis of *o*-quinodimethanes
s. *35,* 676; hydroindoles s. J. Org. Chem. *47,* 3634–43 (1982).

Ring opening by elimination of sulfur dioxide
1,3-Dienes
s. *17,* 968; 2,4-dienones with 2,6-di-*tert*-butyl-4-methylphenol as stabilizer s. K. Kosugi et al., Chem. Letters *1981,* 1341–4.

Ring contraction by elimination of sulfur monoxide
Thiophenes
s. *13,* 895; s.a. C.L. Gajurel, S.R. Vaidya, Indian J. Chem. *19B,* 911–12 (1980).

Sodium hydroxide/Aliquat 336 ←

Methylene compds. from chloromethyl sulfones $CHSO_2CH_2Cl \rightarrow C{=}CH_2$
Ramberg-Bäcklund reaction
s. *20,* 629; with Aliquat 336 as phase transfer catalyst cf. G.D. Hartman, R.D. Hartman, Synthesis *1982,* 504–6.

Potassium hydroxide *KOH*

Divinyl sulfoxide as vinyl carbonium equivalent ←
Vinyl ethers from alcohols

958. $$CH_2{=}CH\text{–}S(O)\text{–}CH{=}CH_2 \xrightarrow{HOPr} CH_2{=}CH\text{–}S(O)\text{–}CH_2CH_2OPr \rightarrow CH_2{=}CHOPr$$

A mixture of the startg. alkoxysulfoxide (prepd. by heating divinyl sulfoxide with propanol at 45–70° in the presence of a little KOH) and KOH heated at 280–90° for 1.5 hrs., and distilled at 70° and 1 mm Hg during 1.5 hrs. → product. Y 71%. F.e.s. B.A. Trofimov et al., Zh. Org. Khim. *18,* 450–1 (1982); **vinyl thioethers** s. Izv. Akad. Nauk SSSR Ser. Khim. *1981,* 1104–8.

n-*Butyllithium (s.a. under Molybdenum peroxide complex)* n-*BuLi*

Ethylene derivs. from 1,3-dithiolane-2-thiones
s. *37,* 690

Lithium diisopropylamide (s.a. under MoO_5.HMPA) $i\text{-}Pr_2NLi$

Ketones from 1-acylaminothioethers RCONC(SR′) → RCOR′
with NaH cf. *36,* 963; with Li diisopropylamide, also from *α*-acylaminosulfones, s. Chem. Pharm. Bull. *29,* 2496–502 (1981).

Lithium diisopropylamide (s.a. under $MoO_5.HMPA$) *i-Pr_2NLi*

Cyclic β-ketocarboxylic acid esters from monothioldicarboxylic acid esters ○

Directed Dieckmann-type cyclization

959. 

Ethyl 5-thioethoxycarbonylvalerate treated with 1.5 equivalents Li diisopropylamide in tetrahydrofuran at –30° during 3 hrs., the mixture acidified, and worked up in the normal procedure → product. Y 74%. Also N- and S-heterocyclic derivs. with NaH s. Y. Yamada et al., Tetrahedron Letters *22,* 1353–4 (1981).

Lithium tetrahydridoaluminate s. under $TiCl_3$ *$LiAlH_4$*

Aliquat 336 s. under NaOH ←

Azodiisobutyronitrile s. under n-*Bu_3SnH* *$Me_2C(CN)N{=}NC(CN)Me_2$*

Titanium *Ti*

Ethylene derivs. from 2-acoxythioethers C(OAc)C(SR) → C:C

with lithium di-*n*-butylcuprate cf. *31,* 967; with Ti (prepd. *in situ* from $TiCl_3$/K), methylene compds. s. S.C. Welch, J.-P. Loh, J. Org. Chem. *46,* 4072–4 (1981).

Titanium trichloride/lithium tetrahydridoaluminate *$TiCl_3/LiAlH_4$*

Ethylene derivs. from 1,1′-disiloxy-1-thioethers C(OSi⋮)–S–C(OSi⋮) → C=C

960. 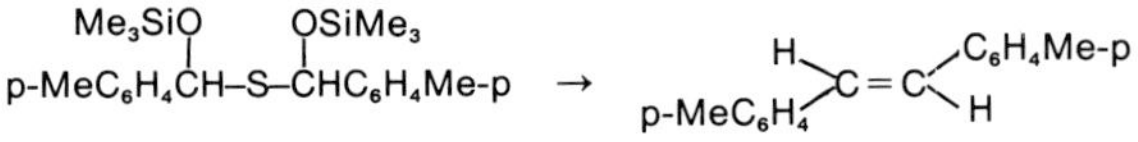

A 4:1 mixture of $TiCl_3/LiAlH_4$ added in small portions to dry tetrahydrofuran at room temp. under N_2, the mixture stirred for 3–4 hrs., a soln. of startg. *a,a′*-bis(trimethylsiloxy)sulfide in the same solvent added, and refluxed for 12 hrs. → product. Y 98% by g.l.c. ((E)-isomer >80%). F.e., **also from sulfido compds.** or 2 aldehyde molecules, s. T.H. Chan et al., Tetrahedron Letters *23,* 837–40 (1982).

*Tri-*n*-butyltin hydride* n-*Bu_3SnH*

1,3-Dienes from β,γ-ethylenesulfones $C(OH)CSO_2R$ → C=C

via 3-ethylene-2-sulfonylalcohols

Geospecific conversion

961. 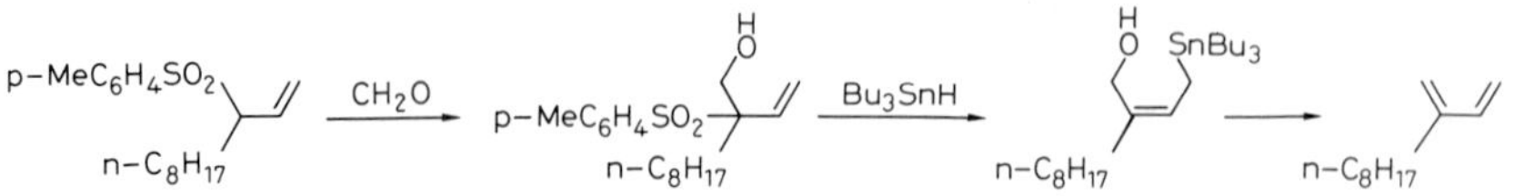

A soln. of *n*-butyllithium in hexane added with stirring under N_2 to a soln. of 3-tolylsulfonylundeca-1-ene and TMEDA in tetrahydrofuran at –70°, after 1 hr. treated with paraformaldehyde, and stirring continued at the same temp. for 1 hr., then at room temp. for 3 hrs. → 3-hydroxymethyl-3-tolylsulfonylundeca-1-ene (Y 95%), in benzene refluxed for 2 hrs. with tri-*n*-butyltin hydride in the presence of a little azodiisobutyronitrile → 2-octyl-1,3-butadiene(Y 92%). F.e.s. Y. Ueno et al., Tetrahedron Letters *22,* 2675–8 (1981).

*Tri-*n*-butyltin hydride/ azodiisobutyronitrile* — n-*Bu_3SnH/ $Me_2C(CN)N{=}NC(CN)Me_2$*

Ethylene derivs. from 2-nitrosulfones — $C(NO_2)CSO_2R \rightarrow C{=}C$

Geospecific radical chain elimination

962. 

Startg. 1,2-nitrosulfone heated with tri-*n*-butyl hydride in the presence of a small amount of a azodiisobutyronitrile in benzene at 80° for 1–2 hrs. → product. Y 86%. (E:Z = 98:2). F.e., **also from 1,2-dinitro compds.,** s. N. Ono et al., Chem. Letters *1981,* 1139–42.

*Molybdenum peroxide complex/ lithium diisopropylamide/*n*-butyllithium* — *MoO_5.HMPA/i-Pr_2NLi/*n*-BuLi*

Indoles from azaheptafulvenes via sulfene adducts ○ ◑

963. 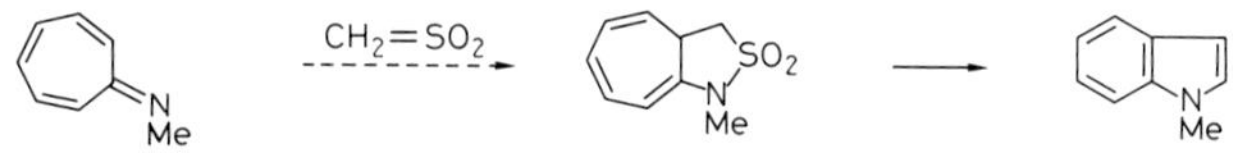

A soln. of 1 equivalent *n*-butyllithium in hexane added by syringe to a soln. of startg. sultam in tetrahydrofuran at –78°, stirred for 1 hr. at the same temp., soln. of MoO_5.HMPA and Li diisopropylamide in tetrahydrofuran/hexamethylphosphoramide added via a double-ended needle, stirring continued for 2 hrs. at –78°, the soln. warmed to room temp., quenched with water, warmed at 45° for 1 hr., and treated with $NaHSO_3$ → 1-methylindole. Y 79%. F.e., **also benzofurans from tropones,** s. B.D. Dean, W.E. Truce, J. Org. Chem. *46,* 3575–6 (1981); **[3+2]-cycloaddition of tropones with sulfenes or ketenes, review,** s. Heterocycles *18,* 343–56 (1982).

Hydrogen chloride — *HCl*

Ring closure with acetals

Modified Pomeranz-Fritsch isoquinoline ring closure ○

s. *29,* 975; f. modifications cf. D.L. Boger, C.E. Brotherton, M.D. Kelley, Tetrahedron *37,* 3977–80 (1981).

N-Condensed 1,2-dihydroisoquinoline ring from N-benzylcyclimmonium salts via 2-benzylamino-1-sulfinylthioethers

964. 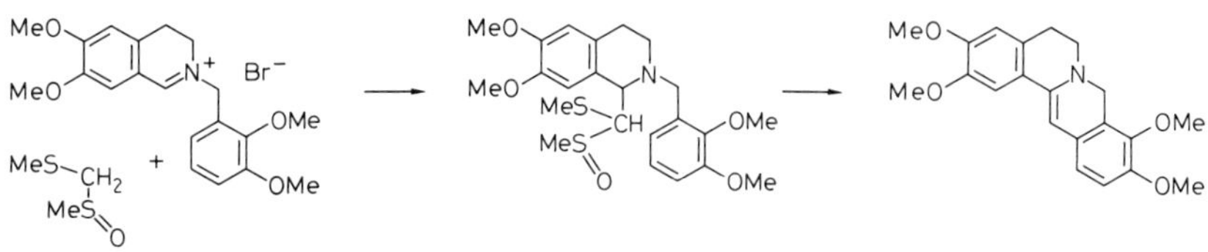

n-Butyllithium in hexane added under N_2 to a soln. of methyl methylthiomethyl sulfoxide in tetrahydrofuran at –35°, after 30 min. added dropwise by syringe to a stirred suspension of 3,4-dihydro-6,7-dimethoxy-N-(2,3-dimethoxybenzyl)isoquinolinium bromide in the same solvent, and stirring continued at –30° for 3 hrs. → intermediate (Y 98%) and concd. HCl heated on a steam bath for 5 min. under N_2 → (+)-tetrahydropalmatine (Y 66%). F.e.s. Z. Kiparissides et al., Can. J. Chem. *58,* 2770–9 (1980).

Remaining Elements ↑ CC⇑Rem

Without additional reagents *w.a.r.*

Hydrocarbons from selenides by elimination of selenium C–Se–C → C–C
Flash vacuum pyrolysis

965. $PhCH_2SeCH_2Ph \rightarrow PhCH_2CH_2Ph + Se$

Sym. 1,2-diarylethanes. Dibenzyl selenide flash vacuum pyrolysed at 600 °C → dibenzyl. Y 80%. F.e.s. T. Otsubo et al., Synth. Commun. *10,* 595–601 (1980); **also ethylene derivs.,** and pyrolysis of diselenides, s. Bull. Chem. Soc. Japan *55,* 182–7 (1982).

α,β-Acetylenenitriles from carboxylic acid chlorides CO–C(=P⋛)CN → C≡CCN
via β-keto-α-phosphoranylidenenitriles
Synthesis with addition of two C-atoms
Perfluoro-α,β-acetylenenitriles

966. $n\text{-}C_3F_7COCl + HC(=PPh_3)CN \rightarrow n\text{-}C_3F_7COC(=PPh_3)CN \rightarrow n\text{-}C_3F_7C{\equiv}CCN$

Heptafluoro-*n*-butyryl chloride added slowly to a stirred, lukewarm soln. of 2 equivalents of cyanomethylenetriphenylphosphorane in benzene, and the mixture stirred at room temp. for 2 hrs. followed by 1 hr. at 80° → heptafluoro-*n*-butyryl(cyano)methylenetriphenylphosphorane (Y 90%), pyrolysed at 220–60° and 10–15 mmHg for 40 min. under N_2 → heptafluoro-2-hexynenitrile (Y 80%). F.e.s. Y.Z. Huang et al., Tetrahedron Letters *22,* 5283–4 (1981).

α-Aminoketones from aldehydes CH(OH)C(N⋜)POPh$_2$ → COCHN⋜
via α-amino-β-hydroxyphosphine oxides
Synthesis with addition of one C-atom

967.

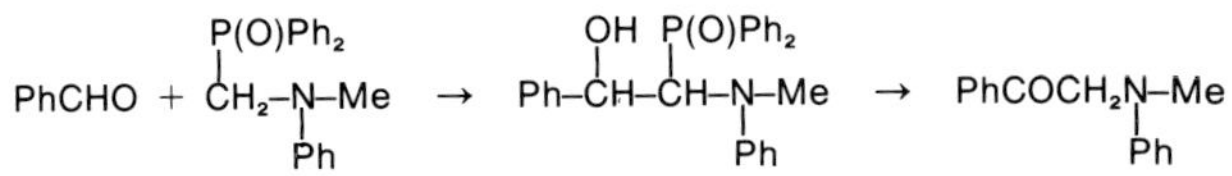

Aminomethyl ketones. Startg. aldehyde adduct heated in ethylene glycol at 140–50° for 3 hrs. → product. Y 82%. F.e. and method s. N.L.J.M. Broekhof, A. van der Gen, Tetrahedron Letters *22,* 2799–802 (1981).

Enol phosphates CHal–C(OSi⋛)PO(OR)$_2$ → C=COPO(OR)$_2$
from β-chloro-α-siloxyphosphonic esters
s. *37,* 793

Tandem siloxy-Cope-Cope rearrangement ←
4,8-Dienals

968.

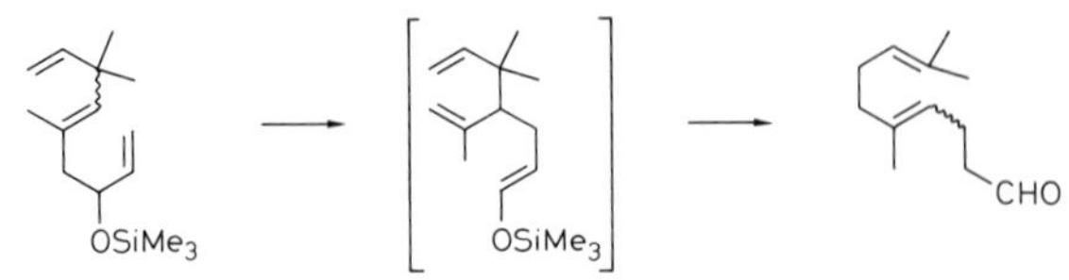

The startg. m. heated neat at 250° → product. Y 86%. K. Mikami et al., J. Org. Chem. *46,* 5447–9 (1981).

Without additional reagents *w.a.r.*

Benzocyclobutenones from *o*-*α*-silylcarboxylic acid chlorides ○

969.

Startg. m. distilled into a horizontal quartz tube loosely packed with quartz chips at 590–610° and 0.01 mmHg → product. Y 73%. F.e.s. B.L. Chenard et al., J. Chem. Soc. Chem. Commun. *1981,* 179–80.

Acetylene derivs. from 1,2,3-selenadiazole ring C
Cyclic acetylene derivs.
s. *34,* 974; s.a. H. Meier et al., Ann. Chem. *1982,* 914–23; formation of highly-strained cycloalkynes *in situ* s. Chem. Ber. *114,* 2382–91 (1981).

Synthesis of *α,β*-ethyleneketones from enoxysilanes and 1,1-dihalides via 2-halogeno-1-siloxycyclopropanes ○

970.

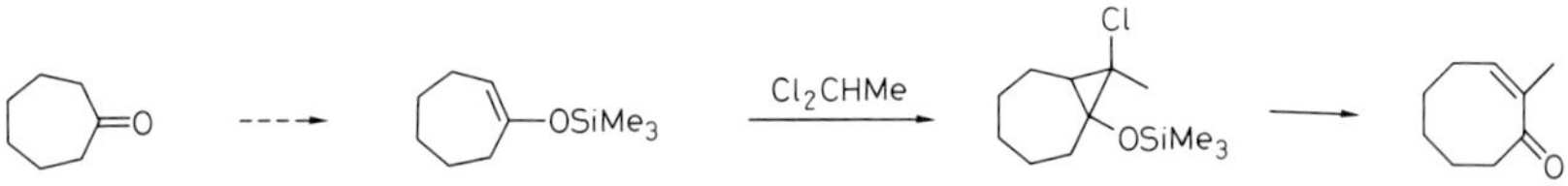

Ring expansion. A soln. of *n*-butyllithium in hexane added during 7 hrs. to a stirred mixture of startg. enoxysilane and 1,1-dichloroethane in ether at −40 to −30°, and allowed to warm to 0° during 2 hrs. → 8-chloro-8-methyl-1-trimethylsilyloxybicyclo[5.1.0]octane (Y 90%) in xylene refluxed for 6 days → 2-methyl-2-cyclooctenone (Y 80%). F.e.s. L. Blanco, P. Amice, J.-M. Conia, Synthesis *1981,* 289–91; *α*-methyl derivs. **and *α,β*-ethylenealdehydes** s. ibid. *1981,* 291–3; ***α,β*-ethylenecarboxylic acid esters from O-silyl O-alkyl keteneacetals** s. ibid. *1982,* 58–60.

Sodium hydride *NaH*

Intramolecular Emmons reaction ○
Cyclopent-2-enones
s. *25,* 693s*36;* bicyclo[3.3.0]oct-1(2)-en-3-ones s. D.K. Klipa, H. Hart, J. Org. Chem. *46,* 2815–6 (1981); M.J. Begley, K. Cooper, G. Pattenden, Tetrahedron *37,* 4503–8 (1981); 4-alkoxy- and 4-amino-isocarbostyril ring with Li diisopropylamide s. K. Akiba, Y. Negishi, N. Inamoto, Chem. Letters *1982,* 349–50.

Potassium hydride *KH*

Ethylene derivs. from 2-hydroxysilanes C(OH)C(Si≡) → C=C
with $HClO_4$ cf. *28,* 964s*32;* with KH, 3-C-methylenesugars, s. F.A. Carey, W.C. Frank, J. Org. Chem. *47,* 3548–50 (1982).

Caesium fluoride *CsF*

Intramolecular heterodiene synthesis via *o*-quinone methidimines ○
N-Condensed 1,2,3,4-tetrahydroquinolines

971.

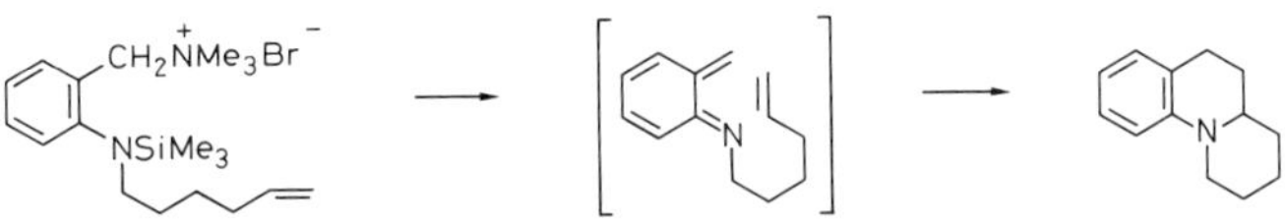

A soln. of [*o*-[(trimethylsilyl)hex-5-enylamino]benzyl]trimethylammonium bromide in acetonitrile added dropwise during 1 hr. to a refluxing suspension of CsF in the same solvent, and heated for a further 1 hr. → benzo[*c*]quinolizidine. Y 58%. F.e. incl. azasteroids, and stereospecificity, s. Y. Ito et al., J. Am. Chem. Soc. *103,* 5250–1 (1981).

Diethylamine s. under O_3 *Et_2NH*

Triethylamine s. under P_2I_4, $MeSO_2Cl$, and Br_2 *Et_3N*

Boron fluoride/acetic acid *$BF_3/MeCOOH$*

Ethylene derivs. from 3-hydroxysilanes ←

972. $Me_3SiCH_2CH(Ph)CMe_2OH \rightarrow CH_2{=}CHCMe_2Ph$

with aryl migration. A mixture of the startg. 3-hydroxysilane and BF_3-acetic acid complex in methylene chloride allowed to react at 0° for 15 min. → product. Y 85%. F.e., also with hydride shift and alkyl migration in lower yield, s. I. Fleming, S.K. Patel, Tetrahedron Letters *22*, 2321–4 (1981).

Terminal ethylene derivs. ←

s. *37*, 601

Dimethyl(methylthio)sulfonium fluoroborate *$[Me_2SSMe]^+BF_4^-$*

Regiospecific ring closures via thionium salts ○

3-Alkylthiocyclohex-1-enes from δ,ε-ethylene-ε-silylmercaptals

973.

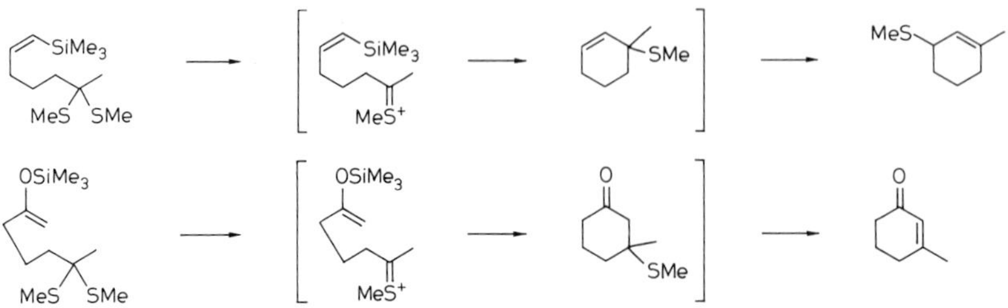

Startg. mercaptal treated with dimethyl(methylthio)sulfonium fluoroborate in methylene chloride at room temp. for 20 hrs. → product. Y 68%. F.e., **also cyclic α,β-ethyleneketones from enoxysilanes** as an **equivalent** of a **directed intramolecular Aldol condensation,** with and without isolation of intermediate cyclic β-(alkylthio)ketones, s. B.M. Trost, E. Murayama, J. Am. Chem. Soc. *103*, 6529–30 (1981).

Maleic anhydride ←

Reactions with nickelacyclopent-3-ene-2,5-diones ←

974.

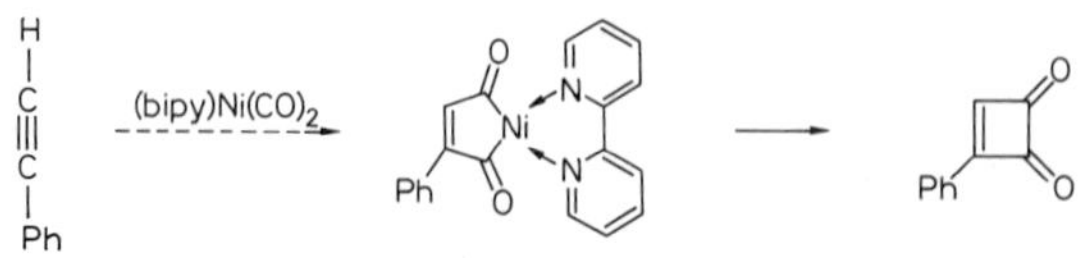

Cyclobutenediones. A stirred suspension of (2,2′-bipyridyl)-2,5-dioxo-3-phenylnickelacyclopentene (prepared from (2,2′-bipyridyl)dicarbonylnickel(0) and phenylacetylene) in tetrahydrofuran treated with a soln. of maleic anhydride in the same solvent at 20°, and the mixture allowed to stand for 6 hrs. → phenylcyclobutenedione. Y 70%. F.e., **also maleic acid anhydrides** by O_2 oxidation, **cyclopentene-1,3-diones from 1,1-dihalides, and *p*-quinols from ethylene derivs.,** s. A. Herrera, H. Hoberg, Synthesis *1981*, 831–3.

Chloramine-T ←
Ethylene derivs. from tellurides CH–CTeR → C=C

975. $n\text{-}C_{10}H_{21}CH_2CH_2TePh \rightarrow n\text{-}C_{10}H_{21}CH=CH_2$

A soln. of 1-dodecyl phenyl telluride and 2 equivalents of chloramine-T trihydrate in tetrahydrofuran refluxed for 40 min. under N_2 → 1-dodecene. Y 78%. F.e.s. T. Otsubo et al., Chem. Letters *1981,* 447–8.

Oxalyl chloride/titanium tetrachloride *$(COCl)_2/TiCl_4$*
Cyclopent-2-enones from γ,δ-ethylene-δ-silylcarboxylic acids ○

976.

4-Spirocyclopent-2-enones. Oxalyl chloride added dropwise by syringe to a soln. of startg. acid in benzene, the mixture stirred at room temp. for 3 hrs., solvent replaced by dichloromethane, cooled to –30°, $TiCl_4$ added, and stirring continued at this temp. for 2 hrs. then at room temp. for 15 hrs. → 9,9-dimethylspiro[4.5]deca-1,6-dien-3-one. Y 82%. F.e.s. S.D. Burke et al., J. Org. Chem. *46,* 2400–2 (1981).

Titanium tetrachloride *$TiCl_4$*
1,2-Dicarbalkoxycyclopropanes from 1,5-dialkoxy-1,5-disiloxy-1,4-dienes ▽
Stereospecific ring closure

977.

A 1 *M* soln. of $TiCl_4$ in dichloromethane added to a stirred soln. of startg. m. in the same solvent at 0°, the mixture stirred at 20° for 1 hr., and quenched with satd. $NaHCO_3$ → product. Y 69% (*trans*-isomer >99%). F.e. and prepn. of startg. m. s. T.H. Chan, I.H.M. Wallace, Tetrahedron Letters *23,* 799–802 (1982).

Stannic chloride *$SnCl_4$*
3-Vinyl-O-heterocyclics ○
s. *37,* 892

Diphosphorus tetraiodide/triethylamine *P_2I_4/Et_3N*
Ketene selenoacetals from carboxylic acid esters $CHC(SeR)_3 \rightarrow C=C(SeR)_2$

978. $n\text{-}C_{10}H_{21}CH_2C(SeMe)_3 \rightarrow n\text{-}C_{10}H_{21}CH=C(SeMe)_2$

1,1,1-Tris(methylseleno)dodecane and triethylamine added at 20° to a stirred suspension of P_2I_4 in methylene chloride, and stirring continued for 1 hr. → 1,1-bis(methylseleno)-1-dodecene. Y 90%. F.e., **also eneselenides from selenoacetals,** S-analogs, and with PI_3 s. J.N. Denis, A. Krief, Tetrahedron Letters *23,* 3407–10 (1982).

...from α-hydroxytriseleno-orthocarboxylic esters $C(OH)C(SeR)_3 \rightarrow C=C(SeR)_2$

979. $n\text{-}C_{10}H_{21}CH(OH)C(SeMe)_3 \rightarrow n\text{-}C_{10}H_{21}CH=C(SeMe)_2$

1,1,1-Tris(methylseleno)-2-hydroxydodecane and triethylamine in methylene chloride added at 0° to a stirred suspension of P_2I_4 in the same solvent, and stirring continued at the same temp. for 30 min. → 1,1-bis(methylseleno)-1-dodecene. Y 85%. F.e. and reagents (PI_3 or $SOCl_2$), **also eneselenides from α-hydroxyselenoacetals** (cf. *37,* 980), and S-analogs with and without organothio group migration, s. J.N. Denis et al., Tetrahedron Letters *22,* 4009–12 (1981).

Ozone/diethylamine *O_3/Et_2NH*

Ethylene derivs. from 2-pyridyl selenides C(SeR)CH → C=C

with H_2O_2 cf. *36,* 979; with O_3 in the presence of Et_2NH, *α,β*-ethyleneoxo compds., s. A. Toshimitsu et al., Tetrahedron Letters *23,* 2105–8 (1982).

Hydrogen peroxide *H_2O_2*

Ethylene derivs. from selenides

s. *29,* 912; 2-ethyleneacylamines from 2-aminoselenides s. A. Toshimitsu et al., J. Chem. Soc. Chem. Commun. *1981,* 546–7; dehydrogenation of *β*-dicarbonyl compds. s. D. Liotta et al., J. Org. Chem. *46,* 2920–3 (1981); J.M. Renga, H.J. Reich, Org. Synth. *59,* 58–65 (1979); Tetrahedron Letters *22,* 3043–6 (1981); *α,β*-ethylenesulfones from 1-sulfonylselenides s. T. Kametani, M. Aizawa, H. Nemoto, Tetrahedron *37,* 2547-53 (1981); *α,β*-ethylenephosphonic esters s. M. Mikolajczyk, S. Grzejszczak, K. Korbacz, Tetrahedron Letters *22,* 3097–100 (1981).

***α,β*-Acetylenesulfones**
from 1-ethylene-2-sulfonylselenides $C(SeAr)=CSO_2R \rightarrow C{\equiv}CSO_2R$

s. *37,* 561

Methanesulfonyl chloride/triethylamine *$MeSO_2Cl/Et_3N$*

Eneselenides from selenoacetals
via α-hydroxyselenoacetals $C(OH)C(SeR)_2 \rightarrow C=CSeR$

980. $(PhSe)_2CH_2 + O{=}CHMe \rightarrow (PhSe)_2CH{-}CH(OH)Me \rightarrow PhSeCH{=}CHMe$

1.0 *M* Li diisopropylamide in tetrahydrofuran added with stirring to a soln. of bis(phenylseleno)methane in the same solvent at –78° under N_2, after 5 min. treated with acetaldehyde, and worked up after a further 10 min. → 1,1-bis(phenylseleno)-2-propanol (Y 92%) in methylene chloride treated at 0° under N_2 with triethylamine followed by dropwise addition of methanesulfonyl chloride, and stirring continued for an additional 15 min. at room temp. → 1-(phenylseleno)propene (Y 84%). F.e. and metalation (cf. Synth. Meth. *34,* 780) s. H.J. Reich, W.W. Willis, P.D. Clark, J. Org. Chem. *46,* 2775–84 (1981).

Bromine/triethylamine *Br_2/Et_3N*

***α,β*-Ethyleneoxo compds. from siloxycyclopropane ring**

s. *31,* 974; *32,* 512; direct method with Br_2/NEt_3 cf. I. Böhm, R. Schulz, H.-U. Reissig, Tetrahedron Letters *23,* 2013–6 (1982).

Hydrogen chloride *HCl*

3(2*H*)-Furanones from γ-siloxy-β-diketones

981.

A soln. of 6-methyl-6-(trimethylsiloxy)-1-heptene-3,5-dione in 50% aq. tetrahydrofuran containing HCl stirred at room temp. under N_2 for 4.5 hrs. → 2,2-dimethyl-5-ethenyl-3(2*H*)-furanone. Y 50%. F.e. and reactions (alkylation and 1,4-addition of mercaptans), **also from α-acoxyketones** with limitations, s. A.B. Smith et al., J. Am. Chem. Soc. *103,* 1501–13 (1981).

Ferric chloride *$FeCl_3$*

Silicon-directed Nazarov reaction
Cyclopent-2-enones from 1-silyl-1,4-dien-3-ones

982. 

A mixture of startg. m. and 1.05 equivalents $FeCl_3$ in methylene chloride allowed to react at –10° for 30 min., and quenched with water → 4,5-dimethylcyclopent-2-enone. Y 95%. – The trimethylsilyl leaving group directs the formation of the double bond to the least substituted position. F.e.s. S.E. Denmark, T.K. Jones, J. Am. Chem. Soc. *104,* 2642–5 (1982).

Palladium(II) acetate *$(MeCOO)_2Pd$*

Cyclic ethyleneketones from siloxydienes
s. *34,* 978; bridged and spirocyclic cycloalkenones s. A.S. Kende, B. Roth, P.J. Sanfilippo, J. Am. Chem. Soc. *104,* 1784–5 (1982).

Carbon ↑ CC⇑C

Without additional reagents *w.a.r.*

Retrodiene scission C

s. *17,* 198s*36;* acetylene derivs. s. B. Tarnchompoo et al., Chem. Letters *1981,* 1241–2; 5-oxabicyclo[2.1.0]pent-2-enes (Dewar furans) s. D. Wirth, D.M. Lemal, J. Am. Chem. Soc. *104,* 847–8 (1982); pyrroles s. J. Leroy, D. Cantacuzene, C. Wakselman, Synthesis *1982,* 313–5; furans s. A.F. Oleinik, E.V. Adamskaya, K.Y. Novitskii, Khim. Geterotsikl. Soedin. *1982,* 453–6.

Enacylamines from Δ^2-5-oxazolones via acylimines
Decarbonylation

983. 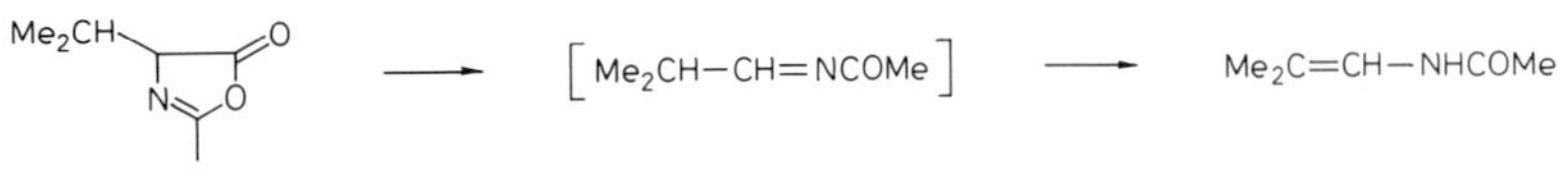

Startg. oxazolone heated during 1–2 hrs. at 100° with pyrolysis of the vapour in a quartz tube at 0.1–0.2 Torr. for 1–2 hrs. at 500° → N-(2-methyl-1-propenyl)acetamide. Y 100%. – The intermediate acylimines are isolable if isomerization is prevented by quaternary C-atoms. F.e.s. S. Jendrzejewski, W. Steglich, Chem. Ber. *114,* 1337–42 (1981).

Sodium hydride *NaH*

Cyclodehydrohalogenation with simultaneous cyclodehydrocyanation ○
N-Condensed isocarbostyril ring from Reissert compds.

984. 

A stirred soln. of 2-(2-chloromethylbenzoyl)-1-cyano-1,2-dihydroisoquinoline in dimethylformamide at 0–5° treated with 50%-NaH under N_2, and the mixture stirred for 2 hrs. → 5,6,13,14-dehydro-8-oxoberbine. Y 80%. F.e.s. S. Veeraraghavan, F.D. Popp, J. Heterocyc. Chem. *18,* 71–3 (1981).

Potassium tert-*butoxide* t-*BuOK*

Cyclopentadienones from 2-β-keto-3(2*H*)-furanones

985.

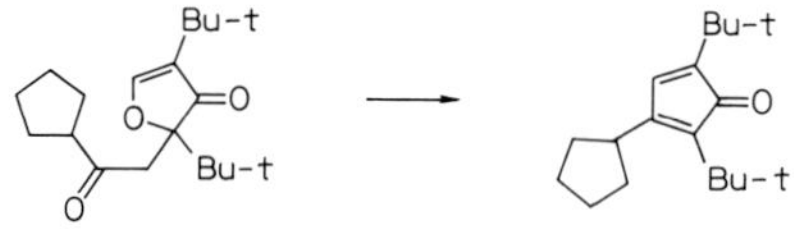

A soln. of equimolar amounts of startg. 5-acylmethyl-2,5-di-*tert*-butyl-4-oxa-2-cyclopentenone and K *tert*-butoxide in *tert*-butanol heated at 70° under N_2 for 2.5 hrs. → 3-alkyl-2,5-di-*tert*-butyl-2,4-cyclopentadienone. Y 91%. F.e.s. A. Nishinaga, K. Nakamura, T. Matsuura, Tetrahedron Letters *22,* 3261–4 (1981); J. Org. Chem. *47,* 1431–5 (1982).

n-*Butyllithium* n-*BuLi*

Pyridines from 1-carbalkoxy-1,4-dihydropyridine-4-phosphonic esters ←
s. *37,* 779

Sodium acetate *MeCOONa*

Decarboxylative ring closure ○
Benzofurans
s. *6,* 896; f.e.s. T. Zawadowski, J. Mazur, Acta Pol. Pharm. *37,* 289 (1980) (Pol.); Chem. Abstr. *95,* 7105 (1981).

Lithium chloride/hexamethylphosphoramide $LiCl/(Me_2N)_3PO$

Intramolecular decarbalkoxylative alkylation
s. *32,* 903s*37*

Sodium chloride *NaCl*

Cyclopropyl ketones from α-acyl-γ-lactones
with NaBr cf. *31,* 984; with NaCl, spirocyclopropanes, s. M. Kawada et al., Chem. Pharm. Bull. *29,* 1900–11, 1912–9 (1981).

Triethylamine/p-*toluenesulfonyl chloride* $Et_3N/TsCl$

Enacylamines from α-acylaminooximes ←
C-Cleavage

986.

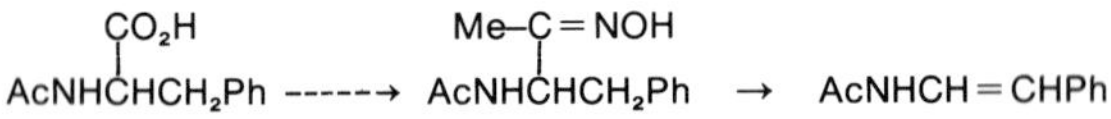

A soln. of startg. oxime in acetonitrile at –10° treated with *p*-toluenesulfonyl chloride and triethylamine for 2 hrs., and allowed to warm to 20° → product. Y 81%. – This is part of an overall synthesis **from α-acylaminocarboxylic acids.** F.e.s. U. Redeker, N. Engel, W. Steglich, Tetrahedron Letters *22,* 4263–4 (1981).

Calcium carbonate $CaCO_3$

Carboxylic acid esters from α-sulfonyloxyketals with aryl migration ←

987.

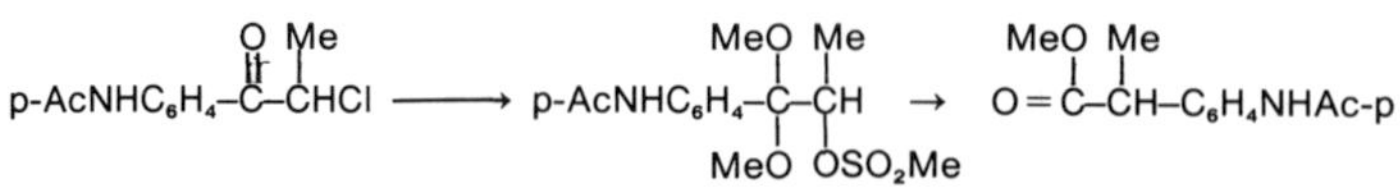

α-Arylcarboxylic acid esters. A soln. of startg. sulfonate in 7:3 methanol water refluxed for 21 hrs. in the presence of $CaCO_3$ → methyl α-(4-acetylaminophenyl)-propionate. Y 93%. – This is the key step in a 3-stage prepn. of carboxylic acid derivs. **from α-halogenoketones.** F.e. and hetaryl analogs s. G. Tsuchihashi, K. Kitajima, S. Mitamura, Tetrahedron Letters *22,* 4305–8 (1981).

Acetic acid *MeCOOH*

Ethylene derivs. from β-lactones ⊂

s. *27,* 981; with acetic acid, inversion of configuration with increasing acidity, s. J. Mulzer, M. Zippel, J. Chem. Soc. Chem. Commun. *1981,* 891–2.

Hexamethylphosphoramide s. under LiCl $(Me_2N)_3PO$

Phenylphosphonic dichloride $PhP(O)Cl_2$

N-Condensed heterocyclics via cyclic azomethinium salts ○

with $POCl_3$ cf. *32,* 454; with $PhP(O)Cl_2$, octahydroindolo[2,3-*a*]quinolizines, s. J. Org. Chem. *46,* 4914–20 (1981).

Oxygen/hydrogen chloride O_2/HCl

Pyridines from 1-carbalkoxy-1,4-dihydropyridines ←

s. *37,* 659

p-*Toluenesulfonyl chloride s. under* Et_3N *TsCl*

Formation of Electron Pair on Nitrogen

Elimination ⇑

Oxygen ↑ ElN⇑O

Zinc/acetic acid *Zn/MeCOOH*

Azomethines from nitrones C=N→O → C=N

with $NaBH_4/NaOH$ cf. *30,* 696; also with Zn/acetic acid s. M. Rai et al., J. Indian Chem. Soc. *59,* 80–1 (1982).

Carbon disulfide CS_2

Deoxygenation of N-oxides ⋜N→O → ⋜N

of quinoline N-oxides s. *25,* 700; of amine oxides s. T. Yoshimura, K. Asada, S. Oae, Bull. Chem. Soc. Japan *55,* 3000–3 (1982).

*2,4-Bis(*p*-methoxyphenyl)-1,3,2,4-dithiaphosphetane-2,4-disulfide* ←

988. $PhCH{=}N(O)C_6H_4\text{–}Cl\text{-}p \rightarrow PhCH{=}N\text{–}C_6H_4\text{–}Cl\text{-}p$

Azomethines from nitrones. A mixture of startg. nitrone and 2,4-bis(*p*-methoxyphenyl)-1,3,2,4-dithiadiphosphetane-2,4-disulfide in benzene stirred at 20° for 30 min. → product. Y 64%. F.e., **also deoxygenation of cyclic** (pyridine) **N-oxides,** s. K.A. Jorgensen et al., Bull. Soc. Chim. Belges *89,* 247–53 (1980).

Molybdenum(IV) oxide complex/triphenylphosphine ←

Deoxygenation of pyridine N-oxides ⊂NO → ⊂N

and of azoxy compds. N=NO → N=N

s. *37,* 994

Nitrogen **ElN⇑N**

Fluoroboric acid s. under Fe *HBF_4*

Iron/fluoroboric acid *Fe/HBF_4*
Diamines from cyclic quaternary hydrazinium dications 
in HCl cf. *34,* 992; in HBF_4, also f. reducing agents, scope and limitations, s. J. Chem. Soc. Perkin Trans. I *1982,* 603–9.

Carbon ↑ **ElN⇑C**

1-Methylimidazole
N-Demethylation of cyclimmonium iodides 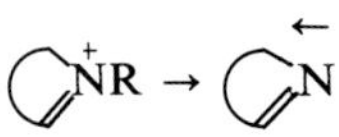

989. 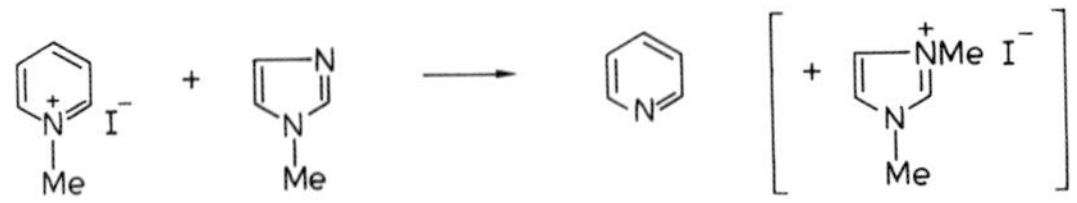

A soln. of 1-methylpyridinium iodide in 1-methylimidazole heated at 160° → pyridine. Y 84%. – **Trans-N-alkylation** is effected if the added base is of low nucleophilicity and is sufficiently different in boiling point to permit adequate separation of product. F.e., also with nicotinic acid, s. L.W. Deady, W.L. Finlayson, Synth. Commun. *10,* 947–50 (1980).

Formation of Electron Pair on Sulfur

Elimination ⇑

Oxygen ↑ **ElS⇑O**

Sodium iodide s. under Me_3SiCl *NaI*

Sodium tetrahydridoborate s.a. under $TiCl_4$ *$NaBH_4$*

Sodium tetrahydridoborate/alumina *$NaBH_4/Al_2O_3$*
Sulfoxides from sulfones via aryloxysulfoxonium salts $\gtrsim SO_2 \rightarrow \gtrsim SO$
s. *36,* 992; **prepn. of sulfoxides, review,** s. J. Drabowicz, M. Mikolajczyk, Org. Prep. Proc. Intern. *14,* 45–89 (1982).

Borane-tetrahydrofuran *$H_3B.THF$*
Thioethers from sulfoxides $\gtrsim SO \rightarrow \gtrsim S$
2-Hydroxythioethers
s. *37,* 612

Bis(trimethylsilyl) sulfide *$(Me_3Si)_2S$*

Bis(trimethylsilyl) selenide *Me_3Si_2Se*

Bis(trimethylsilyl) telluride *$(Me_3Si)_2Te$*

990. 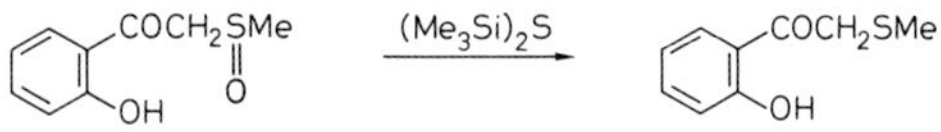

A soln. of startg. sulfoxide in tetrahydrofuran degassed with a slow stream of N_2 for 15 min., hexamethyldisilthiane added in one portion under argon, and stirred at room temp. for 1 hr. → product. Y 90%. F.e., **also selenides and tellurides from selenoxides and telluroxides, respectively,** and with bis(trimethylsilyl) selenide and telluride as reducing agents, s. M.R. Detty, M.D. Seidler, J. Org. Chem. *47*, 1354–6 (1982).

Trimethylsilyl chloride/sodium iodide *Me_3SiCl/NaI*

Disulfides from thiosulfonyl esters SO_2–S → S–S

with NaCN cf. *31*, 991; with trimethylsilyl chloride/sodium iodide s. G. Palumbo et al., Tetrahedron Letters *23*, 2391–4 (1982).

Reviews of trimethylsilyl bromide and iodide as reagents ←

s. *34*, 999s*35*, *36*; of trimethylsilyl iodide s.a. G.A. Olah, S.C. Narang, Tetrahedron *38*, 2225–77 (1982).

Titanium tetrachloride/sodium tetrahydridoborate *$TiCl_4/NaBH_4$*

Thioethers from sulfoxides ⪸SO → ⪸S

s. *37*, 16

Phosphorus pentasulfide *P_2S_5*

Thioketones from sulfines C=S=O → C=S

991. $(p\text{-}Me_2NC_6H_4)_2C=S=O \rightarrow (p\text{-}Me_2NC_6H_4)_2C=S$

A stirred soln. of startg. sulfine in dichloromethane treated with P_4S_{10}, and the mixture stirred at 25° for 1 hr. → bis(4-dimethylaminophenyl)sulfine. Y 95%. F.e., also with thiophosphoryl bromide, s. J.A.M. Kuipers et al., Synthesis *1981*, 295–7.

Trifluoromethanesulfonyl chloride/iodine *CF_3SO_2Cl/I_2*

Thioethers from sulfoxides ⪸SO → ⪸S

992. $Ph_2S=O \rightarrow Ph_2S$

3 Equivalents trifluoromethanesulfonyl chloride, followed by a soln. of I_2 in tetrahydrofuran, added at 0° under argon to a soln. of diphenyl sulfoxide in the same solvent, and stirred at room temp. for 48 hrs. → diphenyl sulfide. Y 96%. F.e.s. N. Fukamiya, M. Okano, T. Aratani, Chem. Ind. *1982*, 199.

Molybdenum(IV) oxide complex/triphenylphosphine ←

s. *37*, 994

Hydrogen iodide *HI*

s. *16*, 983; cyclic thioethers s. N.A. Orlova et al., Izv. Sib. Otd. Akad. Nauk SSSR, Ser. Khim. Nauk *1981*, 125; Chem. Abstr. *96*, 20048 (1982).

Nitrogen ↑ — ElS ⇑ N

Zinc s. under $TiCl_4$ — *Zn*

Titanium tetrachloride/zinc — *$TiCl_4/Zn$*
Thioethers from N-sulfonylsulfilimines — $>S=NSO_2R \rightarrow >S$

993.

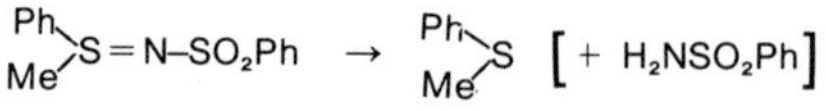

A stirred mixture of startg. sulfilimine and Zn dust in ether treated with $TiCl_4$, and stirring continued for 1 hr. → thioanisole. Y 92%. F.e., also with trifluoroacetic anhydride/NaI, and applications of this method to provide a novel route to **prim. amines from halides via sulfilimines,** s. J. Drabowicz, P. Łyzwa, M. Mikołajczyk, Synthesis *1981,* 890–1.

p-*Toluenesulfonyl nitrite* — *TsNO*
Reductive deimination — $>SO(NH) \rightarrow >SO$
Sulfoxides from sulfoximines
s. *36,* 995; also with *tert*-butyl thionitrate s. Phosphorus-Sulfur *12,* 103–13 (1981).

Halogen ↑ — ElS ⇑ Hal

Acetone — *$(CH_3)_2CO$*
Thioethers from bromosulfonium bromides — $>\overset{+}{S}-Br \rightarrow >S$
s. *17,* 575; **review of structure and reactivity of halogenosulfonium salts** s. G.E. Wilson, Tetrahedron *38,* 2597–625 (1982).

Formation of Electron Pair on Remaining Elements

Elimination ⇑

Oxygen ↑ — ElRem ⇑ O

Hexamethyldisilthiane — *$(Me_3Si)_2S$*
Selenides from selenoxides — $>SeO \rightarrow >Se$
Tellurides from telluroxides — $>TeO \rightarrow >Te$
s. *37,* 990

Titanium tetrachloride/zinc — *$TiCl_4/Zn$*
Arsines from arsine oxides — $\equiv AsO \rightarrow \equiv As$
s. *37,* 994

Deoxygenation with
molybdenum(IV) oxide complex/triphenylphosphine ←

994. $$Ph_2As{=}O \xrightarrow{[Et_2NCS_2]_2MoO_2/PPh_3} Ph_3As \quad [+\ Ph_3P{=}O]$$

A soln. of triphenylarsine oxide in benzene added dropwise with stirring to a little Mo(IV)-oxide complex in the same solvent, the mixture treated with a soln. of triphenylphosphine in the same solvent, and stirring continued at 40° for ca. 20 hrs. → triphenylarsine. Y 84%. F. reaction conditions, **also deoxygenation of pyridine N-oxides, sulfoxides, stibine oxides, and azoxy compds.,** s. X. Lu, J. Sun, X. Tao, Synthesis *1982,* 185–6; arsines from arsine oxides with $TiCl_4$/Zn cf. Y.D. Xing, X.L. Hou, N.Z. Huang, Tetrahedron Letters *22,* 4727–8 (1981).

Carbon ↑ ElRem ⇑C

Without additional reagents *w.a.r.*

Selenides from selenonium salts $>\overset{+}{Se}{-}R \rightarrow\ >Se$

s. *34,* 585; 2-ketoselenides s. N.N. Magdesieva et al., Zh. Org. Khim. *17,* 2552–6 (1981).

Heteropolar Bond

Uptake ⇓

Addition to Nitrogen Het ⇓N

Without additional reagents *w.a.r.*

Nitrilium salts from nitriles $C{\equiv}N \rightarrow C{\equiv}\overset{+}{N}{-}R$

s. *37,* 749

Oxonium fluoroborates $R_3O^+BF_4^-$

5,6-Dihydro-4*H*-1,2-oxazinium salts from 5,6-dihydro-4*H*-1,2-oxazines ←

s. *37,* 209

Addition to Sulfur Het ⇓S

Without additional reagents *w.a.r.*

S-Allylation of thiolactams ←

s. *37,* 704

Trimethylsilyl triflate *$Me_3SiOSO_2CF_3$*

Benzyl- and allyl-sulfonium triflates from thioethers ←

995.

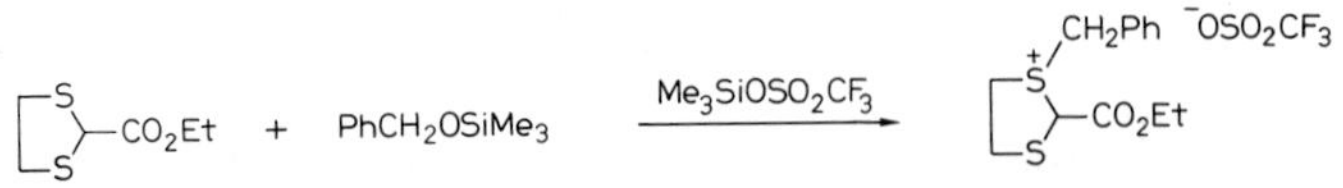

Equimolar amounts of startg. thioether and benzyl trimethylsilyl ether in acetonitrile treated with trimethylsilyl triflate, and the reaction allowed to proceed for 24 hrs. at room temp. → 1-benzyl-2-(carbethoxy)-1,3-dithiolanium triflate. Y 70%. F.e., also allylsulfonium triflates with allyltrimethylsilyl ether or diallyl ether, s. E. Vedejs, J. Eustache, J. Org. Chem. *46,* 3353–4 (1981).

Elimination ⇑

Oxygen ↑ Het ⇑ O

Acetic anhydride *Ac_2O*

1,3-Benzodithiolium salts ←

s. *33,* 998; 2-deuterio-1,3-benzodithiolium salts s. Bull. Chem. Soc. Japan *55,* 2289–90 (1982).

Resolutions Res

Without additional reagents *w.a.r.*

Resolution of stereoisomers

via preferential crystallization of salts ←

s. *5,* 666s*31;* of carboxylic acids with (–)- or (+)-2-amino-2-phenylethanol s. K. Saigo et al., Bull. Chem. Soc. Japan *55,* 1188–90 (1982).

Chromatography ←

Chromatographic resolution on optically active adsorbents ←

reviews s. *5,* 666s*35, 36;* by reverse phase high performance liq. chromatography, resolution of D- and L-aminoacids, s. S. Weinstein, Angew. Chem. Intern. Ed *21,* 218 (1982); of hydrocarbons and sulfoxides **via Pt complexes** cf. J. Am. Chem. Soc. *104,* 1093–5 (1982).

Brucine ←

Resolution of stereoisomers with brucine ←

of halides s. *16,* 990; of 2-acetylene-*tert*-alcohols s. F. Toda, K. Tanaka, H. Ueda, Tetrahedron Letters *22,* 4669–72 (1981).

Via intermediates *v.i.*

Resolution of ketones via β-hydroxysulfoximines ←

996.

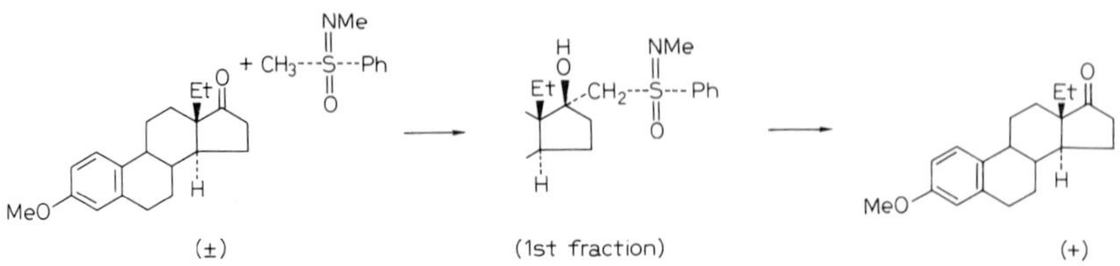

Startg. ketone treated with a soln. of *n*-butyllithium in tetrahydrofuran at 0°, (–)-(R)-methyl phenyl N-methylsulfoximine added at –78°, stirred at the same temp. for

0.5–1 hr., and the mixture separated by medium-pressure or flash liquid chromatography over silica gel using ethyl acetate and hexane as eluent → separated diastereomers [Y 52% (1st fraction); 40% (2nd fraction)], heated in turn under argon at 130° for 5–10 min. → resolved ketone [(+)-enantiomer (Y 86%) from the 1st fraction; (–)-enantiomer (Y 61%) from the 2nd fraction]. – Speed is the main advantage of this method. F.e.s. C.R. Johnson, J.R. Zeller, J. Am. Chem. Soc. *104,* 4021–3 (1982). *β*-hydroxysulfoximines **from oxo compds.,** desulfurization to **optically active alcohols,** s. J. Org. Chem. *47,* 1193–6, 1196–200 (1982).

Index to Volumes 36 and 37

The subject index lists the names of the methods, types of compounds, reagents, etc. For specific compounds and for authors (when a method is not named after them) the reader should consult the indexes of abstract journals. Complex compounds, as those with several functional groups, are referred to under the related simpler compounds, under *special s.,* e.g. aminocarboxylic acids are found under carboxylic acids. Only compounds beginning with letters other than those of the parent or main compounds are as a rule listed under this subentry, and entry numbers are omitted because they are specified where the compound is listed as a main entry. Derivatives used for characterization, identification, or separation are listed under the subentry *derivs.* Methods of synthesis for a given substance are indexed under the name of this substance and the subentry *from,* e.g. carboxylic acids from alcohols, hydrocarbons. Syntheses carried out with a particular starting material are indexed under the starting material and the subentry *startg. m. f.* (starting material for the preparation of...), e.g. alcohols, startg. m. f. carboxylic acids, ketones. Inorganic and organometallic reagents are indexed under the metal.

Classes of compounds may be designated merely by the functional group that is changed during the reaction. Ring compounds may also refer to the corresponding hydrogenated rings, unless the latter are listed specifically. If that is the case, then, as in other cases, *inversion,* customary in the indexes of Chemical Abstracts, has been used. Dihydrofurans, e.g., may be found under Furans, dihydro-. All single letters, which are separated from the name by a hyphen, are ignored in the alphabetic arrangement, e.g. 'O-Acetyl derivs.' are listed under A. Supplementary entries, such as **17**, 935s**36**, can be located by consulting the Index of Supplementary References (see page 557).

Register zu Band 36 und 37

Im alphabetischen Register finden sich als Schlagworte Methoden, Verbindungsklassen, Reagenzien u.dgl.; Einzelverbindungen und Autoren sind, soweit nicht eine Methode nach ihnen benannt ist, in den Registern der Referatenblätter zu suchen. Auf kompliziertere Verbindungen, z.B. mit mehreren funktionellen Gruppen, ist bei den entsprechenden einfacheren Verbindungen unter *special s.* (spezielle siehe) hingewiesen, z.B. bei Carbonsäuren auf Aminocarbonsäuren. In der Regel wird hier jedoch nur auf Verbindungen hingewiesen, die einen anderen Wortanfang haben und deshalb an anderer Stelle im Register stehen. Derivate, die zur Charakterisierung, Identifizierung oder Trennung dienen, sind unter *derivs.* aufgeführt. Methoden zur Synthese eines bestimmten Endprodukts findet man beim Schlagwort dieses Endprodukts unter *from* (aus) registriert, z.B. carboxylic acids from alcohols, hydrocarbons. Synthesen, die sich mit einem bestimmten Ausgangsmaterial ausführen lassen, sind bei dem Schlagwort des Ausgangsmaterials unter *startg. m. f.* (Ausgangsmaterial für die Darstellung von...) zu suchen, z.B. Alcohols, startg. m. f. carboxylic acids, ketones. Anorganische und metallorganische Reagenzien sind unter den entsprechenden Metallen aufgeführt.

Die Bezeichnung der Verbindungsklassen kann sich auf funktionelle Gruppen beschränken, die bei der Reaktion verändert werden. Ringbezeichnungen können sich auch auf die entsprechenden hydrierten Ringe beziehen. Sind hydrierte Ringe aber besonders aufgeführt, dann ist bei ihnen, wie auch in anderen Fällen, von der Inversion Gebrauch gemacht worden, wie sie in den Sachregistern der «Chemical Abstracts» üblich ist. Zum Beispiel stehen Dihydrofurane unter Furans, dihydro-. Griechische Buchstaben und Einzelbuchstaben, die vom eigentlichen Namen durch Bindestrich getrennt sind, werden bei der alphabetischen Anordnung nicht berücksichtigt, z.B. stehen «O-Acetyl-derivs.» unter A. Ergänzungen, wie z.B. **17**, 935s**36**, können mit Hilfe der Tabelle der Ergänzungszitate aufgefunden werden (siehe S. 557).

Deutscher Schlüssel zum Register (Index)

Formula Index of Complex Functional Groups

The elements appear in alphabetical order in the formulae, with the exception of carbon, which is always last. H's and R's are omitted. Hal is used for all halogens.

Formelregister komplexer funktioneller Gruppen

In den Formeln sind die Elemente alphabetisch angeordnet, mit Ausnahme von C, welches stets an letzter Stelle steht. R und H werden nicht berücksichtigt. Für alle Halogene steht das Symbol Hal.

Examples:

1) $RNHCONHNH_2$ — N_3OC
   Semicarbazides
2) $RNHCONHN{=}C$ — N_3OC_2
   Semicarbazones
3) $RSO_2{-}N{=}C(SR')_2$ — NO_2S_3C
   Sulfonyliminodithiocarbonic acid esters
4) $RCON{=}CCl_2$ — Hal_2NOC_2
   Acyliminocarbonyl chlorides
5) $R_2C{=}N{-}S{-}PO(OR')_2$ — NO_3PSC
   Phosphorylthiooximes
6) $RCH{=}C(OR)(NR_2)$ — NOC_2
   1-Alkoxyenamines

Supplementary References in Vols 36–37
Ergänzungszitate in Bd. 36–37

Volume 1

No.	Suppl. Ref. Vol.-Page
23	*37*, 7
85	*36*, 38
147–9	*36*, 73
162	*36*, 82; *37*, 72
171	*36*, 83
182	*36*, 88
203	*36*, 216
277	*37*, 132
278	*37*, 132
319	*36*, 163
344	*37*, 146
403	*36*, 207
451	*37*, 206
479	*36*, 242
488	*37*, 230
489	*37*, 230
529	*36*, 291
550	*36*, 316
591	*37*, 306
660	*36*, 350
680	*37*, 261
716	*37*, 368

Volume 2

No.	Suppl. Ref. Vol.-Page
20	*36*, 6
38	*36*, 11
49	*37*, 9
168	*37*, 60
177/8	*36*, 73
270	*36*, 103
320	*37*, 133
346	*37*, 143
409	*36*, 169
483	*36*, 205
542	*36*, 241
589	*37*, 263
608	*37*, 14
654	*37*, 306
714	*37*, 321
728	*37*, 332
735	*37*, 333
746	*36*, 357
771	*36*, 387
778	*36*, 395
837	*37*, 69

Volume 3

No.	Suppl. Ref. Vol.-Page
42	*36*, 30
188	*36*, 92
251	*37*, 124
255	*37*, 125
324	*36*, 156
429	*37*, 193
441	*36*, 212
447	*37*, 200
485	*37*, 208
493	*36*, 229
523	*37*, 230
552	*36*, 267
583	*36*, 381
585	*36*, 289
645	*36*, 326; *37*, 309
703	*36*, 381
729	*37*, 375

Volume 4

No.	Suppl. Ref. Vol.-Page
2	*36*, 1
60	*36*, 32
102	*36*, 36
149	*36*, 59
154	*37*, 59
173	*37*, 67
182	*37*, 70
187	*36*, 81
213	*37*, 80
298	*37*, 128
311	*37*, 134
314	*36*, 139
326	*37*, 136
392	*37*, 156
405	*36*, 168
465	*37*, 185
475	*36*, 204; *37*, 190
493	*36*, 212
495	*36*, 208
513	*37*, 207
535	*36*, 231
539	*36*, 231
563	*36*, 244
595	*36*, 262
596	*36*, 227
597	*36*, 259
604	*36*, 260
623	*37*, 254
658	*37*, 276
667	*36*, 293; *37*, 285
705	*36*, 316
741	*36*, 327
780	*37*, 333
837	*37*, 378

Volume 5

No.	Suppl. Ref. Vol.-Page
12	*37*, 4
40	*37*, 34
75	*37*, 34
101	*37*, 45
122	*37*, 58
123	*36*, 63
162	*36*, 82

Volume 5 continued

No.	Suppl. Ref. Vol.-Page
213	*36*, 85
216	*36*, 112
372	*36*, 407
409	*36*, 218
454	*37*, 228
477	*37*, 255
483	*37*, 269
518	*36*, 298 (2)
532	*37*, 302
617	*36*, 393
644	*36*, 397
666	*36*, 420, 421
	37, 402 (2)

Volume 6

No.	Suppl. Ref. Vol.-Page
37	*36*, 13
64	*36*, 22
73	*37*, 19
81	*36*, 25
135	*37*, 42
148	*36*, 53
153	*36*, 56
191	*36*, 114
193	*36*, 75
195	*36*, 78
211	*37*, 78
258	*37*, 91
285	*36*, 44
306	*37*, 117
325	*37*, 125
341	*36*, 136
343	*36*, 136
366	*37*, 65
381	*36*, 150
388	*36*, 152
406	*36*, 160
410	*37*, 170
441	*37*, 155
462	*36*, 175
480	*37*, 164
489	*37*, 172
510	*36*, 192; *37*, 180
511	*36*, 193
573	*37*, 199
599	*36*, 224
623	*36*, 234
636	*36*, 238; *37*, 225
637	*36*, 238
656	*36*, 244
694	*36*, 276
696	*37*, 273
752	*36*, 320
820	*36*, 64
828	*36*, 382
874	*36*, 396
896	*37*, 396

Volume 7

No.	Suppl. Ref. Vol.-Page
15	*36*, 4
62	*36*, 9; *37*, 6
78	*36*, 12
91	*36*, 28
95	*36*, 31
105	*36*, 34
112	*37*, 32
125	*36*, 39
161	*36*, 51; *37*, 43
217	*36*, 140; *37*, 135
241	*36*, 83
246	*36*, 88
281	*36*, 97, 100
	37, 86
320	*36*, 110; *37*, 99
325	*36*, 394
339	*36*, 324
417	*36*, 154
437	*37*, 147
459	*37*, 155
493	*36*, 176
503	*36*, 170
529	*36*, 199
583	*36*, 44, 214
600	*37*, 213
634	*37*, 230
638	*36*, 243 (2)
	37, 228
680	*36*, 266
756	*37*, 294
773	*36*, 317
816	*36*, 334; *37*, 315
836	*36*, 349
858	*37*, 369
866	*36*, 385

Volume 8

No.	Suppl. Ref. Vol.-Page
49	*36*, 22
87	*36*, 27
106	*36*, 36
157	*37*, 56
206	*36*, 82; *37*, 75
272	*36*, 69
275	*37*, 90
303	*37*, 103
326	*36*, 118
347	*37*, 190
365	*36*, 129
372	*36*, 133
404	*37*, 135
465	*36*, 118
468	*37*, 147
482	*36*, 162
491	*37*, 157
500	*36*, 178
505	*37*, 164
528	*37*, 224
539	*36*, 189
543	*36*, 190
545	*37*, 179
604	*36*, 212
616	*36*, 215
	37, 203 (2)
618	*37*, 204
635	*37*, 213
643	*37*, 217
657	*37*, 222
662	*36*, 235
666	*36*, 236
667	*36*, 236
670	*37*, 225
676	*36*, 242
691	*36*, 248
692	*36*, 252
695	*37*, 238
736	*36*, 281
782	*36*, 316
865	*36*, 357
888	*36*, 116
889	*36*, 389
930	*37*, 384
944	*36*, 47
952	*36*, 416

No.	Suppl. Ref. Vol.-Page

Volume 9

No.	Suppl. Ref. Vol.-Page
42	*36*, 14; *37*, 9
45	*37*, 85
55	*37*, 14
59	*37*, 15
99	*36*, 32
102	*36*, 32
107	*36*, 34
108	*36*, 39
147	*36*, 42
168	*36*, 49
169	*36*, 49
220	*36*, 65
300	*36*, 95
313	*36*, 97
342	*37*, 44
344	*37*, 79
363	*37*, 73
371	*36*, 121
372	*36*, 122
418	*36*, 149
423	*37*, 142
444	*37*, 33
452	*36*, 157
512	*37*, 164
522	*36*, 179
523	*36*, 178
552	*36*, 190
586	*37*, 188
589	*36*, 234
592	*36*, 204
598	*36*, 205
644	*36*, 223
655	*36*, 407
745	*37*, 255
770	*36*, 288; *37*, 278
794	*37*, 291
810	*36*, 316
851	*37*, 310
860	*36*, 9
888	*37*, 329

Volume 10

No.	Suppl. Ref. Vol.-Page
50	*36*, 19
61	*37*, 20
103	*37*, 40
112	*36*, 129
151	*36*, 74
172	*36*, 98
189	*37*, 90
241	*37*, 116
255	*37*, 126
256	*36*, 131
270	*37*, 134
281	*37*, 137
362	*36*, 175
367	*36*, 176
407	*36*, 203
427	*37*, 197
457	*36*, 221
488	*36*, 243
492	*37*, 235
495	*36*, 253; *37*, 239
498	*36*, 254
499	*37*, 243
506	*36*, 259
508	*36*, 264; *37*, 251
562	*36*, 308
580	*36*, 317
586	*36*, 323
587	*36*, 84, 325
595	*36*, 328
597	*36*, 329
639	*36*, 384; *37*, 272
670	*36*, 393
701	*36*, 419

Volume 11

No.	Suppl. Ref. Vol.-Page
47	*36*, 18
59	*37*, 17
63	*37*, 20
71	*36*, 26
73	*36*, 24
135	*36*, 39; *37*, 34
139	*37*, 37
149	*37*, 41
170	*37*, 50
196	*37*, 65
215	*37*, 69
231	*37*, 197
232	*37*, 78
235	*36*, 166; *37*, 153
259	*36*, 94
284	*37*, 91
293	*36*, 100
294	*36*, 102; *37*, 90
329	*36*, 120
402	*37*, 141
423	*37*, 144
510	*37*, 28
536	*37*, 134
568	*37*, 182
597	*37*, 193
604	*36*, 209
656	*36*, 234
666	*36*, 235
672	*36*, 239
710	*37*, 242
712	*36*, 255
719	*37*, 248
723	*36*, 418
728	*36*, 267
768	*36*, 289
795	*37*, 294
821	*37*, 302
841	*36*, 316
953	*36*, 401

Volume 12

No.	Suppl. Ref. Vol.-Page
16	*36*, 6
42	*36*, 13
64	*36*, 20; *37*, 18
155	*37*, 43
209	*37*, 80
211	*36*, 121
212	*36*, 76
238	*36*, 86
266	*37*, 83
285	*37*, 89
288	*36*, 98, 318
341	*37*, 114
375	*36*, 136, 141
393	*36*, 193
427	*37*, 145
453	*37*, 147
465	*36*, 165 (2)
467	*36*, 165
469	*37*, 150
524	*36*, 180
546	*36*, 192
555	*36*, 194
562	*36*, 197
565	*36*, 199; *37*, 185
582	*36*, 204
585	*36*, 206

Volume 12 continued

No.	Suppl. Ref. Vol.-Page
608	*37*, 202, 203
620	*37*, 207
638	*36*, 225
653	*36*, 231 (2)
	37, 218
671	*37*, 226
679	*36*, 245
687	*36*, 250
700	*36*, 256; *37*, 242
	37, 242
709	*36*, 265
713	*36*, 266
715	*36*, 268
722	*37*, 267
737	*36*, 282; *37*, 274
841	*37*, 324
847	*36*, 340
856	*37*, 333
867	*36*, 357
	37, 342 (2)
878	*37*, 370
881	*36*, 385
884	*36*, 385
959	*36*, 419

Volume 13

No.	Suppl. Ref. Vol.-Page
5	*36*, 3
27	*36*, 11
39	*36*, 15
43	*37*, 6
70	*37*, 19
107	*36*, 34
109	*36*, 34
139	*36*, 47
187	*36*, 239
245	*36*, 169
268	*36*, 98
273	*36*, 96
337	*36*, 129
339	*36*, 56
361	*37*, 133
363	*36*, 138
376	*36*, 142
395	*36*, 151
425	*36*, 156
430	*37*, 152
442	*36*, 154
446	*36*, 143
452	*36*, 316
496	*36*, 178; *37*, 165
501	*37*, 169
511	*36*, 184
513	*37*, 173
529	*36*, 192
530	*37*, 181
535	*37*, 161
548	*37*, 187
550	*36*, 202
561	*36*, 205
604	*37*, 226
608	*36*, 241
624	*37*, 232
651	*36*, 261
667	*36*, 127
668	*36*, 271
681	*37*, 271
690	*37*, 56
693	*36*, 292
732	*36*, 317
752	*36*, 118
835	*36*, 380
870	*36*, 393
884	*36*, 401
895	*37*, 387

Volume 14

No.	Suppl. Ref. Vol.-Page
18	*36*, 64
82	*36*, 254
83	*36*, 26
84	*36*, 24
174	*37*, 58
206	*37*, 38
221	*36*, 84
277	*36*, 97
315	*36*, 114
317	*36*, 207
322	*36*, 117
326	*36*, 119; *37*, 112
334	*36*, 121
382	*37*, 68
393	*36*, 140
402	*37*, 140
442	*36*, 160
450	*36*, 160; *37*, 149
479	*37*, 310
605	*37*, 76
606	*37*, 205
676	*36*, 243, 245
677	*36*, 241, 244
690	*36*, 255
764	*36*, 308
773	*36*, 304
854	*36*, 339
863	*36*, 353; *37*, 334
864	*36*, 353
866	*36*, 343, 383
	37, 324, 375
912	*37*, 374
922	*37*, 377
945	*36*, 395
952	*36*, 396
976	*36*, 408

Volume 15

No.	Suppl. Ref. Vol.-Page
6	*37*, 3
35	*36*, 118
38	*36*, 23
39	*36*, 22; *37*, 19
41	*36*, 23
56	*36*, 28
63	*37*, 25
92	*36*, 38
146	*37*, 12
151	*36*, 62
168	*37*, 374
171	*36*, 75
193	*36*, 86
199	*36*, 90
254	*36*, 116
261	*37*, 111
262	*37*, 113
268	*36*, 123
271	*36*, 122
301	*36*, 138
304	*36*, 64
340	*37*, 145
384	*37*, 165
387	*36*, 11
403	*36*, 187
419	*36*, 197
441	*37*, 202 (2)
470	*37*, 225
531	*37*, 272
555	*36*, 301

No.	Suppl. Ref. Vol.-Page

Volume 15 continued

No.	Suppl. Ref.
569	*36*, 312
583	*36*, 414
586	*36*, 317
589	*36*, 320
593	*36*, 321
600	*36*, 326
611	*36*, 334
634	*37*, 332
652	*37*, 229
671	*36*, 397

Volume 16

No.	Suppl. Ref.
1	*37*, 1
2	*37*, 1
13	*37*, 6
27	*37*, 11
52	*36*, 24; *37*, 20
142	*36*, 50
151	*37*, 47
152	*36*, 65
172	*36*, 59 (2)
181	*37*, 378
184	*37*, 58
194	*36*, 66, 67
	37, 61
196	*36*, 67; *37*, 61
199	*36*, 71
219	*37*, 74
222	*36*, 82
234	*36*, 88
236	*36*, 48, 86
240	*37*, 79
246	*36*, 95
247	*36*, 90; *37*, 80
270	*36*, 99
281	*36*, 102
285	*37*, 93
325	*36*, 121
331	*36*, 121
341	*37*, 77
346	*36*, 89
351	*37*, 124
354	*37*, 124
356	*36*, 129
367	*36*, 14
369	*37*, 212
414	*36*, 152
424	*36*, 161
454	*36*, 162
471	*37*, 157
485	*36*, 174, 176
490	*36*, 177
496	*36*, 178
503	*37*, 161
521	*36*, 189
537	*36*, 192
541	*37*, 181
565	*36*, 388
584	*37*, 196
606	*36*, 217
620	*36*, 221
649	*37*, 216
659	*36*, 240
664	*36*, 241
729	*37*, 262
733	*37*, 272
735	*36*, 281
756	*36*, 301
769	*37*, 1
786	*36*, 316
789	*36*, 316
825	*37*, 305
848	*36*, 396
867	*36*, 348
870	*36*, 349
876	*36*, 355
883	*36*, 357
888	*36*, 281; *37*, 344
926	*36*, 192
931	*36*, 62; *37*, 57
943	*36*, 392
983	*37*, 399
989	*36*, 5
990	*37*, 402

Volume 17

No.	Suppl. Ref.
20	*36*, 9
28	*37*, 9
74	*37*, 22
75	*37*, 22
80	*36*, 26; *37*, 23
82	*36*, 29
83	*36*, 30
112	*36*, 90
117	*36*, 40
153	*37*, 45
155	*36*, 53
161	*36*, 55
169	*36*, 58 (4)
	37, 51 (4)
186	*36*, 69
191	*36*, 70
198	*36*, 412; *37*, 395
236	*37*, 80
259	*37*, 91
276	*36*, 98
288	*37*, 95
292	*37*, 102
330	*37*, 165
333	*36*, 126
338	*36*, 127
345	*36*, 12
370	*36*, 137
381	*36*, 144
403	*37*, 134
424	*36*, 154
433	*37*, 153
471	*36*, 170
482	*37*, 87
526	*37*, 177
546	*36*, 194
549	*37*, 180
553	*36*, 192; *37*, 180
556	*36*, 196
565	*36*, 199
575	*37*, 400
593	*37*, 193
611	*36*, 213
621	*36*, 217
647	*37*, 216
663	*37*, 223
674	*36*, 242
707	*36*, 257
730	*36*, 279
733	*36*, 124
735	*37*, 261
750	*37*, 271
836	*36*, 324
851	*36*, 79
853	*36*, 337
860	*36*, 338
865	*36*, 339
911	*36*, 385
935	*36*, 395
938	*36*, 386
968	*37*, 387

No.	Suppl. Ref. Vol.-Page
Volume 17 continued	
978	*36*, 332
986	*36*, 417
Volume 18	
41	*36*, 16
58	*36*, 20; *37*, 17
66	*37*, 16
76	*36*, 26
88	*36*, 31
106	*36*, 37
109	*36*, 38
142	*37*, 45
176	*36*, 135
191	*37*, 66
193	*36*, 72; *37*, 67
202	*37*, 69
237	*36*, 88; *37*, 77
244	*36*, 93
256	*36*, 95; *37*, 84
262	*37*, 14
297	*37*, 103
301	*36*, 117
302	*36*, 116
306	*37*, 106
338	*36*, 130
354	*37*, 130
369	*36*, 137
381	*36*, 143
386	*36*, 145
391	*37*, 86
430	*36*, 157
434	*36*, 157
442	*36*, 161
448	*36*, 164
465	*37*, 155
488	*36*, 174
502	*36*, 179
522	*36*, 187
542	*36*, 193
543	*37*, 181
544	*36*, 155
550	*37*, 184
555	*36*, 138; *37*, 134
561	*36*, 203
601	*36*, 214
619	*36*, 219
622	*36*, 221
633	*37*, 54
643	*37*, 218
649	*36*, 232; *37*, 221
654	*36*, 234
692	*37*, 231
696	*37*, 237
719	*37*, 250
752	*36*, 290, 295
759	*36*, 281; *37*, 274
776	*36*, 293; *37*, 285
790	*36*, 301
810	*36*, 316
836	*36*, 323
841	*36*, 328
853	*36*, 336
854	*36*, 334
857	*36*, 334
859	*36*, 334; *37*, 317
862	*36*, 333, 336 *37*, 317
872	*37*, 329
880	*36*, 348
888	*36*, 85
891	*37*, 320
904	*36*, 243
940	*36*, 317, 387
947	*37*, 28
960	*36*, 397
Volume 19	
8	*36*, 7; *37*, 4
13	*36*, 9
22	*36*, 14
29	*36*, 17
33	*36*, 167; *37*, 155
51	*36*, 21
56	*37*, 32
72	*37*, 23
73	*37*, 29
81	*36*, 36
88	*36*, 37
112	*36*, 45
125	*37*, 43
132	*37*, 19
155	*36*, 60
187	*36*, 65, 120
200	*36*, 72, 371
200	*37*, 67
206	*36*, 75; *37*, 69
211	*36*, 78
230	*36*, 86
264	*36*, 98
265	*36*, 99
283	*37*, 93
307	*36*, 117
341	*36*, 20
356	*37*, 125
380	*37*, 135
446	*37*, 149
448	*36*, 160; *37*, 148
454	*37*, 149
460	*37*, 149
479	*37*, 154
480	*37*, 144
481	*36*, 167
488	*36*, 170
510	*37*, 164
514	*36*, 160
520	*37*, 163
523	*37*, 203
554	*36*, 190
559	*36*, 193
567	*36*, 90
589	*37*, 189
634	*36*, 216
637	*36*, 219
638	*37*, 208
640	*36*, 223
689	*36*, 241
705	*36*, 251
714	*36*, 418
722	*37*, 244
734	*37*, 254
738	*37*, 256
752	*36*, 116
754	*36*, 273
758a	*36*, 276; *37*, 270
764	*36*, 282
776	*37*, 285
787	*37*, 287
789	*37*, 289
811	*36*, 311
820	*37*, 301
821	*36*, 275
822	*36*, 42
827	*37*, 303
835	*36*, 352
837	*36*, 324
844	*36*, 324

No.	Suppl. Ref. Vol.-Page
Volume 19 continued	
896	*37*, 327
930	*37*, 372
944	*36*, 389
966	*36*, 397

No.	Suppl. Ref. Vol.-Page
Volume 20	
4	*36*, 2
13	*36*, 4
51	*36*, 20
54	*36*, 73
55	*37*, 285
62	*37*, 66
121	*37*, 72
126	*36*, 84
130	*36*, 84
141	*37*, 83
196	*36*, 118
197	*37*, 113
227	*36*, 134
248	*37*, 136
271	*36*, 153; *37*, 143
	37, 177
277	*36*, 85, 324
293	*36*, 157; *37*, 147
294	*37*, 273
308	*36*, 156
311	*37*, 11
379	*37*, 179
381	*37*, 181
417	*37*, 195
420	*36*, 215
443	*37*, 221
448	*36*, 236
459	*36*, 244; *37*, 229
463	*37*, 229
473	*37*, 236
482	*37*, 243
493	*37*, 251
499	*37*, 258
502	*36*, 274
507	*36*, 276
521	*37*, 282
526	*36*, 281
527	*36*, 290
538	*36*, 297
558	*36*, 314
562	*36*, 199
580	*36*, 102
625	*36*, 360
629	*37*, 387
648	*37*, 196
672	*36*, 399
685	*36*, 406
695	*36*, 418

No.	Suppl. Ref. Vol.-Page
Volume 21	
18	*36*, 7, 35
19	*36*, 4
21	*36*, 7
35	*36*, 38
60	*37*, 14
94	*36*, 38
108	*36*, 277
116	*36*, 45
120	*36*, 48; *37*, 42
121	*36*, 48
130	*36*, 52; *37*, 44
133	*36*, 52
158	*37*, 57
174	*37*, 62
191	*37*, 267
198	*36*, 82
216	*36*, 89
217	*37*, 79
285	*37*, 107 (2)
288	*36*, 115
302	*36*, 76
310	*36*, 126
366	*36*, 142
378	*37*, 140
412	*36*, 155
426	*37*, 147
428	*36*, 159
433	*37*, 321
436	*37*, 146
437	*37*, 147
444	*37*, 149
445	*37*, 309
468	*36*, 169
503	*36*, 175; *37*, 163
511	*36*, 184
512	*37*, 174
515	*37*, 174
522	*37*, 158
528	*36*, 190
530	*36*, 190
533	*37*, 180
558	*36*, 197
582	*36*, 204
595	*36*, 207
601	*36*, 210
603	*36*, 212
607	*36*, 215
619	*37*, 209
624	*37*, 211
625a	*36*, 224
627	*36*, 225
634	*36*, 228
639	*37*, 220
641	*37*, 219
642	*36*, 231, 246
676	*37*, 236
729	*36*, 280
731	*37*, 274
737	*37*, 279
744	*36*, 295
764	*37*, 291
858	*36*, 73 (2), 75
867	*37*, 330
894	*36*, 381; *37*, 368
912	*37*, 373
919	*36*, 390
933	*37*, 379
936	*36*, 395
940	*36*, 390
948	*37*, 383
949	*36*, 399

No.	Suppl. Ref. Vol.-Page
Volume 22	
18	*36*, 12
30	*36*, 16
40	*36*, 23
59	*36*, 31
73	*36*, 127
75	*36*, 38
80	*36*, 38
84	*36*, 43
86	*37*, 39
87	*36*, 47
95	*36*, 79
104	*36*, 56; *37*, 48
123	*37*, 157
143	*36*, 66; *37*, 60
153	*36*, 71

No.	Suppl. Ref. Vol.-Page

Volume 22 continued

No.	Suppl. Ref. Vol.-Page
169	*36*, 81
178	*37*, 386
188	*37*, 80
202	*36*, 93
251	*36*, 114
288	*36*, 214
301	*36*, 133; *37*, 129
313	*37*, 133
321	*36*, 139, 276
366	*36*, 17
384	*36*, 87
394	*36*, 161; *37*, 149
399	*37*, 153
406	*36*, 165
409	*37*, 154
420	*36*, 168; *37*, 155
425	*37*, 157
448	*37*, 162
458	*36*, 173
493	*37*, 175
495	*37*, 177
511	*36*, 18
542	*37*, 189
561	*36*, 210
580	*36*, 219
581	*36*, 219
587	*36*, 224
589	*36*, 224
608	*37*, 220
617	*36*, 234, 379
625	*37*, 188
632	*36*, 242
653	*36*, 252
656	*37*, 239
661	*36*, 257
664	*37*, 244
672	*36*, 259
705	*36*, 281; *37*, 274
720	*36*, 278, 289
731	*36*, 297
761	*36*, 280; *37*, 296
763	*36*, 314
776	*36*, 316
782	*37*, 311
786	*36*, 319
791	*36*, 321
793	*37*, 305
835	*36*, 307; *37*, 322
837	*36*, 12, 348
847	*36*, 107
864	*36*, 370
868	*36*, 369
871	*37*, 353
874	*37*, 367
877	*36*, 279; *37*, 348
924	*37*, 378
935	*36*, 397
944	*36*, 399; *37*, 383

Volume 23

No.	Suppl. Ref. Vol.-Page
11	*37*, 6
17	*36*, 154; *37*, 5
29	*37*, 31
36	*36*, 160
48	*36*, 20
51	*37*, 20
58	*37*, 160
62	*36*, 28
63	*37*, 24
74	*36*, 33
85	*36*, 264
92	*37*, 33
114	*37*, 42
139	*36*, 64
150	*37*, 62
186	*37*, 358
223	*37*, 84
248	*36*, 104
258	*37*, 103
260	*36*, 118
291	*36*, 113
296	*36*, 127
310	*37*, 120
365	*36*, 147
369	*36*, 148
380	*37*, 142
409	*37*, 157
415	*36*, 160
424	*37*, 150
428	*36*, 163; *37*, 151
430	*37*, 154
467	*36*, 179
477	*36*, 258; *37*, 240
479	*37*, 12
494	*37*, 179
500	*37*, 181
506	*36*, 315
507	*36*, 191
521	*37*, 154, 301
524	*37*, 186
584	*36*, 230
630	*36*, 250
637	*36*, 254
652	*36*, 260
656	*37*, 248
688	*36*, 280
693	*36*, 288; *37*, 276 *37*, 278
700	*37*, 283
703	*36*, 289
704	*36*, 289
715	*36*, 294
718	*36*, 296
727	*36*, 298
735	*36*, 302
757	*36*, 314
784	*36*, 41
793	*37*, 308
801	*37*, 288
816	*37*, 312
819	*36*, 337; *37*, 318
831	*36*, 337, 352 *37*, 332
832	*36*, 340, 348 *37*, 321
838	*36*, 345
840	*37*, 329
863	*36*, 337
864	*37*, 280
865	*36*, 227
868	*36*, 359
869	*36*, 359 *37*, 344 (2)
879	*36*, 365; *37*, 351
884	*36*, 376
885	*37*, 368
887	*36*, 379
928	*36*, 395
948	*36*, 400, 401 *37*, 384
955	*37*, 385
956	*37*, 385
971	*36*, 408

No.	Suppl. Ref. Vol.-Page
Volume 24	
8	*36*, 6
9	*37*, 10, 12
34	*37*, 10
40	*37*, 10
74	*36*, 33
83	*36*, 46
113	*36*, 57
114	*37*, 243
128	*36*, 63
140	*36*, 66
143	*36*, 70; *37*, 65
149	*36*, 70; *37*, 65
168	*37*, 71
175	*36*, 118; *37*, 108
176	*36*, 82
196	*36*, 91
210	*37*, 87
213	*37*, 88
223	*36*, 174
225	*37*, 162
236	*37*, 100
239	*36*, 114
260	*36*, 119
261	*36*, 116
269	*37*, 25
342	*37*, 138
346	*36*, 147
364	*36*, 153
386	*37*, 145
391	*36*, 34; *37*, 150
401	*36*, 159; *37*, 148
421	*37*, 155
429	*36*, 173
472	*36*, 163
473	*37*, 163
476	*36*, 304
482	*36*, 182
491	*36*, 187
507	*37*, 75
527	*36*, 49
528	*37*, 182
535	*37*, 184
554	*36*, 205
560	*36*, 206
	37, 192 (2)
562	*37*, 192
568	*37*, 194
573	*36*, 209
574	*37*, 201
579	*36*, 214
586	*36*, 247
593	*37*, 212
612	*37*, 222
615a	*36*, 234
634	*37*, 228
636	*37*, 228
663	*36*, 267
674	*37*, 268
688	*37*, 272
714	*36*, 296
718	*37*, 287
733	*36*, 62
745	*36*, 305
764	*36*, 313, 314
765	*36*, 94
784	*36*, 322; *37*, 305
802	*36*, 395; *37*, 41
830	*37*, 322
875	*37*, 348
877	*36*, 365
884	*36*, 376
900	*36*, 382
925	*36*, 386
928	*36*, 394
934	*36*, 395
974	*36*, 408
Volume 25	
16	*36*, 13
19	*36*, 14
32	*37*, 17
37	*37*, 21
43	*36*, 28
51	*36*, 5
52	*37*, 29
62	*37*, 27, 34
69	*36*, 39
82	*37*, 46
88	*36*, 58 (2)
102	*36*, 65
123	*36*, 79; *37*, 71
125	*36*, 80
130	*36*, 82
156	*36*, 98
163a	*37*, 89
167	*36*, 104
185	*36*, 118; *37*, 108
188	*36*, 121
189	*36*, 120; *37*, 113
209	*36*, 131; *37*, 126
233	*36*, 147
238	*36*, 148
244	*36*, 150
255	*36*, 154
257	*36*, 165; *37*, 144
258	*36*, 155
276	*37*, 146
278	*36*, 164; *37*, 152
287	*37*, 151
293	*37*, 157
308	*36*, 176
330	*36*, 175
354	*36*, 195
359	*36*, 197
372	*37*, 187
374	*36*, 202
376	*36*, 204
385	*37*, 194
389	*37*, 196
396	*37*, 202
398	*36*, 214; *37*, 198
410	*37*, 211
421	*36*, 229
456	*36*, 246; *37*, 231
465	*37*, 240
467	*36*, 257
471	*37*, 246
487	*36*, 266
499	*36*, 276
506	*36*, 279; *37*, 272
519	*36*, 295
522	*36*, 296; *37*, 287
573	*36*, 327
581	*37*, 314
584	*36*, 333
591	*37*, 321
599	*36*, 349
605	*37*, 334
618	*37*, 385
622	*36*, 361
628	*36*, 366
634	*36*, 378
636	*36*, 379
637	*36*, 381
656	*36*, 388
693	*36*, 408; *37*, 391
695	*36*, 413
700	*36*, 415; *37*, 397

Volume 27 continued

No.	Suppl. Ref. Vol.-Page
543	*37*, 43
548	*37*, 9
557	*36*, 133
564	*36*, 210
574	*37*, 201
603	*36*, 233
677	*36*, 275
694	*36*, 280
696	*36*, 280; *37*, 272
697	*36*, 280
698	*36*, 280
704	*36*, 414
705	*36*, 283
723	*37*, 285
724	*36*, 294; *37*, 285
726	*36*, 340
756	*36*, 310 (2)
761	*36*, 313
775	*36*, 317, 324
781	*37*, 306
782	*36*, 325
786	*36*, 311
790a	*37*, 309
795	*36*, 324; *37*, 307
833	*36*, 350
835	*36*, 127
837	*36*, 342
839	*37*, 329
843	*36*, 348
851	*36*, 359; *37*, 342
855	*36*, 353
856	*37*, 334
857	*37*, 334
866	*36*, 375
869	*37*, 341
870	*36*, 358
871	*36*, 358; *37*, 342
883	*37*, 270
884	*36*, 373
894	*36*, 365
907	*36*, 312
928	*36*, 391
931	*36*, 389
932	*37*, 379
937	*36*, 389, 391
942	*36*, 393
943	*36*, 386
958	*36*, 398
969	*36*, 402
978	*36*, 194
981	*37*, 397
983	*36*, 412
991	*36*, 415
998	*36*, 421

Volume 28

No.	Suppl. Ref. Vol.-Page
4	*36*, 2 (2)
7	*36*, 4
13	*37*, 5
29	*37*, 15
34	*36*, 24
37	*36*, 32
38	*37*, 23
48	*37*, 27
65	*36*, 47
85	*37*, 57
86	*37*, 77
88	*36*, 89
90	*36*, 63; *37*, 58
113	*36*, 72
119	*36*, 75
144	*36*, 164, 165
145	*37*, 79
157	*36*, 92
170	*36*, 7
182	*36*, 104
204	*36*, 80
251	*36*, 133
265	*36*, 136
266	*36*, 136; *37*, 133
278	*36*, 140
282	*36*, 131
309	*37*, 141
311	*36*, 181
335	*36*, 156
346	*36*, 159
353	*36*, 164 (2)
	37, 152, 153
358	*36*, 157
382	*37*, 160
392	*37*, 163
404	*37*, 171
431	*37*, 179
435	*37*, 180
440	*36*, 193 (2)
	37, 149
450	*37*, 182
453	*36*, 195
454	*37*, 182
469	*36*, 204
473	*36*, 205
481	*36*, 72, 208
483	*37*, 3
497	*37*, 202
507	*37*, 76
511	*36*, 221; *37*, 210
515	*36*, 223
516	*37*, 208
520	*37*, 215
524	*37*, 216
534	*37*, 219
539	*37*, 222
547	*37*, 226
548	*36*, 236
550	*36*, 237
574	*37*, 234
584	*37*, 238
596	*36*, 261
610	*37*, 260
613	*37*, 267
637	*36*, 330
647	*36*, 285
648	*37*, 275, 282
655	*36*, 279, 289
	37, 272, 280
667	*36*, 296; *37*, 287
669	*37*, 288
683	*37*, 291
693	*36*, 175
703	*36*, 313; *37*, 298
704	*36*, 312
720	*36*, 316
721	*37*, 370
726	*36*, 318
733	*37*, 303 (2)
752	*36*, 326
753	*36*, 325
757	*37*, 310
766	*37*, 311
780	*36*, 333
802	*36*, 220
805	*36*, 338
812	*37*, 320
818	*37*, 337
819	*36*, 326, 355
841	*36*, 318
845	*36*, 374
851	*36*, 365

No.	Suppl. Ref. Vol.-Page

Volume 28 continued

No.	Suppl. Ref. Vol.-Page
854	*36*, 367
856	*37*, 249, 353
858	*37*, 355
869	*36*, 379
874	*36*, 381
889	*36*, 384
890	*37*, 371
892	*37*, 383
916	*36*, 393; *37*, 378
930	*36*, 397
934	*37*, 381
945	*36*, 401
964	*37*, 391
965	*36*, 410
991	*36*, 417

Volume 29

No.	Suppl. Ref. Vol.-Page
2	*37*, 4
28	*37*, 12, 122
35	*36*, 21
42	*37*, 22
50	*36*, 28, 30
51	*36*, 31
53	*36*, 31
57	*37*, 27
63	*37*, 30
67	*36*, 36, *37* (2)
68	*37*, 32
76	*36*, 39
77	*37*, 36
79	*37*, 35
80	*37*, 36
84	*36*, 44
97	*36*, 53
101	*37*, 47
106	*36*, 55, 419
107	*36*, 55
117	*36*, 59
119	*36*, 59
132	*36*, 65; *37*, 59
137	*36*, 65
149	*36*, 68
153	*36*, 209
154	*36*, 93
160	*36*, 95
172	*36*, 42, 80
173	*36*, 81
177	*37*, 75
180	*36*, 255
207	*36*, 101
219	*36*, 105
253	*36*, 121
262	*36*, 267; 37, 259
268	*37*, 120
283	*36*, 130
290	*36*, 141
320	*36*, 145
325	*37*, 139
331	*36*, 149
348	*36*, 154
350	*36*, 154
354	*37*, 145
358	*36*, 157
359	*36*, 14
372	*37*, 154
374	*36*, 158
391	*37*, 133
409	*36*, 179
410	*36*, 175; *37*, 86
423	*36*, 180
434	*36*, 169
456	*36*, 193
457	*37*, 180
465	*37*, 184
476	*36*, 199
487	*37*, 190
497	*37*, 193
502	*37*, 194
511	*36*, 215
512	*36*, 216; *37*, 203
530	*36*, 226
547	*36*, 84
556	*36*, 236
558	*37*, 225
576	*36*, 337
581	*36*, 248; *37*, 232
588	*36*, 252
595	*36*, 254
601	*36*, 254
602	*36*, 256
603	*36*, 256; *37*, 242
605	*36*, 259
618	*36*, 267
627	*36*, 272
630	*37*, 264
634	*36*, 275
638	*37*, 269
671	*36*, 287
676	*37*, 281
677	*36*, 290
688	*36*, 298
702	*36*, 305
709	*37*, 295
717	*36*, 313
739	*37*, 26
746	*36*, 321
747	*37*, 305
750	*36*, 18
752	*36*, 323
796	*36*, 363; *37*, 255
	37, 321
807	*36*, 345
808	*36*, 346; *37*, 176
810	*36*, 347; *37*, 328
821	*36*, 219
829	*36*, 108; *37*, 50
830	*36*, 345
836	*36*, 357
846	*36*, 372
854	*36*, 363; *37*, 346
866	*36*, 365
868	*37*, 354
874	*37*, 321
901	*36*, 384
910	*36*, 314
912	*36*, 255, 260
	36, 385, 411
	37, 394
920	*37*, 373
926	*37*, 375
929	*37*, 374
933	*37*, 16
974	*36*, 405
975	*37*, 389
992	*36*, 416

Volume 30

No.	Suppl. Ref. Vol.-Page
1	*36*, 1
4	*36*, 4; *37*, 2
10a	*36*, 12
12	*36*, 5
18	*36*, 24; *37*, 20
27	*36*, 39
39	*36*, 39
41	*36*, 47
43	*36*, 343

No.	Suppl. Ref. Vol.-Page
Volume 31 continued	
633	*36*, 285
634	*37*, 276
638	*36*, 288
639	*37*, 277
641	*37*, 278
642	*36*, 291
658	*37*, 284
664	*36*, 295
	37, 286 (2)
676	*36*, 298
677	*36*, 298
680	*37*, 289
683	*36*, 301
704	*37*, 295
708	*36*, 309; *37*, 298
709	*36*, 310; *37*, 295
720	*36*, 314
722	*36*, 314
728	*37*, 301
771	*36*, 328
789	*36*, 336; *37*, 317
799	*36*, 340, 350
	37, 339
804	*36*, 340
806	*37*, 323
808	*36*, 347
813	*36*, 346
833	*36*, 354
840	*36*, 356
849	*36*, 375
865	*36*, 365; *37*, 348
868	*37*, 361
890	*36*, 375
905	*36*, 384 (2)
915	*37*, 373
919	*36*, 391
920	*37*, 378
930	*37*, 375
932	*36*, 395; *37*, 379
943	*36*, 401
949	*36*, 398
967	*37*, 388
974	*37*, 394
981	*36*, 413
984	*37*, 396
986	*36*, 75
991	*37*, 399
Volume 32	
11	*36*, 9
26	*37*, 18
50	*36*, 20; *37*, 28
52	*36*, 34
56	*36*, 124
64	*37*, 35
85	*37*, 47
90	*37*, 50
95	*36*, 59
96	*37*, 53
99	*37*, 253
110	*36*, 65
148	*36*, 89, 167
159	*36*, 91; *37*, 82
161	*37*, 80
166	*36*, 99
170	*36*, 98
171	*37*, 250
180	*37*, 103
199	*36*, 112
237	*37*, 121
250	*37*, 128
253	*36*, 132
254	*36*, 133
317	*36*, 160; *37*, 147
332	*36*, 163
339	*36*, 169
348	*36*, 168
367	*36*, 173
368	*36*, 174
387	*37*, 168
398	*36*, 184
416	*36*, 191
417	*37*, 179
431	*36*, 197
454	*37*, 397
465	*36*, 205
467	*36*, 205
468	*36*, 206
489	*36*, 208
495	*37*, 204
504	*36*, 177
507	*36*, 220
512	*37*, 394
513	*37*, 205
537	*36*, 236; *37*, 223
569	*37*, 236
575	*37*, 237
576	*36*, 97
593	*37*, 246
599	*37*, 151, 205
601	*36*, 265; *37*, 254
605	*37*, 255
610	*36*, 94
614	*36*, 371
618	*37*, 269
628	*37*, 273
629	*36*, 281
632	*36*, 281
637	*36*, 267
663	*37*, 71
709	*36*, 310
711	*36*, 310; *37*, 296
715	*36*, 312; *37*, 298
	37, 300
726	*37*, 306
780	*36*, 335
796	*36*, 343
800	*36*, 348; *37*, 329
806	*37*, 323
808	*36*, 288
821	*36*, 351
828	*36*, 354; *37*, 336
843	*37*, 344
856	*37*, 348
863	*36*, 365
864	*36*, 365
866	*36*, 378; *37*, 352
872	*36*, 367
876	*37*, 355
881	*37*, 359
886	*36*, 365
887	*36*, 373
888	*36*, 374; *37*, 363
901	*36*, 379
903	*37*, 369
940	*36*, 394
964	*37*, 385
966	*37*, 384
985	*36*, 412
988	*37*, 374
Volume 33	
1	*36*, 1
10	*36*, 6
21	*36*, 7
31	*36*, 175

No.	Suppl. Ref. Vol.-Page
Volume 33 continued	
36	*36*, 16
38	*36*, 17
43	*36*, 20 (2)
	37, 16
44	*37*, 17
46	*36*, 23
47	*36*, 23; *37*, 20
52	*36*, 27
54	*37*, 24
57	*36*, 29
68	*36*, 37
76	*36*, 13; *37*, 7
80	*36*, 43
93	*36*, 52; *37*, 47
100	*36*, 57 (2)
	37, 50
102	*37*, 52
124	*37*, 61
127	*37*, 64
128	*37*, 73
150	*36*, 74
154	*36*, 75
157	*36*, 80
164	*37*, 73
169	*37*, 77
174	*37*, 78
179	*36*, 91
184	*36*, 94
195	*36*, 93, 240
	37, 227
196	*36*, 96
200	*36*, 98
207	*36*, 93
229	*36*, 109
231	*36*, 109
236	*37*, 282
240	*36*, 112
241	*36*, 112; *37*, 101
243	*36*, 116
245	*36*, 117
247	*37*, 36
252	*37*, 111
262	*36*, 122; *37*, 116
265	*37*, 117
283	*37*, 122
311	*36*, 143, 144
329	*36*, 339
332	*37*, 130
335	*37*, 143
339	*36*, 153
342	*36*, 152
352	*36*, 154
357	*36*, 157
360	*37*, 146
363	*36*, 159
369	*36*, 162
374	*36*, 167
379	*36*, 13; *37*, 146
387	*37*, 158
391	*36*, 172
392	*37*, 158
394	*37*, 160
397	*36*, 170
400	*36*, 172
403	*36*, 173; *37*, 163
408	*37*, 162
412	*36*, 176
441	*37*, 177
471	*36*, 207
477	*37*, 195
484	*37*, 199
486	*37*, 204
492	*37*, 203
499	*36*, 219
501	*36*, 221
511	*36*, 229
513	*36*, 229, 363
516	*36*, 231
531	*36*, 203, 241
533	*36*, 234
534	*37*, 225
545	*36*, 239
561	*37*, 169
565	*37*, 233
583	*36*, 254, 269
	36, 411
588	*36*, 254
592	*36*, 258
595	*36*, 260
596	*36*, 261, 411
599	*36*, 259
605	*37*, 151
608	*36*, 340
609	*36*, 268
626	*37*, 271
629	*37*, 271
634	*36*, 187
639	*36*, 285 (2)
639	*37*, 257
641	*37*, 259
654	*37*, 273
658	*36*, 293; *37*, 285
662	*36*, 297; *37*, 287
669	*37*, 58
672	*36*, 300
690	*36*, 306
691	*37*, 293
694	*36*, 308
697	*36*, 308
701	*36*, 311
702	*36*, 311
707	*36*, 312
714	*37*, 300
715	*37*, 349
722	*36*, 320
723	*36*, 318
724	*36*, 318; *37*, 303
727	*37*, 304
735	*36*, 325; *37*, 307
741	*37*, 309
749	*36*, 321; *37*, 50
752	*36*, 330
753	*36*, 335
756	*37*, 313
760	*36*, 332
761	*37*, 314
767	*36*, 334
768	*36*, 336
774	*37*, 318
776	*37*, 272
778	*36*, 338
779	*37*, 324
806	*36*, 356; *37*, 338
	37, 366
828	*37*, 342
830	*37*, 333
849	*37*, 347
855	*36*, 354
858	*36*, 367
865	*36*, 370; *37*, 355
873	*37*, 348
874	*36*, 372, 375
879	*36*, 374
883	*36*, 377
884	*36*, 377
886	*36*, 379
891	*36*, 339, 380
895	*36*, 333
906	*36*, 265
907	*36*, 383

No.	Suppl. Ref. Vol.-Page

Volume 35 continued

No.	Suppl. Ref. Vol.-Page
83	*36*, 79
85	*37*, 198
87	*36*, 80; *37*, 71
88	*37*, 73
94	*36*, 83
98	*36*, 85
105	*37*, 78
119	*36*, 97; *37*, 86
143	*36*, 110
145	*36*, 110
148	*37*, 105
149	*36*, 115
151	*36*, 116
154	*37*, 108
155	*37*, 108
159	*37*, 106
209	*36*, 138; *37*, 134
245	*36*, 180
250	*37*, 150
254	*37*, 153
274	*37*, 163
278	*37*, 164
286	*36*, 337
313	*36*, 417
337	*36*, 186
349	*37*, 192
351	*37*, 195
362	*37*, 110
372	*37*, 108
390	*36*, 241
404	*36*, 247
414	*37*, 236
439	*36*, 271; *37*, 256
440	*37*, 261
441	*37*, 261
443	*37*, 19
471	*37*, 283
473	*37*, 276
476	*36*, 293; *37*, 284
477	*37*, 273
478	*37*, 296
489	*36*, 299
523	*37*, 306
541	*37*, 313
549	*37*, 341
555	*36*, 343
556	*36*, 285; *37*, 275
560	*36*, 344
571	*37*, 333
581	*37*, 330
583	*37*, 332
596	*37*, 345
622	*37*, 362
623	*36*, 373; *37*, 363
627	*36*, 376
631	*37*, 367
650	*36*, 412
664	*36*, 392
670	*36*, 400; *37*, 383
673	*37*, 385
676	*36*, 404; *37*, 387
686	*36*, 409
702	*36*, 418

Volume 36

No.	Suppl. Ref. Vol.-Page
6	*37*, 3
31	*37*, 16
36	*37*, 20
45	*37*, 23
49	*37*, 26
62	*37*, 33
79	*37*, 41
85	*37*, 187
129	*37*, 66
143	*37*, 138
148	*37*, 71
151	*37*, 71
153	*37*, 72
169	*37*, 122
171	*37*, 79
184	*37*, 81
198	*37*, 89
231	*37*, 104
245	*37*, 69
262	*37*, 123
288	*37*, 134
291	*37*, 135
294	*37*, 126
303	*37*, 137
311	*37*, 296
327	*37*, 144
348	*37*, 150
367	*37*, 157
369	*37*, 314
370	*37*, 158
376	*37*, 159
411	*37*, 175
443	*37*, 184
469	*37*, 196
471	*37*, 198
487	*37*, 204
489	*37*, 203
492	*37*, 206
512	*37*, 216
531	*37*, 222
540	*37*, 225
553	*37*, 228, 247
561	*37*, 230
585	*37*, 242
592	*37*, 245
595	*37*, 248
600	*37*, 239, 250
608	*37*, 252
618	*37*, 261
620	*37*, 261
624	*37*, 266
625	*37*, 266
633	*37*, 269
660	*37*, 283
661	*37*, 276
665	*37*, 281
667	*37*, 272
670	*37*, 30
671	*37*, 275
675	*37*, 284
682	*37*, 287
686	*37*, 288
724	*37*, 295
726	*37*, 299
740	*37*, 303
789	*37*, 276
793	*37*, 321
795	*37*, 322
816	*37*, 324
824	*37*, 330
844	*37*, 341
857	*37*, 346
872	*37*, 352
879	*37*, 358
893	*37*, 366
896	*37*, 366
927	*37*, 377
929	*37*, 376
932	*37*, 199
938	*37*, 380
943	*37*, 381
946	*37*, 383
963	*37*, 387
979	*37*, 394
992	*37*, 398
995	*37*, 400